INTRODUCTION TO LINEAR ALGEBRA WITH APPLICATIONS

Jim DeFranza
St. Lawrence University

Dan Gagliardi
SUNY Canton

Mc Graw Hill **Higher Education**

Boston   Burr Ridge, IL   Dubuque, IA   New York   San Francisco   St. Louis
Bangkok   Bogotá   Caracas   Kuala Lumpur   Lisbon   London   Madrid   Mexico City
Milan   Montreal   New Delhi   Santiago   Seoul   Singapore   Sydney   Taipei   Toronto

 **Higher Education**

INTRODUCTION TO LINEAR ALGEBRA WITH APPLICATIONS

Published by McGraw-Hill, a business unit of The McGraw-Hill Companies, Inc., 1221 Avenue of the Americas, New York, NY 10020. Copyright © 2009 by The McGraw-Hill Companies, Inc. All rights reserved. No part of this publication may be reproduced or distributed in any form or by any means, or stored in a database or retrieval system, without the prior written consent of The McGraw-Hill Companies, Inc., including, but not limited to, in any network or other electronic storage or transmission, or broadcast for distance learning.

Some ancillaries, including electronic and print components, may not be available to customers outside the United States.

This book is printed on acid-free paper.

1 2 3 4 5 6 7 8 9 0 DOC/DOC 0 9 8

ISBN 978–0–07–353235–6
MHID 0–07–353235–5

Editorial Director: *Stewart K. Mattson*
Senior Sponsoring Editor: *Elizabeth Covello*
Director of Development: *Kristine Tibbetts*
Developmental Editor: *Michelle Driscoll*
Director of Marketing: *Ryan Blankenship*
Marketing Coordinator: *Sabina Navsariwala*
Project Manager: *Joyce Watters*
Senior Production Supervisor: *Sherry L. Kane*
Senior Media Project Manager: *Tammy Juran*
Designer: *Laurie B. Janssen*
Cover Designer: *Ron Bissell*
(USE) Cover Image: *Royalty-Free/CORBIS*
Senior Photo Research Coordinator: *John C. Leland*
Supplement Producer: *Melissa M. Leick*
Compositor: *Laserwords Private Limited*
Typeface: *10.25/12 Times*
Printer: *R. R. Donnelly Crawfordsville, IN*

Library of Congress Cataloging-in-Publication Data

DeFranza, James, 1950–
  Introduction to linear algebra / James DeFranza, Daniel Gagliardi. —1st ed.
    p. cm.
  Includes index.
  ISBN 978–0–07–353235–6—ISBN 0–07–353235–5 (hard copy : alk. paper)
  1. Algebras, Linear—Textbooks. 2. Algebras, Linear—Problems, exercises, etc. I. Gagliardi, Daniel. II. Title.

  QA184.2.D44   2009
  515′.5—dc22

                                                              2008026020

www.mhhe.com

To Regan, Sara, and David
—JD

To Robin, Zachary, Michael, and Eric
—DG

About the Authors

Jim DeFranza was born in 1950 in Yonkers New York and grew up in Dobbs Ferry New York on the Hudson River. Jim DeFranza is Professor of Mathematics at St. Lawrence University in Canton New York where he has taught undergraduate mathematics for 25 years. St. Lawrence University is a small Liberal Arts College in upstate New York that prides itself in the close interaction that exists between students and faculty. It is this many years of working closely with students that has shaped this text in Linear Algebra and the other texts he has written. He received his Ph.D. in Pure Mathematics from Kent State University in 1979. Dr. DeFranza has coauthored PRECALCULUS, Fourth Edition and two other texts in single variable and multivariable calculus. Dr. DeFranza has also published a dozen research articles in the areas of Sequence Spaces and Classical Summability Theory. Jim is married and has two children David and Sara. Jim and his wife Regan live outside of Canton New York in a 150 year old farm house.

Daniel Gagliardi is an Assistant Professor of Mathematics at SUNY Canton, in Canton New York. Dr. Gagliardi began his career as a software engineer at IBM in East Fishkill New York writing programs to support semiconductor development and manufacturing. He received his Ph.D. in Pure Mathematics from North Carolina State University in 2003 under the supervision of Aloysius Helminck. Dr. Gagliardi's principle area of research is in Symmetric Spaces. In particular, his current work is concerned with developing algorithmic formulations to describe the fine structure (characters and Weyl groups) of local symmetric spaces. Dr. Gagliardi also does research in Graph Theory. His focus there is on the graphical realization of certain types of sequences. In addition to his work as a mathematician, Dr. Gagliardi is an accomplished double bassist and has recently recorded a CD of jazz standards with Author/Pianist Bill Vitek. Dr. Gagliardi lives in northern New York in the picturesque Saint Lawrence River Valley with his wife Robin, and children Zachary, Michael, and Eric.

Contents

Preface

Introduction to Linear Algebra with Applications is an introductory text targeted to second-year or advanced first-year undergraduate students. The organization of this text is motivated by what our experience tells us are the essential concepts that students should master in a one-semester undergraduate linear algebra course. The centerpiece of our philosophy regarding the presentation of the material is that *each topic should be fully developed before the reader moves onto the next*. In addition, there should be a natural connection between topics. We take great care to meet both of these objectives. This allows us to *stay on task* so that each topic can be covered with the depth required before progression to the next logical one. As a result, the reader is prepared for each new unit, and there is no need to repeat a concept in a subsequent chapter when it is utilized.

Linear algebra is taken early in an undergraduate curriculum and yet offers the opportunity to introduce the importance of abstraction, not only in mathematics, but in many other areas where linear algebra is used. Our approach is to take advantage of this opportunity by presenting abstract vector spaces as early as possible. Throughout the text, we are mindful of the difficulties that students at this level have with abstraction and introduce new concepts first through examples which gently illustrate the idea. To motivate the definition of an abstract vector space, and the subtle concept of linear independence, we use addition and scalar multiplication of vectors in Euclidean space. We have strived to create a balance among computation, problem solving, and abstraction. This approach equips students with the necessary skills and problem-solving strategies in an abstract setting that allows for a greater understanding and appreciation for the numerous applications of the subject.

Pedagogical Features

1. **Linear systems, matrix algebra, and determinants:** We have given a streamlined, but complete, discussion of solving linear systems, matrix algebra, determinants, and their connection in Chap. 1. Computational techniques are introduced, and a number of theorems are proved. In this way, students can hone their problem-solving skills while beginning to develop a conceptual sense of the fundamental ideas of linear algebra. Determinants are no longer central in linear algebra, and we believe that in a course at this level, only a few lectures should be devoted to the topic. For this reason we have presented all the essentials on determinants, including their connection to linear systems and matrix inverses, in Chap. 1. This choice also enables us to use determinants as a theoretical tool throughout the text whenever the need arises.

2. **Vectors:** Vectors are introduced in Chap. 1, providing students with a familiar structure to work with as they start to explore the properties which are used later to characterize abstract vector spaces.

3. **Linear independence:** We have found that many students have difficulties with linear combinations and the concept of linear independence. These ideas are fundamental to linear algebra and are essential to almost every topic after linear systems. When students fail to grasp them, the full benefits of the course cannot be realized. In *Introduction to Linear Algebra with Applications* we have devoted Chap. 2 to a careful exposition of linear combinations and linear independence in the context of Euclidean space. This serves several purposes. First, by placing these concepts in a separate chapter their importance in linear algebra is highlighted. Second, an instructor using the text can give exclusive focus to these ideas before applying them to other problems and situations. Third, many of the important ramifications of linear combinations and linear independence are considered in the familiar territory of Euclidean spaces.

4. **Euclidean spaces $\mathbb{R}^n$:** The Euclidean spaces and their algebraic properties are introduced in Chap. 2 and are used as a model for the abstract vectors spaces of Chap. 3. We have found that this approach works well for students with limited exposure to abstraction at this level.

5. **Geometric representations:** Whenever possible, we include figures with geometric representations and interpretations to illuminate the ideas being presented.

6. **New concepts:** New concepts are almost always introduced first through concrete examples. Formal definitions and theorems are then given to describe the situation in general. Additional examples are also provided to further develop the new idea and to explore it in greater depth.

7. **True/false chapter tests:** Each chapter ends with a true/false Chapter Test with approximately 40 questions. These questions are designed to help the student connect concepts and better understand the facts presented in the chapter.

8. **Rigor and intuition:** The approach we have taken attempts to strike a balance between presenting a rigorous development of linear algebra and building intuition. For example, we have chosen to omit the proofs for theorems that are not especially enlightening or that contain excessive computations. When a proof is not present, we include a motivating discussion describing the importance and use of the result and, if possible, the idea behind a proof.

9. **Abstract vector spaces:** We have positioned abstract vector spaces as a central topic within *Introduction to Linear Algebra with Applications* by placing their introduction as early as possible in Chap. 3. We do this to ensure that abstract vector spaces receive the appropriate emphasis. In a typical undergraduate mathematics curriculum, a course on linear algebra is the first time that students are exposed to this level of abstraction. However, Euclidean spaces *still play a central role in our approach* because of their familiarity and since they are so widely used. At the end of this chapter, we include a section on differential equations which underscores the need for the abstract theory of vector spaces.

10. **Section fact summaries:** Each section ends with a summary of the important facts and techniques established in the section. They are written, whenever possible, using nontechnical language and mostly without notation. These summaries are not meant to give a recapitulation of the details and formulas of the section; rather they are designed to give an overview of the main ideas of the section. Our intention is to help students to make connections between the concepts of the section as they survey the topic from a greater vantage point.

Applications

Over the last few decades the applications of linear algebra have mushroomed, increasing not only in their numbers, but also in the diversity of fields to which they apply. Much of this growth is fueled by the power of modern computers and the availability of computer algebra systems used to carry out computations for problems involving large matrices. This impressive power has made linear algebra more relevant than ever. Recently, a consortium of mathematics educators has placed its importance, relative to applications, second only to calculus. Increasingly, universities are offering courses in linear algebra that are specifically geared toward its applications. Whether the intended audience is engineering, economics, science, or mathematics students, the abstract theory is essential to understanding how linear algebra is applied.

In this text our introduction to the applications of linear algebra begins in Sec. 1.8 where we show how linear systems can be used to solve problems related to chemistry, engineering, economics, nutrition, and urban planning. However, many types of applications involve the more sophisticated concepts we develop in the text. These applications require the theoretical notions beyond the basic ideas of Chap. 1, and are presented at the end of a chapter as soon as the required background material is completed. Naturally, we have had to limit the number of applications considered. It is our hope that the topics we have chosen will interest the reader and lead to further inquiry.

Specifically, in Sec. 4.6, we discuss the role of linear algebra in computer graphics. An introduction to the connection between differential equations and linear algebra is given in Secs. 3.5 and 5.3. Markov chains and quadratic forms are examined in Secs. 5.4 and 6.7, respectively. Section 6.5 focuses on the problem of finding approximate solutions to inconsistent linear systems. One of the most familiar applications here is the problem of finding the equation of a line that best fits a set of data points. Finally, in Sec. 6.8 we consider the singular value decomposition of a matrix and its application to data compression.

Technology

Computations are an integral part of any introductory course in mathematics and certainly in linear algebra. To gain mastery of the techniques, we encourage the student to solve as many problems as possible by hand. That said, we also encourage the student to make appropriate use of the available technologies designed to facilitate, or to completely carry out, some of the more tedious computations. For example, it is quite reasonable to use a computer algebra system, such as MAPLE or MATLAB, to row-reduce a large matrix. Our approach in *Introduction to Linear Algebra with Applications* is to assume that some form of technology will be used, but leave the choice to the individual instructor and student. We do not think that it is necessary to include discussions or exercises that use particular software. Note that this text can be used *with or without technology*. The degree to which it is used is left to the discretion of the instructor. From our own experience, we have found that Scientific Notebook,[TM] which offers a front end for LaTeX along with menu access to the computer algebra system MuPad, allows the student to gain experience using technology to carry out computations while learning to write clear mathematics. Another option is to use LaTeX for writing mathematics and a computer algebra system to perform computations.

Another aspect of technology in linear algebra has to do with the accuracy and efficiency of computations. Some applications, such as those related to Internet search engines, involve very large matrices which require extensive processing. Moreover, the accuracy of the results can be affected by computer *roundoff error*. For example, using the characteristic equation to find the eigenvalues of a large matrix is not feasible. Overcoming problems of this kind is extremely important. The field of study known as *numerical linear algebra* is an area of vibrant research for both software engineers and applied mathematicians who are concerned with developing practical solutions. In our text, the fundamental concepts of linear algebra are introduced using simple examples. However, students should be made aware of the computational difficulties that arise when extending these ideas beyond the small matrices used in the illustrations.

Other Features

1. **Chapter openers:** The opening remarks for each chapter describe an application that is directly related to the material in the chapter. These provide additional motivation and emphasize the relevance of the material that is about to be covered.

2. **Writing style:** The writing style is clear, engaging, and easy to follow. Important new concepts are first introduced with examples to help develop the reader's intuition. We limit the use of jargon and provide explanations that are as reader-friendly as possible. Every explanation is crafted with the student in mind. *Introduction to Linear Algebra with Applications* is specifically designed to be a readable text from which a student can learn the fundamental concepts in linear algebra.

3. **Exercise sets:** Exercise sets are organized with routine exercises at the beginning and the more difficult problems toward the end. There is a mix of computational and theoretical exercises with some requiring proof. The early portion of each exercise set tests the student's ability to apply the basic concepts. These exercises are primarily computational, and their solutions follow from the worked examples in the section. The latter portion of each exercise set extends the concepts and techniques by asking the student to construct complete arguments.

4. **Review exercise sets:** The review exercise sets are organized as sample exams with 10 exercises. These exercises tend to have multiple parts, which connect the various techniques and concepts presented in the text. At least one problem in each of these sets presents a new idea in the context of the material of the chapter.

5. **Length:** The length of the text reflects the fact that it is specifically designed for a one-semester course in linear algebra at the undergraduate level.

6. **Appendix:** The appendix contains background material on the algebra of sets, functions, techniques of proof, and mathematical induction. With this feature, the instructor is able to cover, as needed, topics that are typically included in a *Bridge Course* to higher mathematics.

Course Outline

The topics we have chosen for *Introduction to Linear Algebra with Applications* closely follow those commonly covered in a first introductory course. The order in which we present these topics reflects our approach and preferences for emphasis. Nevertheless, we have written the text to be flexible, allowing for some permutations of the order of topics without any loss of consistency. In **Chap. 1** we present all the basic material on linear systems, matrix algebra, determinants, elementary matrices, and the LU decomposition. **Chap. 2** is entirely devoted to a careful exposition of linear combinations and linear independence in $\mathbb{R}^n$. We have found that many students have difficulty with these essential concepts. The addition of this chapter gives us the opportunity to develop all the important ideas in a familiar setting. As mentioned earlier, to emphasize the importance of abstract vector spaces, we have positioned their introduction as early as possible in **Chap. 3**. Also, in Chap. 3 is a discussion of subspaces, bases, and coordinates. Linear transformations between vector spaces are the subject of **Chap. 4**. We give descriptions of the null space and range of a linear transformation at the beginning of the chapter, and later we show that *every finite dimensional vector space, of dimension n, is isomorphic to* $\mathbb{R}^n$. Also, in Chap. 4 we introduce the four fundamental subspaces of a matrix and discuss the action of an $m \times n$ matrix on a vector in $\mathbb{R}^n$. **Chap. 5** is concerned with eigenvalues and eigenvectors. An abundance of examples are given to illustrate the techniques of computing eigenvalues and finding the corresponding eigenvectors. We discuss the algebraic and geometric multiplicities of eigenvalues and give criteria for when a square matrix is diagonalizable. In **Chap. 6**, using $\mathbb{R}^n$ as a model, we show how a geometry can be defined on a vector space by means of an inner product. We also give a description of the Gram-Schmidt process used to find an orthonormal basis for an inner product space and present material on orthogonal complements. At the end of this chapter we discuss the singular value decomposition of an $m \times n$ matrix. The **Appendix** contains a brief summary of some topics found in a *Bridge Course* to higher mathematics. Here we include material on the algebra of sets, functions, techniques of proof, and mathematical induction. Application sections are placed at the end of chapters as soon as the requisite background material has been covered.

Supplements

1. **Instructor solutions manual:** This manual contains detailed solutions to all exercises.

2. **Student solutions manual:** This manual contains detailed solutions to odd-numbered exercises.

3. **Text website www.mhhe.com/defranza:** This website accompanies the text and is available for both students and their instructors. Students will be able to access self-assessment quizzes and extra examples for each section and end of chapter cumulative quizzes. In addition to these assets, instructors will be able to access additional quizzes, sample exams, the end of chapter true/false tests, and the Instructor's Solutions Manual.

Acknowledgments

We would like to give our heartfelt thanks to the many individuals who reviewed the manuscript at various stages of its development. Their thoughtful comments and excellent suggestions have helped us enormously with our efforts to realize our vision of a reader-friendly introductory text on linear algebra.

We would also like to give special thanks to David Meel of Bowling Green State University, Bowling Green, Ohio, for his thorough review of the manuscript and insightful comments that have improved the exposition of the material in the text. We are also grateful to Ernie Stitzinger of North Carolina State University who had the tiring task of checking the complete manuscript for accuracy, including all the exercises. A very special thanks goes to our editors (and facilitators), Liz Covello (Sr. Sponsoring Editor), Michelle Driscoll (Developmental Editor), and Joyce Watters (Project Manager) who have helped us in more ways than we can name, from the inception of this project to its completion. On a personal level, we would like to thank our wives, Regan DeFranza and Robin Gagliardi, for their love and support; and our students at Saint Lawrence University and SUNY Canton who provided the motivation to write the text. Finally, we want to express our gratitude to the staff at McGraw-Hill Higher Education, Inc., for their work in taking our manuscript and producing the text.

List of Reviewers

Marie Aratari, *Oakland Community College*
Cik Azizah, *Universiti Utara Malaysia (UUM)*
Przcmyslaw Bogacki, *Old Dominion University*
Rita Chattopadhyay, *Eastern Michigan University*
Eugene Don, *Queens College*
Lou Giannini, *Curtin University of Technology*
Gregory Gibson, *North Carolina A&T University*
Mark Gockenback, *Michigan Technological University*
Dr. Leong Wah June, *Universiti Putra Malaysia*
Cerry Klein, *University of Missouri–Columbia*
Kevin Knudson, *Mississippi State University*
Hyungiun Ko, *Yonsei University*
Jacob Kogan, *University of Maryland–Baltimore County*
David Meel, *Bowling Green State University*
Martin Nakashima, *California State Poly University–Pomona*
Eugene Spiegel, *University of Connecticut–Storrs*
Dr. Hajar Sulaiman, *Universiti Sains Malaysia (USM)*
Gnana Bhaskar Tenali, *Florida Institute of Technology–Melbourne*
Peter Wolfe, *University of Maryland–College Park*

To The Student

You are probably taking this course early in your undergraduate studies after two or three semesters of calculus, and most likely in your second year. Like calculus, linear algebra is a subject with elegant theory and many diverse applications. However, in this course you will be exposed to abstraction at a much higher level. To help with this transition, some colleges and universities offer a *Bridge Course to Higher Mathematics*. If you have not already taken such a course, this may likely be the first mathematics course where you will be expected to read and understand proofs of theorems, provide proofs of results as part of the exercise sets, and apply the concepts presented. All this is in the context of a specific body of knowledge. If you approach this task with an open mind and a willingness to read the text, some parts perhaps more than once, it will be an exciting and rewarding experience. Whether you are taking this course as part of a mathematics major or because linear algebra is applied in your specific area of study, a clear understanding of the theory is essential for applying the concepts of linear algebra to mathematics or other fields of science. The solved examples and exercises in the text are designed to prepare you for the types of problems you can expect to see in this course and other more advanced courses in mathematics. The organization of the material is based on our philosophy that *each topic should be fully developed before readers move onto the next*. The image of a tree on the front cover of the text is a metaphor for this learning strategy. It is particularly applicable to the study of mathematics. The trunk of the tree represents the material that forms the basis for everything that comes afterward. In our text, this material is contained in Chaps. 1 through 4. All other branches of the tree, representing more advanced topics and applications, extend from the foundational material of the trunk or from the ancillary material of the intervening branches. We have specifically designed our text so that you can read it and learn the concepts of linear algebra in a sequential and thorough manner. If you remain committed to learning this beautiful subject, the rewards will be significant in other courses you may take, and in your professional career. Good luck!

Jim DeFranza
jdefranza@stlawu.edu

Dan Gagliardi
gagliardid@canton.edu

Applications Index

Systems of Linear Equations and Matrices

*I*n the process of *photosynthesis* solar energy is converted into forms that are used by living organisms. The chemical reaction that occurs in the leaves of plants converts carbon dioxide and water to carbohydrates with the release of oxygen. The *chemical equation* of the reaction takes the form

$$aCO_2 + bH_2O \rightarrow cO_2 + dC_6H_{12}O_6$$

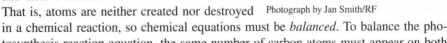

where a, b, c, and d are some positive whole numbers. The **law of conservation of mass** states that the total mass of all substances present before and after a chemical reaction remains the same. That is, atoms are neither created nor destroyed

Photograph by Jan Smith/RF

in a chemical reaction, so chemical equations must be *balanced*. To balance the photosynthesis reaction equation, the same number of carbon atoms must appear on both sides of the equation, so

$$a = 6d$$

The same number of oxygen atoms must appear on both sides, so

$$2a + b = 2c + 6d$$

and the same number of hydrogen atoms must appear on both sides, so

$$2b = 12d$$

This gives us the *system of three linear equations in four variables*

$$\begin{cases} a \quad\qquad -\ 6d = 0 \\ 2a +\ b - 2c -\ 6d = 0 \\ \qquad 2b \quad\ -\ 12d = 0 \end{cases}$$

Any positive integers $a, b, c,$ and d that satisfy all three equations are a *solution* to this system which balances the chemical equation. For example, $a = 6, b = 6, c = 6,$ and $d = 1$ balances the equation.

Many diverse applications are modeled by systems of equations. Systems of equations are also important in mathematics and in particular in linear algebra. In this chapter we develop systematic methods for solving *systems of linear equations*.

1.1 ▶ Systems of Linear Equations

As the introductory example illustrates, many naturally occurring processes are modeled using more than one equation and can require many equations in many variables. For another example, models of the economy contain thousands of equations and thousands of variables. To develop this idea, consider the set of equations

$$\begin{cases} 2x -\ y = 2 \\ \ x + 2y = 6 \end{cases}$$

which is a system of two equations in the common variables x and y. A *solution* to this system consists of values for x and y that simultaneously satisfy each equation. In this example we proceed by solving the first equation for y, so that

$$y = 2x - 2$$

To find the solution, substitute $y = 2x - 2$ into the second equation to obtain

$$x + 2(2x - 2) = 6 \quad\text{and solving for } x \text{ gives}\quad x = 2$$

Substituting $x = 2$ back into the first equation yields $2(2) - y = 2$, so that $y = 2$. Therefore the unique solution to the system is $x = 2, y = 2$. Since both of these equations represent straight lines, a solution exists provided that the lines intersect. These lines intersect at the unique point $(2, 2)$, as shown in Fig. 1(a). A system of equations is **consistent** if there is at least one solution to the system. If there are no solutions, the system is **inconsistent**. In the case of systems of two linear equations with two variables, there are three possibilities:

1. The two lines have different slopes and hence intersect at a unique point, as shown in Fig. 1(a).
2. The two lines are identical (one equation is a nonzero multiple of the other), so there are infinitely many solutions, as shown in Fig. 1(b).

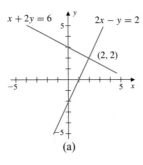

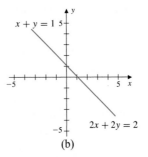

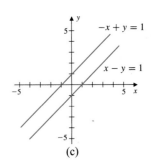

(a)   (b)   (c)

Figure 1

3. The two lines are parallel (have the same slope) and do not intersect, so the system is inconsistent, as shown in Fig. 1(c).

When we are dealing with many variables, the standard method of representing linear equations is to affix subscripts to coefficients and variables. A **linear equation** in the n variables $x_1, x_2, \ldots, x_n$ is an equation of the form

$$a_1 x_1 + a_2 x_2 + \cdots + a_n x_n = b$$

To represent a system of m linear equations in n variables, two subscripts are used for each coefficient. The first subscript indicates the equation number while the second specifies the term of the equation.

DEFINITION 1

System of Linear Equations   A **system of** m **linear equations in** n **variables**, or a **linear system**, is a collection of equations of the form

$$\begin{cases} a_{11}x_1 + a_{12}x_2 + \cdots + a_{1n}x_n = b_1 \\ a_{21}x_1 + a_{22}x_2 + \cdots + a_{2n}x_n = b_2 \\ a_{31}x_1 + a_{32}x_2 + \cdots + a_{3n}x_n = b_3 \\ \quad \vdots \qquad \vdots \qquad \vdots \qquad \vdots \qquad \vdots \\ a_{m1}x_1 + a_{m2}x_2 + \cdots + a_{mn}x_n = b_m \end{cases}$$

This is also referred to as an $m \times n$ linear system.

For example, the collection of equations

$$\begin{cases} -2x_1 + 3x_2 + x_3 - x_4 = -2 \\ x_1 \qquad\quad + x_3 - 4x_4 = 1 \\ 3x_1 - x_2 \qquad\quad - x_4 = 3 \end{cases}$$

is a linear system of three equations in four variables, or a 3×4 linear system.

A **solution** to a linear system with n variables is an ordered sequence $(s_1, s_2, \ldots, s_n)$ such that each equation is satisfied for $x_1 = s_1, x_2 = s_2, \ldots, x_n = s_n$. The **general solution** or **solution set** is the set of all possible solutions.

The Elimination Method

The elimination method, also called *Gaussian elimination*, is an algorithm used to solve linear systems. To describe this algorithm, we first introduce the *triangular form* of a linear system.

An $m \times n$ linear system is in **triangular form** provided that the coefficients $a_{ij} = 0$ whenever $i > j$. In this case we refer to the linear system as a **triangular system**. Two examples of triangular systems are

$$\begin{cases} x_1 - 2x_2 + x_3 = -1 \\ x_2 - 3x_3 = 5 \\ x_3 = 2 \end{cases} \quad \text{and} \quad \begin{cases} x_1 + x_2 - x_3 - x_4 = 2 \\ x_2 - x_3 - 2x_4 = 1 \\ 2x_3 - x_4 = 3 \end{cases}$$

When a linear system is in triangular form, then the solution set can be obtained using a technique called **back substitution**. To illustrate this technique, consider the linear system given by

$$\begin{cases} x_1 - 2x_2 + x_3 = -1 \\ x_2 - 3x_3 = 5 \\ x_3 = 2 \end{cases}$$

From the last equation we see that $x_3 = 2$. Substituting this into the second equation, we obtain $x_2 - 3(2) = 5$, so $x_2 = 11$. Finally, using these values in the first equation, we have $x_1 - 2(11) + 2 = -1$, so $x_1 = 19$. The solution is also written as $(19, 11, 2)$.

DEFINITION 2   **Equivalent Linear Systems**   Two linear systems are **equivalent** if they have the same solutions

---

For example, the system

$$\begin{cases} x_1 - 2x_2 + x_3 = -1 \\ 2x_1 - 3x_2 - x_3 = 3 \\ x_1 - 2x_2 + 2x_3 = 1 \end{cases}$$

has the unique solution $x_1 = 19$, $x_2 = 11$, and $x_3 = 2$, so the linear systems

$$\begin{cases} x_1 - 2x_2 + x_3 = -1 \\ x_2 - 3x_3 = 5 \\ x_3 = 2 \end{cases} \quad \text{and} \quad \begin{cases} x_1 - 2x_2 + x_3 = -1 \\ 2x_1 - 3x_2 - x_3 = 3 \\ x_1 - 2x_2 + 2x_3 = 1 \end{cases}$$

are equivalent.

The next theorem gives three operations that transform a linear system into an equivalent system, and together they can be used to convert any linear system to an equivalent system in triangular form.

THEOREM 1   Let

$$\begin{cases} a_{11}x_1 + a_{12}x_2 + \cdots + a_{1n}x_n = b_1 \\ a_{21}x_1 + a_{22}x_2 + \cdots + a_{2n}x_n = b_2 \\ a_{31}x_1 + a_{32}x_2 + \cdots + a_{3n}x_n = b_3 \\ \quad\vdots \qquad\quad \vdots \qquad\quad \vdots \qquad\quad \vdots \qquad\quad \vdots \\ a_{m1}x_1 + a_{m2}x_2 + \cdots + a_{mn}x_n = b_m \end{cases}$$

be a linear system. Performing any one of the following operations on the linear system produces an equivalent linear system.

1. Interchanging any two equations.
2. Multiplying any equation by a nonzero constant.
3. Adding a multiple of one equation to another.

Proof   Interchanging any two equations does not change the solution of the linear system and therefore yields an equivalent system. If equation i is multiplied by a constant $c \neq 0$, then equation i of the new system is

$$ca_{i1}x_1 + ca_{i2}x_2 + \cdots + ca_{in}x_n = cb_i$$

Let $(s_1, s_2, \ldots, s_n)$ be a solution to the original system. Since

$$a_{i1}s_1 + a_{i2}s_2 + \cdots + a_{in}s_n = b_i, \qquad \text{then} \qquad ca_{i1}s_1 + ca_{i2}s_2 + \cdots + ca_{in}s_n = cb_i$$

Hence $(s_1, s_2, \ldots, s_n)$ is a solution of the new linear system. Consequently, the systems are equivalent.

For part (3) of the theorem, consider the new system obtained by adding c times equation i to equation j of the original system. Thus, equation j of the new system becomes

$$(ca_{i1} + a_{j1})x_1 + (ca_{i2} + a_{j2})x_2 + \cdots + (ca_{in} + a_{jn})x_n = cb_i + b_j$$

or equivalently,

$$c(a_{i1}x_1 + a_{i2}x_2 + \cdots + a_{in}x_n) + (a_{j1}x_1 + a_{j2}x_2 + \cdots + a_{jn}x_n) = cb_i + b_j$$

Now let $(s_1, s_2, \ldots, s_n)$ be a solution for the original system. Then

$$a_{i1}s_1 + a_{i2}s_2 + \cdots + a_{in}s_n = b_i \qquad \text{and} \qquad a_{j1}s_1 + a_{j2}s_2 + \cdots + a_{jn}s_n = b_j$$

Therefore,

$$c(a_{i1}s_1 + a_{i2}s_2 + \cdots + a_{in}s_n) + (a_{j1}s_1 + a_{j2}s_2 + \cdots + a_{jn}s_n) = cb_i + b_j$$

so that $(s_1, s_2, \ldots, s_n)$ is a solution of the modified system and the systems are equivalent.

EXAMPLE 1   Use the elimination method to solve the linear system.

$$\begin{cases} x + y = 1 \\ -x + y = 1 \end{cases}$$

Solution   Adding the first equation to the second gives the equivalent system

$$\begin{cases} x + \ y = 1 \\ \quad 2y = 2 \end{cases}$$

From the second equation, we have $y = 1$. Using back substitution gives $x = 0$.
The graphs of both systems are shown in Fig. 2. Notice that the solution is the same
in both, but that adding the first equation to the second rotates the line $-x + y = 1$
about the point of intersection.

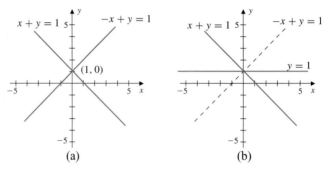

(a)   (b)

Figure 2

Converting a linear system to triangular form often requires many steps. Moreover,
the operations used to convert one linear system to another are not unique and may
not be apparent on inspection. To articulate this process, the notation, for example,

$$(-2) \cdot E_1 + E_3 \longrightarrow E_3$$

will mean *add* -2 *times equation 1 to equation 3, and replace equation 3 with the
result.* The notation $E_i \leftrightarrow E_j$ will be used to indicate that equation i and equation j
are interchanged.

EXAMPLE 2   Solve the linear system.

$$\begin{cases} x + y + \ z = \ \ 4 \\ -x - y + \ z = -2 \\ 2x - y + 2z = \ \ 2 \end{cases}$$

Solution   To convert the system to an equivalent triangular system, we first eliminate the
variable x in the second and third equations to obtain

$$\begin{cases} x + y + z = 4 \\ -x - y + z = -2 \\ 2x - y + 2z = 2 \end{cases} \qquad \begin{aligned} E_1 + E_2 &\rightarrow E_2 \\ -2E_1 + E_3 &\rightarrow E_3 \end{aligned} \longrightarrow \begin{cases} x + y + z = 4 \\ 2z = 2 \\ -3y = -6 \end{cases}$$

Interchanging the second and third equations gives the triangular linear system

$$\begin{cases} x + y + z = 4 \\ 2z = 2 \\ -3y = -6 \end{cases} \qquad E_2 \leftrightarrow E_3 \longrightarrow \begin{cases} x + y + z = 4 \\ -3y = -6 \\ 2z = 2 \end{cases}$$

Using back substitution, we have $z = 1$, $y = 2$, and $x = 4 - y - z = 1$. Therefore, the system is consistent with the unique solution $(1, 2, 1)$.

Recall from solid geometry that the graph of an equation of the form $ax + by + cz = d$ is a plane in three-dimensional space. Hence, the unique solution to the linear system of Example 2 is the point of intersection of three planes, as shown in Fig. 3(a). For another perspective on this, shown in Fig. 3(b) are the lines of the pairwise intersections of the three planes. These lines intersect at a point that is the solution to the 3×3 linear system.

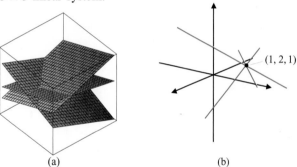

(a)   (b)

Figure 3

Similar to the 2×2 case, the geometry of Euclidean space helps us better understand the possibilities for the general solution of a linear system of three equations in three variables. In particular, the linear system can have a unique solution if the three planes all intersect at a point, as illustrated by Example 2. Alternatively, a 3×3 system can have infinitely many solutions if

1. The three planes are all the same.

2. The three planes intersect in a line (like the pages of a book).

3. Two of the planes are the same with a third plane intersecting them in a line.

For example, the linear system given by

$$\begin{cases} -y + z = 0 \\ y = 0 \\ z = 0 \end{cases}$$

represents three planes whose intersection is the x axis. That is, $z = 0$ is the xy plane, $y = 0$ is the xz plane, and $y = z$ is the plane that cuts through the x axis at a $45°$ angle.

Finally, there are two cases in which a 3×3 linear system has no solutions. First, the linear system has no solutions if at least one of the planes is parallel to, but not the same as, the others. Certainly, when all three planes are parallel, the system has no solutions, as illustrated by the linear system

$$\begin{cases} z = 0 \\ z = 1 \\ z = 2 \end{cases}$$

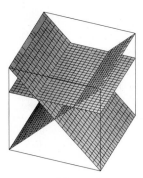

Figure 4

Also, a 3×3 linear system has no solutions, if the lines of the pairwise intersections of the planes are parallel, but not the same, as shown in Fig. 4.

From the previous discussion we see that a 3×3 linear system, like a 2×2 linear system, has no solutions, has a unique solution, or has infinitely many solutions. We will see in Sec. 1.4 that this is the case for linear systems of any size.

In Example 3 we consider a linear system with four variables. Of course the geometric reasoning above cannot be applied to the new situation directly, but provides the motivation for understanding the many possibilities for the solutions to linear systems with several variables.

EXAMPLE 3   Solve the linear system.

$$\begin{cases} 4x_1 - 8x_2 - 3x_3 + 2x_4 = 13 \\ 3x_1 - 4x_2 - x_3 - 3x_4 = 5 \\ 2x_1 - 4x_2 - 2x_3 + 2x_4 = 6 \end{cases}$$

Solution   Since every term of the third equation can be divided evenly by 2, we multiply the third equation by $\frac{1}{2}$. After we do so, the coefficient of x_1 is 1. We then interchange the first and third equations, obtaining

$$\begin{cases} 4x_1 - 8x_2 - 3x_3 + 2x_4 = 13 \\ 3x_1 - 4x_2 - x_3 - 3x_4 = 5 \\ 2x_1 - 4x_2 - 2x_3 + 2x_4 = 6 \end{cases} \quad \frac{1}{2}E_3 \to E_3 \quad \longrightarrow \quad \begin{cases} 4x_1 - 8x_2 - 3x_3 + 2x_4 = 13 \\ 3x_1 - 4x_2 - x_3 - 3x_4 = 5 \\ x_1 - 2x_2 - x_3 + x_4 = 3 \end{cases}$$

$$E_1 \leftrightarrow E_3 \quad \longrightarrow \quad \begin{cases} x_1 - 2x_2 - x_3 + x_4 = 3 \\ 3x_1 - 4x_2 - x_3 - 3x_4 = 5 \\ 4x_1 - 8x_2 - 3x_3 + 2x_4 = 13 \end{cases}$$

Next using the operations $-3E_1 + E_2 \to E_2$ and $-4E_1 + E_3 \to E_3$, we obtain the linear system

$$\begin{cases} x_1 - 2x_2 - x_3 + x_4 = 3 \\ 2x_2 + 2x_3 - 6x_4 = -4 \\ x_3 - 2x_4 = 1 \end{cases}$$

which is an equivalent system in triangular form. Using back substitution, the general solution is

$$x_3 = 2x_4 + 1 \qquad x_2 = x_4 - 3 \qquad x_1 = 3x_4 - 2$$

with x_4 free to assume any real number. It is common in this case to replace x_4 with the parameter t. The general solution can now be written as

$$S = \{(3t - 2, t - 3, 2t + 1, t) \mid t \in \mathbb{R}\}$$

and is called a *one-parameter family* of solutions. The reader can check that $x_1 = 3t - 2$, $x_2 = t - 3$, $x_3 = 2t + 1$, and $x_4 = t$ is a solution for any t by substituting these values in the original equations. A particular solution can be obtained by letting t be a specific value. For example, if $t = 0$, then a particular solution is $(-2, -3, 1, 0)$.

In Example 3, the variable x_4 can assume any real number, giving infinitely many solutions for the linear system. In this case we call x_4 a **free variable**. When a linear system has infinitely many solutions, there can be more than one free variable. In this case, the solution set is an r-parameter family of solutions where r is equal to the number of free variables.

EXAMPLE 4   Solve the linear system.

$$\begin{cases} x_1 - x_2 - 2x_3 - 2x_4 - 2x_5 = 3 \\ 3x_1 - 2x_2 - 2x_3 - 2x_4 - 2x_5 = -1 \\ -3x_1 + 2x_2 + x_3 + x_4 - x_5 = -1 \end{cases}$$

Solution   After performing the operations $E_3 + E_2 \to E_3$ followed by $E_2 - 3E_1 \to E_2$, we have the equivalent system

$$\begin{cases} x_1 - x_2 - 2x_3 - 2x_4 - 2x_5 = 3 \\ x_2 + 4x_3 + 4x_4 + 4x_5 = -10 \\ -x_3 - x_4 - 3x_5 = -2 \end{cases}$$

The variables x_4 and x_5 are both free variables, so to write the solution, let $x_4 = s$ and $x_5 = t$. From the third equation, we have

$$x_3 = 2 - x_4 - 3x_5 = 2 - s - 3t$$

Substitution into the second equation gives

$$\begin{aligned} x_2 &= -10 - 4x_3 - 4x_4 - 4x_5 \\ &= -10 - 4(2 - s - 3t) - 4s - 4t \\ &= -18 + 8t \end{aligned}$$

Finally, substitution into the first equation gives

$$x_1 = 3 + x_2 + 2x_3 + 2x_4 + 2x_5$$
$$= 3 + (-18 + 8t) + 2(2 - s - 3t) + 2s + 2t$$
$$= -11 + 4t$$

The *two-parameter* solution set is therefore given by

$$S = \{(-11 + 4t, -18 + 8t, 2 - s - 3t, s, t) \mid s, t \in \mathbb{R}\}$$

Particular solutions for $s = t = 0$ and $s = 0, t = 1$ are $(-11, -18, 2, 0, 0)$ and $(-7, -10, -1, 0, 1)$, respectively.

EXAMPLE 5   Solve the linear system.

$$\begin{cases} x_1 - x_2 + 2x_3 = 5 \\ 2x_1 + x_2 = 2 \\ x_1 + 8x_2 - x_3 = 3 \\ -x_1 - 5x_2 - 12x_3 = 4 \end{cases}$$

Solution   To convert the linear system to an equivalent triangular system, we will eliminate the first terms in equations 2 through 4, and then the second terms in equations 3 and 4, and then finally the third term in the fourth equation. This is accomplished by using the following operations.

$$\begin{cases} x_1 - x_2 + 2x_3 = 5 \\ 2x_1 + x_2 = 2 \\ x_1 + 8x_2 - x_3 = 3 \\ -x_1 - 5x_2 - 12x_3 = 4 \end{cases} \quad \begin{matrix} -2E_1 + E_2 \to E_2 \\ -E_1 + E_3 \to E_3 \\ E_1 + E_4 \to E_4 \end{matrix} \to \begin{cases} x_1 - x_2 + 2x_3 = 5 \\ 3x_2 - 4x_3 = -8 \\ 9x_2 - 3x_3 = -2 \\ -6x_2 - 10x_3 = 9 \end{cases}$$

$$\begin{matrix} -3E_2 + E_3 \to E_3 \\ 2E_2 + E_4 \to E_4 \end{matrix} \to \begin{cases} x_1 - x_2 + 2x_3 = 5 \\ 3x_2 - 4x_3 = -8 \\ 9x_3 = 22 \\ -18x_3 = -7 \end{cases}$$

$$2E_3 + E_4 \to E_4 \quad \to \begin{cases} x_1 - x_2 + 2x_3 = 5 \\ 3x_2 - 4x_3 = -8 \\ 9x_3 = 22 \\ 0 = -37 \end{cases}$$

The last equation of the final system is an impossibility, so the original linear system is inconsistent and has no solution.

In the previous examples the algorithm for converting a linear system to triangular form is based on using a *leading variable* in an equation to eliminate the same variable in each equation below it. This process can always be used to convert any linear system to triangular form.

EXAMPLE 6

Find the equation of the parabola that passes through the points $(-1, 1)$, $(2, -2)$, and $(3, 1)$. Find the vertex of the parabola.

Solution

The general form of a parabola is given by $y = ax^2 + bx + c$. Conditions on a, b, and c are imposed by substituting the given points into this equation. This gives

$$
\begin{aligned}
1 &= a(-1)^2 + b(-1) + c = a - b + c \\
-2 &= a(2)^2 + b(2) + c = 4a + 2b + c \\
1 &= a(3)^2 + b(3) + c = 9a + 3b + c
\end{aligned}
$$

From these conditions we obtain the linear system

$$
\begin{cases}
a - b + c = 1 \\
4a + 2b + c = -2 \\
9a + 3b + c = 1
\end{cases}
$$

First, with a as the leading variable, we use row operations to eliminate a from equations 2 and 3. In particular, we have

$$
\begin{cases}
a - b + c = 1 \\
4a + 2b + c = -2 \\
9a + 3b + c = 1
\end{cases}
\quad
\begin{aligned}
-4E_1 + E_2 &\to E_2 \\
-9E_1 + E_3 &\to E_3
\end{aligned}
\quad \to \quad
\begin{cases}
a - b + c = 1 \\
6b - 3c = -6 \\
12b - 8c = -8
\end{cases}
$$

Next, with b as the leading variable we, eliminate b from equation 3, so that

$$
\begin{cases}
a - b + c = 1 \\
6b - 3c = -6 \\
12b - 8c = -8
\end{cases}
\quad
-2E_2 + E_3 \to E_3
\quad \to \quad
\begin{cases}
a - b + c = 1 \\
6b - 3c = -6 \\
-2c = 4
\end{cases}
$$

Now, using back substitution on the last system gives $c = -2$, $b = -2$, and $a = 1$. Thus, the parabola we seek is

$$
y = x^2 - 2x - 2
$$

Completing the square gives the parabola in standard form

$$
y = (x - 1)^2 - 3
$$

with vertex $(1, -3)$, as shown in Fig. 5.

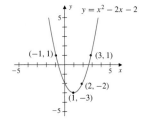

Figure 5

Fact Summary

1. A $m \times n$ linear system has a unique solution, infinitely many solutions, or no solutions.
2. Interchanging any two equations in a linear system does not alter the set of solutions.
3. Multiplying any equation in a linear system by a nonzero constant does not alter the set of solutions.

> **4.** Replacing an equation in a linear system with the sum of the equation and a scalar multiple of another equation does not alter the set of solutions.
>
> **5.** Every linear system can be reduced to an equivalent triangular linear system.

Exercise Set 1.1

1. Consider the linear system

$$\begin{cases} x_1 - x_2 - 2x_3 = 3 \\ -x_1 + 2x_2 + 3x_3 = 1 \\ 2x_1 - 2x_2 - 2x_3 = -2 \end{cases}$$

Perform the operations $E_1 + E_2 \to E_2$ and $-2E_1 + E_3 \to E_3$, and write the new equivalent system. Solve the linear system.

2. Consider the linear system

$$\begin{cases} 2x_1 - 2x_2 - x_3 = -3 \\ x_1 - 3x_2 + x_3 = -2 \\ x_1 - 2x_2 = 2 \end{cases}$$

Perform the operations $E_1 \leftrightarrow E_2$, $-2E_1 + E_2 \to E_2$, $-E_1 + E_3 \to E_3$, $E_2 \leftrightarrow E_3$, and $-4E_2 + E_3 \to E_3$, and write the new equivalent system. Solve the linear system.

3. Consider the linear system

$$\begin{cases} x_1 + 3x_4 = 2 \\ x_1 + x_2 + 4x_4 = 3 \\ 2x_1 + x_3 + 8x_4 = 3 \\ x_1 + x_2 + x_3 + 6x_4 = 2 \end{cases}$$

Perform the operations $-E_1 + E_2 \to E_2$, $-2E_1 + E_3 \to E_3$, $-E_1 + E_4 \to E_4$, $-E_2 + E_4 \to E_4$, and $-E_3 + E_4 \to E_4$, and write the new equivalent system. Solve the linear system.

4. Consider the linear system

$$\begin{cases} x_1 + x_3 = -2 \\ x_1 + x_2 + 4x_3 = -1 \\ 2x_1 + 2x_3 + x_4 = -1 \end{cases}$$

Perform the operations $-E_1 + E_2 \to E_2$ and $-2E_1 + E_3 \to E_3$, and write the new equivalent system. Solve the linear system.

In Exercises 5–18, solve the linear system using the elimination method.

5. $\begin{cases} 2x + 3y = -2 \\ -2x = 0 \end{cases}$

6. $\begin{cases} x + 3y = -1 \\ - y = -1 \end{cases}$

7. $\begin{cases} 4x = 4 \\ -3x + 2y = -3 \end{cases}$

8. $\begin{cases} 2x + 3y = -1 \\ x - y = 0 \end{cases}$

9. $\begin{cases} 3x - 2y = 4 \\ x - \frac{2}{3}y = \frac{4}{3} \end{cases}$

10. $\begin{cases} 3x - 5y = 1 \\ -x + \frac{5}{3}y = -\frac{1}{3} \end{cases}$

11. $\begin{cases} -3x - 2y + 2z = -2 \\ -x - 3y + z = -3 \\ x - 2y + z = -2 \end{cases}$

12. $\begin{cases} x + 3y + z = 2 \\ -2x + 2y - 4z = -1 \\ -y + 3z = 1 \end{cases}$

13. $\begin{cases} -2x - 2y + 2z = 1 \\ x \qquad + 5z = -1 \\ 3x + 2y + 3z = -2 \end{cases}$

14. $\begin{cases} -x + y + 4z = -1 \\ 3x - y + 2z = 2 \\ 2x - 2y - 8z = 2 \end{cases}$

15. $\begin{cases} 3x_1 + 4x_2 + 3x_3 = 0 \\ 3x_1 - 4x_2 + 3x_3 = 4 \end{cases}$

16. $\begin{cases} -2x_1 + x_2 = 2 \\ 3x_1 - x_2 + 2x_3 = 1 \end{cases}$

17. $\begin{cases} x_1 - 2x_2 - 2x_3 - x_4 = -3 \\ -2x_1 + x_2 + x_3 - 2x_4 = -3 \end{cases}$

18. $\begin{cases} 2x_1 + 2x_2 - x_3 = 1 \\ -x_2 + 3x_4 = 2 \end{cases}$

In Exercises 19–22, solve for x, y, and z in terms of a, b, and c.

19. $\begin{cases} -2x + y = a \\ -3x + 2y = b \end{cases}$

20. $\begin{cases} 2x + 3y = a \\ x + y = b \end{cases}$

21. $\begin{cases} 3x + y + 3z = a \\ -x - z = b \\ -x + 2y = c \end{cases}$

22. $\begin{cases} -3x + 2y + z = a \\ x - y - z = b \\ x - y - 2z = c \end{cases}$

In Exercises 23–28, give restrictions on a, b, and c such that the linear system is consistent.

23. $\begin{cases} x - 2y = a \\ -2x + 4y = 2 \end{cases}$

24. $\begin{cases} -x + 3y = a \\ 2x - 6y = 3 \end{cases}$

25. $\begin{cases} x - 2y = a \\ -x + 2y = b \end{cases}$

26. $\begin{cases} 6x - 3y = a \\ -2x + y = b \end{cases}$

27. $\begin{cases} x - 2y + 4z = a \\ 2x + y - z = b \\ 3x - y + 3z = c \end{cases}$

28. $\begin{cases} x - y + 2z = a \\ 2x + 4y - 3z = b \\ 4x + 2y + z = c \end{cases}$

In Exercises 29–32, determine the value of a that makes the system inconsistent.

29. $\begin{cases} x + y = -2 \\ 2x + ay = 3 \end{cases}$

30. $\begin{cases} 2x - y = 4 \\ ax + 3y = 2 \end{cases}$

31. $\begin{cases} x - y = 2 \\ 3x - 3y = a \end{cases}$

32. $\begin{cases} 2x - y = a \\ 6x - 3y = a \end{cases}$

In Exercises 33–36, find an equation in the form $y = ax^2 + bx + c$ for the parabola that passes through the three points. Find the vertex of the parabola.

33. $(0, 0.25)$, $(1, -1.75)$, $(-1, 4.25)$

34. $(0, 2)$, $(-3, -1)$, $(0.5, 0.75)$

35. $(-0.5, -3.25)$, $(1, 2)$, $(2.3, 2.91)$

36. $(0, -2875)$, $(1, -5675)$, $(3, 5525)$

37. Find the point where the three lines $-x + y = 1$, $-6x + 5y = 3$, and $12x + 5y = 39$ intersect. Sketch the lines.

38. Find the point where the four lines $2x + y = 0$, $x + y = -1$, $3x + y = 1$, and $4x + y = 2$ intersect. Sketch the lines.

39. Give an example of a 2×2 linear system that
  a. Has a unique solution
  b. Has infinitely many solutions
  c. Is inconsistent

40. Verify that if $ad - bc \neq 0$, then the system of equations
$$\begin{cases} ax + by = x_1 \\ cx + dy = x_2 \end{cases}$$
has a unique solution.

41. Consider the system

$$\begin{cases} x_1 - x_2 + 3x_3 - x_4 = 1 \\ x_2 - x_3 + 2x_4 = 2 \end{cases}$$

a. Describe the solution set where the variables x_3 and x_4 are free.

b. Describe the solution set where the variables x_2 and x_4 are free.

42. Consider the system

$$\begin{cases} x_1 - x_2 + x_3 - x_4 + x_5 = 1 \\ x_2 - x_4 - x_5 = -1 \\ x_3 - 2x_4 + 3x_5 = 2 \end{cases}$$

a. Describe the solution set where the variables x_4 and x_5 are free.

b. Describe the solution set where the variables x_3 and x_5 are free.

43. Determine the values of k such that the linear system

$$\begin{cases} 9x + ky = 9 \\ kx + y = -3 \end{cases}$$

has

a. No solutions

b. Infinitely many solutions

c. A unique solution

44. Determine the values of k such that the linear system

$$\begin{cases} kx + y + z = 0 \\ x + ky + z = 0 \\ x + y + kz = 0 \end{cases}$$

has

a. A unique solution

b. A one-parameter family of solutions

c. A two-parameter family of solutions

1.2 ▶ Matrices and Elementary Row Operations

In Sec. 1.1 we saw that converting a linear system to an equivalent triangular system provides an algorithm for solving the linear system. The algorithm can be streamlined by introducing *matrices* to represent linear systems.

DEFINITION 1   **Matrix**   An $m \times n$ **matrix** is an array of numbers with m rows and n columns.

For example, the array of numbers

$$\begin{bmatrix} 2 & 3 & -1 & 4 \\ 3 & 1 & 0 & -2 \\ -2 & 4 & 1 & 3 \end{bmatrix}$$

is a 3×4 matrix.

When solving a linear system by the elimination method, only the coefficients of the variables and the constants on the right-hand side are needed to find the solution. The variables are placeholders. Utilizing the structure of a matrix, we can record the coefficients and the constants by using the columns as placeholders for the variables. For example, the coefficients and constants of the linear system

$$\begin{cases} -4x_1 + 2x_2 - 3x_4 = 11 \\ 2x_1 - x_2 - 4x_3 + 2x_4 = -3 \\ 3x_2 - x_4 = 0 \\ -2x_1 + x_4 = 4 \end{cases}$$

can be recorded in matrix form as

$$\left[\begin{array}{cccc|c} -4 & 2 & 0 & -3 & 11 \\ 2 & -1 & -4 & 2 & -3 \\ 0 & 3 & 0 & -1 & 0 \\ -2 & 0 & 0 & 1 & 4 \end{array}\right]$$

This matrix is called the **augmented matrix** of the linear system. Notice that for an $m \times n$ linear system the augmented matrix is $m \times (n + 1)$. The augmented matrix with the last column deleted

$$\left[\begin{array}{cccc} -4 & 2 & 0 & -3 \\ 2 & -1 & -4 & 2 \\ 0 & 3 & 0 & -1 \\ -2 & 0 & 0 & 1 \end{array}\right]$$

is called the **coefficient matrix**. Notice that we always use a 0 to record any missing terms.

The method of elimination on a linear system is equivalent to performing similar operations on the rows of the corresponding augmented matrix. The relationship is illustrated below:

Linear system	Corresponding augmented matrix

$$\begin{cases} x + y - z = 1 \\ 2x - y + z = -1 \\ -x - y + 3z = 2 \end{cases} \qquad \left[\begin{array}{ccc|c} 1 & 1 & -1 & 1 \\ 2 & -1 & 1 & -1 \\ -1 & -1 & 3 & 2 \end{array}\right]$$

Using the operations $-2E_1 + E_2 \to E_2$ and $E_1 + E_3 \to E_3$, we obtain the equivalent triangular system

Using the operations $-2R_1 + R_2 \to R_2$ and $R_1 + R_3 \to R_3$, we obtain the equivalent augmented matrix

$$\begin{cases} x + y - z = 1 \\ -3y + 3z = -3 \\ 2z = 3 \end{cases} \qquad \left[\begin{array}{ccc|c} 1 & 1 & -1 & 1 \\ 0 & -3 & 3 & -3 \\ 0 & 0 & 2 & 3 \end{array}\right]$$

The notation used to describe the operations on an augmented matrix is similar to the notation we introduced for equations. In the example above,

$$-2R_1 + R_2 \longrightarrow R_2$$

means *replace row 2 with −2 times row 1 plus row 2*. Analogous to the triangular form of a linear system, a matrix is in **triangular form** provided that the first nonzero entry for each row of the matrix is to the right of the first nonzero entry in the row above it.

The next theorem is a restatement of Theorem 1 of Sec. 1.1, in terms of operations on the rows of an augmented matrix.

THEOREM 2   Any one of the following operations performed on the augmented matrix, corresponding to a linear system, produces an augmented matrix corresponding to an equivalent linear system.

  1. Interchanging any two rows.
  2. Multiplying any row by a nonzero constant.
  3. Adding a multiple of one row to another.

Solving Linear Systems with Augmented Matrices

The operations in Theorem 2 are called **row operations**. An $m \times n$ matrix A is called **row equivalent** to an $m \times n$ matrix B if B can be obtained from A by a sequence of row operations.

The following steps summarize a process for solving a linear system.

  1. Write the augmented matrix of the linear system.
  2. Use row operations to reduce the augmented matrix to triangular form.
  3. Interpret the final matrix as a linear system (which is equivalent to the original).
  4. Use back substitution to write the solution.

Example 1 illustrates how we can carry out steps 3 and 4.

EXAMPLE 1   Given the augmented matrix, find the solution of the corresponding linear system.

$$\textbf{a.} \begin{bmatrix} 1 & 0 & 0 & | & 1 \\ 0 & 1 & 0 & | & 2 \\ 0 & 0 & 1 & | & 3 \end{bmatrix} \quad \textbf{b.} \begin{bmatrix} 1 & 0 & 0 & 0 & | & 5 \\ 0 & 1 & -1 & 0 & | & 1 \\ 0 & 0 & 0 & 1 & | & 3 \end{bmatrix} \quad \textbf{c.} \begin{bmatrix} 1 & 2 & 1 & -1 & | & 1 \\ 0 & 3 & -1 & 0 & | & 1 \\ 0 & 0 & 0 & 0 & | & 0 \end{bmatrix}$$

Solution

  a. Reading directly from the augmented matrix, we have $x_3 = 3, x_2 = 2$, and $x_1 = 1$. So the system is consistent and has a unique solution.

  b. In this case the solution to the linear system is $x_4 = 3, x_2 = 1 + x_3$, and $x_1 = 5$. So the variable x_3 is free, and the general solution is $S = \{(5, 1 + t, t, 3) \mid t \in \mathbb{R}\}$.

  c. The augmented matrix is equivalent to the linear system

$$\begin{cases} x_1 + 2x_2 + x_3 - x_4 = 1 \\ \quad\quad 3x_2 - x_3 \quad\quad = 1 \end{cases}$$

Using back substitution, we have

$$x_2 = \frac{1}{3}(1 + x_3) \quad\text{and}\quad x_1 = 1 - 2x_2 - x_3 + x_4 = \frac{1}{3} - \frac{5}{3}x_3 + x_4$$

So the variables x_3 and x_4 are free, and the two-parameter solution set is given by

$$S = \left\{ \left. \left(\frac{1}{3} - \frac{5s}{3} + t, \frac{1}{3} + \frac{s}{3}, s, t \right) \right| s, t \in \mathbb{R} \right\}$$

EXAMPLE 2 Write the augmented matrix and solve the linear system.

$$\begin{cases} x - 6y - 4z = -5 \\ 2x - 10y - 9z = -4 \\ -x + 6y + 5z = 3 \end{cases}$$

Solution To solve this system, we write the augmented matrix

$$\left[\begin{array}{ccc|c} 1 & -6 & -4 & -5 \\ 2 & -10 & -9 & -4 \\ -1 & 6 & 5 & 3 \end{array} \right]$$

where we have shaded the entries to eliminate. Using the procedure described above, the augmented matrix is reduced to triangular form as follows:

$$\left[\begin{array}{ccc|c} 1 & -6 & -4 & -5 \\ 2 & -10 & -9 & -4 \\ -1 & 6 & 5 & 3 \end{array} \right] \quad \begin{array}{c} -2R_1 + R_2 \to R_2 \\ R_1 + R_3 \to R_3 \end{array} \quad \longrightarrow \quad \left[\begin{array}{ccc|c} 1 & -6 & -4 & -5 \\ 0 & 2 & -1 & 6 \\ 0 & 0 & 1 & -2 \end{array} \right]$$

The equivalent triangular linear system is

$$\begin{cases} x - 6y - 4z = -5 \\ 2y - z = 6 \\ z = -2 \end{cases}$$

which has the solution $x = -1$, $y = 2$, and $z = -2$.

Echelon Form of a Matrix

In Example 2, the final augmented matrix

$$\left[\begin{array}{ccc|c} 1 & -6 & -4 & -5 \\ 0 & 2 & -1 & 6 \\ 0 & 0 & 1 & -2 \end{array} \right]$$

is in *row echelon form*. The general structure of a matrix in row echelon form is shown in Fig. 1. The height of each step is one row, and the first nonzero term in a row, denoted in Fig. 1 by *, is to the right of the first nonzero term in the previous row. All the terms below the stairs are 0.

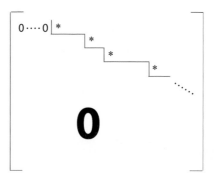

Figure 1

Although, the height of each step in Fig. 1 is one row, a step may extend over several columns. The leading nonzero term in each row is called a **pivot** element. The matrix is in *reduced row echelon form* if, in addition, each pivot is a 1 and all other entries in this column are 0. For example, the reduced row echelon form of the matrix

$$\begin{bmatrix} 1 & -6 & -4 & -5 \\ 0 & 2 & -1 & 6 \\ 0 & 0 & 1 & -2 \end{bmatrix}$$

is given by

$$\begin{bmatrix} 1 & 0 & 0 & -1 \\ 0 & 1 & 0 & 2 \\ 0 & 0 & 1 & -2 \end{bmatrix}$$

Transforming a matrix in row echelon form to reduced row echelon form in effect incorporates back substitution as row operations on the matrix. If we read from the last matrix above, the solution to the corresponding linear system is, as before, $x = -1$, $y = 2$, and $z = -2$.

Here are three additional matrices that *are* in reduced row echelon form

$$\begin{bmatrix} 1 & 0 & 0 & -1 \\ 0 & 1 & 0 & 2 \\ 0 & 0 & 1 & 4 \end{bmatrix} \qquad \begin{bmatrix} 1 & 0 & 0 & 0 \\ 0 & 1 & 0 & 0 \\ 0 & 0 & 1 & 0 \\ 0 & 0 & 0 & 1 \end{bmatrix} \qquad \begin{bmatrix} 1 & -2 & 0 & 1 & -1 \\ 0 & 0 & 1 & -1 & 2 \\ 0 & 0 & 0 & 0 & 0 \end{bmatrix}$$

and two that are *not* in reduced row echelon form

$$\begin{bmatrix} 1 & 3 & 2 & 1 \\ 0 & 1 & -5 & 6 \\ 0 & 1 & 4 & 1 \end{bmatrix} \qquad \begin{bmatrix} 1 & -2 & 1 & 0 \\ 0 & 0 & 2 & 1 \\ 0 & 0 & 0 & 3 \end{bmatrix}$$

In general, for any $m \times n$ matrix in reduced row echelon form, the pivot entries correspond to dependent variables, and the nonpivot entries correspond to independent or free variables. We summarize the previous discussion on row echelon form in the next definition.

DEFINITION 2

Echelon Form   An $m \times n$ matrix is in **row echelon form** if

1. Every row with all 0 entries is below every row with nonzero entries.
2. If rows $1, 2, \ldots, k$ are the rows with nonzero entries and if the leading nonzero entry (**pivot**) in row i occurs in column c_i, for $1, 2, \ldots, k$, then $c_1 < c_2 < \cdots < c_k$.

The matrix is in **reduced row echelon form** if, in addition,

3. The first nonzero entry of each row is a 1.
4. Each column that contains a pivot has all other entries 0.

The process of transforming a matrix to reduced row echelon form is called *Gauss-Jordan elimination*.

EXAMPLE 3

Solve the linear system by transforming the augmented matrix to reduced row echelon form.

$$\begin{cases} x_1 - x_2 - 2x_3 + x_4 = 0 \\ 2x_1 - x_2 - 3x_3 + 2x_4 = -6 \\ -x_1 + 2x_2 + x_3 + 3x_4 = 2 \\ x_1 + x_2 - x_3 + 2x_4 = 1 \end{cases}$$

Solution   The augmented matrix of the linear system is

$$\left[\begin{array}{cccc|c} 1 & -1 & -2 & 1 & 0 \\ 2 & -1 & -3 & 2 & -6 \\ -1 & 2 & 1 & 3 & 2 \\ 1 & 1 & -1 & 2 & 1 \end{array}\right]$$

To transform the matrix into reduced row echelon form, we first use the leading 1 in row 1 as a pivot to eliminate the terms in column 1 of rows 2, 3, and 4. To do this, we use the three row operations

$$-2R_1 + R_2 \rightarrow R_2$$
$$R_1 + R_3 \rightarrow R_3$$
$$-R_1 + R_4 \rightarrow R_4$$

in succession, transforming the matrix

$$\left[\begin{array}{cccc|c} 1 & -1 & -2 & 1 & 0 \\ 2 & -1 & -3 & 2 & -6 \\ -1 & 2 & 1 & 3 & 2 \\ 1 & 1 & -1 & 2 & 1 \end{array}\right] \quad \text{to} \quad \left[\begin{array}{cccc|c} 1 & -1 & -2 & 1 & 0 \\ 0 & 1 & 1 & 0 & -6 \\ 0 & 1 & -1 & 4 & 2 \\ 0 & 2 & 1 & 1 & 1 \end{array}\right]$$

For the second step we use the leftmost 1 in row 2 as the pivot and eliminate the term in column 2 above the pivot, and the two terms below the pivot. The required row operations are

$$R_2 + R_1 \rightarrow R_1$$
$$-R_2 + R_3 \rightarrow R_3$$
$$-2R_2 + R_4 \rightarrow R_4$$

reducing the matrix

$$\begin{bmatrix} 1 & -1 & -2 & 1 & | & 0 \\ 0 & 1 & 1 & 0 & | & -6 \\ 0 & 1 & -1 & 4 & | & 2 \\ 0 & 2 & 1 & 1 & | & 1 \end{bmatrix} \quad \text{to} \quad \begin{bmatrix} 1 & 0 & -1 & 1 & | & -6 \\ 0 & 1 & 1 & 0 & | & -6 \\ 0 & 0 & -2 & 4 & | & 8 \\ 0 & 0 & -1 & 1 & | & 13 \end{bmatrix}$$

Notice that each entry in row 3 is evenly divisible by 2. Therefore, a leading 1 in row 3 is obtained using the operation $-\frac{1}{2}R_3 \rightarrow R_3$, which results in the matrix

$$\begin{bmatrix} 1 & 0 & -1 & 1 & | & -6 \\ 0 & 1 & 1 & 0 & | & -6 \\ 0 & 0 & 1 & -2 & | & -4 \\ 0 & 0 & -1 & 1 & | & 13 \end{bmatrix}$$

Now, by using the leading 1 in row 3 as a pivot, the operations

$$R_3 + R_1 \rightarrow R_1$$
$$-R_3 + R_2 \rightarrow R_2$$
$$R_3 + R_4 \rightarrow R_4$$

row-reduce the matrix

$$\begin{bmatrix} 1 & 0 & -1 & 1 & | & -6 \\ 0 & 1 & 1 & 0 & | & -6 \\ 0 & 0 & 1 & -2 & | & -4 \\ 0 & 0 & -1 & 1 & | & 13 \end{bmatrix} \quad \text{to} \quad \begin{bmatrix} 1 & 0 & 0 & -1 & | & -10 \\ 0 & 1 & 0 & 2 & | & -2 \\ 0 & 0 & 1 & -2 & | & -4 \\ 0 & 0 & 0 & -1 & | & 9 \end{bmatrix}$$

Using the operation $-R_4 \rightarrow R_4$, we change the signs of the entries in row 4 to obtain the matrix

$$\begin{bmatrix} 1 & 0 & 0 & -1 & | & -10 \\ 0 & 1 & 0 & 2 & | & -2 \\ 0 & 0 & 1 & -2 & | & -4 \\ 0 & 0 & 0 & 1 & | & -9 \end{bmatrix}$$

Finally, using the leading 1 in row 4 as the pivot, we eliminate the terms above it in column 4. Specifically, the operations

$$R_4 + R_1 \rightarrow R_1$$
$$-2R_4 + R_2 \rightarrow R_2$$
$$2R_4 + R_3 \rightarrow R_3$$

applied to the last matrix give

$$\begin{bmatrix} 1 & 0 & 0 & 0 & | & -19 \\ 0 & 1 & 0 & 0 & | & 16 \\ 0 & 0 & 1 & 0 & | & -22 \\ 0 & 0 & 0 & 1 & | & -9 \end{bmatrix}$$

which is in reduced row echelon form.

The solution can now be read directly from the reduced matrix, giving us

$$x_1 = -19 \quad x_2 = 16 \quad x_3 = -22 \quad \text{and} \quad x_4 = -9.$$

EXAMPLE 4   Solve the linear system.

$$\begin{cases} 3x_1 - x_2 + x_3 + 2x_4 = -2 \\ x_1 + 2x_2 - x_3 + x_4 = 1 \\ -x_1 - 3x_2 + 2x_3 - 4x_4 = -6 \end{cases}$$

Solution   The linear system in matrix form is

$$\begin{bmatrix} 3 & -1 & 1 & 2 & | & -2 \\ 1 & 2 & -1 & 1 & | & 1 \\ -1 & -3 & 2 & -4 & | & -6 \end{bmatrix}$$

which can be reduced to

$$\begin{bmatrix} 1 & 2 & -1 & 1 & | & 1 \\ 0 & 1 & -1 & 3 & | & 5 \\ 0 & 0 & 1 & -\frac{20}{3} & | & -10 \end{bmatrix}$$

Notice that the system has infinitely many solutions, since from the last row we see that the variable x_4 is a free variable. We can reduce the matrix further, but the solution can easily be found from the echelon form by back substitution, giving us

$$x_3 = -10 + \frac{20}{3}x_4$$

$$x_2 = 5 + x_3 - 3x_4 = 5 + \left(-10 + \frac{20}{3}x_4\right) - 3x_4 = -5 + \frac{11}{3}x_4$$

$$x_1 = 1 - 2x_2 + x_3 - x_4 = 1 - \frac{5}{3}x_4$$

Letting x_4 be the arbitrary parameter t, we see the general solution is

$$S = \left\{ \left(1 - \frac{5t}{3}, -5 + \frac{11t}{3}, -10 + \frac{20t}{3}, t\right) \middle| t \in \mathbb{R} \right\}$$

Example 5 gives an illustration of a reduced matrix for an inconsistent linear system.

EXAMPLE 5   Solve the linear system.

$$\begin{cases} x + y + z = 4 \\ 3x - y - z = 2 \\ x + 3y + 3z = 8 \end{cases}$$

Solution   To solve this system, we reduce the augmented matrix to triangular form. The following steps describe the process.

$$\begin{bmatrix} 1 & 1 & 1 & | & 4 \\ 3 & -1 & -1 & | & 2 \\ 1 & 3 & 3 & | & 8 \end{bmatrix} \quad \begin{array}{c} -3R_1 + R_2 \rightarrow R_2 \\ -R_1 + R_3 \rightarrow R_3 \end{array} \quad \longrightarrow \quad \begin{bmatrix} 1 & 1 & 1 & | & 4 \\ 0 & -4 & -4 & | & -10 \\ 0 & 2 & 2 & | & 4 \end{bmatrix}$$

$$R_2 \leftrightarrow R_3 \quad \longrightarrow \quad \begin{bmatrix} 1 & 1 & 1 & 4 \\ 0 & 2 & 2 & 4 \\ 0 & -4 & -4 & -10 \end{bmatrix}$$

$$\tfrac{1}{2}R_2 \rightarrow R_2 \quad \longrightarrow \quad \begin{bmatrix} 1 & 1 & 1 & 4 \\ 0 & 1 & 1 & 2 \\ 0 & -4 & -4 & -10 \end{bmatrix}$$

$$4R_2 + R_3 \rightarrow R_3 \quad \longrightarrow \quad \begin{bmatrix} 1 & 1 & 1 & 4 \\ 0 & 1 & 1 & 2 \\ 0 & 0 & 0 & -2 \end{bmatrix}$$

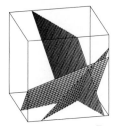

Figure 2

The third row of the last matrix corresponds to the equation $0 = -2$. As this system has no solution, the system is inconsistent. This can also be seen from the fact that the three planes do not have a common intersection, as shown in Fig.2.

In Example 5, each entry in the last row of the reduced coefficient matrix is 0, but the constant term is nonzero and the linear system is inconsistent. The reduced augmented matrix for a consistent linear system can have a row of zeros. However, in this case the term in the last column of this row must also be zero. Example 6 gives an illustration.

EXAMPLE 6   Determine when the augmented matrix represents a consistent linear system.

$$\begin{bmatrix} 1 & 0 & 2 & a \\ 2 & 1 & 5 & b \\ 1 & -1 & 1 & c \end{bmatrix}$$

Solution   The operation $-2R_1 + R_2 \rightarrow R_2$ followed by $-R_1 + R_3 \rightarrow R_3$ and finally followed by $R_2 + R_3 \rightarrow R_3$ reduces the augmented matrix to

$$\begin{bmatrix} 1 & 0 & 2 & a \\ 0 & 1 & 1 & b - 2a \\ 0 & 0 & 0 & b + c - 3a \end{bmatrix}$$

Hence, the corresponding linear system is consistent provided that $b + c - 3a = 0$. That is, the system is consistent for all a, b, and c such that the point (a, b, c) lies on the plane $b + c - 3a = 0$. Notice also that when the system is consistent, the third row will contain all zeros and the variable x_3 is a free variable.

Shown in the following list is an outline that summarizes the process for transforming a matrix to its equivalent reduced row echelon form.

1. If necessary, interchange rows so that the leading nonzero entry of row 1 is the leftmost nonzero entry of the matrix. Then divide each entry of row 1 by the leading entry.

2. Eliminate all other nonzero terms in this leading column.
3. Repeat the first two steps, starting with row 2. Note that the leading entry may not be in column 2.
4. Continue in this way, making sure that the leading entry of each row is a 1 with zeros elsewhere in that column.
5. The leading 1 in any row should be to the right of a leading 1 in the row above it.
6. All rows of zeros are placed at the bottom of the matrix.

We have implicitly assumed in our discussion that every matrix is row equivalent to exactly one matrix in reduced row echelon form. It is an important fact that we will state here as a theorem without proof.

THEOREM 3   The reduced row echelon form of every matrix is unique.

Fact Summary

1. Altering an augmented matrix by interchanging two rows, or multiplying a row by a nonzero constant, or replacing a row with the sum of the same row and a scalar multiple of another row does not alter the set of solutions of the corresponding linear system.
2. If an augmented matrix is row-reduced to triangular form, the coefficient matrix has a row of zeros, and the corresponding augmented term is not zero, then the linear system has no solutions.
3. Every matrix has a unique reduced row echelon form.
4. If the augmented matrix of an $n \times n$ linear system is row-reduced to triangular form and the coefficient matrix has no rows of zeros, then the linear system has a unique solution.
5. If the augmented matrix of an $n \times n$ linear system is row-reduced to triangular form, the coefficient matrix has rows of zeros, and each corresponding augmented term is 0, then the linear system has infinitely many solutions.

Exercise Set 1.2

In Exercises 1–8, write the linear system as an augmented matrix. Do not solve the system.

1. $\begin{cases} 2x - 3y = 5 \\ -x + y = -3 \end{cases}$

2. $\begin{cases} 2x - 2y = 1 \\ 3x \quad\;\; = 1 \end{cases}$

3. $\begin{cases} 2x \quad\;\; - z = 4 \\ x + 4y + z = 2 \\ 4x + y - z = 1 \end{cases}$

4. $\begin{cases} -3x + y + z = 2 \\ \qquad\quad - 4z = 0 \\ -4x + 2y - 3z = 1 \end{cases}$

5. $\begin{cases} 2x_1 \qquad - x_3 = 4 \\ x_1 + 4x_2 + x_3 = 2 \end{cases}$

6. $\begin{cases} 4x_1 + x_2 - 4x_3 = 1 \\ 4x_1 - 4x_2 + 2x_3 = -2 \end{cases}$

7. $\begin{cases} 2x_1 + 4x_2 + 2x_3 + 2x_4 = -2 \\ 4x_1 - 2x_2 - 3x_3 - 2x_4 = 2 \\ x_1 + 3x_2 + 3x_3 - 3x_4 = -4 \end{cases}$

8. $\begin{cases} 3x_1 \qquad\quad - 3x_3 + 4x_4 = -3 \\ -4x_1 + 2x_2 - 2x_3 - 4x_4 = 4 \\ \qquad\quad 4x_2 - 3x_3 + 2x_4 = -3 \end{cases}$

In Exercises 9–20, write the solution of the linear system corresponding to the reduced augmented matrix.

9. $\left[\begin{array}{ccc|c} 1 & 0 & 0 & -1 \\ 0 & 1 & 0 & \frac{1}{2} \\ 0 & 0 & 1 & 0 \end{array}\right]$

10. $\left[\begin{array}{ccc|c} 1 & 0 & 0 & 2 \\ 0 & 1 & 0 & 0 \\ 0 & 0 & 1 & -\frac{2}{3} \end{array}\right]$

11. $\left[\begin{array}{ccc|c} 1 & 0 & 2 & -3 \\ 0 & 1 & -1 & 2 \\ 0 & 0 & 0 & 0 \end{array}\right]$

12. $\left[\begin{array}{ccc|c} 1 & 0 & -\frac{1}{3} & 4 \\ 0 & 1 & 3 & \frac{4}{3} \\ 0 & 0 & 0 & 0 \end{array}\right]$

13. $\left[\begin{array}{ccc|c} 1 & -2 & 0 & -3 \\ 0 & 0 & 1 & 2 \\ 0 & 0 & 0 & 0 \end{array}\right]$

14. $\left[\begin{array}{ccc|c} 1 & 5 & 5 & -1 \\ 0 & 0 & 0 & 0 \\ 0 & 0 & 0 & 0 \end{array}\right]$

15. $\left[\begin{array}{ccc|c} 1 & 0 & 0 & 0 \\ 0 & 1 & 0 & 0 \\ 0 & 0 & 0 & 1 \end{array}\right]$

16. $\left[\begin{array}{ccc|c} 1 & 0 & 0 & 0 \\ 0 & 0 & 1 & 0 \\ 0 & 0 & 0 & 1 \end{array}\right]$

17. $\left[\begin{array}{cccc|c} 1 & 0 & -2 & 5 & 3 \\ 0 & 1 & -1 & 2 & 2 \end{array}\right]$

18. $\left[\begin{array}{cccc|c} 1 & 3 & -3 & 0 & 1 \\ 0 & 0 & 0 & 1 & 4 \end{array}\right]$

19. $\left[\begin{array}{cccc|c} 1 & 0 & 0 & -3 & 1 \\ 0 & 1 & 0 & -1 & 7 \\ 0 & 0 & 1 & 2 & -1 \end{array}\right]$

20. $\left[\begin{array}{cccc|c} 1 & 0 & \frac{2}{5} & 0 & -1 \\ 0 & 1 & -3 & 0 & 1 \\ 0 & 0 & 0 & 1 & \frac{4}{5} \end{array}\right]$

In Exercises 21–28, determine whether the matrices are in reduced row echelon form.

21. $\begin{bmatrix} 1 & 0 & 2 \\ 0 & 1 & 3 \end{bmatrix}$

22. $\begin{bmatrix} 1 & 2 & 0 \\ 0 & 0 & 1 \end{bmatrix}$

23. $\begin{bmatrix} 1 & 2 & 3 \\ 0 & 1 & 2 \\ 0 & 0 & 1 \end{bmatrix}$

24. $\begin{bmatrix} 1 & 2 & 0 \\ 0 & 0 & 2 \\ 0 & 0 & 0 \end{bmatrix}$

25. $\begin{bmatrix} 1 & 2 & 0 & -1 \\ 0 & 0 & 1 & -2 \\ 0 & 0 & 0 & 0 \end{bmatrix}$

26. $\begin{bmatrix} 1 & 0 & -3 & 4 \\ 0 & 1 & 1 & 5 \\ 0 & 0 & 0 & 0 \end{bmatrix}$

27. $\begin{bmatrix} 1 & 0 & 0 & 4 & -1 \\ 0 & 0 & 1 & 5 & 2 \\ 0 & 1 & 0 & 0 & -1 \end{bmatrix}$

28. $\begin{bmatrix} 1 & 1 & 0 & 4 & \frac{2}{3} \\ 0 & 1 & 1 & 5 & 6 \\ 0 & 0 & 0 & 1 & \frac{1}{3} \end{bmatrix}$

In Exercises 29–36, find the reduced row echelon form of the matrix.

29. $\begin{bmatrix} 2 & 3 \\ -2 & 1 \end{bmatrix}$

30. $\begin{bmatrix} -3 & 2 \\ 3 & 3 \end{bmatrix}$

31. $\begin{bmatrix} 3 & 3 & 1 \\ 3 & -1 & 0 \\ -1 & -1 & 2 \end{bmatrix}$

32. $\begin{bmatrix} 0 & 2 & 1 \\ 1 & -3 & -3 \\ 1 & 2 & -3 \end{bmatrix}$

33. $\begin{bmatrix} -4 & 1 & 4 \\ 3 & 4 & -3 \end{bmatrix}$

34. $\begin{bmatrix} -4 & -2 & -1 \\ -2 & -3 & 0 \end{bmatrix}$

35. $\begin{bmatrix} -2 & 2 & -1 & 2 \\ 0 & 3 & 3 & -3 \\ 1 & -4 & 2 & 2 \end{bmatrix}$

36. $\begin{bmatrix} 4 & -3 & -4 & -2 \\ -4 & 2 & 1 & -4 \\ -1 & -3 & 1 & -4 \end{bmatrix}$

In Exercises 37–48, write the linear system as an augmented matrix. Convert the augmented matrix to reduced row echelon form, and find the solution of the linear system.

37. $\begin{cases} x + y = 1 \\ 4x + 3y = 2 \end{cases}$

38. $\begin{cases} -3x + y = 1 \\ 4x + 2y = 0 \end{cases}$

39. $\begin{cases} 3x - 3y = 3 \\ 4x - y - 3z = 3 \\ -2x - 2y = -2 \end{cases}$

40. $\begin{cases} 2x - 4z = 1 \\ 4x + 3y - 2z = 0 \\ 2x + 2z = 2 \end{cases}$

41. $\begin{cases} x + 2y + z = 1 \\ 2x + 3y + 2z = 0 \\ x + y + z = 2 \end{cases}$

42. $\begin{cases} 3x - 2z = -3 \\ -2x + z = -2 \\ -z = 2 \end{cases}$

43. $\begin{cases} 3x_1 + 2x_2 + 3x_3 = -3 \\ x_1 + 2x_2 - x_3 = -2 \end{cases}$

44. $\begin{cases} -3x_2 - x_3 = 2 \\ x_1 + x_3 = -2 \end{cases}$

45. $\begin{cases} -x_1 + 3x_3 + x_4 = 2 \\ 2x_1 + 3x_2 - 3x_3 + x_4 = 2 \\ 2x_1 - 2x_2 - 2x_3 - x_4 = -2 \end{cases}$

46. $\begin{cases} -3x_1 - x_2 + 3x_3 + 3x_4 = -3 \\ x_1 - x_2 + x_3 + x_4 = 3 \\ -3x_1 + 3x_2 - x_3 + 2x_4 = 1 \end{cases}$

47. $\begin{cases} 3x_1 - 3x_2 + x_3 + 3x_4 = -3 \\ x_1 + x_2 - x_3 - 2x_4 = 3 \\ 4x_1 - 2x_2 + x_4 = 0 \end{cases}$

48. $\begin{cases} -3x_1 + 2x_2 - x_3 - 2x_4 = 2 \\ x_1 - x_2 - 3x_4 = 3 \\ 4x_1 - 3x_2 + x_3 - x_4 = 1 \end{cases}$

49. The augmented matrix of a linear system has the form

$$\begin{bmatrix} 1 & 2 & -1 & a \\ 2 & 3 & -2 & b \\ -1 & -1 & 1 & c \end{bmatrix}$$

a. Determine the values of $a, b,$ and c for which the linear system is consistent.

b. Determine the values of $a, b,$ and c for which the linear system is inconsistent.

c. When it is consistent, does the linear system have a unique solution or infinitely many solutions?

d. Give a specific consistent linear system and find one particular solution.

50. The augmented matrix of a linear system has the form

$$\begin{bmatrix} ax & y & 1 \\ 2x & (a-1)y & 1 \end{bmatrix}$$

a. Determine the values of a for which the linear system is consistent.

b. When it is consistent, does the linear system have a unique solution or infinitely many solutions?

c. Give a specific consistent linear system and find one particular solution.

51. The augmented matrix of a linear system has the form

$$\begin{bmatrix} -2 & 3 & 1 & a \\ 1 & 1 & -1 & b \\ 0 & 5 & -1 & c \end{bmatrix}$$

a. Determine the values of a, b, and c for which the linear system is consistent.

b. Determine the values of a, b, and c for which the linear system is inconsistent.

c. When it is consistent, does the linear system have a unique solution or infinitely many solutions?

d. Give a specific consistent linear system and find one particular solution.

52. Give examples to describe all 2×2 reduced row echelon matrices.

1.3 ▶ Matrix Algebra

Mathematics deals with abstractions that are based on natural concepts in concrete settings. For example, we accept the use of numbers and all the algebraic properties that go with them. Numbers can be added and multiplied, and they have properties such as the distributive and associative properties. In some ways matrices can be treated as numbers. For example, we can define addition and multiplication so that algebra can be performed with matrices. This extends the application of matrices beyond just a means for representing a linear system.

Let A be an $m \times n$ matrix. Then each entry of A can be uniquely specified by using the row and column indices of its location, as shown in Fig. 1.

Column j

$\downarrow$

$$\text{Row } i \longrightarrow \begin{bmatrix} a_{11} & \cdots & a_{1j} & \cdots & a_{1n} \\ \vdots & & \vdots & & \vdots \\ a_{i1} & \cdots & a_{ij} & \cdots & a_{in} \\ \vdots & & \vdots & & \vdots \\ a_{m1} & \cdots & a_{mj} & \cdots & a_{mn} \end{bmatrix} = A$$

Figure 1

For example, if

$$A = \begin{bmatrix} -2 & 1 & 4 \\ 5 & 7 & 11 \\ 2 & 3 & 22 \end{bmatrix}$$

then

$$a_{11} = -2 \qquad a_{12} = 1 \qquad a_{13} = 4$$
$$a_{21} = 5 \qquad a_{22} = 7 \qquad a_{23} = 11$$
$$a_{31} = 2 \qquad a_{32} = 3 \qquad a_{33} = 22$$

A **vector** is an $n \times 1$ matrix. The entries of a vector are called its **components**. For a given matrix A, it is convenient to refer to its *row vectors* and its *column vectors*. For example, let

$$A = \begin{bmatrix} 1 & 2 & -1 \\ 3 & 0 & 1 \\ 4 & -1 & 2 \end{bmatrix}$$

Then the column vectors of A are

$$\begin{bmatrix} 1 \\ 3 \\ 4 \end{bmatrix} \qquad \begin{bmatrix} 2 \\ 0 \\ -1 \end{bmatrix} \quad \text{and} \quad \begin{bmatrix} -1 \\ 1 \\ 2 \end{bmatrix}$$

while the row vectors of A, written vertically, are

$$\begin{bmatrix} 1 \\ 2 \\ -1 \end{bmatrix} \qquad \begin{bmatrix} 3 \\ 0 \\ 1 \end{bmatrix} \quad \text{and} \quad \begin{bmatrix} 4 \\ -1 \\ 2 \end{bmatrix}$$

Two $m \times n$ matrices A and B are **equal** if they have the same number of rows and columns and their corresponding entries are equal. Thus, $A = B$ if and only if $a_{ij} = b_{ij}$, for $1 \le i \le m$ and $1 \le j \le n$. Addition and scalar multiplication of matrices are also defined componentwise.

DEFINITION 1

Addition and Scalar Multiplication   If A and B are two $m \times n$ matrices, then the **sum** of the matrices $A + B$ is the $m \times n$ matrix with the ij term given by $a_{ij} + b_{ij}$. The **scalar product** of the matrix A with the real number c, denoted by cA, is the $m \times n$ matrix with the ij term given by ca_{ij}.

EXAMPLE 1

Perform the operations on the matrices

$$A = \begin{bmatrix} 2 & 0 & 1 \\ 4 & 3 & -1 \\ -3 & 6 & 5 \end{bmatrix} \quad \text{and} \quad B = \begin{bmatrix} -2 & 3 & -1 \\ 3 & 5 & 6 \\ 4 & 2 & 1 \end{bmatrix}$$

a. $A + B$ **b.** $2A - 3B$

Solution   **a.** We add the two matrices by adding their corresponding entries, so that

$$A + B = \begin{bmatrix} 2 & 0 & 1 \\ 4 & 3 & -1 \\ -3 & 6 & 5 \end{bmatrix} + \begin{bmatrix} -2 & 3 & -1 \\ 3 & 5 & 6 \\ 4 & 2 & 1 \end{bmatrix}$$

$$= \begin{bmatrix} 2+(-2) & 0+3 & 1+(-1) \\ 4+3 & 3+5 & -1+6 \\ -3+4 & 6+2 & 5+1 \end{bmatrix}$$

$$= \begin{bmatrix} 0 & 3 & 0 \\ 7 & 8 & 5 \\ 1 & 8 & 6 \end{bmatrix}$$

b. To evaluate this expression, we first multiply each entry of the matrix A by 2 and each entry of the matrix B by -3. Then we add the resulting matrices. This gives

$$2A + (-3B) = 2 \begin{bmatrix} 2 & 0 & 1 \\ 4 & 3 & -1 \\ -3 & 6 & 5 \end{bmatrix} + (-3) \begin{bmatrix} -2 & 3 & -1 \\ 3 & 5 & 6 \\ 4 & 2 & 1 \end{bmatrix}$$

$$= \begin{bmatrix} 4 & 0 & 2 \\ 8 & 6 & -2 \\ -6 & 12 & 10 \end{bmatrix} + \begin{bmatrix} 6 & -9 & 3 \\ -9 & -15 & -18 \\ -12 & -6 & -3 \end{bmatrix}$$

$$= \begin{bmatrix} 10 & -9 & 5 \\ -1 & -9 & -20 \\ -18 & 6 & 7 \end{bmatrix}$$

In Example 1(a) reversing the order of the addition of the matrices gives the same result. That is, $A + B = B + A$. This is so because addition of real numbers is commutative. This result holds in general, giving us that matrix addition is also a commutative operation. Some other familiar properties that hold for real numbers also hold for matrices and scalars. These properties are given in Theorem 4.

THEOREM 4   **Properties of Matrix Addition and Scalar Multiplication**   Let A, B, and C be $m \times n$ matrices and c and d be real numbers.

1. $A + B = B + A$
2. $A + (B + C) = (A + B) + C$
3. $c(A + B) = cA + cB$
4. $(c + d)A = cA + dA$
5. $c(dA) = (cd)A$
6. The $m \times n$ matrix with all zero entries, denoted by $\mathbf{0}$, is such that $A + \mathbf{0} = \mathbf{0} + A = A$.
7. For any matrix A, the matrix $-A$, whose components are the negative of each component of A, is such that $A + (-A) = (-A) + A = \mathbf{0}$.

Proof In each case it is sufficient to show that the column vectors of the two matrices agree. We will prove property 2 and leave the others as exercises.

(2) Since the matrices A, B, and C have the same size, the sums $(A + B) + C$ and $A + (B + C)$ are defined and also have the same size. Let $\mathbf{A}_i$, $\mathbf{B}_i$, and $\mathbf{C}_i$ denote the ith column vector of A, B, and C, respectively. Then

$$(\mathbf{A}_i + \mathbf{B}_i) + \mathbf{C}_i = \left(\begin{bmatrix} a_{1i} \\ \vdots \\ a_{mi} \end{bmatrix} + \begin{bmatrix} b_{1i} \\ \vdots \\ b_{mi} \end{bmatrix} \right) + \begin{bmatrix} c_{1i} \\ \vdots \\ c_{mi} \end{bmatrix}$$

$$= \begin{bmatrix} a_{1i} + b_{1i} \\ \vdots \\ a_{mi} + b_{mi} \end{bmatrix} + \begin{bmatrix} c_{1i} \\ \vdots \\ c_{mi} \end{bmatrix} = \begin{bmatrix} (a_{1i} + b_{1i}) + c_{1i} \\ \vdots \\ (a_{mi} + b_{mi}) + c_{mi} \end{bmatrix}$$

Since the components are real numbers, where the associative property of addition holds, we have

$$(\mathbf{A}_i + \mathbf{B}_i) + \mathbf{C}_i = \begin{bmatrix} (a_{1i} + b_{1i}) + c_{1i} \\ \vdots \\ (a_{mi} + b_{mi}) + c_{mi} \end{bmatrix}$$

$$= \begin{bmatrix} a_{1i} + (b_{1i} + c_{1i}) \\ \vdots \\ a_{mi} + (b_{mi} + c_{mi}) \end{bmatrix} = \mathbf{A}_i + (\mathbf{B}_i + \mathbf{C}_i)$$

As this holds for every column vector, the matrices $(A + B) + C$ and $A + (B + C)$ are equal, and we have $(A + B) + C = A + (B + C)$.

Matrix Multiplication

We have defined matrix addition and a scalar multiplication, and we observed that these operations satisfy many of the analogous properties for real numbers. We have not yet considered the *product* of two matrices. Matrix multiplication is more difficult to define and is developed from the *dot product* of two vectors.

DEFINITION 2   **Dot Product of Vectors**   Given two vectors

$$\mathbf{u} = \begin{bmatrix} u_1 \\ u_2 \\ \vdots \\ u_n \end{bmatrix} \qquad \text{and} \qquad \mathbf{v} = \begin{bmatrix} v_1 \\ v_2 \\ \vdots \\ v_n \end{bmatrix}$$

the **dot product** is defined by

$$\mathbf{u} \cdot \mathbf{v} = u_1 v_1 + u_2 v_2 + \cdots + u_n v_n = \sum_{i=1}^{n} u_i v_i$$

Observe that the dot product of two vectors is a scalar. For example,

$$\begin{bmatrix} 2 \\ -3 \\ -1 \end{bmatrix} \cdot \begin{bmatrix} -5 \\ 1 \\ 4 \end{bmatrix} = (2)(-5) + (-3)(1) + (-1)(4) = -17$$

Now to motivate the concept and need for matrix multiplication we first introduce the operation of multiplying a vector by a matrix. As an illustration let

$$B = \begin{bmatrix} 1 & -1 \\ -2 & 1 \end{bmatrix} \qquad \text{and} \qquad \mathbf{v} = \begin{bmatrix} 1 \\ 3 \end{bmatrix}$$

The **product** of B and $\mathbf{v}$, denoted by $B\mathbf{v}$, is a vector, in this case with two components. The first component of $B\mathbf{v}$ is the dot product of the first row vector of B with $\mathbf{v}$, while the second component is the dot product of the second row vector of B with $\mathbf{v}$, so that

$$B\mathbf{v} = \begin{bmatrix} 1 & -1 \\ -2 & 1 \end{bmatrix} \begin{bmatrix} 1 \\ 3 \end{bmatrix} = \begin{bmatrix} (1)(1) + (-1)(3) \\ (-2)(1) + (1)(3) \end{bmatrix} = \begin{bmatrix} -2 \\ 1 \end{bmatrix}$$

Using this operation, the matrix B *transforms* the vector $\mathbf{v} = \begin{bmatrix} 1 \\ 3 \end{bmatrix}$ to the vector $B\mathbf{v} = \begin{bmatrix} -2 \\ 1 \end{bmatrix}$. If $A = \begin{bmatrix} -1 & 2 \\ 0 & 1 \end{bmatrix}$ is another matrix, then the product of A and $B\mathbf{v}$ is given by

$$A(B\mathbf{v}) = \begin{bmatrix} -1 & 2 \\ 0 & 1 \end{bmatrix} \begin{bmatrix} -2 \\ 1 \end{bmatrix} = \begin{bmatrix} 4 \\ 1 \end{bmatrix}$$

The question then arises, is there a single matrix which can be used to transform the original vector $\begin{bmatrix} 1 \\ 3 \end{bmatrix}$ to $\begin{bmatrix} 4 \\ 1 \end{bmatrix}$? To answer this question, let

$$\mathbf{v} = \begin{bmatrix} x \\ y \end{bmatrix} \qquad A = \begin{bmatrix} a_{11} & a_{12} \\ a_{21} & a_{22} \end{bmatrix} \qquad \text{and} \qquad B = \begin{bmatrix} b_{11} & b_{12} \\ b_{21} & b_{22} \end{bmatrix}$$

The product of B and $\mathbf{v}$ is

$$B\mathbf{v} = \begin{bmatrix} b_{11}x + b_{12}y \\ b_{21}x + b_{22}y \end{bmatrix}$$

Now, the product of A and $B\mathbf{v}$ is

$$A(B\mathbf{v}) = \begin{bmatrix} a_{11} & a_{12} \\ a_{21} & a_{22} \end{bmatrix} \begin{bmatrix} b_{11}x + b_{12}y \\ b_{21}x + b_{22}y \end{bmatrix}$$

$$= \begin{bmatrix} a_{11}(b_{11}x + b_{12}y) + a_{12}(b_{21}x + b_{22}y) \\ a_{21}(b_{11}x + b_{12}y) + a_{22}(b_{21}x + b_{22}y) \end{bmatrix}$$

$$= \begin{bmatrix} (a_{11}b_{11} + a_{12}b_{21})x + (a_{11}b_{12} + a_{12}b_{22})y \\ (a_{21}b_{11} + a_{22}b_{21})x + (a_{21}b_{12} + a_{22}b_{22})y \end{bmatrix}$$

$$= \begin{bmatrix} a_{11}b_{11} + a_{12}b_{21} & a_{11}b_{12} + a_{12}b_{22} \\ a_{21}b_{11} + a_{22}b_{21} & a_{21}b_{12} + a_{22}b_{22} \end{bmatrix} \begin{bmatrix} x \\ y \end{bmatrix}$$

Thus, we see that $A(B\mathbf{v})$ is the product of the matrix

$$\begin{bmatrix} a_{11}b_{11} + a_{12}b_{21} & a_{11}b_{12} + a_{12}b_{22} \\ a_{21}b_{11} + a_{22}b_{21} & a_{21}b_{12} + a_{22}b_{22} \end{bmatrix}$$

and the vector $\begin{bmatrix} x \\ y \end{bmatrix}$. We refer to this matrix as the **product** of A and B, denoted by AB, so that

$$A(B\mathbf{v}) = (AB)\mathbf{v}$$

See Fig. 2.

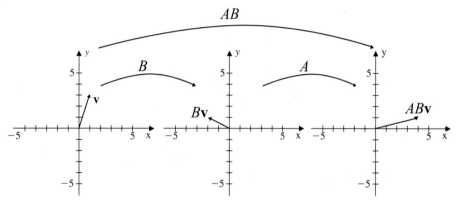

Figure 2

Notice that the product matrix AB is obtained by computing the dot product of each row vector of the matrix A, on the left, with each column vector of the matrix B, on the right. Using the matrices A and B given above, we have

$$AB = \begin{bmatrix} -1 & 2 \\ 0 & 1 \end{bmatrix} \begin{bmatrix} 1 & -1 \\ -2 & 1 \end{bmatrix}$$
$$= \begin{bmatrix} (-1)(1) + (2)(-2) & (-1)(-1) + (2)(1) \\ (0)(1) + (1)(-2) & (0)(-1) + (1)(1) \end{bmatrix} = \begin{bmatrix} -5 & 3 \\ -2 & 1 \end{bmatrix}$$

This matrix transforms the vector $\begin{bmatrix} 1 \\ 3 \end{bmatrix}$ to $\begin{bmatrix} 4 \\ 1 \end{bmatrix}$ in one step. That is,

$$(AB)\mathbf{v} = \begin{bmatrix} -5 & 3 \\ -2 & 1 \end{bmatrix} \begin{bmatrix} 1 \\ 3 \end{bmatrix} = \begin{bmatrix} 4 \\ 1 \end{bmatrix}$$

which was our original aim. The notion of matrices as transformations is taken up again in Chap. 4 where we consider more general transformations of vectors.

For another illustration of the matrix product let

$$A = \begin{bmatrix} 1 & 3 & 0 \\ 2 & 1 & -3 \\ -4 & 6 & 2 \end{bmatrix} \quad \text{and} \quad B = \begin{bmatrix} 3 & -2 & 5 \\ -1 & 4 & -2 \\ 1 & 0 & 3 \end{bmatrix}$$

The entries across the first row of the product matrix AB are obtained from the dot product of the first row vector of A with the first, second, and third column vectors of

B, respectively. The terms in the second row of AB are the dot products of the second row vector of A with the first, second, and third column vectors of B, respectively. Finally, the terms in the third row of AB are the dot products of the third row vector of A again with the first, second, and third column vectors of B, respectively. Thus, the product matrix AB is

$$AB = \begin{bmatrix} (1)(3)+(3)(-1)+(0)(1) & -2+12+0 & 5-6+0 \\ 6-1-3 & -4+4+0 & 10-2-9 \\ -12-6+2 & 8+24+0 & -20-12+6 \end{bmatrix}$$

$$= \begin{bmatrix} 0 & 10 & -1 \\ 2 & 0 & -1 \\ -16 & 32 & -26 \end{bmatrix}$$

In the previous example, the product AB exists since the matrices A and B have the same number of rows and the same number of columns. This condition can be relaxed somewhat. In general, the product of two matrices A and B exists if the number of columns of A is equal to the number of rows of B.

DEFINITION 3   **Matrix Multiplication**   Let A be an $m \times n$ matrix and B an $n \times p$ matrix; then the product AB is an $m \times p$ matrix. The ij term of AB is the dot product of the ith row vector of A with the jth column vector of B, so that

$$(AB)_{ij} = a_{i1}b_{1j} + a_{i2}b_{2j} + \cdots + a_{in}b_{nj} = \sum_{k=1}^{n} a_{ik}b_{kj}$$

It is important to recognize that not all properties of real numbers carry over to properties of matrices. Because matrix multiplication is only defined when the number of columns of the matrix on the left equals the number of rows of the matrix on the right, it is possible for AB to exist with BA being undefined. For example,

$$AB = \begin{bmatrix} 1 & 3 & 0 \\ 2 & 1 & -3 \end{bmatrix} \begin{bmatrix} 3 & -2 & 5 \\ -1 & 4 & -2 \\ 1 & 0 & 3 \end{bmatrix}$$

is defined, but

$$BA = \begin{bmatrix} 3 & -2 & 5 \\ -1 & 4 & -2 \\ 1 & 0 & 3 \end{bmatrix} \begin{bmatrix} 1 & 3 & 0 \\ 2 & 1 & -3 \end{bmatrix}$$

is not. As a result, we cannot interchange the order when multiplying two matrices unless we know beforehand that the matrices *commute*. We say two matrices A and B **commute** when $AB = BA$.

Example 2 illustrates that even when AB and BA are both defined, they might not be equal.

EXAMPLE 2   Verify that the matrices

$$A = \begin{bmatrix} 1 & 0 \\ -1 & 2 \end{bmatrix} \quad \text{and} \quad B = \begin{bmatrix} 0 & 1 \\ 1 & 1 \end{bmatrix}$$

do not satisfy the commutative property for multiplication.

Solution   The products are

$$AB = \begin{bmatrix} 1 & 0 \\ -1 & 2 \end{bmatrix} \begin{bmatrix} 0 & 1 \\ 1 & 1 \end{bmatrix} = \begin{bmatrix} 0 & 1 \\ 2 & 1 \end{bmatrix}$$

and

$$BA = \begin{bmatrix} 0 & 1 \\ 1 & 1 \end{bmatrix} \begin{bmatrix} 1 & 0 \\ -1 & 2 \end{bmatrix} = \begin{bmatrix} -1 & 2 \\ 0 & 2 \end{bmatrix}$$

so that $AB \neq BA$.

In Example 3 we describe all matrices that commute with a particular matrix.

EXAMPLE 3   Find all 2×2 matrices that commute with the matrix

$$A = \begin{bmatrix} 1 & 0 \\ 1 & 1 \end{bmatrix}$$

Solution   We start by letting B denote an arbitrary 2×2 matrix

$$B = \begin{bmatrix} a & b \\ c & d \end{bmatrix}$$

Then the product of matrix A on the left with matrix B on the right is given by

$$AB = \begin{bmatrix} 1 & 0 \\ 1 & 1 \end{bmatrix} \begin{bmatrix} a & b \\ c & d \end{bmatrix} = \begin{bmatrix} a & b \\ a+c & b+d \end{bmatrix}$$

On the other hand,

$$BA = \begin{bmatrix} a & b \\ c & d \end{bmatrix} \begin{bmatrix} 1 & 0 \\ 1 & 1 \end{bmatrix} = \begin{bmatrix} a+b & b \\ c+d & d \end{bmatrix}$$

Setting $AB = BA$, we obtain

$$a = a+b \qquad a+c = c+d \qquad \text{and} \qquad b+d = d$$

so that $b = 0$ and $a = d$. Let S be the set of all 2×2 matrices defined by

$$S = \left\{ \begin{bmatrix} a & 0 \\ c & a \end{bmatrix} \,\middle|\, a, c \in \mathbb{R} \right\}$$

Then each matrix in S commutes with the matrix A.

EXAMPLE 4 Perform the operations on the matrices

$$A = \begin{bmatrix} -3 & 1 \\ 2 & 2 \\ -1 & 5 \end{bmatrix} \qquad B = \begin{bmatrix} -1 & 1 & -1 & 3 \\ 2 & 5 & -3 & 1 \end{bmatrix} \qquad \text{and} \qquad C = \begin{bmatrix} 3 & 2 & -2 & 1 \\ 1 & 6 & -2 & 4 \end{bmatrix}$$

   a. $A(B + C)$ **b.** $AB + AC$

Solution We first notice that the matrix A is 3×2 and both B and C are 2×4, so AB and AC are defined. Also since the matrices B and C have the same number of rows and columns, the matrix $B + C$ is defined, so the expressions in parts (a) and (b) are defined.

  a. We first add the matrices B and C inside the parentheses and then multiply on the left by the matrix A. This gives us

$A(B + C)$

$$= \begin{bmatrix} -3 & 1 \\ 2 & 2 \\ -1 & 5 \end{bmatrix} \left(\begin{bmatrix} -1 & 1 & -1 & 3 \\ 2 & 5 & -3 & 1 \end{bmatrix} + \begin{bmatrix} 3 & 2 & -2 & 1 \\ 1 & 6 & -2 & 4 \end{bmatrix} \right)$$

$$= \begin{bmatrix} -3 & 1 \\ 2 & 2 \\ -1 & 5 \end{bmatrix} \begin{bmatrix} 2 & 3 & -3 & 4 \\ 3 & 11 & -5 & 5 \end{bmatrix}$$

$$= \begin{bmatrix} -3(2)+1(3) & -3(3)+1(11) & -3(-3)+1(-5) & -3(4)+1(5) \\ 2(2)+2(3) & 2(3)+2(11) & 2(-3)+2(-5) & 2(4)+2(5) \\ -1(2)+5(3) & -1(3)+5(11) & -1(-3)+5(-5) & -1(4)+5(5) \end{bmatrix}$$

$$= \begin{bmatrix} -3 & 2 & 4 & -7 \\ 10 & 28 & -16 & 18 \\ 13 & 52 & -22 & 21 \end{bmatrix}$$

  b. In this case we compute AB and AC separately and then add the two resulting matrices. We have

$$AB + AC = \begin{bmatrix} -3 & 1 \\ 2 & 2 \\ -1 & 5 \end{bmatrix} \begin{bmatrix} -1 & 1 & -1 & 3 \\ 2 & 5 & -3 & 1 \end{bmatrix}$$

$$+ \begin{bmatrix} -3 & 1 \\ 2 & 2 \\ -1 & 5 \end{bmatrix} \begin{bmatrix} 3 & 2 & -2 & 1 \\ 1 & 6 & -2 & 4 \end{bmatrix}$$

$$= \begin{bmatrix} 5 & 2 & 0 & -8 \\ 2 & 12 & -8 & 8 \\ 11 & 24 & -14 & 2 \end{bmatrix} + \begin{bmatrix} -8 & 0 & 4 & 1 \\ 8 & 16 & -8 & 10 \\ 2 & 28 & -8 & 19 \end{bmatrix}$$

$$= \begin{bmatrix} -3 & 2 & 4 & -7 \\ 10 & 28 & -16 & 18 \\ 13 & 52 & -22 & 21 \end{bmatrix}$$

Notice that in Example 4 the matrix equation

$$A(B + C) = AB + AC$$

holds. This and other familiar properties involving multiplication and addition of real numbers hold for matrices. They are listed in Theorem 5.

THEOREM 5   **Properties of Matrix Multiplication**   Let A, B, and C be matrices with sizes so that the given expressions are all defined, and let c be a real number.

1. $A(BC) = (AB)C$
2. $c(AB) = (cA)B = A(cB)$
3. $A(B + C) = AB + AC$
4. $(B + C)A = BA + CA$

We have already seen that unlike with real numbers, matrix multiplication does not commute. There are other properties of the real numbers that do not hold for matrices. Recall that if x and y are real numbers such that $xy = 0$, then either $x = 0$ or $y = 0$. This property does not hold for matrices. For example, let

$$A = \begin{bmatrix} 1 & 1 \\ 1 & 1 \end{bmatrix} \quad \text{and} \quad B = \begin{bmatrix} -1 & -1 \\ 1 & 1 \end{bmatrix}$$

Then

$$AB = \begin{bmatrix} 1 & 1 \\ 1 & 1 \end{bmatrix} \begin{bmatrix} -1 & -1 \\ 1 & 1 \end{bmatrix} = \begin{bmatrix} 0 & 0 \\ 0 & 0 \end{bmatrix}$$

Transpose of a Matrix

The *transpose* of a matrix is obtained by interchanging the rows and columns of a matrix.

DEFINITION 4   **Transpose**   If A is an $m \times n$ matrix, the **transpose** of A, denoted by A^t, is the $n \times m$ matrix with ij term

$$(A^t)_{ij} = a_{ji}$$

where $1 \le i \le n$ and $1 \le j \le m$.
For example, the transpose of the matrix

$$A = \begin{bmatrix} 1 & 2 & -3 \\ 0 & 1 & 4 \\ -1 & 2 & 1 \end{bmatrix} \quad \text{is} \quad A^t = \begin{bmatrix} 1 & 0 & -1 \\ 2 & 1 & 2 \\ -3 & 4 & 1 \end{bmatrix}$$

Notice that the row vectors of A become the column vectors of A^t. Theorem 6 gives some properties of the transpose.

THEOREM 6   Suppose A and B are $m \times n$ matrices, C is an $n \times p$ matrix, and c is a scalar.

1. $(A + B)^t = A^t + B^t$
2. $(AC)^t = C^t A^t$
3. $(A^t)^t = A$
4. $(cA)^t = cA^t$

Proof   (2) We start by showing that the products involved are all defined. Since AC is $m \times p$, then $(AC)^t$ is $p \times m$. As C^t is $p \times n$ and A^t is $n \times m$, then $C^t A^t$ is also $p \times m$. So the sizes of the products agree. Now to show that the products are equal, we apply the definitions of multiplication and transpose to obtain

$$(C^t A^t)_{ij} = \sum_{k=1}^{n} (C^t)_{ik}(A^t)_{kj}$$

$$= \sum_{k=1}^{n} c_{ki} a_{jk} = \sum_{k=1}^{n} a_{jk} c_{ki} = (AC)_{ji}$$

$$= ((AC)^t)_{ij}$$

The proofs of parts 1, 3, and 4 are left as exercises.

DEFINITION 5   **Symmetric Matrix**   An $n \times n$ matrix is **symmetric** provided that $A^t = A$.

EXAMPLE 5   Find all 2×2 matrices that are symmetric.

Solution   Let

$$A = \begin{bmatrix} a & b \\ c & d \end{bmatrix}$$

Then A is symmetric if and only if

$$A = \begin{bmatrix} a & b \\ c & d \end{bmatrix} = \begin{bmatrix} a & c \\ b & d \end{bmatrix} = A^t$$

which holds if and only if $b = c$. So a 2×2 matrix is symmetric if and only if the matrix has the form

$$\begin{bmatrix} a & b \\ b & d \end{bmatrix}$$

Fact Summary

Let A, B, and C be matrices.

1. The definitions of matrix addition and scalar multiplication satisfy many of the properties enjoyed by real numbers. This allows algebra to be carried out with matrices.

> **2.** When AB is defined, the ij entry of the product matrix is the dot product of the ith row vector of A with the jth column vector of B.
>
> **3.** Matrix multiplication does not in general commute. Even when AB and BA are both defined, it is possible for $AB \neq BA$.
>
> **4.** The distributive properties hold. That is, $A(B + C) = AB + AC$ and $(B + C)A = BA + CA$.
>
> **5.** $(A + B)^t = A^t + B^t$, $(AB)^t = B^t A^t$, $(A^t)^t = A$, $(cA)^t = cA^t$
>
> **6.** The matrix $A = \begin{bmatrix} a & b \\ c & d \end{bmatrix}$ is symmetric if and only if $b = c$.

Exercise Set 1.3

In Exercises 1–4, use the matrices

$$A = \begin{bmatrix} 2 & -3 \\ 4 & 1 \end{bmatrix} \qquad B = \begin{bmatrix} -1 & 3 \\ -2 & 5 \end{bmatrix}$$

$$C = \begin{bmatrix} 1 & 1 \\ 5 & -2 \end{bmatrix}$$

1. Find $A + B$ and $B + A$.

2. Find $3A - 2B$.

3. Find $(A + B) + C$ and $A + (B + C)$.

4. Find $3(A + B) - 5C$.

In Exercises 5 and 6, use the matrices

$$A = \begin{bmatrix} -3 & -3 & 3 \\ 1 & 0 & 2 \\ 0 & -2 & 3 \end{bmatrix}$$

$$B = \begin{bmatrix} -1 & 3 & 3 \\ -2 & 5 & 2 \\ 1 & 2 & 4 \end{bmatrix}$$

$$C = \begin{bmatrix} -5 & 3 & 9 \\ -3 & 10 & 6 \\ 2 & 2 & 11 \end{bmatrix}$$

5. Find $(A - B) + C$ and $2A + B$.

6. Show that $A + 2B - C = 0$.

In Exercises 7 and 8, use the matrices

$$A = \begin{bmatrix} 3 & 1 \\ -2 & 4 \end{bmatrix} \qquad B = \begin{bmatrix} 2 & 0 \\ 1 & -2 \end{bmatrix}$$

7. Find AB and BA.

8. Show that $3(AB) = A(3B)$.

In Exercises 9 and 10, use the matrices

$$A = \begin{bmatrix} 2 & -3 & -3 \\ -3 & -2 & 0 \end{bmatrix}$$

$$B = \begin{bmatrix} 3 & -1 \\ 2 & -2 \\ 3 & 0 \end{bmatrix}$$

9. Find AB.

10. Find BA.

11. Let

$$A = \begin{bmatrix} -1 & 1 & 1 \\ 3 & -3 & 3 \\ -1 & 2 & 1 \end{bmatrix}$$

and

$$B = \begin{bmatrix} -2 & 3 & -3 \\ 0 & -1 & 2 \\ 3 & -2 & -1 \end{bmatrix}$$

Find AB.

12. Let

$$A = \begin{bmatrix} -2 & -2 & -1 \\ -3 & 2 & 1 \\ 1 & -1 & -1 \end{bmatrix}$$

and

$$B = \begin{bmatrix} 1 & -1 & -2 \\ -2 & -2 & 3 \\ -3 & 1 & -3 \end{bmatrix}$$

Find AB.

In Exercises 13–16, use the matrices

$$A = \begin{bmatrix} -2 & -3 \\ 3 & 0 \end{bmatrix} \qquad B = \begin{bmatrix} 2 & 0 \\ -2 & 0 \end{bmatrix}$$

$$C = \begin{bmatrix} 2 & 0 \\ -1 & -1 \end{bmatrix}$$

13. Find $A(B + C)$.

14. Find $(A + B)C$.

15. Find $2A(B - 3C)$.

16. Find $(A + 2B)(3C)$.

In Exercises 17–24, use the matrices

$$A = \begin{bmatrix} 2 & 0 & -1 \\ 1 & 0 & -2 \end{bmatrix}$$

$$B = \begin{bmatrix} -3 & 1 & 1 \\ -3 & -3 & -2 \end{bmatrix}$$

$$C = \begin{bmatrix} 3 & -1 \\ -1 & -3 \end{bmatrix}$$

Whenever possible, perform the operations. If a computation cannot be made, explain why.

17. $2A^t - B^t$

18. $B^t - 2A$

19. AB^t

20. BA^t

21. $(A^t + B^t)C$

22. $C(A^t + B^t)$

23. $(A^tC)B$

24. $(A^tB^t)C$

25. Let

$$A = \begin{bmatrix} -1 & -2 \\ 1 & 2 \end{bmatrix} \qquad B = \begin{bmatrix} 1 & 3 \\ 2 & -1 \end{bmatrix}$$

$$C = \begin{bmatrix} 7 & 5 \\ -1 & -2 \end{bmatrix}$$

Show that $AB = AC$ and yet $B \neq C$.

26. Let

$$A = \begin{bmatrix} 0 & 2 \\ 0 & 5 \end{bmatrix}$$

Find a 2×2 matrix B that is not the zero matrix, such that AB is the zero matrix.

27. Find all 2×2 matrices of the form

$$A = \begin{bmatrix} a & b \\ 0 & c \end{bmatrix}$$

such that

$$A^2 = AA = \begin{bmatrix} 1 & 0 \\ 0 & 1 \end{bmatrix}$$

28. Let $A = \begin{bmatrix} 2 & 1 \\ 1 & 1 \end{bmatrix}$. Find all matrices of the form

$$M = \begin{bmatrix} a & b \\ c & d \end{bmatrix} \text{ such that } AM = MA.$$

29. Find matrices A and B such that $AB = 0$ but $BA \neq 0$.

30. Show there are no 2×2 matrices A and B such that

$$AB - BA = \begin{bmatrix} 1 & 0 \\ 0 & 1 \end{bmatrix}$$

31. Determine all values of a and b such that

$$\begin{bmatrix} 1 & 2 \\ a & 0 \end{bmatrix}\begin{bmatrix} 3 & b \\ -4 & 1 \end{bmatrix} = \begin{bmatrix} -5 & 6 \\ 12 & 16 \end{bmatrix}$$

32. If A and B are 2×2 matrices, show that the sum of the terms on the diagonal of $AB - BA$ is 0.

33. Let

$$A = \begin{bmatrix} 1 & 0 & 0 \\ 0 & -1 & 0 \\ 0 & 0 & 1 \end{bmatrix}$$

Find the matrix A^{20}.

34. If A and B are $n \times n$ matrices, when does $(A + B)(A - B) = A^2 - B^2$?

35. If the matrices A and B commute, show that $A^2B = BA^2$.

36. Suppose A, B, and C are $n \times n$ matrices and B and C both commute with A.

   a. Show that BC and A commute.

   b. Give specific matrices to show that BC and CB do not have to be equal.

37. Suppose that A is an $n \times n$ matrix. Show that if for each vector $\mathbf{x}$ in $\mathbb{R}^n$, $A\mathbf{x} = \mathbf{0}$, then A is the zero matrix.

38. For each positive integer n, let

$$A_n = \begin{bmatrix} 1 - n & -n \\ n & 1 + n \end{bmatrix}$$

Show that $A_n A_m = A_{n+m}$.

39. Find all 2×2 matrices that satisfy $AA^t = \mathbf{0}$.

40. Suppose that A and B are symmetric matrices. Show that if $AB = BA$, then AB is symmetric.

41. If A is an $m \times n$ matrix, show that AA^t and $A^t A$ are both defined and are both symmetric.

42. An $n \times n$ matrix A is called *idempotent* provided that $A^2 = AA = A$. Suppose that A and B are $n \times n$ idempotent matrices. Show that if $AB = BA$, then the matrix AB is idempotent.

43. An $n \times n$ matrix A is *skew-symmetric* provided $A^t = -A$. Show that if a matrix is skew-symmetric, then the diagonal entries are 0.

44. The trace of an $n \times n$ matrix A is the sum of the diagonal terms, denoted $\mathbf{tr}(A)$.

   a. If A and B are $n \times n$ matrices, show that $\mathbf{tr}(A + B) = \mathbf{tr}(A) + \mathbf{tr}(B)$.

   b. If A is an $n \times n$ matrix and c is a scalar, show that $\mathbf{tr}(cA) = c\,\mathbf{tr}(A)$.

1.4 ▶ The Inverse of a Square Matrix

In the real number system, the number 1 is the multiplicative identity. That is, for any real number a,

$$a \cdot 1 = 1 \cdot a = a$$

We also know that for every number x with $x \neq 0$, there exists the number $\frac{1}{x}$, also written x^{-1}, such that

$$x \cdot \frac{1}{x} = 1$$

We seek a similar relationship for square matrices. For an $n \times n$ matrix A, we can check that the $n \times n$ matrix

$$I = \begin{bmatrix} 1 & 0 & 0 & \cdots & 0 \\ 0 & 1 & 0 & \cdots & 0 \\ 0 & 0 & 1 & \cdots & 0 \\ \vdots & \vdots & \vdots & \ddots & \vdots \\ 0 & 0 & 0 & \cdots & 1 \end{bmatrix}$$

is the **multiplicative identity**. That is, if A is any $n \times n$ matrix, then

$$AI = IA = A$$

This special matrix is called the **identity matrix**. For example, the 2×2, 3×3, and 4×4 identity matrices are, respectively,

$$\begin{bmatrix} 1 & 0 \\ 0 & 1 \end{bmatrix} \qquad \begin{bmatrix} 1 & 0 & 0 \\ 0 & 1 & 0 \\ 0 & 0 & 1 \end{bmatrix} \qquad \text{and} \qquad \begin{bmatrix} 1 & 0 & 0 & 0 \\ 0 & 1 & 0 & 0 \\ 0 & 0 & 1 & 0 \\ 0 & 0 & 0 & 1 \end{bmatrix}$$

DEFINITION 1   **Inverse of a Square Matrix**   Let A be an $n \times n$ matrix. If there exists an $n \times n$ matrix B such that
$$AB = I = BA$$
then the matrix B is a (multiplicative) **inverse** of the matrix A.

EXAMPLE 1   Find an inverse of the matrix
$$A = \begin{bmatrix} 1 & 1 \\ 1 & 2 \end{bmatrix}$$

Solution   In order for a 2×2 matrix $B = \begin{bmatrix} x_1 & x_2 \\ x_3 & x_4 \end{bmatrix}$ to be an inverse of A, matrix B must satisfy
$$\begin{bmatrix} 1 & 1 \\ 1 & 2 \end{bmatrix}\begin{bmatrix} x_1 & x_2 \\ x_3 & x_4 \end{bmatrix} = \begin{bmatrix} x_1 + x_3 & x_2 + x_4 \\ x_1 + 2x_3 & x_2 + 2x_4 \end{bmatrix} = \begin{bmatrix} 1 & 0 \\ 0 & 1 \end{bmatrix}$$

This matrix equation is equivalent to the linear system
$$\begin{cases} x_1 & + x_3 & & = 1 \\ & x_2 & + x_4 & = 0 \\ x_1 & + 2x_3 & & = 0 \\ & x_2 & + 2x_4 & = 1 \end{cases}$$

The augmented matrix and the reduced row echelon form are given by
$$\begin{bmatrix} 1 & 0 & 1 & 0 & | & 1 \\ 0 & 1 & 0 & 1 & | & 0 \\ 1 & 0 & 2 & 0 & | & 0 \\ 0 & 1 & 0 & 2 & | & 1 \end{bmatrix} \rightarrow \begin{bmatrix} 1 & 0 & 0 & 0 & | & 2 \\ 0 & 1 & 0 & 0 & | & -1 \\ 0 & 0 & 1 & 0 & | & -1 \\ 0 & 0 & 0 & 1 & | & 1 \end{bmatrix}$$

Thus, the solution is $x_1 = 2, x_2 = -1, x_3 = -1, x_4 = 1$, and an inverse matrix is
$$B = \begin{bmatrix} 2 & -1 \\ -1 & 1 \end{bmatrix}$$

The reader should verify that $AB = BA = I$.

Theorem 7 establishes the uniqueness, when it exists, of the multiplicative inverse.

THEOREM 7   The inverse of a matrix, if it exists, is unique.

Proof   Assume that the square matrix A has an inverse and that B and C are both inverse matrices of A. That is, $AB = BA = I$ and $AC = CA = I$. We show that $B = C$. Indeed,
$$B = BI = B(AC) = (BA)C = (I)C = C$$

We refer to the unique inverse as *the inverse of* A and denote it by A^{-1}. When the inverse of a matrix A exists, we call A **invertible**. Otherwise, the matrix A is called **noninvertible**.

THEOREM 8

The inverse of the matrix $A = \begin{bmatrix} a & b \\ c & d \end{bmatrix}$ exists if and only if $ad - bc \neq 0$. In this case the inverse is the matrix

$$A^{-1} = \frac{1}{ad - bc} \begin{bmatrix} d & -b \\ -c & a \end{bmatrix}$$

Proof First, assume that $ad - bc \neq 0$, and let

$$B = \begin{bmatrix} x_1 & x_2 \\ x_3 & x_4 \end{bmatrix}$$

We need to find x_1, x_2, x_3, and x_4 such that

$$AB = \begin{bmatrix} 1 & 0 \\ 0 & 1 \end{bmatrix}$$

Taking the product of the two matrices yields

$$\begin{bmatrix} ax_1 + bx_3 & ax_2 + bx_4 \\ cx_1 + dx_3 & cx_2 + dx_4 \end{bmatrix} = \begin{bmatrix} 1 & 0 \\ 0 & 1 \end{bmatrix}$$

which results in the linear system

$$\begin{cases} ax_1 & + bx_3 & & = 1 \\ & ax_2 & & + bx_4 = 0 \\ cx_1 & + dx_3 & & = 0 \\ & cx_2 & & + dx_4 = 1 \end{cases}$$

The augmented matrix of this linear system is given by

$$\begin{bmatrix} a & 0 & b & 0 & | & 1 \\ 0 & a & 0 & b & | & 0 \\ c & 0 & d & 0 & | & 0 \\ 0 & c & 0 & d & | & 1 \end{bmatrix}$$

which reduces to

$$\begin{bmatrix} 1 & 0 & 0 & 0 & | & \frac{d}{ad-bc} \\ 0 & 1 & 0 & 0 & | & -\frac{b}{ad-bc} \\ 0 & 0 & 1 & 0 & | & -\frac{c}{ad-bc} \\ 0 & 0 & 0 & 1 & | & \frac{a}{ad-bc} \end{bmatrix}$$

Since $ad - bc \neq 0$, the inverse of the matrix A is

$$A^{-1} = \begin{bmatrix} \frac{d}{ad-bc} & -\frac{b}{ad-bc} \\ -\frac{c}{ad-bc} & \frac{a}{ad-bc} \end{bmatrix} = \frac{1}{ad - bc} \begin{bmatrix} d & -b \\ -c & a \end{bmatrix}$$

To prove the reverse claim, we use the contrapositive. That is, if $ad - bc = 0$, then the inverse does not exist. An outline of the proof is given in Exercise 41.

To illustrate the use of the formula, let

$$A = \begin{bmatrix} 2 & -1 \\ 1 & 3 \end{bmatrix}$$

then

$$A^{-1} = \frac{1}{6 - (-1)} \begin{bmatrix} 3 & 1 \\ -1 & 2 \end{bmatrix} = \begin{bmatrix} \frac{3}{7} & \frac{1}{7} \\ -\frac{1}{7} & \frac{2}{7} \end{bmatrix}$$

For an example which underscores the necessity of the condition that $ad - bc \neq 0$, we consider the matrix

$$A = \begin{bmatrix} 1 & 1 \\ 1 & 1 \end{bmatrix}$$

Observe that in this case $ad - bc = 1 - 1 = 0$. Now, the matrix A is invertible if there is a $B = \begin{bmatrix} x_1 & x_2 \\ x_3 & x_4 \end{bmatrix}$ such that

$$\begin{bmatrix} 1 & 1 \\ 1 & 1 \end{bmatrix} \begin{bmatrix} x_1 & x_2 \\ x_3 & x_4 \end{bmatrix} = \begin{bmatrix} 1 & 0 \\ 0 & 1 \end{bmatrix}$$

This matrix equation yields the inconsistent system

$$\begin{cases} x_1 & + x_3 & & = 1 \\ & x_2 & & + x_4 = 0 \\ x_1 & + x_3 & & = 0 \\ & x_2 & & + x_4 = 1 \end{cases}$$

Hence, A is not invertible.

To find the inverse of larger square matrices, we extend the method of augmented matrices. Let A be an $n \times n$ matrix. Let B be another $n \times n$ matrix, and let $\mathbf{B}_1, \mathbf{B}_2, \ldots, \mathbf{B}_n$ denote the n column vectors of B. Since $A\mathbf{B}_1, A\mathbf{B}_2, \ldots, A\mathbf{B}_n$ are the column vectors of AB, in order for B to be the inverse of A, we must have

$$A\mathbf{B}_1 = \begin{bmatrix} 1 \\ 0 \\ \vdots \\ 0 \end{bmatrix} \qquad A\mathbf{B}_2 = \begin{bmatrix} 0 \\ 1 \\ \vdots \\ 0 \end{bmatrix} \qquad \ldots \qquad A\mathbf{B}_n = \begin{bmatrix} 0 \\ 0 \\ \vdots \\ 1 \end{bmatrix}$$

That is, the *matrix equations*

$$A\mathbf{x} = \begin{bmatrix} 1 \\ 0 \\ \vdots \\ 0 \end{bmatrix} \qquad A\mathbf{x} = \begin{bmatrix} 0 \\ 1 \\ \vdots \\ 0 \end{bmatrix} \qquad \ldots \qquad A\mathbf{x} = \begin{bmatrix} 0 \\ 0 \\ \vdots \\ 1 \end{bmatrix}$$

must all have unique solutions. But all n linear systems can be solved simultaneously by row-reducing the $n \times 2n$ augmented matrix

$$\begin{bmatrix} a_{11} & a_{12} & \ldots & a_{1n} & 1 & 0 & \ldots & 0 \\ a_{21} & a_{22} & \ldots & a_{2n} & 0 & 1 & \ldots & 0 \\ \vdots & \vdots & \ddots & \vdots & \vdots & \vdots & \ddots & \vdots \\ a_{n1} & a_{n2} & \ldots & a_{nn} & 0 & 0 & \ldots & 1 \end{bmatrix}$$

On the left is the matrix A, and on the right is the matrix I. Then A **will have an inverse if and only if it is row equivalent to the identity matrix.** In this case, each of the linear systems can be solved. If the matrix A does not have an inverse, then the row-reduced matrix on the left will have a row of zeros, indicating at least one of the linear systems does not have a solution.

Example 2 illustrates the procedure.

EXAMPLE 2   Find the inverse of the matrix

$$A = \begin{bmatrix} 1 & 1 & -2 \\ -1 & 2 & 0 \\ 0 & -1 & 1 \end{bmatrix}$$

Solution   To find the inverse of this matrix, place the identity on the right to form the 3×6 matrix

$$\left[\begin{array}{rrr|rrr} 1 & 1 & -2 & 1 & 0 & 0 \\ -1 & 2 & 0 & 0 & 1 & 0 \\ 0 & -1 & 1 & 0 & 0 & 1 \end{array}\right]$$

Now use row operations to reduce the matrix on the left to the identity, while applying the same operations to the matrix on the right. The final result is

$$\left[\begin{array}{rrr|rrr} 1 & 0 & 0 & 2 & 1 & 4 \\ 0 & 1 & 0 & 1 & 1 & 2 \\ 0 & 0 & 1 & 1 & 1 & 3 \end{array}\right]$$

so the inverse matrix is

$$A^{-1} = \begin{bmatrix} 2 & 1 & 4 \\ 1 & 1 & 2 \\ 1 & 1 & 3 \end{bmatrix}$$

The reader should check that $AA^{-1} = A^{-1}A = I$.

EXAMPLE 3   Use the method of Example 2 to determine whether the matrix

$$A = \begin{bmatrix} 1 & -1 & 2 \\ 3 & -3 & 1 \\ 3 & -3 & 1 \end{bmatrix}$$

is invertible.

Solution   Following the procedure described above, we start with the matrix

$$\left[\begin{array}{rrr|rrr} 1 & -1 & 2 & 1 & 0 & 0 \\ 3 & -3 & 1 & 0 & 1 & 0 \\ 3 & -3 & 1 & 0 & 0 & 1 \end{array}\right]$$

After the two row operations $-3R_1 + R_2 \rightarrow R_2$ followed by $-3R_1 + R_3 \rightarrow R_3$, this matrix is reduced to

$$\left[\begin{array}{ccc|ccc} 1 & -1 & 2 & 1 & 0 & 0 \\ 0 & 0 & -5 & -3 & 1 & 0 \\ 0 & 0 & -5 & -3 & 0 & 1 \end{array}\right]$$

Next perform the row operation $-R_2 + R_3 \rightarrow R_3$ to obtain

$$\left[\begin{array}{ccc|ccc} 1 & -1 & 2 & 1 & 0 & 0 \\ 0 & 0 & -5 & -3 & 1 & 0 \\ 0 & 0 & 0 & 0 & -1 & 1 \end{array}\right]$$

The 3×3 matrix of coefficients on the left cannot be reduced to the identity matrix, and therefore, the original matrix does not have an inverse. Also notice that a solution does exist to

$$A\mathbf{x} = \left[\begin{array}{c} 1 \\ 0 \\ 0 \end{array}\right]$$

while solutions to

$$A\mathbf{x} = \left[\begin{array}{c} 0 \\ 1 \\ 0 \end{array}\right] \quad \text{and} \quad A\mathbf{x} = \left[\begin{array}{c} 0 \\ 0 \\ 1 \end{array}\right]$$

do not exist.

The matrix A of Example 3 has two equal rows and cannot be row-reduced to the identity matrix. This is true for any $n \times n$ matrix with two equal rows and provides an alternative method for concluding that such a matrix is not invertible.

Theorem 9 gives a formula for the inverse of the product of invertible matrices.

THEOREM 9   Let A and B be $n \times n$ invertible matrices. Then AB is invertible and

$$(AB)^{-1} = B^{-1}A^{-1}$$

Proof   Using the properties of matrix multiplication, we have

$$(AB)(B^{-1}A^{-1}) = A(BB^{-1})A^{-1} = AIA^{-1} = AA^{-1} = I$$

and

$$(B^{-1}A^{-1})(AB) = B^{-1}(A^{-1}A)B = B^{-1}IB = BB^{-1} = I$$

Since, when it exists, the inverse matrix is unique, we have shown that the inverse of AB is the matrix $B^{-1}A^{-1}$.

| **EXAMPLE 4** | Suppose that B is an invertible matrix and A is any matrix with $AB = BA$. Show that A and B^{-1} commute. |

Solution Since $AB = BA$, we can multiply both sides on the right by B^{-1} to obtain

$$(AB)B^{-1} = (BA)B^{-1}$$

By the associative property of matrix multiplication this last equation can be written as

$$A(BB^{-1}) = BAB^{-1}$$

and since $BB^{-1} = I$, we have

$$A = BAB^{-1}$$

Next we multiply on the left by B^{-1} to obtain

$$B^{-1}A = B^{-1}BAB^{-1} \qquad \text{so} \qquad B^{-1}A = AB^{-1}$$

as required.

Fact Summary

Let A and B denote matrices.

1. The inverse of a matrix, when it exists, is unique.
2. If $A = \begin{bmatrix} a & b \\ c & d \end{bmatrix}$ and $ad - bc \neq 0$, then $A^{-1} = \frac{1}{ad-bc} \begin{bmatrix} d & -b \\ -c & a \end{bmatrix}$.
3. A matrix A is invertible if and only if it is row equivalent to the identity matrix.
4. If A and B are invertible $n \times n$ matrices, then AB is invertible and $(AB)^{-1} = B^{-1}A^{-1}$.

Exercise Set 1.4

In Exercises 1–16, a matrix A is given. Find A^{-1} or indicate that it does not exist. When A^{-1} exists, check your answer by showing that $AA^{-1} = I$.

1. $\begin{bmatrix} 1 & -2 \\ 3 & -1 \end{bmatrix}$

2. $\begin{bmatrix} -3 & 1 \\ 1 & 2 \end{bmatrix}$

3. $\begin{bmatrix} -2 & 4 \\ 2 & -4 \end{bmatrix}$

4. $\begin{bmatrix} 1 & 1 \\ 2 & 2 \end{bmatrix}$

5. $\begin{bmatrix} 0 & 1 & -1 \\ 3 & 1 & 1 \\ 1 & 2 & -1 \end{bmatrix}$

6. $\begin{bmatrix} 0 & 2 & 1 \\ -1 & 0 & 0 \\ 2 & 1 & 1 \end{bmatrix}$

7. $\begin{bmatrix} 3 & -3 & 1 \\ 0 & 0 & 1 \\ -2 & 2 & -1 \end{bmatrix}$

8. $\begin{bmatrix} 1 & 3 & 0 \\ 1 & 2 & 3 \\ 0 & -1 & 3 \end{bmatrix}$

9. $\begin{bmatrix} 3 & 3 & 0 & -3 \\ 0 & 1 & 2 & 0 \\ 0 & 0 & -1 & -1 \\ 0 & 0 & 0 & -2 \end{bmatrix}$

10. $\begin{bmatrix} 1 & 3 & 0 & -3 \\ 0 & 1 & 2 & -3 \\ 0 & 0 & 2 & -2 \\ 0 & 0 & 0 & 2 \end{bmatrix}$

11. $\begin{bmatrix} 1 & 0 & 0 & 0 \\ 2 & 1 & 0 & 0 \\ -3 & -2 & -3 & 0 \\ 0 & 1 & 3 & 3 \end{bmatrix}$

12. $\begin{bmatrix} 1 & 0 & 0 & 0 \\ -2 & 1 & 0 & 0 \\ 1 & -1 & -2 & 0 \\ 2 & -2 & 0 & 2 \end{bmatrix}$

13. $\begin{bmatrix} 2 & -1 & 4 & -5 \\ 0 & -1 & 3 & -1 \\ 0 & 0 & 0 & 2 \\ 0 & 0 & 0 & -1 \end{bmatrix}$

14. $\begin{bmatrix} 3 & 0 & 0 & 0 \\ -6 & 1 & 0 & 0 \\ 2 & -5 & 0 & 0 \\ 1 & -3 & 4 & 2 \end{bmatrix}$

15. $\begin{bmatrix} -1 & 1 & 0 & -1 \\ -1 & 1 & -1 & 0 \\ -1 & 0 & 0 & 0 \\ -2 & 1 & -1 & 1 \end{bmatrix}$

16. $\begin{bmatrix} -2 & -3 & 3 & 0 \\ 2 & 0 & -2 & 0 \\ 2 & 0 & -1 & -1 \\ -2 & 0 & 1 & 1 \end{bmatrix}$

17. Let
$$A = \begin{bmatrix} 2 & 1 \\ 3 & -4 \end{bmatrix} \qquad B = \begin{bmatrix} 1 & 2 \\ -1 & 3 \end{bmatrix}$$
Verify that $AB + A$ can be factored as $A(B + I)$ and $AB + B$ can be factored as $(A + I)B$.

18. If A is an $n \times n$ matrix, write $A^2 + 2A + I$ in factored form.

19. Let
$$A = \begin{bmatrix} 1 & 2 \\ -2 & 1 \end{bmatrix}$$

a. Show that $A^2 - 2A + 5I = 0$.
b. Show that $A^{-1} = \frac{1}{5}(2I - A)$.
c. Show in general that for any square matrix A satisfying $A^2 - 2A + 5I = 0$, the inverse is $A^{-1} = \frac{1}{5}(2I - A)$.

20. Determine those values of λ for which the matrix
$$\begin{bmatrix} 1 & \lambda & 0 \\ 3 & 2 & 0 \\ 1 & 2 & 1 \end{bmatrix}$$
is not invertible.

21. Determine those values of λ for which the matrix
$$\begin{bmatrix} 1 & \lambda & 0 \\ 1 & 3 & 1 \\ 2 & 1 & 1 \end{bmatrix}$$
is not invertible.

22. Determine those values of λ for which the matrix
$$\begin{bmatrix} 2 & \lambda & 1 \\ 3 & 2 & 1 \\ 1 & 2 & 1 \end{bmatrix}$$
is not invertible.

23. Let
$$A = \begin{bmatrix} 1 & \lambda & 0 \\ 1 & 1 & 1 \\ 0 & 0 & 1 \end{bmatrix}$$

a. Determine those values of λ for which A is invertible.

b. For those values found in part (a) find the inverse of A.

24. Determine those values of λ for which the matrix

$$\begin{bmatrix} \lambda & -1 & 0 \\ -1 & \lambda & -1 \\ 0 & -1 & \lambda \end{bmatrix}$$

is invertible.

25. Find 2×2 matrices A and B that are not invertible but $A + B$ is invertible.

26. Find 2×2 matrices A and B that are invertible but $A + B$ is not invertible.

27. If A and B are $n \times n$ matrices and A is invertible, show that

$$(A + B)A^{-1}(A - B) = (A - B)A^{-1}(A + B)$$

28. If $B = PAP^{-1}$, express $B^2, B^3, \ldots, B^k$, where k is any positive integer, in terms of A, P, and P^{-1}.

29. Let A and B be $n \times n$ matrices.

a. Show that if A is invertible and $AB = 0$, then $B = 0$.

b. If A is not invertible, show there is an $n \times n$ matrix B that is not the zero matrix and such that $AB = 0$.

30. Show that if A is symmetric and invertible, then A^{-1} is symmetric.

In Exercises 31–34, the matrices A and B are invertible symmetric matrices and $AB = BA$.

31. Show that AB is symmetric.

32. Show that $A^{-1}B$ is symmetric.

33. Show that AB^{-1} is symmetric.

34. Show that $A^{-1}B^{-1}$ is symmetric.

35. A matrix A is *orthogonal* provided that $A^t = A^{-1}$. Show that the product of two orthogonal matrices is orthogonal.

36. Show the matrix

$$A = \begin{bmatrix} \cos\theta & -\sin\theta \\ \sin\theta & \cos\theta \end{bmatrix}$$

is orthogonal. (See Exercise 35.)

37. a. If A, B, and C are $n \times n$ invertible matrices, show that

$$(ABC)^{-1} = C^{-1}B^{-1}A^{-1}$$

b. Use mathematical induction to show that for all positive integers k, if $A_1, A_2, \ldots, A_k$ are $n \times n$ invertible matrices, then

$$(A_1 A_2 \cdots A_k)^{-1} = A_k^{-1} A_{k-1}^{-1} \cdots A_1^{-1}$$

38. An $n \times n$ matrix A is *diagonal* provided that $a_{ij} = 0$ whenever $i \neq j$. Show that if $a_{nn} \neq 0$ for all n, then A is invertible and the inverse is

$$\begin{bmatrix} \frac{1}{a_{11}} & 0 & 0 & \cdots & 0 \\ 0 & \frac{1}{a_{22}} & 0 & \cdots & 0 \\ \vdots & \vdots & \ddots & \vdots & \vdots \\ 0 & 0 & \cdots & \frac{1}{a_{n-1,n-1}} & 0 \\ 0 & 0 & \cdots & 0 & \frac{1}{a_{nn}} \end{bmatrix}$$

39. Let A be an $n \times n$ invertible matrix. Show that if A is in upper (lower) triangular form, then A^{-1} is also in upper (lower) triangular form.

40. Suppose B is row equivalent to the $n \times n$ invertible matrix A. Show that B is invertible.

41. Show that if $ad - bc = 0$, then $A = \begin{bmatrix} a & b \\ c & d \end{bmatrix}$ is not invertible.

a. Expand the matrix equation

$$\begin{bmatrix} a & b \\ c & d \end{bmatrix} \begin{bmatrix} x_1 & x_2 \\ x_3 & x_4 \end{bmatrix} = \begin{bmatrix} 1 & 0 \\ 0 & 1 \end{bmatrix}$$

b. Show the 2×2 linear system in the variables x_1 and x_3 that is generated in part (a) yields $d = 0$. Similarly, show the system in the variables x_2 and x_4 yields $b = 0$.

c. Use the results of part (b) to conclude that $ad - bc = 0$.

1.5 ▶ **Matrix Equations**

In this section we show how matrix multiplication can be used to write a linear system in terms of matrices and vectors. We can then write a linear system as a single equation, using a matrix and two vectors, which generalizes the linear equation $ax = b$ for real numbers. As we will see, in some cases the linear system can then be solved using algebraic operations similar to the operations used to solve equations involving real numbers.

To illustrate the process, consider the linear system

$$\begin{cases} x - 6y - 4z = -5 \\ 2x - 10y - 9z = -4 \\ -x + 6y + 5z = 3 \end{cases}$$

The matrix of coefficients is given by

$$A = \begin{bmatrix} 1 & -6 & -4 \\ 2 & -10 & -9 \\ -1 & 6 & 5 \end{bmatrix}$$

Now let **x** and **b** be the vectors

$$\mathbf{x} = \begin{bmatrix} x \\ y \\ z \end{bmatrix} \quad \text{and} \quad \mathbf{b} = \begin{bmatrix} -5 \\ -4 \\ 3 \end{bmatrix}$$

Then the original linear system can be rewritten as

$$A\mathbf{x} = \mathbf{b}$$

We refer to this equation as the **matrix form** of the linear system and **x** as the **vector form** of the solution.

In certain cases we can find the solution of a linear system in matrix form directly by matrix multiplication. In particular, if A is invertible, we can multiply both sides of the previous equation on the left by A^{-1}, so that

$$A^{-1}(A\mathbf{x}) = A^{-1}\mathbf{b}$$

Since matrix multiplication is associative, we have

$$\left(A^{-1}A\right)\mathbf{x} = A^{-1}\mathbf{b}$$

therefore,

$$\mathbf{x} = A^{-1}\mathbf{b}$$

For the example above, the inverse of the matrix

$$A = \begin{bmatrix} 1 & -6 & -4 \\ 2 & -10 & -9 \\ -1 & 6 & 5 \end{bmatrix} \quad \text{is} \quad A^{-1} = \begin{bmatrix} 2 & 3 & 7 \\ -\frac{1}{2} & \frac{1}{2} & \frac{1}{2} \\ 1 & 0 & 1 \end{bmatrix}$$

Therefore, the solution to the linear system in vector form is given by

$$\mathbf{x} = A^{-1}\mathbf{b} = \begin{bmatrix} 2 & 3 & 7 \\ -\frac{1}{2} & \frac{1}{2} & \frac{1}{2} \\ 1 & 0 & 1 \end{bmatrix} \begin{bmatrix} -5 \\ -4 \\ 3 \end{bmatrix} = \begin{bmatrix} -1 \\ 2 \\ -2 \end{bmatrix}$$

That is,

$$x = -1 \qquad y = 2 \qquad \text{and} \qquad z = -2$$

We have just seen that if the matrix A has an inverse, then the equation $A\mathbf{x} = \mathbf{b}$ has a unique solution. This fact is recorded in Theorem 10.

THEOREM 10   If the $n \times n$ matrix A is invertible, then for every vector $\mathbf{b}$, with n components, the linear system $A\mathbf{x} = \mathbf{b}$ has the unique solution $\mathbf{x} = A^{-1}\mathbf{b}$.

EXAMPLE 1   Write the linear system in matrix form and solve.

$$\begin{cases} 2x + y = 1 \\ -4x + 3y = 2 \end{cases}$$

Solution   The matrix form of the linear system is given by

$$\begin{bmatrix} 2 & 1 \\ -4 & 3 \end{bmatrix} \begin{bmatrix} x \\ y \end{bmatrix} = \begin{bmatrix} 1 \\ 2 \end{bmatrix}$$

Notice that since $2(3) - (1)(-4) = 10 \neq 0$, the coefficient matrix is invertible. By Theorem 8, of Sec. 1.4, the inverse is

$$\frac{1}{10} \begin{bmatrix} 3 & -1 \\ 4 & 2 \end{bmatrix}$$

Now, by Theorem 10, the solution to the linear system is

$$\mathbf{x} = \frac{1}{10} \begin{bmatrix} 3 & -1 \\ 4 & 2 \end{bmatrix} \begin{bmatrix} 1 \\ 2 \end{bmatrix} = \frac{1}{10} \begin{bmatrix} 1 \\ 8 \end{bmatrix} = \begin{bmatrix} \frac{1}{10} \\ \frac{8}{10} \end{bmatrix}$$

so that

$$x = \frac{1}{10} \qquad \text{and} \qquad y = \frac{8}{10}$$

DEFINITION 1   **Homogeneous Linear System**   A **homogeneous linear system** is a system of the form $A\mathbf{x} = \mathbf{0}$.

The vector $\mathbf{x} = \mathbf{0}$ is always a solution to the homogeneous system $A\mathbf{x} = \mathbf{0}$, and is called the **trivial solution**.

EXAMPLE 2   Let

$$A = \begin{bmatrix} 1 & 2 & 1 \\ 1 & 3 & 0 \\ 1 & 1 & 2 \end{bmatrix} \quad \text{and} \quad \mathbf{x} = \begin{bmatrix} x_1 \\ x_2 \\ x_3 \end{bmatrix}$$

Find all vectors $\mathbf{x}$ such that $A\mathbf{x} = \mathbf{0}$.

Solution   First observe that $\mathbf{x} = \mathbf{0}$ is one solution. To find the general solution, we row-reduce the augmented matrix

$$\begin{bmatrix} 1 & 2 & 1 & 0 \\ 1 & 3 & 0 & 0 \\ 1 & 1 & 2 & 0 \end{bmatrix} \quad \text{to} \quad \begin{bmatrix} 1 & 2 & 1 & 0 \\ 0 & 1 & -1 & 0 \\ 0 & 0 & 0 & 0 \end{bmatrix}$$

From the reduced matrix we see that x_3 is free with $x_2 = x_3$, and $x_1 = -2x_2 - x_3 = -3x_3$. The solution set in *vector form* is given by

$$S = \left\{ \begin{bmatrix} -3t \\ t \\ t \end{bmatrix} \middle| \, t \in \mathbb{R} \right\}.$$

Notice that the trivial solution is also included in S as a particular solution with $t = 0$.

Observe that in Example 2, the coefficient matrix is not row equivalent to I, and hence A is not invertible.

If a homogeneous linear system $A\mathbf{x} = \mathbf{0}$ is such that A is invertible, then by Theorem 10, the only solution is $\mathbf{x} = \mathbf{0}$. In Sec. 1.6 we will show that the converse is also true.

EXAMPLE 3   Show that if $\mathbf{x}$ and $\mathbf{y}$ are distinct solutions to the homogeneous system $A\mathbf{x} = \mathbf{0}$, then $\mathbf{x} + c\mathbf{y}$ is a solution for every real number c.

Solution   Using the algebraic properties of matrices, we have that

$$\begin{aligned} A(\mathbf{x} + c\mathbf{y}) &= A(\mathbf{x}) + A(c\mathbf{y}) \\ &= A\mathbf{x} + cA\mathbf{y} \\ &= \mathbf{0} + c\mathbf{0} \\ &= \mathbf{0} \end{aligned}$$

Hence, $\mathbf{x} + c\mathbf{y}$ is a solution to the homogeneous system.

The result of Example 3 shows that if the homogeneous equation $A\mathbf{x} = \mathbf{0}$ has two distinct solutions, then it has infinitely many solutions. That is, the homogeneous equation $A\mathbf{x} = \mathbf{0}$ either has one solution (the trivial solution) or has infinitely many solutions. The same result holds for the nonhomogeneous equation $A\mathbf{x} = \mathbf{b}$, with $\mathbf{b} \neq \mathbf{0}$. To see this, let $\mathbf{u}$ and $\mathbf{v}$ be distinct solutions to $A\mathbf{x} = \mathbf{b}$ and c a real number. Then

$$
\begin{aligned}
A(\mathbf{v} + c(\mathbf{u} - \mathbf{v})) &= A\mathbf{v} + A(c(\mathbf{u} - \mathbf{v})) \\
&= A\mathbf{v} + cA\mathbf{u} - cA\mathbf{v} \\
&= \mathbf{b} + c\mathbf{b} - c\mathbf{b} = \mathbf{b}
\end{aligned}
$$

These observations are summarized in Theorem 11.

THEOREM 11   If A is an $m \times n$ matrix, then the linear system $A\mathbf{x} = \mathbf{b}$ has no solutions, one solution, or infinitely many solutions.

Fact Summary

Let A be an $m \times n$ matrix.

1. If A is invertible, then for every $n \times 1$ vector $\mathbf{b}$ the matrix equation $A\mathbf{x} = \mathbf{b}$ has a unique solution given by $\mathbf{x} = A^{-1}\mathbf{b}$.
2. If A is invertible, then the only solution to the homogeneous equation $A\mathbf{x} = \mathbf{0}$ is the trivial solution $\mathbf{x} = \mathbf{0}$.
3. If $\mathbf{u}$ and $\mathbf{v}$ are solutions to $A\mathbf{x} = \mathbf{0}$, then the vector $\mathbf{u} + c\mathbf{v}$ is another solution for every scalar c.
4. The linear system $A\mathbf{x} = \mathbf{b}$ has a unique solution, infinitely many solutions, or no solution.

Exercise Set 1.5

In Exercises 1–6, find a matrix A and vectors $\mathbf{x}$ and $\mathbf{b}$ such that the linear system can be written as $A\mathbf{x} = \mathbf{b}$.

1. $\begin{cases} 2x + 3y = -1 \\ -x + 2y = 4 \end{cases}$

2. $\begin{cases} -4x - y = 3 \\ -2x - 5y = 2 \end{cases}$

3. $\begin{cases} 2x - 3y + z = -1 \\ -x - y + 2z = -1 \\ 3x - 2y - 2z = 3 \end{cases}$

4. $\begin{cases} 3y - 2z = 2 \\ -x + 4z = -3 \\ -x - 3z = 4 \end{cases}$

5. $\begin{cases} 4x_1 + 3x_2 - 2x_3 - 3x_4 = -1 \\ -3x_1 - 3x_2 + x_3 = 4 \\ 2x_1 - 3x_2 + 4x_3 - 4x_4 = 3 \end{cases}$

6. $\begin{cases} 3x_2 + x_3 - 2x_4 = -4 \\ 4x_2 - 2x_3 - 4x_4 = 0 \\ x_1 + 3x_2 - 2x_3 = 3 \end{cases}$

In Exercises 7–12, given the matrix A and vectors $\mathbf{x}$ and $\mathbf{b}$, write the equation $A\mathbf{x} = \mathbf{b}$ as a linear system.

7. $A = \begin{bmatrix} 2 & -5 \\ 2 & 1 \end{bmatrix}$ $\mathbf{x} = \begin{bmatrix} x \\ y \end{bmatrix}$ $\mathbf{b} = \begin{bmatrix} 3 \\ 2 \end{bmatrix}$

8. $A = \begin{bmatrix} -2 & 4 \\ 0 & 3 \end{bmatrix}$ $\mathbf{x} = \begin{bmatrix} x \\ y \end{bmatrix}$ $\mathbf{b} = \begin{bmatrix} -1 \\ 1 \end{bmatrix}$

9. $A = \begin{bmatrix} 0 & -2 & 0 \\ 2 & -1 & -1 \\ 3 & -1 & 2 \end{bmatrix}$

$\mathbf{x} = \begin{bmatrix} x \\ y \\ z \end{bmatrix}$ $\mathbf{b} = \begin{bmatrix} 3 \\ 1 \\ -1 \end{bmatrix}$

10. $A = \begin{bmatrix} -4 & -5 & 5 \\ 4 & -1 & 1 \\ -4 & 3 & 5 \end{bmatrix}$

$\mathbf{x} = \begin{bmatrix} x \\ y \\ z \end{bmatrix}$ $\mathbf{b} = \begin{bmatrix} -3 \\ 2 \\ 1 \end{bmatrix}$

11. $A = \begin{bmatrix} 2 & 5 & -5 & 3 \\ 3 & 1 & -2 & -4 \end{bmatrix}$

$\mathbf{x} = \begin{bmatrix} x_1 \\ x_2 \\ x_3 \\ x_4 \end{bmatrix}$ $\mathbf{b} = \begin{bmatrix} 2 \\ 0 \end{bmatrix}$

12. $A = \begin{bmatrix} 0 & -2 & 4 & -2 \\ 2 & 0 & 1 & 1 \\ 1 & 0 & 1 & -2 \end{bmatrix}$

$\mathbf{x} = \begin{bmatrix} x_1 \\ x_2 \\ x_3 \\ x_4 \end{bmatrix}$ $\mathbf{b} = \begin{bmatrix} 4 \\ -3 \\ 1 \end{bmatrix}$

In Exercises 13–16, use the information given to solve the linear system $A\mathbf{x} = \mathbf{b}$.

13.

$A^{-1} = \begin{bmatrix} 2 & 0 & -1 \\ 4 & 1 & 4 \\ 1 & 2 & 4 \end{bmatrix}$

$\mathbf{b} = \begin{bmatrix} 1 \\ -4 \\ 1 \end{bmatrix}$

14.

$A^{-1} = \begin{bmatrix} -4 & 3 & -4 \\ 2 & 2 & 0 \\ 1 & 2 & 4 \end{bmatrix}$

$\mathbf{b} = \begin{bmatrix} 2 \\ 2 \\ -2 \end{bmatrix}$

15.

$A^{-1} = \begin{bmatrix} -3 & -2 & 0 & 3 \\ -1 & 2 & -2 & 3 \\ 0 & 1 & 2 & -3 \\ -1 & 0 & 3 & 1 \end{bmatrix}$

$\mathbf{b} = \begin{bmatrix} 2 \\ -3 \\ 2 \\ 3 \end{bmatrix}$

16.

$A^{-1} = \begin{bmatrix} 3 & 0 & -2 & -2 \\ 2 & 0 & 1 & -1 \\ -3 & -1 & -1 & 1 \\ 2 & -1 & -2 & -3 \end{bmatrix}$

$\mathbf{b} = \begin{bmatrix} 1 \\ -4 \\ 1 \\ 1 \end{bmatrix}$

In Exercises 17–22, solve the linear system by finding the inverse of the coefficient matrix.

17. $\begin{cases} x + 4y = 2 \\ 3x + 2y = -3 \end{cases}$

18. $\begin{cases} 2x - 4y = 4 \\ -2x + 3y = 3 \end{cases}$

19. $\begin{cases} -x \quad\ - z = -1 \\ -3x + y - 3z = 1 \\ x - 3y + 2z = 1 \end{cases}$

20. $\begin{cases} -2x - 2y - z = 0 \\ -x - y \quad\quad = -1 \\ \quad - y + 2z = 2 \end{cases}$

21. $\begin{cases} -\ x_1 - x_2 - 2x_3 +\ x_4 = -1 \\ \quad 2x_1 + x_2 + 2x_3 -\ x_4 = \quad 1 \\ -2x_1 - x_2 - 2x_3 - 2x_4 = \quad 0 \\ -2x_1 - x_2 -\ x_3 -\ x_4 = \quad 0 \end{cases}$

22. $\begin{cases} -x_1 - 2x_2 \qquad\quad +\ x_4 = -3 \\ -x_1 +\ x_2 - 2x_3 +\ x_4 = -2 \\ -x_1 + 2x_2 - 2x_3 +\ x_4 = \quad 3 \\ \qquad\ -2x_2 + 2x_3 - 2x_4 = -1 \end{cases}$

23. Let

$$A = \begin{bmatrix} 1 & -1 \\ 2 & 3 \end{bmatrix}$$

Use the inverse matrix to solve the linear system $A\mathbf{x} = \mathbf{b}$ for the given vector $\mathbf{b}$.

a. $\mathbf{b} = \begin{bmatrix} 2 \\ 1 \end{bmatrix}$

b. $\mathbf{b} = \begin{bmatrix} -3 \\ 2 \end{bmatrix}$

24. Let

$$A = \begin{bmatrix} -1 & 0 & -1 \\ -3 & 1 & -3 \\ 1 & -3 & 2 \end{bmatrix}$$

Use the inverse matrix to solve the linear system $A\mathbf{x} = \mathbf{b}$ for the given vector $\mathbf{b}$.

a. $\mathbf{b} = \begin{bmatrix} -2 \\ 1 \\ 1 \end{bmatrix}$

b. $\mathbf{b} = \begin{bmatrix} 1 \\ -1 \\ 0 \end{bmatrix}$

25. Let

$$A = \begin{bmatrix} -1 & -4 \\ 3 & 12 \\ 2 & 8 \end{bmatrix}$$

Find a nontrivial solution to $A\mathbf{x} = \mathbf{0}$.

26. Let

$$A = \begin{bmatrix} 1 & -2 & 4 \\ 2 & -4 & 8 \\ 3 & -6 & 12 \end{bmatrix}$$

Find a nontrivial solution to $A\mathbf{x} = \mathbf{0}$.

27. Find a nonzero 3×3 matrix A such that the vector

$$\begin{bmatrix} 1 \\ -1 \\ 1 \end{bmatrix}$$

is a solution to $A\mathbf{x} = \mathbf{0}$.

28. Find a nonzero 3×3 matrix A such that the vector

$$\begin{bmatrix} -1 \\ 2 \\ 1 \end{bmatrix}$$

is a solution to $A\mathbf{x} = \mathbf{0}$.

29. Suppose that A is an $n \times n$ matrix and $\mathbf{u}$ and $\mathbf{v}$ are vectors in $\mathbb{R}^n$. Show that if $A\mathbf{u} = A\mathbf{v}$ and $\mathbf{u} \neq \mathbf{v}$, then A is not invertible.

30. Suppose that $\mathbf{u}$ is a solution to $A\mathbf{x} = \mathbf{b}$ and that $\mathbf{v}$ is a solution to $A\mathbf{x} = \mathbf{0}$. Show that $\mathbf{u} + \mathbf{v}$ is a solution to $A\mathbf{x} = \mathbf{b}$.

31. Consider the linear system

$$\begin{cases} 2x + \ y = \quad 1 \\ -x + \ y = -2 \\ \quad x + 2y = -1 \end{cases}$$

a. Write the linear system in matrix form $A\mathbf{x} = \mathbf{b}$ and find the solution.

b. Find a 2×3 matrix C such that $CA = I$. (The matrix C is called a *left inverse.*)

c. Show that the solution to the linear system is given by $\mathbf{x} = C\mathbf{b}$.

32. Consider the linear system

$$\begin{cases} 2x + \ y = \quad 3 \\ -x - \ y = -2 \\ 3x + 2y = \quad 5 \end{cases}$$

a. Write the linear system in matrix form $A\mathbf{x} = \mathbf{b}$ and find the solution.

b. Find a 2×3 matrix C such that $CA = I$.

c. Show that the solution to the linear system is given by $\mathbf{x} = C\mathbf{b}$.

1.6 ▶ Determinants

In Sec. 1.4 we saw that the number $ad - bc$, associated with the 2×2 matrix

$$A = \begin{bmatrix} a & b \\ c & d \end{bmatrix}$$

has special significance. This number is called the *determinant* of A and provides useful information about the matrix. In particular, using this terminology, the matrix A is invertible if and only if the determinant is not equal to 0. In this section the definition of the determinant is extended to larger square matrices. The information provided by the determinant has theoretical value and is used in some applications. In practice, however, the computational difficulty in evaluating the determinant of a very large matrix is significant. For this reason the information desired is generally found by using other more efficient methods.

DEFINITION 1   **Determinant of a 2 × 2 Matrix**   The **determinant** of the matrix $A = \begin{bmatrix} a & b \\ c & d \end{bmatrix}$, denoted by $|A|$ or $\det(A)$, is given by

$$|A| = \det(A) = \begin{vmatrix} a & b \\ c & d \end{vmatrix} = ad - bc$$

Using this terminology a 2×2 matrix is invertible if and only if its determinant is nonzero.

EXAMPLE 1   Find the determinant of the matrix.

   a. $A = \begin{bmatrix} 3 & 1 \\ -2 & 2 \end{bmatrix}$ **b.** $A = \begin{bmatrix} 3 & 5 \\ 4 & 2 \end{bmatrix}$ **c.** $A = \begin{bmatrix} 1 & 0 \\ -3 & 0 \end{bmatrix}$

Solution   **a.** $|A| = \begin{vmatrix} 3 & 1 \\ -2 & 2 \end{vmatrix} = (3)(2) - (1)(-2) = 8$

   b. $|A| = \begin{vmatrix} 3 & 5 \\ 4 & 2 \end{vmatrix} = (3)(2) - (5)(4) = -14$

   c. $|A| = \begin{vmatrix} 1 & 0 \\ -3 & 0 \end{vmatrix} = (1)(0) - (0)(-3) = 0$

Using the determinant of a 2×2 matrix, we now extend this definition to 3×3 matrices.

DEFINITION 2   **Determinant of a 3 × 3 Matrix**   The determinant of the matrix

$$A = \begin{bmatrix} a_{11} & a_{12} & a_{13} \\ a_{21} & a_{22} & a_{23} \\ a_{31} & a_{32} & a_{33} \end{bmatrix}$$

is

$$|A| = a_{11} \begin{vmatrix} a_{22} & a_{23} \\ a_{32} & a_{33} \end{vmatrix} - a_{12} \begin{vmatrix} a_{21} & a_{23} \\ a_{31} & a_{33} \end{vmatrix} + a_{13} \begin{vmatrix} a_{21} & a_{22} \\ a_{31} & a_{32} \end{vmatrix}$$

The computation of the 3×3 determinant takes the form

$$|A| = a_{11} \begin{vmatrix} * & * & * \\ * & a_{22} & a_{23} \\ * & a_{32} & a_{33} \end{vmatrix} - a_{12} \begin{vmatrix} * & * & * \\ a_{21} & * & a_{23} \\ a_{31} & * & a_{33} \end{vmatrix} + a_{13} \begin{vmatrix} * & * & * \\ a_{21} & a_{22} & * \\ a_{31} & a_{32} & * \end{vmatrix}$$

where the first 2×2 determinant is obtained by deleting the first row and first column, the second by deleting the first row and second column, and the third by deleting the first row and third column.

EXAMPLE 2

Find the determinant of the matrix

$$A = \begin{bmatrix} 2 & 1 & -1 \\ 3 & 1 & 4 \\ 5 & -3 & 3 \end{bmatrix}$$

Solution

By Definition 2, the determinant is given by

$$\det(A) = |A| = 2 \begin{vmatrix} 1 & 4 \\ -3 & 3 \end{vmatrix} - 1 \begin{vmatrix} 3 & 4 \\ 5 & 3 \end{vmatrix} + (-1) \begin{vmatrix} 3 & 1 \\ 5 & -3 \end{vmatrix}$$
$$= (2)[3 - (-12)] - (1)(9 - 20) + (-1)(-9 - 5)$$
$$= 30 + 11 + 14$$
$$= 55$$

In Example 2, we found the determinant of a 3×3 matrix by using an *expansion along the first row*. With an adjustment of signs the determinant can be computed by using an expansion along any row. The pattern for the signs is shown in Fig. 1. The expansion along the second row is given by

$$\begin{bmatrix} + & - & + \\ - & + & - \\ + & - & + \end{bmatrix}$$

Figure 1

$$\det(A) = |A| = -3 \begin{vmatrix} 1 & -1 \\ -3 & 3 \end{vmatrix} + 1 \begin{vmatrix} 2 & -1 \\ 5 & 3 \end{vmatrix} - 4 \begin{vmatrix} 2 & 1 \\ 5 & -3 \end{vmatrix}$$
$$= -3(3 - 3) + (6 + 5) - 4(-6 - 5) = 55$$

The 2×2 determinants in this last equation are found from the original matrix by deleting the second row and first column, the second row and second column, and the second row and the third column, respectively. Expansion along the third row is found in a similar way. In this case

$$\det(A) = |A| = 5 \begin{vmatrix} 1 & -1 \\ 1 & 4 \end{vmatrix} - (-3) \begin{vmatrix} 2 & -1 \\ 3 & 4 \end{vmatrix} + 3 \begin{vmatrix} 2 & 1 \\ 3 & 1 \end{vmatrix}$$
$$= 5(4 + 1) + 3(8 + 3) + 3(2 - 3) = 55$$

The determinant can also be computed using expansions along any column in a similar manner. The method used to compute the determinant of a 3×3 matrix can be extended to any square matrix.

DEFINITION 3

Minors and Cofactors of a Matrix   If A is a square matrix, then the **minor** M_{ij}, associated with the entry a_{ij}, is the determinant of the $(n-1) \times (n-1)$ matrix obtained by deleting row i and column j from the matrix A. The **cofactor** of a_{ij} is $C_{ij} = (-1)^{i+j} M_{ij}$.

For the matrix of Example 2, several minors are

$$M_{11} = \begin{vmatrix} 1 & 4 \\ -3 & 3 \end{vmatrix} \qquad M_{12} = \begin{vmatrix} 3 & 4 \\ 5 & 3 \end{vmatrix} \qquad \text{and} \qquad M_{13} = \begin{vmatrix} 3 & 1 \\ 5 & -3 \end{vmatrix}$$

Using the notation of Definition 3, the determinant of A is given by the *cofactor expansion*

$$\begin{aligned} \det(A) &= a_{11}C_{11} + a_{12}C_{12} + a_{13}C_{13} \\ &= 2(-1)^2(15) + 1(-1)^3(-11) - 1(-1)^4(-14) \\ &= 30 + 11 + 14 = 55 \end{aligned}$$

DEFINITION 4

Determinant of a Square Matrix   If A is an $n \times n$ matrix, then

$$\det(A) = a_{11}C_{11} + a_{12}C_{12} + \cdots + a_{1n}C_{1n} = \sum_{k=1}^{n} a_{1k}C_{1k}$$

Similar to the situation for 3×3 matrices, the determinant of any square matrix can be found by expanding along any row or column.

THEOREM 12

Let A be an $n \times n$ matrix. Then the determinant of A equals the cofactor expansion along any row or any column of the matrix. That is, for every $i = 1, \ldots, n$ and $j = 1, \ldots, n$,

$$\det(A) = a_{i1}C_{i1} + a_{i2}C_{i2} + \cdots + a_{in}C_{in} = \sum_{k=1}^{n} a_{ik}C_{ik}$$

and

$$\det(A) = a_{1j}C_{1j} + a_{2j}C_{2j} + \cdots + a_{nj}C_{nj} = \sum_{k=1}^{n} a_{kj}C_{kj}$$

For certain square matrices the computation of the determinant is simplified. One such class of matrices is the square *triangular* matrices.

DEFINITION 5

Triangular Matrices   An $m \times n$ matrix is **upper triangular** if $a_{ij} = 0$, for all $i > j$, and is **lower triangular** if $a_{ij} = 0$, for all $i < j$. A square matrix is a **diagonal** matrix if $a_{ij} = 0$, for all $i \neq j$.

Some examples of upper triangular matrices are

$$\begin{bmatrix} 1 & 1 \\ 0 & 2 \end{bmatrix} \qquad \begin{bmatrix} 2 & -1 & 0 \\ 0 & 0 & 3 \\ 0 & 0 & 2 \end{bmatrix} \qquad \text{and} \qquad \begin{bmatrix} 1 & 1 & 0 & 1 \\ 0 & 0 & 0 & 1 \\ 0 & 0 & 1 & 1 \end{bmatrix}$$

and some examples of lower triangular matrices are

$$\begin{bmatrix} 1 & 0 \\ 1 & 1 \end{bmatrix} \qquad \begin{bmatrix} 2 & 0 & 0 \\ 0 & 1 & 0 \\ 1 & 0 & 2 \end{bmatrix} \qquad \text{and} \qquad \begin{bmatrix} 1 & 0 & 0 & 0 \\ 0 & 0 & 0 & 0 \\ 1 & 3 & 1 & 0 \\ 0 & 1 & 2 & 1 \end{bmatrix}$$

THEOREM 13   If A is an $n \times n$ triangular matrix, then the determinant of A is the product of the terms on the diagonal. That is,

$$\det(A) = a_{11} \cdot a_{22} \cdots a_{nn}$$

Proof   We present the proof for an upper triangular matrix. The proof for a lower triangular matrix is identical. The proof is by induction on n. If $n = 2$, then $\det(A) = a_{11}a_{22} - 0$ and hence is the product of the diagonal terms.

Assume that the result holds for an $n \times n$ triangular matrix. We need to show that the same is true for an $(n + 1) \times (n + 1)$ triangular matrix A. To this end let

$$A = \begin{bmatrix} a_{11} & a_{12} & a_{13} & \cdots & a_{1n} & a_{1,n+1} \\ 0 & a_{22} & a_{23} & \cdots & a_{2n} & a_{2,n+1} \\ 0 & 0 & a_{33} & \cdots & a_{3n} & a_{3,n+1} \\ \vdots & \vdots & \vdots & \ddots & \vdots & \vdots \\ 0 & 0 & 0 & \cdots & a_{nn} & a_{n,n+1} \\ 0 & 0 & 0 & \cdots & 0 & a_{n+1,n+1} \end{bmatrix}$$

Using the cofactor expansion along row $n + 1$, we have

$$\det(A) = (-1)^{(n+1)+(n+1)} a_{n+1,n+1} \begin{vmatrix} a_{11} & a_{12} & a_{13} & \cdots & a_{1n} \\ 0 & a_{22} & a_{23} & \cdots & a_{2n} \\ 0 & 0 & a_{33} & \cdots & a_{3n} \\ \vdots & \vdots & \vdots & \ddots & \vdots \\ 0 & 0 & 0 & \cdots & a_{nn} \end{vmatrix}$$

Since the determinant on the right is $n \times n$ and upper triangular, by the inductive hypothesis

$$\det(A) = (-1)^{2n+2}(a_{n+1,n+1})(a_{11}a_{22} \cdots a_{nn})$$
$$= a_{11}a_{22} \cdots a_{nn}a_{n+1,n+1}$$

Properties of Determinants

Determinants for large matrices can be time-consuming to compute, so any properties of determinants that reduce the number of computations are useful. Theorem 14 shows how row operations affect the determinant.

THEOREM 14 Let A be a square matrix.

1. If two rows of A are interchanged to produce a matrix B, then $\det(B) = -\det(A)$.

2. If a multiple of one row of A is added to another row to produce a matrix B, then $\det(B) = \det(A)$.

3. If a row of A is multiplied by a real number α to produce a matrix B, then $\det(B) = \alpha\det(A)$.

Proof (1) The proof is by induction on n. For the case $n = 2$ let

$$A = \begin{bmatrix} a & b \\ c & d \end{bmatrix}$$

Then $\det(A) = ad - bc$. If the two rows of A are interchanged to give the matrix

$$B = \begin{bmatrix} c & d \\ a & b \end{bmatrix}$$

then $\det(B) = bc - ad = -\det(A)$.

Assume that the result holds for $n \times n$ matrices and A is an $(n + 1) \times (n + 1)$ matrix. Let B be the matrix obtained by interchanging rows i and j of A. Expanding the determinant of A along row i and of B along row j, we have

$$\det(A) = a_{i1}C_{i1} + a_{i2}C_{i2} + \cdots + a_{in}C_{in}$$

and

$$\det(B) = a_{j1}D_{j1} + a_{j2}D_{j2} + \cdots + a_{jn}D_{jn}$$
$$= a_{i1}D_{j1} + a_{i2}D_{j2} + \cdots + a_{in}D_{jn}$$

where C_{ij} and D_{ij} are the cofactors of A and B, respectively. To obtain the result there are two cases. If the signs of the cofactors C_{ij} and D_{ij} are the same, then they differ by one row interchanged. If the signs of the cofactors C_{ij} and D_{ij} are opposite, then they differ by two rows interchanged. In either case, by the inductive hypothesis, we have

$$\det(B) = -\det(A)$$

The proofs of parts 2 and 3 are left as exercises.

We note that in Theorem 14 the same results hold for the similar column operations. To highlight the usefulness of this theorem, recall that by Theorem 13, the determinant of a triangular matrix is the product of the diagonal entries. So an alternative approach to finding the determinant of a matrix A is to row-reduce A to triangular form and apply Theorem 14 to record the effect on the determinant. This method is illustrated in Example 3.

EXAMPLE 3 Find the determinant of the matrix

$$A = \begin{bmatrix} 0 & 1 & 3 & -1 \\ 2 & 4 & -6 & 1 \\ 0 & 3 & 9 & 2 \\ -2 & -4 & 1 & -3 \end{bmatrix}$$

Solution Since column 1 has two zeros, an expansion along this column will involve the fewest computations. Also by Theorem 14, if row 2 is added to row 4, then the determinant is unchanged and

$$\det(A) = \begin{vmatrix} 0 & 1 & 3 & -1 \\ 2 & 4 & -6 & 1 \\ 0 & 3 & 9 & 2 \\ 0 & 0 & -5 & -2 \end{vmatrix}$$

Expansion along the first column gives

$$\det(A) = -2 \begin{vmatrix} 1 & 3 & -1 \\ 3 & 9 & 2 \\ 0 & -5 & -2 \end{vmatrix}$$

We next perform the operation $-3R_1 + R_2 \longrightarrow R_2$, leaving the determinant again unchanged, so that

$$\det(A) = -2 \begin{vmatrix} 1 & 3 & -1 \\ 0 & 0 & 5 \\ 0 & -5 & -2 \end{vmatrix}$$

Now, interchanging the second and third rows gives

$$\det(A) = (-2)(-1) \begin{vmatrix} 1 & 3 & -1 \\ 0 & -5 & -2 \\ 0 & 0 & 5 \end{vmatrix}$$

This last matrix is triangular, thus by Theorem 13,

$$\det(A) = (-2)(-1)[(1)(-5)(5)]$$
$$= -50$$

Theorem 15 lists additional useful properties of the determinant.

THEOREM 15 Let A and B be $n \times n$ matrices and α a real number.

1. The determinant computation is multiplicative. That is,

$$\det(AB) = \det(A)\det(B)$$

2. $\det(\alpha A) = \alpha^n \det(A)$

3. $\det(A^t) = \det(A)$

4. If A has a row (or column) of all zeros, then $\det(A) = 0$.

5. If A has two equal rows (or columns), then $\det(A) = 0$.

6. If A has a row (or column) that is a multiple of another row (or column), then $\det(A) = 0$.

EXAMPLE 4   Let $A = \begin{bmatrix} 1 & 2 \\ 3 & -2 \end{bmatrix}$ and $B = \begin{bmatrix} 1 & -1 \\ 1 & 4 \end{bmatrix}$. Verify Theorem 15, part 1.

Solution   In this case the product is

$$AB = \begin{bmatrix} 3 & 7 \\ 1 & -11 \end{bmatrix}$$

so that $\det(AB) = -33 - 7 = -40$. We also have $\det(A)\det(B) = (-8)(5) = -40$.

Properties of the determinant given in Theorem 15 can be used to establish the connection between the determinant and the invertibility of a square matrix.

THEOREM 16   A square matrix A is invertible if and only if $\det(A) \neq 0$.

Proof   If the matrix A is invertible, then by Theorem 15,

$$1 = \det(I) = \det(AA^{-1}) = \det(A)\det(A^{-1})$$

Since the product of two real numbers is zero if and only if at least one of them is zero, we have $\det(A) \neq 0$ [also $\det(A^{-1}) \neq 0$].

To establish the converse, we will prove the contrapositive statement. Assume that A is not invertible. By the remarks at the end of Sec. 1.4, the matrix A is row equivalent to a matrix R with a row of zeros. Hence, by Theorem 14, there is some real number $k \neq 0$ such that $\det(A) = k\det(R)$, and therefore by Theorem 15, part 4,

$$\det(A) = k\det(R) = k(0) = 0$$

COROLLARY 1   Let A be an invertible matrix. Then

$$\det(A^{-1}) = \frac{1}{\det(A)}$$

Proof   If A is invertible, then as in the proof of Theorem 16, $\det(A) \neq 0$, $\det(A^{-1}) \neq 0$, and

$$\det(A)\det(A^{-1}) = 1$$

Therefore,

$$\det(A^{-1}) = \frac{1}{\det(A)}$$

The final theorem of this section summarizes the connections between inverses, determinants, and linear systems.

THEOREM 17   Let A be a square matrix. Then the following statements are equivalent.

  1. The matrix A is invertible.

  2. The linear system $A\mathbf{x} = \mathbf{b}$ has a unique solution for every vector $\mathbf{b}$.

3. The homogeneous linear system $A\mathbf{x} = \mathbf{0}$ has only the trivial solution.

4. The matrix A is row equivalent to the identity matrix.

5. The determinant of the matrix A is nonzero.

The graph of the equation

$$\frac{(x-h)^2}{a^2} + \frac{(y-k)^2}{b^2} = 1$$

is an ellipse with center (h, k), horizontal axis of length $2a$, and vertical axis of length $2b$.

Determinants can be used to find the equation of a *conic section* passing through specified points. In the 17th century, Johannes Kepler's observations of the orbits of planets about the sun led to the conjecture that these orbits are elliptical. It was Isaac Newton who, later in the same century, proved Kepler's conjecture. The graph of an equation of the form

$$Ax^2 + Bxy + Cy^2 + Dx + Ey + F = 0$$

is a **conic section**. Essentially, the graphs of conic sections are circles, ellipses, hyperbolas, or parabolas.

EXAMPLE 5   An astronomer who wants to determine the approximate orbit of an object traveling about the sun sets up a coordinate system in the plane of the orbit with the sun at the origin. Five observations of the location of the object are then made and are approximated to be $(0, 0.31)$, $(1, 1)$, $(1.5, 1.21)$, $(2, 1.31)$, and $(2.5, 1)$. Use these measurements to find the equation of the ellipse that approximates the orbit.

Solution   We need to find the equation of an ellipse in the form

$$Ax^2 + Bxy + Cy^2 + Dx + Ey + F = 0$$

Each data point must satisfy this equation; for example, since the point $(2, 1.31)$ is on the graph of the conic section,

$$A(2)^2 + B(2)(1.31) + C(1.31)^2 + D(2) + E(1.31) + F = 0$$

so

$$4A + 2.62B + 1.7161C + 2D + 1.31E + F = 0$$

Substituting the five points in the general equation, we obtain the 5×6 linear system (with coefficients rounded to two decimal places)

$$\begin{cases} \quad\quad\quad\quad 0.1C + \quad\quad\quad 0.31E + F = 0 \\ A + \quad B + \quad C + \quad D + \quad E + F = 0 \\ 4A + 2.62B + 1.72C + \quad 2D + 1.31E + F = 0 \\ 2.25A + 1.82B + 1.46C + 1.5D + 1.21E + F = 0 \\ 6.25A + \quad 2.5B + \quad C + 2.5D + \quad E + F = 0 \end{cases}$$

Since the equation $Ax^2 + Bxy + Cy^2 + Dx + Ey + F = 0$ describing the ellipse passing through the five given points has infinitely many solutions, by Theorem 17, we have

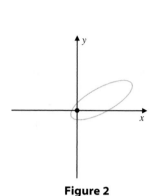

Figure 2

$$\begin{vmatrix} x^2 & xy & y^2 & x & y & 1 \\ 0 & 0 & 0.1 & 0 & 0.31 & 1 \\ 1 & 1 & 1 & 1 & 1 & 1 \\ 4 & 2.62 & 1.72 & 2 & 1.31 & 1 \\ 2.25 & 1.82 & 1.46 & 1.5 & 1.21 & 1 \\ 6.25 & 2.5 & 1 & 2.5 & 1 & 1 \end{vmatrix} = 0$$

Expanding the determinant gives us the equation

$$-0.014868x^2 + 0.0348xy - 0.039y^2 + 0.017238x - 0.003y + 0.00483 = 0$$

The graph of the orbit is shown in Fig. 2.

Cramer's Rule

Determinants can also be used to solve linear systems. To illustrate the technique consider the 2×2 linear system

$$\begin{cases} ax + by = u \\ cx + dy = v \end{cases}$$

with $ad - bc \neq 0$. By Theorem 17, the linear system has a unique solution.

To eliminate the variable y, we multiply the first equation by d and the second equation by b, and then we subtract the two equations. This gives

$$adx + bdy - (bcx + bdy) = du - bv$$

Simplifying, we have

$$(ad - bc)x = du - bv \qquad \text{so that} \qquad x = \frac{du - bv}{ad - bc}$$

Using a similar procedure, we can solve for y.

$$y = \frac{av - cu}{ad - bc}$$

Using determinants, we can write the solution as

$$x = \frac{\begin{vmatrix} u & b \\ v & d \end{vmatrix}}{\begin{vmatrix} a & b \\ c & d \end{vmatrix}} \qquad \text{and} \qquad y = \frac{\begin{vmatrix} a & u \\ c & v \end{vmatrix}}{\begin{vmatrix} a & b \\ c & d \end{vmatrix}}$$

Notice that the solutions for x and y are similar. The denominator for each is the determinant of the coefficient matrix. The determinant in the numerator for x is formed by replacing the first column of the coefficient matrix with the column of constants on the right-hand side of the linear system. The determinant in the numerator for y is formed by replacing the second column of the coefficient matrix with the column of constants. This method of solving a linear system is called *Cramer's rule*.

EXAMPLE 6   Use Cramer's rule to solve the linear system.

$$\begin{cases} 2x + 3y = 2 \\ -5x + 7y = 3 \end{cases}$$

Solution   The determinant of the coefficient matrix is given by

$$\begin{vmatrix} 2 & 3 \\ -5 & 7 \end{vmatrix} = 14 - (-15) = 29$$

and since the determinant is not zero, the system has a unique solution. The solution is given by

$$x = \frac{\begin{vmatrix} 2 & 3 \\ 3 & 7 \end{vmatrix}}{29} = \frac{14 - 9}{29} = \frac{5}{29} \quad \text{and} \quad y = \frac{\begin{vmatrix} 2 & 2 \\ -5 & 3 \end{vmatrix}}{29} = \frac{6 - (-10)}{29} = \frac{16}{29}$$

THEOREM 18

Cramer's Rule   Let A be an $n \times n$ invertible matrix, and let **b** be a column vector with n components. Let A_i be the matrix obtained by replacing the ith column of A with **b**. If $\mathbf{x} = \begin{bmatrix} x_1 \\ x_2 \\ \vdots \\ x_n \end{bmatrix}$ is the unique solution to the linear system $A\mathbf{x} = \mathbf{b}$, then

$$x_i = \frac{\det(A_i)}{\det(A)} \quad \text{for} \quad i = 1, 2, \ldots, n$$

Proof   Let I_i be the matrix obtained by replacing the ith column of the identity matrix with **x**. Then the linear system is equivalent to the matrix equation

$$A I_i = A_i \quad \text{so} \quad \det(A I_i) = \det(A_i)$$

By Theorem 15, part 1, we have

$$\det(A) \det(I_i) = \det(A I_i) = \det(A_i)$$

Since A is invertible, $\det(A) \neq 0$ and hence

$$\det(I_i) = \frac{\det(A_i)}{\det(A)}$$

Expanding along the ith row to find the determinant of I_i gives

$$\det(I_i) = x_i \det(I) = x_i$$

where I is the $(n - 1) \times (n - 1)$ identity. Therefore,

$$x_i = \frac{\det(A_i)}{\det(A)}$$

If a unique solution exists, then Cramer's rule can be used to solve larger square linear systems. Example 7 illustrates the technique for a 3×3 system.

EXAMPLE 7   Solve the linear system.

$$\begin{cases} 2x + 3y - z = 2 \\ 3x - 2y + z = -1 \\ -5x - 4y + 2z = 3 \end{cases}$$

Solution   The determinant of the coefficient matrix is given by

$$\begin{vmatrix} 2 & 3 & -1 \\ 3 & -2 & 1 \\ -5 & -4 & 2 \end{vmatrix} = -11$$

By Cramer's rule the solution to the system is

$$x = -\frac{1}{11} \begin{vmatrix} 2 & 3 & -1 \\ -1 & -2 & 1 \\ 3 & -4 & 2 \end{vmatrix} = -\frac{5}{11}$$

$$y = -\frac{1}{11} \begin{vmatrix} 2 & 2 & -1 \\ 3 & -1 & 1 \\ -5 & 3 & 2 \end{vmatrix} = \frac{36}{11}$$

$$z = -\frac{1}{11} \begin{vmatrix} 2 & 3 & 2 \\ 3 & -2 & -1 \\ -5 & -4 & 3 \end{vmatrix} = \frac{76}{11}$$

The reader should verify this solution by substitution into the original system.

Fact Summary

Let A and B be $n \times n$ matrices.

1. $\det \begin{bmatrix} a & b \\ c & d \end{bmatrix} = ad - bc$.

2. The determinant of A can be computed by expanding along any row or column provided that the signs are adjusted using the pattern

$$\begin{bmatrix} + & - & + & - & \cdots \\ - & + & - & + & \cdots \\ + & - & + & - & \cdots \\ - & + & - & + & \cdots \\ \vdots & \vdots & \vdots & \vdots & \ddots \end{bmatrix}$$

3. The matrix A is invertible if and only if $\det(A) \neq 0$.

4. If A is a triangular matrix, then the determinant of A is the
   diagonal terms.

5. If two rows of A are interchanged, the determinant of the r
   is the negative of the determinant of A.

6. If a multiple of one row of A is added to another row, the determinant is
   not altered.

7. If one row of A is multiplied by a scalar c, the determinant of the resulting
   matrix is c times the determinant of A.

8. $\det(AB) = \det(A)\det(B)$, $\det(cA) = c^n \det(A)$, $\det(A^t) = \det(A)$

9. If A has a row or column of zeros, then $\det(A) = 0$.

10. If one row or column of A is a multiple of another row or column, then
    $\det(A) = 0$.

11. If A is invertible, then $\det(A^{-1}) = \frac{1}{\det(A)}$.

Exercise Set 1.6

In Exercises 1–4, evaluate the determinant of the
matrix by inspection.

1. $\begin{bmatrix} 2 & -40 & 10 \\ 0 & 3 & 12 \\ 0 & 0 & 4 \end{bmatrix}$

2. $\begin{bmatrix} 1 & 2 & 3 \\ 4 & 5 & 6 \\ 1 & 2 & 3 \end{bmatrix}$

3. $\begin{bmatrix} 1 & 0 & 0 & 0 \\ 3 & -1 & 0 & 0 \\ 4 & 2 & 2 & 0 \\ 1 & 1 & 6 & 5 \end{bmatrix}$

4. $\begin{bmatrix} 1 & -1 & 2 \\ 2 & -2 & 4 \\ 1 & 2 & -1 \end{bmatrix}$

In Exercises 5–8, use determinants to decide if the
matrix is invertible.

5. $\begin{bmatrix} 2 & -1 \\ -2 & 2 \end{bmatrix}$

6. $\begin{bmatrix} 1 & 3 \\ 5 & -2 \end{bmatrix}$

7. $\begin{bmatrix} 1 & 0 & 0 \\ 3 & 6 & 0 \\ 0 & 8 & -1 \end{bmatrix}$

8. $\begin{bmatrix} 7 & 2 & 1 \\ 7 & 2 & 1 \\ 3 & 6 & 6 \end{bmatrix}$

9. Answer the questions using the matrix

$$A = \begin{bmatrix} 2 & 0 & 1 \\ 3 & -1 & 4 \\ -4 & 1 & -2 \end{bmatrix}$$

a. Find the determinant of the matrix by using an
   expansion along row 1.

b. Find the determinant of the matrix by using an
   expansion along row 2.

c. Find the determinant of the matrix by using an
   expansion along column 2.

d. Interchange rows 1 and 3 of the matrix, and
   find the determinant of the transformed matrix.

e. Multiply row 1 of the matrix found in part (d)
   by -2, and find the determinant of the new
   matrix. Use the value to find the determinant of
   the original matrix.

f. Replace row 3 of the matrix found in part (e) with the sum of row 3 and -2 times row 1 and find the determinant of the new matrix in two ways. First, use an expansion along row 3 of the new matrix. Second, use the value for the determinant of A that has already been calculated.

g. Does the matrix A have an inverse? Do not try to compute the inverse.

10. Answer the questions using the matrix

$$A = \begin{bmatrix} -1 & 1 & 1 & 2 \\ 3 & -2 & 0 & -1 \\ 0 & 1 & 0 & 1 \\ 3 & 3 & 3 & 3 \end{bmatrix}$$

a. Find the determinant of the matrix by using an expansion along row 4.

b. Find the determinant of the matrix by using an expansion along row 3.

c. Find the determinant of the matrix by using an expansion along column 2.

d. In (a), (b), and (c), which computation do you prefer, and why?

e. Does the matrix A have an inverse? Do not try to compute the inverse.

In Exercises 11–26, find the determinant of the matrix. Specify whether the matrix has an inverse without trying to compute the inverse.

11. $\begin{bmatrix} 5 & 6 \\ -8 & -7 \end{bmatrix}$

12. $\begin{bmatrix} 4 & 3 \\ 9 & 2 \end{bmatrix}$

13. $\begin{bmatrix} -1 & -1 \\ -11 & 5 \end{bmatrix}$

14. $\begin{bmatrix} 1 & 2 \\ 4 & 13 \end{bmatrix}$

15. $\begin{bmatrix} -1 & 4 \\ 1 & -4 \end{bmatrix}$

16. $\begin{bmatrix} 1 & 1 \\ 2 & 2 \end{bmatrix}$

17. $\begin{bmatrix} 5 & -5 & -4 \\ -1 & -3 & 5 \\ -3 & 1 & 3 \end{bmatrix}$

18. $\begin{bmatrix} 3 & -3 & 5 \\ 2 & 4 & -3 \\ -3 & -1 & -5 \end{bmatrix}$

19. $\begin{bmatrix} -3 & 4 & 5 \\ 1 & 1 & 4 \\ -1 & -3 & 4 \end{bmatrix}$

20. $\begin{bmatrix} -2 & -2 & -4 \\ 1 & 1 & 3 \\ -4 & 0 & 4 \end{bmatrix}$

21. $\begin{bmatrix} 1 & -4 & 1 \\ 1 & -2 & 4 \\ 0 & 2 & 3 \end{bmatrix}$

22. $\begin{bmatrix} 1 & 2 & 4 \\ 4 & 0 & 0 \\ 1 & 2 & 4 \end{bmatrix}$

23. $\begin{bmatrix} 2 & -2 & -2 & -2 \\ -2 & 2 & 3 & 0 \\ -2 & -2 & 2 & 0 \\ 1 & -1 & -3 & -1 \end{bmatrix}$

24. $\begin{bmatrix} 1 & -1 & 0 & 0 \\ -3 & -3 & -1 & -1 \\ -1 & -1 & -3 & 2 \\ -1 & -2 & 2 & 1 \end{bmatrix}$

25. $\begin{bmatrix} -1 & 1 & 1 & 0 & 0 \\ 0 & 0 & -1 & 0 & 0 \\ 0 & 0 & 1 & -1 & 0 \\ 0 & 1 & 1 & 0 & 1 \\ 1 & -1 & 1 & 1 & 0 \end{bmatrix}$

26. $\begin{bmatrix} 1 & 0 & -1 & 0 & -1 \\ -1 & -1 & 0 & 0 & -1 \\ 1 & 0 & 0 & 0 & -1 \\ 0 & 1 & 1 & 1 & 0 \\ -1 & 1 & 1 & -1 & 0 \end{bmatrix}$

In Exercises 27–30, let

$$A = \begin{bmatrix} a & b & c \\ d & e & f \\ g & h & i \end{bmatrix}$$

and assume $\det(A) = 10$.

27. Find $\det(3A)$.

28. Find $\det(2A^{-1})$.

29. Find $\det\left[(2A)^{-1}\right]$.

30. Find

$$\det \begin{bmatrix} a & g & d \\ b & h & e \\ c & i & f \end{bmatrix}$$

31. Find x, assuming

$$\det \begin{bmatrix} x^2 & x & 2 \\ 2 & 1 & 1 \\ 0 & 0 & -5 \end{bmatrix} = 0$$

32. Find the determinant of the matrix

$$\begin{bmatrix} 1 & 1 & 1 & 1 & 1 \\ 0 & 1 & 1 & 1 & 1 \\ 1 & 0 & 1 & 1 & 1 \\ 1 & 1 & 0 & 1 & 1 \\ 1 & 1 & 1 & 0 & 1 \end{bmatrix}$$

33. Suppose $a_1 \neq b_1$. Describe the set of all points (x, y) that satisfy the equation

$$\det \begin{bmatrix} 1 & 1 & 1 \\ x & a_1 & b_1 \\ y & a_2 & b_2 \end{bmatrix} = 0$$

34. Use the three systems to answer the questions.

$$(1) \begin{cases} x + y = 3 \\ 2x + 2y = 1 \end{cases} \quad (2) \begin{cases} x + y = 3 \\ 2x + 2y = 6 \end{cases}$$

$$(3) \begin{cases} x + y = 3 \\ 2x - 2y = 1 \end{cases}$$

a. Form the coefficient matrices A, B, and C, respectively, for the three systems.

b. Find $\det(A)$, $\det(B)$, and $\det(C)$. How are they related?

c. Which of the coefficient matrices have inverses?

d. Find all solutions to system (1).

e. Find all solutions to system (2).

f. Find all solutions to system (3).

35. Answer the questions about the linear system.

$$\begin{cases} x - y - 2z = 3 \\ -x + 2y + 3z = 1 \\ 2x - 2y - 2z = -2 \end{cases}$$

a. Form the coefficient matrix A for the linear system.

b. Find $\det(A)$.

c. Does the system have a unique solution? Explain.

d. Find all solutions to the system.

36. Answer the questions about the linear system.

$$\begin{cases} x + 3y - 2z = -1 \\ 2x + 5y + z = 2 \\ 2x + 6y - 4z = -2 \end{cases}$$

a. Form the coefficient matrix A for the linear system.

b. Find $\det(A)$.

c. Does the system have a unique solution? Explain.

d. Find all solutions to the system.

37. Answer the questions about the linear system.

$$\begin{cases} -x - z = -1 \\ 2x + 2z = 1 \\ x - 3y - 3z = 1 \end{cases}$$

a. Form the coefficient matrix A for the linear system.

b. Find $\det(A)$.

c. Does the system have a unique solution? Explain.

d. Find all solutions for the system.

In Exercises 38–43, use the fact that the graph of the general equation

$$Ax^2 + Bxy + Cy^2 + Dx + Ey + F = 0$$

is essentially a parabola, circle, ellipse, or hyperbola.

38. a. Find the equation of the parabola in the form

$$Ax^2 + Dx + Ey + F = 0$$

that passes through the points $(0, 3)$, $(1, 1)$, and $(4, -2)$.

b. Sketch the graph of the parabola.

39. a. Find the equation of the parabola in the form

$$Cy^2 + Dx + Ey + F = 0$$

that passes through the points $(-2, -2)$, $(3, 2)$, and $(4, -3)$.

b. Sketch the graph of the parabola.

40. a. Find the equation of the circle in the form

$$A(x^2 + y^2) + Dx + Ey + F = 0$$

that passes through the points $(-3, -3)$, $(-1, 2)$, and $(3, 0)$.

b. Sketch the graph of the circle.

41. a. Find the equation of the hyperbola in the form

$$Ax^2 + Cy^2 + Dx + Ey + F = 0$$

that passes through the points $(0, -4)$, $(0, 4)$, $(1, -2)$, and $(2, 3)$.

b. Sketch the graph of the hyperbola.

42. a. Find the equation of the ellipse in the form

$$Ax^2 + Cy^2 + Dx + Ey + F = 0$$

that passes through the points $(-3, 2)$, $(-1, 3)$, $(1, -1)$, and $(4, 2)$.

b. Sketch the graph of the ellipse.

43. a. Find the equation of the ellipse in the form

$$Ax^2 + Bxy + Cy^2 + Dx + Ey + F = 0$$

that passes through the points $(-1, 0)$, $(0, 1)$, $(1, 0)$, $(2, 2)$, and $(3, 1)$.

b. Sketch the graph of the ellipse.

In Exercises 44–51, use Cramer's rule to solve the linear system.

44. $\begin{cases} 2x + 3y = 4 \\ 2x + 2y = 4 \end{cases}$

45. $\begin{cases} 5x - 5y = 7 \\ 2x - 3y = 6 \end{cases}$

46. $\begin{cases} 2x + 5y = 4 \\ 4x + y = 3 \end{cases}$

47. $\begin{cases} -9x - 4y = 3 \\ -7x + 5y = -10 \end{cases}$

48. $\begin{cases} -10x - 7y = -12 \\ 12x + 11y = 5 \end{cases}$

49. $\begin{cases} -x - 3y = 4 \\ -8x + 4y = 3 \end{cases}$

50. $\begin{cases} -2x + y - 4z = -8 \\ - 4y + z = 3 \\ 4x - z = -8 \end{cases}$

51. $\begin{cases} 2x + 3y + 2z = -2 \\ -x - 3y - 8z = -2 \\ -3x + 2y - 7z = 2 \end{cases}$

52. An $n \times n$ matrix is *skew-symmetric* provided $A^t = -A$. Show that if A is skew-symmetric and n is an odd positive integer, then A is not invertible.

53. If A is a 3×3 matrix, show that $\det(A) = \det(A^t)$.

54. If A is an $n \times n$ upper triangular matrix, show that $\det(A) = \det(A^t)$.

1.7 ▶ Elementary Matrices and *LU* Factorization

In Sec. 1.2 we saw how the linear system $A\mathbf{x} = \mathbf{b}$ can be solved by using Gaussian elimination on the corresponding augmented matrix. Recall that the idea there was to use row operations to transform the coefficient matrix to row echelon form. The upper triangular form of the resulting matrix made it easy to find the solution by using back substitution. (See Example 1 of Sec. 1.2.) In a similar manner, if an augmented matrix is reduced to lower triangular form, then *forward substitution* can be used to find the solution of the corresponding linear system. For example, starting from the first equation of the linear system

$$\begin{cases} x_1 & = 3 \\ -x_1 + x_2 & = -1 \\ 2x_1 - x_2 + x_3 & = 5 \end{cases}$$

we obtain the solution $x_1 = 3$, $x_2 = 2$, and $x_3 = 1$. Thus, from a computational perspective, to find the solution of a linear system, it is desirable that the corresponding matrix be either upper or lower triangular.

In this section we show how, in certain cases, an $m \times n$ matrix A can be written as $A = LU$, where L is a lower triangular matrix and U is an upper triangular matrix. We call this an ***LU* factorization of** A. For example, an LU factorization of the matrix $\begin{bmatrix} -3 & -2 \\ 3 & 4 \end{bmatrix}$ is given by

$$\begin{bmatrix} -3 & -2 \\ 3 & 4 \end{bmatrix} = \begin{bmatrix} -1 & 0 \\ 1 & 2 \end{bmatrix} \begin{bmatrix} 3 & 2 \\ 0 & 1 \end{bmatrix}$$

with $L = \begin{bmatrix} -1 & 0 \\ 1 & 2 \end{bmatrix}$ and $U = \begin{bmatrix} 3 & 2 \\ 0 & 1 \end{bmatrix}$. We also show in this section that when such a factorization of A exists, a process that involves both forward and back substitution can be used to find the solution to the linear system $A\mathbf{x} = \mathbf{b}$.

Elementary Matrices

As a first step we describe an alternative method for carrying out row operations using *elementary matrices*.

DEFINITION 1　　**Elementary Matrix**　An **elementary matrix** is any matrix that can be obtained from the identity matrix by performing a single elementary row operation.

As an illustration, the elementary matrix E_1 is formed by interchanging the first and third rows of the 3×3 identity matrix I, that is,

$$E_1 = \begin{bmatrix} 0 & 0 & 1 \\ 0 & 1 & 0 \\ 1 & 0 & 0 \end{bmatrix}$$

Corresponding to the three row operations given in Theorem 2 of Sec. 1.2, there are three types of elementary matrices. For example, as we have just seen, E_1 is derived from I by means of the row operation $R_1 \leftrightarrow R_3$ which interchanges the first and third rows. Also, the row operation $kR_1 + R_2 \longrightarrow R_2$ applied to I yields the elementary matrix

$$E_2 = \begin{bmatrix} 1 & 0 & 0 \\ k & 1 & 0 \\ 0 & 0 & 1 \end{bmatrix}$$

Next, if $c \neq 0$, the row operation $cR_2 \longrightarrow R_2$ performed on I produces the matrix

$$E_3 = \begin{bmatrix} 1 & 0 & 0 \\ 0 & c & 0 \\ 0 & 0 & 1 \end{bmatrix}$$

Using any row operation, we can construct larger elementary matrices from larger identity matrices in a similar manner.

We now show how elementary matrices can be used to perform row operations. To illustrate the process, let A be the 3×3 matrix given by

$$A = \begin{bmatrix} 1 & 2 & 3 \\ 4 & 5 & 6 \\ 7 & 8 & 9 \end{bmatrix}$$

Multiplying A by the matrix E_1, defined above, we obtain

$$E_1 A = \begin{bmatrix} 7 & 8 & 9 \\ 4 & 5 & 6 \\ 1 & 2 & 3 \end{bmatrix}$$

Observe that $E_1 A$ is the result of interchanging the first and third rows of A. Theorem 19 gives the situation in general.

THEOREM 19   Let A be an $m \times n$ matrix and E the elementary matrix obtained from the $m \times m$ identity matrix I by a single row operation $\mathcal{R}$. Denote by $\mathcal{R}(A)$ the result of performing the row operation on A. Then $\mathcal{R}(A) = EA$.

By repeated application of Theorem 19, a sequence of row operations can be performed on a matrix A by successively multiplying A by the corresponding elementary matrices. Specifically, let E_i be the elementary matrix corresponding to the row operation $\mathcal{R}_i$ with $1 \leq i \leq k$. Then

$$\mathcal{R}_k \cdots \mathcal{R}_2 (\mathcal{R}_1(A)) = E_k \cdots E_2 E_1 A$$

EXAMPLE 1   Let A be the matrix

$$A = \begin{bmatrix} 1 & 2 & -1 \\ 3 & 5 & 0 \\ -1 & 1 & 1 \end{bmatrix}$$

Use elementary matrices to perform the row operations $\mathcal{R}_1$: $R_2 - 3R_1 \longrightarrow R_2$, $\mathcal{R}_2$: $R_3 + R_1 \longrightarrow R_3$, and $\mathcal{R}_3$: $R_3 + 3R_2 \longrightarrow R_3$.

Solution   The elementary matrices corresponding to these row operations are given by

$$E_1 = \begin{bmatrix} 1 & 0 & 0 \\ -3 & 1 & 0 \\ 0 & 0 & 1 \end{bmatrix} \qquad E_2 = \begin{bmatrix} 1 & 0 & 0 \\ 0 & 1 & 0 \\ 1 & 0 & 1 \end{bmatrix} \qquad E_3 = \begin{bmatrix} 1 & 0 & 0 \\ 0 & 1 & 0 \\ 0 & 3 & 1 \end{bmatrix}$$

respectively, so that

$$E_3 E_2 E_1 A = \begin{bmatrix} 1 & 2 & -1 \\ 0 & -1 & 3 \\ 0 & 0 & 9 \end{bmatrix}$$

The reader should check that the matrix on the right-hand side is equal to the result of performing the row operations above on A in the order given.

The Inverse of an Elementary Matrix

An important property of elementary matrices is that they are invertible.

THEOREM 20

Let E be an $n \times n$ elementary matrix. Then E is invertible. Moreover, its inverse is also an elementary matrix.

Proof Let E be an elementary matrix. To show that E is invertible, we compute its determinant and apply Theorem 17 of Sec. 1.6. There are three cases depending on the form of E. First, if E is derived from I by an interchange of two rows, then $\det(E) = -\det(I) = -1$. Second, if E is the result of multiplying one row of I by a nonzero scalar c, then $\det(E) = c \det(I) = c \neq 0$. Third, if E is formed by adding a multiple of one row of I to another row, then $\det(E) = \det(I) = 1$. In either case, $\det(E) \neq 0$ and hence E is invertible. To show that E^{-1} is an elementary matrix, we use the algorithm of Sec. 1.4 to compute the inverse. In this case starting with the $n \times 2n$ augmented matrix

$$[E \mid I]$$

we reduce the elementary matrix on the left (to I) by applying the reverse operation used to form E, obtaining

$$\left[I \mid E^{-1} \right]$$

That E^{-1} is also an elementary matrix follows from the fact that the reverse of each row operation is also a row operation.

As an illustration of the above theorem, let $\mathcal{R}$ be the row operation $2R_2 + R_1 \longrightarrow R_1$, which says to *add 2 times row 2 to row 1*. The corresponding elementary matrix is given by

$$E = \begin{bmatrix} 1 & 2 & 0 \\ 0 & 1 & 0 \\ 0 & 0 & 1 \end{bmatrix}$$

Since $\det(E) = 1$, then E is invertible with

$$E^{-1} = \begin{bmatrix} 1 & -2 & 0 \\ 0 & 1 & 0 \\ 0 & 0 & 1 \end{bmatrix}$$

Observe that E^{-1} corresponds to the row operation $\mathcal{R}_2: -2R_2 + R_1 \longrightarrow R_1$ which says to *subtract 2 times row 2 from row 1*, reversing the original row operation $\mathcal{R}$.

Recall from Sec. 1.2 that an $m \times n$ matrix A is row equivalent to an $m \times n$ matrix B if B can be obtained from A by a finite sequence of row operations. Theorem 21 gives a restatement of this fact in terms of elementary matrices.

THEOREM 21   Let A and B be $m \times n$ matrices. The matrix A is row equivalent to B if and only if there are elementary matrices $E_1, E_2, \ldots, E_k$ such that $B = E_k E_{k-1} \cdots E_2 E_1 A$.

In light of Theorem 21, if A is row equivalent to B, then B is row equivalent to A. Indeed, if A is row equivalent to B, then

$$B = E_k E_{k-1} \cdots E_2 E_1 A$$

for some elementary matrices $E_1, E_2, \ldots, E_k$. Successively multiplying both sides of this equation by $E_k^{-1}, E_{k-1}^{-1}, \ldots,$ and E_1^{-1}, we obtain

$$A = E_1^{-1} \cdots E_{k-1}^{-1} E_k^{-1} B$$

Since each of the matrices $E_1^{-1}, E_2^{-1}, \ldots, E_k^{-1}$ is an elementary matrix, B is row equivalent to A.

Theorem 22 uses elementary matrices to provide a characterization of invertible matrices.

THEOREM 22   An $n \times n$ matrix A is invertible if and only if it can be written as the product of elementary matrices.

Proof   First assume that there are elementary matrices $E_1, E_2, \ldots, E_k$ such that

$$A = E_1 E_2 \cdots E_{k-1} E_k$$

We claim that the matrix $B = E_k^{-1} \cdots E_2^{-1} E_1^{-1}$ is the inverse of A. To show this, we multiply both sides of $A = E_1 E_2 \cdots E_{k-1} E_k$ by B to obtain

$$BA = (E_k^{-1} \cdots E_2^{-1} E_1^{-1})A = (E_k^{-1} \cdots E_2^{-1} E_1^{-1})(E_1 E_2 \cdots E_{k-1} E_k) = I$$

establishing the claim. On the other hand, suppose that A is invertible. In Sec. 1.4, we showed that A is row equivalent to the identity matrix. So by Theorem 21, there are elementary matrices $E_1, E_2, \ldots, E_k$ such that $I = E_k E_{k-1} \cdots E_2 E_1 A$. Consequently, $A = E_1^{-1} \cdots E_k^{-1} I$. Since $E_1^{-1}, \ldots, E_k^{-1}$ and I are all elementary matrices, A is the product of elementary matrices as desired.

LU Factorization

There are many reasons why it is desirable to obtain an *LU* factorization of a matrix. For example, suppose that A is an $m \times n$ matrix and $\mathbf{b}_i$, with $1 \leq i \leq k$, is a collection of vectors in $\mathbb{R}^n$, which represent *outputs* for the linear systems $A\mathbf{x} = \mathbf{b}_i$. Finding *input* vectors $\mathbf{x}_i$ requires that we solve k linear systems. However, since the matrix A is the

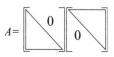

$$A = \begin{array}{cc} \boxed{\begin{matrix} & 0 \\ & \end{matrix}} & \boxed{\begin{matrix} & \\ 0 & \end{matrix}} \end{array}$$

Figure 1

same for each linear system, the process is greatly simplified if A is replaced with its *LU* factorization. The details for solving a linear system using an *LU* factorization are presented later in this section. If A is an $n \times n$ matrix with an *LU* factorization given by $A = LU$, then L and U are also $n \times n$. See Fig. 1. Then by Theorem 13 of Sec. 1.6, the determinant of A is given by

$$\det(A) = (\ell_{11} \cdots \ell_{nn})(u_{11} \cdots u_{nn})$$

where ℓ_{ii} and u_{ii} are the diagonal entries of L and U, respectively. If this determinant is not zero, then by Theorem 9 of Sec. 1.4 the inverse of the matrix A is given by

$$A^{-1} = (LU)^{-1} = U^{-1}L^{-1}$$

To describe the process of obtaining an *LU* factorization of an $m \times n$ matrix A, suppose that A can be reduced to an upper triangular matrix by a sequence of row operations which correspond to lower triangular elementary matrices. That is, there exist lower triangular elementary matrices $L_1, L_2, \ldots, L_k$ such that

$$L_k L_{k-1} \cdots L_1 A = U$$

Since each of the matrices L_i with $1 \le i \le k$ is invertible, we have

$$A = L_1^{-1} L_2^{-1} \cdots L_k^{-1} U$$

By Theorem 20, $L_1^{-1}, L_2^{-1}, \ldots,$ and L_k^{-1} are elementary matrices. They are also lower triangular. Now let $L = L_1^{-1} L_2^{-1} \cdots L_k^{-1}$. Observe that L is lower triangular as it is the product of lower triangular matrices. The desired factorization is thus given by $A = LU$.

EXAMPLE 2   Find an *LU* factorization of the matrix

$$A = \begin{bmatrix} 3 & 6 & -3 \\ 6 & 15 & -5 \\ -1 & -2 & 6 \end{bmatrix}$$

Solution   Observe that A can be row-reduced to an upper triangular matrix by means of the row operations $\mathcal{R}_1 : \frac{1}{3}R_1 \longrightarrow R_1$, $\mathcal{R}_2 : -6R_1 + R_2 \longrightarrow R_2$, and $\mathcal{R}_3 : R_1 + R_3 \longrightarrow R_3$. The corresponding elementary matrices are therefore given by

$$E_1 = \begin{bmatrix} \frac{1}{3} & 0 & 0 \\ 0 & 1 & 0 \\ 0 & 0 & 1 \end{bmatrix} \qquad E_2 = \begin{bmatrix} 1 & 0 & 0 \\ -6 & 1 & 0 \\ 0 & 0 & 1 \end{bmatrix} \qquad E_3 = \begin{bmatrix} 1 & 0 & 0 \\ 0 & 1 & 0 \\ 1 & 0 & 1 \end{bmatrix}$$

respectively, so that

$$E_3 E_2 E_1 A = \begin{bmatrix} 1 & 0 & 0 \\ 0 & 1 & 0 \\ 1 & 0 & 1 \end{bmatrix} \begin{bmatrix} 1 & 0 & 0 \\ -6 & 1 & 0 \\ 0 & 0 & 1 \end{bmatrix} \begin{bmatrix} \frac{1}{3} & 0 & 0 \\ 0 & 1 & 0 \\ 0 & 0 & 1 \end{bmatrix} \begin{bmatrix} 3 & 6 & -3 \\ 6 & 15 & -5 \\ -1 & -2 & 6 \end{bmatrix}$$

$$= \begin{bmatrix} 1 & 2 & -1 \\ 0 & 3 & 1 \\ 0 & 0 & 5 \end{bmatrix} = U$$

An LU factorization of A is then given by $A = \left(E_1^{-1} E_2^{-1} E_3^{-1} \right) U$, so that

$$
\begin{bmatrix} 3 & 6 & -3 \\ 6 & 15 & -5 \\ -1 & -2 & 6 \end{bmatrix} = \begin{bmatrix} 3 & 0 & 0 \\ 6 & 1 & 0 \\ -1 & 0 & 1 \end{bmatrix} \begin{bmatrix} 1 & 2 & -1 \\ 0 & 3 & 1 \\ 0 & 0 & 5 \end{bmatrix}
$$

From the remarks preceding Example 2, we see that to use the above procedure with success, there are limitations on the matrix A. Specifically, A must be reducible to upper triangular form without any row interchanges. This will ensure that the elementary matrices used in the elimination process will all be lower triangular. Theorem 23 summarizes these results.

THEOREM 23   Let A be an $m \times n$ matrix that can be reduced to the upper triangular matrix U without row interchanges by means of the $m \times m$ lower triangular matrices L_1, $L_2, \ldots L_k$. If $L = L_1^{-1} L_2^{-1} \cdots L_k^{-1}$, then A has the LU factorization $A = LU$.

A simple example of a matrix that cannot be reduced to upper triangular form without interchanges is given by $P = \begin{bmatrix} 0 & 1 \\ 1 & 0 \end{bmatrix}$. This matrix does not have an LU factorization. (See Exercise 29.)

As stated in Theorem 23, it is not necessary for A to be square, as shown in Example 3.

EXAMPLE 3   Find an LU factorization of the matrix

$$
A = \begin{bmatrix} 1 & -3 & -2 & 0 \\ 1 & -2 & 1 & -1 \\ 2 & -4 & 3 & 2 \end{bmatrix}
$$

Solution   Observe that A can be reduced to the upper triangular matrix $U = \begin{bmatrix} 1 & -3 & -2 & 0 \\ 0 & 1 & 3 & -1 \\ 0 & 0 & 1 & 4 \end{bmatrix}$

by means of the elementary matrices

$$
E_1 = \begin{bmatrix} 1 & 0 & 0 \\ -1 & 1 & 0 \\ 0 & 0 & 1 \end{bmatrix} \qquad E_2 = \begin{bmatrix} 1 & 0 & 0 \\ 0 & 1 & 0 \\ -2 & 0 & 1 \end{bmatrix} \qquad E_3 = \begin{bmatrix} 1 & 0 & 0 \\ 0 & 1 & 0 \\ 1 & -2 & 1 \end{bmatrix}
$$

Hence,

$$A = \left(E_1^{-1} E_2^{-1} E_3^{-1} \right) U$$

$$= \left(\begin{bmatrix} 1 & 0 & 0 \\ 1 & 1 & 0 \\ 0 & 0 & 1 \end{bmatrix} \begin{bmatrix} 1 & 0 & 0 \\ 0 & 1 & 0 \\ 2 & 0 & 1 \end{bmatrix} \begin{bmatrix} 1 & 0 & 0 \\ 0 & 1 & 0 \\ 0 & 2 & 1 \end{bmatrix} \right) \begin{bmatrix} 1 & -3 & -2 & 0 \\ 0 & 1 & 3 & -1 \\ 0 & 0 & 1 & 4 \end{bmatrix}$$

$$= \begin{bmatrix} 1 & 0 & 0 \\ 1 & 1 & 0 \\ 2 & 2 & 1 \end{bmatrix} \begin{bmatrix} 1 & -3 & -2 & 0 \\ 0 & 1 & 3 & -1 \\ 0 & 0 & 1 & 4 \end{bmatrix} = LU$$

Solving a Linear System Using *LU* Factorization

We now turn our attention to the process of solving a linear system by using an *LU* factorization. To illustrate the procedure, consider the linear system $A\mathbf{x} = \mathbf{b}$ with

$$\mathbf{b} = \begin{bmatrix} 3 \\ 11 \\ 9 \end{bmatrix}$$ and A the matrix of Example 2. By using the *LU* factorization of A found in Example 2,

$$A\mathbf{x} = \begin{bmatrix} 3 & 6 & -3 \\ 6 & 15 & -5 \\ -1 & -2 & 6 \end{bmatrix} \begin{bmatrix} x_1 \\ x_2 \\ x_3 \end{bmatrix} = \begin{bmatrix} 3 \\ 11 \\ 9 \end{bmatrix}$$

can be written equivalently as

$$LU\mathbf{x} = \begin{bmatrix} 3 & 0 & 0 \\ 6 & 1 & 0 \\ -1 & 0 & 1 \end{bmatrix} \begin{bmatrix} 1 & 2 & -1 \\ 0 & 3 & 1 \\ 0 & 0 & 5 \end{bmatrix} \begin{bmatrix} x_1 \\ x_2 \\ x_3 \end{bmatrix} = \begin{bmatrix} 3 \\ 11 \\ 9 \end{bmatrix}$$

To solve this equation efficiently, we define the vector $\mathbf{y} = \begin{bmatrix} y_1 \\ y_2 \\ y_3 \end{bmatrix}$ by the equation $U\mathbf{x} = \mathbf{y}$, so that

$$\begin{bmatrix} 1 & 2 & -1 \\ 0 & 3 & 1 \\ 0 & 0 & 5 \end{bmatrix} \begin{bmatrix} x_1 \\ x_2 \\ x_3 \end{bmatrix} = \begin{bmatrix} y_1 \\ y_2 \\ y_3 \end{bmatrix}$$

Making this substitution in the linear system $L(U\mathbf{x}) = \mathbf{b}$ gives

$$\begin{bmatrix} 3 & 0 & 0 \\ 6 & 1 & 0 \\ -1 & 0 & 1 \end{bmatrix} \begin{bmatrix} y_1 \\ y_2 \\ y_3 \end{bmatrix} = \begin{bmatrix} 3 \\ 11 \\ 9 \end{bmatrix}$$

Using forward substitution, we solve the system $L\mathbf{y} = \mathbf{b}$ for $\mathbf{y}$, obtaining $y_1 = 1$, $y_2 = 5$, and $y_3 = 10$. Next we solve the linear system $U\mathbf{x} = \mathbf{y}$. That is,

$$\begin{bmatrix} 1 & 2 & -1 \\ 0 & 3 & 1 \\ 0 & 0 & 5 \end{bmatrix} \begin{bmatrix} x_1 \\ x_2 \\ x_3 \end{bmatrix} = \begin{bmatrix} 1 \\ 5 \\ 10 \end{bmatrix}$$

Using back substitution, we obtain $x_3 = 2$, $x_2 = 1$, and $x_1 = 1$.

The following steps summarize the procedure for solving the linear system $A\mathbf{x} = \mathbf{b}$ when A admits an LU factorization.

1. Use Theorem 23 to write the linear system $A\mathbf{x} = \mathbf{b}$ as $L(U\mathbf{x}) = \mathbf{b}$.
2. Define the vector $\mathbf{y}$ by means of the equation $U\mathbf{x} = \mathbf{y}$.
3. Use forward substitution to solve the system $L\mathbf{y} = \mathbf{b}$ for $\mathbf{y}$.
4. Use back substitution to solve the system $U\mathbf{x} = \mathbf{y}$ for $\mathbf{x}$. Note that $\mathbf{x}$ is the solution to the original linear system.

PLU Factorization

We have seen that a matrix A has an LU factorization provided that it can be row-reduced without interchanging rows. We conclude this section by noting that when row interchanges are required to reduce A, a factorization is still possible. In this case the matrix A can be factored as $A = PLU$, where P is a *permutation matrix*, that is, a matrix that results from interchanging rows of the identity matrix. As an illustration, let

$$A = \begin{bmatrix} 0 & 2 & -2 \\ 1 & 4 & 3 \\ 1 & 2 & 0 \end{bmatrix}$$

The matrix A can be reduced to

$$U = \begin{bmatrix} 1 & 2 & 0 \\ 0 & 2 & 3 \\ 0 & 0 & -5 \end{bmatrix}$$

by means of the row operations $\mathcal{R}_1\colon\ R_1 \leftrightarrow R_3$, $\mathcal{R}_2\colon\ -R_1 + R_2 \longrightarrow R_2$, and $\mathcal{R}_3\colon\ -R_2 + R_3 \longrightarrow R_3$. The corresponding elementary matrices are given by

$$E_1 = \begin{bmatrix} 0 & 0 & 1 \\ 0 & 1 & 0 \\ 1 & 0 & 0 \end{bmatrix} \qquad E_2 = \begin{bmatrix} 1 & 0 & 0 \\ -1 & 1 & 0 \\ 0 & 0 & 1 \end{bmatrix} \qquad \text{and} \qquad E_3 = \begin{bmatrix} 1 & 0 & 0 \\ 0 & 1 & 0 \\ 0 & -1 & 1 \end{bmatrix}$$

Observe that the elementary matrix E_1 is a permutation matrix while E_2 and E_3 are lower triangular. Hence,

$$A = E_1^{-1} \left(E_2^{-1} E_3^{-1} \right) U$$

$$= \begin{bmatrix} 0 & 0 & 1 \\ 0 & 1 & 0 \\ 1 & 0 & 0 \end{bmatrix} \begin{bmatrix} 1 & 0 & 0 \\ 1 & 1 & 0 \\ 0 & 1 & 1 \end{bmatrix} \begin{bmatrix} 1 & 2 & 0 \\ 0 & 2 & 3 \\ 0 & 0 & -5 \end{bmatrix}$$

$$= PLU$$

Fact Summary

1. A row operation on a matrix A can be performed by multiplying A by an elementary matrix.
2. An elementary matrix is invertible, and the inverse is an elementary matrix.
3. An $n \times n$ matrix A is invertible if and only if it is the product of elementary matrices.
4. An $m \times n$ matrix A has an LU factorization if it can be reduced to an upper triangular matrix with no row interchanges.
5. If $A = LU$, then L is invertible.
6. An LU factorization of A provides an efficient method for solving $A\mathbf{x} = \mathbf{b}$.

Exercise Set 1.7

In Exercises 1–4:

a. Find the 3×3 elementary matrix E that performs the row operation.

b. Compute EA, where

$$A = \begin{bmatrix} 1 & 2 & 1 \\ 3 & 1 & 2 \\ 1 & 1 & -4 \end{bmatrix}$$

1. $2R_1 + R_2 \longrightarrow R_2$

2. $R_1 \leftrightarrow R_2$

3. $-3R_2 + R_3 \longrightarrow R_3$

4. $-R_1 + R_3 \longrightarrow R_3$

In Exercises 5–10:

    a. Find the elementary matrices required to reduce A to the identity.

b. Write A as the product of elementary matrices.

5. $A = \begin{bmatrix} 1 & 3 \\ -2 & 4 \end{bmatrix}$

6. $A = \begin{bmatrix} -2 & 5 \\ 2 & 5 \end{bmatrix}$

7. $A = \begin{bmatrix} 1 & 2 & -1 \\ 2 & 5 & 3 \\ 1 & 2 & 0 \end{bmatrix}$

8. $A = \begin{bmatrix} -1 & 1 & 1 \\ 3 & 1 & 0 \\ -2 & 1 & 1 \end{bmatrix}$

9. $A = \begin{bmatrix} 0 & 1 & 1 \\ 1 & 2 & 3 \\ 0 & 1 & 0 \end{bmatrix}$

10. $A = \begin{bmatrix} 0 & 0 & 0 & 1 \\ 0 & 0 & 1 & 0 \\ 0 & 1 & 0 & 0 \\ 1 & 0 & 0 & 0 \end{bmatrix}$

In Exercises 11–16, find the LU factorization of the matrix A.

11. $A = \begin{bmatrix} 1 & -2 \\ -3 & 7 \end{bmatrix}$

12. $A = \begin{bmatrix} 3 & 9 \\ \frac{1}{2} & 1 \end{bmatrix}$

13. $A = \begin{bmatrix} 1 & 2 & 1 \\ 2 & 5 & 5 \\ -3 & -6 & -2 \end{bmatrix}$

14. $A = \begin{bmatrix} 1 & 1 & 1 \\ -1 & 0 & -4 \\ 2 & 2 & 3 \end{bmatrix}$

15. $A = \begin{bmatrix} 1 & \frac{1}{2} & -3 \\ 1 & \frac{3}{2} & 1 \\ -1 & -1 & 4 \end{bmatrix}$

16. $A = \begin{bmatrix} 1 & -2 & 1 & 3 \\ -2 & 5 & -3 & -7 \\ 1 & -2 & 2 & 8 \\ 3 & -6 & 3 & 10 \end{bmatrix}$

In Exercises 17–22, solve the linear system by using LU factorization.

17. $\begin{cases} -2x + y = -1 \\ 4x - y = 5 \end{cases}$

18. $\begin{cases} 3x - 2y = 2 \\ -6x + 5y = -\frac{7}{2} \end{cases}$

19. $\begin{cases} x + 4y - 3z = 0 \\ -x - 3y + 5z = -3 \\ 2x + 8y - 5z = 1 \end{cases}$

20. $\begin{cases} x - 2y + z = -1 \\ 2x - 3y + 6z = 8 \\ -2x + 4y - z = 4 \end{cases}$

21. $\begin{cases} x - 2y + 3z + w = 5 \\ x - y + 5z + 3w = 6 \\ 2x - 4y + 7z + 3w = 14 \\ -x + y - 5z - 2w = -8 \end{cases}$

22. $\begin{cases} x + 2y + 2z - w = 5 \\ y + z - w = -2 \\ -x - 2y - z + 4w = 1 \\ 2x + 2y + 2z + 2w = 1 \end{cases}$

In Exercises 23 and 24, find the PLU factorization of the matrix A.

23. $A = \begin{bmatrix} 0 & 1 & -1 \\ 2 & -1 & 0 \\ 1 & -3 & 2 \end{bmatrix}$

24. $A = \begin{bmatrix} 0 & 0 & 1 \\ 2 & 1 & 1 \\ 1 & 0 & -3 \end{bmatrix}$

In Exercises 25–28, find the inverse of the matrix A by using an LU factorization.

25. $A = \begin{bmatrix} 1 & 4 \\ -3 & -11 \end{bmatrix}$

26. $A = \begin{bmatrix} 1 & 7 \\ 2 & 20 \end{bmatrix}$

27. $A = \begin{bmatrix} 2 & 1 & -1 \\ 2 & 2 & -2 \\ 2 & 2 & 1 \end{bmatrix}$

28. $A = \begin{bmatrix} -3 & 2 & 1 \\ 3 & -1 & 1 \\ -3 & 1 & 0 \end{bmatrix}$

29. Show directly that the matrix $A = \begin{bmatrix} 0 & 1 \\ 1 & 0 \end{bmatrix}$ does not have an LU factorization.

30. Let A, B, and C be $m \times n$ matrices. Show that if A is row equivalent to B and B is row equivalent to C, then A is row equivalent to C.

31. Show that if A and B are $n \times n$ invertible matrices, then A and B are row equivalent.

32. Suppose that A is an $n \times n$ matrix with an LU factorization, $A = LU$.

   a. What can be said about the diagonal entries of L?

   b. Express $\det(A)$ in terms of the entries of L and U.

   c. Show that A can be row-reduced to U using only replacement operations.

1.8 ▶ Applications of Systems of Linear Equations

In the opening to this chapter we introduced linear systems by describing their connection to the process of photosynthesis. In this section we enlarge the scope of the applications we consider and show how linear systems are used to model a wide variety of problems.

Balancing Chemical Equations

Recall from the introduction to this chapter that a chemical equation is balanced if there are the same number of atoms, of each element, on both sides of the equation. Finding the number of molecules needed to balance a chemical equation involves solving a linear system.

EXAMPLE 1   Propane is a common gas used for cooking and home heating. Each molecule of propane is comprised of 3 atoms of carbon and 8 atoms of hydrogen, written as C_3H_8. When propane burns, it combines with oxygen gas, O_2, to form carbon dioxide, CO_2, and water, H_2O. Balance the chemical equation

$$C_3H_8 + O_2 \longrightarrow CO_2 + H_2O$$

that describes this process.

Solution   We need to find whole numbers $x_1, x_2, x_3,$ and x_4, so that the equation

$$x_1C_3H_8 + x_2O_2 \longrightarrow x_3CO_2 + x_4H_2O$$

is balanced. Equating the number of carbon, hydrogen, and oxygen atoms on both sides of this equation yields the linear system

$$\begin{cases} 3x_1 \quad - \quad x_3 \quad\quad = 0 \\ 8x_1 \quad\quad\quad - 2x_4 = 0 \\ \quad 2x_2 - 2x_3 - \quad x_4 = 0 \end{cases}$$

Solving this system, we obtain the solution set

$$S = \left\{ \left(\frac{1}{4}t, \frac{5}{4}t, \frac{3}{4}t, t \right) \,\middle|\, t \in \mathbb{R} \right\}$$

Since whole numbers are required to balance the chemical equation, particular solutions are obtained by letting $t = 0, 4, 8, \ldots.$ For example, if $t = 8$, then $x_1 = 2, x_2 = 10, x_3 = 6,$ and $x_4 = 8.$ The corresponding balanced equation is given by

$$2C_3H_8 + 10O_2 \longrightarrow 6CO_2 + 8H_2O$$

Network Flow

To study the flow of traffic through city streets, urban planners use mathematical models called *directed graphs* or *digraphs*. In these models, edges and points are used to represent streets and intersections, respectively. Arrows are used to indicate the direction of traffic. To balance a traffic network, we assume that the *outflow* of each intersection is equal to the *inflow*, and that the total flow into the network is equal to the total flow out.

EXAMPLE 2 Partial traffic flow information, given by average hourly volume, is known about a network of five streets, as shown in Fig. 1. Complete the flow pattern for the network.

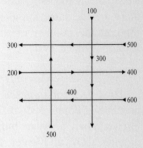

Figure 1

Solution To complete the traffic model, we need to find values for the eight unknown flows, as shown in Fig. 2.

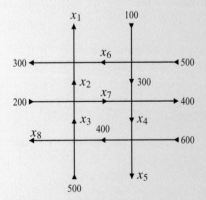

Figure 2

Our assumptions about the intersections give us the set of linear equations

$$\begin{cases} x_2 + x_6 & = 300 + x_1 \\ 100 + 500 & = x_6 + 300 \\ 200 + x_3 & = x_2 + x_7 \\ 300 + x_7 & = 400 + x_4 \\ 400 + 500 & = x_3 + x_8 \\ x_4 + 600 & = 400 + x_5 \end{cases}$$

Moreover, balancing the total flow into the network with the total flow out gives us the additional equation

$$500 + 600 + 500 + 200 + 100 = 400 + x_5 + x_8 + 300 + x_1$$

The final linear system is

$$\begin{cases} -x_1 + x_2 & + x_6 & = 300 \\ & x_6 & = 300 \\ x_2 - x_3 & + x_7 & = 200 \\ & -x_4 & + x_7 & = 100 \\ & x_3 & + x_8 = 900 \\ & -x_4 + x_5 & = 200 \\ x_1 & + x_5 & + x_8 = 1200 \end{cases}$$

The solution is given by

$$x_1 = 1100 - s - t \qquad x_2 = 1100 - s - t \qquad x_3 = 900 - t \qquad x_4 = -100 + s$$
$$x_5 = 100 + s \qquad x_6 = 300 \qquad x_7 = s \qquad x_8 = t$$

Notice that x_7 and x_8 are free variables. However, to obtain particular solutions, we must choose numbers for s and t that produce positive values for each x_i in the system (otherwise we will have traffic going in the wrong direction!) For example, $s = 400$ and $t = 300$ give a viable solution.

Nutrition

Designing a healthy diet involves selecting foods from different groups that, when combined in the proper amounts, satisfy certain nutritional requirements.

EXAMPLE 3   Table 1 gives the amount, in milligrams (mg), of vitamin A, vitamin C, and calcium contained in 1 gram (g) of four different foods. For example, food 1 has 10 mg of vitamin A, 50 mg of vitamin C, and 60 mg of calcium per gram of food. Suppose that a dietician wants to prepare a meal that provides 200 mg of vitamin A, 250 mg of vitamin C, and 300 mg of calcium. How much of each food should be used?

Table 1

	Food 1	Food 2	Food 3	Food 4
Vitamin A	10	30	20	10
Vitamin C	50	30	25	10
Calcium	60	20	40	25

Solution   Let $x_1, x_2, x_3,$ and x_4 denote the amounts of foods 1 through 4, respectively. The amounts for each of the foods needed to satisfy the dietician's requirement can be found by solving the linear system

$$\begin{cases} 10x_1 + 30x_2 + 20x_3 + 10x_4 = 200 \\ 50x_1 + 30x_2 + 25x_3 + 10x_4 = 250 \\ 60x_1 + 20x_2 + 40x_3 + 25x_4 = 300 \end{cases}$$

Rounded to two decimal places, the solution to the linear system is given by

$$x_1 = 0.63 + 0.11t \qquad x_2 = 3.13 + 0.24t$$
$$x_3 = 5 - 0.92t \qquad x_4 = t$$

Observe that each of these values must be nonnegative. Hence, particular solutions can be found by choosing nonnegative values of t such that

$$0 \le 5 - 0.92t$$

Isolating t gives

$$t \le \frac{5}{0.92} \approx 5.4$$

Economic Input-Output Models

Constructing models of the economy is another application of linear systems. In a real economy there are tens of thousands of goods and services. By focusing on specific sectors of the economy the Leontief input-output model gives a method for describing a simplified, but useful model of a real economy. For example, consider an economy for which the outputs are services, raw materials, and manufactured goods. Table 2 provides the inputs needed per unit of output.

Table 2

	Services	Raw materials	Manufacturing
Services	0.04	0.05	0.02
Raw materials	0.03	0.04	0.04
Manufactured goods	0.02	0.3	0.2

Here to provide \$1.00 worth of service, the service sector requires \$0.04 worth of services, \$0.05 worth of raw materials, and \$0.02 worth of manufactured goods. The data in Table 2 are recorded in the matrix

$$A = \begin{bmatrix} 0.04 & 0.05 & 0.02 \\ 0.03 & 0.04 & 0.04 \\ 0.02 & 0.3 & 0.2 \end{bmatrix}$$

This matrix is called the **input-output** matrix. The **demand vector D** gives the total demand on the three sectors, in billions of dollars, and the **production vector x**, also in billions of dollars, contains the production level information for each sector. Each component of $A\mathbf{x}$ represents the level of production that is used by the corresponding sector and is called the **internal demand**.

As an example, suppose that the production vector is

$$\mathbf{x} = \begin{bmatrix} 200 \\ 100 \\ 150 \end{bmatrix}$$

Then the the internal demand is given by

$$A\mathbf{x} = \begin{bmatrix} 0.04 & 0.05 & 0.02 \\ 0.03 & 0.04 & 0.04 \\ 0.02 & 0.3 & 0.2 \end{bmatrix} \begin{bmatrix} 200 \\ 100 \\ 150 \end{bmatrix} = \begin{bmatrix} 16 \\ 16 \\ 64 \end{bmatrix}$$

This result means that the service sector requires $16 billion of services, raw materials, and manufactured goods. It also means that the external demand cannot exceed $184 billion of services, $84 billion of raw materials, and $86 billion in manufactured goods.

Alternatively, suppose that the external demand **D** is given. We wish to find a level of production for each sector such that the internal and external demands are met. Thus, to balance the economy, **x** must satisfy

$$\mathbf{x} - A\mathbf{x} = \mathbf{D}$$

that is,

$$(I - A)\mathbf{x} = \mathbf{D}$$

When $I - A$ is invertible, then

$$\mathbf{x} = (I - A)^{-1}\mathbf{D}$$

EXAMPLE 4   Suppose that the external demand for services, raw materials, and manufactured goods in the economy described in Table 2 is given by

$$\mathbf{D} = \begin{bmatrix} 300 \\ 500 \\ 600 \end{bmatrix}$$

Find the levels of production that balance the economy.

Solution   From the discussion above we have that the production vector **x** must satisfy

$$(I - A)\mathbf{x} = \mathbf{D}$$

that is,

$$\begin{bmatrix} 0.96 & -0.05 & -0.02 \\ -0.03 & 0.96 & -0.04 \\ -0.02 & -0.3 & 0.8 \end{bmatrix} \begin{bmatrix} x_1 \\ x_2 \\ x_3 \end{bmatrix} = \begin{bmatrix} 300 \\ 500 \\ 600 \end{bmatrix}$$

Since the matrix on the left is invertible, the production vector **x** can be found by multiplying both sides by the inverse. Thus,

$$\begin{bmatrix} x_1 \\ x_2 \\ x_3 \end{bmatrix} = \begin{bmatrix} 1.04 & 0.06 & 0.03 \\ 0.03 & 1.06 & 0.05 \\ 0.04 & 0.4 & 1.27 \end{bmatrix} \begin{bmatrix} 300 \\ 500 \\ 600 \end{bmatrix}$$

$$\approx \begin{bmatrix} 360 \\ 569 \\ 974 \end{bmatrix}$$

So the service sector must produce approximately $360 billion worth of services, the raw material sector must produce approximately $569 billion worth of raw materials, and the manufacturing sector must produce approximately $974 billion worth of manufactured goods.

Exercise Set 1.8

In Exercises 1–4, use the smallest possible positive integers to balance the chemical equation.

1. When subjected to heat, aluminium reacts with copper oxide to produce copper metal and aluminium oxide according to the equation

$$Al_3 + CuO \longrightarrow Al_2O_3 + Cu$$

   Balance the chemical equation.

2. When sodium thiosulfate solution is mixed with brown iodine solution, the mixture becomes colorless as the iodine is converted to colorless sodium iodide according to the equation

$$I_2 + Na_2S_2O_3 \longrightarrow NaI + Na_2S_4O_6$$

   Balance the chemical equation.

3. Cold remedies such as Alka-Seltzer use the reaction of sodium bicarbonate with citric acid in solution to produce a fizz (carbon dioxide gas). The reaction produces sodium citrate, water, and carbon dioxide according to the equation

$$NaHCO_3 + C_6H_8O_7 \longrightarrow$$
$$Na_3C_6H_5O_7 + H_2O + CO_2$$

   Balance the chemical equation. For every 100 mg of sodium bicarbonate, how much citric acid should be used? What mass of carbon dioxide will be produced?

4. Balance the chemical equation

$$MnS + As_2Cr_{10}O_{35} + H_2SO_4 \longrightarrow$$
$$HMnO_4 + AsH_3 + CrS_3O_{12} + H_2O$$

5. Find the traffic flow pattern for the network in the figure. Flow rates are in cars per hour. Give one specific solution.

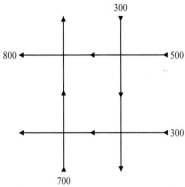

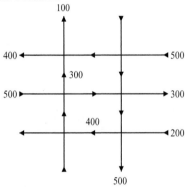

6. Find the traffic flow pattern for the network in the figure. Flow rates are in cars per hour. Give one specific solution.

7. Find the traffic flow pattern for the network in the figure. Flow rates are in cars per half-hour. What is the current status of the road labeled x_5?

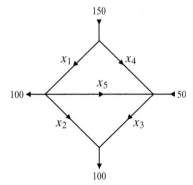

8. Find the traffic flow pattern for the network in the figure. Flow rates are in cars per half-hour. What is the smallest possible value for x_8?

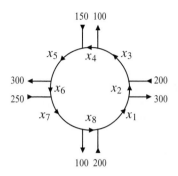

9. The table lists the number of milligrams of vitamin A, vitamin B, vitamin C, and niacin contained in 1 g of four different foods. A dietician wants to prepare a meal that provides 250 mg of vitamin A, 300 mg of vitamin B, 400 mg of vitamin C, and 70 mg of niacin. Determine how many grams of each food must be included, and describe any limitations on the quantities of each food that can be used.

	Group 1	Group 2	Group 3	Group 4
Vitamin A	20	30	40	10
Vitamin B	40	20	35	20
Vitamin C	50	40	10	30
Niacin	5	5	10	5

10. The table lists the amounts of sodium, potassium, carbohydrates, and fiber in a single serving of three food groups. Also listed are the daily recommended amounts based on a 2000-calorie diet. Is it possible to prepare a diet using the three food groups alone that meets the recommended amounts?

	Group 1	Group 2	Group 3	Requirement
Sodium (mg)	200	400	300	2400
Potassium (mg)	300	500	400	3500
Carbohydrates (g)	40	50	20	300
Fiber (g)	5	3	2	25

11. An economy is divided into three sectors as described in the table. Each entry represents the number of units required by the sector to produce 1 unit of output.

	Services	Raw materials	Manufacturing
Services	0.02	0.04	0.05
Raw materials	0.03	0.02	0.04
Manufacturing	0.03	0.3	0.1

a. Write the input-output matrix A for the economy.

b. If the levels of production, in billions, of the three sectors of the economy are 300, 150, and 200, respectively, find the internal demand vector for the economy. What is the total external demand that can be met by the three sectors?

c. Find the inverse of the matrix $I - A$.

d. If the external demands on the three sectors are 350, 400, and 600, respectively, determine the levels of production that balance the economy.

12. Economies are, in general, very complicated with many sectors. The input-output matrix A is based on grouping the different industries and services into 10 separate sectors. If the external demands to the sectors are given in the vector **D**, determine the levels of production that balance the economy.

$$A = \begin{bmatrix}
0.041 & 0.032 & 0.018 & 0.041 & 0.009 & 0.002 & 0.039 & 0.048 & 0.04 & 0.021 \\
0.023 & 0.037 & 0.046 & 0.011 & 0.004 & 0.024 & 0.041 & 0.006 & 0.004 & 0.007 \\
0.018 & 0.03 & 0.039 & 0.05 & 0.038 & 0.011 & 0.049 & 0.001 & 0.028 & 0.047 \\
.034 & 0.005 & 0.034 & 0.039 & 0.023 & 0.007 & 0.009 & 0.023 & 0.05 & 0.006 \\
0.022 & 0.019 & 0.021 & 0.009 & 0.007 & 0.035 & 0.044 & 0.023 & 0.019 & 0.019 \\
0.044 & 0.005 & 0.02 & 0.006 & 0.013 & 0.005 & 0.032 & 0.016 & 0.047 & 0.02 \\
0.018 & 0.001 & 0.049 & 0.011 & 0.043 & 0.003 & 0.024 & 0.047 & 0.027 & 0.042 \\
0.026 & 0.004 & 0.03 & 0.015 & 0.044 & 0.021 & 0.01 & 0.004 & 0.011 & 0.044 \\
0.01 & 0.011 & 0.039 & 0.025 & 0.005 & 0.029 & 0.024 & 0.023 & 0.021 & 0.042 \\
0.048 & 0.03 & 0.019 & 0.045 & 0.044 & 0.033 & 0.014 & 0.03 & 0.042 & 0.05
\end{bmatrix} \quad D = \begin{bmatrix} 45 \\ 10 \\ 11 \\ 17 \\ 48 \\ 32 \\ 42 \\ 21 \\ 34 \\ 40 \end{bmatrix}$$

13. The table contains estimates for national health care in billions of dollars.

Year	Dollars (billions)
1965	30
1970	80
1975	120
1980	250
1985	400
1990	690

a. Make a scatter plot of the data.

b. Use the 1970, 1980, and 1990 data to write a system of equations that can be used to find a parabola that approximates the data.

c. Solve the system found in part (b).

d. Plot the parabola along with the data points.

e. Use the model found in part (c) to predict an estimate for national health care spending in 2010.

14. The number of cellular phone subscribers worldwide from 1985 to 2002 is given in the table. Use the data from 1985, 1990, and 2000 to fit a parabola to the data points. Use the quadratic function to predict the number of cellular phone subscribers expected in 2010.

Year	Cellular Phone Subscribers(millions)
1985	1
1990	11
2000	741
2001	955
2002	1155

In Exercises 15–18, use the *power* of a matrix to solve the problems. That is, for a matrix A, the nth power is

$$A^n = \underbrace{A \cdot A \cdot A \cdots A}_{n \text{ times}}$$

15. Demographers are interested in the movement of populations or groups of populations from one region to another. Suppose each year it is estimated that 90 percent of the people of a city remain in the city, 10 percent move to the suburbs, 92 percent of the suburban population remain in the suburbs, and 8 percent move to the city.

a. Write a 2×2 *transition* matrix that describes the percentage of the populations that move from city to city (remain in the city), city to suburbs, suburbs to suburbs (remain in the suburbs), and suburbs to city.

b. If in the year 2002 the population of a city was 1,500,000 and of the suburbs was 600,000, write a matrix product that gives a 2×1 vector containing the populations in the city and in the suburbs in the year 2003. Multiply the matrices to find the populations.

c. If in the year 2002 the population of a city was 1,500,000 and of the suburbs was 600,000, write a matrix product that gives a 2×1 vector containing the populations in the city and in

the suburbs in the year 2004. Multiply the matrices to find the populations.

d. Give a matrix product in terms of powers of the matrix found in part (a) for the size of the city and suburban populations in any year after 2002.

16. To study the spread of a disease, a medical researcher infects 200 laboratory mice of a population of 1000. The researcher estimates that it is likely that 80 percent of the infected mice will recover in a week and 20 percent of healthy mice will contract the disease in the same week.

a. Write a 2×2 matrix that describes the percentage of the population that transition from healthy to healthy, healthy to infected, infected to infected, and infected to healthy.

b. Determine the number of healthy and infected mice after the first week.

c. Determine the number of healthy and infected mice after the second week.

d. Determine the number of healthy and infected mice after six weeks.

17. In a population of 50,000 there are 20,000 nonsmokers, 20,000 smokers of one pack or less a day, and 10,000 smokers of more than one pack a day. During any month it is likely that only 10 percent of the nonsmokers will become smokers of one pack or less a day and the rest will remain nonsmokers, 20 percent of the smokers of a pack or less will quit smoking, 30 percent will increase their smoking to more than one pack a day, 30 percent of the heavy smokers will remain smokers but decrease their smoking to one pack or less, and 10 percent will go cold turkey and quit. After one month what part of the population is in each category? After two months how many are in each category? After one year how many are in each category?

18. An entrepreneur has just formed a new company to compete with the established giant in the market. She hired an advertising firm to develop a campaign to introduce her product to the market. The advertising blitz seems to be working, and in any given month 2 percent of the consumers switch from the time-honored product to the new improved version, but at the same time 5 percent of those using the new product decide to switch back to the old established brand. How long will it take for the new company to acquire 20 percent of the consumers?

In Exercises 19 and 20, the figure shows an electrical network. In an electrical network, current is measured in amperes, resistance in ohms, and the product of current and resistance in volts. Batteries are represented using two parallel line segments of unequal length, and it is understood the current flows out of the terminal denoted by the longer line segment. Resistance is denoted using a sawtooth. To analyze an electrical network requires Kirchhoff's laws, which state all current flowing into a junction, denoted using a black dot, must flow out and the sum of the products of current I and resistance R around a closed path (a loop) is equal to the total voltage in the path.

19. a. Apply Kirchhoff's first law to either junction to write an equation involving I_1, I_2, and I_3.

   b. Apply Kirchhoff's second law to the two loops to write two linear equations.

   c. Solve the system of equations from parts (a) and (b) to find the currents I_1, I_2, and I_3.

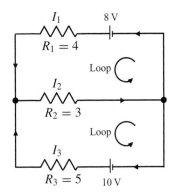

20. a. Apply Kirchhoff's first law to the four junctions to write four equations involving currents.

   b. Apply Kirchhoff's second law to the three loops to write three linear equations.

   c. Solve the system of equations from parts (a) and (b) to find the currents I_1, I_2, I_3, I_4, I_5, and I_6.

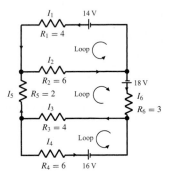

In Exercises 21 and 22, use the fact that if a plate has reached a thermal equilibrium, then the temperature at a grid point, not on the boundary of the plate, is the average of the temperatures of the four closest grid points. The temperatures are equal at each point on a boundary, as shown in the figure. Estimate the temperature at each interior grid point.

21.

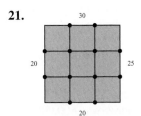

22.

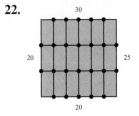

Review Exercises for Chapter 1

1. Consider the linear system

$$\begin{cases} x + y + 2z + w = 3 \\ -x + z + 2w = 1 \\ 2x + 2y + w = -2 \\ x + y + 2z + 3w = 5 \end{cases}$$

a. Define the coefficient matrix A for the linear system.

b. Find $\det(A)$.

c. Is the linear system consistent? Explain.

d. Find all solutions to $A\mathbf{x} = \mathbf{0}$.

e. Is the matrix A invertible? If yes, then find the inverse.

f. Solve the linear system.

2. The augmented matrix of a linear system has the form

$$\begin{bmatrix} 1 & -1 & 2 & 1 & a \\ -1 & 3 & 1 & 1 & b \\ 3 & -5 & 5 & 1 & c \\ 2 & -2 & 4 & 2 & d \end{bmatrix}$$

a. Can you decide by inspection whether the determinant of the coefficient matrix is 0? Explain.

b. Can you decide by inspection whether the linear system has a unique solution for every choice of $a, b, c,$ and d? Explain.

c. Determine the values of $a, b, c,$ and d for which the linear system is consistent.

d. Determine the values of $a, b, c,$ and d for which the linear system is inconsistent.

e. Does the linear system have a unique solution or infinitely many solutions?

f. If $a = 2, b = 1, c = -1,$ and $d = 4$, describe the solution set for the linear system.

3. Find all idempotent matrices of the form

$$\begin{bmatrix} a & b \\ 0 & c \end{bmatrix}$$

4. Let S denote the set of all 2×2 matrices. Find all matrices $\begin{bmatrix} a & b \\ c & d \end{bmatrix}$ that will commute with **every** matrix in S.

5. Let A and B be 2×2 matrices.

a. Show that the sum of the terms on the main diagonal of $AB - BA$ is 0.

b. If M is a 2×2 matrix and the sum of the main diagonal entries is 0, show there is a constant c such that

$$M^2 = cI$$

c. If $A, B,$ and C are 2×2 matrices, then use parts (a) and (b) to show that

$$(AB - BA)^2 C = C(AB - BA)^2$$

6. Find the traffic flow pattern for the network in the figure. Flow rates are in cars per hour. Give one specific solution.

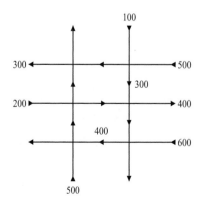

7. a. Explain why the matrix

$$A = \begin{bmatrix} 1 & 1 & 1 & 1 & 1 \\ 0 & 1 & 1 & 1 & 1 \\ 0 & 0 & 1 & 1 & 1 \\ 0 & 0 & 0 & 1 & 1 \\ 0 & 0 & 0 & 0 & 1 \end{bmatrix}$$

is invertible.

b. Determine the maximum number of 1's that can be added to A such that the resulting matrix is invertible.

8. Show that if A is invertible, then A^t is invertible and $(A^t)^{-1} = (A^{-1})^t$.

9. A matrix A is *skew-symmetric* provided $A^t = -A$.

a. Let A be an $n \times n$ matrix and define

$$B = A + A^t \quad \text{and} \quad C = A - A^t$$

Show that B is symmetric and C is skew-symmetric.

b. Show that every $n \times n$ matrix can be written as the sum of a symmetric and a skew-symmetric matrix.

10. Suppose $\mathbf{u}$ and $\mathbf{v}$ are solutions to the linear system $A\mathbf{x} = \mathbf{b}$. Show that if scalars α and β satisfy $\alpha + \beta = 1$, then $\alpha\mathbf{u} + \beta\mathbf{v}$ is also a solution to the linear system $A\mathbf{x} = \mathbf{b}$.

Chapter 1: Chapter Test

In Exercises 1–45, determine whether the statement is true or false.

1. A 2×2 linear system has one solution, no solutions, or infinitely many solutions.

2. A 3×3 linear system has no solutions, one solution, two solutions, three solutions, or infinitely many solutions.

3. If A and B are $n \times n$ matrices with no zero entries, then $AB \neq \mathbf{0}$.

4. Homogeneous linear systems always have at least one solution.

5. If A is an $n \times n$ matrix, then $A\mathbf{x} = \mathbf{0}$ has a nontrivial solution if and only if the matrix A has an inverse.

6. If A and B are $n \times n$ matrices and $A\mathbf{x} = B\mathbf{x}$ for every $n \times 1$ matrix $\mathbf{x}$, then $A = B$.

7. If A, B, and C are invertible $n \times n$ matrices, then $(ABC)^{-1} = A^{-1}B^{-1}C^{-1}$.

8. If A is an invertible $n \times n$ matrix, then the linear system $A\mathbf{x} = \mathbf{b}$ has a unique solution.

9. If A and B are $n \times n$ invertible matrices and $AB = BA$, then A commutes with B^{-1}.

10. If A and B commute, then $A^2B = BA^2$.

11. The matrix

$$\begin{bmatrix} 1 & -2 & 3 & 1 & 0 \\ 0 & -1 & 4 & 3 & 2 \\ 0 & 0 & 3 & 5 & -2 \\ 0 & 0 & 0 & 0 & 4 \\ 0 & 0 & 0 & 0 & 6 \end{bmatrix}$$

does not have an inverse.

12. Interchanging two rows of a matrix changes the sign of its determinant.

13. Multiplying a row of a matrix by a nonzero constant results in the determinant being multiplied by the same nonzero constant.

14. If two rows of a matrix are equal, then the determinant of the matrix is 0.

15. Performing the operation $aR_i + R_j \to R_j$ on a matrix multiplies the determinant by the constant a.

16. If $A = \begin{bmatrix} 1 & 2 \\ 4 & 6 \end{bmatrix}$, then $A^2 - 7A = 2I$.

17. If A and B are invertible matrices, then $A + B$ is an invertible matrix.

18. If A and B are invertible matrices, then AB is an invertible matrix.

19. If A is an $n \times n$ matrix and A does not have an inverse, then the linear system $A\mathbf{x} = \mathbf{b}$ is inconsistent.

20. The linear system

$$\begin{bmatrix} 1 & 2 & 3 \\ 6 & 5 & 4 \\ 0 & 0 & 0 \end{bmatrix} \begin{bmatrix} x \\ y \\ z \end{bmatrix} = \begin{bmatrix} 1 \\ 2 \\ 3 \end{bmatrix}$$

is inconsistent.

21. The inverse of the matrix

$$\begin{bmatrix} 2 & -1 \\ 3 & 1 \end{bmatrix} \quad \text{is} \quad \begin{bmatrix} 1 & 1 \\ -3 & 2 \end{bmatrix}$$

22. The matrix

$$\begin{bmatrix} 2 & -1 \\ 4 & -2 \end{bmatrix}$$

does not have an inverse.

23. If the $n \times n$ matrix A is idempotent and invertible, then $A = I$.

24. If A and B commute, then A^t and B^t commute.

25. If A is an $n \times n$ matrix and $\det(A) = 3$, then $\det(A^t A) = 9$.

In Exercises 26–32, use the linear system

$$\begin{cases} 2x + 2y = 3 \\ x - y = 1 \end{cases}$$

26. The coefficient matrix is

$$A = \begin{bmatrix} 2 & 2 \\ 1 & -1 \end{bmatrix}$$

27. The coefficient matrix A has determinant

$$\det(A) = 0$$

28. The linear system has a unique solution.

29. The only solution to the linear system is $x = -7/4$ and $y = -5/4$.

30. The inverse of the coefficient matrix A is

$$A^{-1} = \begin{bmatrix} \frac{1}{4} & \frac{1}{2} \\ \frac{1}{4} & -\frac{1}{2} \end{bmatrix}$$

31. The linear system is equivalent to the matrix equation

$$\begin{bmatrix} 2 & 2 \\ 1 & -1 \end{bmatrix} \begin{bmatrix} x \\ y \end{bmatrix} = \begin{bmatrix} 3 \\ 1 \end{bmatrix}$$

32. The solution to the system is given by the matrix equation

$$\begin{bmatrix} x \\ y \end{bmatrix} = \begin{bmatrix} \frac{1}{4} & \frac{1}{2} \\ \frac{1}{4} & -\frac{1}{2} \end{bmatrix} \begin{bmatrix} 3 \\ 1 \end{bmatrix}$$

In Exercises 33–36, use the linear system

$$\begin{cases} x_1 + 2x_2 - 3x_3 = 1 \\ 2x_1 + 5x_2 - 8x_3 = 4 \\ -2x_1 - 4x_2 + 6x_3 = -2 \end{cases}$$

33. The determinant of the coefficient matrix is

$$\begin{vmatrix} 5 & -8 \\ -4 & 6 \end{vmatrix} + \begin{vmatrix} 2 & -8 \\ -2 & 6 \end{vmatrix} + \begin{vmatrix} 2 & 5 \\ -2 & -4 \end{vmatrix}$$

34. The determinant of the coefficient matrix is 0.

35. A solution to the linear system is $x_1 = -4$, $x_2 = 0$, and $x_3 = -1$.

36. The linear system has infinitely many solutions, and the general solution is given by x_3 is free, $x_2 = 2 + 2x_3$, and $x_1 = -3 - x_3$.

In Exercises 37–41, use the matrix

$$A = \begin{bmatrix} -1 & -2 & 1 & 3 \\ 1 & 0 & 1 & -1 \\ 2 & 1 & 2 & 1 \end{bmatrix}$$

37. After the operation $R_1 \longleftrightarrow R_2$ is performed, the matrix becomes

$$\begin{bmatrix} 1 & 0 & 1 & -1 \\ -1 & -2 & 1 & 3 \\ 2 & 1 & 2 & 1 \end{bmatrix}$$

38. After the operation $-2R_1 + R_3 \longrightarrow R_3$ is performed on the matrix found in Exercise 37, the matrix becomes

$$\begin{bmatrix} 1 & 0 & 1 & -1 \\ -1 & -2 & 1 & 3 \\ 0 & -2 & 0 & -3 \end{bmatrix}$$

39. The matrix A is row equivalent to

$$\begin{bmatrix} 1 & 0 & 1 & -1 \\ 0 & -2 & 2 & 2 \\ 0 & 0 & 1 & 4 \end{bmatrix}$$

40. The reduced row echelon form of A is

$$\begin{bmatrix} 1 & 0 & 0 & -5 \\ 0 & 1 & 0 & 3 \\ 0 & 0 & 1 & 4 \end{bmatrix}$$

41. If A is viewed as the augmented matrix of a linear system, then the solution to the linear system is $x = -5$, $y = 3$, and $z = 4$.

In Exercises 42–45, use the matrices

$$A = \begin{bmatrix} 1 & 1 & 2 \\ -2 & 3 & 1 \\ 4 & 0 & -3 \end{bmatrix}$$

$$B = \begin{bmatrix} 1 & 2 & 1 \\ -1 & 3 & 2 \end{bmatrix}$$

42. The matrix products AB and BA are both defined.

43. The matrix expression $-2BA + 3B$ simplifies to a 2×3 matrix.

44. The matrix expression $-2BA + 3B$ equals

$$\begin{bmatrix} -3 & -5 & 3 \\ -5 & 7 & 16 \end{bmatrix}$$

45. The matrix A^2 is

$$\begin{bmatrix} 7 & 4 & -3 \\ -4 & 7 & -4 \\ -8 & 4 & 17 \end{bmatrix}$$

2

Linear Combinations and Linear Independence

*I*n the broadest sense a *signal* is any time-varying quantity. The motion of a particle through space, for example, can be thought of as a signal. A seismic disturbance is detected as signals from within the earth. Sound caused by the vibration of a string is a signal, radio waves are signals, and a digital picture with colors represented numerically also can be considered a signal. A video signal is a sequence of images. Signals represented using real numbers are called *continuous* while others that use integers are called *discrete*. A compact disc contains discrete signals representing sound. Some signals are periodic; that is, the *waveform* or shape of the signal repeats at regular intervals. The *period* of a wave is the time it takes for one cycle of the wave, and the *frequency* is the number of cycles that occur per unit of time. If the period of a wave

is $2T$, then the frequency is $F = \frac{1}{2T}$. Every periodic motion is the mixture of sine and cosine waves with frequencies proportional to a common frequency, called the *fundamental frequency*. A signal with period $2T$ is a mixture of the functions

$$1, \cos \frac{\pi x}{T}, \sin \frac{\pi x}{T}, \cos \frac{2\pi x}{T}, \sin \frac{2\pi x}{T}, \cos \frac{3\pi x}{T}, \sin \frac{3\pi x}{T}, \ldots$$

and for any n, the signal can be approximated by the *fundamental* set

$$1, \cos \frac{\pi x}{T}, \sin \frac{\pi x}{T}, \cos \frac{2\pi x}{T}, \sin \frac{2\pi x}{T}, \ldots, \cos \frac{n\pi x}{T}, \sin \frac{n\pi x}{T}$$

The approximation obtained from the sum of the elements of the fundamental set with appropriate coefficients, or *weights*, has the form

$$a_0 + a_1 \cos \frac{\pi x}{T} + b_1 \sin \frac{\pi x}{T} + a_2 \cos \frac{2\pi x}{T}$$

$$+ b_2 \sin \frac{2\pi x}{T} + \cdots + a_n \cos \frac{n\pi x}{T} + b_n \sin \frac{n\pi x}{T}$$

This sum is called a *linear combination* of the elements of the fundamental set. A *square wave* on the interval $[-\pi, \pi]$ along with the approximations

$$\frac{4}{\pi} \sin x, \quad \frac{4}{\pi} \sin x + \frac{4}{3\pi} \sin 3x, \quad \frac{4}{\pi} \sin x + \frac{4}{3\pi} \sin 3x + \frac{4}{5\pi} \sin 5x$$

and

$$\frac{4}{\pi} \sin x + \frac{4}{3\pi} \sin 3x + \frac{4}{5\pi} \sin 5x + \frac{4}{7\pi} \sin 7x$$

are shown in Fig. 1. As more terms are added, the approximations become better.

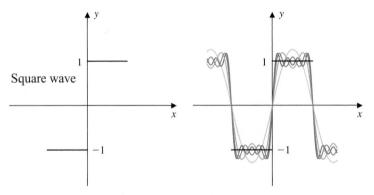

Figure 1

In Chap. 1 we defined a vector, with n entries, as an $n \times 1$ matrix. Vectors are used not only in mathematics, but in virtually every branch of science. In this chapter we study sets of vectors and analyze their additive properties. The concepts presented here, in the context of vectors, are fundamental to the study of linear algebra. In Chap. 3, we extend these concepts to *abstract vector spaces*, including spaces of functions as described in the opening example.

2.1 ▶ Vectors in $\mathbb{R}^n$

Euclidean 2-space, denoted by $\mathbb{R}^2$, is the set of all vectors with two entries, that is,

$$\mathbb{R}^2 = \left\{ \begin{bmatrix} x_1 \\ x_2 \end{bmatrix} \middle| \ x_1, x_2 \text{ are real numbers} \right\}$$

Similarly Euclidean 3-space, denoted by $\mathbb{R}^3$, is the set of all vectors with three entries, that is,

$$\mathbb{R}^3 = \left\{ \begin{bmatrix} x_1 \\ x_2 \\ x_3 \end{bmatrix} \,\middle|\, x_1, x_2, x_3 \text{ are real numbers} \right\}$$

In general, Euclidean n-space consists of vectors with n entries.

DEFINITION 1

Vectors in $\mathbb{R}^n$ Euclidean n-space, denoted by $\mathbb{R}^n$, or simply n-space, is defined by

$$\mathbb{R}^n = \left\{ \begin{bmatrix} x_1 \\ x_2 \\ \vdots \\ x_n \end{bmatrix} \,\middle|\, x_i \in \mathbb{R}, \text{ for } i = 1, 2, \ldots, n \right\}$$

The entries of a vector are called the **components** of the vector.

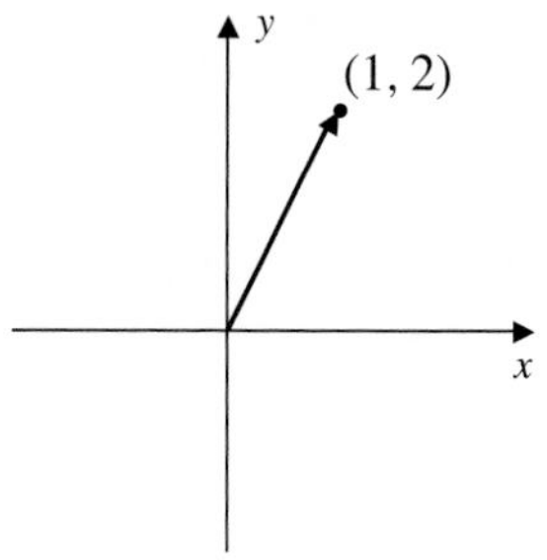

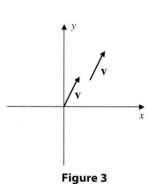

Figure 2

Geometrically, in $\mathbb{R}^2$ and $\mathbb{R}^3$ a vector is a directed line segment from the origin to the point whose coordinates are equal to the components of the vector. For example, the vector in $\mathbb{R}^2$ given by

$$\mathbf{v} = \begin{bmatrix} 1 \\ 2 \end{bmatrix}$$

is the directed line segment from the origin $(0, 0)$ to the point $(1, 2)$, as shown in Fig. 2. The point $(0, 0)$ is the **initial point**, and the point $(1, 2)$ is the **terminal point**. The **length** of a vector is the length of the line segment from the initial point to the terminal point. For example, the length of $\mathbf{v} = \begin{bmatrix} 1 \\ 2 \end{bmatrix}$ is $\sqrt{1^2 + 2^2} = \sqrt{5}$. A vector is unchanged if it is relocated elsewhere in the plane, provided that the length and direction remain unchanged. For example, the directed line segments between $(0, 0)$ and $(1, 2)$ and between $(2, 2)$ and $(3, 4)$ are both representations of the same vector $\mathbf{v} = \begin{bmatrix} 1 \\ 2 \end{bmatrix}$. See Fig. 3. When the initial point of a vector is the origin, we say that the vector is in **standard position**.

Since vectors are matrices, two vectors are **equal** provided that their corresponding components are equal. The operations of addition and scalar multiplication are defined componentwise as they are for matrices.

Figure 3

DEFINITION 2

Addition and Scalar Multiplication of Vectors   Let $\mathbf{u}$ and $\mathbf{v}$ be vectors in $\mathbb{R}^n$ and c a scalar.

1. The **sum** of $\mathbf{u}$ and $\mathbf{v}$ is

$$\mathbf{u} + \mathbf{v} = \begin{bmatrix} u_1 \\ u_2 \\ \vdots \\ u_n \end{bmatrix} + \begin{bmatrix} v_1 \\ v_2 \\ \vdots \\ v_n \end{bmatrix} = \begin{bmatrix} u_1 + v_1 \\ u_2 + v_2 \\ \vdots \\ u_n + v_n \end{bmatrix}$$

2. The **scalar product** of c and $\mathbf{u}$ is

$$c\mathbf{u} = c \begin{bmatrix} u_1 \\ u_2 \\ \vdots \\ u_n \end{bmatrix} = \begin{bmatrix} cu_1 \\ cu_2 \\ \vdots \\ cu_n \end{bmatrix}$$

These algebraic definitions of vector addition and scalar multiplication agree with the standard geometric definitions. Two vectors $\mathbf{u}$ and $\mathbf{v}$ are added according to the *parallelogram rule*, as shown in Fig. 4(a). The vector $c\mathbf{u}$ is a *scaling* of the vector $\mathbf{u}$. In Fig. 4(b) are examples of scaling a vector with $0 < c < 1$ and $c > 1$. In addition, if $c < 0$, then the vector $c\mathbf{u}$ is reflected through the origin, as shown in Fig. 4(b). The difference $\mathbf{u} - \mathbf{v} = \mathbf{u} + (-\mathbf{v})$ is the vector shown in Fig. 4(c). As shown in Fig. 4(c), it is common to draw the difference vector $\mathbf{u} - \mathbf{v}$ from the terminal point of $\mathbf{v}$ to the terminal point of $\mathbf{u}$.

EXAMPLE 1   Let

$$\mathbf{u} = \begin{bmatrix} 1 \\ -2 \\ 3 \end{bmatrix} \qquad \mathbf{v} = \begin{bmatrix} -1 \\ 4 \\ 3 \end{bmatrix} \qquad \text{and} \qquad \mathbf{w} = \begin{bmatrix} 4 \\ 2 \\ 6 \end{bmatrix}$$

Find $(2\mathbf{u} + \mathbf{v}) - 3\mathbf{w}$.

Solution   Using the componentwise definitions of addition and scalar multiplication, we have

$$(2\mathbf{u} + \mathbf{v}) - 3\mathbf{w} = \left(2 \begin{bmatrix} 1 \\ -2 \\ 3 \end{bmatrix} + \begin{bmatrix} -1 \\ 4 \\ 3 \end{bmatrix} \right) - 3 \begin{bmatrix} 4 \\ 2 \\ 6 \end{bmatrix}$$

$$= \left(\begin{bmatrix} 2 \\ -4 \\ 6 \end{bmatrix} + \begin{bmatrix} -1 \\ 4 \\ 3 \end{bmatrix} \right) + \begin{bmatrix} -12 \\ -6 \\ -18 \end{bmatrix}$$

$$= \begin{bmatrix} 1 \\ 0 \\ 9 \end{bmatrix} + \begin{bmatrix} -12 \\ -6 \\ -18 \end{bmatrix} = \begin{bmatrix} -11 \\ -6 \\ -9 \end{bmatrix}$$

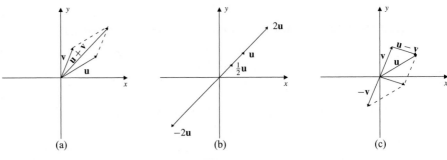

(a)   (b)   (c)

Figure 4

Vectors in $\mathbb{R}^n$, being matrices with n rows and 1 column, enjoy all the algebraic properties of matrices that we saw in Chap. 1.

EXAMPLE 2   Show that vector addition is commutative.

Solution   If $\mathbf{u}$ and $\mathbf{v}$ are vectors in $\mathbb{R}^n$, then

$$\mathbf{u} + \mathbf{v} = \begin{bmatrix} u_1 \\ u_2 \\ \vdots \\ u_n \end{bmatrix} + \begin{bmatrix} v_1 \\ v_2 \\ \vdots \\ v_n \end{bmatrix} = \begin{bmatrix} u_1 + v_1 \\ u_2 + v_2 \\ \vdots \\ u_n + v_n \end{bmatrix}$$

Since addition for real numbers is commutative,

$$\mathbf{u} + \mathbf{v} = \begin{bmatrix} u_1 + v_1 \\ u_2 + v_2 \\ \vdots \\ u_n + v_n \end{bmatrix} = \begin{bmatrix} v_1 + u_1 \\ v_2 + u_2 \\ \vdots \\ v_n + u_n \end{bmatrix} = \mathbf{v} + \mathbf{u}$$

The **zero vector** in $\mathbb{R}^n$ is the vector with each component equal to 0, that is,

$$\mathbf{0} = \begin{bmatrix} 0 \\ 0 \\ \vdots \\ 0 \end{bmatrix}$$

Hence, for any vector $\mathbf{v}$ in $\mathbb{R}^n$, we have $\mathbf{v} + \mathbf{0} = \mathbf{v}$. Recall that for any real number a there is a unique number $-a$ such that $a + (-a) = 0$. This enables us to define the **additive inverse** of any vector $\mathbf{v}$ as the vector

$$-\mathbf{v} = \begin{bmatrix} -v_1 \\ -v_2 \\ \vdots \\ -v_n \end{bmatrix}$$

so that $\mathbf{v} + (-\mathbf{v}) = \mathbf{0}$.

Theorem 1 summarizes the essential algebraic properties of vectors in $\mathbb{R}^n$. These properties serve as a model for the structure we will require for the abstract vector spaces of Chap. 3. The first of these properties was proved in Example 2. The remaining justifications are left as exercises.

THEOREM 1   Let $\mathbf{u}$, $\mathbf{v}$, and $\mathbf{w}$ be vectors in $\mathbb{R}^n$, and let c and d be scalars. The following algebraic properties hold.

1. Commutative property: $\mathbf{u} + \mathbf{v} = \mathbf{v} + \mathbf{u}$
2. Associative property: $(\mathbf{u} + \mathbf{v}) + \mathbf{w} = \mathbf{u} + (\mathbf{v} + \mathbf{w})$

3. Additive identity: The vector $\mathbf{0}$ satisfies $\mathbf{0} + \mathbf{u} = \mathbf{u} + \mathbf{0} = \mathbf{u}$.

4. Additive inverse: For every vector $\mathbf{u}$, the vector $-\mathbf{u}$ satisfies
$\mathbf{u} + (-\mathbf{u}) = -\mathbf{u} + \mathbf{u} = \mathbf{0}$.

5. $c(\mathbf{u} + \mathbf{v}) = c\mathbf{u} + c\mathbf{v}$

6. $(c + d)\mathbf{u} = c\mathbf{u} + d\mathbf{u}$

7. $c(d\mathbf{u}) = (cd)\mathbf{u}$

8. $(1)\mathbf{u} = \mathbf{u}$

By the associative property, the vector sum $\mathbf{u}_1 + \mathbf{u}_2 + \cdots + \mathbf{u}_n$ can be computed unambiguously, without the need for parentheses. This will be important in Sec. 2.2.

EXAMPLE 3   Let

$$\mathbf{u} = \begin{bmatrix} 1 \\ -1 \end{bmatrix} \qquad \mathbf{v} = \begin{bmatrix} 2 \\ 3 \end{bmatrix} \qquad \text{and} \qquad \mathbf{w} = \begin{bmatrix} 4 \\ -3 \end{bmatrix}$$

Verify that the associative property holds for these three vectors. Also verify that for any scalars c and d, $c(d\mathbf{u}) = (cd)\mathbf{u}$.

Solution   To verify the associative property, we have

$$(\mathbf{u} + \mathbf{v}) + \mathbf{w} = \left(\begin{bmatrix} 1 \\ -1 \end{bmatrix} + \begin{bmatrix} 2 \\ 3 \end{bmatrix} \right) + \begin{bmatrix} 4 \\ -3 \end{bmatrix} = \begin{bmatrix} 1+2 \\ -1+3 \end{bmatrix} + \begin{bmatrix} 4 \\ -3 \end{bmatrix}$$

$$= \begin{bmatrix} 3 \\ 2 \end{bmatrix} + \begin{bmatrix} 4 \\ -3 \end{bmatrix} = \begin{bmatrix} 7 \\ -1 \end{bmatrix}$$

and

$$\mathbf{u} + (\mathbf{v} + \mathbf{w}) = \begin{bmatrix} 1 \\ -1 \end{bmatrix} + \left(\begin{bmatrix} 2 \\ 3 \end{bmatrix} + \begin{bmatrix} 4 \\ -3 \end{bmatrix} \right)$$

$$= \begin{bmatrix} 1 \\ -1 \end{bmatrix} + \begin{bmatrix} 6 \\ 0 \end{bmatrix} = \begin{bmatrix} 7 \\ -1 \end{bmatrix}$$

Hence, $(\mathbf{u} + \mathbf{v}) + \mathbf{w} = \mathbf{u} + (\mathbf{v} + \mathbf{w})$.

For the second verification, we have

$$c(d\mathbf{u}) = c \left(d \begin{bmatrix} 1 \\ -1 \end{bmatrix} \right) = c \begin{bmatrix} d \\ -d \end{bmatrix} = \begin{bmatrix} cd \\ -cd \end{bmatrix} = cd \begin{bmatrix} 1 \\ -1 \end{bmatrix} = (cd)\mathbf{u}$$

The properties given in Theorem 1 can be used to establish other useful properties of vectors in $\mathbb{R}^n$. For example, if $\mathbf{u} \in \mathbb{R}^n$ and c is a scalar, then

$$0\mathbf{u} = 0 \begin{bmatrix} u_1 \\ u_2 \\ \vdots \\ u_n \end{bmatrix} = \begin{bmatrix} 0 \\ 0 \\ \vdots \\ 0 \end{bmatrix} = \mathbf{0} \qquad \text{and} \qquad c\mathbf{0} = \mathbf{0}$$

We also have the property that $(-1)\mathbf{u} = -\mathbf{u}$. That is, the scalar product of -1 with $\mathbf{u}$ is the additive inverse of $\mathbf{u}$. In the case of real numbers, the statement $xy = 0$ is equivalent to $x = 0$ or $y = 0$. A similar property holds for scalar multiplication. That is, if $c\mathbf{u} = \mathbf{0}$, then either $c = 0$ or $\mathbf{u} = \mathbf{0}$. To see this, let

$$\begin{bmatrix} cu_1 \\ cu_2 \\ \vdots \\ cu_n \end{bmatrix} = \begin{bmatrix} 0 \\ 0 \\ \vdots \\ 0 \end{bmatrix}$$

so that $cu_1 = 0, cu_2 = 0, \ldots, cu_n = 0$. If $c = 0$, then the conclusion holds. Otherwise, $u_1 = u_2 = \cdots = u_n = 0$, that is, $\mathbf{u} = \mathbf{0}$.

Fact Summary

1. The definitions of vector addition and scalar multiplication in $\mathbb{R}^n$ agree with the definitions for matrices in general and satisfy all the algebraic properties of matrices.

2. The zero vector, whose components are all 0, is the additive identity for vectors in $\mathbb{R}^n$. The additive inverse of a vector $\mathbf{v}$, denoted by $-\mathbf{v}$, is obtained by negating each component of $\mathbf{v}$.

3. For vectors in $\mathbb{R}^2$ and $\mathbb{R}^3$ vector addition agrees with the standard parallelogram law. Multiplying such a vector by a positive scalar changes the length of the vector but not the direction. If the scalar is negative, the vector is reflected through the origin.

Exercise Set 2.1

In Exercises 1–6, use the vectors

$$\mathbf{u} = \begin{bmatrix} 1 \\ -2 \\ 3 \end{bmatrix} \qquad \mathbf{v} = \begin{bmatrix} -2 \\ 4 \\ 0 \end{bmatrix}$$

$$\mathbf{w} = \begin{bmatrix} 2 \\ 1 \\ -1 \end{bmatrix}$$

1. Find $\mathbf{u} + \mathbf{v}$ and $\mathbf{v} + \mathbf{u}$.

2. Find $(\mathbf{u} + \mathbf{v}) + \mathbf{w}$ and $\mathbf{u} + (\mathbf{v} + \mathbf{w})$.

3. Find $\mathbf{u} - 2\mathbf{v} + 3\mathbf{w}$.

4. Find $-\mathbf{u} + \frac{1}{2}\mathbf{v} - 2\mathbf{w}$.

5. Find $-3(\mathbf{u} + \mathbf{v}) - \mathbf{w}$.

6. Find $2\mathbf{u} - 3(\mathbf{v} - 2\mathbf{w})$.

In Exercises 7–10, use the vectors

$$\mathbf{u} = \begin{bmatrix} 1 \\ -2 \\ 3 \\ 0 \end{bmatrix} \qquad \mathbf{v} = \begin{bmatrix} 3 \\ 2 \\ -1 \\ 1 \end{bmatrix}$$

7. Find $-2(\mathbf{u} + 3\mathbf{v}) + 3\mathbf{u}$.

8. Find $3\mathbf{u} - 2\mathbf{v}$.

9. If x_1 and x_2 are real scalars, verify that
$(x_1 + x_2)\mathbf{u} = x_1\mathbf{u} + x_2\mathbf{u}$.

10. If x_1 is a real scalar, verify that
$$x_1(\mathbf{u} + \mathbf{v}) = x_1\mathbf{u} + x_1\mathbf{v}.$$

In Exercises 11–14, let

$$\mathbf{e}_1 = \begin{bmatrix} 1 \\ 0 \\ 0 \end{bmatrix} \qquad \mathbf{e}_2 = \begin{bmatrix} 0 \\ 1 \\ 0 \end{bmatrix}$$

$$\mathbf{e}_3 = \begin{bmatrix} 0 \\ 0 \\ 1 \end{bmatrix}$$

Write the given vector in terms of the vectors $\mathbf{e}_1, \mathbf{e}_2,$ and $\mathbf{e}_3$.

11. $\mathbf{v} = \begin{bmatrix} 2 \\ 4 \\ 1 \end{bmatrix}$

12. $\mathbf{v} = \begin{bmatrix} -1 \\ 3 \\ 2 \end{bmatrix}$

13. $\mathbf{v} = \begin{bmatrix} 0 \\ 3 \\ -2 \end{bmatrix}$

14. $\mathbf{v} = \begin{bmatrix} -1 \\ 0 \\ \frac{1}{2} \end{bmatrix}$

In Exercises 15 and 16, find $\mathbf{w}$ such that
$-\mathbf{u} + 3\mathbf{v} - 2\mathbf{w} = \mathbf{0}$.

15. $\mathbf{u} = \begin{bmatrix} 1 \\ 4 \\ 2 \end{bmatrix} \qquad \mathbf{v} = \begin{bmatrix} -2 \\ 2 \\ 0 \end{bmatrix}$

16. $\mathbf{u} = \begin{bmatrix} -2 \\ 0 \\ 1 \end{bmatrix} \qquad \mathbf{v} = \begin{bmatrix} 2 \\ -3 \\ 4 \end{bmatrix}$

In Exercises 17–24, write the vector equation as an equivalent linear system and then solve the system. Explain what the solution to the linear system implies about the vector equation.

17. $c_1 \begin{bmatrix} 1 \\ -2 \end{bmatrix} + c_2 \begin{bmatrix} 3 \\ -2 \end{bmatrix} = \begin{bmatrix} -2 \\ -1 \end{bmatrix}$

18. $c_1 \begin{bmatrix} 2 \\ 5 \end{bmatrix} + c_2 \begin{bmatrix} -1 \\ -2 \end{bmatrix} = \begin{bmatrix} 0 \\ 5 \end{bmatrix}$

19. $c_1 \begin{bmatrix} 1 \\ 2 \end{bmatrix} + c_2 \begin{bmatrix} -1 \\ -2 \end{bmatrix} = \begin{bmatrix} 3 \\ 1 \end{bmatrix}$

20. $c_1 \begin{bmatrix} -1 \\ 3 \end{bmatrix} + c_2 \begin{bmatrix} 2 \\ -6 \end{bmatrix} = \begin{bmatrix} -1 \\ 1 \end{bmatrix}$

21. $c_1 \begin{bmatrix} -4 \\ 4 \\ 3 \end{bmatrix} + c_2 \begin{bmatrix} 0 \\ 3 \\ -1 \end{bmatrix} + c_3 \begin{bmatrix} -5 \\ 1 \\ -5 \end{bmatrix} = \begin{bmatrix} -3 \\ -3 \\ 4 \end{bmatrix}$

22. $c_1 \begin{bmatrix} 0 \\ -1 \\ 1 \end{bmatrix} + c_2 \begin{bmatrix} 1 \\ 1 \\ 0 \end{bmatrix} + c_3 \begin{bmatrix} 1 \\ 1 \\ -1 \end{bmatrix} = \begin{bmatrix} -1 \\ 0 \\ -1 \end{bmatrix}$

23. $c_1 \begin{bmatrix} -1 \\ 0 \\ 1 \end{bmatrix} + c_2 \begin{bmatrix} -1 \\ 1 \\ 1 \end{bmatrix} + c_3 \begin{bmatrix} 1 \\ -1 \\ -1 \end{bmatrix} = \begin{bmatrix} -1 \\ 0 \\ 2 \end{bmatrix}$

24. $c_1 \begin{bmatrix} -1 \\ 2 \\ 4 \end{bmatrix} + c_2 \begin{bmatrix} 0 \\ 2 \\ 4 \end{bmatrix} + c_3 \begin{bmatrix} 2 \\ 1 \\ 2 \end{bmatrix} = \begin{bmatrix} 6 \\ 7 \\ 3 \end{bmatrix}$

In Exercises 25–28, find all vectors $\begin{bmatrix} a \\ b \end{bmatrix}$, so that the vector equation can be solved.

25. $c_1 \begin{bmatrix} 1 \\ -1 \end{bmatrix} + c_2 \begin{bmatrix} 2 \\ 1 \end{bmatrix} = \begin{bmatrix} a \\ b \end{bmatrix}$

26. $c_1 \begin{bmatrix} 1 \\ 1 \end{bmatrix} + c_2 \begin{bmatrix} -1 \\ 1 \end{bmatrix} = \begin{bmatrix} a \\ b \end{bmatrix}$

27. $c_1 \begin{bmatrix} 1 \\ -1 \end{bmatrix} + c_2 \begin{bmatrix} 2 \\ -2 \end{bmatrix} = \begin{bmatrix} a \\ b \end{bmatrix}$

28. $c_1 \begin{bmatrix} 3 \\ 1 \end{bmatrix} + c_2 \begin{bmatrix} 6 \\ 2 \end{bmatrix} = \begin{bmatrix} a \\ b \end{bmatrix}$

In Exercises 29–32, find all vectors
$$\mathbf{v} = \begin{bmatrix} a \\ b \\ c \end{bmatrix}$$
so that the vector equation $c_1\mathbf{v}_1 + c_2\mathbf{v}_2 + c_3\mathbf{v}_3 = \mathbf{v}$ can be solved.

29. $\mathbf{v}_1 = \begin{bmatrix} 1 \\ 0 \\ 1 \end{bmatrix}$ $\mathbf{v}_2 = \begin{bmatrix} 0 \\ 1 \\ 1 \end{bmatrix}$

$\mathbf{v}_3 = \begin{bmatrix} 2 \\ 1 \\ 0 \end{bmatrix}$

30. $\mathbf{v}_1 = \begin{bmatrix} 1 \\ 1 \\ 1 \end{bmatrix}$ $\mathbf{v}_2 = \begin{bmatrix} 0 \\ 1 \\ 0 \end{bmatrix}$

$\mathbf{v}_3 = \begin{bmatrix} 1 \\ 1 \\ 0 \end{bmatrix}$

31. $\mathbf{v}_1 = \begin{bmatrix} 1 \\ 1 \\ -1 \end{bmatrix}$ $\mathbf{v}_2 = \begin{bmatrix} 2 \\ 1 \\ 1 \end{bmatrix}$

$\mathbf{v}_3 = \begin{bmatrix} 3 \\ 2 \\ 0 \end{bmatrix}$

32. $\mathbf{v}_1 = \begin{bmatrix} -1 \\ 0 \\ 2 \end{bmatrix}$ $\mathbf{v}_2 = \begin{bmatrix} 1 \\ -2 \\ 8 \end{bmatrix}$

$\mathbf{v}_3 = \begin{bmatrix} 1 \\ -1 \\ 3 \end{bmatrix}$

In Exercises 33–39, verify the indicated vector property of Theorem 1 for vectors in $\mathbb{R}^n$.

33. Property 2.

34. Property 3.

35. Property 4.

36. Property 5.

37. Property 6.

38. Property 7.

39. Property 8.

40. Prove that the zero vector in $\mathbb{R}^n$ is unique.

2.2 ▶ Linear Combinations

In three-dimensional Euclidean space $\mathbb{R}^3$ the *coordinate vectors* that define the three axes are the vectors

$$\mathbf{e}_1 = \begin{bmatrix} 1 \\ 0 \\ 0 \end{bmatrix} \qquad \mathbf{e}_2 = \begin{bmatrix} 0 \\ 1 \\ 0 \end{bmatrix} \qquad \text{and} \qquad \mathbf{e}_3 = \begin{bmatrix} 0 \\ 0 \\ 1 \end{bmatrix}$$

Every vector in $\mathbb{R}^3$ can then be obtained from these three coordinate vectors, for example, the vector

$$\mathbf{v} = \begin{bmatrix} 2 \\ 3 \\ 3 \end{bmatrix} = 2 \begin{bmatrix} 1 \\ 0 \\ 0 \end{bmatrix} + 3 \begin{bmatrix} 0 \\ 1 \\ 0 \end{bmatrix} + 3 \begin{bmatrix} 0 \\ 0 \\ 1 \end{bmatrix}$$

Geometrically, the vector $\mathbf{v}$ is obtained by adding scalar multiples of the coordinate vectors, as shown in Fig. 1. The vectors $\mathbf{e}_1, \mathbf{e}_2$, and $\mathbf{e}_3$ are not unique in this respect. For example, the vector $\mathbf{v}$ can also be written as a combination of the vectors

$$\mathbf{v}_1 = \begin{bmatrix} 1 \\ 1 \\ 1 \end{bmatrix} \qquad \mathbf{v}_2 = \begin{bmatrix} 0 \\ 1 \\ 1 \end{bmatrix} \qquad \text{and} \qquad \mathbf{v}_3 = \begin{bmatrix} -1 \\ 1 \\ 1 \end{bmatrix}$$

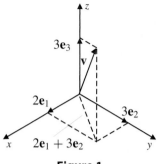

Figure 1

that is,

$$3\mathbf{v}_1 - \mathbf{v}_2 + \mathbf{v}_3 = \begin{bmatrix} 2 \\ 3 \\ 3 \end{bmatrix}$$

A vector written as a combination of other vectors using addition and scalar multiplication is called a *linear combination*. Combining vectors in this manner plays a central role in describing Euclidean spaces and, as we will see in Chap. 3, in describing abstract vector spaces.

DEFINITION 1   **Linear Combination**   Let $S = \{\mathbf{v}_1, \mathbf{v}_2, \ldots, \mathbf{v}_k\}$ be a set of vectors in $\mathbb{R}^n$, and let $c_1, c_2, \ldots, c_k$ be scalars. An expression of the form

$$c_1\mathbf{v}_1 + c_2\mathbf{v}_2 + \cdots + c_k\mathbf{v}_k = \sum_{i=1}^{k} c_i\mathbf{v}_i$$

is called a **linear combination** of the vectors of S. Any vector $\mathbf{v}$ that can be written in this form is also called a linear combination of the vectors of S.

In Example 1 we show how linear systems are used to decide if a vector is a linear combination of a set of vectors.

EXAMPLE 1   Determine whether the vector

$$\mathbf{v} = \begin{bmatrix} -1 \\ 1 \\ 10 \end{bmatrix}$$

is a linear combination of the vectors

$$\mathbf{v}_1 = \begin{bmatrix} 1 \\ 0 \\ 1 \end{bmatrix} \qquad \mathbf{v}_2 = \begin{bmatrix} -2 \\ 3 \\ -2 \end{bmatrix} \qquad \text{and} \qquad \mathbf{v}_3 = \begin{bmatrix} -6 \\ 7 \\ 5 \end{bmatrix}$$

Solution   The vector $\mathbf{v}$ is a linear combination of the vectors $\mathbf{v}_1$, $\mathbf{v}_2$, and $\mathbf{v}_3$ if there are scalars c_1, c_2, and c_3, such that

$$\mathbf{v} = \begin{bmatrix} -1 \\ 1 \\ 10 \end{bmatrix} = c_1\mathbf{v}_1 + c_2\mathbf{v}_2 + c_3\mathbf{v}_3$$

$$= c_1 \begin{bmatrix} 1 \\ 0 \\ 1 \end{bmatrix} + c_2 \begin{bmatrix} -2 \\ 3 \\ -2 \end{bmatrix} + c_3 \begin{bmatrix} -6 \\ 7 \\ 5 \end{bmatrix}$$

$$= \begin{bmatrix} c_1 - 2c_2 - 6c_3 \\ 3c_2 + 7c_3 \\ c_1 - 2c_2 + 5c_3 \end{bmatrix}$$

Equating components gives the linear system

$$\begin{cases} c_1 - 2c_2 - 6c_3 = -1 \\ 3c_2 + 7c_3 = 1 \\ c_1 - 2c_2 + 5c_3 = 10 \end{cases}$$

To solve this linear system, we reduce the augmented matrix

$$\begin{bmatrix} 1 & -2 & -6 & -1 \\ 0 & 3 & 7 & 1 \\ 1 & -2 & 5 & 10 \end{bmatrix} \quad \text{to} \quad \begin{bmatrix} 1 & 0 & 0 & 1 \\ 0 & 1 & 0 & -2 \\ 0 & 0 & 1 & 1 \end{bmatrix}$$

From the last matrix, we see that the linear system is consistent with the unique solution

$$c_1 = 1 \qquad c_2 = -2 \qquad \text{and} \qquad c_3 = 1$$

Using these scalars, we can write $\mathbf{v}$ as the linear combination

$$\mathbf{v} = \begin{bmatrix} -1 \\ 1 \\ 10 \end{bmatrix} = 1 \begin{bmatrix} 1 \\ 0 \\ 1 \end{bmatrix} + (-2) \begin{bmatrix} -2 \\ 3 \\ -2 \end{bmatrix} + 1 \begin{bmatrix} -6 \\ 7 \\ 5 \end{bmatrix}$$

The case for which a vector is not a linear combination of a set of vectors is illustrated in Example 2.

EXAMPLE 2   Determine whether the vector

$$\mathbf{v} = \begin{bmatrix} -5 \\ 11 \\ -7 \end{bmatrix}$$

is a linear combination of the vectors

$$\mathbf{v}_1 = \begin{bmatrix} 1 \\ -2 \\ 2 \end{bmatrix} \qquad \mathbf{v}_2 = \begin{bmatrix} 0 \\ 5 \\ 5 \end{bmatrix} \qquad \text{and} \qquad \mathbf{v}_3 = \begin{bmatrix} 2 \\ 0 \\ 8 \end{bmatrix}$$

Solution   The vector $\mathbf{v}$ is a linear combination of the vectors $\mathbf{v}_1$, $\mathbf{v}_2$, and $\mathbf{v}_3$ if there are scalars c_1, c_2, and c_3, such that

$$\begin{bmatrix} -5 \\ 11 \\ -7 \end{bmatrix} = c_1 \begin{bmatrix} 1 \\ -2 \\ 2 \end{bmatrix} + c_2 \begin{bmatrix} 0 \\ 5 \\ 5 \end{bmatrix} + c_3 \begin{bmatrix} 2 \\ 0 \\ 8 \end{bmatrix}$$

The augmented matrix corresponding to this equation is given by

$$\begin{bmatrix} 1 & 0 & 2 & | & -5 \\ -2 & 5 & 0 & | & 11 \\ 2 & 5 & 8 & | & -7 \end{bmatrix}$$

Reducing the augmented matrix

$$\begin{bmatrix} 1 & 0 & 2 & | & -5 \\ -2 & 5 & 0 & | & 11 \\ 2 & 5 & 8 & | & -7 \end{bmatrix} \quad \text{to} \quad \begin{bmatrix} 1 & 0 & 2 & | & -5 \\ 0 & 5 & 4 & | & 1 \\ 0 & 0 & 0 & | & 2 \end{bmatrix}$$

shows that the linear system is inconsistent. Therefore, the vector $\mathbf{v}$ cannot be written as a linear combination of the three vectors $\mathbf{v}_1$, $\mathbf{v}_2$, and $\mathbf{v}_3$.

To see this geometrically, first observe that the vector $\mathbf{v}_3$ is a linear combination of $\mathbf{v}_1$ and $\mathbf{v}_2$. That is,

$$\mathbf{v}_3 = 2\mathbf{v}_1 + \frac{4}{5}\mathbf{v}_2$$

Therefore, any linear combination of the three vectors $\mathbf{v}_1$, $\mathbf{v}_2$, and $\mathbf{v}_3$, is just a linear combination of $\mathbf{v}_1$ and $\mathbf{v}_2$. Specifically,

$$c_1\mathbf{v}_1 + c_2\mathbf{v}_2 + c_3\mathbf{v}_3 = c_1\mathbf{v}_1 + c_2\mathbf{v}_2 + c_3\left(2\mathbf{v}_1 + \frac{4}{5}\mathbf{v}_2\right)$$

$$= (c_1 + 2c_3)\mathbf{v}_1 + \left(c_2 + \frac{4}{5}c_3\right)\mathbf{v}_2$$

The set of all vectors that are linear combinations of $\mathbf{v}_1$ and $\mathbf{v}_2$ is a plane in $\mathbb{R}^3$, which does not contain the vector $\mathbf{v}$, as shown in Fig. 2.

Figure 2

In $\mathbb{R}^n$ the **coordinate vectors** are the n vectors given by

$$\mathbf{e}_1 = \begin{bmatrix} 1 \\ 0 \\ \vdots \\ 0 \end{bmatrix} \qquad \mathbf{e}_2 = \begin{bmatrix} 0 \\ 1 \\ \vdots \\ 0 \end{bmatrix} \quad \cdots \quad \mathbf{e}_n = \begin{bmatrix} 0 \\ 0 \\ \vdots \\ 1 \end{bmatrix}$$

These vectors can also be defined by the equations

$$(\mathbf{e}_k)_i = \begin{cases} 1 & \text{if } i = k \\ 0 & \text{if } i \neq k \end{cases}$$

where $1 \leq k \leq n$.

An important property of the coordinate vectors is that *every vector in $\mathbb{R}^n$ can be written as a linear combination of the coordinate vectors*. Indeed, for any vector $\mathbf{v}$ in $\mathbb{R}^n$, let the scalars be the components of the vector, so that

$$\mathbf{v} = \begin{bmatrix} v_1 \\ v_2 \\ \vdots \\ v_n \end{bmatrix} = v_1 \begin{bmatrix} 1 \\ 0 \\ \vdots \\ 0 \end{bmatrix} + v_2 \begin{bmatrix} 0 \\ 1 \\ \vdots \\ 0 \end{bmatrix} + \cdots + v_n \begin{bmatrix} 0 \\ 0 \\ \vdots \\ 1 \end{bmatrix}$$

$$= v_1 \mathbf{e}_1 + v_2 \mathbf{e}_2 + \cdots + v_n \mathbf{e}_n$$

Linear combinations of more abstract objects can also be formed, as illustrated in Example 3 using 2×2 matrices. This type of construction is used extensively in Chap. 3 when we consider abstract vector spaces.

EXAMPLE 3   Show that the matrix

$$A = \begin{bmatrix} 1 & 1 \\ 1 & 0 \end{bmatrix}$$

is a linear combination of the matrices

$$M_1 = \begin{bmatrix} 1 & 0 \\ 0 & 1 \end{bmatrix} \qquad M_2 = \begin{bmatrix} 0 & 1 \\ 1 & 1 \end{bmatrix} \qquad \text{and} \qquad M_3 = \begin{bmatrix} 1 & 1 \\ 1 & 1 \end{bmatrix}$$

Solution   Similar to the situation with vectors, we must find scalars c_1, c_2, and c_3 such that

$$c_1 M_1 + c_2 M_2 + c_3 M_3 = A$$

that is,

$$c_1 \begin{bmatrix} 1 & 0 \\ 0 & 1 \end{bmatrix} + c_2 \begin{bmatrix} 0 & 1 \\ 1 & 1 \end{bmatrix} + c_3 \begin{bmatrix} 1 & 1 \\ 1 & 1 \end{bmatrix} = \begin{bmatrix} 1 & 1 \\ 1 & 0 \end{bmatrix}$$

After performing the scalar multiplication and addition, we obtain

$$\begin{bmatrix} c_1 + c_3 & c_2 + c_3 \\ c_2 + c_3 & c_1 + c_2 + c_3 \end{bmatrix} = \begin{bmatrix} 1 & 1 \\ 1 & 0 \end{bmatrix}$$

Equating corresponding entries gives the linear system

$$\begin{cases} c_1 + c_3 = 1 \\ c_2 + c_3 = 1 \\ c_1 + c_2 + c_3 = 0 \end{cases}$$

This system is consistent with solution $c_1 = -1, c_2 = -1$, and $c_3 = 2$. Thus, the matrix A is a linear combination of the matrices M_1, M_2, and M_3.

EXAMPLE 4   Consider the homogeneous equation $A\mathbf{x} = \mathbf{0}$. Show that if $\mathbf{x}_1, \mathbf{x}_2, \ldots, \mathbf{x}_n$ are solutions of the equation, then every linear combination $c_1\mathbf{x}_1 + c_2\mathbf{x}_2 + \cdots + c_n\mathbf{x}_n$ is also a solution of the equation.

Solution   Since $\mathbf{x}_1, \mathbf{x}_2, \ldots, \mathbf{x}_n$ are solutions of the matrix equation, we have

$$A\mathbf{x}_1 = \mathbf{0} \qquad A\mathbf{x}_2 = \mathbf{0} \qquad \ldots \qquad A\mathbf{x}_n = \mathbf{0}$$

Then using the algebraic properties of matrices, we have

$$\begin{aligned} A(c_1\mathbf{x}_1 + c_2\mathbf{x}_2 + \cdots + c_n\mathbf{x}_n) &= A(c_1\mathbf{x}_1) + A(c_2\mathbf{x}_2) + \cdots + A(c_n\mathbf{x}_n) \\ &= c_1(A\mathbf{x}_1) + c_2(A\mathbf{x}_2) + \cdots + c_n(A\mathbf{x}_n) \\ &= c_1\mathbf{0} + c_2\mathbf{0} + \cdots + c_n\mathbf{0} \\ &= \mathbf{0} \end{aligned}$$

The result of Example 4 is an extension of the one given in Example 3 of Sec. 1.5.

Vector Form of a Linear System

We have already seen that a linear system with m equations and n variables

$$\begin{cases} a_{11}x_1 + a_{12}x_2 + \cdots + a_{1n}x_n = b_1 \\ a_{21}x_1 + a_{22}x_2 + \cdots + a_{2n}x_n = b_2 \\ \vdots \qquad \vdots \qquad\qquad \vdots \qquad \vdots \\ a_{m1}x_1 + a_{m2}x_2 + \cdots + a_{mn}x_n = b_m \end{cases}$$

can be written in matrix form as $A\mathbf{x} = \mathbf{b}$, where A is the $m \times n$ coefficient matrix, $\mathbf{x}$ is the vector in $\mathbb{R}^n$ of variables, and $\mathbf{b}$ is the vector in $\mathbb{R}^m$ of constants. If we use the column vectors of the coefficient matrix A, then the matrix equation can be written in the equivalent form

$$
x_1 \begin{bmatrix} a_{11} \\ a_{21} \\ \vdots \\ a_{m1} \end{bmatrix} + x_2 \begin{bmatrix} a_{12} \\ a_{22} \\ \vdots \\ a_{m2} \end{bmatrix} + \cdots + x_n \begin{bmatrix} a_{1n} \\ a_{2n} \\ \vdots \\ a_{mn} \end{bmatrix} = \begin{bmatrix} b_1 \\ b_2 \\ \vdots \\ b_m \end{bmatrix}
$$

This last equation is called the **vector form** of a linear system. This equation can also be written as

$$
x_1 \mathbf{A}_1 + x_2 \mathbf{A}_2 + \cdots + x_n \mathbf{A}_n = \mathbf{b}
$$

where $\mathbf{A}_i$ denotes the ith column vector of the matrix A. Observe that this equation is consistent whenever the vector $\mathbf{b}$ can be written as a linear combination of the column vectors of A.

THEOREM 2   The linear system $A\mathbf{x} = \mathbf{b}$ is consistent if and only if the vector $\mathbf{b}$ can be expressed as a linear combination of the column vectors of A.

Matrix Multiplication

Before concluding this section, we comment on how linear combinations can be used to describe the product of two matrices. Let A be an $m \times n$ matrix and B an $n \times p$ matrix. If $\mathbf{B}_i$ is the ith column vector of B, then the ith column vector of the product AB is given by

$$
\begin{aligned}
A\mathbf{B}_i &= \begin{bmatrix} a_{11} & a_{12} & \cdots & a_{1n} \\ a_{21} & a_{22} & \cdots & a_{2n} \\ \vdots & \vdots & \ddots & \vdots \\ a_{m1} & a_{m2} & \cdots & a_{mn} \end{bmatrix} \begin{bmatrix} b_{1i} \\ b_{2i} \\ \vdots \\ b_{ni} \end{bmatrix} \\[2em]
&= \begin{bmatrix} a_{11}b_{1i} + a_{12}b_{2i} + \cdots + a_{1n}b_{ni} \\ a_{21}b_{1i} + a_{22}b_{2i} + \cdots + a_{2n}b_{ni} \\ \vdots & \vdots & \vdots \\ a_{m1}b_{1i} + a_{m2}b_{2i} + \cdots + a_{mn}b_{ni} \end{bmatrix} \\[2em]
&= \begin{bmatrix} a_{11}b_{1i} \\ a_{21}b_{1i} \\ \vdots \\ a_{m1}b_{1i} \end{bmatrix} + \begin{bmatrix} a_{12}b_{2i} \\ a_{22}b_{2i} \\ \vdots \\ a_{m2}b_{2i} \end{bmatrix} + \cdots + \begin{bmatrix} a_{1n}b_{ni} \\ a_{2n}b_{ni} \\ \vdots \\ a_{mn}b_{ni} \end{bmatrix} \\[2em]
&= b_{1i}\mathbf{A}_1 + b_{2i}\mathbf{A}_2 + \cdots + b_{ni}\mathbf{A}_n
\end{aligned}
$$

Since for $i = 1, 2, \ldots, p$ the product $A\mathbf{B}_i$ is the ith column vector of AB, each column vector of AB is a linear combination of the column vectors of A.

> **Fact Summary**
>
> 1. Every vector in $\mathbb{R}^n$ is a linear combination of the coordinate vectors $e_1, e_2, \ldots, e_n$.
> 2. If $x_1, x_2, \ldots, x_k$ are all solutions to the homogeneous equation $Ax = 0$, then so is every linear combination of these vectors.
> 3. The linear system $Ax = b$ can be written in the equivalent vector form as $x_1A_1 + x_2A_2 + \cdots + x_nA_n = b$. The left side is a linear combination of the column vectors of A.
> 4. The linear system $Ax = b$ is consistent if and only if b is a linear combination of the column vectors of A.

Exercise Set 2.2

In Exercises 1–6, determine whether the vector v is a linear combination of the vectors v_1 and v_2.

1. $v = \begin{bmatrix} -4 \\ 11 \end{bmatrix}$ $v_1 = \begin{bmatrix} 1 \\ 1 \end{bmatrix}$

$v_2 = \begin{bmatrix} -2 \\ 3 \end{bmatrix}$

2. $v = \begin{bmatrix} 13 \\ -2 \end{bmatrix}$ $v_1 = \begin{bmatrix} -1 \\ 2 \end{bmatrix}$

$v_2 = \begin{bmatrix} 3 \\ 0 \end{bmatrix}$

3. $v = \begin{bmatrix} 1 \\ 1 \end{bmatrix}$ $v_1 = \begin{bmatrix} -2 \\ 4 \end{bmatrix}$

$v_2 = \begin{bmatrix} 3 \\ -6 \end{bmatrix}$

4. $v = \begin{bmatrix} 3 \\ 2 \end{bmatrix}$ $v_1 = \begin{bmatrix} 1 \\ -2 \end{bmatrix}$

$v_2 = \begin{bmatrix} \frac{1}{2} \\ -1 \end{bmatrix}$

5. $v = \begin{bmatrix} -3 \\ 10 \\ 10 \end{bmatrix}$ $v_1 = \begin{bmatrix} -2 \\ 3 \\ 4 \end{bmatrix}$

$v_2 = \begin{bmatrix} 1 \\ 4 \\ 2 \end{bmatrix}$

6. $v = \begin{bmatrix} -2 \\ 6 \\ 8 \end{bmatrix}$ $v_1 = \begin{bmatrix} 3 \\ 4 \\ -1 \end{bmatrix}$

$v_2 = \begin{bmatrix} 2 \\ 7 \\ 3 \end{bmatrix}$

In Exercises 7–12, determine whether the vector v is a linear combination of the vectors v_1, v_2, and v_3.

7. $v = \begin{bmatrix} 2 \\ 8 \\ 2 \end{bmatrix}$ $v_1 = \begin{bmatrix} 2 \\ -2 \\ 0 \end{bmatrix}$

$v_2 = \begin{bmatrix} 3 \\ 0 \\ -3 \end{bmatrix}$ $v_3 = \begin{bmatrix} -2 \\ 0 \\ -1 \end{bmatrix}$

8. $\mathbf{v} = \begin{bmatrix} 5 \\ -4 \\ -7 \end{bmatrix}$ $\mathbf{v}_1 = \begin{bmatrix} 1 \\ -1 \\ 0 \end{bmatrix}$

$\mathbf{v}_2 = \begin{bmatrix} -2 \\ -1 \\ -1 \end{bmatrix}$ $\mathbf{v}_3 = \begin{bmatrix} 3 \\ -1 \\ -3 \end{bmatrix}$

9. $\mathbf{v} = \begin{bmatrix} -1 \\ 1 \\ 5 \end{bmatrix}$ $\mathbf{v}_1 = \begin{bmatrix} 1 \\ 2 \\ -1 \end{bmatrix}$

$\mathbf{v}_2 = \begin{bmatrix} -1 \\ -1 \\ 3 \end{bmatrix}$ $\mathbf{v}_3 = \begin{bmatrix} 0 \\ 1 \\ 2 \end{bmatrix}$

10. $\mathbf{v} = \begin{bmatrix} -3 \\ 5 \\ 5 \end{bmatrix}$ $\mathbf{v}_1 = \begin{bmatrix} -3 \\ 2 \\ 1 \end{bmatrix}$

$\mathbf{v}_2 = \begin{bmatrix} 1 \\ 4 \\ 1 \end{bmatrix}$ $\mathbf{v}_3 = \begin{bmatrix} -1 \\ 10 \\ 3 \end{bmatrix}$

11. $\mathbf{v} = \begin{bmatrix} 3 \\ -17 \\ 17 \\ 7 \end{bmatrix}$ $\mathbf{v}_1 = \begin{bmatrix} 2 \\ -3 \\ 4 \\ 1 \end{bmatrix}$

$\mathbf{v}_2 = \begin{bmatrix} 1 \\ 6 \\ -1 \\ 2 \end{bmatrix}$ $\mathbf{v}_3 = \begin{bmatrix} -1 \\ -1 \\ 2 \\ 3 \end{bmatrix}$

12. $\mathbf{v} = \begin{bmatrix} 6 \\ 3 \\ 3 \\ 7 \end{bmatrix}$ $\mathbf{v}_1 = \begin{bmatrix} 2 \\ 3 \\ 4 \\ 5 \end{bmatrix}$

$\mathbf{v}_2 = \begin{bmatrix} 1 \\ -1 \\ 2 \\ 3 \end{bmatrix}$ $\mathbf{v}_3 = \begin{bmatrix} 3 \\ 1 \\ -3 \\ 1 \end{bmatrix}$

In Exercises 13–16, find all the ways that $\mathbf{v}$ can be written as a linear combination of the given vectors.

13. $\mathbf{v} = \begin{bmatrix} 3 \\ 0 \end{bmatrix}$ $\mathbf{v}_1 = \begin{bmatrix} 3 \\ 1 \end{bmatrix}$

$\mathbf{v}_2 = \begin{bmatrix} 0 \\ -1 \end{bmatrix}$ $\mathbf{v}_3 = \begin{bmatrix} -1 \\ 2 \end{bmatrix}$

14. $\mathbf{v} = \begin{bmatrix} -1 \\ -1 \end{bmatrix}$ $\mathbf{v}_1 = \begin{bmatrix} 1 \\ -1 \end{bmatrix}$

$\mathbf{v}_2 = \begin{bmatrix} -2 \\ -1 \end{bmatrix}$ $\mathbf{v}_3 = \begin{bmatrix} 3 \\ 0 \end{bmatrix}$

15. $\mathbf{v} = \begin{bmatrix} 0 \\ -1 \\ -3 \end{bmatrix}$ $\mathbf{v}_1 = \begin{bmatrix} 0 \\ 1 \\ 1 \end{bmatrix}$

$\mathbf{v}_2 = \begin{bmatrix} -2 \\ -1 \\ 2 \end{bmatrix}$ $\mathbf{v}_3 = \begin{bmatrix} -2 \\ -3 \\ -1 \end{bmatrix}$

$\mathbf{v}_4 = \begin{bmatrix} 2 \\ -1 \\ -2 \end{bmatrix}$

16. $\mathbf{v} = \begin{bmatrix} -3 \\ -3 \\ 1 \end{bmatrix}$ $\mathbf{v}_1 = \begin{bmatrix} -1 \\ -1 \\ 2 \end{bmatrix}$

$\mathbf{v}_2 = \begin{bmatrix} 0 \\ -1 \\ -1 \end{bmatrix}$ $\mathbf{v}_3 = \begin{bmatrix} 0 \\ -1 \\ -2 \end{bmatrix}$

$\mathbf{v}_4 = \begin{bmatrix} -3 \\ -1 \\ -2 \end{bmatrix}$

In Exercises 17–20, determine if the matrix M is a linear combination of the matrices M_1, M_2, and M_3.

17. $M = \begin{bmatrix} -2 & 4 \\ 4 & 0 \end{bmatrix}$

$M_1 = \begin{bmatrix} 1 & 2 \\ 1 & -1 \end{bmatrix}$ $M_2 = \begin{bmatrix} -2 & 3 \\ 1 & 4 \end{bmatrix}$

$M_3 = \begin{bmatrix} -1 & 3 \\ 2 & 1 \end{bmatrix}$

18. $M = \begin{bmatrix} 2 & 3 \\ 1 & 2 \end{bmatrix}$

$M_1 = \begin{bmatrix} 2 & 2 \\ 1 & 1 \end{bmatrix}$ $M_2 = \begin{bmatrix} -1 & 1 \\ 2 & 1 \end{bmatrix}$

$M_3 = \begin{bmatrix} 1 & 2 \\ 3 & 1 \end{bmatrix}$

19. $M = \begin{bmatrix} 2 & 1 \\ -1 & 2 \end{bmatrix}$

$M_1 = \begin{bmatrix} 2 & 2 \\ -1 & 3 \end{bmatrix}$ $M_2 = \begin{bmatrix} 3 & -1 \\ 2 & -2 \end{bmatrix}$

$M_3 = \begin{bmatrix} 3 & -1 \\ 2 & 2 \end{bmatrix}$

20. $M = \begin{bmatrix} 2 & 1 \\ 3 & 4 \end{bmatrix}$

$M_1 = \begin{bmatrix} 1 & 0 \\ 0 & -1 \end{bmatrix}$ $M_2 = \begin{bmatrix} 0 & 1 \\ 0 & 0 \end{bmatrix}$

$M_3 = \begin{bmatrix} 0 & 0 \\ 0 & 1 \end{bmatrix}$

21. Let $A = \begin{bmatrix} 1 & 3 \\ -2 & 1 \end{bmatrix}$ and $\mathbf{x} = \begin{bmatrix} 2 \\ -1 \end{bmatrix}$. Write the product $A\mathbf{x}$ as a linear combination of the column vectors of A.

22. Let $A = \begin{bmatrix} 1 & 2 & -1 \\ 2 & 3 & 4 \\ -3 & 2 & 1 \end{bmatrix}$ and $\mathbf{x} = \begin{bmatrix} -1 \\ -1 \\ 3 \end{bmatrix}$.

Write the product $A\mathbf{x}$ as a linear combination of the column vectors of A.

23. Let $A = \begin{bmatrix} -1 & -2 \\ 3 & 4 \end{bmatrix}$ and $B = \begin{bmatrix} 3 & 2 \\ 2 & 5 \end{bmatrix}$. Write each column vector of AB as a linear combination of the column vectors of A.

24. Let $A = \begin{bmatrix} 2 & 0 & -1 \\ 1 & -1 & 4 \\ -4 & 3 & 1 \end{bmatrix}$ and

$B = \begin{bmatrix} 3 & 2 & 1 \\ -2 & 1 & 0 \\ 2 & -1 & 1 \end{bmatrix}$. Write each column

vector of AB as a linear combination of the column vectors of A.

In Exercises 25 and 26, write the polynomial $p(x)$, if possible, as a linear combination of the polynomials

$$1 + x \qquad \text{and} \qquad x^2$$

25. $p(x) = 2x^2 - 3x - 1$

26. $p(x) = -x^2 + 3x + 3$

In Exercises 27 and 28, write the polynomial $p(x)$, if possible, as a linear combination of the polynomials

$$1 + x, -x, x^2 + 1 \qquad \text{and} \qquad 2x^3 - x + 1$$

27. $p(x) = x^3 - 2x + 1$

28. $p(x) = x^3$

29. Describe all vectors in $\mathbb{R}^3$ that can be written as a linear combination of the vectors

$$\begin{bmatrix} 1 \\ 2 \\ -1 \end{bmatrix} \qquad \begin{bmatrix} 3 \\ 7 \\ -2 \end{bmatrix} \quad \text{and} \quad \begin{bmatrix} 1 \\ 3 \\ 0 \end{bmatrix}$$

30. Describe all 2×2 matrices that can be written as a linear combination of the matrices

$$\begin{bmatrix} 1 & 0 \\ 0 & 0 \end{bmatrix} \qquad \begin{bmatrix} 0 & 1 \\ 1 & 0 \end{bmatrix} \quad \text{and} \quad \begin{bmatrix} 0 & 0 \\ 0 & 1 \end{bmatrix}$$

31. If $\mathbf{v} = \mathbf{v}_1 + \mathbf{v}_2 + \mathbf{v}_3 + \mathbf{v}_4$ and $\mathbf{v}_4 = \mathbf{v}_1 - 2\mathbf{v}_2 + 3\mathbf{v}_3$, write $\mathbf{v}$ as a linear combination of $\mathbf{v}_1$, $\mathbf{v}_2$, and $\mathbf{v}_3$.

32. If $\mathbf{v} = \mathbf{v}_1 + \mathbf{v}_2 + \mathbf{v}_3 + \mathbf{v}_4$ and $\mathbf{v}_2 = 2\mathbf{v}_1 - 4\mathbf{v}_3$, write $\mathbf{v}$ as a linear combination of $\mathbf{v}_1$, $\mathbf{v}_3$, and $\mathbf{v}_4$.

33. Suppose that the vector $\mathbf{v}$ is a linear combination of the vectors $\mathbf{v}_1, \mathbf{v}_2, \ldots, \mathbf{v}_n$, and $c_1\mathbf{v}_1 + c_2\mathbf{v}_2 + \cdots + c_n\mathbf{v}_n = \mathbf{0}$, with $c_1 \neq 0$. Show that $\mathbf{v}$ is a linear combination of $\mathbf{v}_2, \ldots, \mathbf{v}_n$.

34. Suppose that the vector $\mathbf{v}$ is a linear combination of the vectors $\mathbf{v}_1, \mathbf{v}_2, \ldots, \mathbf{v}_n$, and $\mathbf{w}_1, \mathbf{w}_2, \ldots, \mathbf{w}_m$, are another m vectors. Show that $\mathbf{v}$ is a linear combination of $\mathbf{v}_1, \mathbf{v}_2, \ldots, \mathbf{v}_n, \mathbf{w}_1, \mathbf{w}_2, \ldots, \mathbf{w}_m$.

35. Let S_1 be the set of all linear combinations of the vectors $\mathbf{v}_1, \mathbf{v}_2, \ldots, \mathbf{v}_k$ in $\mathbb{R}^n$, and S_2 be the set of all linear combinations of the vectors $\mathbf{v}_1, \mathbf{v}_2, \ldots,$

$\mathbf{v}_k, c\mathbf{v}_k$, where c is a nonzero scalar. Show that $S_1 = S_2$.

36. Let S_1 be the set of all linear combinations of the vectors $\mathbf{v}_1, \mathbf{v}_2, \ldots, \mathbf{v}_k$ in $\mathbb{R}^n$, and S_2 be the set of all linear combinations of the vectors $\mathbf{v}_1, \mathbf{v}_2, \ldots, \mathbf{v}_k, \mathbf{v}_1 + \mathbf{v}_2$. Show that $S_1 = S_2$.

37. Suppose that $A\mathbf{x} = \mathbf{b}$ is a 3×3 linear system that is consistent. If there is a scalar c such that $\mathbf{A}_3 = c\mathbf{A}_1$, then show that the linear system has infinitely many solutions.

38. Suppose that $A\mathbf{x} = \mathbf{b}$ is a 3×3 linear system that is consistent. If $\mathbf{A}_3 = \mathbf{A}_1 + \mathbf{A}_2$, then show that the linear system has infinitely many solutions.

39. The equation

$$2y'' - 3y' + y = 0$$

is an example of a *differential equation*. Show that $y = f(x) = e^x$ and $y = g(x) = e^{\frac{1}{2}x}$ are solutions to the equation. Then show that any linear combination of $f(x)$ and $g(x)$ is another solution to the differential equation.

2.3 ▶ Linear Independence

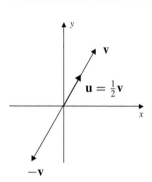

Figure 1

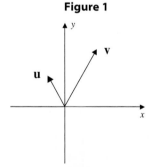

Figure 2

In Sec. 2.2 we saw that given a set S of vectors in $\mathbb{R}^n$, it is not always possible to express every vector in $\mathbb{R}^n$ as a linear combination of vectors from S. At the other extreme, there are infinitely many different subsets S such that the collection of all linear combinations of vectors from S is $\mathbb{R}^n$. For example, the collection of all linear combinations of the set of coordinate vectors $S = \{\mathbf{e}_1, \ldots, \mathbf{e}_n\}$ is $\mathbb{R}^n$, but so is the collection of linear combinations of $T = \{\mathbf{e}_1, \ldots, \mathbf{e}_n, \mathbf{e}_1 + \mathbf{e}_2\}$. In this way S and T both *generate* $\mathbb{R}^n$. To characterize those minimal sets S that generate $\mathbb{R}^n$, we require the concept of *linear independence*. As motivation let two vectors $\mathbf{u}$ and $\mathbf{v}$ in $\mathbb{R}^2$ lie on the same line, as shown in Fig. 1. Thus, there is a nonzero scalar c such that

$$\mathbf{u} = c\mathbf{v}$$

This condition can also be written as

$$\mathbf{u} - c\mathbf{v} = \mathbf{0}$$

In this case we say that the vectors $\mathbf{u}$ and $\mathbf{v}$ are *linearly dependent*. Evidently we have that two vectors $\mathbf{u}$ and $\mathbf{v}$ are linearly dependent provided that the zero vector is a nontrivial (not both scalars 0) linear combination of the vectors. On the other hand, the vectors shown in Fig. 2 are not linearly dependent. This concept is generalized to sets of vectors in $\mathbb{R}^n$.

DEFINITION 1 **Linearly Indpendent and Linearly Dependent** The set of vectors $S = \{\mathbf{v}_1, \mathbf{v}_2, \ldots, \mathbf{v}_m\}$ in $\mathbb{R}^n$ is **linearly independent** provided that the only solution to the equation

$$c_1\mathbf{v}_1 + c_2\mathbf{v}_2 + \cdots + c_m\mathbf{v}_m = \mathbf{0}$$

is the trivial solution $c_1 = c_2 = \cdots = c_m = 0$. If the above linear combination has a nontrivial solution, then the set S is called **linearly dependent**.

For example, the set of coordinate vectors

$$S = \{\mathbf{e}_1, \ldots, \mathbf{e}_n\}$$

in $\mathbb{R}^n$ is linearly independent.

EXAMPLE 1   Determine whether the vectors

$$\mathbf{v}_1 = \begin{bmatrix} 1 \\ 0 \\ 1 \\ 2 \end{bmatrix} \qquad \mathbf{v}_2 = \begin{bmatrix} 0 \\ 1 \\ 1 \\ 2 \end{bmatrix} \qquad \text{and} \qquad \mathbf{v}_3 = \begin{bmatrix} 1 \\ 1 \\ 1 \\ 3 \end{bmatrix}$$

are linearly independent or linearly dependent.

Solution   We seek solutions to the vector equation

$$c_1 \begin{bmatrix} 1 \\ 0 \\ 1 \\ 2 \end{bmatrix} + c_2 \begin{bmatrix} 0 \\ 1 \\ 1 \\ 2 \end{bmatrix} + c_3 \begin{bmatrix} 1 \\ 1 \\ 1 \\ 3 \end{bmatrix} = \begin{bmatrix} 0 \\ 0 \\ 0 \\ 0 \end{bmatrix}$$

From this we obtain the linear system

$$\begin{cases} c_1 + c_3 = 0 \\ c_2 + c_3 = 0 \\ c_1 + c_2 + c_3 = 0 \\ 2c_1 + 2c_2 + 3c_3 = 0 \end{cases}$$

Subtracting the first equation from the third equation gives $c_2 = 0$. Then, from equation 2, we have $c_3 = 0$ and from equation 1 we have $c_1 = 0$. Hence, the only solution to the linear system is the trivial solution $c_1 = c_2 = c_3 = 0$. Therefore, the vectors are linearly independent.

EXAMPLE 2   Determine whether the vectors

$$\mathbf{v}_1 = \begin{bmatrix} 1 \\ 0 \\ 2 \end{bmatrix} \qquad \mathbf{v}_2 = \begin{bmatrix} -1 \\ 1 \\ 2 \end{bmatrix} \qquad \mathbf{v}_3 = \begin{bmatrix} -2 \\ 3 \\ 1 \end{bmatrix} \qquad \mathbf{v}_4 = \begin{bmatrix} 2 \\ 1 \\ 1 \end{bmatrix}$$

are linearly independent.

Solution   As in Example 1, we need to solve

$$c_1 \begin{bmatrix} 1 \\ 0 \\ 2 \end{bmatrix} + c_2 \begin{bmatrix} -1 \\ 1 \\ 2 \end{bmatrix} + c_3 \begin{bmatrix} -2 \\ 3 \\ 1 \end{bmatrix} + c_4 \begin{bmatrix} 2 \\ 1 \\ 1 \end{bmatrix} = \begin{bmatrix} 0 \\ 0 \\ 0 \end{bmatrix}$$

This leads to the homogeneous linear system

$$\begin{cases} c_1 - c_2 - 2c_3 + 2c_4 = 0 \\ c_2 + 3c_3 + c_4 = 0 \\ 2c_1 + 2c_2 + c_3 + c_4 = 0 \end{cases}$$

with solution set given by

$$S = \{(-2t, 2t, -t, t) \mid t \in \mathbb{R}\}$$

Since the linear system has infinitely many solutions, the set of vectors $\{v_1, v_2, v_3, v_4\}$ is linearly dependent.

In Example 2, we verified that the set of vectors $\{v_1, v_2, v_3, v_4\}$ is linearly dependent. Observe further that the vector v_4 is a linear combination of v_1, v_2, and v_3, that is,

$$v_4 = 2v_1 - 2v_2 + v_3$$

In Theorem 3 we establish that any finite collection of vectors in $\mathbb{R}^n$, where the number of vectors exceeds n, is linearly dependent.

THEOREM 3   Let $S = \{v_1, v_2, \ldots, v_n\}$ be a set of n nonzero vectors in $\mathbb{R}^m$. If $n > m$, then the set S is linearly dependent.

Proof   Let A be the $m \times n$ matrix with column vectors the vectors of S so that

$$A_i = v_i \qquad \text{for} \qquad i = 1, 2, \ldots, n$$

In this way we have

$$c_1 v_1 + c_2 v_2 + \cdots + c_n v_n = 0$$

in matrix form, is the homogeneous linear system

$$A\mathbf{c} = \mathbf{0} \quad \text{' where} \qquad \mathbf{c} = \begin{bmatrix} c_1 \\ c_2 \\ \vdots \\ c_n \end{bmatrix}$$

As A is not square with $n > m$, there is at least one free variable. Thus, the solution is not unique and $S = \{v_1, \ldots, v_n\}$ is linearly dependent.

Notice that from Theorem 3, any set of three or more vectors in $\mathbb{R}^2$, four or more vectors in $\mathbb{R}^3$, five or more vectors in $\mathbb{R}^4$, and so on, is linearly dependent. This theorem does not address the case for which $n \leq m$. In this case, a set of n vectors in $\mathbb{R}^m$ may be either linearly independent or linearly dependent.

The notions of linear independence and dependence can be generalized to include other objects, as illustrated in Example 3.

EXAMPLE 3   Determine whether the matrices
$$M_1 = \begin{bmatrix} 1 & 0 \\ 3 & 2 \end{bmatrix} \qquad M_2 = \begin{bmatrix} -1 & 2 \\ 3 & 2 \end{bmatrix} \qquad \text{and} \qquad M_3 = \begin{bmatrix} 5 & -6 \\ -3 & -2 \end{bmatrix}$$
are linearly independent.

Solution   Solving the equation
$$c_1 \begin{bmatrix} 1 & 0 \\ 3 & 2 \end{bmatrix} + c_2 \begin{bmatrix} -1 & 2 \\ 3 & 2 \end{bmatrix} + c_3 \begin{bmatrix} 5 & -6 \\ -3 & -2 \end{bmatrix} = \begin{bmatrix} 0 & 0 \\ 0 & 0 \end{bmatrix}$$
is equivalent to solving
$$\begin{bmatrix} c_1 - c_2 + 5c_3 & 2c_2 - 6c_3 \\ 3c_1 + 3c_2 - 3c_3 & 2c_1 + 2c_2 - 2c_3 \end{bmatrix} = \begin{bmatrix} 0 & 0 \\ 0 & 0 \end{bmatrix}$$
Equating corresponding entries gives the linear system
$$\begin{cases} c_1 - c_2 + 5c_3 = 0 \\ 2c_2 - 6c_3 = 0 \\ 3c_1 + 3c_2 - 3c_3 = 0 \\ 2c_1 + 2c_2 - 2c_3 = 0 \end{cases}$$
The augmented matrix of the linear system is
$$\left[\begin{array}{ccc|c} 1 & -1 & 5 & 0 \\ 0 & 2 & -6 & 0 \\ 3 & 3 & -3 & 0 \\ 2 & 2 & -2 & 0 \end{array} \right] \qquad \text{which reduces to} \qquad \left[\begin{array}{ccc|c} 1 & 0 & 2 & 0 \\ 0 & 1 & -3 & 0 \\ 0 & 0 & 0 & 0 \\ 0 & 0 & 0 & 0 \end{array} \right]$$
Therefore, the solution set is
$$S = \{(-2t, 3t, t) \mid t \in \mathbb{R}\}$$
Since the original equation has infinitely many solutions, the matrices are linearly dependent.

Criteria to determine if a set of vectors is linearly independent or dependent are extremely useful. The next several theorems give situations where such a determination can be made.

THEOREM 4   If a set of vectors $S = \{v_1, v_2, \ldots, v_n\}$ contains the zero vector, then S is linearly dependent.

Proof   Suppose that the vector $v_k = 0$, for some index k, with $1 \le k \le n$. Setting $c_1 = c_2 = \cdots = c_{k-1} = 0$, $c_k = 1$, and $c_{k+1} = c_{k+2} = \cdots = c_n = 0$, we have
$$0v_1 + \cdots + 0v_{k-1} + 1v_k + 0v_{k+1} + \cdots + 0v_n = 0$$
which shows that the set of vectors is linearly dependent.

THEOREM 5   A set of nonzero vectors is linearly dependent if and only if at least one of the vectors is a linear combination of other vectors in the set.

Proof   Let $S = \{\mathbf{v}_1, \mathbf{v}_2, \ldots, \mathbf{v}_n\}$ be a set of nonzero vectors that is linearly dependent. Then there are scalars $c_1, c_2, \ldots, c_n$, not all 0, with

$$c_1\mathbf{v}_1 + c_2\mathbf{v}_2 + \cdots + c_n\mathbf{v}_n = \mathbf{0}$$

Suppose that $c_k \neq 0$ for some index k. Then solving the previous equation for the vector $\mathbf{v}_k$, we have

$$\mathbf{v}_k = -\frac{c_1}{c_k}\mathbf{v}_1 - \cdots - \frac{c_{k-1}}{c_k}\mathbf{v}_{k-1} - \frac{c_{k+1}}{c_k}\mathbf{v}_{k+1} - \cdots - \frac{c_n}{c_k}\mathbf{v}_n$$

Conversely, let $\mathbf{v}_k$ be such that

$$\mathbf{v}_k = c_1\mathbf{v}_1 + c_2\mathbf{v}_2 + \cdots + c_{k-1}\mathbf{v}_{k-1} + c_{k+1}\mathbf{v}_{k+1} + \cdots + c_n\mathbf{v}_n$$

Then

$$c_1\mathbf{v}_1 + c_2\mathbf{v}_2 + \cdots + c_{k-1}\mathbf{v}_{k-1} - \mathbf{v}_k + c_{k+1}\mathbf{v}_{k+1} + \cdots + c_n\mathbf{v}_n = \mathbf{0}$$

Since the coefficient of $\mathbf{v}_k$ is -1, the linear system has a nontrivial solution. Hence, the set S is linearly dependent.

As an illustration, let S be the set of vectors

$$S = \left\{ \begin{bmatrix} 1 \\ 3 \\ 1 \end{bmatrix}, \begin{bmatrix} -1 \\ 2 \\ 1 \end{bmatrix}, \begin{bmatrix} 2 \\ 6 \\ 2 \end{bmatrix} \right\}$$

Notice that the third vector is twice the first vector, that is,

$$\begin{bmatrix} 2 \\ 6 \\ 2 \end{bmatrix} = 2\begin{bmatrix} 1 \\ 3 \\ 1 \end{bmatrix}$$

Thus, by Theorem 5, the set S is linearly dependent.

EXAMPLE 4   Verify that the vectors

$$\mathbf{v}_1 = \begin{bmatrix} -1 \\ 0 \end{bmatrix} \qquad \mathbf{v}_2 = \begin{bmatrix} 0 \\ 1 \end{bmatrix} \qquad \text{and} \qquad \mathbf{v}_3 = \begin{bmatrix} 1 \\ 0 \end{bmatrix}$$

are linearly dependent. Then show that not every vector can be written as a linear combination of the others.

Solution   By Theorem 3, any three vectors in $\mathbb{R}^2$ are linearly dependent. Now, observe that $\mathbf{v}_1$ and $\mathbf{v}_3$ are linear combinations of the other two vectors, that is,

$$\mathbf{v}_1 = 0\mathbf{v}_2 - \mathbf{v}_3 \qquad \text{and} \qquad \mathbf{v}_3 = 0\mathbf{v}_2 - \mathbf{v}_1$$

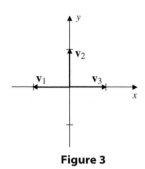

Figure 3

However, $\mathbf{v}_2$ cannot be written as a linear combination of $\mathbf{v}_1$ and $\mathbf{v}_3$. To see this, notice that the equation

$$a\mathbf{v}_1 + b\mathbf{v}_3 = \mathbf{v}_2$$

is equivalent to the inconsistent linear system

$$\begin{cases} -a + b & = 0 \\ 0 & = 1 \end{cases}$$

As shown in Fig. 3, any linear combination of the vectors $\mathbf{v}_1$ and $\mathbf{v}_3$ is a vector that is along the x axis. Therefore, $\mathbf{v}_2$ is not a linear combination of $\mathbf{v}_1$ and $\mathbf{v}_3$.

THEOREM 6

1. If a set of vectors S is linearly independent, then any subset of S is also a linearly independent set of vectors.
2. If a set of vectors T is linearly dependent and S is a set of vectors that contains T, then S is also a linearly dependent set of vectors.

Proof (1) Let T be a subset of S. Reorder and relabel the vectors of S, if necessary, so that $T = \{\mathbf{v}_1, \ldots, \mathbf{v}_k\}$ and $S = \{\mathbf{v}_1, \ldots, \mathbf{v}_k, \mathbf{v}_{k+1}, \ldots, \mathbf{v}_m\}$. Consider the equation

$$c_1\mathbf{v}_1 + c_2\mathbf{v}_2 + \cdots + c_k\mathbf{v}_k = \mathbf{0}$$

Next let $c_{k+1} = c_{k+2} = \cdots = c_m = 0$, and consider the linear combination

$$c_1\mathbf{v}_1 + c_2\mathbf{v}_2 + \cdots + c_k\mathbf{v}_k + 0\mathbf{v}_{k+1} + \cdots + 0\mathbf{v}_m = \mathbf{0}$$

Since S is linearly independent, $c_1 = c_2 = \cdots = c_k = 0$ and hence T is linearly independent.

(2) Let $T = \{\mathbf{v}_1, \ldots, \mathbf{v}_k\}$ and suppose that $T \subset S$. Label the vectors of S that are not in T as $\mathbf{v}_{k+1}, \ldots, \mathbf{v}_m$. Since T is linearly dependent, there are scalars $c_1, \ldots, c_k$, not all 0, such that

$$c_1\mathbf{v}_1 + c_2\mathbf{v}_2 + \cdots + c_k\mathbf{v}_k = \mathbf{0}$$

Then $c_1, c_2, \ldots, c_k, c_{k+1} = c_{k+2} = \cdots = c_m = 0$ is a collection of m scalars, not all 0, with

$$c_1\mathbf{v}_1 + c_2\mathbf{v}_2 + \cdots + c_k\mathbf{v}_k + 0\mathbf{v}_{k+1} + \cdots + 0\mathbf{v}_m = \mathbf{0}$$

Consequently, S is linearly dependent.

Given a set of vectors $S = \{\mathbf{v}_1, \ldots, \mathbf{v}_n\}$ and an arbitrary vector not in S, we have seen that it may or may not be possible to write $\mathbf{v}$ as a linear combination of S. We have also seen that sometimes $\mathbf{v}$ can be written as a linear combination of the vectors of S in infinitely many ways. That this cannot happen for a linearly independent set is the content of Theorem 7.

THEOREM 7   Let $S = \{\mathbf{v}_1, \mathbf{v}_2, \ldots, \mathbf{v}_n\}$ be a linearly independent set. Suppose that there are scalars $c_1, c_2, \ldots, c_n$ such that

$$\mathbf{v} = \sum_{k=1}^{n} c_k \mathbf{v}_k$$

Then the scalars are unique.

Proof To prove the result, let $\mathbf{v}$ be written as

$$\mathbf{v} = \sum_{k=1}^{n} c_k \mathbf{v}_k \qquad \text{and as} \qquad \mathbf{v} = \sum_{k=1}^{n} d_k \mathbf{v}_k$$

Then

$$\mathbf{0} = \mathbf{v} - \mathbf{v} = \sum_{k=1}^{n} c_k \mathbf{v}_k - \sum_{k=1}^{n} d_k \mathbf{v}_k$$

$$= \sum_{k=1}^{n} (c_k - d_k) \mathbf{v}_k$$

Since the set of vectors S is linearly independent, the only solution to this last equation is the trivial one. That is,

$$c_1 - d_1 = 0, c_2 - d_2 = 0, \ldots, c_n - d_n = 0, \quad \text{or} \quad c_1 = d_1, c_2 = d_2, \ldots, c_n = d_n$$

Linear Systems

At the end of Sec. 2.2, in Theorem 2, we made the observation that a linear system $A\mathbf{x} = \mathbf{b}$ is consistent if and only if the vector $\mathbf{b}$ is a linear combination of the column vectors of the matrix A. Theorem 8 gives criteria for when the solution is unique.

THEOREM 8   Let $A\mathbf{x} = \mathbf{b}$ be a consistent $m \times n$ linear system. The solution is unique if and only if the column vectors of A are linearly independent.

Proof First we prove that the condition is necessary. Suppose that the column vectors $\mathbf{A}_1, \mathbf{A}_2, \ldots, \mathbf{A}_n$ are linearly independent, and let

$$\mathbf{c} = \begin{bmatrix} c_1 \\ c_2 \\ \vdots \\ c_n \end{bmatrix} \qquad \text{and} \qquad \mathbf{d} = \begin{bmatrix} d_1 \\ d_2 \\ \vdots \\ d_n \end{bmatrix}$$

be solutions to the linear system. In vector form, we have

$$c_1\mathbf{A}_1 + c_2\mathbf{A}_2 + \cdots + c_n\mathbf{A}_n = \mathbf{b} \qquad \text{and} \qquad d_1\mathbf{A}_1 + d_2\mathbf{A}_2 + \cdots + d_n\mathbf{A}_n = \mathbf{b}$$

By Theorem 7, $c_1 = d_1, c_2 = d_2, \ldots, c_n = d_n$. Hence, $\mathbf{c} = \mathbf{d}$ and the solution to the linear system is unique.

   To prove the sufficiency, we will prove the contrapositive statement. Let $\mathbf{v}$ be a solution to the linear system $A\mathbf{x} = \mathbf{b}$, and assume that the column vectors of A are linearly dependent. Then there are scalars $c_1, c_2, \ldots, c_n$, not all 0, such that

$$c_1\mathbf{A}_1 + c_2\mathbf{A}_2 + \cdots + c_n\mathbf{A}_n = \mathbf{0}$$

that is, if $\mathbf{c} = \begin{bmatrix} c_1 \\ c_2 \\ \vdots \\ c_n \end{bmatrix}$, then $A\mathbf{c} = \mathbf{0}$. Since matrix multiplication satisfies the distributive property,

$$A(\mathbf{v} + \mathbf{c}) = A\mathbf{v} + A\mathbf{c} = \mathbf{b} + \mathbf{0} = \mathbf{b}$$

Therefore, the vector $\mathbf{v} + \mathbf{c}$ is another solution to the linear system, and the solution is not unique. This completes the proof of the contrapositive statement. Therefore, we have shown that if the solution is unique, then the column vectors of A are linearly independent.

   Theorem 8 provides another way of establishing Theorem 11 of Sec. 1.5 that a linear system has no solutions, one solution, or infinitely many solutions.

Linear Independence and Determinants

In Chap. 1, we established that a square matrix A is invertible if and only if $\det(A) \neq 0$. (See Theorem 16 of Sec. 1.6.) This is equivalent to the statement that the linear system $A\mathbf{x} = \mathbf{b}$ has a unique solution for every $\mathbf{b}$ if and only if $\det(A) \neq 0$. This gives an alternative method for showing that a set of vectors is linearly independent. Specifically, if A is a square matrix, then by Theorem 8, the column vectors of A are linearly independent if and only if $\det(A) \neq 0$.

EXAMPLE 5   Let

$$S = \left\{ \begin{bmatrix} 1 \\ 0 \\ 3 \end{bmatrix}, \begin{bmatrix} 1 \\ 2 \\ 4 \end{bmatrix}, \begin{bmatrix} 1 \\ 4 \\ 5 \end{bmatrix} \right\}$$

Determine whether the set S is linearly independent.

Solution   Let A be the matrix whose column vectors are the vectors of S; that is,

$$A = \begin{bmatrix} 1 & 1 & 1 \\ 0 & 2 & 4 \\ 3 & 4 & 5 \end{bmatrix}$$

The determinant of A can be found by expanding along the first column, so

$$\det(A) = 1 \begin{vmatrix} 2 & 4 \\ 4 & 5 \end{vmatrix} - 0 \begin{vmatrix} 1 & 1 \\ 4 & 5 \end{vmatrix} + 3 \begin{vmatrix} 1 & 1 \\ 2 & 4 \end{vmatrix}$$

$$= -6 - 0 + 3(2) = 0$$

Therefore, by the previous remarks S is linearly dependent.

The final theorem summarizes the connections that have thus far been established concerning solutions to a linear system, linear independence, invertibility of matrices, and determinants.

THEOREM 9   Let A be a square matrix. Then the following statements are equivalent.

1. The matrix A is invertible.

2. The linear system $Ax = b$ has a unique solution for every vector b.

3. The homogeneous linear system $Ax = 0$ has only the trivial solution.

4. The matrix A is row equivalent to the identity matrix.

5. The determinant of the matrix A is nonzero.

6. The column vectors of A are linearly independent.

Fact Summary

Let S be a set of m vectors in $\mathbb{R}^n$.

1. If $m > n$, then S is linearly dependent.

2. If the zero vector is in S, then S is linearly dependent.

3. If u and v are in S and there is a scalar c such that $u = cv$, then S is linearly dependent.

4. If any vector in S is a linear combination of other vectors in S, then S is linearly dependent.

5. If S is linearly independent and T is a subset of S, then T is linearly independent.

6. If T is linearly dependent and T is a subset of S, then S is linearly dependent.

7. If $S = \{v_1, \ldots, v_m\}$ is linearly independent and $v = c_1 v_1 + \cdots + c_m v_m$, then the set of scalars $c_1, \ldots, c_m$ is uniquely determined.

8. The linear system $Ax = b$ has a unique solution if and only if the column vectors of A are linearly independent.

9. If A is a square matrix, then the column vectors of A are linearly independent if and only if $\det(A) \neq 0$.

Exercise Set 2.3

In Exercises 1–10, determine whether the given
vectors are linearly independent.

1. $v_1 = \begin{bmatrix} -1 \\ 1 \end{bmatrix}$ $v_2 = \begin{bmatrix} 2 \\ -3 \end{bmatrix}$

2. $v_1 = \begin{bmatrix} 2 \\ -4 \end{bmatrix}$ $v_2 = \begin{bmatrix} 1 \\ 2 \end{bmatrix}$

3. $v_1 = \begin{bmatrix} 1 \\ -4 \end{bmatrix}$ $v_2 = \begin{bmatrix} -2 \\ 8 \end{bmatrix}$

4. $v_1 = \begin{bmatrix} 1 \\ 0 \end{bmatrix}$ $v_2 = \begin{bmatrix} 0 \\ -1 \end{bmatrix}$

$v_3 = \begin{bmatrix} 1 \\ -1 \end{bmatrix}$

5. $v_1 = \begin{bmatrix} -1 \\ 2 \\ 1 \end{bmatrix}$ $v_2 = \begin{bmatrix} -2 \\ 2 \\ 3 \end{bmatrix}$

6. $v_1 = \begin{bmatrix} 4 \\ 2 \\ -6 \end{bmatrix}$ $v_2 = \begin{bmatrix} -2 \\ -1 \\ 3 \end{bmatrix}$

7. $v_1 = \begin{bmatrix} -4 \\ 4 \\ -1 \end{bmatrix}$ $v_2 = \begin{bmatrix} -5 \\ 3 \\ 3 \end{bmatrix}$

$v_3 = \begin{bmatrix} 3 \\ -5 \\ 5 \end{bmatrix}$

8. $v_1 = \begin{bmatrix} 3 \\ -3 \\ -1 \end{bmatrix}$ $v_2 = \begin{bmatrix} -1 \\ 2 \\ -2 \end{bmatrix}$

$v_3 = \begin{bmatrix} -1 \\ 3 \\ 1 \end{bmatrix}$

9. $v_1 = \begin{bmatrix} 3 \\ -1 \\ -1 \\ 2 \end{bmatrix}$ $v_2 = \begin{bmatrix} 1 \\ 0 \\ 2 \\ 1 \end{bmatrix}$

$v_3 = \begin{bmatrix} 3 \\ -1 \\ 0 \\ 1 \end{bmatrix}$

10. $v_1 = \begin{bmatrix} -2 \\ -4 \\ 1 \\ 1 \end{bmatrix}$ $v_2 = \begin{bmatrix} 3 \\ -4 \\ 0 \\ 4 \end{bmatrix}$

$v_3 = \begin{bmatrix} -1 \\ -12 \\ 2 \\ 6 \end{bmatrix}$

In Exercises 11–14, determine whether the matrices
are linearly independent.

11. $M_1 = \begin{bmatrix} 3 & 3 \\ 2 & 1 \end{bmatrix}$ $M_2 = \begin{bmatrix} 0 & 1 \\ 0 & 0 \end{bmatrix}$

$M_3 = \begin{bmatrix} 1 & -1 \\ -1 & -2 \end{bmatrix}$

12. $M_1 = \begin{bmatrix} -1 & 2 \\ 1 & 1 \end{bmatrix}$ $M_2 = \begin{bmatrix} 1 & 4 \\ 0 & 1 \end{bmatrix}$

$M_3 = \begin{bmatrix} 2 & 2 \\ -1 & 0 \end{bmatrix}$

13. $M_1 = \begin{bmatrix} 1 & -2 \\ -2 & -2 \end{bmatrix}$ $M_2 = \begin{bmatrix} 0 & -1 \\ 2 & 2 \end{bmatrix}$

$M_3 = \begin{bmatrix} -1 & 1 \\ -2 & 2 \end{bmatrix}$ $M_4 = \begin{bmatrix} 1 & 1 \\ -1 & -2 \end{bmatrix}$

14. $M_1 = \begin{bmatrix} 0 & -1 \\ -1 & 1 \end{bmatrix}$ $M_2 = \begin{bmatrix} -2 & -1 \\ 1 & -1 \end{bmatrix}$

$M_3 = \begin{bmatrix} 2 & 0 \\ -1 & 2 \end{bmatrix}$ $M_4 = \begin{bmatrix} -2 & 2 \\ 2 & -1 \end{bmatrix}$

In Exercises 15–18, explain, without solving a linear system, why the set of vectors is linearly dependent.

15. $\mathbf{v}_1 = \begin{bmatrix} -1 \\ 4 \end{bmatrix}$ $\mathbf{v}_2 = \begin{bmatrix} \frac{1}{2} \\ -2 \end{bmatrix}$

16. $\mathbf{v}_1 = \begin{bmatrix} 2 \\ -1 \end{bmatrix}$ $\mathbf{v}_2 = \begin{bmatrix} -1 \\ -2 \end{bmatrix}$

$\mathbf{v}_3 = \begin{bmatrix} 1 \\ 3 \end{bmatrix}$

17. $\mathbf{v}_1 = \begin{bmatrix} 1 \\ -6 \\ 2 \end{bmatrix}$ $\mathbf{v}_2 = \begin{bmatrix} 0 \\ 0 \\ 0 \end{bmatrix}$

$\mathbf{v}_3 = \begin{bmatrix} 4 \\ 7 \\ 1 \end{bmatrix}$

18. $\mathbf{v}_1 = \begin{bmatrix} 1 \\ 0 \\ -2 \end{bmatrix}$ $\mathbf{v}_2 = \begin{bmatrix} 1 \\ -2 \\ 1 \end{bmatrix}$

$\mathbf{v}_3 = \begin{bmatrix} 2 \\ -2 \\ -1 \end{bmatrix}$

In Exercises 19 and 20, explain, without solving a linear system, why the column vectors of the matrix A are linearly dependent.

19. a. $A = \begin{bmatrix} -1 & 2 & 5 \\ 3 & -6 & 3 \\ 2 & -4 & 3 \end{bmatrix}$

b. $A = \begin{bmatrix} 2 & 1 & 3 \\ 1 & 0 & 1 \\ -1 & 1 & 0 \end{bmatrix}$

20. a. $A = \begin{bmatrix} 4 & -1 & 2 & 6 \\ 5 & -2 & 0 & 2 \\ -2 & 4 & -3 & -2 \end{bmatrix}$

b. $A = \begin{bmatrix} -1 & 2 & 3 \\ 2 & 3 & 1 \\ -1 & -1 & 0 \\ 3 & 5 & 2 \end{bmatrix}$

21. Determine the values of a such that the vectors

$$\begin{bmatrix} 1 \\ 2 \\ 1 \end{bmatrix} \quad \begin{bmatrix} -1 \\ 0 \\ 1 \end{bmatrix} \quad \begin{bmatrix} 2 \\ a \\ 4 \end{bmatrix}$$

are linearly independent.

22. Determine the values of a such that the matrices

$$\begin{bmatrix} 1 & 2 \\ 0 & 1 \end{bmatrix} \quad \begin{bmatrix} 1 & 0 \\ 1 & 0 \end{bmatrix} \quad \begin{bmatrix} 1 & -4 \\ a & -2 \end{bmatrix}$$

are linearly independent.

23. Let

$$\mathbf{v}_1 = \begin{bmatrix} 1 \\ 1 \\ 1 \end{bmatrix} \quad \mathbf{v}_2 = \begin{bmatrix} 1 \\ 2 \\ 3 \end{bmatrix} \quad \mathbf{v}_3 = \begin{bmatrix} 1 \\ 1 \\ 2 \end{bmatrix}$$

a. Show that the vectors are linearly independent.

b. Find the unique scalars c_1, c_2, c_3 such that the vector

$$\mathbf{v} = \begin{bmatrix} 2 \\ 1 \\ 3 \end{bmatrix}$$

can be written as

$$\mathbf{v} = c_1\mathbf{v}_1 + c_2\mathbf{v}_2 + c_3\mathbf{v}_3$$

24. Let

$$M_1 = \begin{bmatrix} 1 & 0 \\ -1 & 0 \end{bmatrix} \quad M_2 = \begin{bmatrix} 1 & 1 \\ 1 & 0 \end{bmatrix}$$

$$M_3 = \begin{bmatrix} 0 & 1 \\ 1 & 1 \end{bmatrix}$$

a. Show that the matrices are linearly independent.

b. Find the unique scalars c_1, c_2, c_3 such that the matrix

$$M = \begin{bmatrix} 3 & 5 \\ 4 & 3 \end{bmatrix}$$

can be written as

$$M = c_1M_1 + c_2M_2 + c_3M_3$$

c. Show that the matrix

$$M = \begin{bmatrix} 0 & 3 \\ 3 & 1 \end{bmatrix}$$

cannot be written as a linear combination of M_1, M_2, and M_3.

In Exercises 25 and 26, for the given matrix A determine if the linear system $A\mathbf{x} = \mathbf{b}$ has a unique solution.

25. $A = \begin{bmatrix} 1 & 2 & 0 \\ -1 & 0 & 3 \\ 2 & 1 & 2 \end{bmatrix}$

26. $A = \begin{bmatrix} 3 & 2 & 4 \\ 1 & -1 & 4 \\ 0 & 2 & -4 \end{bmatrix}$

In Exercises 27–30, determine whether the set of polynomials is linearly independent or linearly dependent. A set of polynomials $S = \{p_1(x), p_2(x), \ldots, p_n(x)\}$ is linearly independent provided

$$c_1 p_1(x) + c_2 p_2(x) + \cdots + c_n p_n(x) = 0$$

for all x implies that

$$c_1 = c_2 = \cdots = c_n = 0$$

27. $p_1(x) = 1 \; p_2(x) = -2 + 4x^2$
$p_3(x) = 2x \; p_4(x) = -12x + 8x^3$

28. $p_1(x) = 1 \; p_2(x) = x$
$p_3(x) = 5 + 2x - x^2$

29. $p_1(x) = 2 \; p_2(x) = x \; p_3(x) = x^2$
$p_4(x) = 3x - 1$

30. $p_1(x) = x^3 - 2x^2 + 1 \; p_2(x) = 5x$
$p_3(x) = x^2 - 4 \; p_4(x) = x^3 + 2x$

In Exercises 31–34, show that the set of functions is linearly independent on the interval $[0, 1]$. A set of functions $S = \{f_1(x), f_2(x), \ldots, f_n(x)\}$ is linearly independent on the interval $[a, b]$ provided

$$c_1 f_1(x) + c_2 f_2(x) + \cdots + c_n f_n(x) = 0$$

for all $x \in [a, b]$ implies that

$$c_1 = c_2 = \cdots = c_n = 0$$

31. $f_1(x) = \cos \pi x \; f_2(x) = \sin \pi x$

32. $f_1(x) = e^x \; f_2(x) = e^{-x}$
$f_3(x) = e^{2x}$

33. $f_1(x) = x \; f_2(x) = x^2 \; f_3(x) = e^x$

34. $f_1(x) = x \; f_2(x) = e^x$
$f_3(x) = \sin \pi x$

35. Verify that two vectors $\mathbf{u}$ and $\mathbf{v}$ in $\mathbb{R}^n$ are linearly dependent if and only if one is a scalar multiple of the other.

36. Suppose that $S = \{\mathbf{v}_1, \mathbf{v}_2, \mathbf{v}_3\}$ is linearly independent and

$$\mathbf{w_1} = \mathbf{v}_1 + \mathbf{v}_2 + \mathbf{v}_3 \qquad \mathbf{w_2} = \mathbf{v}_2 + \mathbf{v}_3$$

and

$$\mathbf{w_3} = \mathbf{v}_3$$

Show that $T = \{\mathbf{w_1}, \mathbf{w_2}, \mathbf{w_3}\}$ is linearly independent.

37. Suppose that $S = \{\mathbf{v}_1, \mathbf{v}_2, \mathbf{v}_3\}$ is linearly independent and

$$\mathbf{w_1} = \mathbf{v}_1 + \mathbf{v}_2 \qquad \mathbf{w_2} = \mathbf{v}_2 - \mathbf{v}_3$$

and

$$\mathbf{w_3} = \mathbf{v}_2 + \mathbf{v}_3$$

Show that $T = \{\mathbf{w_1}, \mathbf{w_2}, \mathbf{w_3}\}$ is linearly independent.

38. Suppose that $S = \{\mathbf{v}_1, \mathbf{v}_2, \mathbf{v}_3\}$ is linearly independent and

$$\mathbf{w_1} = \mathbf{v}_2 \qquad \mathbf{w_2} = \mathbf{v}_1 + \mathbf{v}_3$$

and

$$\mathbf{w_3} = \mathbf{v}_1 + \mathbf{v}_2 + \mathbf{v}_3$$

Determine whether the set $T = \{\mathbf{w_1}, \mathbf{w_2}, \mathbf{w_3}\}$ is linearly independent or linearly dependent.

39. Suppose that the set $S = \{\mathbf{v}_1, \mathbf{v}_2\}$ is linearly independent. Show that if $\mathbf{v}_3$ cannot be written as a linear combination of $\mathbf{v}_1$ and $\mathbf{v}_2$, then $\{\mathbf{v}_1, \mathbf{v}_2, \mathbf{v}_3\}$ is linearly independent.

40. Let $S = \{\mathbf{v}_1, \mathbf{v}_2, \mathbf{v}_3\}$, where $\mathbf{v}_3 = \mathbf{v}_1 + \mathbf{v}_2$.

a. Write $\mathbf{v}_1$ as a linear combination of the vectors in S in three different ways.

b. Find all scalars c_1, c_2, and c_3 such that
$$\mathbf{v}_1 = c_1\mathbf{v}_1 + c_2\mathbf{v}_2 + c_3\mathbf{v}_3.$$

41. Show that if the column vectors of an $m \times n$ matrix $\mathbf{A}_1, \ldots, \mathbf{A}_n$ are linearly independent, then
$$\{\mathbf{x} \in \mathbb{R}^n \mid A\mathbf{x} = \mathbf{0}\} = \{\mathbf{0}\}$$
.

42. Let $\mathbf{v}_1, \ldots, \mathbf{v}_k$ be linearly independent vectors in $\mathbb{R}^n$, and suppose A is an invertible $n \times n$ matrix. Define vectors $\mathbf{w}_i = A\mathbf{v}_i$, for $i = 1, \ldots, k$. Show that the vectors $\mathbf{w}_1, \ldots, \mathbf{w}_k$ are linearly independent. Show, using a 2×2 matrix, that the requirement of invertibility is necessary.

Review Exercises for Chapter 2

1. If $ad - bc \neq 0$, show that the vectors
$$\begin{bmatrix} a \\ b \end{bmatrix} \quad \text{and} \quad \begin{bmatrix} c \\ d \end{bmatrix}$$
are linearly independent. Suppose that $ad - bc = 0$. What can you say about the two vectors?

2. Suppose that $S = \{\mathbf{v}_1, \mathbf{v}_2, \mathbf{v}_3\}$ is a linearly independent set of vectors in $\mathbb{R}^n$. Show that $T = \{\mathbf{v}_1, \mathbf{v}_2, \mathbf{v}_1 + \mathbf{v}_2 + \mathbf{v}_3\}$ is also linearly independent.

3. Determine for which nonzero values of a the vectors
$$\begin{bmatrix} a^2 \\ 0 \\ 1 \end{bmatrix} \quad \begin{bmatrix} 0 \\ a \\ 2 \end{bmatrix} \quad \text{and} \quad \begin{bmatrix} 1 \\ 0 \\ 1 \end{bmatrix}$$
are linearly independent.

4. Let
$$S = \left\{ \begin{bmatrix} 2s - t \\ s \\ t \\ s \end{bmatrix} \middle| s, t \in \mathbb{R} \right\}$$

a. Find two vectors in $\mathbb{R}^4$ so that all vectors in S can be written as a linear combination of the two vectors.

b. Are the vectors found in part (a) linearly independent?

5. Let
$$\mathbf{v}_1 = \begin{bmatrix} 1 \\ 0 \\ 2 \end{bmatrix} \quad \text{and} \quad \mathbf{v}_2 = \begin{bmatrix} 1 \\ 1 \\ 1 \end{bmatrix}$$

a. Is $S = \{\mathbf{v}_1, \mathbf{v}_2\}$ linearly independent?

b. Find a vector $\begin{bmatrix} a \\ b \\ c \end{bmatrix}$ that cannot be written as a linear combination of $\mathbf{v}_1$ and $\mathbf{v}_2$.

c. Describe all vectors in $\mathbb{R}^3$ that can be written as a linear combination of $\mathbf{v}_1$ and $\mathbf{v}_2$.

d. Let
$$\mathbf{v}_3 = \begin{bmatrix} 1 \\ 0 \\ 0 \end{bmatrix}$$

Is $T = \{\mathbf{v}_1, \mathbf{v}_2, \mathbf{v}_3\}$ linearly independent or linearly dependent?

e. Describe all vectors in $\mathbb{R}^3$ that can be written as a linear combination of $\mathbf{v}_1, \mathbf{v}_2$, and $\mathbf{v}_3$.

6. Let
$$\mathbf{v}_1 = \begin{bmatrix} 1 \\ -1 \\ 1 \end{bmatrix} \quad \mathbf{v}_2 = \begin{bmatrix} 2 \\ 1 \\ 1 \end{bmatrix}$$
$$\mathbf{v}_3 = \begin{bmatrix} 0 \\ 2 \\ 1 \end{bmatrix} \quad \text{and} \quad \mathbf{v}_4 = \begin{bmatrix} -2 \\ 2 \\ 1 \end{bmatrix}$$

a. Show that $S = \{\mathbf{v}_1, \mathbf{v}_2, \mathbf{v}_3, \mathbf{v}_4\}$ is linearly dependent.

b. Show that $T = \{\mathbf{v}_1, \mathbf{v}_2, \mathbf{v}_3\}$ is linearly independent.

c. Show that $\mathbf{v}_4$ can be written as a linear combination of $\mathbf{v}_1, \mathbf{v}_2$, and $\mathbf{v}_3$.

d. How does the set of all linear combinations of vectors in S compare with the set of all linear combinations of vectors in T?

7. Consider the linear system

$$\begin{cases} x + y + 2z + w = 3 \\ -x + z + 2w = 1 \\ 2x + 2y + w = -2 \\ x + y + 2z + 3w = 5 \end{cases}$$

a. Write the linear system in the matrix form $A\mathbf{x} = \mathbf{b}$.

b. Find the determinant of the coefficient matrix A.

c. Are the column vectors of A linearly independent?

d. Without solving the linear system, determine whether it has a unique solution.

e. Solve the linear system.

8. Let

$$M_1 = \begin{bmatrix} 1 & 0 \\ -1 & 1 \end{bmatrix} \qquad M_2 = \begin{bmatrix} 1 & 1 \\ 2 & 1 \end{bmatrix}$$

$$M_3 = \begin{bmatrix} 0 & 1 \\ 1 & 0 \end{bmatrix}$$

a. Show that the set $\{M_1, M_2, M_3\}$ is linearly independent.

b. Find the unique scalars c_1, c_2, and c_3 such that

$$\begin{bmatrix} 1 & -1 \\ 2 & 1 \end{bmatrix} = c_1 M_1 + c_2 M_2 + c_3 M_3$$

c. Can the matrix

$$\begin{bmatrix} 1 & -1 \\ 1 & 2 \end{bmatrix}$$

be written as a linear combination of M_1, M_2, and M_3?

d. Describe all matrices $\begin{bmatrix} a & b \\ c & d \end{bmatrix}$ that can be written as a linear combination of M_1, M_2, and M_3.

9. Let

$$A = \begin{bmatrix} 1 & 3 & 2 \\ 2 & -1 & 3 \\ 1 & 1 & -1 \end{bmatrix}$$

a. Write the linear system $A\mathbf{x} = \mathbf{b}$ in vector form.

b. Compute $\det(A)$. What can you conclude as to whether the linear system is consistent or inconsistent?

c. Are the column vectors of A linearly independent?

d. Without solving the linear system, does the system have a unique solution? Give two reasons.

10. Two vectors in $\mathbb{R}^n$ are perpendicular provided their dot product is 0. Suppose $S = \{\mathbf{v}_1, \mathbf{v}_2, \ldots, \mathbf{v}_n\}$ is a set of nonzero vectors which are pairwise perpendicular. Follow the steps to show S is linearly independent.

a. Show that for any vector $\mathbf{v}$ the dot product satisfies $\mathbf{v} \cdot \mathbf{v} \geq 0$.

b. Show that for any vector $\mathbf{v} \neq \mathbf{0}$ the dot product satisfies $\mathbf{v} \cdot \mathbf{v} > 0$.

c. Show that for all vectors $\mathbf{u}, \mathbf{v}$, and $\mathbf{w}$ the dot product satisfies

$$\mathbf{u} \cdot (\mathbf{v} + \mathbf{w}) = \mathbf{u} \cdot \mathbf{v} + \mathbf{u} \cdot \mathbf{w}$$

d. Consider the equation

$$c_1 \mathbf{v}_1 + c_2 \mathbf{v}_2 + \cdots + c_n \mathbf{v}_n = \mathbf{0}$$

Use the dot product of $\mathbf{v}_i$, for each $1 \leq i \leq n$, with the expression on the left of the previous equation to show that $c_i = 0$, for each $1 \leq i \leq n$.

Chapter 2: Chapter Test

In Exercises 1–33, determine whether the statement is true or false.

1. Every vector in $\mathbb{R}^3$ can be written as a linear combination of

$$\begin{bmatrix} 1 \\ 0 \\ 0 \end{bmatrix} \quad \begin{bmatrix} 0 \\ 1 \\ 0 \end{bmatrix} \quad \begin{bmatrix} 0 \\ 0 \\ 1 \end{bmatrix}$$

2. Every 2×2 matrix can be written as a linear combination of

$$\begin{bmatrix} 1 & 0 \\ 0 & 0 \end{bmatrix} \quad \begin{bmatrix} 0 & 1 \\ 0 & 0 \end{bmatrix} \quad \begin{bmatrix} 0 & 0 \\ 1 & 0 \end{bmatrix}$$

3. Every 2×2 matrix can be written as a linear combination of

$$\begin{bmatrix} 1 & 0 \\ 0 & 0 \end{bmatrix} \quad \begin{bmatrix} 0 & 1 \\ 0 & 0 \end{bmatrix} \quad \begin{bmatrix} 0 & 0 \\ 1 & 0 \end{bmatrix}$$

$$\begin{bmatrix} 0 & 0 \\ 0 & 1 \end{bmatrix}$$

In Exercises 4–8, use the vectors

$$\mathbf{v}_1 = \begin{bmatrix} 1 \\ 0 \\ 1 \end{bmatrix} \quad \mathbf{v}_2 = \begin{bmatrix} 2 \\ 1 \\ 0 \end{bmatrix}$$

$$\mathbf{v}_3 = \begin{bmatrix} 4 \\ 3 \\ -1 \end{bmatrix}$$

4. The set $S = \{\mathbf{v}_1, \mathbf{v}_2, \mathbf{v}_3\}$ is linearly independent.

5. There are scalars c_1 and c_2 so that $\mathbf{v}_3 = c_1\mathbf{v}_1 + c_2\mathbf{v}_2$.

6. The vector $\mathbf{v}_2$ can be written as a linear combination of $\mathbf{v}_1$ and $\mathbf{v}_3$.

7. The vector $\mathbf{v}_1$ can be written as a linear combination of $\mathbf{v}_2$ and $\mathbf{v}_3$.

8. If $\mathbf{v}_1$, $\mathbf{v}_2$, and $\mathbf{v}_3$ are the column vectors of a 3×3 matrix A, then the linear system $A\mathbf{x} = \mathbf{b}$ has a unique solution for all vectors $\mathbf{b}$ in $\mathbb{R}^3$.

9. The polynomial $p(x) = 3 + x$ can be written as a linear combination of $q_1(x) = 1 + x$ and $q_2(x) = 1 - x - x^2$.

10. The set

$$S = \left\{ \begin{bmatrix} 1 \\ -1 \\ 3 \end{bmatrix}, \begin{bmatrix} 1 \\ 0 \\ 1 \end{bmatrix}, \begin{bmatrix} 2 \\ -2 \\ 6 \end{bmatrix}, \begin{bmatrix} 0 \\ 1 \\ 0 \end{bmatrix} \right\}$$

is linearly independent.

In Exercises 11–14, use the matrices

$$M_1 = \begin{bmatrix} 1 & -1 \\ 0 & 0 \end{bmatrix} \quad M_2 = \begin{bmatrix} 0 & 0 \\ 1 & 0 \end{bmatrix}$$

$$M_3 = \begin{bmatrix} 0 & 0 \\ 0 & 1 \end{bmatrix} \quad M_4 = \begin{bmatrix} 2 & -1 \\ 1 & 3 \end{bmatrix}$$

11. The set $S = \{M_1, M_2, M_3, M_4\}$ is linearly independent.

12. The set $T = \{M_1, M_2, M_3\}$ is linearly independent.

13. The set of all linear combinations of matrices in S is equal to the set of all linear combinations of matrices in T.

14. Every matrix that can be written as a linear combination of the matrices in T has the form

$$\begin{bmatrix} x & -x \\ y & z \end{bmatrix}$$

15. The vectors

$$\begin{bmatrix} s \\ 0 \\ 0 \end{bmatrix} \qquad \begin{bmatrix} 1 \\ s \\ 1 \end{bmatrix} \qquad \begin{bmatrix} 0 \\ 1 \\ s \end{bmatrix}$$

are linearly independent if and only if $s = 0$ or $s = 1$.

In Exercises 16–19, use the vectors

$$\mathbf{v}_1 = \begin{bmatrix} 1 \\ 1 \end{bmatrix} \qquad \mathbf{v}_2 = \begin{bmatrix} 3 \\ -1 \end{bmatrix}$$

16. The set $S = \{\mathbf{v}_1, \mathbf{v}_2\}$ is linearly independent.

17. Every vector in $\mathbb{R}^2$ can be written as a linear combination of $\mathbf{v}_1$ and $\mathbf{v}_2$.

18. If the column vectors of a matrix A are $\mathbf{v}_1$ and $\mathbf{v}_2$, then $\det(A) = 0$.

19. If $\mathbf{b}$ is in $\mathbb{R}^2$ and $c_1\mathbf{v}_1 + c_2\mathbf{v}_2 = \mathbf{b}$, then

$$\begin{bmatrix} c_1 \\ c_2 \end{bmatrix} = A^{-1}\mathbf{b}$$

where A is the 2×2 matrix with column vectors $\mathbf{v}_1$ and $\mathbf{v}_2$.

20. The column vectors of the matrix

$$\begin{bmatrix} \cos\theta & \sin\theta \\ -\sin\theta & \cos\theta \end{bmatrix}$$

are linearly independent.

21. If $\mathbf{v}_1$ and $\mathbf{v}_2$ are linearly independent vectors in $\mathbb{R}^n$ and $\mathbf{v}_3$ cannot be written as a scalar multiple of $\mathbf{v}_1$, then $\mathbf{v}_1$, $\mathbf{v}_2$, and $\mathbf{v}_3$ are linearly independent.

22. If $S = \{\mathbf{v}_1, \mathbf{v}_2, \ldots, \mathbf{v}_m\}$ is a set of nonzero vectors in $\mathbb{R}^n$ that are linearly dependent, then every vector in S can be written as a linear combination of the others.

23. If $\mathbf{v}_1$ and $\mathbf{v}_2$ are in $\mathbb{R}^3$, then the matrix with column vectors $\mathbf{v}_1$, $\mathbf{v}_2$, and $\mathbf{v}_1 + \mathbf{v}_2$ has a nonzero determinant.

24. If $\mathbf{v}_1$ and $\mathbf{v}_2$ are linearly independent, $\mathbf{v}_1$, $\mathbf{v}_2$, and $\mathbf{v}_1 + \mathbf{v}_2$ are also linearly independent.

25. If the set S contains the zero vector, then S is linearly dependent.

26. The column vectors of an $n \times n$ invertible matrix can be linearly dependent.

27. If A is an $n \times n$ matrix with linearly independent column vectors, then the row vectors of A are also linearly independent.

28. If the row vectors of a nonsquare matrix are linearly independent, then the column vectors are also linearly independent.

29. If $\mathbf{v}_1$, $\mathbf{v}_2$, $\mathbf{v}_3$, and $\mathbf{v}_4$ are in $\mathbb{R}^4$ and $\{\mathbf{v}_1, \mathbf{v}_2, \mathbf{v}_3\}$ is linearly dependent, then $\{\mathbf{v}_1, \mathbf{v}_2, \mathbf{v}_3, \mathbf{v}_4\}$ is linearly dependent.

30. If $\mathbf{v}_1$, $\mathbf{v}_2$, $\mathbf{v}_3$, and $\mathbf{v}_4$ are in $\mathbb{R}^4$ and $\{\mathbf{v}_1, \mathbf{v}_2, \mathbf{v}_3\}$ is linearly independent, then $\{\mathbf{v}_1, \mathbf{v}_2, \mathbf{v}_3, \mathbf{v}_4\}$ is linearly independent.

31. If $\mathbf{v}_1$, $\mathbf{v}_2$, $\mathbf{v}_3$, and $\mathbf{v}_4$ are in $\mathbb{R}^4$ and $\{\mathbf{v}_1, \mathbf{v}_2, \mathbf{v}_3, \mathbf{v}_4\}$ is linearly independent, then $\{\mathbf{v}_1, \mathbf{v}_2, \mathbf{v}_3\}$ is linearly independent.

32. If $\mathbf{v}_1$, $\mathbf{v}_2$, $\mathbf{v}_3$, and $\mathbf{v}_4$ are in $\mathbb{R}^4$ and $\{\mathbf{v}_1, \mathbf{v}_2, \mathbf{v}_3, \mathbf{v}_4\}$ is linearly dependent, then $\{\mathbf{v}_1, \mathbf{v}_2, \mathbf{v}_3\}$ is linearly dependent.

33. If $S = \{\mathbf{v}_1, \mathbf{v}_2, \ldots, \mathbf{v}_5\}$ is a subset of $\mathbb{R}^4$, then S is linearly dependent.

3

Vector Spaces

When a digital signal is sent through space (sometimes across millions of miles), errors in the signal are bound to occur. In response to the need for reliable information, mathematicians and scientists from a variety of disciplines have developed ways to improve the quality of these transmissions. One obvious method is to send messages repeatedly to increase the likelihood of receiving them correctly. This, however, is time-consuming and limits the number of messages that can be sent. An innovative methodology developed by Richard Hamming in 1947 involves embedding in the transmission a means for error detection and self-correction. One of Hamming's coding schemes, known as Hamming's (7,4) code, uses binary vectors (vectors consisting of 1s and 0s) with seven components. Some of these vectors are identified as *codewords* depending on the configuration of the 1s and 0s within it. To decide if the binary vector

© Brand X Pictures/PunchStock/RF

$$\mathbf{b} = \begin{bmatrix} b_1 \\ b_2 \\ \vdots \\ b_7 \end{bmatrix}$$

is a codeword, a test using matrix multiplication is performed. The matrix given by

$$C = \begin{bmatrix} 1 & 1 & 1 & 0 & 1 & 0 & 0 \\ 0 & 1 & 1 & 1 & 0 & 1 & 0 \\ 1 & 0 & 1 & 1 & 0 & 0 & 1 \end{bmatrix}$$

is called the *check matrix*. To carry out the test, we compute the product of C and $\mathbf{b}$, using modulo 2 arithmetic, where an even result corresponds to a 0 and an odd result corresponds to a 1. This product produces a binary vector with three components called the *syndrome* vector given by

$$C\mathbf{b} = \mathbf{s}$$

A binary vector $\mathbf{b}$ is a **codeword** if the syndrome vector $\mathbf{s} = \mathbf{0}$. Put another way, $\mathbf{b}$ is a codeword if it is a solution to the homogeneous equation $C\mathbf{b} \equiv \mathbf{0}$ (mod 2). For example, the vector

$$\mathbf{u} = \begin{bmatrix} 1 \\ 1 \\ 0 \\ 0 \\ 0 \\ 1 \\ 1 \end{bmatrix}$$

is a codeword since

$$C\mathbf{u} = \begin{bmatrix} 1 & 1 & 1 & 0 & 1 & 0 & 0 \\ 0 & 1 & 1 & 1 & 0 & 1 & 0 \\ 1 & 0 & 1 & 1 & 0 & 0 & 1 \end{bmatrix} \begin{bmatrix} 1 \\ 1 \\ 0 \\ 0 \\ 0 \\ 1 \\ 1 \end{bmatrix} = \begin{bmatrix} 2 \\ 2 \\ 2 \end{bmatrix} \equiv \begin{bmatrix} 0 \\ 0 \\ 0 \end{bmatrix} \text{ (mod 2)}$$

whereas the vector

$$\mathbf{v} = \begin{bmatrix} 1 \\ 1 \\ 1 \\ 0 \\ 0 \\ 0 \\ 0 \end{bmatrix} \quad \text{is not since} \quad C\mathbf{v} = \begin{bmatrix} 3 \\ 2 \\ 2 \end{bmatrix} \equiv \begin{bmatrix} 1 \\ 0 \\ 0 \end{bmatrix} \text{ (mod 2)}$$

With this ingenious strategy the recipient of a legitimate codeword can safely assume that the vector is free from errors. On the other hand, if the vector received is not a codeword, an algorithm involving the syndrome vector can be applied to restore it to the original. In the previous example the fifth digit of $\mathbf{v}$ was altered during the transmission. The intended vector is given by

$$\mathbf{v}^* = \begin{bmatrix} 1 \\ 1 \\ 1 \\ 0 \\ 1 \\ 0 \\ 0 \end{bmatrix}$$

Hamming's (7,4) code is classified as a *linear code* since the sum of any two codewords is also a codeword. To see this, observe that if $\mathbf{u}$ and $\mathbf{v}$ are codewords, then the sum $\mathbf{u} + \mathbf{v}$ is also a codeword since

$$C(\mathbf{u} + \mathbf{v}) = C\mathbf{u} + C\mathbf{v} = \mathbf{0} + \mathbf{0} = \mathbf{0} \ (\text{mod } 2)$$

It also has the property that every codeword can be written as a linear combination of a few key codewords.

In this chapter we will see how the set of all linear combinations of a set of vectors forms a *vector space*. The set of codewords in the chapter opener is an example.

3.1 ▶ Definition of a Vector Space

In Chap. 2 we defined a natural *addition* and *scalar multiplication* on vectors in $\mathbb{R}^n$ as generalizations of the same operations on real numbers. With respect to these operations, we saw in Theorem 1 of Sec. 2.1 that sets of vectors satisfy many of the familiar algebraic properties enjoyed by numbers. In this section we use these properties as axioms to generalize the concept of a vector still further. In particular, we consider as vectors any class of objects with definitions for addition and scalar multiplication that satisfy the properties of this theorem. In this way our new concept of a vector will include vectors in $\mathbb{R}^n$ but many new kinds as well.

DEFINITION 1

Vector Space   A set V is called a **vector space** over the real numbers provided that there are two operations—addition, denoted by $\oplus$, and scalar multiplication, denoted by $\odot$—that satisfy all the following axioms. The axioms must hold for all vectors $\mathbf{u}, \mathbf{v}$, and $\mathbf{w}$ in V and all scalars c and d in $\mathbb{R}$.

1. The sum $\mathbf{u} \oplus \mathbf{v}$ is in V.   Closed under addition
2. $\mathbf{u} \oplus \mathbf{v} = \mathbf{v} \oplus \mathbf{u}$ Addition is commutative
3. $(\mathbf{u} \oplus \mathbf{v}) \oplus \mathbf{w} = \mathbf{u} \oplus (\mathbf{v} \oplus \mathbf{w})$ Addition is associative

4. There exists a vector $\mathbf{0} \in V$ such that for   Additive identity
   every vector $\mathbf{u} \in V$, $\mathbf{0} \oplus \mathbf{u} = \mathbf{u} \oplus \mathbf{0} = \mathbf{u}$.

5. For every vector $\mathbf{u} \in V$, there exists a vec-   Additive inverse
   tor, denoted by $-\mathbf{u}$, such that $\mathbf{u} \oplus (-\mathbf{u}) = -\mathbf{u} \oplus \mathbf{u} = \mathbf{0}$.

6. The scalar product $c \odot \mathbf{u}$ is in V.   Closed under scalar
   multiplication

7. $c \odot (\mathbf{u} \oplus \mathbf{v}) = (c \odot \mathbf{u}) \oplus (c \odot \mathbf{v})$

8. $(c + d) \odot \mathbf{u} = (c \odot \mathbf{u}) \oplus (d \odot \mathbf{u})$

9. $c \odot (d \odot \mathbf{u}) = (cd) \odot \mathbf{u}$

10 $1 \odot \mathbf{u} = \mathbf{u}$

In this section (and elsewhere when necessary) we use the special symbols $\oplus$ and $\odot$ of the previous definition to distinguish vector addition and scalar multiplication from ordinary addition and multiplication of real numbers. We also will point out that for general vector spaces the set of scalars can be chosen from any *field*. In this text, unless otherwise stated, we chose scalars from the set of real numbers.

EXAMPLE 1   **Euclidean Vector Spaces**   The set $V = \mathbb{R}^n$ with the standard operations of addition and scalar multiplication is a vector space.

Solution   Axioms 2 through 5 and 7 through 10 are shown to hold in Theorem 1 of Sec. 2.1. The fact that $\mathbb{R}^n$ is closed under addition and scalar multiplication is a direct consequence of how these operations are defined. The Euclidean vector spaces $\mathbb{R}^n$ are the prototypical vector spaces on which the general theory of vector spaces is built.

EXAMPLE 2   **Vector Spaces of Matrices**   Show that the set $V = M_{m \times n}$ of all $m \times n$ matrices is a vector space over the scalar field $\mathbb{R}$, with $\oplus$ and $\odot$ defined componentwise.

Solution   Since addition of matrices is componentwise, the sum of two $m \times n$ matrices is another $m \times n$ matrix as is a scalar times an $m \times n$ matrix. Thus, the closure axioms (axioms 1 and 6) are satisfied. We also have that $1 \odot A = A$. The other seven axioms are given in Theorem 4 of Sec. 1.3.

When we are working with more abstract sets of objects, the operations of addition and scalar multiplication can be defined in nonstandard ways. The result is not always a vector space. This is illustrated in the next several examples.

EXAMPLE 3   Let $V = \mathbb{R}$. Define addition and scalar multiplication by

$$\mathbf{a} \oplus \mathbf{b} = 2a + 2b \qquad \text{and} \qquad k \odot \mathbf{a} = ka$$

Show that addition is commutative but not associative.

Solution   Since the usual addition of real numbers (on the right-hand side) is commutative,

$$\mathbf{a} \oplus \mathbf{b} = 2a + 2b$$
$$= 2b + 2a$$
$$= \mathbf{b} \oplus \mathbf{a}$$

Thus, the operation $\oplus$ is commutative.

To determine whether addition is associative, we evaluate and compare the expressions

$$(\mathbf{a} \oplus \mathbf{b}) \oplus \mathbf{c} \qquad \text{and} \qquad \mathbf{a} \oplus (\mathbf{b} \oplus \mathbf{c})$$

In this case, we have

$$(\mathbf{a} \oplus \mathbf{b}) \oplus \mathbf{c} = (2a + 2b) \oplus c \qquad \text{and} \qquad \mathbf{a} \oplus (\mathbf{b} \oplus \mathbf{c}) = a \oplus (2b + 2c)$$
$$= 2(2a + 2b) + 2c \qquad\qquad\qquad\qquad = 2a + 2(2b + 2c)$$
$$= 4a + 4b + 2c \qquad\qquad\qquad\qquad\quad = 2a + 4b + 4c$$

We see that the two final expressions are not equal for all choices of a, b, and c. Therefore, the associative property is not upheld, and V is not a vector space.

EXAMPLE 4   Let $V = \mathbb{R}$. Define addition and scalar multiplication by

$$\mathbf{a} \oplus \mathbf{b} = a^b \qquad \text{and} \qquad k \odot \mathbf{a} = ka$$

Show that V is not a vector space.

Solution   In this case

$$\mathbf{a} \oplus \mathbf{b} = a^b \qquad \text{and} \qquad \mathbf{b} \oplus \mathbf{a} = b^a$$

Since $a^b \neq b^a$ for all choices of a and b, the commutative property of addition is not upheld, and V is not a vector space.

In Example 5 we show that familiar sets with nonstandard definitions for addition and scalar multiplication can be vector spaces.

| EXAMPLE 5 | Let $V = \{(a, b) \mid a, b \in \mathbb{R}\}$. Let $\mathbf{v} = (v_1, v_2)$ and $\mathbf{w} = (w_1, w_2)$. Define |

$$(v_1, v_2) \oplus (w_1, w_2) = (v_1 + w_1 + 1, v_2 + w_2 + 1) \qquad \text{and}$$
$$c \odot (v_1, v_2) = (cv_1 + c - 1, cv_2 + c - 1)$$

Verify that V is a vector space.

Solution   First observe that since the result of addition or scalar multiplication is an ordered pair, V is closed under addition and scalar multiplication. Since addition of real numbers is commutative and associative, axioms 2 and 3 hold for the $\oplus$ defined here. Now an element $\mathbf{w} \in V$ is the additive identity provided that for all $\mathbf{v} \in V$

$$\mathbf{v} \oplus \mathbf{w} = \mathbf{v} \qquad \text{or} \qquad (v_1 + w_1 + 1, v_2 + w_2 + 1) = (v_1, v_2)$$

Equating components gives

$$v_1 + w_1 + 1 = v_1 \qquad \text{and} \qquad v_2 + w_2 + 1 = v_2 \quad \text{so}$$
$$w_1 = -1 \qquad \text{and} \qquad w_2 = -1$$

This establishes the existence of an additive identity. Specifically, $\mathbf{0} = (-1, -1)$, so axiom 4 holds.

To show that each element $\mathbf{v}$ in V has an additive inverse, we must find a vector $\mathbf{w}$ such that

$$\mathbf{v} \oplus \mathbf{w} = \mathbf{0} = (-1, -1)$$

Since $\mathbf{v} \oplus \mathbf{w} = (v_1 + w_1 + 1, v_2 + w_2 + 1)$, this last equation requires that

$$v_1 + w_1 + 1 = -1 \qquad \text{and} \qquad v_2 + w_2 + 1 = -1 \qquad \text{so that}$$
$$w_1 = -v_1 - 2 \qquad \text{and} \qquad w_2 = -v_2 - 2$$

Thus, for any element $\mathbf{v} = (v_1, v_2)$ in V, we have $-\mathbf{v} = (-v_1 - 2, -v_2 - 2)$. The remaining axioms all follow from the similar properties of the real numbers.

A **polynomial of degree** n is an expression of the form

$$p(x) = a_0 + a_1 x + a_2 x^2 + \cdots + a_{n-1} x^{n-1} + a_n x^n$$

where $a_0, \ldots, a_n$ are real numbers and $a_n \neq 0$. The degree of the **zero polynomial** is undefined since it can be written as $p(x) = 0 x^n$ for any positive integer n. Polynomials comprise one of the most basic sets of functions and have many applications in mathematics.

| EXAMPLE 6 | **Vector Space of Polynomials**   Let n be a fixed positive integer. Denote by $\mathcal{P}_n$ the set of all polynomials of degree n or less. Define addition by adding like terms. That is, if |

$$p(x) = a_0 + a_1 x + a_2 x^2 + \cdots + a_{n-1} x^{n-1} + a_n x^n$$

and

$$q(x) = b_0 + b_1 x + b_2 x^2 + \cdots + b_{n-1} x^{n-1} + b_n x^n$$

then

$$p(x) \oplus q(x) = (a_0 + b_0) + (a_1 + b_1)x + (a_2 + b_2)x^2 + \cdots + (a_n + b_n)x^n$$

If c is a scalar, then scalar multiplication is defined by

$$c \odot p(x) = ca_0 + ca_1 x + ca_2 x^2 + \cdots + ca_{n-1} x^{n-1} + ca_n x^n$$

Verify that $V = \mathcal{P}_n \cup \{\mathbf{0}\}$ is a real vector space, where $\mathbf{0}$ is the zero polynomial.

Solution   Since the sum of two polynomials of degree n or less is another polynomial of degree n or less, with the same holding for scalar multiplication, the set V is closed under addition and scalar multiplication. The zero vector is just the zero polynomial, and the additive inverse of $p(x)$ is given by

$$-p(x) = -a_0 - a_1 x - a_2 x^2 - \cdots - a_{n-1} x^{n-1} - a_n x^n$$

The remaining axioms are consequences of the properties of real numbers. For example,

$$\begin{aligned} p(x) \oplus q(x) &= (a_0 + b_0) + (a_1 + b_1)x + (a_2 + b_2)x^2 + \cdots + (a_n + b_n)x^n \\ &= (b_0 + a_0) + (b_1 + a_1)x + (b_2 + a_2)x^2 + \cdots + (b_n + a_n)x^n \\ &= q(x) \oplus p(x) \end{aligned}$$

In the sequel we will use $\mathcal{P}_n$ to denote the vector space of polynomials of degree n or less along with the zero polynomial.

The condition *degree n or less* cannot be replaced with all polynomials of *degree equal to n*. The latter set is not closed under addition. For example, the polynomials $x^2 - 2x + 1$ and $-x^2 + 3x + 4$ are both polynomials of degree 2, but the sum is $x + 5$, which has degree equal to 1.

EXAMPLE 7   **Vector Space of Real-Valued Functions**   Let V be the set of real-valued functions defined on a common domain given by the interval $[a, b]$. For all f and g in V and $c \in \mathbb{R}$, define addition and scalar multiplication, respectively, by

$$(f \oplus g)(x) = f(x) + g(x) \qquad \text{and} \qquad (c \odot f)(x) = cf(x)$$

for each x in $[a, b]$. Show that V is a real vector space.

Solution   Since the pointwise sum of two functions with domain $[a, b]$ is another function with domain $[a, b]$, the set V is closed under addition. Similarly, the set V is closed under scalar multiplication.

To show that addition in V is commutative, let f and g be functions in V. Then

$$(f \oplus g)(x) = f(x) + g(x) = g(x) + f(x) = (g \oplus f)(x)$$

Addition is also associative since for any functions f, g, and h in V, we have

$$\begin{aligned}(f \oplus (g \oplus h))(x) &= f(x) + (g \oplus h)(x) \\ &= f(x) + g(x) + h(x) \\ &= (f \oplus g)(x) + h(x) \\ &= ((f \oplus g) \oplus h)(x)\end{aligned}$$

The zero element of V, denoted by $\mathbf{0}$, is the function that is 0 for all real numbers in $[a, b]$. We have that $\mathbf{0}$ is the additive identity on V since

$$(f \oplus \mathbf{0})(x) = f(x) + \mathbf{0}(x) = f(x)$$

Next we let c and d be real numbers and let f be an element of V. The distributive property of real numbers gives us

$$\begin{aligned}(c + d) \odot f(x) &= (c + d)f(x) = cf(x) + df(x) \\ &= (c \odot f)(x) \oplus (d \odot f)(x)\end{aligned}$$

so $(c + d) \odot f = (c \odot f) \oplus (d \odot f)$, establishing property 8.

The other properties follow in a similar manner.

The set of **complex numbers**, denoted by $\mathbb{C}$, is defined by

$$\mathbb{C} = \{a + bi \mid a, b \in \mathbb{R}\}$$

where i satisfies

$$i^2 = -1 \qquad \text{or equivalently} \qquad i = \sqrt{-1}$$

The set of complex numbers is an *algebraic extension* of the real numbers, which it contains as a subset. For every complex number $z = a + bi$, the real number a is called the **real part** of z and the real number b the **imaginary part** of z.

With the appropriate definitions of addition and scalar multiplication, the set of complex numbers $\mathbb{C}$ is a vector space.

EXAMPLE 8   **Vector Space of Complex Numbers**   Let $\mathbf{z} = a + bi$ and $\mathbf{w} = c + di$ be elements of $\mathbb{C}$ and α a real number. Define vector addition on $\mathbb{C}$ by

$$\mathbf{z} \oplus \mathbf{w} = (a + bi) + (c + di) = (a + c) + (b + d)i$$

and scalar multiplication by

$$\alpha \odot \mathbf{z} = \alpha \odot (a + bi) = \alpha a + (\alpha b)i$$

Verify that $\mathbb{C}$ is a vector space.

Solution   For each element $\mathbf{z} = a + bi$ in $\mathbb{C}$, associate the vector in $\mathbb{R}^2$ whose components are the real and imaginary parts of $\mathbf{z}$. That is, let

$$\mathbf{z} = a + bi \longleftrightarrow \begin{bmatrix} a \\ b \end{bmatrix}$$

Observe that addition and scalar multiplication in $\mathbb{C}$ correspond to those same operations in $\mathbb{R}^2$. In this way $\mathbb{C}$ and $\mathbb{R}^2$ have the same algebraic structure. Since $\mathbb{R}^2$ is a vector space, so is $\mathbb{C}$.

In Example 8, we showed that $\mathbb{C}$ is a vector space over the real numbers. It is also possible to show that $\mathbb{C}$ is a vector space over the complex scalars. We leave the details to the reader.

Example 9 is from analytic geometry.

EXAMPLE 9   Let a, b, and c be fixed real numbers. Let V be the set of points in three-dimensional Euclidean space that lie on the plane P given by

$$ax + by + cz = 0$$

Define addition and scalar multiplication on V coordinatewise. Verify that V is a vector space.

Solution   To show that V is closed under addition, let $\mathbf{u} = (u_1, u_2, u_3)$ and $\mathbf{v} = (v_1, v_2, v_3)$ be points in V. The vectors $\mathbf{u}$ and $\mathbf{v}$ are in V provided that

$$au_1 + bu_2 + cu_3 = 0 \qquad \text{and} \qquad av_1 + bv_2 + cv_3 = 0$$

Now by definition

$$\mathbf{u} \oplus \mathbf{v} = (u_1 + v_1, u_2 + v_2, u_3 + v_3)$$

We know that $\mathbf{u} \oplus \mathbf{v}$ is in V since

$$\begin{aligned} a(u_1 + v_1) + b(u_2 + v_2) + c(u_3 + v_3) &= au_1 + av_1 + bu_2 + bv_2 + cu_3 + cv_3 \\ &= (au_1 + bu_2 + cu_3) + (av_1 + bv_2 + cv_3) \\ &= 0 \end{aligned}$$

Similarly, V is closed under scalar multiplication since for any scalar α, we have

$$\alpha \odot \mathbf{u} = (\alpha u_1, \alpha u_2, \alpha u_3)$$

and

$$a(\alpha u_1) + b(\alpha u_2) + c(\alpha u_3) = \alpha(au_1 + bu_2 + cu_3) = \alpha(0) = 0$$

In this case the zero vector is $(0, 0, 0)$, which is also on the plane P. Since the addition and scalar multiplication defined on V are the analogous operations defined on the vector space $\mathbb{R}^3$, the remaining axioms are satisfied for elements of V as well.

We conclude this section by showing that some familiar algebraic properties of $\mathbb{R}^n$ extend to abstract vector spaces.

THEOREM 1   In a vector space V, additive inverses are unique.

Proof   Let $\mathbf{u}$ be an element of V. Suppose that $\mathbf{v}$ and $\mathbf{w}$ are elements of V and both are additive inverses of $\mathbf{u}$. We show that $\mathbf{v} = \mathbf{w}$. Since

$$\mathbf{u} \oplus \mathbf{v} = \mathbf{0} \qquad \text{and} \qquad \mathbf{u} \oplus \mathbf{w} = \mathbf{0}$$

axioms 4, 3, and 2 give

$$\mathbf{v} = \mathbf{v} \oplus \mathbf{0} = \mathbf{v} \oplus (\mathbf{u} \oplus \mathbf{w}) = (\mathbf{v} \oplus \mathbf{u}) \oplus \mathbf{w} = \mathbf{0} \oplus \mathbf{w} = \mathbf{w}$$

establishing the result.

THEOREM 2   Let V be a vector space, $\mathbf{u}$ a vector in V, and c a real number.

1. $0 \odot \mathbf{u} = \mathbf{0}$
2. $c \odot \mathbf{0} = \mathbf{0}$
3. $(-1) \odot \mathbf{u} = -\mathbf{u}$
4. If $c \odot \mathbf{u} = \mathbf{0}$, then either $c = 0$ or $\mathbf{u} = \mathbf{0}$.

Proof   (1) By axiom 8, we have

$$0 \odot \mathbf{u} = (0 + 0) \odot \mathbf{u} = (0 \odot \mathbf{u}) \oplus (0 \odot \mathbf{u})$$

Adding the inverse $-(0 \odot \mathbf{u})$ to both sides of the preceding equation gives the result.

(2) By axiom 4, we know that $\mathbf{0} \oplus \mathbf{0} = \mathbf{0}$. Combining this with axiom 7 gives

$$c \odot \mathbf{0} = c \odot (\mathbf{0} \oplus \mathbf{0}) = (c \odot \mathbf{0}) \oplus (c \odot \mathbf{0})$$

Again adding the inverse $-(c \odot \mathbf{0})$ to both sides of the last equation gives the result.

(3) By axioms 10 and 8 and part 1 of this theorem,

$$\begin{aligned} \mathbf{u} \oplus (-1) \odot \mathbf{u} &= (1 \odot \mathbf{u}) \oplus [(-1) \odot \mathbf{u}] \\ &= (1 - 1) \odot \mathbf{u} \\ &= 0 \odot \mathbf{u} \\ &= \mathbf{0} \end{aligned}$$

Thus, $(-1) \odot \mathbf{u}$ is an additive inverse of $\mathbf{u}$. Since $-\mathbf{u}$ is by definition the additive inverse of $\mathbf{u}$ and by Theorem 1 additive inverses are unique, we have $(-1) \odot \mathbf{u} = -\mathbf{u}$.

(4) Let $c \odot \mathbf{u} = \mathbf{0}$. If $c = 0$, then the conclusion holds. Suppose that $c \neq 0$. Then multiply both sides of

$$c \odot \mathbf{u} = \mathbf{0}$$

by $\dfrac{1}{c}$ and apply part 2 of this theorem to obtain

$$\frac{1}{c} \odot (c \odot \mathbf{u}) = \mathbf{0} \qquad \text{so that} \qquad 1 \odot \mathbf{u} = \mathbf{0}$$

and hence $\mathbf{u} = \mathbf{0}$.

Fact Summary

1. To determine whether a set V with addition and scalar multiplication defined on V is a vector space requires verification of the 10 vector space axioms.
2. The Euclidean space $\mathbb{R}^n$ and the set of matrices $M_{m \times n}$, with the standard componentwise operations, are vector spaces. The set of polynomials of degree n or less with termwise operations is a vector space.
3. In all vector spaces, additive inverses are unique. Also

$$0 \odot \mathbf{u} = \mathbf{0} \qquad c \odot \mathbf{0} = \mathbf{0} \qquad \text{and} \qquad (-1) \odot \mathbf{u} = -\mathbf{u}$$

In addition if $c \odot \mathbf{u} = \mathbf{0}$, then either the scalar c is the number 0 or the vector $\mathbf{u}$ is the zero vector.

Exercise Set 3.1

In Exercises 1–4, let $V = \mathbb{R}^3$. Show that V with the given operations for $\oplus$ and $\odot$ is not a vector space.

1.
$$\begin{bmatrix} x_1 \\ y_1 \\ z_1 \end{bmatrix} \oplus \begin{bmatrix} x_2 \\ y_2 \\ z_2 \end{bmatrix} = \begin{bmatrix} x_1 - x_2 \\ y_1 - y_2 \\ z_1 - z_2 \end{bmatrix}$$

$$c \odot \begin{bmatrix} x_1 \\ y_1 \\ z_1 \end{bmatrix} = \begin{bmatrix} cx_1 \\ cy_1 \\ cz_1 \end{bmatrix}$$

2.
$$\begin{bmatrix} x_1 \\ y_1 \\ z_1 \end{bmatrix} \oplus \begin{bmatrix} x_2 \\ y_2 \\ z_2 \end{bmatrix} = \begin{bmatrix} x_1 + x_2 - 1 \\ y_1 + y_2 - 1 \\ z_1 + z_2 - 1 \end{bmatrix}$$

$$c \odot \begin{bmatrix} x_1 \\ y_1 \\ z_1 \end{bmatrix} = \begin{bmatrix} cx_1 \\ cy_1 \\ cz_1 \end{bmatrix}$$

3.
$$\begin{bmatrix} x_1 \\ y_1 \\ z_1 \end{bmatrix} \oplus \begin{bmatrix} x_2 \\ y_2 \\ z_2 \end{bmatrix} = \begin{bmatrix} 2x_1 + 2x_2 \\ 2y_1 + 2y_2 \\ 2z_1 + 2z_2 \end{bmatrix}$$

$$c \odot \begin{bmatrix} x_1 \\ y_1 \\ z_1 \end{bmatrix} = \begin{bmatrix} cx_1 \\ cy_1 \\ cz_1 \end{bmatrix}$$

4.
$$\begin{bmatrix} x_1 \\ y_1 \\ z_1 \end{bmatrix} \oplus \begin{bmatrix} x_2 \\ y_2 \\ z_2 \end{bmatrix} = \begin{bmatrix} x_1 + x_2 \\ y_1 + y_2 \\ z_1 + z_2 \end{bmatrix}$$

$$c \odot \begin{bmatrix} x_1 \\ y_1 \\ z_1 \end{bmatrix} = \begin{bmatrix} c + x_1 \\ y_1 \\ z_1 \end{bmatrix}$$

5. Write out all 10 vector space axioms to show $\mathbb{R}^2$ with the standard componentwise operations is a vector space.

6. Write out all 10 vector space axioms to show that $M_{2 \times 2}$ with the standard componentwise operations is a vector space.

7. Let $V = \mathbb{R}^2$ and define addition as the standard componentwise addition and define scalar multiplication by

$$c \odot \begin{bmatrix} x \\ y \end{bmatrix} = \begin{bmatrix} x + c \\ y \end{bmatrix}$$

Show that V is not a vector space.

8. Let

$$V = \left\{ \begin{bmatrix} a \\ b \\ 1 \end{bmatrix} \Bigg| \ a, b \in \mathbb{R} \right\}$$

a. With the standard componentwise operations show that V is not a vector space.

b. If addition and scalar multiplication are defined componentwise only on the first two components and the third is always 1, show that V is a vector space.

9. Let $V = \mathbb{R}^2$ and define

$$\begin{bmatrix} a \\ b \end{bmatrix} \oplus \begin{bmatrix} c \\ d \end{bmatrix} = \begin{bmatrix} a + 2c \\ b + 2d \end{bmatrix}$$

$$c \odot \begin{bmatrix} a \\ b \end{bmatrix} = \begin{bmatrix} ca \\ cb \end{bmatrix}$$

Determine whether V is a vector space.

10. Let

$$V = \left\{ \begin{bmatrix} t \\ -3t \end{bmatrix} \Bigg| \ t \in \mathbb{R} \right\}$$

and let addition and scalar multiplication be the standard operations on vectors. Determine whether V is a vector space.

11. Let

$$V = \left\{ \begin{bmatrix} t + 1 \\ 2t \end{bmatrix} \Bigg| \ t \in \mathbb{R} \right\}$$

and let addition and scalar multiplication be the standard operations on vectors. Determine whether V is a vector space.

12. Let

$$V = \left\{ \begin{bmatrix} a & b \\ c & 0 \end{bmatrix} \Bigg| \ a, b, c \in \mathbb{R} \right\}$$

and let addition and scalar multiplication be the standard componentwise operations. Determine whether V is a vector space.

13. Let

$$V = \left\{ \begin{bmatrix} a & b \\ c & 1 \end{bmatrix} \Bigg| \ a, b, c \in \mathbb{R} \right\}$$

a. If addition and scalar multiplication are the standard componentwise operations, show that V is not a vector space.

b. Define

$$\begin{bmatrix} a & b \\ c & 1 \end{bmatrix} \oplus \begin{bmatrix} d & e \\ f & 1 \end{bmatrix} = \begin{bmatrix} a + d & b + e \\ c + f & 1 \end{bmatrix}$$

and

$$k \odot \begin{bmatrix} a & b \\ c & 1 \end{bmatrix} = \begin{bmatrix} ka & kb \\ kc & 1 \end{bmatrix}$$

Show that V is a vector space.

In Exercises 14–19, let V be the set of 2×2 matrices with the standard (componentwise) definitions for vector addition and scalar multiplication. Determine whether V is a vector space. If V is not a vector space, show that at least one of the 10 axioms does not hold.

14. Let V be the set of all skew-symmetric matrices, that is, the set of all matrices such that $A^t = -A$.

15. Let V be the set of all upper triangular matrices.

16. Let V be the set of all real symmetric matrices, that is, the set of all matrices such that $A^t = A$.

17. Let V be the set of all invertible matrices.

18. Let V be the set of all idempotent matrices.

19. Let B be a fixed matrix, and let V be the set of all matrices A such that $AB = \mathbf{0}$.

20. Let

$$V = \left\{ \begin{bmatrix} a & b \\ c & -a \end{bmatrix} \Bigg| \ a, b, c \in \mathbb{R} \right\}$$

and define addition and scalar multiplication as the standard componentwise operations. Determine whether V is a vector space.

21. Let V denote the set of 2×2 invertible matrices. Define

$$A \oplus B = AB \qquad c \odot A = cA$$

a. Determine the additive identity and additive inverse.

b. Show that V is not a vector space.

22. Let
$$V = \left\{ \begin{bmatrix} t \\ 1+t \end{bmatrix} \;\middle|\; t \in \mathbb{R} \right\}$$

Define
$$\begin{bmatrix} t_1 \\ 1+t_1 \end{bmatrix} \oplus \begin{bmatrix} t_2 \\ 1+t_2 \end{bmatrix}$$
$$= \begin{bmatrix} t_1 + t_2 \\ 1 + t_1 + t_2 \end{bmatrix}$$
$$c \odot \begin{bmatrix} t \\ 1+t \end{bmatrix} = \begin{bmatrix} ct \\ 1+ct \end{bmatrix}$$

a. Find the additive identity and inverse.

b. Show that V is a vector space.

c. Verify that $0 \odot \mathbf{v} = \mathbf{0}$ for all $\mathbf{v}$.

23. Let
$$V = \left\{ \begin{bmatrix} 1+t \\ 2-t \\ 3+2t \end{bmatrix} \;\middle|\; t \in \mathbb{R} \right\}$$

Define
$$\begin{bmatrix} 1+t_1 \\ 2-t_1 \\ 3+2t_1 \end{bmatrix} \oplus \begin{bmatrix} 1+t_2 \\ 2-t_2 \\ 3+2t_2 \end{bmatrix}$$
$$= \begin{bmatrix} 1+(t_1+t_2) \\ 2-(t_1+t_2) \\ 3+(2t_1+2t_2) \end{bmatrix}$$
$$c \odot \begin{bmatrix} 1+t \\ 2-t \\ 3+2t \end{bmatrix} = \begin{bmatrix} 1+ct \\ 2-ct \\ 3+2ct \end{bmatrix}$$

a. Find the additive identity and inverse.

b. Show that V is a vector space.

c. Verify $0 \odot \mathbf{v} = \mathbf{0}$ for all $\mathbf{v}$.

24. Let
$$\mathbf{u} = \begin{bmatrix} 1 \\ 0 \\ 1 \end{bmatrix} \qquad \mathbf{v} = \begin{bmatrix} 2 \\ -1 \\ 1 \end{bmatrix}$$
and
$$S = \{a\mathbf{u} + b\mathbf{v} \mid a, b \in \mathbb{R}\}$$

Show that S with the standard componentwise operations is a vector space.

25. Let $\mathbf{v}$ be a vector in $\mathbb{R}^n$, and let
$$S = \{\mathbf{v}\}$$
Define $\oplus$ and $\odot$ by
$$\mathbf{v} \oplus \mathbf{v} = \mathbf{v} \qquad c \odot \mathbf{v} = \mathbf{v}$$
Show that S is a vector space.

26. Let
$$S = \left\{ \begin{bmatrix} x \\ y \\ z \end{bmatrix} \;\middle|\; 3x - 2y + z = 0 \right\}$$

Show that S with the standard componentwise operations is a vector space.

27. Let S be the set of all vectors
$$\begin{bmatrix} x \\ y \\ z \end{bmatrix}$$
in $\mathbb{R}^3$ such that $x + y - z = 0$ and $2x - 3y + 2z = 0$. Show that S with the standard componentwise operations is a vector space.

28. Let
$$V = \left\{ \begin{bmatrix} \cos t \\ \sin t \end{bmatrix} \;\middle|\; t \in \mathbb{R} \right\}$$
and define
$$\begin{bmatrix} \cos t_1 \\ \sin t_1 \end{bmatrix} \oplus \begin{bmatrix} \cos t_2 \\ \sin t_2 \end{bmatrix}$$
$$= \begin{bmatrix} \cos(t_1 + t_2) \\ \sin(t_1 + t_2) \end{bmatrix}$$
$$c \odot \begin{bmatrix} \cos t \\ \sin t \end{bmatrix} = \begin{bmatrix} \cos ct \\ \sin ct \end{bmatrix}$$

a. Determine the additive identity and additive inverse.

b. Show that V is a vector space.

c. Show that if $\oplus$ and $\odot$ are the standard componentwise operations, then V is not a vector space.

29. Let V be the set of all real-valued functions defined on $\mathbb{R}$ with the standard operations that satisfy $f(0) = 1$. Determine whether V is a vector space.

30. Let V be the set of all real-valued functions defined on $\mathbb{R}$.

Define $f \oplus g$ by

$$(f \oplus g)(x) = f(x) + g(x)$$

and define $c \odot f$ by

$$(c \odot f)(x) = f(x + c)$$

Determine whether V is a vector space.

31. Let $f(x) = x^3$ defined on $\mathbb{R}$ and let

$$V = \{f(x + t) \mid t \in \mathbb{R}\}$$

Define

$$f(x + t_1) \oplus f(x + t_2) = f(x + t_1 + t_2)$$

$$c \odot f(x + t) = f(x + ct)$$

a. Determine the additive identity and additive inverses.

b. Show that V is a vector space.

3.2 ▶ Subspaces

Many interesting examples of vector spaces are subsets of a given vector space V that are vector spaces in their own right. For example, the xy plane in $\mathbb{R}^3$ given by

$$\left\{ \begin{bmatrix} x \\ y \\ 0 \end{bmatrix} \;\middle|\; x, y \in \mathbb{R} \right\}$$

is a subset of $\mathbb{R}^3$. It is also a vector space with the same standard componentwise operations defined on $\mathbb{R}^3$. Another example of a *subspace* of a vector space is given in Example 9 of Sec. 3.1. The determination as to whether a subset of a vector space is itself a vector space is simplified since many of the required properties are *inherited* from the parent space.

DEFINITION 1   **Subspace**   A **subspace** W of a vector space V is a nonempty subset that is itself a vector space with respect to the inherited operations of vector addition and scalar multiplication on V.

The first requirement for a subset $W \subseteq V$ to be a subspace is that W be closed under the operations of V. For example, let V be the vector space $\mathbb{R}^2$ with the standard definitions of addition and scalar multiplication. Let $W \subseteq \mathbb{R}^2$ be the subset defined by

$$W = \left\{ \begin{bmatrix} a \\ 0 \end{bmatrix} \;\middle|\; a \in \mathbb{R} \right\}$$

Observe that the sum of any two vectors in W is another vector in W, since

$$\begin{bmatrix} a \\ 0 \end{bmatrix} \oplus \begin{bmatrix} b \\ 0 \end{bmatrix} = \begin{bmatrix} a + b \\ 0 \end{bmatrix}$$

In this way we say that W is closed under addition. The subset W is also closed under scalar multiplication since for any real number c,

$$c \odot \begin{bmatrix} a \\ 0 \end{bmatrix} = \begin{bmatrix} ca \\ 0 \end{bmatrix}$$

which is again in W.

On the other hand, the subset

$$W = \left\{ \begin{bmatrix} a \\ 1 \end{bmatrix} \middle| a \in \mathbb{R} \right\}$$

is not closed under addition, since

$$\begin{bmatrix} a \\ 1 \end{bmatrix} \oplus \begin{bmatrix} b \\ 1 \end{bmatrix} = \begin{bmatrix} a + b \\ 2 \end{bmatrix}$$

which is not in W. See Fig. 1. The subset W is also not closed under scalar multiplication since

$$c \odot \begin{bmatrix} a \\ 1 \end{bmatrix} = \begin{bmatrix} ca \\ c \end{bmatrix}$$

which is not in W for all values of $c \neq 1$.

Now let us suppose that a nonempty subset W is closed under both of the operations on V. To determine whether W is a subspace, we must show that each of the remaining vector space axioms hold. Fortunately, our task is simplified as most of these properties are inherited from the vector space V. For example, to show that the commutative property holds in W, let $\mathbf{u}$ and $\mathbf{v}$ be vectors in W. Since $\mathbf{u}$ and $\mathbf{v}$ are also in V, then

$$\mathbf{u} \oplus \mathbf{v} = \mathbf{v} \oplus \mathbf{u}$$

Similarly, any three vectors in W satisfy the associative property, as this property is also inherited from V. To show that W contains the zero vector, let $\mathbf{w}$ be any vector in W. Since W is closed under scalar multiplication, $0 \odot \mathbf{w} \in W$. Now, by Theorem 2 of Sec. 3.1, we have $0 \odot \mathbf{w} = \mathbf{0}$. Thus, $\mathbf{0} \in W$. Similarly, for any $\mathbf{w} \in W$,

$$(-1) \odot \mathbf{w} = -\mathbf{w}$$

is also in W. All the other vector space properties, axioms 7 through 10, are inherited from V. This shows that W is a subspace of V. Conversely, if W is a subspace of V, then it is necessarily closed under addition and scalar multiplication. This proves Theorem 3.

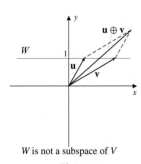

W is not a subspace of V

Figure 1

THEOREM 3   Let W be a nonempty subset of the vector space V. Then W is a subspace of V if and only if W is closed under addition and scalar multiplication.

By Theorem 3, the first of the examples above with

$$W = \left\{ \begin{bmatrix} a \\ 0 \end{bmatrix} \middle| a \in \mathbb{R} \right\}$$

is a subspace of $\mathbb{R}^2$ while the second subset

$$W = \left\{ \left[\begin{array}{c} a \\ 1 \end{array} \right] \,\middle|\, a \in \mathbb{R} \right\}$$

is not.

For any vector space V the subset $W = \{\mathbf{0}\}$, consisting of only the zero vector, is a subspace of V, called the **trivial subspace**. We also have that any vector space V, being a subset of itself, is a subspace.

EXAMPLE 1   Let

$$W = \left\{ \left[\begin{array}{c} a \\ a+1 \end{array} \right] \,\middle|\, a \in \mathbb{R} \right\}$$

be a subset of the vector space $V = \mathbb{R}^2$ with the standard definitions of addition and scalar multiplication. Determine whether W is a subspace of V.

Solution   In light of Theorem 3, we check to see if W is closed under addition and scalar multiplication. Let

$$\mathbf{u} = \left[\begin{array}{c} u \\ u+1 \end{array} \right] \quad \text{and} \quad \mathbf{v} = \left[\begin{array}{c} v \\ v+1 \end{array} \right]$$

be vectors in W. Adding the vectors gives

$$\mathbf{u} \oplus \mathbf{v} = \left[\begin{array}{c} u \\ u+1 \end{array} \right] \oplus \left[\begin{array}{c} v \\ v+1 \end{array} \right]$$

$$= \left[\begin{array}{c} u+v \\ u+v+2 \end{array} \right]$$

This last vector is not in the required form since

$$u+v+2 \neq u+v+1$$

and hence we see that $\mathbf{u} \oplus \mathbf{v}$ is not in W. Thus, W is not a subspace of V.

It is sometimes easy to show that a subset W of a vector space is not a subspace. In particular, if $\mathbf{0} \notin W$ or the additive inverse of a vector is not in W, then W is not a subspace. In Example 1, W is not a subspace since it does not contain the zero vector.

EXAMPLE 2   The **trace** of a square matrix is the sum of the entries on the diagonal. Let $M_{2 \times 2}$ be the vector space of 2×2 matrices with the standard operations for addition and scalar multiplication, and let W be the subset of all 2×2 matrices with trace 0, that is,

$$W = \left\{ \left[\begin{array}{cc} a & b \\ c & d \end{array} \right] \,\middle|\, a+d = 0 \right\}$$

Show that W is a subspace of $M_{2 \times 2}$.

Solution   Let

$$\mathbf{w}_1 = \begin{bmatrix} a_1 & b_1 \\ c_1 & d_1 \end{bmatrix} \qquad \text{and} \qquad \mathbf{w}_2 = \begin{bmatrix} a_2 & b_2 \\ c_2 & d_2 \end{bmatrix}$$

be matrices in W, so that $a_1 + d_1 = 0$ and $a_2 + d_2 = 0$. The sum of the two matrices is

$$\mathbf{w}_1 \oplus \mathbf{w}_2 = \begin{bmatrix} a_1 & b_1 \\ c_1 & d_1 \end{bmatrix} \oplus \begin{bmatrix} a_2 & b_2 \\ c_2 & d_2 \end{bmatrix} = \begin{bmatrix} a_1 + a_2 & b_1 + b_2 \\ c_1 + c_2 & d_1 + d_2 \end{bmatrix}$$

Since the trace of $\mathbf{w}_1 \oplus \mathbf{w}_2$ is

$$(a_1 + a_2) + (d_1 + d_2) = (a_1 + d_1) + (a_2 + d_2) = 0$$

then W is closed under addition. Also, for any scalar c,

$$c \odot \mathbf{w}_1 = c \odot \begin{bmatrix} a_1 & b_1 \\ c_1 & d_1 \end{bmatrix} = \begin{bmatrix} ca_1 & cb_1 \\ cc_1 & cd_1 \end{bmatrix}$$

The trace of this matrix is $ca_1 + cd_1 = c(a_1 + d_1) = 0$. Thus, W is also closed under scalar multiplication. Therefore, W is a subspace of $M_{2 \times 2}$.

EXAMPLE 3   Let W be the subset of $V = M_{n \times n}$ consisting of all symmetric matrices. Let the operations of addition and scalar multiplication on V be the standard operations. Show that W is a subspace of V.

Solution   Recall from Sec. 1.3 that a matrix A is symmetric provided that $A^t = A$. Let A and B be matrices in W and c be a real number. By Theorem 6 of Sec. 1.3,

$$(A \oplus B)^t = A^t \oplus B^t = A \oplus B \qquad \text{and} \qquad (c \odot A)^t = c \odot A^t = c \odot A$$

Thus, W is closed under addition and scalar multiplication, and consequently, by Theorem 3, W is a subspace.

EXAMPLE 4   Let $V = M_{n \times n}$ with the standard operations and W be the subset of V consisting of all idempotent matrices. Determine whether W is a subspace.

Solution   Recall that a matrix A is idempotent provided that $A^2 = A$ (See Exercise 42 of Sec. 1.3.) Let A be an element of W, so that $A^2 = A$. Then

$$(c \odot A)^2 = (cA)^2 = c^2 A^2 = c^2 A = c^2 \odot A$$

so that

$$(c \odot A)^2 = c \odot A \qquad \text{if and only if} \qquad c^2 = c$$

Since this is not true for all values of c, then W is not closed under scalar multiplication and is not a subspace.

The two closure criteria for a subspace can be combined into one as stated in Theorem 4.

THEOREM 4

A nonempty subset W of a vector space V is a subspace of V if and only if for each pair of vectors $\mathbf{u}$ and $\mathbf{v}$ in W and each scalar c, the vector $\mathbf{u} \oplus (c \odot \mathbf{v})$ is in W.

Proof   Let W be a nonempty subset of V, and suppose that $\mathbf{u} \oplus (c \odot \mathbf{v})$ belongs to W for all vectors $\mathbf{u}$ and $\mathbf{v}$ in W and all scalars c. By Theorem 3 it suffices to show that W is closed under addition and scalar multiplication. Suppose that $\mathbf{u}$ and $\mathbf{v}$ are in W; then $\mathbf{u} \oplus (1 \odot \mathbf{v}) = \mathbf{u} \oplus \mathbf{v}$ is in W, so that W is closed under addition. Next, since W is nonempty, let $\mathbf{u}$ be any vector in W. Then $\mathbf{0} = \mathbf{u} \oplus [(-1) \odot \mathbf{u}]$, so that the zero vector is in W. Now, if c is any scalar, then $c \odot \mathbf{u} = \mathbf{0} \oplus (c \odot \mathbf{u})$ and hence is in W. Therefore, W is also closed under scalar multiplication.

Conversely, if W is a subspace with $\mathbf{u}$ and $\mathbf{v}$ in W, and c a scalar, then since W is closed under addition and scalar multiplication, we know that $\mathbf{u} \oplus (c \odot \mathbf{v})$ is in W.

EXAMPLE 5

Let W be the subset of $\mathbb{R}^3$ defined by

$$W = \left\{ \begin{bmatrix} 3t \\ 0 \\ -2t \end{bmatrix} \, \middle| \, t \in \mathbb{R} \right\}$$

Use Theorem 4 to show that W is a subspace.

Solution   Let $\mathbf{u}$ and $\mathbf{v}$ be vectors in W and c be a real number. Then there are real numbers p and q such that

$$\mathbf{u} \oplus (c \odot \mathbf{v}) = \begin{bmatrix} 3p \\ 0 \\ -2p \end{bmatrix} \oplus \left(c \odot \begin{bmatrix} 3q \\ 0 \\ -2q \end{bmatrix} \right)$$

$$= \begin{bmatrix} 3(p+cq) \\ 0 \\ -2(p+cq) \end{bmatrix}$$

As this vector is in W, by Theorem 4, W is a subspace.

Alternatively, the set W can be written as

$$W = \left\{ t \begin{bmatrix} 3 \\ 0 \\ -2 \end{bmatrix} \, \middle| \, t \in \mathbb{R} \right\}$$

which is a line through the origin in $\mathbb{R}^3$.

We now consider what happens when subspaces are combined. In particular, let W_1 and W_2 be subspaces of a vector space V. Then the intersection $W_1 \cap W_2$ is also a subspace of V. To show this, let **u** and **v** be elements of $W_1 \cap W_2$ and let c be a scalar. Since W_1 and W_2 are both subspaces, then by Theorem 4, $\mathbf{u} \oplus (c \odot \mathbf{v})$ is in W_1 and is in W_2, and hence is in the intersection. Applying Theorem 4 again, we have that $W_1 \cap W_2$ is a subspace.

The extension to an arbitrary number of subspaces is stated in Theorem 5.

THEOREM 5

The intersection of any collection of subspaces of a vector space is a subspace of the vector space.

Example 6 shows that the union of two subspaces need not be a subspace.

EXAMPLE 6   Let W_1 and W_2 be the subspaces of $\mathbb{R}^2$ with the standard operations given by

$$W_1 = \left\{ \begin{bmatrix} x \\ 0 \end{bmatrix} \,\middle|\, x \in \mathbb{R} \right\} \qquad \text{and} \qquad W_2 = \left\{ \begin{bmatrix} 0 \\ y \end{bmatrix} \,\middle|\, y \in \mathbb{R} \right\}$$

Show that $W_1 \cup W_2$ is not a subspace.

Solution   The subspaces W_1 and W_2 consist of all vectors that lie on the x axis and the y axis, respectively. Their union is the collection of all vectors that lie on either axis and is given by

$$W_1 \cup W_2 = \left\{ \begin{bmatrix} x \\ y \end{bmatrix} \,\middle|\, x = 0 \text{ or } y = 0 \right\}$$

This set is not closed under addition since

$$\begin{bmatrix} 1 \\ 0 \end{bmatrix} \oplus \begin{bmatrix} 0 \\ 1 \end{bmatrix} = \begin{bmatrix} 1 \\ 1 \end{bmatrix}$$

which is not in $W_1 \cup W_2$, as shown in Fig. 2.

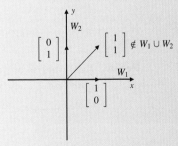

Figure 2

Span of a Set of Vectors

Subspaces of a vector space can be constructed by collecting all linear combinations of a set of vectors from the space. These subspaces are used to analyze certain properties of the vector space. A linear combination is defined in abstract vector spaces exactly as it is defined in $\mathbb{R}^n$ in Chap. 2.

DEFINITION 2   **Linear Combination**   Let $S = \{\mathbf{v}_1, \mathbf{v}_2, \ldots, \mathbf{v}_k\}$ be a set of vectors in a vector space V, and let $c_1, c_2, \ldots, c_k$ be scalars. A **linear combination** of the vectors of S is an expression of the form

$$(c_1 \odot \mathbf{v}_1) \oplus (c_2 \odot \mathbf{v}_2) \oplus \cdots \oplus (c_k \odot \mathbf{v}_k)$$

When the operations of vector addition and scalar multiplication are clear, we will drop the use of the symbols $\oplus$ and $\odot$. For example, the linear combination given in Definition 2 will be written as

$$c_1 \mathbf{v}_1 + c_2 \mathbf{v}_2 + \cdots + c_k \mathbf{v}_k = \sum_{i=1}^{k} c_i \mathbf{v}_i$$

Unless otherwise stated, the operations on the vector spaces $\mathbb{R}^n$, $M_{m \times n}$, $\mathcal{P}_n$, and their subspaces are the standard operations. Care is still needed when interpreting expressions defining linear combinations to distinguish between vector space operations and addition and multiplication of real numbers.

DEFINITION 3   **Span of a Set of Vectors**   Let V be a vector space and let $S = \{\mathbf{v}_1, \ldots, \mathbf{v}_n\}$ be a (finite) set of vectors in V. The **span** of S, denoted by **span**(S), is the set

$$\mathbf{span}(S) = \{c_1 \mathbf{v}_1 + c_2 \mathbf{v}_2 + \cdots + c_n \mathbf{v}_n \mid c_1, c_2, \ldots, c_n \in \mathbb{R}\}$$

PROPOSITION 1   If $S = \{\mathbf{v}_1, \mathbf{v}_2, \ldots, \mathbf{v}_n\}$ is a set of vectors in a vector space V, then **span**(S) is a subspace.

Proof   Let $\mathbf{u}$ and $\mathbf{w}$ be vectors in **span**(S) and c a scalar. Then there are scalars $c_1, \ldots, c_n$ and $d_1, \ldots, d_n$ such that

$$\mathbf{u} + c\mathbf{w} = (c_1 \mathbf{v}_1 + \cdots + c_n \mathbf{v}_n) + c(d_1 \mathbf{v}_1 + \cdots + d_n \mathbf{v}_n)$$
$$= (c_1 + cd_1)\mathbf{v}_1 + \cdots + (c_n + cd_n)\mathbf{v}_n$$

Therefore, $\mathbf{u} + c\mathbf{w}$ is in **span**(S), and hence the span is a subspace.

EXAMPLE 7   Let S be the subset of the vector space $\mathbb{R}^3$ defined by

$$S = \left\{ \begin{bmatrix} 2 \\ -1 \\ 0 \end{bmatrix}, \begin{bmatrix} 1 \\ 3 \\ -2 \end{bmatrix}, \begin{bmatrix} 1 \\ 1 \\ 4 \end{bmatrix} \right\}$$

Show that

$$\mathbf{v} = \begin{bmatrix} -4 \\ 4 \\ -6 \end{bmatrix}$$

is in **span**(S).

Solution   To determine if $\mathbf{v}$ is in the span of S, we consider the equation

$$c_1 \begin{bmatrix} 2 \\ -1 \\ 0 \end{bmatrix} + c_2 \begin{bmatrix} 1 \\ 3 \\ -2 \end{bmatrix} + c_3 \begin{bmatrix} 1 \\ 1 \\ 4 \end{bmatrix} = \begin{bmatrix} -4 \\ 4 \\ -6 \end{bmatrix}$$

Solving this linear system, we obtain

$$c_1 = -2 \qquad c_2 = 1 \qquad \text{and} \qquad c_3 = -1$$

This shows that $\mathbf{v}$ is a linear combination of the vectors in S and is thus in **span**(S).

The span of a single nonzero vector in $\mathbb{R}^n$ is a line through the origin, and the span of two linearly independent vectors is a plane through the origin as shown in Fig. 3.

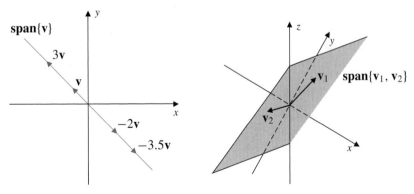

Figure 3

Since every line through the origin in $\mathbb{R}^2$ and $\mathbb{R}^3$, and every plane through the origin in $\mathbb{R}^3$, can be written as the span of vectors, these sets are subspaces.

EXAMPLE 8   Let

$$S = \left\{ \begin{bmatrix} 1 & 0 \\ 0 & 0 \end{bmatrix}, \begin{bmatrix} 0 & 1 \\ 1 & 0 \end{bmatrix}, \begin{bmatrix} 0 & 0 \\ 0 & 1 \end{bmatrix} \right\}$$

Show that the span of S is the subspace of $M_{2 \times 2}$ of all symmetric matrices.

Solution   Recall that a 2×2 matrix is symmetric provided that it has the form

$$\begin{bmatrix} a & b \\ b & c \end{bmatrix}$$

Since any matrix in **span**(S) has the form

$$a \begin{bmatrix} 1 & 0 \\ 0 & 0 \end{bmatrix} + b \begin{bmatrix} 0 & 1 \\ 1 & 0 \end{bmatrix} + c \begin{bmatrix} 0 & 0 \\ 0 & 1 \end{bmatrix} = \begin{bmatrix} a & b \\ b & c \end{bmatrix}$$

span(S) is the collection of all 2×2 symmetric matrices.

EXAMPLE 9   Show that

$$\mathbf{span} \left\{ \begin{bmatrix} 1 \\ 1 \\ 1 \end{bmatrix}, \begin{bmatrix} 1 \\ 0 \\ 2 \end{bmatrix}, \begin{bmatrix} 1 \\ 1 \\ 0 \end{bmatrix} \right\} = \mathbb{R}^3$$

Solution   Let

$$\mathbf{v} = \begin{bmatrix} a \\ b \\ c \end{bmatrix}$$

be an arbitrary element of $\mathbb{R}^3$. The vector $\mathbf{v}$ is in **span**(S) provided that there are scalars c_1, c_2, and c_3 such that

$$c_1 \begin{bmatrix} 1 \\ 1 \\ 1 \end{bmatrix} + c_2 \begin{bmatrix} 1 \\ 0 \\ 2 \end{bmatrix} + c_3 \begin{bmatrix} 1 \\ 1 \\ 0 \end{bmatrix} = \begin{bmatrix} a \\ b \\ c \end{bmatrix}$$

This linear system in matrix form is given by

$$\left[\begin{array}{ccc|c} 1 & 1 & 1 & a \\ 1 & 0 & 1 & b \\ 1 & 2 & 0 & c \end{array} \right]$$

After row-reducing, we obtain

$$\left[\begin{array}{ccc|c} 1 & 0 & 0 & -2a + 2b + c \\ 0 & 1 & 0 & a - b \\ 0 & 0 & 1 & 2a - b - c \end{array} \right]$$

From this final augmented matrix the original system is consistent, having solution $c_1 = -2a + 2b + c$, $c_2 = a - b$, and $c_3 = 2a - b - c$, for all choices of a, b, and c. Thus, every vector in $\mathbb{R}^3$ can be written as a linear combination of the three given vectors. Hence, the span of the three vectors is all of $\mathbb{R}^3$.

EXAMPLE 10   Show that

$$\text{span} \left\{ \begin{bmatrix} -1 \\ 2 \\ 1 \end{bmatrix}, \begin{bmatrix} 4 \\ 1 \\ -3 \end{bmatrix}, \begin{bmatrix} -6 \\ 3 \\ 5 \end{bmatrix} \right\} \neq \mathbb{R}^3$$

Solution   We approach this problem in the same manner as in Example 9. In this case, however, the resulting linear system is not always consistent. We can see this by reducing the augmented matrix

$$\begin{bmatrix} -1 & 4 & -6 & | & a \\ 2 & 1 & 3 & | & b \\ 1 & -3 & 5 & | & c \end{bmatrix} \quad \text{to} \quad \begin{bmatrix} -1 & 4 & -6 & | & a \\ 0 & 9 & -9 & | & b + 2a \\ 0 & 0 & 0 & | & c + \frac{7}{9}a - \frac{1}{9}b \end{bmatrix}$$

This last augmented matrix shows that the original system is consistent only if $7a - b + 9c = 0$. This is the equation of a plane in 3-space, and hence the span is not all of $\mathbb{R}^3$. See Fig. 4.

Notice that the solution to the equation $7a - b + 9c = 0$ can be written in parametric form by letting $b = s$, $c = t$, and $a = \frac{1}{7}s - \frac{9}{7}t$, so that

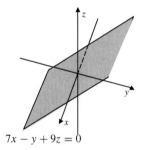

$$\text{span} \left\{ \begin{bmatrix} -1 \\ 2 \\ 1 \end{bmatrix}, \begin{bmatrix} 4 \\ 1 \\ -3 \end{bmatrix}, \begin{bmatrix} -6 \\ 3 \\ 5 \end{bmatrix} \right\} = \left\{ s \begin{bmatrix} \frac{1}{7} \\ 1 \\ 0 \end{bmatrix} + t \begin{bmatrix} -\frac{9}{7} \\ 0 \\ 1 \end{bmatrix} \,\middle|\, s, t \in \mathbb{R} \right\}$$

$7x - y + 9z = 0$

Figure 4

In this way, we see that the span is the subspace of all linear combinations of two linearly independent vectors, highlighting the geometric interpretation of the solution as a plane.

With Examples 9 and 10 we have completed the groundwork for the notion of a *basis*, which is central to linear algebra and is the subject of Sec. 3.3. Specifically, in Example 9, we saw that the set of vectors

$$S = \{\mathbf{v}_1, \ \mathbf{v}_2, \ \mathbf{v}_3\} = \left\{ \begin{bmatrix} 1 \\ 1 \\ 1 \end{bmatrix}, \begin{bmatrix} 1 \\ 0 \\ 2 \end{bmatrix}, \begin{bmatrix} 1 \\ 1 \\ 0 \end{bmatrix} \right\}$$

spans $\mathbb{R}^3$. These vectors are also linearly independent. To see this, observe that the matrix

$$A = \begin{bmatrix} 1 & 1 & 1 \\ 1 & 0 & 1 \\ 1 & 2 & 0 \end{bmatrix}$$

whose column vectors are the vectors of S, is row equivalent to the 3×3 identity matrix, as seen in the solution to Example 9. [Another way of showing that S is linearly independent is to observe that $\det(A) = 1 \neq 0$.] Consequently, by Theorem 7 of Sec. 2.3, we have that every vector in $\mathbb{R}^3$ can be written in only one way as a linear combination of the vectors of S.

On the other hand, the span of the set of vectors

$$S' = \{\mathbf{v}_1', \mathbf{v}_2', \mathbf{v}_3'\} = \left\{ \begin{bmatrix} -1 \\ 2 \\ 1 \end{bmatrix}, \begin{bmatrix} 4 \\ 1 \\ -3 \end{bmatrix}, \begin{bmatrix} -6 \\ 3 \\ 5 \end{bmatrix} \right\}$$

of Example 10 is a plane passing through the origin. Hence, not every vector in $\mathbb{R}^3$ can be written as a linear combination of the vectors in S'. As we expect, these vectors are linearly dependent since

$$\det\left(\begin{bmatrix} -1 & 4 & -6 \\ 2 & 1 & 3 \\ 1 & -3 & 5 \end{bmatrix} \right) = 0$$

In particular, $\mathbf{v}_3' = 2\mathbf{v}_1' - \mathbf{v}_2'$. The vectors $\mathbf{v}_1'$ and $\mathbf{v}_2'$ are linearly independent vectors which span the plane shown in Fig. 4, but not $\mathbb{R}^3$.

To pursue these notions a bit further, there are many sets of vectors which span $\mathbb{R}^3$. For example, the set

$$B = \{\mathbf{e}_1, \mathbf{e}_2, \mathbf{e}_3, \mathbf{v}\} = \left\{ \begin{bmatrix} 1 \\ 0 \\ 0 \end{bmatrix}, \begin{bmatrix} 0 \\ 1 \\ 0 \end{bmatrix}, \begin{bmatrix} 0 \\ 0 \\ 1 \end{bmatrix}, \begin{bmatrix} 1 \\ 2 \\ 3 \end{bmatrix} \right\}$$

spans $\mathbb{R}^3$, but by Theorem 3 of Sec. 2.3 must necessarily be linearly dependent. The ideal case, in terms of minimizing the number of vectors, is illustrated in Example 9 where the three linearly independent vectors of S span $\mathbb{R}^3$. In Sec. 3.3 we will see that S is a basis for $\mathbb{R}^3$, and that every basis for $\mathbb{R}^3$ consists of exactly three linearly independent vectors.

EXAMPLE 11   Show that the set of matrices

$$S = \left\{ \begin{bmatrix} -1 & 0 \\ 2 & 1 \end{bmatrix}, \begin{bmatrix} 1 & 1 \\ 1 & 0 \end{bmatrix} \right\}$$

does not span $M_{2\times2}$. Describe **span**(S).

Solution   The equation

$$c_1 \begin{bmatrix} -1 & 0 \\ 2 & 1 \end{bmatrix} + c_2 \begin{bmatrix} 1 & 1 \\ 1 & 0 \end{bmatrix} = \begin{bmatrix} a & b \\ c & d \end{bmatrix}$$

is equivalent to the linear system

$$\begin{cases} -c_1 + c_2 & = a \\ c_2 & = b \\ 2c_1 + c_2 & = c \\ c_1 & = d \end{cases}$$

From these equations we see that

$$c_1 = d \qquad \text{and} \qquad c_2 = b$$

which gives

$$a = b - d \qquad \text{and} \qquad c = b + 2d$$

Therefore,

$$\mathbf{span}(S) = \left\{ \begin{bmatrix} b-d & b \\ b+2d & d \end{bmatrix} \ \middle|\ b, d \in \mathbb{R} \right\}$$

EXAMPLE 12   Show that the set of polynomials

$$S = \{x^2 + 2x + 1, x^2 + 2, x\}$$

spans the vector space $\mathcal{P}_2$.

Solution   An arbitrary vector in $\mathcal{P}_2$ can be written in the form $ax^2 + bx + c$. To determine whether the $\mathbf{span}(S) = \mathcal{P}_2$, we consider the equation

$$c_1(x^2 + 2x + 1) + c_2(x^2 + 2) + c_3 x = ax^2 + bx + c$$

which simplifies to

$$(c_1 + c_2)x^2 + (2c_1 + c_3)x + (c_1 + 2c_2) = ax^2 + bx + c$$

Since two polynomials are equal if and only if the coefficients of like terms are equal, equating coefficients in the previous equation gives, in matrix form, the linear system

$$\begin{bmatrix} 1 & 1 & 0 & a \\ 2 & 0 & 1 & b \\ 1 & 2 & 0 & c \end{bmatrix}$$

This matrix reduces to

$$\begin{bmatrix} 1 & 0 & 0 & 2a - c \\ 0 & 1 & 0 & -a + c \\ 0 & 0 & 1 & -4a + b + 2c \end{bmatrix}$$

Hence, the linear system has the unique solution $c_1 = 2a - c$, $c_2 = -a + c$, and $c_3 = -4a + b + 2c$, for all a, b, and c. Therefore, $\mathbf{span}(S) = \mathcal{P}_2$.

The Null Space and Column Space of a Matrix

Two special subspaces associated with every matrix A are the *null space* and *column space* of the matrix.

DEFINITION 4

Null Space and Column Space   Let A be an $m \times n$ matrix.

1. The **null space** of A, denoted by $N(A)$, is the set of all vectors in $\mathbb{R}^n$ such that $A\mathbf{x} = \mathbf{0}$.
2. The **column space** of A, denoted by $\mathbf{col}(A)$, is the set of all linear combinations of the column vectors of A.

Observe that $N(A)$ is a subset of $\mathbb{R}^n$ and $\mathbf{col}(A)$ is a subset of $\mathbb{R}^m$. Moreover, by Proposition 1, $\mathbf{col}(A)$ is a subspace of $\mathbb{R}^m$. Using this terminology, we give a restatement of Theorem 2 of Sec. 2.2.

THEOREM 6

Let A be an $m \times n$ matrix. The linear system $A\mathbf{x} = \mathbf{b}$ is consistent if and only if $\mathbf{b}$ is in the column space of A.

EXAMPLE 13   Let

$$A = \begin{bmatrix} 1 & -1 & -2 \\ -1 & 2 & 3 \\ 2 & -2 & -2 \end{bmatrix} \quad \text{and} \quad \mathbf{b} = \begin{bmatrix} 3 \\ 1 \\ -2 \end{bmatrix}$$

a. Determine whether $\mathbf{b}$ is in $\mathbf{col}(A)$.
b. Find $N(A)$.

Solution

a. By Theorem 6, the vector $\mathbf{b}$ is in $\mathbf{col}(A)$ if and only if there is a vector $\mathbf{x}$ such that $A\mathbf{x} = \mathbf{b}$. The corresponding augmented matrix is given by

$$\begin{bmatrix} 1 & -1 & -2 & 3 \\ -1 & 2 & 3 & 1 \\ 2 & -2 & -2 & -2 \end{bmatrix} \quad \text{which reduces to} \quad \begin{bmatrix} 1 & 0 & 0 & 3 \\ 0 & 1 & 0 & 8 \\ 0 & 0 & 1 & -4 \end{bmatrix}$$

Hence, the linear system $A\mathbf{x} = \mathbf{b}$ is consistent, and vector $\mathbf{b}$ is in $\mathbf{col}(A)$. Specifically,

$$\begin{bmatrix} 3 \\ 1 \\ -2 \end{bmatrix} = 3 \begin{bmatrix} 1 \\ -1 \\ 2 \end{bmatrix} + 8 \begin{bmatrix} -1 \\ 2 \\ -2 \end{bmatrix} - 4 \begin{bmatrix} -2 \\ 3 \\ -2 \end{bmatrix}$$

b. To find the null space of A, we solve the homogeneous equation $A\mathbf{x} = \mathbf{0}$. The corresponding augmented matrix for this linear system is the same as in part (a), except for the right column that consists of three zeros. Consequently, the only solution is the trivial solution and hence $N(A) = \{\mathbf{0}\}$.

In Theorem 7 we show that the null space of a matrix also is a subspace.

THEOREM 7

Let A be an $m \times n$ matrix. Then the null space of A is a subspace of $\mathbb{R}^n$.

Proof   The null space of A is nonempty since $\mathbf{0}$ is in $N(A)$. That is, $A\mathbf{0} = \mathbf{0}$. Now let $\mathbf{u}$ and $\mathbf{v}$ be vectors in $N(A)$ and c a scalar. Then

$$\begin{aligned} A(\mathbf{u} + c\mathbf{v}) &= A\mathbf{u} + A(c\mathbf{v}) \\ &= A\mathbf{u} + cA(\mathbf{v}) \\ &= \mathbf{0} + c\mathbf{0} = \mathbf{0} \end{aligned}$$

Hence, $\mathbf{u} + c\mathbf{v}$ is in $N(A)$, and therefore by Theorem 4, $N(A)$ is a subspace.

Fact Summary

Let V be a vector space and W a nonempty subset of V.

1. To verify that W is a subspace of V, show that $\mathbf{u} \oplus c \odot \mathbf{v}$ is in W for any $\mathbf{u}$ and $\mathbf{v}$ in W and any scalar c.
2. The span of a set of vectors from V is a subspace.
3. The span of a single nonzero vector in $\mathbb{R}^2$ or $\mathbb{R}^3$ is a line that passes through the origin. The span of two linearly independent vectors in $\mathbb{R}^3$ is a plane that passes through the origin. These sets are subspaces.
4. The intersection of subspaces is a subspace. The union of two subspaces may not be a subspace.
5. If A is an $m \times n$ matrix, the null space of A is a subspace of $\mathbb{R}^n$ and the column space of A is a subspace of $\mathbb{R}^m$.
6. The linear system $A\mathbf{x} = \mathbf{b}$ is consistent if and only if $\mathbf{b}$ is in the column space of A.

Exercise Set 3.2

In Exercises 1–6, determine whether the subset S of $\mathbb{R}^2$ is a subspace. If S is not a subspace, find vectors $\mathbf{u}$ and $\mathbf{v}$ in S such that $\mathbf{u} + \mathbf{v}$ is not in S; or a vector $\mathbf{u}$ and a scalar c such that $c\mathbf{u}$ is not in S.

1. $S = \left\{ \begin{bmatrix} 0 \\ y \end{bmatrix} \;\middle|\; y \in \mathbb{R} \right\}$

2. $S = \left\{ \begin{bmatrix} x \\ y \end{bmatrix} \;\middle|\; xy \geq 0 \right\}$

3. $S = \left\{ \begin{bmatrix} x \\ y \end{bmatrix} \;\middle|\; xy \leq 0 \right\}$

4. $S = \left\{ \begin{bmatrix} x \\ y \end{bmatrix} \;\middle|\; x^2 + y^2 \leq 1 \right\}$

5. $S = \left\{ \begin{bmatrix} x \\ 2x - 1 \end{bmatrix} \;\middle|\; x \in \mathbb{R} \right\}$

6. $S = \left\{ \begin{bmatrix} x \\ 3x \end{bmatrix} \;\middle|\; x \in \mathbb{R} \right\}$

In Exercises 7–10, determine whether the subset S of $\mathbb{R}^3$ is a subspace.

7. $S = \left\{ \begin{bmatrix} x_1 \\ x_2 \\ x_3 \end{bmatrix} \;\middle|\; x_1 + x_3 = -2 \right\}$

8. $S = \left\{ \begin{bmatrix} x_1 \\ x_2 \\ x_3 \end{bmatrix} \;\middle|\; x_1 x_2 x_3 = 0 \right\}$

9. $S = \left\{ \begin{bmatrix} s - 2t \\ s \\ t + s \end{bmatrix} \;\middle|\; s, t \in \mathbb{R} \right\}$

10. $S = \left\{ \begin{bmatrix} x_1 \\ 2 \\ x_3 \end{bmatrix} \;\middle|\; x_1, x_3 > 0 \right\}$

In Exercises 11–18, determine whether the subset S of $M_{2\times 2}$ is a subspace.

11. Let S be the set of all symmetric matrices.

12. Let S be the set of all idempotent matrices.

13. Let S be the set of all invertible matrices.

14. Let S be the set of all skew-symmetric matrices.

15. Let S be the set of all upper triangular matrices.

16. Let S be the set of all diagonal matrices.

17. Let S be the set of all matrices with $a_{22} = 0$.

18. Let S be the set of all matrices with $a_{11} + a_{22} = 0$.

In Exercises 19–24, determine whether the subset S of $\mathcal{P}_5$ is a subspace.

19. Let S be the set of all polynomials with degree equal to 3.

20. Let S be the set of all polynomials with even degree.

21. Let S be the set of all polynomials such that $p(0) = 0$.

22. Let S be the set of all polynomials of the form $p(x) = ax^2$.

23. Let S be the set of all polynomials of the form $p(x) = ax^2 + 1$.

24. Let S be the set of all polynomials of degree less than or equal to 4.

In Exercises 25 and 26, determine if the vector $\mathbf{v}$ is in the span of

$$S = \left\{ \begin{bmatrix} 1 \\ 1 \\ 0 \end{bmatrix}, \begin{bmatrix} -1 \\ -1 \\ 1 \end{bmatrix}, \begin{bmatrix} -1 \\ 2 \\ 0 \end{bmatrix} \right\}$$

25. $\mathbf{v} = \begin{bmatrix} 1 \\ -1 \\ 1 \end{bmatrix}$

26. $\mathbf{v} = \begin{bmatrix} -2 \\ 7 \\ -3 \end{bmatrix}$

In Exercises 27 and 28, determine if the matrix M is in the span of

$$S = \left\{ \begin{bmatrix} 1 & 1 \\ 0 & -1 \end{bmatrix}, \begin{bmatrix} 0 & 1 \\ 2 & 1 \end{bmatrix}, \begin{bmatrix} 1 & -1 \\ -4 & -3 \end{bmatrix} \right\}$$

27. $M = \begin{bmatrix} -2 & 1 \\ 6 & 5 \end{bmatrix}$

28. $M = \begin{bmatrix} 1 & 1 \\ 2 & -3 \end{bmatrix}$

In Exercises 29 and 30, determine if the polynomial $p(x)$ is in the span of

$$S = \{1 + x, x^2 - 2, 3x\}$$

29. $p(x) = 2x^2 - 6x - 11$

30. $3x^2 - x - 4$

In Exercises 31–36, give an explicit description of the span of S.

31. $S = \left\{ \begin{bmatrix} 2 \\ -1 \\ -2 \end{bmatrix}, \begin{bmatrix} 1 \\ 3 \\ -1 \end{bmatrix} \right\}$

32. $S = \left\{ \begin{bmatrix} 1 \\ 1 \\ 2 \end{bmatrix}, \begin{bmatrix} 2 \\ 3 \\ 1 \end{bmatrix}, \begin{bmatrix} 1 \\ 2 \\ -1 \end{bmatrix} \right\}$

33. $\left\{ \begin{bmatrix} 1 & 2 \\ 1 & 0 \end{bmatrix}, \begin{bmatrix} 1 & -1 \\ 0 & 1 \end{bmatrix} \right\}$

34. $S = \left\{ \begin{bmatrix} 1 & 0 \\ 1 & 0 \end{bmatrix}, \begin{bmatrix} 1 & 1 \\ -1 & -1 \end{bmatrix}, \begin{bmatrix} 0 & -1 \\ 1 & 1 \end{bmatrix} \right\}$

35. $S = \{x, (x + 1)^2, x^2 + 3x + 1\}$

36. $S = \{x^2 - 4, 2 - x, x^2 + x + 2\}$

In Exercises 37–40, a subset S of $\mathbb{R}^3$ is given.
  a. Find **span**(S).
  b. Is S linearly independent?

37. $S = \left\{ \begin{bmatrix} 2 \\ 1 \\ -1 \end{bmatrix}, \begin{bmatrix} 3 \\ 0 \\ -2 \end{bmatrix} \right\}$

38. $S = \left\{ \begin{bmatrix} 1 \\ 1 \\ 2 \end{bmatrix}, \begin{bmatrix} 0 \\ -1 \\ 1 \end{bmatrix}, \begin{bmatrix} 2 \\ 5 \\ 1 \end{bmatrix} \right\}$

39. $S = \left\{ \begin{bmatrix} 3 \\ 3 \\ 2 \end{bmatrix}, \begin{bmatrix} 0 \\ 1 \\ 0 \end{bmatrix}, \begin{bmatrix} 1 \\ -1 \\ -1 \end{bmatrix} \right\}$

40. $S = \left\{ \begin{bmatrix} 1 \\ 2 \\ 1 \end{bmatrix}, \begin{bmatrix} -1 \\ 0 \\ 3 \end{bmatrix}, \begin{bmatrix} 0 \\ 1 \\ 1 \end{bmatrix}, \begin{bmatrix} 2 \\ 1 \\ 1 \end{bmatrix} \right\}$

41. Let

$$S = \left\{ \begin{bmatrix} 1 \\ 2 \\ 2 \end{bmatrix}, \begin{bmatrix} -1 \\ 3 \\ -1 \end{bmatrix}, \begin{bmatrix} 1 \\ 2 \\ -1 \end{bmatrix}, \right.$$
$$\left. \begin{bmatrix} 0 \\ 6 \\ 1 \end{bmatrix}, \begin{bmatrix} -3 \\ 4 \\ 5 \end{bmatrix} \right\}$$

  a. Find **span**(S).
  b. Is S linearly independent?
  c. Let

$$T = \left\{ \begin{bmatrix} 1 \\ 2 \\ 2 \end{bmatrix}, \begin{bmatrix} -1 \\ 3 \\ -1 \end{bmatrix}, \begin{bmatrix} 1 \\ 2 \\ -1 \end{bmatrix}, \right.$$
$$\left. \begin{bmatrix} -3 \\ 4 \\ 5 \end{bmatrix} \right\}$$

  Is **span**(T) = $\mathbb{R}^3$? Is T linearly independent?
  d. Let

$$H = \left\{ \begin{bmatrix} 1 \\ 2 \\ 2 \end{bmatrix}, \begin{bmatrix} -1 \\ 3 \\ -1 \end{bmatrix}, \begin{bmatrix} 1 \\ 2 \\ -1 \end{bmatrix} \right\}$$

  Is **span**(H) = $\mathbb{R}^3$? Is H linearly independent?

42. Let

$$S = \left\{ \begin{bmatrix} 2 & -3 \\ 0 & 0 \end{bmatrix}, \begin{bmatrix} 1 & 1 \\ 1 & 0 \end{bmatrix}, \begin{bmatrix} -3 & 1 \\ 1 & 0 \end{bmatrix} \right\}$$

  a. Find **span**(S).
  b. Is S linearly independent?
  c. Let

$$T = \left\{ \begin{bmatrix} 2 & -3 \\ 0 & 0 \end{bmatrix}, \begin{bmatrix} 1 & 1 \\ 1 & 0 \end{bmatrix}, \begin{bmatrix} -3 & 1 \\ 1 & 0 \end{bmatrix}, \right.$$
$$\left. \begin{bmatrix} 0 & 0 \\ 0 & 1 \end{bmatrix} \right\}$$

  Is **span**(T) = $M_{2\times2}$? Is T linearly independent?

43. Let

$$S = \left\{1, x - 3, x^2 + 2x, 2x^2 + 3x + 5\right\}$$

a. Find **span**(S).

b. Is S linearly independent?

c. Show that $2x^2 + 3x + 5$ is a linear combination of the other three polynomials in S.

d. Let $T = \left\{1, x - 3, x^2 + 2x, x^3\right\}$. Is T linearly independent? Is **span**(T) = $\mathcal{P}_3$?

44. Let

$$S = \left\{ \begin{bmatrix} 2s - t \\ s \\ t \\ -s \end{bmatrix} \;\middle|\; s, t \in \mathbb{R} \right\}$$

a. Show that S is a subspace of $\mathbb{R}^4$.

b. Find two vectors that span S.

c. Are the two vectors found in part (b) linearly independent?

d. Is $S = \mathbb{R}^4$?

45. Let

$$S = \left\{ \begin{bmatrix} -s \\ s - 5t \\ 3t + 2s \end{bmatrix} \;\middle|\; s, t \in \mathbb{R} \right\}$$

a. Show that S is a subspace of $\mathbb{R}^3$.

b. Find a set of vectors that span S.

c. Are the two vectors found in part (b) linearly independent?

d. Is $S = \mathbb{R}^3$?

46. Let
$$S = \textbf{span} \left\{ \begin{bmatrix} 1 & 0 \\ 0 & 1 \end{bmatrix}, \begin{bmatrix} 1 & 0 \\ 1 & 0 \end{bmatrix}, \begin{bmatrix} 0 & 1 \\ 1 & 1 \end{bmatrix} \right\}$$

a. Describe the subspace S.

b. Is $S = M_{2 \times 2}$?

c. Are the three matrices that generate S linearly independent?

47. Let A be a 2 × 3 matrix and let

$$S = \left\{ \mathbf{x} \in \mathbb{R}^3 \;\middle|\; A\mathbf{x} = \begin{bmatrix} 1 \\ 2 \end{bmatrix} \right\}$$

Is S a subspace? Explain.

48. Let A be an $m \times n$ matrix and let

$$S = \left\{ \mathbf{x} \in \mathbb{R}^n \;\middle|\; A\mathbf{x} = \mathbf{0} \right\}$$

Is S a subspace? Explain.

49. Let A be a fixed $n \times n$ matrix and let

$$S = \{ B \in M_{n \times n} \;\mid\; AB = BA \}$$

Is S a subspace? Explain.

50. Suppose S and T are subspaces of a vector space V. Define

$$S + T = \{ \mathbf{u} + \mathbf{v} \mid \mathbf{u} \in S, \mathbf{v} \in T \}$$

Show that $S + T$ is a subspace of V.

51. Let $S = \textbf{span}(\{\mathbf{u}_1, \mathbf{u}_2, \ldots \mathbf{u}_m\})$ and $T = \textbf{span}(\{\mathbf{v}_1, \mathbf{v}_2, \ldots \mathbf{v}_n\})$ be subspaces of a vector space V. Show that

$$S + T = \textbf{span}(\{\mathbf{u}_1, \ldots \mathbf{u}_m, \mathbf{v}_1, \ldots \mathbf{v}_n\})$$

(See Exercise 50.)

52. Let

$$S = \left\{ \begin{bmatrix} x & -x \\ y & z \end{bmatrix} \;\middle|\; x, y, z \in \mathbb{R} \right\}$$

and

$$T = \left\{ \begin{bmatrix} a & b \\ -a & c \end{bmatrix} \;\middle|\; a, b, c \in \mathbb{R} \right\}$$

a. Show that S and T are subspaces.

b. Describe all matrices in $S + T$.

(See Exercises 50 and 51.)

3.3 ▶ Basis and Dimension

In Sec. 2.3 we introduced the notion of linear independence and its connection to the minimal sets that can be used to generate or span $\mathbb{R}^n$. In this section we explore this connection further and see how to determine whether a spanning set is minimal.

This leads to the concept of a basis for an abstract vector space. As a first step, we generalize the concept of linear independence to abstract vector spaces introduced in Sec. 3.1.

DEFINITION 1

Linear Independence and Linear Dependence   The set of vectors $S = \{v_1, v_2, \ldots, v_m\}$ in a vector space V is called **linearly independent** provided that the only solution to the equation

$$c_1 v_1 + c_2 v_2 + \cdots + c_m v_m = \mathbf{0}$$

is the trivial solution $c_1 = c_2 = \cdots = c_m = 0$. If the equation has a nontrivial solution, then the set S is called **linearly dependent**.

EXAMPLE 1

Let

$$v_1 = \begin{bmatrix} 1 \\ 0 \\ -1 \end{bmatrix} \qquad v_2 = \begin{bmatrix} 0 \\ 2 \\ 2 \end{bmatrix} \qquad v_3 = \begin{bmatrix} -3 \\ 4 \\ 7 \end{bmatrix}$$

and let $W = \mathbf{span}\{v_1, v_2, v_3\}$.

a. Show that v_3 is a linear combination of v_1 and v_2.

b. Show that $\mathbf{span}\{v_1, v_2\} = W$.

c. Show that v_1 and v_2 are linearly independent.

Solution

a. To solve the vector equation

$$c_1 \begin{bmatrix} 1 \\ 0 \\ -1 \end{bmatrix} + c_2 \begin{bmatrix} 0 \\ 2 \\ 2 \end{bmatrix} = \begin{bmatrix} -3 \\ 4 \\ 7 \end{bmatrix}$$

we row-reduce the corresponding augmented matrix for the linear system to obtain

$$\begin{bmatrix} 1 & 0 & -3 \\ 0 & 2 & 4 \\ -1 & 2 & 7 \end{bmatrix} \longrightarrow \begin{bmatrix} 1 & 0 & -3 \\ 0 & 1 & 2 \\ 0 & 0 & 0 \end{bmatrix}$$

The solution to the vector equation above is $c_1 = -3$ and $c_2 = 2$, therefore

$$v_3 = -3 v_1 + 2 v_2$$

Notice that the vector v_3 lies in the plane spanned by v_1 and v_2, as shown in Fig. 1.

Figure 1

b. From part (a) an element of $W = \{c_1 v_1 + c_2 v_2 + c_3 v_3 \mid c_1, c_2, c_3 \in \mathbb{R}\}$ can be written in the form

$$c_1 v_1 + c_2 v_2 + c_3 v_3 = c_1 v_1 + c_2 v_2 + c_3(-3 v_1 + 2 v_2)$$
$$= (c_1 - 3 c_3) v_1 + (c_2 + 2 c_3) v_2$$

and therefore, every vector in W is a linear combination of $\mathbf{v}_1$ and $\mathbf{v}_2$. As a result, the vector $\mathbf{v}_3$ is not needed to generate W, so that $\mathbf{span}\{\mathbf{v}_1, \mathbf{v}_2\} = W$.

c. Since neither vector is a scalar multiple of the other, the vectors $\mathbf{v}_1$ and $\mathbf{v}_2$ are linearly independent.

In Example 1, we were able to reduce the number of linearly dependent vectors that span W to a linearly independent set of vectors which also spans W. We accomplished this by eliminating the vector $\mathbf{v}_3$ from the set, which, as we saw in the solution, is a linear combination of the vectors $\mathbf{v}_1$ and $\mathbf{v}_2$ and hence does not affect the span. Theorem 8 gives a general description of the process.

THEOREM 8   Let $\mathbf{v}_1, \ldots, \mathbf{v}_n$ be vectors in a vector space V, and let $W = \mathbf{span}\{\mathbf{v}_1, \ldots, \mathbf{v}_n\}$. If $\mathbf{v}_n$ is a linear combination of $\mathbf{v}_1, \ldots, \mathbf{v}_{n-1}$, then

$$W = \mathbf{span}\{\mathbf{v}_1, \ldots, \mathbf{v}_{n-1}\}$$

Proof   If $\mathbf{v}$ is in $\mathbf{span}\{\mathbf{v}_1, \ldots, \mathbf{v}_{n-1}\}$, then there are scalars $c_1, c_2, \ldots, c_{n-1}$ such that $\mathbf{v} = c_1\mathbf{v}_1 + \cdots + c_{n-1}\mathbf{v}_{n-1}$. Then $\mathbf{v} = c_1\mathbf{v}_1 + \cdots + c_{n-1}\mathbf{v}_{n-1} + 0\mathbf{v}_n$, so that $\mathbf{v}$ is also in $\mathbf{span}\{\mathbf{v}_1, \ldots, \mathbf{v}_n\}$. Therefore,

$$\mathbf{span}\{\mathbf{v}_1, \ldots, \mathbf{v}_{n-1}\} \subseteq \mathbf{span}\{\mathbf{v}_1, \ldots, \mathbf{v}_n\}$$

Conversely, if $\mathbf{v}$ is in $\mathbf{span}\{\mathbf{v}_1, \ldots, \mathbf{v}_n\}$, then there are scalars $c_1, \ldots, c_n$ such that $\mathbf{v} = c_1\mathbf{v}_1 + \cdots + c_n\mathbf{v}_n$. Also, since $\mathbf{v}_n$ is a linear combination of $\mathbf{v}_1, \ldots, \mathbf{v}_{n-1}$, there are scalars $d_1, \ldots, d_{n-1}$ such that $\mathbf{v}_n = d_1\mathbf{v}_1 + \cdots + d_{n-1}\mathbf{v}_{n-1}$. Then

$$\begin{aligned}
\mathbf{v} &= c_1\mathbf{v}_1 + \cdots + c_{n-1}\mathbf{v}_{n-1} + c_n\mathbf{v}_n \\
&= c_1\mathbf{v}_1 + \cdots + c_{n-1}\mathbf{v}_{n-1} + c_n(d_1\mathbf{v}_1 + \cdots + d_{n-1}\mathbf{v}_{n-1}) \\
&= (c_1 + c_nd_1)\mathbf{v}_1 + \cdots + (c_{n-1} + c_nd_{n-1})\mathbf{v}_{n-1}
\end{aligned}$$

so that $\mathbf{v} \in \mathbf{span}\{\mathbf{v}_1, \ldots, \mathbf{v}_{n-1}\}$ and $\mathbf{span}\{\mathbf{v}_1, \ldots, \mathbf{v}_n\} \subseteq \mathbf{span}\{\mathbf{v}_1, \ldots, \mathbf{v}_{n-1}\}$. Therefore,

$$W = \mathbf{span}\{\mathbf{v}_1, \ldots, \mathbf{v}_n\} = \mathbf{span}\{\mathbf{v}_1, \ldots, \mathbf{v}_{n-1}\}$$

EXAMPLE 2   Compare the column spaces of the matrices

$$A = \begin{bmatrix} 1 & 0 & -1 & 1 \\ 2 & 0 & 1 & 7 \\ 1 & 1 & 2 & 7 \\ 3 & 4 & 1 & 5 \end{bmatrix} \quad \text{and} \quad B = \begin{bmatrix} 1 & 0 & -1 & 1 & 2 \\ 2 & 0 & 1 & 7 & -1 \\ 1 & 1 & 2 & 7 & 1 \\ 3 & 4 & 1 & 5 & -2 \end{bmatrix}$$

Solution   By using the methods presented in Chap. 2 it can be shown that the the column vectors of the matrix A are linearly independent. Since the column vectors of B consist of a set of five vectors in $\mathbb{R}^4$, by Theorem 3 of Sec. 2.3, the vectors are linearly dependent. In addition, the first four column vectors of B are the same as the linearly independent column vectors of A, hence by Theorem 5 of Sec. 2.3 the fifth column vector of B must be a linear combination of the other four vectors. Finally by Theorem 8, we know that $\mathbf{col}(A) = \mathbf{col}(B)$.

As a consequence of Theorem 8, a set of vectors $\{\mathbf{v}_1, \ldots, \mathbf{v}_n\}$ such that $V = \mathbf{span}\{\mathbf{v}_1, \ldots, \mathbf{v}_n\}$ is minimal, in the sense of the number of spanning vectors, when they are linearly independent. We also saw in Chap. 2 that when a vector in $\mathbb{R}^n$ can be written as a linear combination of vectors from a linearly independent set, then the representation is unique. The same result holds for abstract vector spaces.

THEOREM 9   If $B = \{\mathbf{v}_1, \mathbf{v}_2, \ldots, \mathbf{v}_m\}$ is a linearly independent set of vectors in a vector space V, then every vector in $\mathbf{span}(B)$ can be written uniquely as a linearly combination of vectors from B.

Motivated by these ideas, we now define what we mean by a basis of a vector space.

DEFINITION 2   **Basis for a Vector Space**   A subset B of a vector space V is a **basis** for V provided that

1. B is a linearly independent set of vectors in V
2. $\mathbf{span}(B) = V$

As an example, the set of coordinate vectors

$$S = \{\mathbf{e}_1, \ldots, \mathbf{e}_n\}$$

spans $\mathbb{R}^n$ and is linearly independent, so that S is a basis for $\mathbb{R}^n$. This particular basis is called the **standard basis** for $\mathbb{R}^n$. In Example 3 we give a basis for $\mathbb{R}^3$, which is not the standard basis.

EXAMPLE 3   Show that the set

$$B = \left\{ \begin{bmatrix} 1 \\ 1 \\ 0 \end{bmatrix}, \begin{bmatrix} 1 \\ 1 \\ 1 \end{bmatrix}, \begin{bmatrix} 0 \\ 1 \\ -1 \end{bmatrix} \right\}$$

is a basis for $\mathbb{R}^3$.

Solution   First, to show that S spans $\mathbb{R}^3$, we must show that the equation

$$c_1 \begin{bmatrix} 1 \\ 1 \\ 0 \end{bmatrix} + c_2 \begin{bmatrix} 1 \\ 1 \\ 1 \end{bmatrix} + c_3 \begin{bmatrix} 0 \\ 1 \\ -1 \end{bmatrix} = \begin{bmatrix} a \\ b \\ c \end{bmatrix}$$

has a solution for every choice of a, b, and c in $\mathbb{R}$. To solve this linear system, we reduce the corresponding augmented matrix

$$\begin{bmatrix} 1 & 1 & 0 & a \\ 1 & 1 & 1 & b \\ 0 & 1 & -1 & c \end{bmatrix} \quad \text{to} \quad AB = \begin{bmatrix} 1 & 0 & 0 & 2a - b - c \\ 0 & 1 & 0 & -a + b + c \\ 0 & 0 & 1 & -a + b \end{bmatrix}$$

Therefore, $c_1 = 2a - b - c$, $c_2 = -a + b + c$, and $c_3 = -a + b$. For example, suppose that $\mathbf{v} = \begin{bmatrix} 1 \\ 2 \\ 3 \end{bmatrix}$; then

$$c_1 = 2(1) - 2 - 3 = -3$$
$$c_2 = -1 + 2 + 3 = 4$$
$$c_3 = -1 + 2 = 1$$

so that

$$-3 \begin{bmatrix} 1 \\ 1 \\ 0 \end{bmatrix} + 4 \begin{bmatrix} 1 \\ 1 \\ 1 \end{bmatrix} + \begin{bmatrix} 0 \\ 1 \\ -1 \end{bmatrix} = \begin{bmatrix} 1 \\ 2 \\ 3 \end{bmatrix}$$

Since the linear system is consistent for all choices of a, b, and c, we know that **span**$(B) = \mathbb{R}^3$.

To show that B is linearly independent, we compute the determinant of the matrix whose column vectors are the vectors of B, that is,

$$\begin{vmatrix} 1 & 1 & 0 \\ 1 & 1 & 1 \\ 0 & 1 & -1 \end{vmatrix} = 1$$

Since this determinant is nonzero, by Theorem 9 of Sec. 2.3 the set B is linearly independent. Therefore, B is a basis for $\mathbb{R}^3$. Alternatively, to show that B is linearly independent, notice from the reduced matrix above that

$$A = \begin{bmatrix} 1 & 1 & 0 \\ 1 & 1 & 1 \\ 0 & 1 & -1 \end{bmatrix}$$

is row equivalent to I. Again by Theorem 9 of Sec. 2.3, B is linearly independent.

As we have already illustrated in the examples above, bases for a vector space are not unique. For example, consider the standard basis $B = \{\mathbf{e}_1, \mathbf{e}_2, \mathbf{e}_3\}$ for $\mathbb{R}^3$. Another basis for $\mathbb{R}^3$ is given by $B' = \{2\mathbf{e}_1, \mathbf{e}_2, \mathbf{e}_3\}$, where we have simply multiplied the first vector by 2.

Theorem 10 generalizes this idea, showing that an infinite family of bases can be derived from a given basis for a vector space by scalar multiplication.

THEOREM 10   Let $B = \{\mathbf{v}_1, \ldots, \mathbf{v}_n\}$ be a basis for a vector space V and c a nonzero scalar. Then $B_c = \{c\mathbf{v}_1, \mathbf{v}_2, \ldots, \mathbf{v}_n\}$ is a basis.

Proof   If $\mathbf{v}$ is an element of the vector space V, then since B is a basis there are scalars $c_1, \ldots, c_n$ such that $\mathbf{v} = c_1\mathbf{v}_1 + c_2\mathbf{v}_2 + \cdots + c_n\mathbf{v}_n$. But since $c \neq 0$, we can also write

$$\mathbf{v} = \frac{c_1}{c}(c\mathbf{v}_1) + c_2\mathbf{v}_2 + \cdots + c_n\mathbf{v}_n$$

so that $\mathbf{v}$ is a linear combination of the vectors in B_c. Thus, $\mathbf{span}(B_c) = V$. To show that B_c is linearly independent, consider the equation

$$c_1(c\mathbf{v}_1) + c_2\mathbf{v}_2 + \cdots + c_n\mathbf{v}_n = \mathbf{0}$$

By vector space axiom 9 we can write this as

$$(c_1 c)(\mathbf{v}_1) + c_2\mathbf{v}_2 + \cdots + c_n\mathbf{v}_n = \mathbf{0}$$

Now, since B is linearly independent, the only solution to the previous equation is the trivial solution

$$c_1 c = 0 \qquad c_2 = 0 \qquad \ldots \qquad c_n = 0$$

Since $c \neq 0$, then $c_1 = 0$. Therefore, B_c is linearly independent and hence is a basis.

EXAMPLE 4   Let W be the subspace of $M_{2\times2}$ of matrices with trace equal to 0, and let

$$S = \left\{ \begin{bmatrix} 1 & 0 \\ 0 & -1 \end{bmatrix}, \begin{bmatrix} 0 & 1 \\ 0 & 0 \end{bmatrix}, \begin{bmatrix} 0 & 0 \\ 1 & 0 \end{bmatrix} \right\}$$

Show that S is a basis for W.

Solution   In Example 2 of Sec. 3.2 we showed that W is a subspace of $M_{2\times2}$. To show that $\mathbf{span}(S) = W$, first recall that a matrix

$$A = \begin{bmatrix} a & b \\ c & d \end{bmatrix}$$

has trace 0 if and only if $a + d = 0$, so that A has the form

$$A = \begin{bmatrix} a & b \\ c & -a \end{bmatrix}$$

Since for every such matrix

$$\begin{bmatrix} a & b \\ c & -a \end{bmatrix} = a \begin{bmatrix} 1 & 0 \\ 0 & -1 \end{bmatrix} + b \begin{bmatrix} 0 & 1 \\ 0 & 0 \end{bmatrix} + c \begin{bmatrix} 0 & 0 \\ 1 & 0 \end{bmatrix}$$

then **span**$(S) = W$. We also know that S is linearly independent since the linear system

$$c_1 \begin{bmatrix} 1 & 0 \\ 0 & -1 \end{bmatrix} + c_2 \begin{bmatrix} 0 & 1 \\ 0 & 0 \end{bmatrix} + c_3 \begin{bmatrix} 0 & 0 \\ 1 & 0 \end{bmatrix} = \begin{bmatrix} 0 & 0 \\ 0 & 0 \end{bmatrix}$$

is equivalent to

$$\begin{bmatrix} c_1 & c_2 \\ c_3 & -c_1 \end{bmatrix} = \begin{bmatrix} 0 & 0 \\ 0 & 0 \end{bmatrix}$$

which has only the trivial solution $c_1 = c_2 = c_3 = 0$. Thus, S is a basis for W.

Similar to the situation for $\mathbb{R}^n$, there is a natural set of matrices in $M_{m \times n}$ that comprise a standard basis. Let $\mathbf{e}_{ij}$ be the matrix with a one in the ij position, and 0s elsewhere. The set $S = \{\mathbf{e}_{ij} \mid 1 \leq i \leq m, 1 \leq j \leq n\}$ is the **standard basis for** $M_{m \times n}$. For example, the standard basis for $M_{2 \times 2}$ is

$$S = \left\{ \begin{bmatrix} 1 & 0 \\ 0 & 0 \end{bmatrix}, \begin{bmatrix} 0 & 1 \\ 0 & 0 \end{bmatrix}, \begin{bmatrix} 0 & 0 \\ 1 & 0 \end{bmatrix}, \begin{bmatrix} 0 & 0 \\ 0 & 1 \end{bmatrix} \right\}$$

EXAMPLE 5   Determine whether

$$B = \left\{ \begin{bmatrix} 1 & 3 \\ 2 & 1 \end{bmatrix}, \begin{bmatrix} -1 & 2 \\ 1 & 0 \end{bmatrix}, \begin{bmatrix} 0 & 1 \\ 0 & -4 \end{bmatrix} \right\}$$

is a basis for $M_{2 \times 2}$.

Solution   Let

$$A = \begin{bmatrix} a & b \\ c & d \end{bmatrix}$$

be an arbitrary matrix in $M_{2 \times 2}$. To see if A is in the span of S, we consider the equation

$$c_1 \begin{bmatrix} 1 & 3 \\ 2 & 1 \end{bmatrix} + c_2 \begin{bmatrix} -1 & 2 \\ 1 & 0 \end{bmatrix} + c_3 \begin{bmatrix} 0 & 1 \\ 0 & -4 \end{bmatrix} = \begin{bmatrix} a & b \\ c & d \end{bmatrix}$$

The augmented matrix corresponding to this equation is given by

$$\begin{bmatrix} 1 & -1 & 0 & a \\ 3 & 2 & 1 & b \\ 2 & 1 & 0 & c \\ 1 & 0 & -4 & d \end{bmatrix}$$

After row-reducing, we obtain

$$\begin{bmatrix} 1 & -1 & 0 & a \\ 0 & 5 & 1 & -3a + b \\ 0 & 0 & -3 & -a - 3b + 5c \\ 0 & 0 & 0 & a + 4b - 7c + d \end{bmatrix}$$

Observe that the above linear system is consistent only if $a + 4b - 7c + d = 0$. Hence, B does not span $M_{2 \times 2}$, and therefore is not a basis.

Notice that in Example 5 the set B is linearly independent, but the three mat do not span the set of all 2×2 matrices. We will see that the minimal number of matrices required to span $M_{2 \times 2}$ is four.

Another vector space we have already considered is $\mathcal{P}_n$, the vector space of polynomials of degree less than or equal to n. The **standard basis for** $\mathcal{P}_n$ is the set

$$B = \{1, x, x^2, \ldots, x^n\}$$

Indeed, if $p(x) = a_0 + a_1x + a_2x^2 + \cdots + a_nx^n$ is any polynomial in $\mathcal{P}_n$, then it is a linear combination of the vectors in B, so $\mathbf{span}(B) = \mathcal{P}_n$. To show that B is linearly independent, suppose that

$$c_0 + c_1x + c_2x^2 + \cdots + c_nx^n = 0$$

for all real numbers x. We can write this equation as

$$c_0 + c_1x + c_2x^2 + \cdots + c_nx^n = 0 + 0x + 0x^2 + \cdots + 0x^n$$

Since two polynomials are identical if and only if the coefficients of like terms are equal, then $c_1 = c_2 = c_3 = \cdots = c_n = 0$.

Another basis for $\mathcal{P}_2$ is given in Example 6.

EXAMPLE 6

Show that $B = \{x + 1, x - 1, x^2\}$ is a basis for $\mathcal{P}_2$.

Solution

Let $ax^2 + bx + c$ be an arbitrary polynomial in $\mathcal{P}_2$. To verify that B spans $\mathcal{P}_2$, we must show that scalars c_1, c_2, and c_3 can be found such that

$$c_1(x + 1) + c_2(x - 1) + c_3x^2 = ax^2 + bx + c$$

for every choice of a, b, and c. Collecting like terms on the left-hand side gives

$$c_3x^2 + (c_1 + c_2)x + (c_1 - c_2) = ax^2 + bx + c$$

Equating coefficients on both sides, we obtain $c_3 = a$, $c_1 + c_2 = b$ and $c_1 - c_2 = c$. This linear system has the unique solution

$$c_1 = \tfrac{1}{2}(b + c) \qquad c_2 = \tfrac{1}{2}(b - c) \qquad c_3 = a$$

Therefore, $\mathbf{span}(B) = \mathcal{P}_2$. To show linear independence, we consider the equation

$$c_1(x + 1) + c_2(x - 1) + c_3x^2 = 0 + 0x + 0x^2$$

Observe that the solution above holds for all choices of a, b, and c, so that $c_1 = c_2 = c_3 = 0$. Therefore, the set B is also linearly independent and hence is a basis.

Another way of showing that the set B of Example 6 is a basis for $\mathcal{P}_2$ is to show that the polynomials of the standard basis can be written as linear combinations of the polynomials in B. Specifically, we have

$$1 = \tfrac{1}{2}(x + 1) - \tfrac{1}{2}(x - 1) \qquad x = \tfrac{1}{2}(x + 1) + \tfrac{1}{2}(x - 1)$$

and x^2 is already in B.

Dimension

We have already seen in Theorem 3 of Sec. 2.3 that any set of m vectors from $\mathbb{R}^n$, with $m > n$, must necessarily be linearly dependent. Hence, any basis of $\mathbb{R}^n$ contains at most n vectors. It can also be shown that any linearly independent set of m vectors, with $m < n$, does not span $\mathbb{R}^n$. For example, as we have already seen, two linearly independent vectors in $\mathbb{R}^3$ span a plane. Hence, any basis of $\mathbb{R}^n$ must contain exactly n vectors. The number n, an invariant of $\mathbb{R}^n$, is called the *dimension* of $\mathbb{R}^n$. Theorem 11 shows that this holds for abstract vector spaces.

THEOREM 11   If a vector space V has a basis with n vectors, then every basis has n vectors.

Proof   Let $B = \{\mathbf{v}_1, \mathbf{v}_2, \ldots, \mathbf{v}_n\}$ be a basis for V, and let $T = \{\mathbf{u}_1, \mathbf{u}_2, \ldots, \mathbf{u}_m\}$ be a subset of V with $m > n$. We claim that T is linearly dependent. To establish this result, observe that since B is a basis, then every vector in T can be written as a linear combination of the vectors from B. That is,

$$\mathbf{u}_1 = \lambda_{11}\mathbf{v}_1 + \lambda_{12}\mathbf{v}_2 + \cdots + \lambda_{1n}\mathbf{v}_n$$
$$\mathbf{u}_2 = \lambda_{21}\mathbf{v}_1 + \lambda_{22}\mathbf{v}_2 + \cdots + \lambda_{2n}\mathbf{v}_n$$
$$\vdots$$
$$\mathbf{u}_m = \lambda_{m1}\mathbf{v}_1 + \lambda_{m2}\mathbf{v}_2 + \cdots + \lambda_{mn}\mathbf{v}_n$$

Now consider the equation

$$c_1\mathbf{u}_1 + c_2\mathbf{u}_2 + \cdots + c_m\mathbf{u}_m = \mathbf{0}$$

Using the equations above, we can write this last equation in terms of the basis vectors. After collecting like terms, we obtain

$$(c_1\lambda_{11} + c_2\lambda_{21} + \cdots + c_m\lambda_{m1})\mathbf{v}_1$$
$$+ (c_1\lambda_{12} + c_2\lambda_{22} + \cdots + c_m\lambda_{m2})\mathbf{v}_2$$
$$\vdots$$
$$+ (c_1\lambda_{1n} + c_2\lambda_{2n} + \cdots + c_m\lambda_{mn})\mathbf{v}_n = \mathbf{0}$$

Since B is a basis, it is linearly independent, hence

$$c_1\lambda_{11} + c_2\lambda_{21} + \cdots + c_m\lambda_{m1} = 0$$
$$c_1\lambda_{12} + c_2\lambda_{22} + \cdots + c_m\lambda_{m2} = 0$$
$$\vdots$$
$$c_1\lambda_{1n} + c_2\lambda_{2n} + \cdots + c_m\lambda_{mn} = 0$$

This last linear system is not square with n equations in the m variables $c_1, \ldots, c_m$. Since $m > n$, by Theorem 3 of Sec. 2.3 the linear system has a nontrivial solution, and hence T is linearly dependent.

Now, suppose that $T = \{\mathbf{u}_1, \mathbf{u}_2, \ldots, \mathbf{u}_m\}$ is another basis for the vector space V. By the result we just established it must be the case that $m \leq n$. But by the same reasoning, the number of vectors in the basis B also cannot exceed the number of vectors in T, so $n \leq m$. Consequently $n = m$ as desired.

We can now give a definition for the dimension of an abstract vector space.

DEFINITION 3 **Dimension of a Vector Space** The **dimension** of the vector space V, denoted by $\dim(V)$, is the number of vectors in any basis of V.

For example, since the standard bases for $\mathbb{R}^n$, $M_{2 \times 2}$, $M_{m \times n}$, and $\mathcal{P}_n$ are

$$\{\mathbf{e}_1, \mathbf{e}_2, \ldots, \mathbf{e}_n\}$$
$$\{\mathbf{e}_{11}, \mathbf{e}_{12}, \mathbf{e}_{21}, \mathbf{e}_{22}\}$$
$$\{\mathbf{e}_{ij} \mid 1 \leq i \leq m, i \leq j \leq n\}$$
$$\{1, x, x^2, \ldots, x^n\}$$

respectively, we have

$$\dim(\mathbb{R}^n) = n \qquad \dim(M_{2 \times 2}) = 4 \qquad \dim(M_{m \times n}) = mn \qquad \dim(\mathcal{P}_n) = n + 1$$

We call a vector space V **finite dimensional** if there exists a basis for V with a finite number of vectors. If such a basis does not exist, then V is called **infinite dimensional**. The trivial vector space $V = \{\mathbf{0}\}$ is considered finite dimensional, with $\dim(V) = 0$, even though it does not have a basis. In this text our focus is on finite dimensional vector spaces, although infinite dimensional vector spaces arise naturally in many areas of science and mathematics.

To determine whether a set of n vectors from a vector space of dimension n is or is not a basis, it is sufficient to verify either that the set spans the vector space or that the set is linearly independent.

THEOREM 12 Suppose that V is a vector space with $\dim(V) = n$.

1. If $S = \{\mathbf{v}_1, \mathbf{v}_2, \ldots, \mathbf{v}_n\}$ is linearly independent, then $\mathbf{span}(S) = V$ and S is a basis.
2. If $S = \{\mathbf{v}_1, \mathbf{v}_2, \ldots, \mathbf{v}_n\}$ and $\mathbf{span}(S) = V$, then S is linearly independent and S is a basis.

Proof (1) Suppose that S is linearly independent, and let $\mathbf{v}$ be any vector in V. If $\mathbf{v}$ is in S, then $\mathbf{v}$ is in $\mathbf{span}(S)$. Now suppose that $\mathbf{v}$ is not in S. As in the proof of

Theorem 11, the set $\{\mathbf{v}, \mathbf{v}_1, \mathbf{v}_2, \ldots, \mathbf{v}_n\}$ is linearly dependent. Thus, there are scalars $c_1, \ldots, c_n, c_{n+1}$, not all zero, such that

$$c_1\mathbf{v}_1 + c_2\mathbf{v}_2 + \cdots + c_n\mathbf{v}_n + c_{n+1}\mathbf{v} = \mathbf{0}$$

Observe that $c_{n+1} \neq 0$, since if it were, then S would be linearly dependent, violating the hypothesis that it is linearly independent. Solving for $\mathbf{v}$ gives

$$\mathbf{v} = -\frac{c_1}{c_{n+1}}\mathbf{v}_1 - \frac{c_2}{c_{n+1}}\mathbf{v}_2 - \cdots - \frac{c_n}{c_{n+1}}\mathbf{v}_n$$

As $\mathbf{v}$ was chosen arbitrarily, every vector in V is in **span**(S) and therefore $V = $ **span**(S).

(2) (Proof by contradiction) Assume that S is linearly dependent. Then by Theorem 5 of Sec. 2.3 one of the vectors in S can be written as a linear combination of the other vectors. We can eliminate this vector from S without changing the span. We continue this process until we arrive at a linearly independent spanning set with less than n elements. This contradicts the fact that the dimension of V is n.

EXAMPLE 7   Determine whether

$$B = \left\{ \begin{bmatrix} 1 \\ 0 \\ 1 \end{bmatrix}, \begin{bmatrix} 1 \\ 1 \\ 0 \end{bmatrix}, \begin{bmatrix} 0 \\ 0 \\ 1 \end{bmatrix} \right\}$$

is a basis for $\mathbb{R}^3$.

Solution   Since $\dim(\mathbb{R}^3) = 3$, the set B is a basis if it is linearly independent. Let

$$A = \begin{bmatrix} 1 & 1 & 0 \\ 0 & 1 & 0 \\ 1 & 0 & 1 \end{bmatrix}$$

be the matrix whose column vectors are the vectors of B. The determinant of this matrix is 1, so that by Theorem 9 of Sec. 2.3 the set B is linearly independent and hence, by Theorem 12, is a basis. We can also show that B is a basis by showing that B spans $\mathbb{R}^3$.

Finding a Basis

In Sec. 3.2, we saw that the span of a nonempty set of vectors $S = \{\mathbf{v}_1, \ldots, \mathbf{v}_m\}$ is a subspace. We then ask whether S is a basis for this subspace (or a vector space). From Theorem 12, this is equivalent to determining whether S is linearly independent. When the vectors $\mathbf{v}_1, \ldots, \mathbf{v}_m$ are in $\mathbb{R}^n$, as in Example 7, form the matrix A with ith column vector equal to $\mathbf{v}_i$. By Theorem 2 of Sec. 1.2, if B is the row echelon matrix obtained from reducing A, then $A\mathbf{x} = \mathbf{0}$ if and only if $B\mathbf{x} = \mathbf{0}$. Now if the column vectors of A are linearly dependent, then there are scalars $c_1, \ldots, c_m$, not all 0, such that $c_1\mathbf{v}_1 + \cdots + c_m\mathbf{v}_m = \mathbf{0}$. To express this in matrix form, let

$$\mathbf{c} = \begin{bmatrix} c_1 \\ c_2 \\ \vdots \\ c_m \end{bmatrix}$$

Then $A\mathbf{c} = \mathbf{0} = B\mathbf{c}$. Hence, the column vectors of A and B are both linearly dependent or linearly independent. Observe that the column vectors of B associated with the pivots are linearly independent since none of the vectors can be a linear combination of the column vectors that come before. Therefore, by the previous remarks, the corresponding column vectors of A are also linearly independent. By Theorem 12, these same column vectors form a basis for $\mathbf{col}(A)$. When choosing vectors for a basis of $\mathbf{col}(A)$, we must select the column vectors in A corresponding to the pivot column vectors of B, and not the pivot column vectors of B. For example, the row-reduced echelon form of the matrix

$$A = \begin{bmatrix} 1 & 0 & 1 \\ 0 & 0 & 0 \\ 0 & 1 & 1 \end{bmatrix} \qquad \text{is the matrix} \qquad B = \begin{bmatrix} 1 & 0 & 1 \\ 0 & 1 & 1 \\ 0 & 0 & 0 \end{bmatrix}$$

However, the column spaces of A and B are different. In this case $\mathbf{col}(A)$ is the xz plane and $\mathbf{col}(B)$ is the xy plane with

$$\mathbf{col}(A) = \mathbf{span}\{\mathbf{v}_1, \mathbf{v}_2\} = \mathbf{span}\left\{ \begin{bmatrix} 1 \\ 0 \\ 0 \end{bmatrix}, \begin{bmatrix} 0 \\ 0 \\ 1 \end{bmatrix} \right\}$$

and

$$\mathbf{col}(B) = \mathbf{span}\{\mathbf{w}_1, \mathbf{w}_2\} = \mathbf{span}\left\{ \begin{bmatrix} 1 \\ 0 \\ 0 \end{bmatrix}, \begin{bmatrix} 0 \\ 1 \\ 0 \end{bmatrix} \right\}$$

respectively. See Fig. 2.

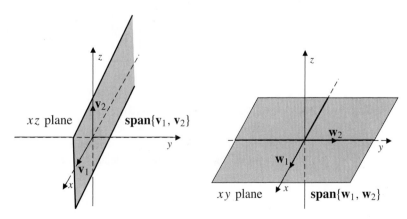

Figure 2

The details of these observations are made clearer by considering a specific example. Let

$$S = \{\mathbf{v}_1, \mathbf{v}_2, \mathbf{v}_3, \mathbf{v}_4, \mathbf{v}_5\} = \left\{ \begin{bmatrix} 1 \\ 1 \\ 0 \end{bmatrix}, \begin{bmatrix} 1 \\ 0 \\ 1 \end{bmatrix}, \begin{bmatrix} 2 \\ 1 \\ 2 \end{bmatrix}, \begin{bmatrix} 2 \\ 1 \\ 1 \end{bmatrix}, \begin{bmatrix} 3 \\ 1 \\ 3 \end{bmatrix} \right\}$$

We begin by considering the equation

$$c_1 \begin{bmatrix} 1 \\ 1 \\ 0 \end{bmatrix} + c_2 \begin{bmatrix} 1 \\ 0 \\ 1 \end{bmatrix} + c_3 \begin{bmatrix} 2 \\ 1 \\ 2 \end{bmatrix} + c_4 \begin{bmatrix} 2 \\ 1 \\ 1 \end{bmatrix} + c_5 \begin{bmatrix} 3 \\ 1 \\ 3 \end{bmatrix} = \begin{bmatrix} 0 \\ 0 \\ 0 \end{bmatrix}$$

To solve this system, we reduce the corresponding augmented matrix to reduced echelon form. That is,

$$\begin{bmatrix} 1 & 1 & 2 & 2 & 3 & | & 0 \\ 1 & 0 & 1 & 1 & 1 & | & 0 \\ 0 & 1 & 2 & 1 & 3 & | & 0 \end{bmatrix} \quad \text{reduces to} \quad \begin{bmatrix} 1 & 0 & 0 & 1 & 0 & | & 0 \\ 0 & 1 & 0 & 1 & 1 & | & 0 \\ 0 & 0 & 1 & 0 & 1 & | & 0 \end{bmatrix}$$

In the general solution, the variables c_1, c_2, and c_3 are the dependent variables corresponding to the leading ones in the reduced matrix, while c_4 and c_5 are free. Thus, the solution is given by

$$S = \{(-s, -s - t, -t, s, t) \mid s, t \in \mathbb{R}\}$$

Now to find a basis for **span**(S), we substitute these values into the original vector equation to obtain

$$-s \begin{bmatrix} 1 \\ 1 \\ 0 \end{bmatrix} + (-s - t) \begin{bmatrix} 1 \\ 0 \\ 1 \end{bmatrix} + (-t) \begin{bmatrix} 2 \\ 1 \\ 2 \end{bmatrix} + s \begin{bmatrix} 2 \\ 1 \\ 1 \end{bmatrix} + t \begin{bmatrix} 3 \\ 1 \\ 3 \end{bmatrix} = \begin{bmatrix} 0 \\ 0 \\ 0 \end{bmatrix}$$

We claim that *each of the vectors corresponding to a free variable is a linear combination of the others*. To establish the claim in this case, let $s = 1$ and $t = 0$. The above vector equation now becomes

$$- \begin{bmatrix} 1 \\ 1 \\ 0 \end{bmatrix} - \begin{bmatrix} 1 \\ 0 \\ 1 \end{bmatrix} + \begin{bmatrix} 2 \\ 1 \\ 1 \end{bmatrix} = \begin{bmatrix} 0 \\ 0 \\ 0 \end{bmatrix}$$

that is,

$$-\mathbf{v}_1 - \mathbf{v}_2 + \mathbf{v}_4 = \mathbf{0}$$

Thus, $\mathbf{v}_4$ is a linear combination of $\mathbf{v}_1$ and $\mathbf{v}_2$. Also, to see that $\mathbf{v}_5$ is a linear combination of $\mathbf{v}_1$, $\mathbf{v}_2$, and $\mathbf{v}_3$, we let $s = 0$ and $t = 1$.

In light of Theorem 8 we eliminate $\mathbf{v}_4$ and $\mathbf{v}_5$ from S to obtain $S' = \{\mathbf{v}_1, \mathbf{v}_2, \mathbf{v}_3\}$. Observe that S' is linearly independent since each of these vectors corresponds to a column with a leading 1. Thus, the equation

$$c_1 \begin{bmatrix} 1 \\ 1 \\ 0 \end{bmatrix} + c_2 \begin{bmatrix} 1 \\ 0 \\ 1 \end{bmatrix} + c_3 \begin{bmatrix} 2 \\ 1 \\ 2 \end{bmatrix} = \begin{bmatrix} 0 \\ 0 \\ 0 \end{bmatrix}$$

has only the trivial solution.

We summarize the procedure for finding a basis for the span of a set of vectors.

Given a set $S = \{\mathbf{v}_1, \mathbf{v}_2, \mathbf{v}_3, \ldots, \mathbf{v}_n\}$ to find a basis for **span**(S):

1. Form a matrix A whose column vectors are $\mathbf{v}_1, \mathbf{v}_2, \ldots, \mathbf{v}_n$.

2. Reduce A to row echelon form.

3. The vectors from S that correspond to the columns of the reduced matrix with the leading 1s are a basis for **span**(S).

In Example 8 we use the process described above to show how to obtain a basis from a spanning set.

EXAMPLE 8   Let

$$S = \left\{ \begin{bmatrix} 1 \\ 0 \\ 1 \end{bmatrix}, \begin{bmatrix} 0 \\ 1 \\ 1 \end{bmatrix}, \begin{bmatrix} 1 \\ 1 \\ 2 \end{bmatrix}, \begin{bmatrix} 1 \\ 2 \\ 1 \end{bmatrix}, \begin{bmatrix} -1 \\ 1 \\ -2 \end{bmatrix} \right\}$$

Find a basis for the span of S.

Solution   Start by constructing the matrix whose column vectors are the vectors in S. We reduce the matrix

$$\begin{bmatrix} 1 & 0 & 1 & 1 & -1 \\ 0 & 1 & 1 & 2 & 1 \\ 1 & 1 & 2 & 1 & -2 \end{bmatrix} \quad \text{to} \quad \begin{bmatrix} 1 & 0 & 1 & 0 & -2 \\ 0 & 1 & 1 & 0 & -1 \\ 0 & 0 & 0 & 1 & 1 \end{bmatrix}$$

Observe that the leading 1s in the reduced matrix are in columns 1, 2, and 4. Therefore, a basis B for **span**(S) is given by $\{\mathbf{v}_1, \mathbf{v}_2, \mathbf{v}_4\}$, that is,

$$B = \left\{ \begin{bmatrix} 1 \\ 0 \\ 1 \end{bmatrix}, \begin{bmatrix} 0 \\ 1 \\ 1 \end{bmatrix}, \begin{bmatrix} 1 \\ 2 \\ 1 \end{bmatrix} \right\}$$

A set of vectors in a vector space that is not a basis can be expanded to a basis by using Theorem 13.

THEOREM 13   Suppose that $S = \{\mathbf{v}_1, \mathbf{v}_2, \ldots, \mathbf{v}_n\}$ is a linearly independent subset of a vector space V. If $\mathbf{v}$ is a vector in V that is not in **span**(S), then $T = \{\mathbf{v}, \mathbf{v}_1, \mathbf{v}_2, \ldots, \mathbf{v}_n\}$ is linearly independent.

Proof   To show that T is linearly independent, we consider the equation

$$c_1\mathbf{v}_1 + c_2\mathbf{v}_2 + \cdots + c_n\mathbf{v}_n + c_{n+1}\mathbf{v} = \mathbf{0}$$

If $c_{n+1} \neq 0$, then we can solve for the vector $\mathbf{v}$ in terms of the vectors of S, contrary to the hypothesis that $\mathbf{v}$ is not in the span of S. Thus, $c_{n+1} = 0$ and the starting equation is equivalent to

$$c_1\mathbf{v}_1 + c_2\mathbf{v}_2 + \cdots + c_n\mathbf{v}_n = \mathbf{0}$$

Since S is linearly independent, then

$$c_1 = 0 \qquad c_2 = 0 \qquad \ldots \qquad c_n = 0$$

Hence, T is linearly independent.

An alternative method for expanding a set of vectors in $\mathbb{R}^n$ to a basis is to add the coordinate vectors to the set and then *trim* the resulting set to a basis. This technique is illustrated in Example 9.

EXAMPLE 9   Find a basis for $\mathbb{R}^4$ that contains the vectors

$$\mathbf{v}_1 = \begin{bmatrix} 1 \\ 0 \\ 1 \\ 0 \end{bmatrix} \qquad \text{and} \qquad \mathbf{v}_2 = \begin{bmatrix} -1 \\ 1 \\ -1 \\ 0 \end{bmatrix}$$

Solution   Notice that the set $\{\mathbf{v}_1, \mathbf{v}_2\}$ is linearly independent. However, it cannot span $\mathbb{R}^4$ since $\dim(\mathbb{R}^4) = 4$. To find a basis, form the set $S = \{\mathbf{v}_1, \mathbf{v}_2, \mathbf{e}_1, \mathbf{e}_2, \mathbf{e}_3, \mathbf{e}_4\}$. Since $\mathbf{span}\{\mathbf{e}_1, \mathbf{e}_2, \mathbf{e}_3, \mathbf{e}_4\} = \mathbb{R}^4$, we know that $\mathbf{span}(S) = \mathbb{R}^4$. Now proceed as in Example 8 by reducing the matrix

$$\begin{bmatrix} 1 & -1 & 1 & 0 & 0 & 0 \\ 0 & 1 & 0 & 1 & 0 & 0 \\ 1 & -1 & 0 & 0 & 1 & 0 \\ 0 & 0 & 0 & 0 & 0 & 1 \end{bmatrix}$$

to reduced row echelon form

$$\begin{bmatrix} 1 & 0 & 0 & 1 & 1 & 0 \\ 0 & 1 & 0 & 1 & 0 & 0 \\ 0 & 0 & 1 & 0 & -1 & 0 \\ 0 & 0 & 0 & 0 & 0 & 1 \end{bmatrix}$$

Observe that the pivot columns are 1, 2, 3, and 6. A basis is therefore given by the set of vectors $\{\mathbf{v}_1, \mathbf{v}_2, \mathbf{e}_1, \mathbf{e}_4\}$.

The following useful corollary results from repeated application of Theorem 13.

COROLLARY 1   Let $S = \{\mathbf{v}_1, \mathbf{v}_2, \ldots, \mathbf{v}_r\}$ be a linearly independent set of vectors in an n-dimensional vector space V with $r < n$. Then S can be expanded to a basis for V. That is, there are vectors $\{\mathbf{v}_{r+1}, \mathbf{v}_{r+2}, \ldots, \mathbf{v}_n\}$ so that $\{\mathbf{v}_1, \mathbf{v}_2, \ldots, \mathbf{v}_r, \mathbf{v}_{r+1}, \ldots, \mathbf{v}_n\}$ is a basis for V.

Fact Summary

Let V be a vector space with $\dim(V) = n$.

1. There are finite sets of vectors that span V. The set of vectors can be linearly independent or linearly dependent. A basis is a linearly independent set of vectors that spans V.
2. Every nontrivial vector space has infinitely many bases.
3. Every basis of V has n elements.
4. The standard basis for $\mathbb{R}^n$ is the n coordinate vectors $\mathbf{e}_1, \mathbf{e}_2, \ldots, \mathbf{e}_n$.
5. The standard basis for $M_{2\times 2}$ consists of the four matrices

$$\begin{bmatrix} 1 & 0 \\ 0 & 0 \end{bmatrix} \quad \begin{bmatrix} 0 & 1 \\ 0 & 0 \end{bmatrix} \quad \begin{bmatrix} 0 & 0 \\ 1 & 0 \end{bmatrix} \quad \begin{bmatrix} 0 & 0 \\ 0 & 1 \end{bmatrix}$$

6. The standard basis for $\mathcal{P}_n$ is $\{1, x, x^2, \ldots, x^n\}$.
7. If $\mathbf{v}_n$ is a linear combination of $\mathbf{v}_1, \mathbf{v}_2, \ldots, \mathbf{v}_{n-1}$, then

$$\text{span}\{\mathbf{v}_1, \mathbf{v}_2, \ldots, \mathbf{v}_{n-1}\} = \text{span}\{\mathbf{v}_1, \mathbf{v}_2, \ldots, \mathbf{v}_{n-1}, \mathbf{v}_n\}$$

8. $\dim(\mathbb{R}^n) = n$, $\dim(M_{m\times n}) = mn$, $\dim(\mathcal{P}_n) = n + 1$
9. If a set B of n vectors of V is linearly independent, then B is a basis for V.
10. If the span of a set B of n vectors is V, then B is a basis for V.
11. Every linearly independent subset of V can be expanded to a basis for V.
12. If S is a set of vectors in $\mathbb{R}^n$, a basis can always be found for $\text{span}(S)$ from the vectors of S.

Exercise Set 3.3

In Exercises 1–6, explain why the set S is not a basis for the vector space V.

1. $S = \left\{ \begin{bmatrix} 2 \\ 1 \\ 3 \end{bmatrix}, \begin{bmatrix} 0 \\ -1 \\ 1 \end{bmatrix} \right\}$ $V = \mathbb{R}^3$

2. $S = \left\{ \begin{bmatrix} 2 \\ 1 \end{bmatrix}, \begin{bmatrix} 1 \\ 0 \end{bmatrix}, \begin{bmatrix} 8 \\ -3 \end{bmatrix} \right\}$ $V = \mathbb{R}^2$

3. $S = \left\{ \begin{bmatrix} 1 \\ 0 \\ 1 \end{bmatrix}, \begin{bmatrix} -1 \\ 1 \\ 0 \end{bmatrix}, \begin{bmatrix} 0 \\ 1 \\ 1 \end{bmatrix} \right\}$ $V = \mathbb{R}^3$

4. $S = \{2, x, x^3 + 2x^2 - 1\}$ $V = \mathcal{P}_3$

5. $S = \{x, x^2, x^2 + 2x, x^3 - x + 1\}$ $V = \mathcal{P}_3$

6. $S = \left\{ \begin{bmatrix} 1 & 0 \\ 0 & 1 \end{bmatrix}, \begin{bmatrix} 0 & 1 \\ 0 & 0 \end{bmatrix}, \right.$

$\left. \begin{bmatrix} 0 & 0 \\ 1 & 0 \end{bmatrix}, \begin{bmatrix} 2 & -3 \\ 1 & 2 \end{bmatrix} \right\}$ $V = M_{2\times 2}$

In Exercises 7–12, show that S is basis for the vector space V.

7. $S = \left\{ \begin{bmatrix} 1 \\ 1 \end{bmatrix}, \begin{bmatrix} -1 \\ 2 \end{bmatrix} \right\}$ $V = \mathbb{R}^2$

8. $S = \left\{ \begin{bmatrix} -1 \\ 3 \end{bmatrix}, \begin{bmatrix} 1 \\ -1 \end{bmatrix} \right\}$ $V = \mathbb{R}^2$

9. $S = \left\{ \begin{bmatrix} 1 \\ -1 \\ 1 \end{bmatrix}, \begin{bmatrix} 0 \\ -2 \\ -3 \end{bmatrix}, \begin{bmatrix} 0 \\ 2 \\ -2 \end{bmatrix} \right\}$

$V = \mathbb{R}^3$

10. $S = \left\{ \begin{bmatrix} -1 \\ -1 \\ 0 \end{bmatrix}, \begin{bmatrix} 2 \\ -1 \\ -3 \end{bmatrix}, \begin{bmatrix} 1 \\ 1 \\ 2 \end{bmatrix} \right\}$

$V = \mathbb{R}^3$

11. $S = \left\{ \begin{bmatrix} 1 & 0 \\ 1 & 0 \end{bmatrix}, \begin{bmatrix} 1 & 1 \\ -1 & 0 \end{bmatrix}, \right.$

$\left. \begin{bmatrix} 0 & 1 \\ -1 & 2 \end{bmatrix}, \begin{bmatrix} 1 & 0 \\ 0 & 1 \end{bmatrix} \right\}$

$V = M_{2 \times 2}$

12. $S = \{x^2 + 1, x + 2, -x^2 + x\}$ $V = \mathcal{P}_2$

In Exercises 13–18, determine whether S is a basis for the vector space V.

13. $S = \left\{ \begin{bmatrix} -1 \\ 2 \\ 1 \end{bmatrix}, \begin{bmatrix} 1 \\ 0 \\ 1 \end{bmatrix}, \begin{bmatrix} 1 \\ 1 \\ 1 \end{bmatrix} \right\}$ $V = \mathbb{R}^3$

14. $S = \left\{ \begin{bmatrix} 2 \\ -2 \\ 1 \end{bmatrix}, \begin{bmatrix} 5 \\ 1 \\ 2 \end{bmatrix}, \begin{bmatrix} 3 \\ 1 \\ 1 \end{bmatrix} \right\}$ $V = \mathbb{R}^3$

15. $S = \left\{ \begin{bmatrix} 1 \\ 1 \\ -1 \\ 1 \end{bmatrix}, \begin{bmatrix} 2 \\ 1 \\ 3 \\ 1 \end{bmatrix}, \begin{bmatrix} 2 \\ 4 \\ 2 \\ 5 \end{bmatrix}, \begin{bmatrix} -1 \\ 2 \\ 0 \\ 3 \end{bmatrix} \right\}$

$V = \mathbb{R}^4$

16. $S = \left\{ \begin{bmatrix} -1 \\ 1 \\ 0 \\ 1 \end{bmatrix}, \begin{bmatrix} 2 \\ 1 \\ -1 \\ 2 \end{bmatrix}, \begin{bmatrix} 1 \\ 3 \\ 1 \\ -1 \end{bmatrix}, \begin{bmatrix} 2 \\ 1 \\ 1 \\ 2 \end{bmatrix} \right\}$

$V = \mathbb{R}^4$

17. $S = \{1, 2x^2 + x + 2, -x^2 + x\}$ $V = \mathcal{P}_2$

18. $S = \left\{ \begin{bmatrix} 1 & 0 \\ 0 & 0 \end{bmatrix}, \begin{bmatrix} 1 & 1 \\ 0 & 0 \end{bmatrix}, \right.$

$\left. \begin{bmatrix} -2 & 1 \\ 1 & 1 \end{bmatrix}, \begin{bmatrix} 0 & 0 \\ 0 & 2 \end{bmatrix} \right\}$

$V = M_{2 \times 2}$

In Exercises 19–24, find a basis for the subspace S of the vector space V. Specify the dimension of S.

19. $S = \left\{ \begin{bmatrix} s + 2t \\ -s + t \\ t \end{bmatrix} \,\middle|\, s, t \in \mathbb{R} \right\}$ $V = \mathbb{R}^3$

20. $S = \left\{ \begin{bmatrix} a & a + d \\ a + d & d \end{bmatrix} \,\middle|\, a, d \in \mathbb{R} \right\}$ $V = M_{2 \times 2}$

21. Let S be the subspace of $V = M_{2 \times 2}$ consisting of all 2×2 symmetric matrices.

22. Let S be the subspace of $V = M_{2 \times 2}$ consisting of all 2×2 skew-symmetric matrices.

23. $S = \{p(x) \mid p(0) = 0\}$ $V = \mathcal{P}_2$

24. $S = \{p(x) \mid p(0) = 0, p(1) = 0\}$ $V = \mathcal{P}_3$

In Exercises 25–30, find a basis for the **span**(S) as a subspace of $\mathbb{R}^3$.

25. $S = \left\{ \begin{bmatrix} 2 \\ 2 \\ -1 \end{bmatrix}, \begin{bmatrix} 2 \\ 0 \\ -2 \end{bmatrix}, \begin{bmatrix} 1 \\ 2 \\ 1 \end{bmatrix} \right\}$

26. $S = \left\{ \begin{bmatrix} -2 \\ 1 \\ 3 \end{bmatrix}, \begin{bmatrix} 4 \\ -1 \\ 2 \end{bmatrix}, \begin{bmatrix} 2 \\ 0 \\ 5 \end{bmatrix} \right\}$

27. $S = \left\{ \begin{bmatrix} 2 \\ -3 \\ 0 \end{bmatrix}, \begin{bmatrix} 0 \\ 2 \\ 2 \end{bmatrix}, \begin{bmatrix} -1 \\ -1 \\ 0 \end{bmatrix}, \begin{bmatrix} 2 \\ 3 \\ -1 \end{bmatrix} \right\}$

28. $S = \left\{ \begin{bmatrix} -2 \\ 0 \\ 2 \end{bmatrix}, \begin{bmatrix} 1 \\ 0 \\ -3 \end{bmatrix}, \begin{bmatrix} -3 \\ -3 \\ -2 \end{bmatrix}, \begin{bmatrix} 1 \\ 2 \\ -2 \end{bmatrix} \right\}$

29. $S = \left\{ \begin{bmatrix} 2 \\ -3 \\ 0 \end{bmatrix}, \begin{bmatrix} 0 \\ 2 \\ 2 \end{bmatrix}, \begin{bmatrix} 2 \\ -1 \\ 2 \end{bmatrix}, \begin{bmatrix} 4 \\ 0 \\ 4 \end{bmatrix} \right\}$

30. $S = \left\{ \begin{bmatrix} 2 \\ 2 \\ 0 \end{bmatrix}, \begin{bmatrix} 1 \\ -1 \\ 0 \end{bmatrix}, \begin{bmatrix} 0 \\ 2 \\ 2 \end{bmatrix}, \begin{bmatrix} 2 \\ 3 \\ 1 \end{bmatrix} \right\}$

In Exercises 31–36, find a basis for the vector space V that contains the given vectors.

31. $S = \left\{ \begin{bmatrix} 2 \\ -1 \\ 3 \end{bmatrix}, \begin{bmatrix} 1 \\ 0 \\ 2 \end{bmatrix} \right\}$ $V = \mathbb{R}^3$

32. $S = \left\{ \begin{bmatrix} -1 \\ 1 \\ 3 \end{bmatrix}, \begin{bmatrix} 1 \\ 1 \\ 1 \end{bmatrix} \right\} V = \mathbb{R}^3$

33. $S = \left\{ \begin{bmatrix} 1 \\ -1 \\ 2 \\ 4 \end{bmatrix}, \begin{bmatrix} 3 \\ 1 \\ 1 \\ 2 \end{bmatrix} \right\} V = \mathbb{R}^4$

34. $S = \left\{ \begin{bmatrix} -1 \\ 1 \\ 1 \\ -1 \end{bmatrix}, \begin{bmatrix} 1 \\ -3 \\ -1 \\ 2 \end{bmatrix}, \begin{bmatrix} 1 \\ -2 \\ -1 \\ 3 \end{bmatrix} \right\} V = \mathbb{R}^4$

35. $S = \left\{ \begin{bmatrix} -1 \\ 1 \\ 3 \end{bmatrix}, \begin{bmatrix} 1 \\ 1 \\ 1 \end{bmatrix} \right\} V = \mathbb{R}^3$

36. $S = \left\{ \begin{bmatrix} 2 \\ 2 \\ -1 \end{bmatrix}, \begin{bmatrix} -1 \\ -1 \\ 3 \end{bmatrix} \right\} V = \mathbb{R}^3$

37. Find a basis for the subspace of $M_{n \times n}$ consisting of all diagonal matrices.

38. Show that if $S = \{\mathbf{v}_1, \mathbf{v}_2, \dots, \mathbf{v}_n\}$ is a basis for the vector space V and c is a nonzero scalar, then $S' = \{c\mathbf{v}_1, c\mathbf{v}_2, \dots, c\mathbf{v}_n\}$ is also a basis for V.

39. Show that if $S = \{\mathbf{v}_1, \mathbf{v}_2, \dots, \mathbf{v}_n\}$ is a basis for $\mathbb{R}^n$ and A is an $n \times n$ invertible matrix, then $S' = \{A\mathbf{v}_1, A\mathbf{v}_2, \dots, A\mathbf{v}_n\}$ is also a basis.

40. Find a basis for the subspace
$S = \{\mathbf{x} \in \mathbb{R}^4 \mid A\mathbf{x} = \mathbf{0}\}$ of $\mathbb{R}^4$ where

$$A = \begin{bmatrix} 3 & 3 & 1 & 3 \\ -1 & 0 & -1 & -1 \\ 2 & 0 & 2 & 1 \end{bmatrix}$$

41. Suppose that V is a vector space with $\dim(V) = n$. Show that if H is a subspace of V and $\dim(H) = n$, then $H = V$.

42. Let S and T be the subspaces of $\mathcal{P}_3$ defined by

$$S = \{p(x) \mid p(0) = 0\}$$

and

$$T = \{q(x) \mid q(1) = 0\}$$

Find $\dim(S)$, $\dim(T)$, and $\dim(S \cap T)$.

43. Let

$$W = \left\{ \begin{bmatrix} 2s + t + 3r \\ 3s - t + 2r \\ s + t + 2r \end{bmatrix} \middle| s, t, r \in \mathbb{R} \right\}$$

Find $\dim(W)$.

44. Let S and T be the subspaces of $\mathbb{R}^4$ defined by

$$S = \left\{ \begin{bmatrix} s \\ t \\ 0 \\ 0 \end{bmatrix} \middle| s, t \in \mathbb{R} \right\}$$

and

$$T = \left\{ \begin{bmatrix} 0 \\ s \\ t \\ 0 \end{bmatrix} \middle| s, t \in \mathbb{R} \right\}$$

Find $\dim(S)$, $\dim(T)$, and $\dim(S \cap T)$.

3.4 ▶ Coordinates and Change of Basis

From our earliest experiences with Euclidean space we have used rectangular coordinates, (or xy coordinates), to specify the location of a point in the plane. Equivalently, these coordinates describe a vector in standard position which terminates at the point. Equipped with our knowledge of linear combinations, we now understand these xy coordinates to be the scalar multiples required to express the vector as a linear combination of the standard basis vectors $\mathbf{e}_1$ and $\mathbf{e}_2$. For example, the vector $\mathbf{v} = \begin{bmatrix} 2 \\ 3 \end{bmatrix}$, with xy coordinates $(2, 3)$, can be written as

$$\mathbf{v} = 2\mathbf{e}_1 + 3\mathbf{e}_2$$

as shown in Fig. 1(a). This point (or vector) can also be specified relative to another pair of linearly independent vectors, describing an $x'y'$ coordinate system. For example, since

$$\begin{bmatrix} 2 \\ 3 \end{bmatrix} = \tfrac{5}{2} \begin{bmatrix} 1 \\ 1 \end{bmatrix} + \tfrac{1}{2} \begin{bmatrix} -1 \\ 1 \end{bmatrix}$$

the $x'y'$ coordinates of **v** are given by $\left(\tfrac{5}{2}, \tfrac{1}{2} \right)$. See Fig. 1(b).

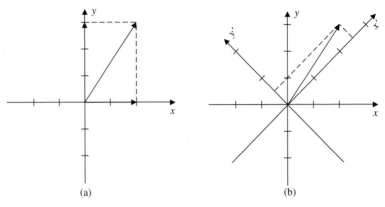

(a)   (b)

Figure 1

In this section we generalize this concept to abstract vector spaces. Let V be a vector space with basis $B = \{\mathbf{v}_1, \mathbf{v}_2, \ldots, \mathbf{v}_n\}$. From Theorem 7 of Sec. 2.3, every vector **v** in V can be written uniquely as a linear combination of the vectors of B. That is, there are unique scalars $c_1, c_2, \ldots, c_n$ such that

$$\mathbf{v} = c_1\mathbf{v}_1 + c_2\mathbf{v}_2 + \cdots + c_n\mathbf{v}_n$$

It is tempting to associate the list of scalars $\{c_1, c_2, \ldots, c_n\}$ with the *coordinates* of the vector **v**. However, changing the order of the basis vectors in B will change the order of the scalars. For example, the sets

$$B = \left\{ \begin{bmatrix} 1 \\ 0 \end{bmatrix}, \begin{bmatrix} 0 \\ 1 \end{bmatrix} \right\}$$

and

$$B' = \left\{ \begin{bmatrix} 0 \\ 1 \end{bmatrix}, \begin{bmatrix} 1 \\ 0 \end{bmatrix} \right\}$$

are both bases for $\mathbb{R}^2$. Then the list of scalars associated with the vector $\begin{bmatrix} 1 \\ 2 \end{bmatrix}$ is $\{1, 2\}$ relative to B but is $\{2, 1\}$ relative to B'. To remove this ambiguity, we introduce the notion of an *ordered basis*.

DEFINITION 1   **Ordered Basis**   An **ordered basis** of a vector space V is a fixed sequence of linearly independent vectors that span V.

DEFINITION 2   **Coordinates**   Let $B = \{\mathbf{v}_1, \mathbf{v}_2, \dots, \mathbf{v}_n\}$ be an ordered basis for the vector space V. Let $\mathbf{v}$ be a vector in V, and let $c_1, c_2, \dots, c_n$ be the unique scalars such that

$$\mathbf{v} = c_1\mathbf{v}_1 + c_2\mathbf{v}_2 + \cdots + c_n\mathbf{v}_n$$

Then $c_1, c_2, \dots, c_n$ are called the **coordinates of $\mathbf{v}$ relative to** B. In this case we write

$$[\mathbf{v}]_B = \begin{bmatrix} c_1 \\ c_2 \\ \vdots \\ c_n \end{bmatrix}$$

and refer to the vector $[\mathbf{v}]_B$ as the **coordinate vector of $\mathbf{v}$ relative to** B.

In $\mathbb{R}^n$ the coordinates of a vector relative to the standard basis $B = \{\mathbf{e}_1, \mathbf{e}_2, \dots, \mathbf{e}_n\}$ are simply the components of the vector. Similarly, the coordinates of a polynomial $p(x) = a_0 + a_1 x + a_2 x^2 + \cdots + a_n x^n$ in $\mathcal{P}_n$ relative to the standard basis $\{1, x, x^2, \dots, x^n\}$ are the coefficients of the polynomial.

EXAMPLE 1   Let $V = \mathbb{R}^2$ and B be the ordered basis

$$B = \left\{ \begin{bmatrix} 1 \\ 1 \end{bmatrix}, \begin{bmatrix} -1 \\ 1 \end{bmatrix} \right\}$$

Find the coordinates of the vector $\mathbf{v} = \begin{bmatrix} 1 \\ 5 \end{bmatrix}$ relative to B.

Solution   The coordinates c_1 and c_2 are found by writing $\mathbf{v}$ as a linear combination of the two vectors in B. That is, we solve the equation

$$c_1 \begin{bmatrix} 1 \\ 1 \end{bmatrix} + c_2 \begin{bmatrix} -1 \\ 1 \end{bmatrix} = \begin{bmatrix} 1 \\ 5 \end{bmatrix}$$

In this case $c_1 = 3$ and $c_2 = 2$. We therefore have that the coordinate vector of $\mathbf{v} = \begin{bmatrix} 1 \\ 5 \end{bmatrix}$ relative to B is

$$[\mathbf{v}]_B = \begin{bmatrix} 3 \\ 2 \end{bmatrix}$$

EXAMPLE 2   Let $V = \mathcal{P}_2$ and B be the ordered basis

$$B = \left\{ 1, x - 1, (x - 1)^2 \right\}$$

Find the coordinates of $p(x) = 2x^2 - 2x + 1$ relative to B.

Solution   We must find $c_1, c_2,$ and c_3 such that

$$c_1(1) + c_2(x - 1) + c_3(x - 1)^2 = 2x^2 - 2x + 1$$

Expanding the left-hand side and collecting like terms give

$$c_3x^2 + (c_2 - 2c_3)x + (c_1 - c_2 + c_3) = 2x^2 - 2x + 1$$

Equating the coefficients of like terms gives the linear system

$$\begin{cases} c_1 - c_2 + c_3 = 1 \\ \phantom{c_1 - {}} c_2 - 2c_3 = -2 \\ \phantom{c_1 - c_2 + {}} c_3 = 2 \end{cases}$$

The unique solution to this system is $c_1 = 1$, $c_2 = 2$, and $c_3 = 2$, so that

$$[\mathbf{v}]_B = \begin{bmatrix} 1 \\ 2 \\ 2 \end{bmatrix}$$

EXAMPLE 3   Let W be the subspace of all symmetric matrices in the vector space $M_{2 \times 2}$. Let

$$B = \left\{ \begin{bmatrix} 1 & 0 \\ 0 & 0 \end{bmatrix}, \begin{bmatrix} 0 & 1 \\ 1 & 0 \end{bmatrix}, \begin{bmatrix} 0 & 0 \\ 0 & 1 \end{bmatrix} \right\}$$

Show that B is a basis for W and find the coordinates of

$$\mathbf{v} = \begin{bmatrix} 2 & 3 \\ 3 & 5 \end{bmatrix}$$

relative to B.

Solution   In Example 8 of Sec. 3.2, we showed that B spans W. The matrices in B are also linearly independent and hence are a basis for W. Observe that $\mathbf{v}$ can be written as

$$2 \begin{bmatrix} 1 & 0 \\ 0 & 0 \end{bmatrix} + 3 \begin{bmatrix} 0 & 1 \\ 1 & 0 \end{bmatrix} + 5 \begin{bmatrix} 0 & 0 \\ 0 & 1 \end{bmatrix} = \begin{bmatrix} 2 & 3 \\ 3 & 5 \end{bmatrix}$$

Then relative to the ordered basis B, the coordinate vector of $\mathbf{v}$ is

$$[\mathbf{v}]_B = \begin{bmatrix} 2 \\ 3 \\ 5 \end{bmatrix}$$

Change of Basis

Many problems in applied mathematics are made easier by changing from one basis of a vector space to another. To simplify our explanation of the process, we will consider a vector space V with $\dim(V) = 2$ and show how to change from coordinates relative to one basis for V to another basis for V.

Let V be a vector space of dimension 2 and let

$$B = \{\mathbf{v}_1, \mathbf{v}_2\} \quad \text{and} \quad B' = \{\mathbf{v}_1', \mathbf{v}_2'\}$$

be ordered bases for V. Now let $\mathbf{v}$ be a vector in V, and suppose that the coordinates of $\mathbf{v}$ relative to B are given by

$$[\mathbf{v}]_B = \begin{bmatrix} x_1 \\ x_2 \end{bmatrix} \quad \text{that is} \quad \mathbf{v} = x_1\mathbf{v}_1 + x_2\mathbf{v}_2$$

To determine the coordinates of $\mathbf{v}$ relative to B', we first write $\mathbf{v}_1$ and $\mathbf{v}_2$ in terms of the vectors $\mathbf{v}_1'$ and $\mathbf{v}_2'$. Since B' is a basis, there are scalars a_1, a_2, b_1, and b_2 such that

$$\mathbf{v}_1 = a_1\mathbf{v}_1' + a_2\mathbf{v}_2'$$
$$\mathbf{v}_2 = b_1\mathbf{v}_1' + b_2\mathbf{v}_2'$$

Then $\mathbf{v}$ can be written as

$$\mathbf{v} = x_1(a_1\mathbf{v}_1' + a_2\mathbf{v}_2') + x_2(b_1\mathbf{v}_1' + b_2\mathbf{v}_2')$$

Collecting the coefficients of $\mathbf{v}_1'$ and $\mathbf{v}_2'$ gives

$$\mathbf{v} = (x_1a_1 + x_2b_1)\mathbf{v}_1' + (x_1a_2 + x_2b_2)\mathbf{v}_2'$$

so that the coordinates of $\mathbf{v}$ relative to the basis B' are given by

$$[\mathbf{v}]_{B'} = \begin{bmatrix} x_1a_1 + x_2b_1 \\ x_1a_2 + x_2b_2 \end{bmatrix}$$

Now by rewriting the vector on the right-hand side as a matrix product, we have

$$[\mathbf{v}]_{B'} = \begin{bmatrix} a_1 & b_1 \\ a_2 & b_2 \end{bmatrix} \begin{bmatrix} x_1 \\ x_2 \end{bmatrix} = \begin{bmatrix} a_1 & b_1 \\ a_2 & b_2 \end{bmatrix} [\mathbf{v}]_B$$

Notice that the column vectors of the matrix are the coordinate vectors $[\mathbf{v}_1]_{B'}$ and $[\mathbf{v}_2]_{B'}$. The matrix

$$\begin{bmatrix} a_1 & b_1 \\ a_2 & b_2 \end{bmatrix}$$

is called the **transition matrix** from B to B' and is denoted by $[I]_B^{B'}$, so that

$$[\mathbf{v}]_{B'} = [I]_B^{B'}[\mathbf{v}]_B$$

EXAMPLE 4   Let $V = \mathbb{R}^2$ with bases

$$B = \left\{ \begin{bmatrix} 1 \\ 1 \end{bmatrix}, \begin{bmatrix} 1 \\ -1 \end{bmatrix} \right\} \quad \text{and} \quad B' = \left\{ \begin{bmatrix} 2 \\ -1 \end{bmatrix}, \begin{bmatrix} -1 \\ 1 \end{bmatrix} \right\}$$

a. Find the transition matrix from B to B'.

b. Let $[\mathbf{v}]_B = \begin{bmatrix} 3 \\ -2 \end{bmatrix}$ and find $[\mathbf{v}]_{B'}$.

Solution   **a.** By denoting the vectors in B by $\mathbf{v}_1$ and $\mathbf{v}_2$ and those in B' by $\mathbf{v}'_1$ and $\mathbf{v}'_2$, the column vectors of the transition matrix are $[\mathbf{v}_1]_{B'}$ and $[\mathbf{v}_2]_{B'}$. These coordinate vectors are found from the equations

$$c_1 \begin{bmatrix} 2 \\ -1 \end{bmatrix} + c_2 \begin{bmatrix} -1 \\ 1 \end{bmatrix} = \begin{bmatrix} 1 \\ 1 \end{bmatrix} \quad \text{and} \quad d_1 \begin{bmatrix} 2 \\ -1 \end{bmatrix} + d_2 \begin{bmatrix} -1 \\ 1 \end{bmatrix} = \begin{bmatrix} 1 \\ -1 \end{bmatrix}$$

Solving these equations gives $c_1 = 2$ and $c_2 = 3$, and $d_1 = 0$ and $d_2 = -1$, so that

$$[\mathbf{v}_1]_{B'} = \begin{bmatrix} 2 \\ 3 \end{bmatrix} \quad \text{and} \quad [\mathbf{v}_2]_{B'} = \begin{bmatrix} 0 \\ -1 \end{bmatrix}$$

Therefore, the transition matrix is

$$[I]_B^{B'} = \begin{bmatrix} 2 & 0 \\ 3 & -1 \end{bmatrix}$$

b. Since

$$[\mathbf{v}]_{B'} = [I]_B^{B'} [\mathbf{v}]_B$$

then

$$[\mathbf{v}]_{B'} = \begin{bmatrix} 2 & 0 \\ 3 & -1 \end{bmatrix} \begin{bmatrix} 3 \\ -2 \end{bmatrix} = \begin{bmatrix} 6 \\ 11 \end{bmatrix}$$

Observe that the same vector, relative to the different bases, is obtained from the coordinates $[\mathbf{v}]_B$ and $[\mathbf{v}]_{B'}$. That is,

$$3 \begin{bmatrix} 1 \\ 1 \end{bmatrix} - 2 \begin{bmatrix} 1 \\ -1 \end{bmatrix} = \begin{bmatrix} 1 \\ 5 \end{bmatrix} = 6 \begin{bmatrix} 2 \\ -1 \end{bmatrix} + 11 \begin{bmatrix} -1 \\ 1 \end{bmatrix}$$

The procedure to find the transition matrix between two bases of a vector space of dimension 2 can be generalized to $\mathbb{R}^n$ and other vector spaces of finite dimension. The result is stated in Theorem 14.

THEOREM 14   Let V be a vector space of dimension n with ordered bases

$$B = \{\mathbf{v}_1, \mathbf{v}_2, \dots, \mathbf{v}_n\} \quad \text{and} \quad B' = \{\mathbf{v}'_1, \mathbf{v}'_2, \dots, \mathbf{v}'_n\}$$

Then the transition matrix from B to B' is given by

$$[I]_B^{B'} = \left[\begin{bmatrix} \\ \mathbf{v}_1 \\ \end{bmatrix}_{B'} \begin{bmatrix} \\ \mathbf{v}_2 \\ \end{bmatrix}_{B'} \cdots \begin{bmatrix} \\ \mathbf{v}_n \\ \end{bmatrix}_{B'} \right]$$

Moreover, a change of coordinates is carried out by

$$[\mathbf{v}]_{B'} = [I]_B^{B'} [\mathbf{v}]_B$$

In Example 5 we use the result of Theorem 14 to change from one basis of $\mathcal{P}_2$ to another.

EXAMPLE 5   Let $V = \mathcal{P}_2$ with bases

$$B = \{1, x, x^2\} \quad \text{and} \quad B' = \{1, x+1, x^2+x+1\}$$

a. Find the transition matrix $[I]_B^{B'}$.

b. Let $p(x) = 3 - x + 2x^2$ and find $[p(x)]_{B'}$.

Solution   **a.** To find the first column vector of the transition matrix, we must find scalars a_1, a_2, and a_3 such that

$$a_1(1) + a_2(x+1) + a_3(x^2+x+1) = 1$$

By inspection we see that the solution is $a_1 = 1$, $a_2 = 0$, and $a_3 = 0$. Therefore,

$$[1]_{B'} = \begin{bmatrix} 1 \\ 0 \\ 0 \end{bmatrix}$$

The second and third column vectors of the transition matrix can be found by solving the equations

$$b_1(1) + b_2(x+1) + b_3(x^2+x+1) = x$$

and

$$c_1(1) + c_2(x+1) + c_3(x^2+x+1) = x^2$$

respectively. The solutions are given by $b_1 = -1$, $b_2 = 1$, and $b_3 = 0$, and $c_1 = 0$, $c_2 = -1$, and $c_3 = 1$. Hence, the transition matrix is

$$[I]_B^{B'} = \begin{bmatrix} 1 & -1 & 0 \\ 0 & 1 & -1 \\ 0 & 0 & 1 \end{bmatrix}$$

b. The basis B is the standard basis for $\mathcal{P}_2$, so the coordinate vector of $p(x) = 3 - x + 2x^2$ relative to B is given by

$$[p(x)]_{B'} = \begin{bmatrix} 3 \\ -1 \\ 2 \end{bmatrix}$$

Hence,

$$[p(x)]_{B'} = \begin{bmatrix} 1 & -1 & 0 \\ 0 & 1 & -1 \\ 0 & 0 & 1 \end{bmatrix} \begin{bmatrix} 3 \\ -1 \\ 2 \end{bmatrix} = \begin{bmatrix} 4 \\ -3 \\ 2 \end{bmatrix}$$

Notice that $3 - x + 2x^2 = 4(1) - 3(x + 1) + 2(x^2 + x + 1)$.

EXAMPLE 6

Let $B = \{e_1, e_2\}$ be the standard ordered basis for $\mathbb{R}^2$, B' be the ordered basis given by

$$B' = \{v_1', v_2'\} = \left\{ \begin{bmatrix} -1 \\ 1 \end{bmatrix}, \begin{bmatrix} 1 \\ 1 \end{bmatrix} \right\}$$

and let $v = \begin{bmatrix} 3 \\ 4 \end{bmatrix}$.

a. Find the transition matrix from B to B'.

b. Find $[v]_{B'}$.

c. Write the vector v as a linear combination of e_1 and e_2 and also as a linear combination of v_1' and v_2'.

d. Show the results of part (c) graphically.

Solution

a. The transition matrix from B to B' is computed by solving the equations

$$c_1 \begin{bmatrix} -1 \\ 1 \end{bmatrix} + c_2 \begin{bmatrix} 1 \\ 1 \end{bmatrix} = \begin{bmatrix} 1 \\ 0 \end{bmatrix} \quad \text{and} \quad d_1 \begin{bmatrix} -1 \\ 1 \end{bmatrix} + d_2 \begin{bmatrix} 1 \\ 1 \end{bmatrix} = \begin{bmatrix} 0 \\ 1 \end{bmatrix}$$

That is, we must solve the linear systems

$$\begin{cases} -c_1 + c_2 = 1 \\ c_1 + c_2 = 0 \end{cases} \quad \text{and} \quad \begin{cases} -d_1 + d_2 = 0 \\ d_1 + d_2 = 1 \end{cases}$$

The unique solutions are given by $c_1 = -\frac{1}{2}, c_2 = \frac{1}{2}$ and $d_1 = \frac{1}{2}, d_2 = \frac{1}{2}$. The transition matrix is then given by

$$[I]_B^{B'} = \begin{bmatrix} -\frac{1}{2} & \frac{1}{2} \\ \frac{1}{2} & \frac{1}{2} \end{bmatrix}$$

b. Since B is the standard basis, the coordinates of v relative to B are $[v]_B = \begin{bmatrix} 3 \\ 4 \end{bmatrix}$. By Theorem 14, the coordinates of v relative to B' are given by

$$[v]_{B'} = \begin{bmatrix} -\frac{1}{2} & \frac{1}{2} \\ \frac{1}{2} & \frac{1}{2} \end{bmatrix} \begin{bmatrix} 3 \\ 4 \end{bmatrix} = \begin{bmatrix} \frac{1}{2} \\ \frac{7}{2} \end{bmatrix}$$

c. Using the coordinates of $\mathbf{v}$ relative to the two bases, we have

$$3\begin{bmatrix} 1 \\ 0 \end{bmatrix} + 4\begin{bmatrix} 0 \\ 1 \end{bmatrix} = \mathbf{v} = \frac{1}{2}\mathbf{v}'_1 + \frac{7}{2}\mathbf{v}'_2$$

d. The picture given in Fig. 2 shows the location of the terminal point $(3, 4)$ of the vector $\mathbf{v}$ relative to the $\mathbf{e}_1\mathbf{e}_2$ axes and the $\mathbf{v}'_1\mathbf{v}'_2$ axes.

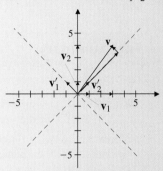

Figure 2

Inverse of a Transition Matrix

A fact that will be useful in Chap. 4 is that the transition matrix $[I]_B^{B'}$ between bases B and B' of a finite dimensional vector space is invertible. Moreover, the transition matrix from B' to B is the inverse of $[I]_B^{B'}$. To see this, suppose that V is a vector space of dimension n with ordered bases

$$B = \{\mathbf{v}_1, \mathbf{v}_2, \ldots, \mathbf{v}_n\} \qquad \text{and} \qquad B' = \{\mathbf{v}'_1, \mathbf{v}'_2, \ldots, \mathbf{v}'_n\}$$

To show that $[I]_B^{B'}$ is invertible, let $\mathbf{x} \in \mathbb{R}^n$ be such that

$$[I]_B^{B'}\mathbf{x} = \mathbf{0}$$

Observe that the left-hand side of this equation in vector form is $x_1[\mathbf{v}_1]_{B'} + \cdots + x_n[\mathbf{v}_n]_{B'}$. Since B is a basis, then the vectors $\mathbf{v}_i$ for $1 \leq i \leq n$ are linearly independent. Hence, so are the vectors $[\mathbf{v}_1]_{B'}, \cdots, [\mathbf{v}_n]_{B'}$. Therefore, $x_1 = x_2 = \cdots = x_n = 0$. Since the only solution to the homogeneous equation $[I]_B^{B'}\mathbf{x} = \mathbf{0}$ is the trivial solution, then by Theorem 17 of Sec. 1.6, the matrix $[I]_B^{B'}$ is invertible. Moreover, by Theorem 14, since

$$[\mathbf{v}]_{B'} = [I]_B^{B'}[\mathbf{v}]_B \qquad \text{we know that} \qquad ([I]_B^{B'})^{-1}[\mathbf{v}]_{B'} = [\mathbf{v}]_B$$

and therefore

$$[I]_{B'}^{B} = ([I]_B^{B'})^{-1}$$

The previous observations are summarized in Theorem 15.

THEOREM 15   Let V be a vector space of dimension n with ordered bases

$$B = \{\mathbf{v}_1, \mathbf{v}_2, \ldots, \mathbf{v}_n\} \qquad \text{and} \qquad B' = \{\mathbf{v}'_1, \mathbf{v}'_2, \ldots, \mathbf{v}'_n\}$$

Then the transition matrix $[I]_B^{B'}$ from B to B' is invertible and

$$[I]_{B'}^{B} = ([I]_B^{B'})^{-1}$$

Fact Summary

Let V be a vector space with $\dim(V) = n$.

1. In $\mathbb{R}^n$, the coordinates of a vector with respect to the standard basis are the components of the vector.

2. Given any two ordered bases for V, a transition matrix can be used to change the coordinates of a vector relative to one basis to the coordinates relative to the other basis.

3. If B and B' are two ordered bases for V, the transition matrix from B to B' is the matrix $[I]_B^{B'}$ whose column vectors are the coordinates of the basis vectors of B relative to the basis B'. Also

$$[\mathbf{v}]_{B'} = [I]_B^{B'}[\mathbf{v}]_B$$

4. If B and B' are two ordered bases for V, the transition matrix from B to B' is invertible and the inverse matrix is the transition matrix from B' to B. That is, $([I]_B^{B'})^{-1} = [I]_{B'}^{B}$

Exercise Set 3.4

In Exercises 1–8, find the coordinates of the vector $\mathbf{v}$ relative to the ordered basis B.

1. $B = \left\{ \begin{bmatrix} 3 \\ 1 \end{bmatrix}, \begin{bmatrix} -2 \\ 2 \end{bmatrix} \right\}$ $\mathbf{v} = \begin{bmatrix} 8 \\ 0 \end{bmatrix}$

2. $B = \left\{ \begin{bmatrix} -2 \\ 4 \end{bmatrix}, \begin{bmatrix} -1 \\ 1 \end{bmatrix} \right\}$ $\mathbf{v} = \begin{bmatrix} -2 \\ 1 \end{bmatrix}$

3. $B = \left\{ \begin{bmatrix} 1 \\ -1 \\ 2 \end{bmatrix}, \begin{bmatrix} 3 \\ -1 \\ 1 \end{bmatrix}, \begin{bmatrix} 1 \\ 0 \\ 2 \end{bmatrix} \right\}$

$\mathbf{v} = \begin{bmatrix} 2 \\ -1 \\ 9 \end{bmatrix}$

4. $B = \left\{ \begin{bmatrix} 2 \\ 2 \\ 1 \end{bmatrix}, \begin{bmatrix} 1 \\ 0 \\ 2 \end{bmatrix}, \begin{bmatrix} 0 \\ 0 \\ 1 \end{bmatrix} \right\}$

$\mathbf{v} = \begin{bmatrix} 0 \\ 1 \\ \frac{1}{2} \end{bmatrix}$

5. $B = \{1, x - 1, x^2\}$ $\mathbf{v} = p(x) = -2x^2 + 2x + 3$

6. $B = \{x^2 + 2x + 2, 2x + 3, -x^2 + x + 1\}$
$\mathbf{v} = p(x) = -3x^2 + 6x + 8$

7. $B = \left\{ \begin{bmatrix} 1 & -1 \\ 0 & 0 \end{bmatrix}, \begin{bmatrix} 0 & 1 \\ 1 & 0 \end{bmatrix}, \right.$

$\left. \begin{bmatrix} 1 & 0 \\ 0 & -1 \end{bmatrix}, \begin{bmatrix} 1 & 0 \\ -1 & 0 \end{bmatrix} \right\}$

$\mathbf{v} = \begin{bmatrix} 1 & 3 \\ -2 & 2 \end{bmatrix}$

8. $B = \left\{ \begin{bmatrix} 1 & -1 \\ 0 & 1 \end{bmatrix}, \begin{bmatrix} 0 & 1 \\ 0 & 2 \end{bmatrix}, \right.$

$\left. \begin{bmatrix} 1 & -1 \\ 1 & 0 \end{bmatrix}, \begin{bmatrix} 1 & 1 \\ 0 & 3 \end{bmatrix} \right\}$

$\mathbf{v} = \begin{bmatrix} 2 & -2 \\ 1 & 3 \end{bmatrix}$

In Exercises 9–12, find the coordinates of the vector $\mathbf{v}$ relative to the two ordered bases B_1 and B_2.

9. $B_1 = \left\{ \begin{bmatrix} -3 \\ 1 \end{bmatrix}, \begin{bmatrix} 2 \\ 2 \end{bmatrix} \right\}$

$B_2 = \left\{ \begin{bmatrix} 2 \\ 1 \end{bmatrix}, \begin{bmatrix} 0 \\ 1 \end{bmatrix} \right\}$ $\mathbf{v} = \begin{bmatrix} 1 \\ 0 \end{bmatrix}$

10. $B_1 = \left\{ \begin{bmatrix} -2 \\ 1 \\ 0 \end{bmatrix}, \begin{bmatrix} 1 \\ 0 \\ 2 \end{bmatrix}, \begin{bmatrix} -1 \\ 1 \\ 0 \end{bmatrix} \right\}$

$B_2 = \left\{ \begin{bmatrix} 1 \\ 0 \\ 1 \end{bmatrix}, \begin{bmatrix} 0 \\ 1 \\ 2 \end{bmatrix}, \begin{bmatrix} -1 \\ -1 \\ 0 \end{bmatrix} \right\}$

$\mathbf{v} = \begin{bmatrix} -2 \\ 2 \\ 2 \end{bmatrix}$

11. $B_1 = \{x^2 - x + 1, x^2 + x + 1, 2x^2\}$

$B_2 = \{2x^2 + 1, -x^2 + x + 2, x + 3\}$

$\mathbf{v} = p(x) = x^2 + x + 3$

12. $B_1 = \left\{ \begin{bmatrix} 1 & 0 \\ 0 & 2 \end{bmatrix}, \begin{bmatrix} 1 & -1 \\ 1 & 0 \end{bmatrix}, \right.$

$\left. \begin{bmatrix} 0 & 1 \\ 1 & 0 \end{bmatrix}, \begin{bmatrix} 2 & 0 \\ -1 & 1 \end{bmatrix} \right\}$

$B_2 = \left\{ \begin{bmatrix} 3 & -1 \\ 0 & 1 \end{bmatrix}, \begin{bmatrix} -1 & 0 \\ 1 & 1 \end{bmatrix}, \right.$

$\left. \begin{bmatrix} 0 & 0 \\ 1 & 0 \end{bmatrix}, \begin{bmatrix} 0 & -1 \\ 1 & 1 \end{bmatrix} \right\}$

$\mathbf{v} = \begin{bmatrix} 0 & 0 \\ 3 & 1 \end{bmatrix}$

In Exercises 13–18, find the transition matrix between the ordered bases B_1 and B_2; then given $[\mathbf{v}]_{B_1}$, find $[\mathbf{v}]_{B_2}$.

13. $B_1 = \left\{ \begin{bmatrix} 1 \\ 1 \end{bmatrix}, \begin{bmatrix} -1 \\ 1 \end{bmatrix} \right\}$

$B_2 = \left\{ \begin{bmatrix} 1 \\ 0 \end{bmatrix}, \begin{bmatrix} 0 \\ 1 \end{bmatrix} \right\}$ $[\mathbf{v}]_{B_1} = \begin{bmatrix} 2 \\ 3 \end{bmatrix}$

14. $B_1 = \left\{ \begin{bmatrix} -2 \\ 1 \end{bmatrix}, \begin{bmatrix} 1 \\ 2 \end{bmatrix} \right\}$

$B_2 = \left\{ \begin{bmatrix} 1 \\ 3 \end{bmatrix}, \begin{bmatrix} 2 \\ 0 \end{bmatrix} \right\}$ $[\mathbf{v}]_{B_1} = \begin{bmatrix} 1 \\ -1 \end{bmatrix}$

15. $B_1 = \left\{ \begin{bmatrix} 1 \\ 2 \\ 1 \end{bmatrix}, \begin{bmatrix} 0 \\ 1 \\ 1 \end{bmatrix}, \begin{bmatrix} 0 \\ 1 \\ 0 \end{bmatrix} \right\}$

$B_2 = \left\{ \begin{bmatrix} 0 \\ 1 \\ 0 \end{bmatrix}, \begin{bmatrix} -1 \\ 1 \\ -1 \end{bmatrix}, \begin{bmatrix} 2 \\ 1 \\ -1 \end{bmatrix} \right\}$

$[\mathbf{v}]_{B_1} = \begin{bmatrix} -1 \\ 0 \\ 2 \end{bmatrix}$

16. $B_1 = \left\{ \begin{bmatrix} -2 \\ 1 \\ 0 \end{bmatrix}, \begin{bmatrix} 1 \\ 0 \\ 2 \end{bmatrix}, \begin{bmatrix} -1 \\ 1 \\ 0 \end{bmatrix} \right\}$

$B_2 = \left\{ \begin{bmatrix} 1 \\ 0 \\ 1 \end{bmatrix}, \begin{bmatrix} 0 \\ 1 \\ 2 \end{bmatrix}, \begin{bmatrix} -1 \\ -1 \\ 0 \end{bmatrix} \right\}$

$$[\mathbf{v}]_{B_1} = \begin{bmatrix} 2 \\ 1 \\ 1 \end{bmatrix}$$

17. $B_1 = \{1, x, x^2\}$ $B_2 = \{x^2, 1, x\}$,

$$[\mathbf{v}]_{B_1} = \begin{bmatrix} 2 \\ 3 \\ 5 \end{bmatrix}$$

18. $B_1 = \{x^2 - 1, 2x^2 + x + 1, -x + 1\}$
    $B_2 = \{(x-1)^2, x+2, (x+1)^2\}$

$$[\mathbf{v}]_{B_1} = \begin{bmatrix} 1 \\ 1 \\ 2 \end{bmatrix}$$

19. Let $B = \{\mathbf{v}_1, \mathbf{v}_2, \mathbf{v}_3\}$ be the ordered basis of $\mathbb{R}^3$ consisting of

$$\mathbf{v}_1 = \begin{bmatrix} -1 \\ 1 \\ 1 \end{bmatrix} \quad \mathbf{v}_2 = \begin{bmatrix} 1 \\ 0 \\ 1 \end{bmatrix}$$

$$\mathbf{v}_3 = \begin{bmatrix} -1 \\ 1 \\ 0 \end{bmatrix}$$

Find the coordinates of the vector

$$\mathbf{v} = \begin{bmatrix} a \\ b \\ c \end{bmatrix}$$

relative to the ordered basis B.

20. Let $B = \{\mathbf{v}_1, \mathbf{v}_2, \mathbf{v}_3, \mathbf{v}_4\}$ be the basis of $\mathbb{R}^4$ consisting of

$$\mathbf{v}_1 = \begin{bmatrix} 1 \\ 0 \\ 1 \\ 0 \end{bmatrix} \quad \mathbf{v}_2 = \begin{bmatrix} 0 \\ -1 \\ 1 \\ -1 \end{bmatrix}$$

$$\mathbf{v}_3 = \begin{bmatrix} 0 \\ -1 \\ -1 \\ 0 \end{bmatrix} \quad \mathbf{v}_4 = \begin{bmatrix} -1 \\ 0 \\ 0 \\ -1 \end{bmatrix}$$

Find the coordinates of the vector

$$\mathbf{v} = \begin{bmatrix} a \\ b \\ c \\ d \end{bmatrix}$$

relative to the ordered basis B.

21. Let

$$B_1 = \left\{ \begin{bmatrix} 1 \\ 0 \\ 0 \end{bmatrix}, \begin{bmatrix} 0 \\ 1 \\ 0 \end{bmatrix}, \begin{bmatrix} 0 \\ 0 \\ 1 \end{bmatrix} \right\}$$

be the standard ordered basis for $\mathbb{R}^3$ and let

$$B_2 = \left\{ \begin{bmatrix} 0 \\ 1 \\ 0 \end{bmatrix}, \begin{bmatrix} 1 \\ 0 \\ 0 \end{bmatrix}, \begin{bmatrix} 0 \\ 0 \\ 1 \end{bmatrix} \right\}$$

be a second ordered basis.

a. Find the transition matrix from the ordered basis B_1 to the ordered basis B_2.

b. Find the coordinates of the vector

$$\mathbf{v} = \begin{bmatrix} 1 \\ 2 \\ 3 \end{bmatrix}$$

relative to the ordered basis B_2.

22. Let

$$B_1 = \left\{ \begin{bmatrix} 2 \\ 2 \end{bmatrix}, \begin{bmatrix} 1 \\ -2 \end{bmatrix} \right\}$$

and

$$B_2 = \left\{ \begin{bmatrix} -1 \\ 2 \end{bmatrix}, \begin{bmatrix} 3 \\ 0 \end{bmatrix} \right\}$$

be two ordered bases for $\mathbb{R}^2$.

a. Find $[I]_{B_1}^{B_2}$

b. Find $[I]_{B_2}^{B_1}$

c. Show that $\left([I]_{B_1}^{B_2} \right)^{-1} = [I]_{B_2}^{B_1}$

23. Let
$$S = \left\{ \begin{bmatrix} 1 \\ 0 \end{bmatrix}, \begin{bmatrix} 0 \\ 1 \end{bmatrix} \right\}$$
be the standard ordered basis for $\mathbb{R}^2$ and let
$$B = \left\{ \begin{bmatrix} 1 \\ 0 \end{bmatrix}, \begin{bmatrix} -\frac{1}{2} \\ \frac{1}{2} \end{bmatrix} \right\}$$
be a second ordered basis.

a. Find $[I]_S^B$

b. Find the coordinates of
$$\begin{bmatrix} 1 \\ 2 \end{bmatrix} \begin{bmatrix} 1 \\ 4 \end{bmatrix} \begin{bmatrix} 4 \\ 2 \end{bmatrix} \begin{bmatrix} 4 \\ 4 \end{bmatrix}$$
relative to the ordered basis B.

c. Draw the rectangle in the plane with vertices $(1, 2)$, $(1, 4)$, $(4, 1)$, and $(4, 4)$.

d. Draw the polygon in the plane with vertices given by the coordinates found in part (b).

24. Fix a real number θ and define the transition matrix from the standard ordered basis S on $\mathbb{R}^2$ to a second ordered basis B by
$$[I]_S^B = \begin{bmatrix} \cos\theta & -\sin\theta \\ \sin\theta & \cos\theta \end{bmatrix}$$

a. If $[\mathbf{v}]_S = \begin{bmatrix} x \\ y \end{bmatrix}$, then find $[\mathbf{v}]_B$.

b. Draw the rectangle in the plane with vertices
$$\begin{bmatrix} 0 \\ 0 \end{bmatrix} \begin{bmatrix} 0 \\ 1 \end{bmatrix} \begin{bmatrix} 1 \\ 0 \end{bmatrix} \begin{bmatrix} 1 \\ 1 \end{bmatrix}$$

c. Let $\theta = \frac{\pi}{2}$. Draw the rectangle in the plane with vertices the coordinates of the vectors, given in part (b), relative to the ordered basis B.

25. Suppose that $B_1 = \{\mathbf{u}_1, \mathbf{u}_2, \mathbf{u}_3\}$ and $B_2 = \{\mathbf{v}_1, \mathbf{v}_2, \mathbf{v}_3\}$ are ordered bases for a vector space V such that $\mathbf{u}_1 = -\mathbf{v}_1 + 2\mathbf{v}_2$, $\mathbf{u}_2 = -\mathbf{v}_1 + 2\mathbf{v}_2 - \mathbf{v}_3$, and $\mathbf{u}_3 = -\mathbf{v}_2 + \mathbf{v}_3$.

a. Find the transition matrix $[I]_{B_1}^{B_2}$

b. Find $[2\mathbf{u}_1 - 3\mathbf{u}_2 + \mathbf{u}_3]_{B_2}$

3.5 ▶ **Application: Differential Equations**

Differential equations arise naturally in virtually every branch of science and technology. They are used extensively by scientists and engineers to solve problems concerning growth, motion, vibrations, forces, or any problem involving the rates of change of variable quantities. Not surprisingly, mathematicians have devoted a great deal of effort to developing methods for solving differential equations. As it turns out, linear algebra is highly useful to these efforts. However, linear algebra also makes it possible to attain a deeper understanding of the theoretical foundations of these equations and their solutions. In this section and in Sec. 5.3 we give a brief introduction to the connection between linear algebra and differential equations.

As a first step, let y be a function of a single variable x. An equation that involves x, y, y', y'', ..., $y^{(n)}$, where n is a fixed positive integer, is called an **ordinary differential equation of order** n. We will henceforth drop the qualifier *ordinary* since none of the equations we investigate will involve partial derivatives. Also, for obvious reasons we will narrow the scope of our discussion and consider only equations of a certain type.

The Exponential Model

One of the simplest kinds of differential equations is the first-order equation given by

$$y' = ky$$

where k is a real number. This equation is used to model quantities that exhibit exponential growth or decay and is based on the assumption that the rate of change of the quantity present at any time t is directly proportional to the quantity present at time t. A **solution** to a differential equation is a function $y = f(t)$ that satisfies the equation, that is, results in an identity when substituted for y in the original equation. To solve this equation, we write it as

$$\frac{y'}{y} = k$$

and integrate both sides of the equation with respect to the independent variable to obtain

$$\ln y = \int \frac{y'}{y}\, dt = \int k\, dt = kt + A$$

Solving for y gives

$$y = e^{\ln y} = e^{kt+A} = e^A e^{kt} = Ce^{kt}$$

where C is an arbitrary constant.

As an illustration, consider the differential equation $y' = 3y$. Then any function of the form $y(t) = Ce^{3t}$ is a solution. Since the parameter C in the solution is arbitrary, the solution produces a family of functions all of which satisfy the differential equation. For this reason $y(t) = Ce^{3t}$ is called the **general solution** to $y' = 3y$.

In certain cases a physical constraint imposes a condition on the solution that allows for the identification of a **particular solution**. If, for example, in the previous problem it is required that $y = 2$ when $t = 0$, then $2 = Ce^{3(0)}$, so that $C = 2$. This is called an **initial condition**. A differential equation together with an initial condition is called an **initial-value problem**. The solution to the previous initial-value problem is given by

$$y(t) = 2e^{3t}$$

From a linear algebra perspective we can think of the general solution to the differential equation $y' = ky$ as the span, over $\mathbb{R}$, of the vector e^{kt} which describes a one-dimensional subspace of the vector space of differentiable functions on the real line.

Second-Order Differential Equations with Constant Coefficients

We now extend the differential equation of the previous subsection to second-order and consider equations of the form

$$y'' + ay' + by = 0$$

Motivated by the solution to the exponential model, we check to see if there are any solutions of the form $y = e^{rx}$, for some real number r. After computing the first and second derivatives $y' = re^{rx}$ and $y'' = r^2 e^{rx}$, we see that $y = e^{rx}$ is a solution of the second-order equation if and only if

$$r^2 e^{rx} + are^{rx} + be^{rx} = 0$$

that is,

$$e^{rx}(r^2 + ar + b) = 0$$

Since $e^{rx} > 0$ for every choice of r and x, we know e^{rx} is a solution of $y'' + ay' + by = 0$ if and only if

$$r^2 + ar + b = 0$$

This equation is called the **auxiliary equation**. As this equation is quadratic there are three possibilities for the roots r_1 and r_2. This in turn yields three possible variations for the solution of the differential equation. The auxiliary equation can have two distinct real roots, one real root, or two distinct complex roots. These cases are considered in order.

Case 1 The roots r_1 and r_2 are real and distinct.
In this case there are two solutions, given by

$$y_1(x) = e^{r_1 x} \qquad \text{and} \qquad y_2(x) = e^{r_2 x}$$

EXAMPLE 1   Find two distinct solutions to the differential equation $y'' - 3y' + 2y = 0$.

Solution   Let $y = e^{rx}$. Since the auxiliary equation $r^2 - 3r + 2 = (r-1)(r-2) = 0$ has the distinct real roots $r_1 = 1$ and $r_2 = 2$, two distinct solutions for the differential equation are

$$y_1(x) = e^x \qquad \text{and} \qquad y_2(x) = e^{2x}$$

Case 2 There is one repeated root r. Although the auxiliary equation has only one root, there are still two distinct solutions, given by

$$y_1(x) = e^{rx} \qquad \text{and} \qquad y_2(x) = xe^{rx}$$

EXAMPLE 2   Find two distinct solutions to the differential equation $y'' - 2y' + y = 0$.

Solution   Let $y = e^{rx}$. Since the auxiliary equation $r^2 - 2r + 1 = (r-1)^2 = 0$ has the repeated root $r = 1$, two distinct solutions of the differential equation are

$$y_1(x) = e^x \qquad \text{and} \qquad y_2(x) = xe^x$$

Case 3 The auxiliary equation has distinct complex (conjugate) roots given by $r_1 = \alpha + \beta i$ and $r_2 = \alpha - \beta i$. In this case the solutions are

$$y_1(x) = e^{\alpha x} \cos \beta x \qquad \text{and} \qquad y_2(x) = e^{\alpha x} \sin \beta x$$

| EXAMPLE 3 | Find two distinct solutions to the differential equation $y'' - 2y' + 5y = 0$. |

Solution   Let $y = e^{rx}$, so the auxiliary equation corresponding to $y'' - 2y' + 5y = 0$ is given by $r^2 - 2r + 5 = 0$. Applying the quadratic formula gives the complex roots $r_1 = 1 + 2i$ and $r_2 = 1 - 2i$. The two solutions to the differential equation are then given by

$$y_1(x) = e^x \cos 2x \qquad \text{and} \qquad y_2(x) = e^x \sin 2x$$

In what follows we require Theorem 16 on existence and uniqueness for second-order linear differential equations. A proof can be found in any text on ordinary differential equations.

| THEOREM 16 | Let $p(x)$, $q(x)$, and $f(x)$ be continuous functions on the interval I. If x_0 is in I, then the initial-value problem |

$$y'' + p(x)y' + q(x)y = f(x) \qquad y(x_0) = y_0 \qquad y'(x_0) = y_0'$$

has a unique solution on I.

Fundamental Sets of Solutions

With solutions in hand for each one of these cases, we now consider the question as to whether there are other solutions to equations of this type, and if so, how they can be described. The simple (but elegant) answer, to which the remainder of this section is devoted, is found by using linear algebra. We will see that in each case the functions $y_1(x)$ and $y_2(x)$ form a basis for the vector space of solutions to the equation $y'' + ay' + by = 0$. Accordingly, every solution $y(x)$ to this equation can be written as a linear combination $y(x) = c_1 y_1(x) + c_2 y_2(x)$.

Toward this end, for a positive integer $n \geq 0$, let $V = C^{(n)}(I)$ be the vector space of all functions that are n times differentiable on the real interval I. If $n = 0$, then $C^{(0)}(I)$ denotes the set of all continuous functions on I. We first show that the solution set to the differential equation $y'' + ay' + by = 0$ is a subspace of $V = C^{(2)}(I)$.

| THEOREM 17 | **Superposition Principle** Suppose that $y_1(x)$ and $y_2(x)$ are functions in $C^{(2)}(I)$. If $y_1(x)$ and $y_2(x)$ are solutions to the differential equation $y'' + ay' + by = 0$ and c is any scalar, then $y_1(x) + cy_2(x)$ is also a solution. |

Proof   Since $y_1(x)$ and $y_2(x)$ are both solutions, then

$$y_1''(x) + ay_1'(x) + by_1(x) = 0 \qquad \text{and} \qquad y_2''(x) + ay_2'(x) + by_2(x) = 0$$

Now to show that $y(x) = y_1(x) + cy_2(x)$ is a solution to the differential equation, observe that

$$y'(x) = y_1'(x) + cy_2'(x) \qquad \text{and} \qquad y''(x) = y_1''(x) + cy_2''(x)$$

Substituting the values for y, y', and y'' in the differential equation and rearranging the terms gives

$$\begin{aligned}
y_1''(x) + cy_2''(x) &+ a[y_1'(x) + cy_2'(x)] + b[y_1(x) + cy_2(x)] \\
&= y_1''(x) + cy_2''(x) + ay_1'(x) + acy_2'(x) + by_1(x) + bcy_2(x) \\
&= [y_1''(x) + ay_1'(x) + by_1(x)] + c[y_2''(x) + ay_2'(x) + by_2(x)] \\
&= 0 + 0 = 0
\end{aligned}$$

Let S be the set of solutions to the differential equation $y'' + ay' + by = 0$. By the superposition principle above and by Theorem 4 of Sec. 3.2, we know that S is a subspace of $C^{(2)}(I)$.

To analyze the algebraic structure of S, we recall from Exercise 31 of Sec. 2.3 that a set of functions $U = \{f_1(x), f_2(x), \ldots, f_n(x)\}$ is linearly independent on an interval I if and only if

$$c_1 f_1(x) + c_2 f_2(x) + \cdots + c_n f_n(x) = 0$$

for all $x \in I$ implies that $c_1 = c_2 = \cdots = c_n = 0$. Theorem 18 provides a useful test to decide whether two functions are linearly independent on an interval.

THEOREM 18

Wronskian   Let $f(x)$ and $g(x)$ be differentiable functions on an interval I. Define the function $W[f, g]$ on I by

$$W[f, g](x) = \begin{vmatrix} f(x) & g(x) \\ f'(x) & g'(x) \end{vmatrix} = f(x)g'(x) - f'(x)g(x)$$

If $W[f, g](x_0)$ is nonzero for some x_0 in I, then $f(x)$ and $g(x)$ are linearly independent on I.

Proof   Consider the equation

$$c_1 f(x) + c_2 g(x) = 0$$

Taking derivatives of both sides, we obtain

$$c_1 f'(x) + c_2 g'(x) = 0$$

Taken together, these equations form a linear system of two equations in the two variables c_1 and c_2. Observe that the determinant of the corresponding coefficient matrix is $W[f, g](x)$. Hence, if $W[f, g](x)$ is nonzero for some $x_0 \in I$, then by

Theorem 17 of Sec. 1.6 we know that $c_1 = c_2 = 0$. Accordingly, $f(x)$ and $g(x)$ are linearly independent.

The function $W[f, g]$ of Theorem 18 is called the **Wronskian** of f and g. The Wronskian, and the result of Theorem 18, can be extended to any finite set of functions that have continuous derivatives up to order n.

If y_1 and y_2 are solutions to the differential equation $y'' + ay' + by = 0$, then *Abel's formula* for the Wronskian gives the next result.

THEOREM 19   Let $y_1(x)$ and $y_2(x)$ be solutions to the differential equation $y'' + ay' + by = 0$. The functions y_1 and y_2 are linearly independent if and only if $W[y_1, y_2](x) \neq 0$ for all x in I.

At this point we are now ready to show that any two linearly independent solutions to the differential equation $y'' + ay' + by = 0$ span the subspace of solutions.

THEOREM 20   **Fundamental Set of Solutions**   Suppose that $y_1(x)$ and $y_2(x)$ are two linearly independent solutions, on the interval I, to the differential equation

$$y'' + ay' + by = 0$$

Then every solution can be written as a linear combination of $y_1(x)$ and $y_2(x)$.

Proof   Let $y(x)$ be a particular solution to the initial-value problem

$$y'' + ay' + by = 0 \quad \text{with} \quad y(x_0) = y_0 \quad \text{and} \quad y'(x_0) = y_0'$$

for some x_0 in I. We claim that there exist real numbers c_1 and c_2 such that

$$y(x) = c_1 y_1(x) + c_2 y_2(x)$$

Differentiating both sides of this equation gives

$$y'(x) = c_1 y_1'(x) + c_2 y_2'(x)$$

Now substituting x_0 into both of these equations and using the initial conditions above, we obtain the linear system of two equations in the two variables c_1 and c_2 given by

$$\begin{cases} c_1 y_1(x_0) + c_2 y_2(x_0) &= y_0 \\ c_1 y_1'(x_0) + c_2 y_2'(x_0) &= y_0' \end{cases}$$

Observe that the determinant of the coefficient matrix is the Wronskian $W[y_1, y_2](x_0)$. Since $y_1(x)$ and $y_2(x)$ are linearly independent, then by Theorem 19, the determinant of the coefficient matrix is nonzero. Consequently, by Theorem 17 of Sec. 1.6, there exist unique numbers c_1 and c_2 that provide a solution for the linear system. Define the function g by

$$g(x) = c_1 y_1(x) + c_2 y_2(x)$$

Then $g(x)$ is also a solution to the original initial-value problem. By the uniqueness part of Theorem 16,

$$y(x) = g(x) = c_1 y_1(x) + c_2 y_2(x)$$

as claimed.

The linearly independent solutions $y_1(x)$ and $y_2(x)$ of Theorem 20 are called a **fundamental set of solutions**. In light of this theorem, the fundamental set $\{y_1(x), y_2(x)\}$ is a basis for the subspace S of solutions to $y'' + ay' + by = 0$. As there are two of them, $\dim(S) = 2$.

We now return to the specific cases for the solutions to $y'' + ay' + by = 0$. Recall that for case 1 we obtained the two solutions

$$y_1(x) = e^{r_1 x} \qquad \text{and} \qquad y_2(x) = e^{r_2 x}$$

with $r_1 \neq r_2$. To show that these functions form a fundamental set, we compute the Wronskian, so that

$$W[y_1, y_2](x) = \begin{vmatrix} e^{r_1 x} & e^{r_2 x} \\ r_1 e^{r_1 x} & r_2 e^{r_2 x} \end{vmatrix}$$
$$= r_2(e^{r_1 x} e^{r_2 x}) - r_1(e^{r_1 x} e^{r_2 x})$$
$$= r_2 e^{(r_1 + r_2)x} - r_1 e^{(r_1 + r_2)x}$$
$$= e^{(r_1 + r_2)x}(r_2 - r_1)$$

Since the exponential function is always greater than 0 and r_1 and r_2 are distinct, the Wronskian is nonzero for all x, and therefore the functions are linearly independent. Hence, $\{e^{r_1 x}, e^{r_2 x}\}$ is a fundamental set, and every solution $y(x)$ to a problem of this type has the form

$$y(x) = c_1 e^{r_1 x} + c_2 e^{r_2 x}$$

for scalars c_1 and c_2.

For case 2, the Wronskian is given by

$$W[e^{rx}, xe^{rx}] = e^{2rx}$$

Since e^{2rx} is never zero, $\{e^{rx}, xe^{rx}\}$ is a fundamental set of solutions for problems of this type.

Finally, for case 3 the Wronskian is given by

$$W[e^{\alpha x} \cos \beta x, e^{\alpha x} \sin \beta x] = \beta e^{2\alpha x}$$

so that $\{e^{\alpha x} \cos \beta x, e^{\alpha x} \sin \beta x\}$ is a fundamental set as long as β is not zero. If $\beta = 0$, then the differential equation becomes $y'' + ay' = 0$ which reduces to case 1.

There are many physical applications of second-order differential equations with constant coefficients. Two important areas are in mechanical and electrical oscillations.

A fundamental problem in mechanics is the motion of an object on a vibrating spring. The motion of the object is described by the solution of an initial-value problem of the form

$$my'' + cy' + ky = f(x) \qquad y(0) = A \qquad y'(0) = B$$

where m is the mass of the object attached to the spring, c is the *damping coefficient*, k is the stiffness of the spring, and $f(x)$ represents some external force. If there are no external forces acting on the system, then $f(x) = 0$.

EXAMPLE 4   Let the mass of an object attached to a spring be $m = 1$, and the spring constant $k = 4$. Solve the three initial-value problems describing the position of the object attached to the spring with no external forces; initial conditions $y(0) = 2$, $y'(0) = 0$; and damping coefficients c equaling 2, 4, and 5.

Solution   The differential equation describing the position of the object is given by

$$y'' + cy' + 4y = 0$$

When $c = 2$, the auxiliary equation for $y'' + 2y' + 4y = 0$ is

$$r^2 + 2r + 4 = 0$$

Since the roots are the complex values $r_1 = -1 + \sqrt{3}i$ and $r_2 = -1 - \sqrt{3}i$, the general solution for the differential equation is

$$y(x) = e^{-x} \left[c_1 \cos(\sqrt{3}x) + c_2 \sin(\sqrt{3}x) \right]$$

From the initial conditions, we have

$$y(x) = 2e^{-x} \left[\cos(\sqrt{3}x) + \frac{\sqrt{3}}{3} \sin(\sqrt{3}x) \right]$$

When $c = 4$, the auxiliary equation for $y'' + 4y' + 4y = 0$ is

$$r^2 + 4r + 4 = (r + 2)^2 = 0$$

Since there is one repeated real root, the general solution for the differential equation is

$$y(x) = c_1 e^{-2x} + c_2 x e^{-2x}$$

From the initial conditions,

$$y(x) = 2e^{-2x}(2x + 1)$$

When $c = 5$, the auxiliary equation for $y'' + 5y' + 4y = 0$ is

$$r^2 + 5r + 4 = (r + 1)(r + 4) = 0$$

Since there are two distinct real roots, the general solution for the differential equation is

$$y(x) = c_1 e^{-x} + c_2 e^{-4x}$$

From the initial conditions,

$$y(x) = \tfrac{2}{3}(4e^{-x} - e^{-4x})$$

The graphs of the solutions are shown in Fig. 1.

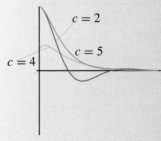

Figure 1

Exercise Set 3.5

In Exercises 1–4, find the general solution to the differential equation.

a. Find two distinct solutions to the homogeneous differential equation.

b. Show that the two solutions from part (a) are linearly independent.

c. Write the general solution.

1. $y'' - 5y' + 6y = 0$

2. $y'' + 3y' + 2y = 0$

3. $y'' + 4y' + 4y = 0$

4. $y'' - 4y' + 5y = 0$

In Exercises 5 and 6, find the solution to the initial-value problem.

5. $y'' - 2y' + y = 0 \quad y(0) = 1 \quad y'(0) = 3$

6. $y'' - 3y' + 2y = 0 \quad y(1) = 0 \quad y'(1) = 1$

7. Consider the the nonhomogeneous differential equation given by

$$y'' - 4y' + 3y = g(x) \qquad \text{where}$$
$$g(x) = 3x^2 + x + 2$$

a. Find the general solution to the associated homogeneous differential equation for which $g(x) = 0$. This is called the *complementary solution* and is denoted by $y_c(x)$.

b. Assume there exists a *particular solution* denoted $y_p(x)$ to the nonhomogeneous equation of the form

$$y_p(x) = ax^2 + bx + c$$

Substitute $y_p(x)$ into the differential equation to find conditions on the coefficients $a, b,$ and c.

c. Verify that $y(x) = y_c(x) + y_p(x)$ is a solution to the differential equation.

8. Consider the nonhomogeneous differential equation given by

$$y'' + 4y' + 3y = g(x) \qquad \text{where}$$
$$g(x) = 3 \sin 2x$$

a. Find the general solution to the associated homogeneous differential equation for which $g(x) = 0$.

b. Assume there exists a particular solution to the nonhomogeneous equation of the form

$$y_p(x) = A \cos 2x + B \sin 2x$$

Substitute $y_p(x)$ into the differential equation to find conditions on the coefficients A and B.

c. Verify that $y_c(x) + y_p(x)$ is a solution to the differential equation.

9. Let w be the weight of an object attached to a spring, g the constant acceleration due to gravity of 32 ft/s^2, k the spring constant, and d the distance in feet that the spring is stretched by the weight. Then the mass of the object is $m = \frac{w}{g}$ and $k = \frac{w}{d}$. Suppose that a 2-lb weight stretches a spring by 6-in. Find the equation of the motion of the weight if the object is pulled down by 3-in and then released. Notice that this system is *undamped*; that is, the damping coefficient is 0.

10. Suppose an 8-lb object is attached to a spring with a spring constant of 4 lb/ft and that the damping force on the system is twice the velocity. Find the equation of the motion if the object is pulled down 1-ft and given an upward velocity of 2 ft/s.

Review Exercises for Chapter 3

1. Determine for which values of k the vectors

$$\begin{bmatrix} 1 \\ -2 \\ 0 \\ 2 \end{bmatrix} \quad \begin{bmatrix} 0 \\ 1 \\ -1 \\ 3 \end{bmatrix} \quad \begin{bmatrix} 0 \\ 0 \\ 1 \\ 4 \end{bmatrix} \quad \begin{bmatrix} 2 \\ 3 \\ 4 \\ k \end{bmatrix}$$

form a basis for $\mathbb{R}^4$.

2. For which values of $a, b, c, d, e,$ and f are the vectors

$$\begin{bmatrix} a \\ 0 \\ 0 \end{bmatrix} \quad \begin{bmatrix} b \\ c \\ 0 \end{bmatrix} \quad \begin{bmatrix} d \\ e \\ f \end{bmatrix}$$

a basis for $\mathbb{R}^3$?

3. Let

$$S = \left\{ \begin{bmatrix} a - b & a \\ b + c & a - c \end{bmatrix} \,\middle|\, a, b, c \in \mathbb{R} \right\}$$

a. Show that S is a subspace of $M_{2\times2}$.

b. Is $\begin{bmatrix} 5 & 3 \\ -2 & 3 \end{bmatrix}$ in S?

c. Find a basis B for S.

d. Give a 2×2 matrix that is not in S.

4. Let $S = \{p(x) = a + bx + cx^2 \mid a + b + c = 0\}$.

a. Show that S is a subspace of $\mathcal{P}_2$.

b. Find a basis for S. Specify the dimension of S.

5. Suppose that $S = \{\mathbf{v}_1, \mathbf{v}_2, \mathbf{v}_3\}$ is a basis for a vector space V.

a. Determine whether the set $T = \{\mathbf{v}_1, \mathbf{v}_1 + \mathbf{v}_2, \mathbf{v}_1 + \mathbf{v}_2 + \mathbf{v}_3\}$ is a basis for V.

b. Determine whether the set $W = \{-\mathbf{v}_2 + \mathbf{v}_3, 3\mathbf{v}_1 + 2\mathbf{v}_2 + \mathbf{v}_3, \mathbf{v}_1 - \mathbf{v}_2 + 2\mathbf{v}_3\}$ is a basis for V.

6. Let $S = \{\mathbf{v}_1, \mathbf{v}_2, \mathbf{v}_3\}$, where

$$\mathbf{v}_1 = \begin{bmatrix} 1 \\ -3 \\ 1 \\ 1 \end{bmatrix} \quad \mathbf{v}_2 = \begin{bmatrix} 2 \\ -1 \\ 1 \\ 1 \end{bmatrix}$$

$$\mathbf{v}_3 = \begin{bmatrix} 4 \\ -7 \\ 3 \\ 3 \end{bmatrix}$$

a. Explain why the set S is not a basis for $\mathbb{R}^4$.

b. Show that $\mathbf{v}_3$ is a linear combination of $\mathbf{v}_1$ and $\mathbf{v}_2$.

c. Find the dimension of the span of the set S.

d. Find a basis B for $\mathbb{R}^4$ that contains the vectors $\mathbf{v}_1$ and $\mathbf{v}_2$.

e. Show that

$$T = \left\{ \begin{bmatrix} 1 \\ 2 \\ -1 \\ 1 \end{bmatrix}, \begin{bmatrix} 1 \\ 0 \\ 1 \\ 0 \end{bmatrix}, \begin{bmatrix} 1 \\ 0 \\ 1 \\ -1 \end{bmatrix}, \begin{bmatrix} 1 \\ 2 \\ 1 \\ -1 \end{bmatrix} \right\}$$

is a basis for $\mathbb{R}^4$.

f. Find the transition matrix from the ordered basis B to the ordered basis T.

g. Use the matrix found in part (f) to find the transition matrix from the ordered basis T to the ordered basis B.

h. If

$$[\mathbf{v}]_B = \begin{bmatrix} 1 \\ 3 \\ -2 \\ 5 \end{bmatrix}$$

find the coordinates of $\mathbf{v}$ relative to the ordered basis T.

i. If

$$[\mathbf{v}]_T = \begin{bmatrix} -2 \\ 13 \\ -5 \\ -1 \end{bmatrix}$$

find the coordinates of $\mathbf{v}$ relative to the ordered basis B.

7. Suppose $\mathbf{span}\{\mathbf{v}_1, \ldots, \mathbf{v}_n\} = V$ and

$$c_1\mathbf{v}_1 + c_2\mathbf{v}_2 + \cdots + c_n\mathbf{v}_n = \mathbf{0}$$

with $c_1 \neq 0$. Show that $\mathbf{span}\{\mathbf{v}_2, \ldots, \mathbf{v}_n\} = V$.

8. Let $V = M_{2 \times 2}$.

a. Give a basis for V and find its dimension. Let S be the set of all matrices of the form

$$\begin{bmatrix} a & b \\ c & a \end{bmatrix}$$

and let T be the set of all matrices of the form

$$\begin{bmatrix} x & y \\ y & z \end{bmatrix}$$

b. Show that S and T are subspaces of the vector space V.

c. Give bases for S and T and specify their dimensions.

d. Give a description of the matrices in $S \cap T$. Find a basis for $S \cap T$ and give its dimension.

9. Let

$$\mathbf{u} = \begin{bmatrix} u_1 \\ u_2 \end{bmatrix} \quad \text{and} \quad \mathbf{v} = \begin{bmatrix} v_1 \\ v_2 \end{bmatrix}$$

such that

$$\mathbf{u} \cdot \mathbf{v} = 0 \quad \text{and} \quad \sqrt{u_1^2 + u_2^2} = 1 = \sqrt{v_1^2 + v_2^2}$$

a. Show that $B = \{\mathbf{u}, \mathbf{v}\}$ is a basis for $\mathbb{R}^2$.

b. Find the coordinates of the vector $\mathbf{w} = \begin{bmatrix} x \\ y \end{bmatrix}$ relative to the ordered basis B.

10. Let c be a fixed scalar and let

$$p_1(x) = 1 \qquad p_2(x) = x + c$$
$$p_3(x) = (x + c)^2$$

a. Show that $B = \{p_1(x), p_2(x), p_3(x)\}$ is a basis for $\mathcal{P}_2$.

b. Find the coordinates of $f(x) = a_0 + a_1x + a_2x^2$ relative to the ordered basis B.

Chapter 3: Chapter Test

In Exercises 1–35, determine whether the statement is true or false.

1. If $V = \mathbb{R}$ and addition and scalar multiplication are defined as

$$x \oplus y = x + 2y \qquad c \odot x = x + c$$

then V is a vector space.

2. The set

$$S = \left\{ \begin{bmatrix} 1 \\ 3 \\ 1 \end{bmatrix}, \begin{bmatrix} 2 \\ 1 \\ -1 \end{bmatrix}, \begin{bmatrix} 0 \\ 4 \\ 3 \end{bmatrix} \right\}$$

is a basis for $\mathbb{R}^3$.

3. A line in $\mathbb{R}^3$ is a subspace of dimension 1.

4. The set

$$S = \left\{ \begin{bmatrix} 2 & 1 \\ 0 & 1 \end{bmatrix}, \begin{bmatrix} 3 & 0 \\ 2 & 1 \end{bmatrix}, \begin{bmatrix} 1 & 0 \\ 2 & 0 \end{bmatrix} \right\}$$

is a basis for $M_{2 \times 2}$.

5. The set

$$\left\{ \begin{bmatrix} 1 \\ 2 \\ 3 \end{bmatrix}, \begin{bmatrix} 0 \\ 1 \\ 2 \end{bmatrix}, \begin{bmatrix} -2 \\ 0 \\ 1 \end{bmatrix} \right\}$$

is a basis for $\mathbb{R}^3$ if and only if

$$\det \begin{bmatrix} 1 & 0 & -2 \\ 2 & 1 & 0 \\ 3 & 2 & 1 \end{bmatrix} \neq 0$$

6. The set

$$S = \left\{ \begin{bmatrix} x \\ y \end{bmatrix} \middle| y \leq 0 \right\}$$

is a subspace of $\mathbb{R}^2$.

7. The set

$$S = \{ A \in M_{2 \times 2} \mid \det(A) = 0 \}$$

is a subspace of $M_{2 \times 2}$.

8. The set

$$\{2, 1 + x, 2 - 3x^2, x^2 - x + 1\}$$

is a basis for $\mathcal{P}_3$.

9. The set

$$\{x^3 - 2x^2 + 1, x^2 - 4, x^3 + 2x, 5x\}$$

is a basis for $\mathcal{P}_3$.

10. The dimension of the subspace

$$S = \left\{ \begin{bmatrix} s + 2t \\ t - s \\ s \end{bmatrix} \middle| s, t \in \mathbb{R} \right\}$$

of $\mathbb{R}^3$ is 2.

11. If

$$S = \left\{ \begin{bmatrix} 1 \\ 4 \end{bmatrix}, \begin{bmatrix} 2 \\ 1 \end{bmatrix} \right\}$$

and

$$T = \left\{ \begin{bmatrix} 1 \\ 4 \end{bmatrix}, \begin{bmatrix} 2 \\ 1 \end{bmatrix}, \begin{bmatrix} 3 \\ 5 \end{bmatrix} \right\}$$

then $\mathbf{span}(S) = \mathbf{span}(T)$.

12. The set

$$S = \left\{ \begin{bmatrix} 2a \\ a \\ 0 \end{bmatrix} \middle| a \in \mathbb{R} \right\}$$

is a subspace of $\mathbb{R}^3$ of dimension 1.

13. If $S = \{\mathbf{v}_1, \mathbf{v}_2, \mathbf{v}_3\}$ and $T = \{\mathbf{v}_1, \mathbf{v}_2, \mathbf{v}_3, \mathbf{v}_1 + \mathbf{v}_2\}$, then $\mathbf{span}(S) = \mathbf{span}(T)$.

14. If $S = \{\mathbf{v}_1, \mathbf{v}_2, \mathbf{v}_3\}$, $T = \{\mathbf{v}_1, \mathbf{v}_2, \mathbf{v}_3, \mathbf{v}_1 + \mathbf{v}_2\}$, and S is a basis, then T is not a basis.

15. If $\{\mathbf{v}_1, \mathbf{v}_2, \mathbf{v}_3\}$ is a basis for a vector space V and $\mathbf{w}_1 = \mathbf{v}_1 + 2\mathbf{v}_2 + \mathbf{v}_3$, $\mathbf{w}_2 = \mathbf{v}_1 + \mathbf{v}_2 + \mathbf{v}_3$, $\mathbf{w}_3 = \mathbf{v}_1 - \mathbf{v}_2 - \mathbf{v}_3$, then $W = \{\mathbf{w}_1, \mathbf{w}_2, \mathbf{w}_3\}$ is also a basis for V.

16. If V is a vector space of dimension n and S is a set of vectors that span V, then the number of vectors in S is less than or equal to n.

17. If V is a vector space of dimension n, then any set of $n - 1$ vectors is linearly dependent.

18. If S and T are subspaces of a vector space V, then $S \cup T$ is a subspace of V.

19. If $S = \{\mathbf{v}_1, \ldots, \mathbf{v}_n\}$ is a linearly independent set of vectors in $\mathbb{R}^n$, then S is a basis.

20. If A is a 3×3 matrix and for every vector

$$\mathbf{b} = \begin{bmatrix} a \\ b \\ c \end{bmatrix} \quad \text{the linear system } A\mathbf{x} = \mathbf{b} \text{ has a}$$

solution, then the column vectors of A span $\mathbb{R}^3$.

21. If an $n \times n$ matrix is invertible, then the column vectors form a basis for $\mathbb{R}^n$.

22. If a vector space has bases S and T and the number of elements of S is n, then the number of elements of T is also n.

23. In a vector space V, if

$$\mathbf{span}\{\mathbf{v}_1, \mathbf{v}_2, \ldots, \mathbf{v}_n\} = V$$

and $\mathbf{w}_1, \mathbf{w}_2, \ldots, \mathbf{w}_m$ are any elements of V, then

$$\mathbf{span}\{\mathbf{v}_1, \mathbf{v}_2, \ldots, \mathbf{v}_n, \mathbf{w}_1, \mathbf{w}_2, \ldots, \mathbf{w}_m\} = V.$$

24. If V is a vector space of dimension n and H is a subspace of dimension n, then $H = V$.

25. If B_1 and B_2 are bases for the vector space V, then the transition matrix from B_1 to B_2 is the inverse of the transition matrix from B_2 to B_1.

In Exercises 26–29, use the bases of $\mathbb{R}^2$

$$B_1 = \left\{ \begin{bmatrix} 1 \\ -1 \end{bmatrix}, \begin{bmatrix} 0 \\ 2 \end{bmatrix} \right\}$$

and

$$B_2 = \left\{ \begin{bmatrix} 1 \\ 1 \end{bmatrix}, \begin{bmatrix} 3 \\ -1 \end{bmatrix} \right\}$$

26. The coordinates of $\begin{bmatrix} 1 \\ 0 \end{bmatrix}$, relative to B_1, are

$$\begin{bmatrix} 1 \\ 1 \end{bmatrix}.$$

27. The coordinates of $\begin{bmatrix} 1 \\ 0 \end{bmatrix}$ relative to B_2 are

$$\frac{1}{4} \begin{bmatrix} 1 \\ 1 \end{bmatrix}.$$

28. The transition matrix from B_1 to B_2 is

$$[I]_{B_1}^{B_2} = \frac{1}{2} \begin{bmatrix} -1 & 3 \\ 1 & -1 \end{bmatrix}$$

29. The transition matrix from B_2 to B_1 is

$$[I]_{B_2}^{B_1} = \begin{bmatrix} 1 & 3 \\ 1 & 1 \end{bmatrix}$$

In Exercises 30–35, use the bases of $\mathcal{P}_3$,

$$B_1 = \{1, x, x^2, x^3\}$$

and

$$B_2 = \{x, x^2, 1, x^3\}$$

30. $[x^3 + 2x^2 - x]_{B_1} = \begin{bmatrix} 1 \\ 2 \\ -1 \end{bmatrix}$

31. $[x^3 + 2x^2 - x]_{B_1} = \begin{bmatrix} 0 \\ -1 \\ 2 \\ 1 \end{bmatrix}$

32. $[x^3 + 2x^2 - x]_{B_2} = \begin{bmatrix} 0 \\ -1 \\ 2 \\ 1 \end{bmatrix}$

33. $[x^3 + 2x^2 - x]_{B_2} = \begin{bmatrix} -1 \\ 2 \\ 0 \\ 1 \end{bmatrix}$

34. $[(1 + x)^2 - 3(x^2 + x - 1) + x^3]_{B_2}$

$$= \begin{bmatrix} 4 \\ -1 \\ -2 \\ 1 \end{bmatrix}$$

35. The transition matrix from B_1 to B_2 is

$$[I]_{B_1}^{B_2} = \begin{bmatrix} 0 & 1 & 0 & 0 \\ 0 & 0 & 1 & 0 \\ 1 & 0 & 0 & 0 \\ 0 & 0 & 0 & 1 \end{bmatrix}$$

Linear Transformations

A critical component in the design of an airplane is the airflow over the *wing*. Four forces that act on an aircraft, and need to be considered in its design, are *lift*, the force of gravity, *thrust*, and *drag*. Lift and drag are aerodynamic forces that are generated by the movement of the aircraft through the air. During take-off, thrust from the engines must overcome drag. Lift, created by the rush of air over the wing, must overcome the force of gravity before the airplane can fly. Mathematical models developed by aeronautical engineers simulate the behavior of an aircraft in flight. These models involve linear systems with millions of equations and variables. As we saw in Chap. 1, linear algebra provides systematic methods for solving these equations. Another use of linear algebra in the design process of an airplane is in modeling the movement of the aircraft

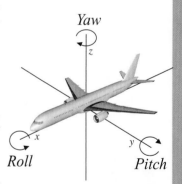

through space. To check the feasibility of their designs, aeronautical engineers use *computer graphics* to visualize simulations of the aircraft in flight. Three control parameters which affect the position of an aircraft are *pitch*, *roll*, and *yaw*. The pitch measures the fore and aft tilt of an airplane, relative to the earth, while the roll measures the tilt from side to side. Together these give the *attitude* of the aircraft. Using the figure above, the pitch is a rotation about the *y* axis, while a roll is a rotation about the *x* axis. The yaw measures the rotation about the *z* axis, and when combined with the pitch, gives the *heading*. During a simulation, the attitude and heading of the aircraft can be changed by applying a *transformation* to its coordinates relative to a predefined center of equilibrium. As we shall see in this chapter, such a transformation can be represented by matrix multiplication. Specifically, if the angles of rotation for pitch, roll, and yaw are given by θ, φ, and ψ, respectively, then the matrix representations for these transformations are given by

$$
\begin{bmatrix} \cos\theta & 0 & -\sin\theta \\ 0 & 1 & 0 \\ \sin\theta & 0 & \cos\theta \end{bmatrix} \quad \begin{bmatrix} 1 & 0 & 0 \\ 0 & \cos\varphi & -\sin\varphi \\ 0 & \sin\varphi & \cos\varphi \end{bmatrix} \quad \text{and} \quad \begin{bmatrix} \cos\psi & \sin\psi & 0 \\ -\sin\psi & \cos\psi & 0 \\ 0 & 0 & 1 \end{bmatrix}
$$

This type of transformation is a *linear map* between vector spaces, in this case from $\mathbb{R}^3$ to $\mathbb{R}^3$. The generation and manipulation of computer graphics are one of many applications that require the linear transformations which are introduced in this chapter.

Due to their wide applicability *linear transformations* on vector spaces are of general interest and are the subject of this chapter. As functions between vector spaces, they are special since they preserve the additive structure of linear combinations. That is, the image of a linear combination under a linear transformation is also a linear combination in the range. In this chapter we investigate the connection between linear transformations and matrices, showing that every linear transformation between finite dimensional vector spaces can be written as a matrix multiplication.

4.1 ▶ Linear Transformations

In mathematics every line of inquiry ultimately leads to a description of some set and functions on that set. One may metaphorically refer to elements of the set as *nouns* and functions that operate on elements as *verbs*. In linear algebra the sets are vector spaces, which we discussed in Chap. 3, and linear transformations on vector spaces are the functions.

If V and W are vector spaces, then a **mapping** T from V to W is a function that assigns to each vector $\mathbf{v}$ in V a unique vector $\mathbf{w}$ in W. In this case we say that T maps V into W, and we write $T: V \longrightarrow W$. For each $\mathbf{v}$ in V the vector $\mathbf{w} = T(\mathbf{v})$ in W is the **image** of $\mathbf{v}$ under T.

EXAMPLE 1 Define a mapping $T: \mathbb{R}^2 \longrightarrow \mathbb{R}^2$ by

$$
T\left(\begin{bmatrix} x \\ y \end{bmatrix}\right) = \begin{bmatrix} x+y \\ x-y \end{bmatrix}
$$

a. Find the image of the coordinate vectors $\mathbf{e}_1$ and $\mathbf{e}_2$ under the mapping T.

b. Give a description of all vectors in $\mathbb{R}^2$ that are mapped to the zero vector.

c. Show that the mapping T satisfies

$$T(\mathbf{u} + \mathbf{v}) = T(\mathbf{u}) + T(\mathbf{v}) \qquad (\textit{preserves} \text{ vector space addition})$$

and

$$T(c\mathbf{v}) = cT(\mathbf{v}) \qquad (\textit{preserves} \text{ scalar multiplication})$$

for all vectors $\mathbf{u}$ and $\mathbf{v}$ in V and all scalars c in $\mathbb{R}$.

Solution   **a.** Since $\mathbf{e}_1 = \begin{bmatrix} 1 \\ 0 \end{bmatrix}$ and $\mathbf{e}_2 = \begin{bmatrix} 0 \\ 1 \end{bmatrix}$, we have

$$T(\mathbf{e}_1) = \begin{bmatrix} 1+0 \\ 1-0 \end{bmatrix} = \begin{bmatrix} 1 \\ 1 \end{bmatrix} \quad \text{and} \quad T(\mathbf{e}_2) = \begin{bmatrix} 1 \\ -1 \end{bmatrix}$$

b. To answer this question, we solve

$$T\left(\begin{bmatrix} x \\ y \end{bmatrix}\right) = \begin{bmatrix} x+y \\ x-y \end{bmatrix} = \begin{bmatrix} 0 \\ 0 \end{bmatrix}$$

This leads to the linear system

$$\begin{cases} x+y = 0 \\ x-y = 0 \end{cases}$$

where the unique solution is $x = y = 0$. Thus, the only vector that is mapped by T to $\begin{bmatrix} 0 \\ 0 \end{bmatrix}$ is the zero vector $\begin{bmatrix} 0 \\ 0 \end{bmatrix}$.

c. To show that the mapping T preserves vector space addition, let

$$\mathbf{u} = \begin{bmatrix} u_1 \\ u_2 \end{bmatrix} \quad \text{and} \quad \mathbf{v} = \begin{bmatrix} v_1 \\ v_2 \end{bmatrix}$$

Then

$$\begin{aligned}
T(\mathbf{u} + \mathbf{v}) &= T\left(\begin{bmatrix} u_1 \\ u_2 \end{bmatrix} + \begin{bmatrix} v_1 \\ v_2 \end{bmatrix}\right) \\
&= T\left(\begin{bmatrix} u_1 + v_1 \\ u_2 + v_2 \end{bmatrix}\right) \\
&= \begin{bmatrix} (u_1 + v_1) + (u_2 + v_2) \\ (u_1 + v_1) - (u_2 + v_2) \end{bmatrix} \\
&= \begin{bmatrix} u_1 + u_2 \\ u_1 - u_2 \end{bmatrix} + \begin{bmatrix} v_1 + v_2 \\ v_1 - v_2 \end{bmatrix} \\
&= T\left(\begin{bmatrix} u_1 \\ u_2 \end{bmatrix}\right) + T\left(\begin{bmatrix} v_1 \\ v_2 \end{bmatrix}\right) \\
&= T(\mathbf{u}) + T(\mathbf{v})
\end{aligned}$$

We also have

$$\begin{aligned}
T(c\mathbf{u}) &= T\left(\begin{bmatrix} cu_1 \\ cu_2 \end{bmatrix}\right) \\
&= \begin{bmatrix} cu_1 + cu_2 \\ cu_1 - cu_2 \end{bmatrix} = c\begin{bmatrix} u_1 + u_2 \\ u_1 - u_2 \end{bmatrix} \\
&= cT(\mathbf{u})
\end{aligned}$$

A mapping T between vector spaces V and W that satisfies the two properties, as in Example 1,

$$T(\mathbf{u} + \mathbf{v}) = T(\mathbf{u}) + T(\mathbf{v}) \quad \text{and} \quad T(c\mathbf{u}) = cT(\mathbf{u})$$

is called a *linear transformation* from V into W. Notice that the operations of addition and scalar multiplication on the left-hand side of each equation refer to operations in the vector space V, and on the right-hand side refer to operations in the vector space W.

Definition 1 combines the two requirements for the linearity of T into one statement.

DEFINITION 1   **Linear Transformation**   Let V and W be vector spaces. The mapping $T: V \rightarrow W$ is called a **linear transformation** if and only if

$$T(c\mathbf{u} + \mathbf{v}) = cT(\mathbf{u}) + T(\mathbf{v})$$

for every choice of $\mathbf{u}$ and $\mathbf{v}$ in V and scalars c in $\mathbb{R}$. In the case for which $V = W$, then T is called a **linear operator**.

The mapping T defined in Example 1 is a linear operator on $\mathbb{R}^2$. In Example 2 we show how matrices can be used to define linear transformations.

EXAMPLE 2   Let A be an $m \times n$ matrix. Define a mapping $T: \mathbb{R}^n \rightarrow \mathbb{R}^m$ by

$$T(\mathbf{x}) = A\mathbf{x}$$

a. Show that T is a linear transformation.
b. Let A be the 2×3 matrix

$$A = \begin{bmatrix} 1 & 2 & -1 \\ -1 & 3 & 2 \end{bmatrix}$$

Find the images of

$$\begin{bmatrix} 1 \\ 1 \\ 1 \end{bmatrix} \quad \text{and} \quad \begin{bmatrix} 7 \\ -1 \\ 5 \end{bmatrix}$$

under the mapping $T: \mathbb{R}^3 \rightarrow \mathbb{R}^2$ with $T(\mathbf{x}) = A\mathbf{x}$.

Solution   **a.** By Theorem 5 of Sec. 1.3, for all vectors $\mathbf{u}$ and $\mathbf{v}$ in $\mathbb{R}^n$ and all scalars c in $\mathbb{R}$,

$$A(c\mathbf{u} + \mathbf{v}) = cA\mathbf{u} + A\mathbf{v}$$

Therefore,

$$T(c\mathbf{u} + \mathbf{v}) = cT(\mathbf{u}) + T(\mathbf{v})$$

b. Since T is defined by matrix multiplication, we have

$$T\left(\begin{bmatrix} 1 \\ 1 \\ 1 \end{bmatrix}\right) = \begin{bmatrix} 1 & 2 & -1 \\ -1 & 3 & 2 \end{bmatrix} \begin{bmatrix} 1 \\ 1 \\ 1 \end{bmatrix} = \begin{bmatrix} 2 \\ 4 \end{bmatrix}$$

and

$$T\left(\begin{bmatrix} 7 \\ -1 \\ 5 \end{bmatrix}\right) = \begin{bmatrix} 1 & 2 & -1 \\ -1 & 3 & 2 \end{bmatrix} \begin{bmatrix} 7 \\ -1 \\ 5 \end{bmatrix} = \begin{bmatrix} 0 \\ 0 \end{bmatrix}$$

Later in this chapter, in Sec. 4.4, we show that every linear transformation between finite dimensional vector spaces can be represented by a matrix. In Examples 1 and 2, we have discussed some of the algebraic properties of linear transformations. In Example 3 we consider the action of a linear transformation from a geometric perspective.

EXAMPLE 3   Define a linear transformation $T: \mathbb{R}^3 \longrightarrow \mathbb{R}^2$ by

$$T\left(\begin{bmatrix} x \\ y \\ z \end{bmatrix}\right) = \begin{bmatrix} x \\ y \end{bmatrix}$$

a. Discuss the action of T on a vector in $\mathbb{R}^3$, and give a geometric interpretation of the equation

$$T\left(\begin{bmatrix} 1 \\ 0 \\ 1 \end{bmatrix} + \begin{bmatrix} 0 \\ 1 \\ 1 \end{bmatrix}\right) = T\left(\begin{bmatrix} 1 \\ 0 \\ 1 \end{bmatrix}\right) + T\left(\begin{bmatrix} 0 \\ 1 \\ 1 \end{bmatrix}\right)$$

b. Find the image of the set

$$S_1 = \left\{ t \begin{bmatrix} 1 \\ 2 \\ 1 \end{bmatrix} \middle| t \in \mathbb{R} \right\}$$

c. Find the image of the set

$$S_2 = \left\{ \begin{bmatrix} x \\ y \\ 3 \end{bmatrix} \middle| x, y \in \mathbb{R} \right\}$$

d. Describe the set

$$S_3 = \left\{ \begin{bmatrix} x \\ 0 \\ z \end{bmatrix} \middle| x, z \in \mathbb{R} \right\}$$

and find its image.

Solution   **a.** The linear transformation T gives the *projection*, or shadow, of a vector in 3-space to its image in the xy plane. Let

$$\mathbf{v}_1 = \begin{bmatrix} 1 \\ 0 \\ 1 \end{bmatrix} \qquad \mathbf{v}_2 = \begin{bmatrix} 0 \\ 1 \\ 1 \end{bmatrix} \qquad \text{and} \qquad \mathbf{v}_3 = \mathbf{v}_1 + \mathbf{v}_2 = \begin{bmatrix} 1 \\ 1 \\ 2 \end{bmatrix}$$

The images of these vectors are shown in Fig. 1. We see from the figure that $T(\mathbf{v}_3) = \begin{bmatrix} 1 \\ 1 \end{bmatrix}$ is equal to the vector sum $T(\mathbf{v}_1) + T(\mathbf{v}_2) = \begin{bmatrix} 1 \\ 0 \end{bmatrix} + \begin{bmatrix} 0 \\ 1 \end{bmatrix}$, as desired.

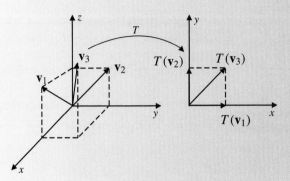

Figure 1

b. The set S_1 is a line in 3-space with direction vector $\begin{bmatrix} 1 \\ 2 \\ 1 \end{bmatrix}$. By the definition of T we have

$$T(S_1) = \left\{ t \begin{bmatrix} 1 \\ 2 \end{bmatrix} \middle| t \in \mathbb{R} \right\}$$

which is a line in $\mathbb{R}^2$ through the origin with slope 2.

c. The set S_2 is a plane in 3-space 3 units above and parallel to the xy plane. In this case,

$$T(S_2) = \left\{ \begin{bmatrix} x \\ y \end{bmatrix} \middle| x, y \in \mathbb{R} \right\}$$

Thus, the image of S_2 is the entire xy plane, which from the description of T as a projection is the result we expect.

d. The set S_3 is the xz plane. Here we have

$$T(S_3) = \left\{ \begin{bmatrix} x \\ 0 \end{bmatrix} \middle| x \in \mathbb{R} \right\}$$

which is just the x axis. Again, this is the expected result, given our description of T.

In Example 4 we use the derivative of a function to define a linear transformation between vector spaces of polynomials.

EXAMPLE 4   Define a mapping $T: \mathcal{P}_3 \longrightarrow \mathcal{P}_2$ by

$$T(p(x)) = p'(x)$$

where $p'(x)$ is the derivative of $p(x)$.

   a. Show that T is a linear transformation.

   b. Find the image of the polynomial $p(x) = 3x^3 + 2x^2 - x + 2$.

   c. Describe the polynomials in $\mathcal{P}_3$ that are mapped to the zero vector of $\mathcal{P}_2$.

Solution   First observe that if $p(x)$ is in $\mathcal{P}_3$, then it has the form

$$p(x) = ax^3 + bx^2 + cx + d$$

so that

$$T(p(x)) = p'(x) = 3ax^2 + 2bx + c$$

Since $p'(x)$ is in $\mathcal{P}_2$, then T is a map from $\mathcal{P}_3$ into $\mathcal{P}_2$.

   a. To show that T is linear, let $p(x)$ and $q(x)$ be polynomials of degree 3 or less, and let k be a scalar. Recall from calculus that the derivative of a sum is the sum of the derivatives, and that the derivative of a scalar times a function is the scalar times the derivative of the function. Consequently,

$$\begin{aligned} T(kp(x) + q(x)) &= \frac{d}{dx}(kp(x) + q(x)) \\ &= \frac{d}{dx}(kp(x)) + \frac{d}{dx}(q(x)) \\ &= kp'(x) + q'(x) \\ &= kT(p(x)) + T(q(x)) \end{aligned}$$

   Therefore, the mapping T is a linear transformation.

   b. The image of the polynomial $p(x) = 3x^3 + 2x^2 - x + 2$ is

$$T(p(x)) = \frac{d}{dx}(3x^3 + 2x^2 - x + 2) = 9x^2 + 4x - 1$$

   c. The only functions in $\mathcal{P}_3$ with derivative equal to zero are the constant polynomials $p(x) = c$, where c is a real number.

PROPOSITION 1   Let V and W be vector spaces, and let $T: V \rightarrow W$ be a linear transformation. Then $T(\mathbf{0}) = \mathbf{0}$.

Proof   Since $T(\mathbf{0}) = T(\mathbf{0} + \mathbf{0})$ and T is a linear transformation, we know that $T(\mathbf{0}) = T(\mathbf{0} + \mathbf{0}) = T(\mathbf{0}) + T(\mathbf{0})$. Subtracting $T(\mathbf{0})$ from both sides of the last equation gives $T(\mathbf{0}) = \mathbf{0}$.

| EXAMPLE 5 | Define a mapping $T: \mathbb{R}^2 \longrightarrow \mathbb{R}^2$ by |

$$T\left(\begin{bmatrix} x \\ y \end{bmatrix}\right) = \left(\begin{bmatrix} e^x \\ e^y \end{bmatrix}\right)$$

Determine whether T is a linear transformation.

Solution   Since

$$T(\mathbf{0}) = T\left(\begin{bmatrix} 0 \\ 0 \end{bmatrix}\right) = \begin{bmatrix} e^0 \\ e^0 \end{bmatrix} = \begin{bmatrix} 1 \\ 1 \end{bmatrix}$$

by the contrapositive of Proposition 1, we know that T is not a linear transformation.

| EXAMPLE 6 | Define a mapping $T: M_{m \times n} \longrightarrow M_{n \times m}$ by |

$$T(A) = A^t$$

Show that the mapping is a linear transformation.

Solution   By Theorem 6 of Sec. 1.3, we have

$$T(A + B) = (A + B)^t = A^t + B^t = T(A) + T(B)$$

Also by this same theorem,

$$T(cA) = (cA)^t = cA^t = cT(A)$$

Thus, T is a linear transformation.

| EXAMPLE 7 | **Coordinates**   Let V be a vector space with $\dim(V) = n$, and $B = \{\mathbf{v}_1, \mathbf{v}_2, \dots, \mathbf{v}_n\}$ an ordered basis for V. Let $T: V \longrightarrow \mathbb{R}^n$ be the map that sends a vector $\mathbf{v}$ in V to its coordinate vector in $\mathbb{R}^n$ relative to B. That is, |

$$T(\mathbf{v}) = [\mathbf{v}]_B$$

It was shown in Sec. 3.4 that this map is well defined, that is, the coordinate vector of $\mathbf{v}$ relative to B is unique. Show that the map T is also a linear transformation.

Solution   Let $\mathbf{u}$ and $\mathbf{v}$ be vectors in V and let k be a scalar. Since B is a basis, there are unique sets of scalars $c_1, \dots, c_n$ and $d_1, \dots, d_n$ such that

$$\mathbf{u} = c_1\mathbf{v}_1 + \cdots + c_n\mathbf{v}_n \qquad \text{and} \qquad \mathbf{v} = d_1\mathbf{v}_1 + \cdots + d_n\mathbf{v}_n$$

Applying T to the vector $k\mathbf{u} + \mathbf{v}$ gives

$$T(k\mathbf{u} + \mathbf{v}) = T((kc_1 + d_1)\mathbf{v}_1 + \cdots + (kc_n + d_n)\mathbf{v}_n)$$

$$= \begin{bmatrix} kc_1 + d_1 \\ kc_2 + d_2 \\ \vdots \\ kc_n + d_n \end{bmatrix} = k \begin{bmatrix} c_1 \\ c_2 \\ \vdots \\ c_n \end{bmatrix} + \begin{bmatrix} d_1 \\ d_2 \\ \vdots \\ d_n \end{bmatrix}$$

$$= kT(\mathbf{u}) + T(\mathbf{v})$$

Therefore, we have shown that the mapping T is a linear transformation.

As mentioned earlier, when $T: V \longrightarrow W$ is a linear transformation, then the structure of V is preserved when it is mapped into W. Specifically, the image of a linear combination of vectors, under a linear map, is equal to a linear combination of the image vectors with the same coefficients. To see this, let V and W be vector spaces and $T: V \to W$ be a linear transformation. Then by repeated application of Definition 1, we have

$$T(c_1\mathbf{v}_1 + c_2\mathbf{v}_2 + \cdots + c_n\mathbf{v}_n) = T(c_1\mathbf{v}_1) + \cdots + T(c_n\mathbf{v}_n)$$
$$= c_1T(\mathbf{v}_1) + c_2T(\mathbf{v}_2) + \cdots + c_nT(\mathbf{v}_n)$$

The fact that a linear transformation T between vector spaces V and W preserves linear combinations is useful in evaluating T when its action on the vectors of a basis for V is known. This is illustrated in Example 8.

EXAMPLE 8 Let $T: \mathbb{R}^3 \to \mathbb{R}^2$ be a linear transformation, and let B be the standard basis for $\mathbb{R}^3$. If

$$T(\mathbf{e}_1) = \begin{bmatrix} 1 \\ 1 \end{bmatrix} \qquad T(\mathbf{e}_2) = \begin{bmatrix} -1 \\ 2 \end{bmatrix} \qquad \text{and} \qquad T(\mathbf{e}_3) = \begin{bmatrix} 0 \\ 1 \end{bmatrix}$$

find $T(\mathbf{v})$, where

$$\mathbf{v} = \begin{bmatrix} 1 \\ 3 \\ 2 \end{bmatrix}$$

Solution To find the image of the vector $\mathbf{v}$, we first write the vector as a linear combination of the basis vectors. In this case

$$\mathbf{v} = \mathbf{e}_1 + 3\mathbf{e}_2 + 2\mathbf{e}_3$$

Applying T to this linear combination and using the linearity properties of T, we have

$$T(\mathbf{v}) = T(\mathbf{e}_1 + 3\mathbf{e}_2 + 2\mathbf{e}_3)$$
$$= T(\mathbf{e}_1) + 3T(\mathbf{e}_2) + 2T(\mathbf{e}_3)$$
$$= \begin{bmatrix} 1 \\ 1 \end{bmatrix} + 3\begin{bmatrix} -1 \\ 2 \end{bmatrix} + 2\begin{bmatrix} 0 \\ 1 \end{bmatrix}$$
$$= \begin{bmatrix} -2 \\ 9 \end{bmatrix}$$

EXAMPLE 9   Let $T: \mathbb{R}^3 \longrightarrow \mathbb{R}^3$ be a linear operator and B a basis for $\mathbb{R}^3$ given by

$$B = \left\{ \begin{bmatrix} 1 \\ 1 \\ 1 \end{bmatrix}, \begin{bmatrix} 1 \\ 2 \\ 3 \end{bmatrix}, \begin{bmatrix} 1 \\ 1 \\ 2 \end{bmatrix} \right\}$$

If

$$T\left(\begin{bmatrix} 1 \\ 1 \\ 1 \end{bmatrix}\right) = \begin{bmatrix} 1 \\ 1 \\ 1 \end{bmatrix} \qquad T\left(\begin{bmatrix} 1 \\ 2 \\ 3 \end{bmatrix}\right) = \begin{bmatrix} -1 \\ -2 \\ -3 \end{bmatrix} \qquad T\left(\begin{bmatrix} 1 \\ 1 \\ 2 \end{bmatrix}\right) = \begin{bmatrix} 2 \\ 2 \\ 4 \end{bmatrix}$$

find

$$T\left(\begin{bmatrix} 2 \\ 3 \\ 6 \end{bmatrix}\right)$$

Solution   Since B is a basis for $\mathbb{R}^3$, there are (unique) scalars c_1, c_2, and c_3 such that

$$c_1\begin{bmatrix} 1 \\ 1 \\ 1 \end{bmatrix} + c_2\begin{bmatrix} 1 \\ 2 \\ 3 \end{bmatrix} + c_3\begin{bmatrix} 1 \\ 1 \\ 2 \end{bmatrix} = \begin{bmatrix} 2 \\ 3 \\ 6 \end{bmatrix}$$

Solving this equation, we obtain $c_1 = -1, c_2 = 1$, and $c_3 = 2$. Hence,

$$T\left(\begin{bmatrix} 2 \\ 3 \\ 6 \end{bmatrix}\right) = T\left(-1\begin{bmatrix} 1 \\ 1 \\ 1 \end{bmatrix} + \begin{bmatrix} 1 \\ 2 \\ 3 \end{bmatrix} + 2\begin{bmatrix} 1 \\ 1 \\ 2 \end{bmatrix}\right)$$

By the linearity of T, we have

$$T\left(\begin{bmatrix} 2 \\ 3 \\ 6 \end{bmatrix}\right) = (-1)T\left(\begin{bmatrix} 1 \\ 1 \\ 1 \end{bmatrix}\right) + T\left(\begin{bmatrix} 1 \\ 2 \\ 3 \end{bmatrix}\right) + 2T\left(\begin{bmatrix} 1 \\ 1 \\ 2 \end{bmatrix}\right)$$
$$= -\begin{bmatrix} 1 \\ 1 \\ 1 \end{bmatrix} + \begin{bmatrix} -1 \\ -2 \\ -3 \end{bmatrix} + 2\begin{bmatrix} 2 \\ 2 \\ 4 \end{bmatrix}$$
$$= \begin{bmatrix} 2 \\ 1 \\ 4 \end{bmatrix}$$

Operations with Linear Transformations

Linear transformations can be combined by using a natural addition and scalar multiplication to produce new linear transformations. For example, let $S, T: \mathbb{R}^2 \rightarrow \mathbb{R}^2$ be defined by

$$S\left(\begin{bmatrix} x \\ y \end{bmatrix}\right) = \begin{bmatrix} x + y \\ -x \end{bmatrix} \quad \text{and} \quad T\left(\begin{bmatrix} x \\ y \end{bmatrix}\right) = \begin{bmatrix} 2x - y \\ x + 3y \end{bmatrix}$$

We then define

$$(S + T)(\mathbf{v}) = S(\mathbf{v}) + T(\mathbf{v}) \quad \text{and} \quad (cS)(\mathbf{v}) = c(S(\mathbf{v}))$$

To illustrate this definition, let $\mathbf{v} = \begin{bmatrix} 2 \\ -1 \end{bmatrix}$; then

$$(S + T)(\mathbf{v}) = S(\mathbf{v}) + T(\mathbf{v}) = \begin{bmatrix} 2 + (-1) \\ -2 \end{bmatrix} + \begin{bmatrix} 2(2) - (-1) \\ 2 + 3(-1) \end{bmatrix} = \begin{bmatrix} 6 \\ -3 \end{bmatrix}$$

For scalar multiplication let $c = 3$. Then

$$(3T)(\mathbf{v}) = 3T(\mathbf{v}) = 3\begin{bmatrix} 5 \\ -1 \end{bmatrix} = \begin{bmatrix} 15 \\ -3 \end{bmatrix}$$

In Theorem 1 we show that these operations on linear transformations produce linear transformations.

THEOREM 1   Let V and W be vector spaces and let $S, T: V \rightarrow W$ be linear transformations. The function $S + T$ defined by

$$(S + T)(\mathbf{v}) = S(\mathbf{v}) + T(\mathbf{v})$$

is a linear transformation from V into W. If c is any scalar, the function cS defined by

$$(cS)(\mathbf{v}) = cS(\mathbf{v})$$

is a linear transformation from V into W.

Proof   Let $\mathbf{u}, \mathbf{v} \in V$ and let d be any scalar. Then

$$(S + T)(d\mathbf{u} + \mathbf{v}) = S(d\mathbf{u} + \mathbf{v}) + T(d\mathbf{u} + \mathbf{v})$$
$$= S(d\mathbf{u}) + S(\mathbf{v}) + T(d\mathbf{u}) + T(\mathbf{v})$$
$$= dS(\mathbf{u}) + S(\mathbf{v}) + dT(\mathbf{u}) + T(\mathbf{v})$$
$$= d(S(\mathbf{u}) + T(\mathbf{u})) + S(\mathbf{v}) + T(\mathbf{v})$$
$$= d(S + T)(\mathbf{u}) + (S + T)(\mathbf{v})$$

so that $S + T$ is a linear transformation. Also

$$
\begin{aligned}
(cS)(d\mathbf{u} + \mathbf{v}) &= c(S(d\mathbf{u} + \mathbf{v})) \\
&= c(S(d\mathbf{u}) + S(\mathbf{v})) \\
&= c(dS(\mathbf{u}) + S(\mathbf{v})) \\
&= (cd)S(\mathbf{u}) + cS(\mathbf{v}) \\
&= d(cS)(\mathbf{u}) + (cS)(\mathbf{v})
\end{aligned}
$$

so that cS is a linear transformation.

Using the sum of two linear transformations and the scalar product defined above, the set of all linear transformations between two given vector spaces is itself a vector space, denoted by $£(U, V)$. Verification of this is left to Exercise 45 at the end of this section.

As we saw in Example 2, every $m \times n$ matrix A defines a linear map from $\mathbb{R}^n$ to $\mathbb{R}^m$. Also, if B is an $n \times p$ matrix, then B defines a linear map from $\mathbb{R}^p$ to $\mathbb{R}^n$. The product matrix AB, which is an $m \times p$ matrix, then defines a linear transformation from $\mathbb{R}^p$ to $\mathbb{R}^m$. As we shall see (in Sec. 4.4), this map corresponds to the composition of the maps defined by A and B. The desire for this correspondence is what motivated the definition of matrix multiplication given in Sec. 1.3.

THEOREM 2   Let U, V, and W be vector spaces. If $T: V \to U$ and $S: U \to W$ are linear transformations, then the composition map $S{\circ}T: V \to W$, defined by

$$
(S{\circ}T)(\mathbf{v}) = S(T(\mathbf{v}))
$$

is a linear transformation. (See Fig. 2.)

Proof   To show that $S{\circ}T$ is a linear transformation, let $\mathbf{v}_1$ and $\mathbf{v}_2$ be vectors in V and c a scalar. Applying $S{\circ}T$ to $c\mathbf{v}_1 + \mathbf{v}_2$, we obtain

$$
\begin{aligned}
(S{\circ}T)(c\mathbf{v}_1 + \mathbf{v}_2) &= S(T(c\mathbf{v}_1 + \mathbf{v}_2)) \\
&= S(cT(\mathbf{v}_1) + T(\mathbf{v}_2)) \\
&= S(cT(\mathbf{v}_1)) + S(T(\mathbf{v}_2)) \\
&= cS(T(\mathbf{v}_1)) + S(T(\mathbf{v}_2)) \\
&= c(S{\circ}T)(\mathbf{v}_1) + (S{\circ}T)(\mathbf{v}_2)
\end{aligned}
$$

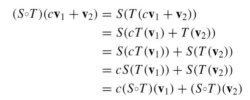

Figure 2

This shows that $S{\circ}T$ is a linear transformation.

In the case of all linear operators on a vector space V, denoted by $£(V, V)$, the operations of addition and scalar multiplication make $£(V, V)$ a vector space. If, in addition, we define a product on $£(V, V)$ by

$$
ST(\mathbf{v}) = (S{\circ}T)(\mathbf{v})
$$

then the product satisfies the necessary properties making $£(V, V)$ a *linear algebra*.

Fact Summary

Let V, W, and Z be vector spaces and S and T functions from V into W.

1. The function T is a linear transformation provided that for all $\mathbf{u}, \mathbf{v}$ in V and all scalars c, $T(c\mathbf{u} + \mathbf{v}) = cT(\mathbf{u}) + T(\mathbf{v})$.

2. If A is an $m \times n$ matrix and T is defined by $T(\mathbf{x}) = A\mathbf{x}$, then T is a linear transformation from $\mathbb{R}^n$ into $\mathbb{R}^m$.

3. If T is a linear transformation, then the zero vector in V is mapped to the zero vector in W, that is, $T(\mathbf{0}) = \mathbf{0}$.

4. If $B = \{\mathbf{v}_1, \mathbf{v}_2, \ldots, \mathbf{v}_n\}$ is an ordered basis for V and $W = \mathbb{R}^n$, then the coordinate mapping $T(\mathbf{v}) = [\mathbf{v}]_B$ is a linear transformation.

5. If $\{\mathbf{v}_1, \mathbf{v}_2, \ldots, \mathbf{v}_n\}$ is a set of vectors in V and T is a linear transformation, then

$$T(c_1\mathbf{v}_1 + c_2\mathbf{v}_2 + \cdots + c_n\mathbf{v}_n) = c_1T(\mathbf{v}_1) + c_2T(\mathbf{v}_2) + \cdots + c_nT(\mathbf{v}_n)$$

for all scalars $c_1, \ldots, c_n$.

6. If S and T are linear transformations and c is a scalar, then $S + T$ and cT are linear transformations.

7. If $T\colon V \longrightarrow W$ is a linear transformation and $L\colon W \longrightarrow Z$ is a linear transformation, then $L{\circ}T\colon V \longrightarrow Z$ is a linear transformation.

Exercise Set 4.1

In Exercises 1–6, determine whether the function $T\colon \mathbb{R}^2 \to \mathbb{R}^2$ is a linear transformation.

1. $T\left(\begin{bmatrix} x \\ y \end{bmatrix}\right) = \begin{bmatrix} y \\ x \end{bmatrix}$

2. $T\left(\begin{bmatrix} x \\ y \end{bmatrix}\right) = \begin{bmatrix} x + y \\ x - y + 2 \end{bmatrix}$

3. $T\left(\begin{bmatrix} x \\ y \end{bmatrix}\right) = \begin{bmatrix} x \\ y^2 \end{bmatrix}$

4. $T\left(\begin{bmatrix} x \\ y \end{bmatrix}\right) = \begin{bmatrix} 2x - y \\ x + 3y \end{bmatrix}$

5. $T\left(\begin{bmatrix} x \\ y \end{bmatrix}\right) = \begin{bmatrix} x \\ 0 \end{bmatrix}$

6. $T\left(\begin{bmatrix} x \\ y \end{bmatrix}\right) = \begin{bmatrix} \frac{x+y}{2} \\ \frac{x+y}{2} \end{bmatrix}$

In Exercises 7–16, determine whether the function is a linear transformation between vector spaces.

7. $T\colon \mathbb{R} \to \mathbb{R}, T(x) = x^2$

8. $T\colon \mathbb{R} \to \mathbb{R}, T(x) = -2x$

9. $T\colon \mathbb{R}^2 \to \mathbb{R}, T\left(\begin{bmatrix} x \\ y \end{bmatrix}\right) = x^2 + y^2$

10. $T\colon \mathbb{R}^3 \to \mathbb{R}^2,$

$$T\left(\begin{bmatrix} x \\ y \\ z \end{bmatrix}\right) = \begin{bmatrix} x \\ y \end{bmatrix}$$

11. $T\colon \mathbb{R}^3 \to \mathbb{R}^3,$

$$T\left(\begin{bmatrix} x \\ y \\ z \end{bmatrix}\right) = \begin{bmatrix} x + y - z \\ 2xy \\ x + z + 1 \end{bmatrix}$$

12. $T: \mathbb{R}^3 \to \mathbb{R}^3$,

$$T\left(\begin{bmatrix} x \\ y \\ z \end{bmatrix}\right) = \begin{bmatrix} \cos x \\ \sin y \\ \sin x + \sin z \end{bmatrix}$$

13. $T: \mathcal{P}_3 \to \mathcal{P}_3$,

$$T(p(x)) = 2p''(x) - 3p'(x) + p(x)$$

14. $T: \mathcal{P}_n \to \mathcal{P}_n$,

$$T(p(x)) = p(x) + x$$

15. $T: M_{2\times2} \to \mathbb{R}, T(A) = \det(A)$

16. $T: M_{2\times2} \to M_{2\times2}, T(A) = A + A^t$

In Exercises 17–20, a function $T: V \to W$ between vector spaces and two vectors **u** and **v** in V are given.

    a. Find $T(\mathbf{u})$ and $T(\mathbf{v})$.

    b. Is $T(\mathbf{u} + \mathbf{v}) = T(\mathbf{u}) + T(\mathbf{v})$?

    c. Is T a linear transformation?

17. Define $T: \mathbb{R}^2 \to \mathbb{R}^2$ by

$$T\left(\begin{bmatrix} x \\ y \end{bmatrix}\right) = \begin{bmatrix} -x \\ y \end{bmatrix}$$

Let

$$\mathbf{u} = \begin{bmatrix} -2 \\ 3 \end{bmatrix} \quad \mathbf{v} = \begin{bmatrix} 2 \\ -2 \end{bmatrix}$$

18. Define $T: \mathcal{P}_2 \to \mathcal{P}_2$ by

$$T(p(x)) = p''(x) - 2p'(x) + p(x)$$

Let

$$\mathbf{u} = x^2 - 3x + 1 \quad \mathbf{v} = -x - 1$$

19. Define $T: \mathcal{P}_3 \to \mathbb{R}^2$ by

$$T(ax^3 + bx^2 + cx + d) = \begin{bmatrix} -a - b + 1 \\ c + d \end{bmatrix}$$

Let

$$\mathbf{u} = -x^3 + 2x^2 - x + 1 \quad \mathbf{v} = x^2 - 1$$

20. Define $T: \mathbb{R}^3 \to \mathbb{R}^2$ by

$$T\left(\begin{bmatrix} x \\ y \\ z \end{bmatrix}\right) = \begin{bmatrix} x^2 - 1 \\ y + z \end{bmatrix}$$

Let

$$\mathbf{u} = \begin{bmatrix} 1 \\ 2 \\ 3 \end{bmatrix} \quad \mathbf{v} = \begin{bmatrix} -\frac{1}{2} \\ -1 \\ 1 \end{bmatrix}$$

21. If $T: \mathbb{R}^2 \to \mathbb{R}^2$ is a linear operator and

$$T\left(\begin{bmatrix} 1 \\ 0 \end{bmatrix}\right) = \begin{bmatrix} 2 \\ 3 \end{bmatrix}$$

$$T\left(\begin{bmatrix} 0 \\ 1 \end{bmatrix}\right) = \begin{bmatrix} -1 \\ 4 \end{bmatrix}$$

then find $T\left(\begin{bmatrix} 1 \\ -3 \end{bmatrix}\right)$.

22. If $T: \mathbb{R}^3 \to \mathbb{R}^3$ is a linear operator and

$$T\left(\begin{bmatrix} 1 \\ 0 \\ 0 \end{bmatrix}\right) = \begin{bmatrix} 1 \\ -1 \\ 0 \end{bmatrix}$$

$$T\left(\begin{bmatrix} 0 \\ 1 \\ 0 \end{bmatrix}\right) = \begin{bmatrix} 2 \\ 0 \\ 1 \end{bmatrix}$$

$$T\left(\begin{bmatrix} 0 \\ 0 \\ 1 \end{bmatrix}\right) = \begin{bmatrix} 1 \\ -1 \\ 1 \end{bmatrix}$$

then find $T\left(\begin{bmatrix} 1 \\ 7 \\ 5 \end{bmatrix}\right)$.

23. If $T: \mathcal{P}_2 \to \mathcal{P}_2$ is a linear operator and

$$T(1) = 1 + x \quad T(x) = 2 + x^2$$
$$T(x^2) = x - 3x^2$$

then find $T(-3 + x - x^2)$.

24. If $T: M_{2\times2} \to M_{2\times2}$ is a linear operator and

$$T(\mathbf{e}_{11}) = \begin{bmatrix} 0 & 1 \\ 0 & 0 \end{bmatrix}$$

$$T(\mathbf{e}_{12}) = \begin{bmatrix} 1 & 0 \\ 0 & -1 \end{bmatrix}$$

$$T(\mathbf{e}_{21}) = \begin{bmatrix} 1 & 1 \\ 0 & 0 \end{bmatrix}$$

$$T(\mathbf{e}_{22}) = \begin{bmatrix} 0 & 0 \\ 2 & 0 \end{bmatrix}$$

then find

$$T\left(\begin{bmatrix} 2 & 1 \\ -1 & 3 \end{bmatrix}\right)$$

25. Suppose that $T: \mathbb{R}^2 \rightarrow \mathbb{R}^2$ is a linear operator such that

$$T\left(\begin{bmatrix} 1 \\ 1 \end{bmatrix}\right) = \begin{bmatrix} 2 \\ -1 \end{bmatrix}$$

$$T\left(\begin{bmatrix} -1 \\ 0 \end{bmatrix}\right) = \begin{bmatrix} 2 \\ -1 \end{bmatrix}$$

Is it possible to determine $T\left(\begin{bmatrix} 3 \\ 7 \end{bmatrix}\right)$? If so, find it; and if not, explain why.

26. Define a linear operator $T: \mathbb{R}^3 \rightarrow \mathbb{R}^3$ by $T(\mathbf{u}) = A\mathbf{u}$, where

$$A = \begin{bmatrix} 1 & 2 & 3 \\ 2 & 1 & 3 \\ 1 & 3 & 2 \end{bmatrix}$$

a. Find $T(\mathbf{e}_1)$, $T(\mathbf{e}_2)$, and $T(\mathbf{e}_3)$.
b. Find $T(3\mathbf{e}_1 - 4\mathbf{e}_2 + 6\mathbf{e}_3)$.

27. Suppose that $T: \mathcal{P}_2 \rightarrow \mathcal{P}_2$ is a linear operator such that

$$T(x^2) = 2x - 1 \qquad T(-3x) = x^2 - 1$$
$$T(-x^2 + 3x) = 2x^2 - 2x + 1$$

a. Is it possible to determine $T(2x^2 - 3x + 2)$? If so, find it; and if not, explain why.
b. Is it possible to determine $T(3x^2 - 4x)$? If so, find it; and if not, explain why.

28. Suppose that $T: \mathbb{R}^3 \rightarrow \mathbb{R}^3$ is a linear operator such that

$$T\left(\begin{bmatrix} 1 \\ 0 \\ 0 \end{bmatrix}\right) = \begin{bmatrix} -1 \\ 2 \\ 3 \end{bmatrix}$$

$$T\left(\begin{bmatrix} 1 \\ 1 \\ 0 \end{bmatrix}\right) = \begin{bmatrix} 2 \\ -2 \\ 1 \end{bmatrix}$$

$$T\left(\begin{bmatrix} 1 \\ 3 \\ 0 \end{bmatrix}\right) = \begin{bmatrix} 8 \\ -10 \\ -3 \end{bmatrix}$$

a. Find

$$T\left(\begin{bmatrix} 2 \\ -5 \\ 0 \end{bmatrix}\right)$$

b. Is it possible to determine $T(\mathbf{v})$ for all vectors $\mathbf{v}$ in $\mathbb{R}^3$? Explain.

29. Define a linear operator $T: \mathbb{R}^2 \rightarrow \mathbb{R}^2$ by

$$T\left(\begin{bmatrix} x \\ y \end{bmatrix}\right) = \begin{bmatrix} -x \\ -y \end{bmatrix}$$

a. Find a matrix A such that $T(\mathbf{v}) = A\mathbf{v}$.
b. Find $T(\mathbf{e}_1)$ and $T(\mathbf{e}_2)$.

30. Define a linear transformation $T: \mathbb{R}^2 \rightarrow \mathbb{R}^3$ by

$$T\left(\begin{bmatrix} x \\ y \end{bmatrix}\right) = \begin{bmatrix} x - 2y \\ 3x + y \\ 2y \end{bmatrix}$$

a. Find a matrix A such that $T(\mathbf{v}) = A\mathbf{v}$.
b. Find $T(\mathbf{e}_1)$ and $T(\mathbf{e}_2)$.

31. Define $T: \mathbb{R}^3 \rightarrow \mathbb{R}^2$ by

$$T\left(\begin{bmatrix} x \\ y \\ z \end{bmatrix}\right) = \begin{bmatrix} x + y \\ x - y \end{bmatrix}$$

Find all vectors that are mapped to $\mathbf{0}$.

32. Define $T: \mathbb{R}^3 \rightarrow \mathbb{R}^2$ by

$$T\left(\begin{bmatrix} x \\ y \\ z \end{bmatrix}\right) = \begin{bmatrix} x + 2y + z \\ -x + 5y + z \end{bmatrix}$$

Find all vectors that are mapped to $\mathbf{0}$.

33. Define $T: \mathbb{R}^3 \rightarrow \mathbb{R}^3$ by

$$T\left(\begin{bmatrix} x \\ y \\ z \end{bmatrix}\right) = \begin{bmatrix} x - y + 2z \\ 2x + 3y - z \\ -x + 2y - 2z \end{bmatrix}$$

a. Find all vectors in $\mathbb{R}^3$ that are mapped to the zero vector.

b. Let $\mathbf{w} = \begin{bmatrix} 7 \\ -6 \\ -9 \end{bmatrix}$. Determine whether there is

a vector $\mathbf{v}$ in $\mathbb{R}^3$ such that $T(\mathbf{v}) = \mathbf{w}$.

34. Define $T: \mathcal{P}_2 \to \mathcal{P}_2$ by

$$T(p(x)) = p'(x) - p(0)$$

a. Find all vectors that are mapped to $\mathbf{0}$.

b. Find two polynomials $p(x)$ and $q(x)$ such that $T(p(x)) = T(q(x)) = 6x - 3$.

c. Is T a linear operator?

35. Suppose $T_1: V \to \mathbb{R}$ and $T_2: V \to \mathbb{R}$ are linear transformations. Define $T: V \to \mathbb{R}^2$ by

$$T(\mathbf{v}) = \begin{bmatrix} T_1(\mathbf{v}) \\ T_2(\mathbf{v}) \end{bmatrix}$$

Show that T is a linear transformation.

36. Define $T: M_{n \times n} \to \mathbb{R}$ by $T(A) = \mathbf{tr}(A)$. Show that T is a linear transformation.

37. Suppose that B is a fixed $n \times n$ matrix. Define $T: M_{n \times n} \to M_{n \times n}$ by $T(A) = AB - BA$. Show that T is a linear operator.

38. Define $T: \mathbb{R} \to \mathbb{R}$ by $T(x) = mx + b$. Determine when T is a linear operator.

39. Define $T: C^{(0)}[0, 1] \to \mathbb{R}$ by

$$T(f) = \int_0^1 f(x)\,dx$$

for each function f in $C^{(0)}[0, 1]$.

a. Show that T is a linear operator.

b. Find $T(2x^2 - x + 3)$.

40. Suppose that $T: V \to W$ is a linear transformation and $T(\mathbf{u}) = \mathbf{w}$. If $T(\mathbf{v}) = \mathbf{0}$, then find $T(\mathbf{u} + \mathbf{v})$.

41. Suppose that $T: \mathbb{R}^n \to \mathbb{R}^m$ is a linear transformation and $\{\mathbf{v}, \mathbf{w}\}$ is a linearly independent subset of $\mathbb{R}^n$. If $\{T(\mathbf{v}), T(\mathbf{w})\}$ is linearly dependent, show that $T(\mathbf{u}) = \mathbf{0}$ has a nontrivial solution.

42. Suppose that $T: V \to V$ is a linear operator and $\{\mathbf{v}_1, \ldots, \mathbf{v}_n\}$ is linearly dependent. Show that $\{T(\mathbf{v}_1), \ldots, T(\mathbf{v}_n)\}$ is linearly dependent.

43. Let $S = \{\mathbf{v}_1, \mathbf{v}_2, \mathbf{v}_3\}$ be a linearly independent subset of $\mathbb{R}^3$. Find a linear operator $T: \mathbb{R}^3 \to \mathbb{R}^3$, such that $\{T(\mathbf{v}_1), T(\mathbf{v}_2), T(\mathbf{v}_3)\}$ is linearly dependent.

44. Suppose that $T_1: V \to V$ and $T_2: V \to V$ are linear operators and $\{\mathbf{v}_1, \ldots, \mathbf{v}_n\}$ is a basis for V. If $T_1(\mathbf{v}_i) = T_2(\mathbf{v}_i)$, for each $i = 1, 2, \ldots, n$, show that $T_1(\mathbf{v}) = T_2(\mathbf{v})$ for all $\mathbf{v}$ in V.

45. Verify that $\pounds(U, V)$ is a vector space.

4.2 ▶ The Null Space and Range

In Sec. 3.2, we defined the null space of an $m \times n$ matrix to be the subspace of $\mathbb{R}^n$ of all vectors $\mathbf{x}$ with $A\mathbf{x} = \mathbf{0}$. We also defined the column space of A as the subspace of $\mathbb{R}^m$ of all linear combinations of the column vectors of A. In this section we extend these ideas to linear transformations.

DEFINITION 1   **Null Space and Range**   Let V and W be vector spaces. For a linear transformation $T: V \longrightarrow W$ the **null space** of T, denoted by $N(T)$, is defined by

$$N(T) = \{\mathbf{v} \in V \mid T(\mathbf{v}) = \mathbf{0}\}$$

The **range** of T, denoted by $R(T)$, is defined by

$$R(T) = \{T(\mathbf{v}) \mid \mathbf{v} \in V\}$$

The null space of a linear transformation is then the set of all vectors in V that are mapped to the zero vector, with the range being the set of all images of the mapping, as shown in Fig. 1.

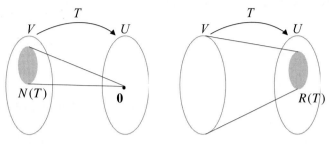

Figure 1

In Theorem 3 we see that the null space and the range of a linear transformation are both subspaces.

THEOREM 3   Let V and W be vector spaces and $T: V \longrightarrow W$ a linear transformation.

1. The null space of T is a subspace of V.
2. The range of T is a subspace of W.

Proof   (1) Let $\mathbf{v}_1$ and $\mathbf{v}_2$ be in $N(T)$, so that $T(\mathbf{v}_1) = \mathbf{0}$ and $T(\mathbf{v}_2) = \mathbf{0}$. If c is a scalar, then using the linearity of T, we have

$$T(c\mathbf{v}_1 + \mathbf{v}_2) = cT(\mathbf{v}_1) + T(\mathbf{v}_2) = c\mathbf{0} + \mathbf{0} = \mathbf{0}$$

Thus, $c\mathbf{v}_1 + \mathbf{v}_2$ is in $N(T)$, and by Theorem 4 of Sec. 3.2, $N(T)$ is a subspace of V.

(2) Let $\mathbf{w}_1$ and $\mathbf{w}_2$ be in $R(T)$. Then there are vectors $\mathbf{v}_1$ and $\mathbf{v}_2$ in V such that $T(\mathbf{v}_1) = \mathbf{w}_1$ and $T(\mathbf{v}_2) = \mathbf{w}_2$. Then for any scalar c,

$$T(c\mathbf{v}_1 + \mathbf{v}_2) = cT(\mathbf{v}_1) + T(\mathbf{v}_2) = c\mathbf{w}_1 + \mathbf{w}_2$$

so that $c\mathbf{w}_1 + \mathbf{w}_2$ is in $R(T)$ and hence $R(T)$ is a subspace of W.

EXAMPLE 1   Define the linear transformation $T: \mathbb{R}^4 \longrightarrow \mathbb{R}^3$ by

$$T\left(\begin{bmatrix} a \\ b \\ c \\ d \end{bmatrix}\right) = \left(\begin{bmatrix} a+b \\ b-c \\ a+d \end{bmatrix}\right)$$

a. Find a basis for the null space of T and its dimension.

   b. Give a description of the range of T.

   c. Find a basis for the range of T and its dimension.

Solution   **a.** The null space of T is found by setting each component of the image vector equal to 0. This yields the linear system

$$\begin{cases} a + b & = 0 \\ b - c & = 0 \\ a & + d = 0 \end{cases}$$

This linear system has infinitely many solutions, given by

$$S = \left\{ \begin{bmatrix} -t \\ t \\ t \\ t \end{bmatrix} \,\middle|\, t \in \mathbb{R} \right\}$$

Hence,

$$N(T) = \mathbf{span} \left\{ \begin{bmatrix} -1 \\ 1 \\ 1 \\ 1 \end{bmatrix} \right\}$$

A basis for $N(T)$ consists of the one vector

$$\begin{bmatrix} -1 \\ 1 \\ 1 \\ 1 \end{bmatrix}$$

Consequently, $\dim(N(T)) = 1$.

   b. Observe that any vector in the range can be written as

$$a \begin{bmatrix} 1 \\ 0 \\ 1 \end{bmatrix} + b \begin{bmatrix} 1 \\ 1 \\ 0 \end{bmatrix} + c \begin{bmatrix} 0 \\ -1 \\ 0 \end{bmatrix} + d \begin{bmatrix} 0 \\ 0 \\ 1 \end{bmatrix}$$

for some real numbers $a, b, c,$ and d. Therefore,

$$R(T) = \mathbf{span} \left\{ \begin{bmatrix} 1 \\ 0 \\ 1 \end{bmatrix}, \begin{bmatrix} 1 \\ 1 \\ 0 \end{bmatrix}, \begin{bmatrix} 0 \\ -1 \\ 0 \end{bmatrix}, \begin{bmatrix} 0 \\ 0 \\ 1 \end{bmatrix} \right\}$$

   c. Since the range is a subspace of $\mathbb{R}^3$, it has dimension less than or equal to 3. Consequently, the four vectors found to span the range in part (b) are linearly dependent and do not form a basis. To find a basis for $R(T)$, we use the trimming procedure given in Sec. 3.3 and reduce the matrix

$$\begin{bmatrix} 1 & 1 & 0 & 0 \\ 0 & 1 & -1 & 0 \\ 1 & 0 & 0 & 1 \end{bmatrix} \quad \text{to} \quad \begin{bmatrix} 1 & 0 & 0 & 1 \\ 0 & 1 & 0 & -1 \\ 0 & 0 & 1 & -1 \end{bmatrix}$$

Since the reduced matrix has pivots in the first three columns, a basis for the range of T is

$$B = \left\{ \begin{bmatrix} 1 \\ 0 \\ 1 \end{bmatrix}, \begin{bmatrix} 1 \\ 1 \\ 0 \end{bmatrix}, \begin{bmatrix} 0 \\ -1 \\ 0 \end{bmatrix} \right\}$$

Therefore, $\dim(R(T)) = 3$. Observe that B also spans $\mathbb{R}^3$, so that $R(T) = \mathbb{R}^3$.

EXAMPLE 2

Define the linear transformation $T: \mathcal{P}_4 \longrightarrow \mathcal{P}_3$, by

$$T(p(x)) = p'(x)$$

Find the null space and range of T.

Solution   Recall that the derivative of a constant polynomial is 0. Since these are the only polynomials for which the derivative is 0, we know that $N(T)$ is the set of constant polynomials in $\mathcal{P}_4$. We claim that the range of T is all of $\mathcal{P}_3$. To see this, let $q(x) = ax^3 + bx^2 + cx + d$ be an arbitrary element of $\mathcal{P}_3$. A polynomial $p(x)$ whose derivative is $q(x)$ is found by using the antiderivative. That is, to find $p(x)$, we integrate $q(x)$ to obtain

$$p(x) = \int q(x)\, dx = \int (ax^3 + bx^2 + cx + d)\, dx = \frac{a}{4}x^4 + \frac{b}{3}x^3 + \frac{c}{2}x^2 + dx + e$$

which is an element of $\mathcal{P}_4$, with $p'(x) = q(x)$. This shows that for every polynomial $q(x)$ in $\mathcal{P}_3$ there is a polynomial $p(x)$ in $\mathcal{P}_4$ such that $T(p(x)) = q(x)$, giving that the range of T is all of $\mathcal{P}_3$.

In Sec. 4.1, we saw that the image of an arbitrary vector $\mathbf{v} \in V$ can be computed if the image $T(\mathbf{v}_i)$ is known for each vector $\mathbf{v}_i$ in a basis for V. This leads to Theorem 4.

THEOREM 4

Let V and W be finite dimensional vector spaces and $B = \{\mathbf{v}_1, \mathbf{v}_2, \ldots, \mathbf{v}_n\}$ a basis for V. If $T: V \to W$ is a linear transformation, then

$$R(T) = \mathbf{span}\{T(\mathbf{v}_1), T(\mathbf{v}_2), \ldots, T(\mathbf{v}_n)\}$$

Proof   To show that the two sets are equal, we will show that each is a subset of the other. First, if $\mathbf{w}$ is in $R(T)$, then there is a vector $\mathbf{v}$ in V such that $T(\mathbf{v}) = \mathbf{w}$. Now, since B is a basis for V, there are scalars $c_1, \ldots, c_n$ with

$$\mathbf{v} = c_1\mathbf{v}_1 + c_2\mathbf{v}_2 + \cdots + c_n\mathbf{v}_n$$

so that

$$T(\mathbf{v}) = T(c_1\mathbf{v}_1 + c_2\mathbf{v}_2 + \cdots + c_n\mathbf{v}_n)$$

From the linearity of T, we have

$$\mathbf{w} = T(\mathbf{v}) = c_1 T(\mathbf{v}_1) + c_2 T(\mathbf{v}_2) + \cdots + c_n T(\mathbf{v}_n)$$

As $\mathbf{w}$ is a linear combination of $T(\mathbf{v}_1), T(\mathbf{v}_2), \ldots, T(\mathbf{v}_n)$, then $\mathbf{w} \in \mathbf{span}\{T(\mathbf{v}_1),$ $T(\mathbf{v}_2), \ldots, T(\mathbf{v}_n)\}$. Since this is true for all $\mathbf{w}$ in $R(T)$, then

$$R(T) \subset \mathbf{span}\{T(\mathbf{v}_1), T(\mathbf{v}_2), \ldots, T(\mathbf{v}_n)\}$$

On the other hand, suppose that $\mathbf{w} \in \mathbf{span}\{T(\mathbf{v}_1), T(\mathbf{v}_2), \ldots, T(\mathbf{v}_n)\}$. Then there are scalars $c_1, \ldots, c_n$ with

$$\mathbf{w} = c_1 T(\mathbf{v}_1) + c_2 T(\mathbf{v}_2) + \cdots + c_n T(\mathbf{v}_n)$$
$$= T(c_1 \mathbf{v}_1 + c_2 \mathbf{v}_2 + \cdots + c_n \mathbf{v}_n)$$

Therefore, $\mathbf{w}$ is the image under T of $c_1 \mathbf{v}_1 + c_2 \mathbf{v}_2 + \cdots + c_n \mathbf{v}_n$, which is an element of V. Therefore, $\mathbf{span}\{T(\mathbf{v}_1), T(\mathbf{v}_2), \ldots, T(\mathbf{v}_n)\} \subset R(T)$.

EXAMPLE 3   Let $T: \mathbb{R}^3 \to \mathbb{R}^3$ be a linear operator and $B = \{\mathbf{v}_1, \mathbf{v}_2, \mathbf{v}_3\}$ a basis for $\mathbb{R}^3$. Suppose that

$$T(\mathbf{v}_1) = \begin{bmatrix} 1 \\ 1 \\ 0 \end{bmatrix} \qquad T(\mathbf{v}_2) = \begin{bmatrix} 1 \\ 0 \\ -1 \end{bmatrix} \qquad T(\mathbf{v}_3) = \begin{bmatrix} 2 \\ 1 \\ -1 \end{bmatrix}$$

a. Is $\begin{bmatrix} 1 \\ 2 \\ 1 \end{bmatrix}$ in $R(T)$?

b. Find a basis for $R(T)$.

c. Find the null space $N(T)$.

Solution   **a.** From Theorem 4, the vector $\mathbf{w} = \begin{bmatrix} 1 \\ 2 \\ 1 \end{bmatrix}$ is in $R(T)$ if there are scalars c_1, c_2, and c_3 such that

$$c_1 T(\mathbf{v}_1) + c_2 T(\mathbf{v}_2) + c_3 T(\mathbf{v}_3) = \begin{bmatrix} 1 \\ 2 \\ 1 \end{bmatrix}$$

that is,

$$c_1 \begin{bmatrix} 1 \\ 1 \\ 0 \end{bmatrix} + c_2 \begin{bmatrix} 1 \\ 0 \\ -1 \end{bmatrix} + c_2 \begin{bmatrix} 2 \\ 1 \\ -1 \end{bmatrix} = \begin{bmatrix} 1 \\ 2 \\ 1 \end{bmatrix}$$

The set of solutions to this linear system is given by $S = \{(2 - t, -1 - t, t) \mid t \in \mathbb{R}\}$. In particular, if $t = 0$, then a solution is $c_1 = 2, c_2 = -1$, and $c_3 = 0$. Thus, $\mathbf{w} \in R(T)$.

b. To find a basis for $R(T)$, we row-reduce the matrix

$$\begin{bmatrix} 1 & 1 & 2 \\ 1 & 0 & 1 \\ 0 & -1 & -1 \end{bmatrix} \quad \text{to obtain} \quad \begin{bmatrix} 1 & 0 & 1 \\ 0 & 1 & 1 \\ 0 & 0 & 0 \end{bmatrix}$$

Since the leading 1s are in columns 1 and 2, a basis for $R(T)$ is given by

$$R(T) = \textbf{span} \left\{ \begin{bmatrix} 1 \\ 1 \\ 0 \end{bmatrix}, \begin{bmatrix} 1 \\ 0 \\ -1 \end{bmatrix} \right\}$$

Observe that since the range is spanned by two linearly independent vectors, $R(T)$ is a plane in $\mathbb{R}^3$, as shown in Fig. 2.

c. Since B is a basis for $\mathbb{R}^3$, the null space is the set of all vectors $c_1 \mathbf{v}_1 + c_2 \mathbf{v}_2 + c_3 \mathbf{v}_3$ such that

$$c_1 T(\mathbf{v}_1) + c_2 T(\mathbf{v}_2) + c_3 T(\mathbf{v}_3) = \begin{bmatrix} 0 \\ 0 \\ 0 \end{bmatrix}$$

By using the reduced matrix

$$\begin{bmatrix} 1 & 0 & 1 \\ 0 & 1 & 1 \\ 0 & 0 & 0 \end{bmatrix}$$

from part (b), the null space consists of all vectors such that $c_1 = -c_3$, $c_2 = -c_3$, and c_3 is any real number. That is,

$$N(T) = \textbf{span} \, \{-\mathbf{v}_1 - \mathbf{v}_2 + \mathbf{v}_3\}$$

which is a line in $\mathbb{R}^3$. See Fig. 2.

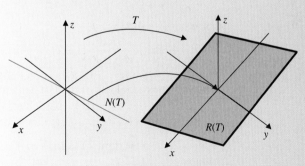

Figure 2

Notice that in Example 3 we have

$$\dim(\mathbb{R}^3) = \dim(R(T)) + \dim(N(T))$$

In Theorem 5 we establish this fundamental result for all linear transformations between finite dimensional vector spaces.

THEOREM 5   Let V and W be finite dimensional vector spaces. If $T: V \longrightarrow W$ is a linear transformation, then

$$\dim(V) = \dim(R(T)) + \dim(N(T))$$

Proof   Suppose that $\dim(V) = n$. To establish the result, we consider three cases.

First, suppose that $\dim(N(T)) = \dim(V) = n$. In this case, the image of every vector in V is the zero vector (in W), so that $R(T) = \{\mathbf{0}\}$. Since the dimension of the vector space containing only the zero element is 0, the result holds.

Now suppose $1 \le r = \dim(N(T)) < n$. Let $\{\mathbf{v}_1, \mathbf{v}_2, \ldots, \mathbf{v}_r\}$ be a basis for $N(T)$. By Corollary 1 of Sec. 3.3, there are $n - r$ vectors $\{\mathbf{v}_{r+1}, \mathbf{v}_{r+2}, \ldots, \mathbf{v}_n\}$, such that $\{\mathbf{v}_1, \mathbf{v}_2, \ldots, \mathbf{v}_r, \mathbf{v}_{r+1}, \ldots, \mathbf{v}_n\}$ is a basis for V. We claim that $S = \{T(\mathbf{v}_{r+1}), T(\mathbf{v}_{r+2}), \ldots, T(\mathbf{v}_n)\}$ is a basis for $R(T)$. By Theorem 4, we have

$$R(T) = \mathbf{span}\{T(\mathbf{v}_1), T(\mathbf{v}_2), \ldots, T(\mathbf{v}_r), T(\mathbf{v}_{r+1}), \ldots, T(\mathbf{v}_n)\}$$

Since $T(\mathbf{v}_1) = T(\mathbf{v}_2) = \cdots = T(\mathbf{v}_r) = \mathbf{0}$, each vector in $R(T)$ is a linear combination of $T(\mathbf{v}_{r+1}), \ldots, T(\mathbf{v}_n)$ and hence $R(T) = \mathbf{span}(S)$. To show that S is linearly independent, we consider the equation

$$c_{r+1}T(\mathbf{v}_{r+1}) + c_{r+2}T(\mathbf{v}_{r+2}) + \cdots + c_n T(\mathbf{v}_n) = \mathbf{0}$$

We need to show that $c_{r+1} = c_{r+2} = \cdots = c_n = 0$. Since T is linear, the previous equation can be written as

$$T(c_{r+1}\mathbf{v}_{r+1} + c_{r+2}\mathbf{v}_{r+2} + \cdots + c_n\mathbf{v}_n) = \mathbf{0}$$

From this last equation, we have $c_{r+1}\mathbf{v}_{r+1} + c_{r+2}\mathbf{v}_{r+2} + \cdots + c_n\mathbf{v}_n$ is in $N(T)$. However, since $\{\mathbf{v}_1, \mathbf{v}_2, \ldots, \mathbf{v}_r\}$ is a basis for $N(T)$, there are scalars $c_1, c_2, \ldots, c_r$ such that

$$c_{r+1}\mathbf{v}_{r+1} + c_{r+2}\mathbf{v}_{r+2} + \cdots + c_n\mathbf{v}_n = c_1\mathbf{v}_1 + c_2\mathbf{v}_2 + \cdots + c_r\mathbf{v}_r$$

that is,

$$-c_1\mathbf{v}_1 - c_2\mathbf{v}_2 - \cdots - c_r\mathbf{v}_r + c_{r+1}\mathbf{v}_{r+1} + c_{r+2}\mathbf{v}_{r+2} + \cdots + c_n\mathbf{v}_n = \mathbf{0}$$

Now, since $\{\mathbf{v}_1, \mathbf{v}_2, \ldots, \mathbf{v}_r, \mathbf{v}_{r+1}, \ldots, \mathbf{v}_n\}$ is a basis for V and hence linearly independent, the coefficients of the last equation must all be 0, that is, $c_1 = c_2 = \cdots = c_r = c_{r+1} = \cdots = c_n = 0$. In particular, $c_{r+1} = c_{r+2} = \cdots = c_n = 0$. Hence, the $n - r$ vectors $T(\mathbf{v}_{r+1}), \ldots, T(\mathbf{v}_n)$ are a basis for $R(T)$. Consequently,

$$n = \dim(V) = (n - r) + r = \dim(R(T)) + \dim(N(T))$$

Finally, suppose that $N(T) = \{\mathbf{0}\}$, so that $\dim(N(T)) = 0$. If $\{\mathbf{v}_1, \ldots, \mathbf{v}_n\}$ is a basis for V, then by Theorem 4 we have

$$R(T) = \mathbf{span}\{T(\mathbf{v}_1), \ldots, T(\mathbf{v}_n)\}$$

A similar argument to the one above shows that $\{T(\mathbf{v}_1), \ldots, T(\mathbf{v}_n)\}$ is linearly independent. Thus, $\dim(R(T)) = n = \dim(V)$, and the result also holds in this case.

EXAMPLE 4   Define a linear transformation $T\colon \mathcal{P}_4 \longrightarrow \mathcal{P}_2$ by

$$T(p(x)) = p''(x)$$

Find the dimension of the range of T, and give a description of the range.

Solution   Let $B = \{1, x, x^2, x^3, x^4\}$ be the standard basis for $\mathcal{P}_4$. Since $p(x)$ is in $N(T)$ if and only if its degree is 0 or 1, the null space is the subspace of $\mathcal{P}_4$ consisting of polynomials with degree 1 or less. Hence, $\{1, x\}$ is a basis for $N(T)$, and $\dim(N(T)) = 2$. Since $\dim(\mathcal{P}_4) = 5$, by Theorem 5 we have

$$2 + \dim(R(T)) = 5 \qquad \text{so} \qquad \dim(R(T)) = 3$$

Then as in the proof of Theorem 5, we have

$$\left\{T(x^2), T(x^3), T(x^4)\right\} = \left\{2, 6x, 12x^2\right\}$$

is a basis for $R(T)$. Observe that $R(T)$ is just the subspace $\mathcal{P}_2$ of $\mathcal{P}_4$.

Matrices

In Sec. 3.2 we defined the column space of a matrix A, denoted by **col**(A), as the span of its column vectors. We also defined the null space of the $m \times n$ matrix A as the set of all vectors $\mathbf{x}$ in $\mathbb{R}^n$ such that $A\mathbf{x} = \mathbf{0}$. We further examine these notions here in the context of linear transformations. In particular, let A be an $m \times n$ matrix and let $T\colon \mathbb{R}^n \longrightarrow \mathbb{R}^m$ be the linear transformation defined by

$$T(\mathbf{v}) = A\mathbf{v}$$

This last equation can be written in vector form as

$$T(\mathbf{v}) = v_1 \mathbf{A}_1 + v_2 \mathbf{A}_2 + \cdots + v_n \mathbf{A}_n$$

where $\mathbf{A}_i$ are the column vectors of A, and v_i are the components of $\mathbf{v}$ for $1 \le i \le n$. In this way we see that the range of T, which is a subspace of $\mathbb{R}^m$, is equal to the column space of A, that is,

$$R(T) = \mathbf{col}(A)$$

The dimension of the column space of A is called the **column rank** of A. We also have

$$N(T) = \{\mathbf{v} \in \mathbb{R}^n \mid A\mathbf{v} = \mathbf{0}\} = N(A)$$

The dimension of $N(A)$ is called the **nullity** of A. Applying Theorem 5, we have

$$\textbf{column rank}(A) + \textbf{nullity}(A) = n$$

Another subspace of $\mathbb{R}^n$ associated with the matrix A is the row space of A, denoted by **row**(A), and is the span of the row vectors of A. Since the transpose operation maps the row vectors of A to the column vectors of A^t, the row space of A is the same as the column space of A^t, that is,

$$\textbf{row}(A) = \textbf{col}(A^t)$$

By using the algorithm for finding a basis, given in Sec. 3.3, a basis for **col**(A) can be found by row reduction. In particular, the columns with the leading 1s in the row-reduced form of A correspond to the column vectors of A needed for a basis of **col**(A). Hence, the column rank of A is equal to the number of leading 1s in the row-reduced form of A. On the other hand, row-reducing A eliminates row vectors that are linear combinations of the others, so that the nonzero row vectors of the reduced form of A form a basis for **row**(A). Hence, the row rank is equal to the number of leading 1s in the reduced form of A. We have now established Theorem 6.

THEOREM 6   The row rank and the column rank of a matrix A are equal.

We can now define the **rank** of a matrix A as $\dim(\textbf{row}(A))$ or $\dim(\textbf{col}(A))$. Again by Theorem 5, we have

$$\textbf{rank}(A) + \textbf{nullity}(A) = n$$

Linear Systems

When the nullity of a matrix A is known, the above formula can sometimes be used to determine whether the linear system $A\mathbf{x} = \mathbf{b}$ is consistent. For example, suppose that a linear system consists of 20 equations each with 22 variables. Further suppose that a basis for the null space of the 20×22 coefficient matrix consists of two vectors. That is, every solution to the homogeneous linear system $A\mathbf{x} = \mathbf{0}$ is a linear combination of two linearly independent vectors in $\mathbb{R}^{22}$. Then **nullity**(A) $= 2$, so that

$$\dim(\textbf{col}(A)) = \textbf{rank}(A) = 22 - \textbf{nullity}(A) = 20$$

But the only subspace of $\mathbb{R}^{20}$ with dimension 20 is $\mathbb{R}^{20}$ itself. Hence, **col**(A) $= \mathbb{R}^{20}$, and consequently every vector **b** in $\mathbb{R}^{20}$ is a linear combination of the columns of A. That is, the linear system $A\mathbf{x} = \mathbf{b}$ is consistent for every vector **b** in $\mathbb{R}^{20}$. In general, if A is an $m \times n$ matrix, **nullity**(A) $= r$, and $\dim(\textbf{col}) = n - r = m$, then the linear system $A\mathbf{x} = \mathbf{b}$ is consistent for every vector **b** in $\mathbb{R}^{m}$.

We now add several more items to the list of equivalences given in Theorem 9 of Sec. 2.3, connecting solutions of the linear system $A\mathbf{x} = \mathbf{b}$ and properties of the coefficient matrix A.

THEOREM 7   Let A be an $n \times n$ matrix. Then the following statements are equivalent.

1. The matrix A is invertible.

2. The linear system $A\mathbf{x} = \mathbf{b}$ has a unique solution for every vector **b**.

3. The homogeneous linear system $A\mathbf{x} = \mathbf{0}$ has only the trivial solution.

4. The matrix A is row equivalent to the identity matrix.

5. The determinant of the matrix A is nonzero.

6. The column vectors of A are linearly independent.

7. The column vectors of A span $\mathbb{R}^n$.

8. The column vectors of A are a basis for $\mathbb{R}^n$.

9. $\mathbf{rank}(A) = n$

10. $R(A) = \mathbf{col}(A) = \mathbb{R}^n$

11. $N(A) = \{\mathbf{0}\}$

12. $\mathbf{row}(A) = \mathbb{R}^n$

13. The number of pivot columns of the reduced row echelon form of A is n.

Fact Summary

Let V and W be vector spaces and T a linear transformation from V into W.

1. The null space $N(T)$ is a subspace of V, and the the range $R(T)$ is a subspace of W.

2. If $B = \{\mathbf{v}_1, \ldots, \mathbf{v}_n\}$ is a basis for V, then

$$R(T) = \mathbf{span}\{T(\mathbf{v}_1), \ldots, \mathbf{T}(\mathbf{v}_n)\}$$

3. If V and W are finite dimensional vector spaces, then

$$\dim(V) = \dim(R(T)) + \dim(N(T))$$

4. If A is an $m \times n$ matrix, then

$$\mathbf{rank}(A) + \mathbf{nullity}(A) = n$$

5. If A is an $m \times n$ matrix, then the rank of A is the number of leading 1s in the row-reduced form of A.

6. If A is an $n \times n$ invertible matrix, in addition to Theorem 9 of Sec. 2.3, we know that $\mathbf{rank}(A) = n$, $R(A) = \mathbf{col}(A) = \mathbb{R}^n$, $N(A) = \{\mathbf{0}\}$, and the number of leading 1s in the row echelon form of A is n.

Exercise Set 4.2

In Exercises 1–4, define a linear operator $T: \mathbb{R}^2 \to \mathbb{R}^2$ by

$$T\left(\begin{bmatrix} x \\ y \end{bmatrix}\right) = \begin{bmatrix} x - 2y \\ -2x + 4y \end{bmatrix}$$

Determine whether the vector $\mathbf{v}$ is in $N(T)$.

1. $\mathbf{v} = \begin{bmatrix} 0 \\ 0 \end{bmatrix}$

2. $\mathbf{v} = \begin{bmatrix} 2 \\ 1 \end{bmatrix}$

3. $\mathbf{v} = \begin{bmatrix} 1 \\ 3 \end{bmatrix}$

4. $\mathbf{v} = \begin{bmatrix} \frac{1}{2} \\ \frac{1}{4} \end{bmatrix}$

In Exercises 5–8, define a linear operator
$T\colon \mathcal{P}_3 \to \mathcal{P}_3$ by
$$T(p(x)) = xp''(x)$$

Determine whether the polynomial $p(x)$ is in $N(T)$.

 5. $p(x) = x^2 - 3x + 1$

6. $p(x) = 5x + 2$

7. $p(x) = 1 - x^2$

8. $p(x) = 3$

In Exercises 9–12, define a linear operator
$T\colon \mathbb{R}^3 \to \mathbb{R}^3$ by
$$T\left(\begin{bmatrix} x \\ y \\ z \end{bmatrix}\right) = \begin{bmatrix} x + 2z \\ 2x + y + 3z \\ x - y + 3z \end{bmatrix}$$

Determine whether the vector $\mathbf{v}$ is in $R(T)$.

9. $\mathbf{v} = \begin{bmatrix} 1 \\ 3 \\ 0 \end{bmatrix}$

10. $\mathbf{v} = \begin{bmatrix} 2 \\ 3 \\ 4 \end{bmatrix}$

11. $\mathbf{v} = \begin{bmatrix} -1 \\ 1 \\ -2 \end{bmatrix}$

12. $\mathbf{v} = \begin{bmatrix} -2 \\ -5 \\ -1 \end{bmatrix}$

In Exercises 13–16, define a linear transformation
$T\colon M_{2\times 2} \to M_{3\times 2}$ by
$$T\begin{bmatrix} a & b \\ c & d \end{bmatrix} = \begin{bmatrix} a + c & b + d \\ -a + 2c & -b + 2d \\ 2a & 2b \end{bmatrix}$$

Determine whether the matrix A is in $R(T)$.

 13. $A = \begin{bmatrix} -1 & -1 \\ -5 & -2 \\ 2 & 0 \end{bmatrix}$

14. $A = \begin{bmatrix} 1 & 2 \\ 3 & -3 \\ -2 & 2 \end{bmatrix}$

 15. $A = \begin{bmatrix} 1 & 0 \\ 2 & 1 \\ 4 & 0 \end{bmatrix}$

16. $A \doteq \begin{bmatrix} 4 & 1 \\ -1 & 5 \\ 6 & -2 \end{bmatrix}$

In Exercises 17–24, find a basis for the null space of
the linear transformation T.

17. $T\colon \mathbb{R}^2 \to \mathbb{R}^2$,
$$T\left(\begin{bmatrix} x \\ y \end{bmatrix}\right) = \begin{bmatrix} 3x + y \\ y \end{bmatrix}$$

18. $T\colon \mathbb{R}^2 \to \mathbb{R}^2$,
$$T\left(\begin{bmatrix} x \\ y \end{bmatrix}\right) = \begin{bmatrix} -x + y \\ x - y \end{bmatrix}$$

19. $T\colon \mathbb{R}^3 \to \mathbb{R}^3$,
$$T\left(\begin{bmatrix} x \\ y \\ z \end{bmatrix}\right) = \begin{bmatrix} x + 2z \\ 2x + y + 3z \\ x - y + 3z \end{bmatrix}$$

20. $T\colon \mathbb{R}^3 \to \mathbb{R}^3$,
$$T\left(\begin{bmatrix} x \\ y \\ z \end{bmatrix}\right) = \begin{bmatrix} -2x + 2y + 2z \\ 3x + 5y + z \\ 2y + z \end{bmatrix}$$

21. $T\colon \mathbb{R}^3 \to \mathbb{R}^3$,
$$T\left(\begin{bmatrix} x \\ y \\ z \end{bmatrix}\right) = \begin{bmatrix} x - 2y - z \\ -x + 2y + z \\ 2x - 4y - 2z \end{bmatrix}$$

22. $T\colon \mathbb{R}^4 \to \mathbb{R}^3$,
$$T\left(\begin{bmatrix} x \\ y \\ z \\ w \end{bmatrix}\right) = \begin{bmatrix} x + y - z + w \\ 2x + y + 4z + w \\ 3x + y + 9z \end{bmatrix}$$

23. $T: \mathcal{P}_2 \to \mathbb{R}$,
$$T(p(x)) = p(0)$$

24. $T: \mathcal{P}_2 \to \mathcal{P}_2$,
$$T(p(x)) = p''(x)$$

In Exercises 25–30, find a basis for the range of the linear transformation T.

25. $T: \mathbb{R}^3 \to \mathbb{R}^3$,
$$T(\mathbf{v}) = \begin{bmatrix} 1 & 1 & 2 \\ 0 & 1 & -1 \\ 2 & 0 & 1 \end{bmatrix} \mathbf{v}$$

26. $T: \mathbb{R}^5 \to \mathbb{R}^3$,
$$T(\mathbf{v}) = \begin{bmatrix} 1 & -2 & -3 & 1 & 5 \\ 3 & -1 & 1 & 0 & 4 \\ 1 & 1 & 3 & 1 & 2 \end{bmatrix} \mathbf{v}$$

27. $T: \mathbb{R}^3 \to \mathbb{R}^3$,
$$T\left(\begin{bmatrix} x \\ y \\ z \end{bmatrix}\right) = \begin{bmatrix} x \\ y \\ 0 \end{bmatrix}$$

28. $T: \mathbb{R}^3 \to \mathbb{R}^3$,
$$T\left(\begin{bmatrix} x \\ y \\ z \end{bmatrix}\right) = \begin{bmatrix} x - y + 3z \\ x + y + z \\ -x + 3y - 5z \end{bmatrix}$$

29. $T: \mathcal{P}_3 \to \mathcal{P}_3$,
$$T(p(x)) = p''(x) + p'(x) + p(0)$$

30. $T: \mathcal{P}_2 \to \mathcal{P}_2$,
$$T(ax^2 + bx + c) = (a + b)x^2 + cx + (a + b)$$

31. Let $T: \mathbb{R}^3 \to \mathbb{R}^3$ be a linear operator and $B = \{\mathbf{v}_1, \mathbf{v}_2, \mathbf{v}_3\}$ a basis for $\mathbb{R}^3$. Suppose
$$T(\mathbf{v}_1) = \begin{bmatrix} -2 \\ 1 \\ 1 \end{bmatrix} \quad T(\mathbf{v}_2) = \begin{bmatrix} 0 \\ 1 \\ -1 \end{bmatrix}$$
$$T(\mathbf{v}_3) = \begin{bmatrix} -2 \\ 2 \\ 0 \end{bmatrix}$$

a. Determine whether
$$\mathbf{w} = \begin{bmatrix} -6 \\ 5 \\ 0 \end{bmatrix}$$
is in the range of T.

b. Find a basis for $R(T)$.

c. Find $\dim(N(T))$.

32. Let $T: \mathbb{R}^3 \to \mathbb{R}^3$ be a linear operator and $B = \{\mathbf{v}_1, \mathbf{v}_2, \mathbf{v}_3\}$ a basis for $\mathbb{R}^3$. Suppose
$$T(\mathbf{v}_1) = \begin{bmatrix} -1 \\ 2 \\ 1 \end{bmatrix} \quad T(\mathbf{v}_2) = \begin{bmatrix} 0 \\ 5 \\ 0 \end{bmatrix}$$
$$T(\mathbf{v}_3) = \begin{bmatrix} -1 \\ -1 \\ 2 \end{bmatrix}$$

a. Determine whether
$$\mathbf{w} = \begin{bmatrix} -2 \\ 1 \\ 2 \end{bmatrix}$$
is in the range of T.

b. Find a basis for $R(T)$.

c. Find $\dim(N(T))$.

33. Let $T: \mathcal{P}_2 \to \mathcal{P}_2$ be defined by
$$T(ax^2 + bx + c) = ax^2 + (a - 2b)x + b$$

a. Determine whether $p(x) = 2x^2 - 4x + 6$ is in the range of T.

b. Find a basis for $R(T)$.

34. Let $T: \mathcal{P}_2 \to \mathcal{P}_2$ be defined by
$$T(ax^2 + bx + c) = cx^2 + bx - b$$

a. Determine whether $p(x) = x^2 - x - 2$ is in the range of T.

b. Find a basis for $R(T)$.

35. Find a linear transformation $T: \mathbb{R}^3 \to \mathbb{R}^2$ such that $R(T) = \mathbb{R}^2$.

36. Find a linear operator $T: \mathbb{R}^2 \to \mathbb{R}^2$ such that $R(T) = N(T)$.

37. Define a linear operator $T: \mathcal{P}_n \to \mathcal{P}_n$ by

$$T(p(x)) = p'(x)$$

a. Describe the range of T.

b. Find $\dim(R(T))$.

c. Find $\dim(N(T))$.

38. Define a linear operator $T: \mathcal{P}_n \to \mathcal{P}_n$ by

$$T(p(x)) = \frac{d^k}{dx^k}(p(x))$$

where $1 \leq k \leq n$. Show $\dim(N(T)) = k$.

39. Suppose $T: \mathbb{R}^4 \to \mathbb{R}^6$ is a linear transformation.

a. If $\dim(N(T)) = 2$, then find $\dim(R(T))$.

b. If $\dim(R(T)) = 3$, then find $\dim(N(T))$.

40. Show that if $T: V \to V$ is a linear operator such that $R(T) = N(T)$, then $\dim(V)$ is even.

41. Let

$$A = \begin{bmatrix} 1 & 0 \\ 0 & -1 \end{bmatrix}$$

Define $T: M_{2 \times 2} \to M_{2 \times 2}$ by

$$T(B) = AB - BA$$

Find a basis for the null space of T.

42. Define $T: M_{n \times n} \to M_{n \times n}$ by $T(A) = A^t$. Show that $R(T) = M_{n \times n}$.

43. Define $T: M_{n \times n} \to M_{n \times n}$ by $T(A) = A + A^t$.

a. Find $R(T)$.

b. Find $N(T)$.

44. Define $T: M_{n \times n} \to M_{n \times n}$ by $T(A) = A - A^t$.

a. Find $R(T)$.

b. Find $N(T)$.

45. Let A be a fixed $n \times n$ matrix, and define $T: M_{n \times n} \to M_{n \times n}$ by $T(B) = AB$. When does $R(T) = M_{n \times n}$?

46. Let A be a fixed $n \times n$ diagonal matrix, and define $T: \mathbb{R}^n \to \mathbb{R}^n$ by $T(\mathbf{v}) = A\mathbf{v}$.

a. Show $\dim(R(T))$ is the number of nonzero entries on the diagonal of A.

b. Find $\dim(N(T))$. How is it related to the diagonal terms of the matrix A?

4.3 ▶ Isomorphisms

Many of the vector spaces that we have discussed are, from an algebraic perspective, the *same*. In this section we show how an *isomorphism*, which is a special kind of linear transformation, can be used to establish a correspondence between two vector spaces. Essential to this discussion are the concepts of *one-to-one* and *onto* mappings. For a more detailed description see App. A, Sec. A.2.

DEFINITION 1

One-to-One and Onto Let V and W be vector spaces and $T: V \to W$ a mapping.

1. The mapping T is called **one-to-one** (or **injective**) if $\mathbf{u} \neq \mathbf{v}$ implies that $T(\mathbf{u}) \neq T(\mathbf{v})$. That is, distinct elements of V must have distinct images in W.
2. The mapping T is called **onto** (or **surjective**) if $T(V) = W$. That is, the range of T is W.

A mapping is called **bijective** if it is both injective and surjective.

When we are trying to show that a mapping is one-to-one, a useful equivalent formulation comes from the contrapositive statement. That is, T is one-to-one if $T(\mathbf{u}) = T(\mathbf{v})$ implies that $\mathbf{u} = \mathbf{v}$. To show that a mapping is onto, we must show that if $\mathbf{w}$ is an arbitrary element of W, then there is some element $\mathbf{v} \in V$ with $T(\mathbf{v}) = \mathbf{w}$.

EXAMPLE 1 Let $T: \mathbb{R}^2 \to \mathbb{R}^2$ be the mapping defined by $T(\mathbf{v}) = A\mathbf{v}$, with

$$A = \begin{bmatrix} 1 & 1 \\ -1 & 0 \end{bmatrix}$$

Show that T is one-to-one and onto.

Solution First, to show that T is one-to-one, let

$$\mathbf{u} = \begin{bmatrix} u_1 \\ u_2 \end{bmatrix} \quad \text{and} \quad \mathbf{v} = \begin{bmatrix} v_1 \\ v_2 \end{bmatrix}$$

Then

$$T(\mathbf{u}) = \begin{bmatrix} 1 & 1 \\ -1 & 0 \end{bmatrix} \begin{bmatrix} u_1 \\ u_2 \end{bmatrix} = \begin{bmatrix} u_1 + u_2 \\ -u_1 \end{bmatrix}$$

and

$$T(\mathbf{v}) = \begin{bmatrix} 1 & 1 \\ -1 & 0 \end{bmatrix} \begin{bmatrix} v_1 \\ v_2 \end{bmatrix} = \begin{bmatrix} v_1 + v_2 \\ -v_1 \end{bmatrix}$$

Now if $T(\mathbf{u}) = T(\mathbf{v})$, then

$$\begin{bmatrix} u_1 + u_2 \\ -u_1 \end{bmatrix} = \begin{bmatrix} v_1 + v_2 \\ -v_1 \end{bmatrix}$$

Equating the second components gives $u_1 = v_1$, and using this when equating the first components gives $u_2 = v_2$. Thus, $\mathbf{u} = \mathbf{v}$, establishing that the mapping is one-to-one.

Next, to show that T is onto, let $\mathbf{w} = \begin{bmatrix} a \\ b \end{bmatrix}$ be an arbitrary vector in $\mathbb{R}^2$. We must show that there is a vector $\mathbf{v} = \begin{bmatrix} v_1 \\ v_2 \end{bmatrix}$ in $\mathbb{R}^2$ such that

$$T(\mathbf{v}) = \begin{bmatrix} 1 & 1 \\ -1 & 0 \end{bmatrix} \begin{bmatrix} v_1 \\ v_2 \end{bmatrix} = \begin{bmatrix} a \\ b \end{bmatrix}$$

Applying the inverse of A to both sides of this equation, we have

$$\begin{bmatrix} v_1 \\ v_2 \end{bmatrix} = \begin{bmatrix} 0 & -1 \\ 1 & 1 \end{bmatrix} \begin{bmatrix} a \\ b \end{bmatrix} \quad \text{that is} \quad \begin{bmatrix} v_1 \\ v_2 \end{bmatrix} = \begin{bmatrix} -b \\ a+b \end{bmatrix}$$

Thus, T is onto. For example, let $\mathbf{w} = \begin{bmatrix} 1 \\ 2 \end{bmatrix}$; then using the above formula for the preimage, we have $\mathbf{v} = \begin{bmatrix} -2 \\ 1+2 \end{bmatrix} = \begin{bmatrix} -2 \\ 3 \end{bmatrix}$. As verification, observe that

$$T(\mathbf{v}) = \begin{bmatrix} 1 & 1 \\ -1 & 0 \end{bmatrix} \begin{bmatrix} -2 \\ 3 \end{bmatrix} = \begin{bmatrix} 1 \\ 2 \end{bmatrix}$$

An alternative argument is to observe that the column vectors of A are linearly independent and hence are a basis for $\mathbb{R}^2$. Therefore, the range of T being the column space of A is all of $\mathbb{R}^2$.

Theorem 8 gives a useful way to determine whether a linear transforation is one-to-one.

THEOREM 8

The linear transformation $T: V \longrightarrow W$ is one-to-one if and only if the null space of T consists of only the zero vector of V.

Proof   First suppose that T is one-to-one. We claim that $N(T) = \{0\}$. To show this, let $\mathbf{v}$ be any vector in the null space of T, so that $T(\mathbf{v}) = \mathbf{0}$. We also have, by Proposition 1 of Sec. 4.1, $T(\mathbf{0}) = \mathbf{0}$. Since T is one-to-one, then $\mathbf{v} = \mathbf{0}$, so only the zero vector is mapped to the zero vector.

Now suppose that $N(T) = \{0\}$ and

$$T(\mathbf{u}) = T(\mathbf{v})$$

Subtracting $T(\mathbf{v})$ from both sides of the last equation and using the linearity of T, we obtain

$$T(\mathbf{u}) - T(\mathbf{v}) = \mathbf{0} \qquad \text{so that} \qquad T(\mathbf{u} - \mathbf{v}) = \mathbf{0}$$

Thus, $\mathbf{u} - \mathbf{v} \in N(T)$. Since the null space consists of only the zero vector, $\mathbf{u} - \mathbf{v} = \mathbf{0}$, that is, $\mathbf{u} = \mathbf{v}$.

EXAMPLE 2

Define a linear operator $T: \mathbb{R}^2 \longrightarrow \mathbb{R}^2$ by

$$T\left(\begin{bmatrix} x \\ y \end{bmatrix}\right) = \begin{bmatrix} 2x - 3y \\ 5x + 2y \end{bmatrix}$$

Use Theorem 8 to show that T is one-to-one.

Solution   The vector $\begin{bmatrix} x \\ y \end{bmatrix}$ is in the null space of T if and only if

$$\begin{cases} 2x - 3y = 0 \\ 5x + 2y = 0 \end{cases}$$

This linear system has the unique solution $x = y = 0$. Thus, $N(T) = \{0\}$ and hence by Theorem 8, T is one-to-one.

The mapping of Example 2 can alternatively be defined by using the matrix

$$A = \begin{bmatrix} 2 & -3 \\ 5 & 2 \end{bmatrix}$$

so that $T(\mathbf{x}) = A\mathbf{x}$. Since $\det(A) \neq 0$, then A is invertible. This allows us to show that the map is also onto. Indeed, if $\mathbf{b}$ is any vector in $\mathbb{R}^2$, then $\mathbf{x} = A^{-1}\mathbf{b}$ is the vector in the domain that is the preimage of $\mathbf{b}$, so that T is onto.

In Theorem 4 of Sec. 4.2, we showed that if $T: V \longrightarrow W$ is a linear transformation between vector spaces, and $B = \{\mathbf{v}_1, \ldots, \mathbf{v}_n\}$ is a basis for V, then $R(T) = \text{span}\{T(\mathbf{v}_1), \ldots, T(\mathbf{v}_n)\}$. If, in addition, the transformation is one-to-one, then the spanning vectors are also a basis for the range, as given in Theorem 9.

THEOREM 9

Suppose that $T: V \to W$ is a linear transformation and $B = \{\mathbf{v}_1, \ldots, \mathbf{v}_n\}$ is a basis for V. If T is one-to-one, then $\{T(\mathbf{v}_1), \ldots, T(\mathbf{v}_n)\}$ is a basis for $R(T)$.

Proof By Theorem 4 of Sec. 4.2, we know that the $\mathbf{span}\{T(\mathbf{v}_1), \ldots, T(\mathbf{v}_n)\} = R(T)$, so it suffices to show that $\{T(\mathbf{v}_1), \ldots, T(\mathbf{v}_n)\}$ is linearly independent. To do so, we consider the equation

$$c_1 T(\mathbf{v}_1) + c_2 T(\mathbf{v}_2) + \cdots + c_n T(\mathbf{v}_n) = \mathbf{0}$$

which is equivalent to

$$T(c_1\mathbf{v}_1 + c_2\mathbf{v}_2 + \cdots + c_n\mathbf{v}_n) = \mathbf{0}$$

Since T is one-to-one, the null space consists of only the zero vector of V, so that

$$c_1\mathbf{v}_1 + c_2\mathbf{v}_2 + \cdots + c_n\mathbf{v}_n = \mathbf{0}$$

Finally, since B is a basis for V, it is linearly independent; hence

$$c_1 = c_2 = \cdots = c_n = 0$$

Therefore, $\{T(\mathbf{v}_1), \ldots, T(\mathbf{v}_n)\}$ is linearly independent.

We note that in Theorem 9 if T is also onto, then $\{T(\mathbf{v}_1), \ldots, T(\mathbf{v}_n)\}$ is a basis for W.

We are now ready to define an isomorphism on vector spaces.

DEFINITION 2

Isomorphism Let V and W be vector spaces. A linear transformation $T: V \longrightarrow W$ that is both one-to-one and onto is called an **isomorphism**. In this case the vector spaces V and W are said to be **isomorphic**.

Proposition 2 builds on the remarks following Example 2 and gives a useful characterization of linear transformations defined by a matrix that are isomorphisms.

PROPOSITION 2

Let A be an $n \times n$ matrix and $T: \mathbb{R}^n \longrightarrow \mathbb{R}^n$ be the mapping defined by $T(\mathbf{x}) = A\mathbf{x}$. Then T is an isomorphism if and only if A is invertible.

Proof Let A be invertible and $\mathbf{b}$ be any vector in $\mathbb{R}^n$. Then $\mathbf{x} = A^{-1}\mathbf{b}$ is the preimage of $\mathbf{b}$. Thus, the mapping T is onto. To show that T is one-to-one, observe that by Theorem 10 of Sec. 1.5 the equation $A\mathbf{x} = \mathbf{0}$ has only the solution $\mathbf{x} = A^{-1}\mathbf{0} = \mathbf{0}$. Thus, by Theorem 8, the mapping T is one-to-one and hence is an isomorphism from $\mathbb{R}^n$ onto $\mathbb{R}^n$.

Conversely, suppose that T is an isomorphism. Then $T: \mathbb{R}^n \longrightarrow \mathbb{R}^n$ is onto, with the column space of A being $\mathbb{R}^n$. Hence, by Theorem 7 of Sec. 4.2 the matrix A is invertible.

Theorem 10 is of fundamental importance to the study of finite dimensional vector spaces and is the main result of the section.

THEOREM 10

If V is a vector space with $\dim(V) = n$, then V and $\mathbb{R}^n$ are isomorphic.

Proof Let $B = \{\mathbf{v}_1, \ldots, \mathbf{v}_n\}$ be an ordered basis for V. Let $T: V \longrightarrow \mathbb{R}^n$ be the coordinate transformation defined by $T(\mathbf{v}) = [\mathbf{v}]_B$, first introduced in Example 7 of Sec. 4.1. We claim that T is an isomorphism. First, to show that T is one-to-one, suppose that $T(\mathbf{v}) = \mathbf{0}$. Since B is a basis, there are unique scalars $c_1, \ldots, c_n$ such that

$$\mathbf{v} = c_1 \mathbf{v}_1 + \cdots + c_n \mathbf{v}_n$$

Thus,

$$T(\mathbf{v}) = [\mathbf{v}]_B = \begin{bmatrix} c_1 \\ c_2 \\ \vdots \\ c_n \end{bmatrix} = \begin{bmatrix} 0 \\ 0 \\ \vdots \\ 0 \end{bmatrix}$$

so that $c_1 = c_2 = \cdots = c_n = 0$ and $\mathbf{v} = \mathbf{0}$. Therefore, $N(T) = \{\mathbf{0}\}$, and by Theorem 8, T is one-to-one.

Now, to show that T is onto, let

$$\mathbf{w} = \begin{bmatrix} k_1 \\ k_2 \\ \vdots \\ k_n \end{bmatrix}$$

be a vector in $\mathbb{R}^n$. Define $\mathbf{v}$ in V by $\mathbf{v} = k_1 \mathbf{v}_1 + \cdots + k_n \mathbf{v}_n$. Observe that $T(\mathbf{v}) = \mathbf{w}$ and hence T is onto. Therefore, the linear transformation T is an isomorphism, and V and $\mathbb{R}^n$ are isomorphic vector spaces.

So far in our experience we have seen that $\dim(\mathcal{P}_2) = 3$ and $\dim(S_{2\times 2}) = 3$, where $S_{2\times 2}$ is the vector space of 2×2 symmetric matrices. Consequently, by Theorem 10, the vector spaces $\mathcal{P}_2$ and $S_{2\times 2}$ are both isomorphic to $\mathbb{R}^3$, where the isomorphism is the coordinate map between the standard bases. Next we show that in fact all vector spaces of dimension n are isomorphic to one another. To do so, we first require the notion of the inverse of a linear transformation.

DEFINITION 3

Inverse of a Linear Transformation   Let V and W be vector spaces and $T: V \longrightarrow W$ a one-to-one linear transformation. The mapping $T^{-1}: R(T) \longrightarrow V$, defined by

$$T^{-1}(\mathbf{w}) = \mathbf{v} \qquad \text{if and only if} \qquad T(\mathbf{v}) = \mathbf{w}$$

is called the **inverse** of T. If T is onto, then T^{-1} is defined on all of W.

By Theorem 4 of Sec. A.2, if T is one-to-one, then the inverse map is well defined. Indeed, let $\mathbf{u}$ and $\mathbf{v}$ be vectors in V such that $T^{-1}(\mathbf{w}) = \mathbf{u}$ and $T^{-1}(\mathbf{w}) = \mathbf{v}$. Applying T gives $T(T^{-1}(\mathbf{w})) = T(\mathbf{u})$ and $T(T^{-1}(\mathbf{w})) = T(\mathbf{v})$, so that $T(\mathbf{u}) = T(\mathbf{v})$. Since T is one-to-one, we have $\mathbf{u} = \mathbf{v}$.

The inverse map of a one-to-one linear transformation is also a linear transformation, as we now show.

PROPOSITION 3 Let V and W be vector spaces and $T: V \longrightarrow W$ a one-to-one linear transformation. Then the mapping $T^{-1}: R(T) \longrightarrow V$ is also a linear transformation.

Proof Let $\mathbf{w}_1$ and $\mathbf{w}_2$ be vectors in $R(T)$, and let c be a scalar. Also let $\mathbf{v}_1$ and $\mathbf{v}_2$ be vectors in V with $T^{-1}(\mathbf{w}_1) = \mathbf{v}_1$ and $T^{-1}(\mathbf{w}_2) = \mathbf{v}_2$. Since T is linear,

$$T(c\mathbf{v}_1 + \mathbf{v}_2) = cT(\mathbf{v}_1) + T(\mathbf{v}_2)$$
$$= c\mathbf{w}_1 + \mathbf{w}_2$$

Hence,

$$T^{-1}(c\mathbf{w}_1 + \mathbf{w}_2) = c\mathbf{v}_1 + \mathbf{v}_2$$
$$= cT^{-1}(\mathbf{w}_1) + T^{-1}(\mathbf{w}_2)$$

Consequently, T^{-1} is a linear transformation.

Proposition 4 shows that the inverse transformation of an isomorphism defined by matrix multiplication can be written using the inverse of the matrix. The proof is left as an exercise.

PROPOSITION 4 Let A be an $n \times n$ invertible matrix and $T: \mathbb{R}^n \longrightarrow \mathbb{R}^n$ the linear transformation defined by $T(\mathbf{x}) = A\mathbf{x}$. Then $T^{-1}(\mathbf{x}) = A^{-1}\mathbf{x}$.

EXAMPLE 3 Let $T: \mathbb{R}^2 \to \mathbb{R}^2$ be the mapping of Example 1 with $T(\mathbf{v}) = A\mathbf{v}$, where

$$A = \begin{bmatrix} 1 & 1 \\ -1 & 0 \end{bmatrix}$$

Verify that the inverse map $T^{-1}: \mathbb{R}^2 \longrightarrow \mathbb{R}^2$ is given by $T^{-1}(\mathbf{w}) = A^{-1}\mathbf{w}$, where

$$A^{-1} = \begin{bmatrix} 0 & -1 \\ 1 & 1 \end{bmatrix}$$

Solution Let $\mathbf{v} = \begin{bmatrix} v_1 \\ v_2 \end{bmatrix}$ be a vector in $\mathbb{R}^2$. Then

$$\mathbf{w} = T(\mathbf{v}) = \begin{bmatrix} v_1 + v_2 \\ -v_1 \end{bmatrix}$$

Applying A^{-1} to $\mathbf{w}$, we obtain

$$\begin{bmatrix} 0 & -1 \\ 1 & 1 \end{bmatrix} \begin{bmatrix} v_1 + v_2 \\ -v_1 \end{bmatrix} = \begin{bmatrix} v_1 \\ v_2 \end{bmatrix} = T^{-1}(\mathbf{w})$$

THEOREM 11   If V and W are vector spaces of dimension n, then V and W are isomorphic.

Proof   By Theorem 10, there are isomorphisms $T_1: V \longrightarrow \mathbb{R}^n$ and $T_2: W \longrightarrow \mathbb{R}^n$, as shown in Fig. 1. Let $\phi = T_2^{-1} \circ T_1: V \longrightarrow W$. To show that ϕ is linear, we first note that T_2^{-1} is linear by Proposition 3. Next by Theorem 2 of Sec. 4.1, the composition $T_2^{-1} \circ T_1$ is linear. Finally, by Theorem 4 of Sec. A.2, the mapping ϕ is one-to-one and onto and is therefore a vector space isomorphism.

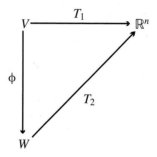

$$\phi = T_2^{-1} \circ T_1: V \longrightarrow W$$

Figure 1

EXAMPLE 4   Find an explicit isomorphism from $\mathcal{P}_2$ onto the vector space of 2×2 symmetric matrices $S_{2\times 2}$.

Solution   To use the method given in the proof of Theorem 11, first let

$$B_1 = \{1, x, x^2\} \quad \text{and} \quad B_2 = \left\{ \begin{bmatrix} 0 & 0 \\ 0 & 1 \end{bmatrix}, \begin{bmatrix} 0 & 1 \\ 1 & 0 \end{bmatrix}, \begin{bmatrix} 1 & 0 \\ 0 & 0 \end{bmatrix} \right\}$$

be ordered bases for $\mathcal{P}_2$ and $S_{2\times 2}$, respectively. Let T_1 and T_2 be the respective coordinate maps from $\mathcal{P}_2$ and $S_{2\times 2}$ into $\mathbb{R}^3$. Then

$$T_1(ax^2 + bx + c) = \begin{bmatrix} c \\ b \\ a \end{bmatrix} \quad \text{and} \quad T_2 \left(\begin{bmatrix} a & b \\ b & c \end{bmatrix} \right) = \begin{bmatrix} c \\ b \\ a \end{bmatrix}$$

Observe that $T_2^{-1} \colon \mathbb{R}^3 \longrightarrow S_{2\times 2}$ maps the vector

$$\begin{bmatrix} c \\ b \\ a \end{bmatrix} \qquad \text{to the symmetric matrix} \qquad \begin{bmatrix} a & b \\ b & c \end{bmatrix}$$

Thus, the desired isomorphism is given by $(T_2^{-1} \circ T_1) \colon \mathcal{P}_2 \longrightarrow S_{2\times 2}$ with

$$(T_2^{-1} \circ T_1)(ax^2 + bx + c) = \begin{bmatrix} a & b \\ b & c \end{bmatrix}$$

For example,

$$(T_2^{-1} \circ T_1)(x^2 - x + 2) = T_2^{-1}(T_1(x^2 - x + 2)) = T_2^{-1}\left(\begin{bmatrix} 2 \\ -1 \\ 1 \end{bmatrix} \right) = \begin{bmatrix} 1 & -1 \\ -1 & 2 \end{bmatrix}$$

Fact Summary

Let V and W be vector spaces and T a linear transformation from V into W.

1. The mapping T is one-to-one if and only if the null space of T consists of only the zero vector.
2. If $\{\mathbf{v}_1, \ldots, \mathbf{v}_n\}$ is a basis for V and T is one-to-one, then $S = \{T(\mathbf{v}_1), \ldots, T(\mathbf{v}_n)\}$ is a basis for the range of T. If T is also onto, then S is a basis for W.
3. Every vector space of dimension n is isomorphic to the Euclidean space $\mathbb{R}^n$.
4. If T is one-to-one, then T^{-1} is also a linear transformation.
5. If V and W are both of dimension n, then they are isomorphic.
6. Let A be an $n \times n$ matrix and $T(\mathbf{x}) = A\mathbf{x}$. Then the mapping T is an isomorphism if and only if A is invertible.
7. If A is an invertible matrix and $T(\mathbf{x}) = A\mathbf{x}$, then $T^{-1}(\mathbf{x}) = A^{-1}\mathbf{x}$.

Exercise Set 4.3

In Exercises 1–6, determine whether the linear transformation is one-to-one.

1. $T \colon \mathbb{R}^2 \to \mathbb{R}^2$,
$$T\left(\begin{bmatrix} x \\ y \end{bmatrix} \right) = \begin{bmatrix} 4x - y \\ x \end{bmatrix}$$

2. $T \colon \mathbb{R}^2 \to \mathbb{R}^2$,
$$T\left(\begin{bmatrix} x \\ y \end{bmatrix} \right) = \begin{bmatrix} \frac{1}{2}x + \frac{1}{2}y \\ \frac{1}{2}x + \frac{1}{2}y \end{bmatrix}$$

3. $T \colon \mathbb{R}^3 \to \mathbb{R}^3$,
$$T\left(\begin{bmatrix} x \\ y \\ z \end{bmatrix} \right) = \begin{bmatrix} x + y - z \\ y \\ y - z \end{bmatrix}$$

4. $T \colon \mathbb{R}^3 \to \mathbb{R}^3$,
$$T\left(\begin{bmatrix} x \\ y \\ z \end{bmatrix} \right) = \begin{bmatrix} \frac{2}{3}x - \frac{2}{3}y - \frac{2}{3}z \\ -\frac{2}{3}x - \frac{1}{3}y - \frac{1}{3}z \\ -\frac{2}{3}x - \frac{4}{3}y - \frac{1}{3}z \end{bmatrix}$$

5. $T: \mathcal{P}_2 \to \mathcal{P}_2$,
$$T(p(x)) = p'(x) - p(x)$$

6. $T: \mathcal{P}_2 \to \mathcal{P}_3$,
$$T(p(x)) = xp(x)$$

In Exercises 7–10, determine whether the linear transformation is onto.

7. $T: \mathbb{R}^2 \to \mathbb{R}^2$,
$$T\left(\begin{bmatrix} x \\ y \end{bmatrix}\right) = \begin{bmatrix} 3x - y \\ x + y \end{bmatrix}$$

8. $T: \mathbb{R}^2 \to \mathbb{R}^2$,
$$T\left(\begin{bmatrix} x \\ y \end{bmatrix}\right) = \begin{bmatrix} -2x + y \\ x - \frac{1}{2}y \end{bmatrix}$$

9. $T: \mathbb{R}^3 \to \mathbb{R}^3$,
$$T\left(\begin{bmatrix} x \\ y \\ z \end{bmatrix}\right) = \begin{bmatrix} x - y + 2z \\ y - z \\ 2z \end{bmatrix}$$

10. $T: \mathbb{R}^3 \to \mathbb{R}^3$,
$$T\left(\begin{bmatrix} x \\ y \\ z \end{bmatrix}\right) = \begin{bmatrix} 2x + 3y - z \\ -x + y + 3z \\ x + 4y + 2z \end{bmatrix}$$

In Exercises 11–14, $T: \mathbb{R}^2 \to \mathbb{R}^2$ is a linear operator. Determine whether the set $\{T(\mathbf{e}_1), T(\mathbf{e}_2)\}$ is a basis for $\mathbb{R}^2$.

11. $T: \mathbb{R}^2 \to \mathbb{R}^2$,
$$T\left(\begin{bmatrix} x \\ y \end{bmatrix}\right) = \begin{bmatrix} -x - 2y \\ 3x \end{bmatrix}$$

12. $T: \mathbb{R}^2 \to \mathbb{R}^2$,
$$T\left(\begin{bmatrix} x \\ y \end{bmatrix}\right) = \begin{bmatrix} x \\ -3x \end{bmatrix}$$

13. $T: \mathbb{R}^2 \to \mathbb{R}^2$,
$$T\left(\begin{bmatrix} x \\ y \end{bmatrix}\right) = \begin{bmatrix} 3x - y \\ -3x - y \end{bmatrix}$$

14. $T: \mathbb{R}^2 \to \mathbb{R}^2$,
$$T\left(\begin{bmatrix} x \\ y \end{bmatrix}\right) = \begin{bmatrix} \frac{1}{10}x + \frac{1}{5}y \\ \frac{1}{5}x + \frac{2}{5}y \end{bmatrix}$$

In Exercises 15–18, $T: \mathbb{R}^3 \to \mathbb{R}^3$ is a linear operator. Determine whether the set $\{T(\mathbf{e}_1), T(\mathbf{e}_2), T(\mathbf{e}_3)\}$ is a basis for $\mathbb{R}^3$.

15. $T: \mathbb{R}^3 \to \mathbb{R}^3$,
$$T\left(\begin{bmatrix} x \\ y \\ z \end{bmatrix}\right) = \begin{bmatrix} -x - y + 2z \\ y - z \\ 5z \end{bmatrix}$$

16. $T: \mathbb{R}^3 \to \mathbb{R}^3$,
$$T\left(\begin{bmatrix} x \\ y \\ z \end{bmatrix}\right) = \begin{bmatrix} 2x + 3y - z \\ 2x + 6y + 3z \\ 4x + 9y + 2z \end{bmatrix}$$

17. $T: \mathbb{R}^3 \to \mathbb{R}^3$,
$$T\left(\begin{bmatrix} x \\ y \\ z \end{bmatrix}\right) = \begin{bmatrix} 4x - 2y + z \\ 2x + z \\ 2x - y + \frac{3}{2}z \end{bmatrix}$$

18. $T: \mathbb{R}^3 \to \mathbb{R}^3$,
$$T\left(\begin{bmatrix} x \\ y \\ z \end{bmatrix}\right) = \begin{bmatrix} x - y + 2z \\ -x + 2y - z \\ -y + 5z \end{bmatrix}$$

In Exercises 19 and 20, $T: \mathcal{P}_2 \to \mathcal{P}_2$ is a linear operator. Determine whether the set $\{T(1), T(x), T(x^2)\}$ is a basis for $\mathcal{P}_2$.

19. $T(ax^2 + bx + c) = (a + b + c)x^2 + (a + b)x + a$

20. $T(p(x)) = xp'(x)$

In Exercises 21–24, let $T: V \to V$ be the linear operator defined by $T(\mathbf{v}) = A\mathbf{v}$.

a. Show that T is an isomorphism.

b. Find A^{-1}.

c. Show directly that $T^{-1}(\mathbf{w}) = A^{-1}\mathbf{w}$ for all $\mathbf{w} \in V$.

21. $T\left(\begin{bmatrix} x \\ y \end{bmatrix}\right) = \begin{bmatrix} 1 & 0 \\ -2 & -3 \end{bmatrix}\begin{bmatrix} x \\ y \end{bmatrix}$

22. $T\left(\begin{bmatrix} x \\ y \end{bmatrix}\right) = \begin{bmatrix} -2 & 3 \\ -1 & -1 \end{bmatrix} \begin{bmatrix} x \\ y \end{bmatrix}$

23. $T\left(\begin{bmatrix} x \\ y \\ z \end{bmatrix}\right) = \begin{bmatrix} -2 & 0 & 1 \\ 1 & -1 & -1 \\ 0 & 1 & 0 \end{bmatrix} \begin{bmatrix} x \\ y \\ z \end{bmatrix}$

24. $T\left(\begin{bmatrix} x \\ y \\ z \end{bmatrix}\right) = \begin{bmatrix} 2 & -1 & 1 \\ -1 & 1 & -1 \\ 0 & 1 & 0 \end{bmatrix} \begin{bmatrix} x \\ y \\ z \end{bmatrix}$

In Exercises 25–28, determine whether the matrix mapping $T: V \to V$ is an isomorphism.

25. $T\left(\begin{bmatrix} x \\ y \end{bmatrix}\right) = \begin{bmatrix} -3 & 1 \\ 1 & -3 \end{bmatrix} \begin{bmatrix} x \\ y \end{bmatrix}$

26. $T\left(\begin{bmatrix} x \\ y \end{bmatrix}\right) = \begin{bmatrix} -3 & 1 \\ -3 & 1 \end{bmatrix} \begin{bmatrix} x \\ y \end{bmatrix}$

27. $T\left(\begin{bmatrix} x \\ y \\ z \end{bmatrix}\right) = \begin{bmatrix} 0 & -1 & -1 \\ 2 & 0 & 2 \\ 1 & 1 & -3 \end{bmatrix} \begin{bmatrix} x \\ y \\ z \end{bmatrix}$

28. $T\left(\begin{bmatrix} x \\ y \\ z \end{bmatrix}\right) = \begin{bmatrix} 1 & 3 & 0 \\ -1 & -2 & -3 \\ 0 & -1 & 3 \end{bmatrix} \begin{bmatrix} x \\ y \\ z \end{bmatrix}$

29. Show that $T: M_{n \times n} \to M_{n \times n}$ defined by

$$T(A) = A^t$$

is an isomorphism.

30. Show that $T: \mathcal{P}_3 \to \mathcal{P}_3$ defined by

$$T(p(x)) = p'''(x) + p''(x) + p'(x) + p(x)$$

is an isomorphism.

31. Let A be an $n \times n$ invertible matrix. Show that $T: M_{n \times n} \to M_{n \times n}$ defined by

$$T(B) = ABA^{-1}$$

is an isomorphism.

32. Find an isomorphism from $M_{2 \times 2}$ onto $\mathbb{R}^4$.

33. Find an isomorphism from $\mathbb{R}^4$ onto $\mathcal{P}_3$.

34. Find an isomorphism from $M_{2 \times 2}$ onto $\mathcal{P}_3$.

35. Let

$$V = \left\{ \begin{bmatrix} x \\ y \\ z \end{bmatrix} \,\middle|\, x + 2y - z = 0 \right\}$$

Find an isomorphism from V onto $\mathbb{R}^2$.

36. Let

$$V = \left\{ \begin{bmatrix} a & b \\ c & -a \end{bmatrix} \,\middle|\, a, b, c \in \mathbb{R} \right\}$$

Find an isomorphism from $\mathcal{P}_2$ onto V.

37. Suppose $T: \mathbb{R}^3 \to \mathbb{R}^3$ is an isomorphism. Show that T takes lines through the origin to lines through the origin and planes through the origin to planes through the origin.

4.4 ▶ Matrix Representation of a Linear Transformation

Matrices have played an important role in our study of linear algebra. In this section we establish the connection between matrices and linear transformations. To illustrate the idea, recall from Sec. 4.1 that given any $m \times n$ matrix A, we can define a linear transformation $T: \mathbb{R}^n \longrightarrow \mathbb{R}^m$ by

$$T(\mathbf{v}) = A\mathbf{v}$$

In Example 8 of Sec. 4.1, we showed how a linear transformation $T: \mathbb{R}^3 \longrightarrow \mathbb{R}^2$ is completely determined by the images of the coordinate vectors $\mathbf{e}_1, \mathbf{e}_2$, and $\mathbf{e}_3$ of $\mathbb{R}^3$.

The key was to recognize that a vector $\mathbf{v} = \begin{bmatrix} v_1 \\ v_2 \\ v_3 \end{bmatrix}$ can be written as

$$\mathbf{v} = v_1\mathbf{e}_1 + v_2\mathbf{e}_2 + v_3\mathbf{e}_3 \qquad \text{so that} \qquad T(\mathbf{v}) = v_1 T(\mathbf{e}_1) + v_2 T(\mathbf{e}_2) + v_3 T(\mathbf{e}_3)$$

In that example, T was defined so that

$$T(\mathbf{e}_1) = \begin{bmatrix} 1 \\ 1 \end{bmatrix}, T(\mathbf{e}_2) = \begin{bmatrix} -1 \\ 2 \end{bmatrix} \qquad T(\mathbf{e}_3) = \begin{bmatrix} 0 \\ 1 \end{bmatrix}$$

Now let A be the 2×3 matrix whose column vectors are $T(\mathbf{e}_1)$, $T(\mathbf{e}_2)$, and $T(\mathbf{e}_3)$. Then

$$T(\mathbf{v}) = \begin{bmatrix} 1 & -1 & 0 \\ 1 & 2 & 1 \end{bmatrix} \mathbf{v} = A\mathbf{v}$$

That is, the linear transformation T is given by a matrix product. In general, if $T \colon \mathbb{R}^n \longrightarrow \mathbb{R}^m$ is a linear transformation, then it is possible to write

$$T(\mathbf{v}) = A\mathbf{v}$$

where A is the $m \times n$ matrix whose jth column vector is $T(\mathbf{e}_j)$ for $j = 1, 2, \ldots, n$. The matrix A is called the **matrix representation of T relative to the standard bases** of $\mathbb{R}^n$ and $\mathbb{R}^m$.

In this section we show that *every linear transformation between finite dimensional vector spaces can be written as a matrix multiplication.* Specifically, let V and W be finite dimensional vector spaces with fixed ordered bases B and B', respectively. If $T \colon V \longrightarrow W$ is a linear transformation, then there exists a matrix A such that

$$[T(\mathbf{v})]_{B'} = A[\mathbf{v}]_B$$

In the case for which $V = \mathbb{R}^n$, $W = \mathbb{R}^m$, and B and B' are, respectively, the standard bases, the last equation is equivalent to

$$T(\mathbf{v}) = A\mathbf{v}$$

as above. We now present the details.

Let V and W be vector spaces with ordered bases $B = \{\mathbf{v}_1, \mathbf{v}_2, \ldots, \mathbf{v}_n\}$ and $B' = \{\mathbf{w}_1, \mathbf{w}_2, \ldots, \mathbf{w}_m\}$, respectively, and let $T \colon V \longrightarrow W$ be a linear transformation. Now let $\mathbf{v}$ be any vector in V and let

$$[\mathbf{v}]_B = \begin{bmatrix} c_1 \\ c_2 \\ \vdots \\ c_n \end{bmatrix}$$

be the coordinate vector of $\mathbf{v}$ relative to the basis B. Thus,

$$\mathbf{v} = c_1\mathbf{v}_1 + c_2\mathbf{v}_2 + \cdots + c_n\mathbf{v}_n$$

Applying T to both sides of this last equation gives

$$T(\mathbf{v}) = T(c_1\mathbf{v}_1 + c_2\mathbf{v}_2 + \cdots + c_n\mathbf{v}_n)$$
$$= c_1 T(\mathbf{v}_1) + c_2 T(\mathbf{v}_2) + \cdots + c_n T(\mathbf{v}_n)$$

Note that for each $i = 1, 2, \ldots, n$ the vector $T(\mathbf{v}_i)$ is in W. Thus, there are unique scalars a_{ij} with $1 \leq i \leq m$ and $1 \leq j \leq n$ such that

$$T(\mathbf{v}_1) = a_{11}\mathbf{w}_1 + a_{21}\mathbf{w}_2 + \cdots + a_{m1}\mathbf{w}_m$$
$$T(\mathbf{v}_2) = a_{12}\mathbf{w}_1 + a_{22}\mathbf{w}_2 + \cdots + a_{m2}\mathbf{w}_m$$
$$\vdots$$
$$T(\mathbf{v}_n) = a_{1n}\mathbf{w}_1 + a_{2n}\mathbf{w}_2 + \cdots + a_{mn}\mathbf{w}_m$$

Thus, the coordinate vectors relative to the ordered basis B' are given by

$$[T(\mathbf{v}_i)]_{B'} = \begin{bmatrix} a_{1i} \\ a_{2i} \\ \vdots \\ a_{mi} \end{bmatrix} \qquad \text{for } i = 1, 2, \ldots, n$$

Recall from Example 7 of Sec. 4.1 that the coordinate map defines a linear transformation. Thus, the coordinate vector of $T(\mathbf{v})$ relative to B' can be written in vector form as

$$[T(\mathbf{v})]_{B'} = c_1 \begin{bmatrix} a_{11} \\ a_{21} \\ \vdots \\ a_{m1} \end{bmatrix} + c_2 \begin{bmatrix} a_{12} \\ a_{22} \\ \vdots \\ a_{m2} \end{bmatrix} + \cdots + c_n \begin{bmatrix} a_{1n} \\ a_{2n} \\ \vdots \\ a_{mn} \end{bmatrix}$$

or in matrix form as

$$[T(\mathbf{v})]_{B'} = \begin{bmatrix} a_{11} & a_{12} & \cdots & a_{1n} \\ a_{21} & a_{22} & \cdots & a_{2n} \\ \vdots & \vdots & \vdots & \vdots \\ a_{m1} & a_{m2} & \cdots & a_{mn} \end{bmatrix} \begin{bmatrix} c_1 \\ c_2 \\ \vdots \\ c_n \end{bmatrix}$$

The matrix on the right-hand side of the last equation is denoted by $[T]_B^{B'}$, with

$$[T]_B^{B'} = \begin{bmatrix} \begin{bmatrix} T(\mathbf{v}_1) \end{bmatrix}_{B'} & \begin{bmatrix} T(\mathbf{v}_2) \end{bmatrix}_{B'} & \cdots & \begin{bmatrix} T(\mathbf{v}_n) \end{bmatrix}_{B'} \end{bmatrix}$$

We call $[T]_B^{B'}$ the **matrix of T relative to B and B'**. In the case for which $T: V \to V$ is a linear operator and B is a fixed ordered basis for V, the matrix representation for the mapping T is denoted by $[T]_B$.

The preceding discussion is summarized in Theorem 12.

THEOREM 12   Let V and W be finite dimensional vector spaces with ordered bases $B = \{\mathbf{v}_1, \mathbf{v}_2, \ldots \mathbf{v}_n\}$ and $B' = \{\mathbf{w}_1, \mathbf{w}_2, \ldots \mathbf{w}_m\}$, respectively, and let $T: V \longrightarrow W$ be a linear transformation. Then the matrix $[T]_B^{B'}$ is the matrix representation for T relative to the bases B and B'. Moreover, the coordinates of $T(v)$ relative to B' are given by

$$[T(\mathbf{v})]_{B'} = [T]_B^{B'} [\mathbf{v}]_B$$

Suppose that in Theorem 12 the vector spaces V and W are the same, B and B' are two different ordered bases for V, and $T: V \longrightarrow V$ is the identity operator, that is, $T(\mathbf{v}) = \mathbf{v}$ for all $\mathbf{v}$ in V. Then $[T]_B^{B'}$ is the change of bases matrix $[I]_B^{B'}$, given in Sec. 3.4.

EXAMPLE 1   Define the linear operator $T: \mathbb{R}^3 \longrightarrow \mathbb{R}^3$ by

$$T\left(\begin{bmatrix} x \\ y \\ z \end{bmatrix}\right) = \begin{bmatrix} x \\ -y \\ z \end{bmatrix}$$

a. Find the matrix of T relative to the standard basis for $\mathbb{R}^3$.

b. Use the result of part (a) to find

$$T\left(\begin{bmatrix} 1 \\ 1 \\ 2 \end{bmatrix}\right)$$

Solution   **a.** Let $B = \{\mathbf{e}_1, \mathbf{e}_2, \mathbf{e}_3\}$ be the standard basis for $\mathbb{R}^3$. Since

$$[T(\mathbf{e}_1)]_B = \begin{bmatrix} 1 \\ 0 \\ 0 \end{bmatrix} \qquad [T(\mathbf{e}_2)]_B = \begin{bmatrix} 0 \\ -1 \\ 0 \end{bmatrix} \qquad [T(\mathbf{e}_3)]_B = \begin{bmatrix} 0 \\ 0 \\ 1 \end{bmatrix}$$

then

$$[T]_B = \begin{bmatrix} 1 & 0 & 0 \\ 0 & -1 & 0 \\ 0 & 0 & 1 \end{bmatrix}$$

b. Since B is the standard basis for $\mathbb{R}^3$, the coordinates of any vector are given by its components. In this case, with

$$\mathbf{v} = \begin{bmatrix} 1 \\ 1 \\ 2 \end{bmatrix} \qquad \text{then} \qquad [\mathbf{v}]_B = \begin{bmatrix} 1 \\ 1 \\ 2 \end{bmatrix}$$

Thus, by Theorem 12,

$$T(\mathbf{v}) = [T(\mathbf{v})]_B = \begin{bmatrix} 1 & 0 & 0 \\ 0 & -1 & 0 \\ 0 & 0 & 1 \end{bmatrix}\begin{bmatrix} 1 \\ 1 \\ 2 \end{bmatrix} = \begin{bmatrix} 1 \\ -1 \\ 2 \end{bmatrix}$$

Notice that the action of T is a reflection through the xz plane, as shown in Fig. 1.

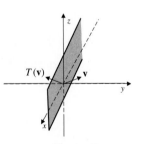

Figure 1

The following steps summarize the process for finding the matrix representation of a linear transformation $T: V \longrightarrow W$ relative to the ordered bases B and B'.

1. For the given basis $B = \{\mathbf{v}_1, \mathbf{v}_2, \dots, \mathbf{v}_n\}$, find $T(\mathbf{v}_1), T(\mathbf{v}_2), \dots, T(\mathbf{v}_n)$.
2. Find the coordinates of $T(\mathbf{v}_1), T(\mathbf{v}_2), \dots, T(\mathbf{v}_n)$ relative to the basis $B' = \{\mathbf{w}_1, \mathbf{w}_2, \dots, \mathbf{w}_m\}$ of W. That is, find $[T(\mathbf{v}_1)]_{B'}, [T(\mathbf{v}_2)]_{B'}, \dots, [T(\mathbf{v}_n)]_{B'}$.
3. Define the $m \times n$ matrix $[T]_B^{B'}$ with ith column vector equal to $[T(\mathbf{v}_i)]_{B'}$.
4. Compute $[\mathbf{v}]_B$.
5. Compute the coordinates of $T(\mathbf{v})$ relative to B' by

$$[T(\mathbf{v})]_{B'} = [T]_B^{B'}[\mathbf{v}]_B = \begin{bmatrix} c_1 \\ c_2 \\ \vdots \\ c_m \end{bmatrix}$$

6. Then $T(\mathbf{v}) = c_1\mathbf{w}_1 + c_2\mathbf{w}_2 + \cdots + c_m\mathbf{w}_m$.

EXAMPLE 2

Let $T: \mathbb{R}^2 \longrightarrow \mathbb{R}^3$ be the linear transformation defined by

$$T(\mathbf{v}) = T\left(\begin{bmatrix} x_1 \\ x_2 \end{bmatrix}\right) = \begin{bmatrix} x_2 \\ x_1 + x_2 \\ x_1 - x_2 \end{bmatrix}$$

and let

$$B = \left\{ \begin{bmatrix} 1 \\ 2 \end{bmatrix}, \begin{bmatrix} 3 \\ 1 \end{bmatrix} \right\} \qquad B' = \left\{ \begin{bmatrix} 1 \\ 0 \\ 0 \end{bmatrix}, \begin{bmatrix} 1 \\ 1 \\ 0 \end{bmatrix}, \begin{bmatrix} 1 \\ 1 \\ 1 \end{bmatrix} \right\}$$

be ordered bases for $\mathbb{R}^2$ and $\mathbb{R}^3$, respectively.

   a. Find the matrix $[T]_B^{B'}$.

   b. Let $\mathbf{v} = \begin{bmatrix} -3 \\ -2 \end{bmatrix}$. Find $T(\mathbf{v})$ directly and then use the matrix found in part (a).

Solution

   a. We first apply T to the basis vectors of B, which gives

$$T\left(\begin{bmatrix} 1 \\ 2 \end{bmatrix}\right) = \begin{bmatrix} 2 \\ 3 \\ -1 \end{bmatrix} \quad \text{and} \quad T\left(\begin{bmatrix} 3 \\ 1 \end{bmatrix}\right) = \begin{bmatrix} 1 \\ 4 \\ 2 \end{bmatrix}$$

Next we find the coordinates of each of these vectors relative to the basis B'. That is, we find scalars such that

$$a_1 \begin{bmatrix} 1 \\ 0 \\ 0 \end{bmatrix} + a_2 \begin{bmatrix} 1 \\ 1 \\ 0 \end{bmatrix} + a_3 \begin{bmatrix} 1 \\ 1 \\ 1 \end{bmatrix} = \begin{bmatrix} 2 \\ 3 \\ -1 \end{bmatrix}$$

and

$$b_1 \begin{bmatrix} 1 \\ 0 \\ 0 \end{bmatrix} + b_2 \begin{bmatrix} 1 \\ 1 \\ 0 \end{bmatrix} + b_3 \begin{bmatrix} 1 \\ 1 \\ 1 \end{bmatrix} = \begin{bmatrix} 1 \\ 4 \\ 2 \end{bmatrix}$$

The solution to the first linear system is

$$a_1 = -1 \qquad a_2 = 4 \qquad a_3 = -1$$

and the solution to the second system is

$$b_1 = -3 \qquad b_2 = 2 \qquad b_3 = 2$$

Thus,

$$[T]_B^{B'} = \begin{bmatrix} -1 & -3 \\ 4 & 2 \\ -1 & 2 \end{bmatrix}$$

b. Using the definition of T directly, we have

$$T\left(\begin{bmatrix} -3 \\ -2 \end{bmatrix} \right) = \begin{bmatrix} -2 \\ -3 - 2 \\ -3 + 2 \end{bmatrix} = \begin{bmatrix} -2 \\ -5 \\ -1 \end{bmatrix}$$

Now, to use the matrix found in part (a), we need to find the coordinates of $\mathbf{v}$ relative to B. Observe that the solution to the equation

$$a_1 \begin{bmatrix} 1 \\ 2 \end{bmatrix} + a_2 \begin{bmatrix} 3 \\ 1 \end{bmatrix} = \begin{bmatrix} -3 \\ -2 \end{bmatrix} \qquad \text{is} \qquad a_1 = -\frac{3}{5} \qquad a_2 = -\frac{4}{5}$$

Thus, the coordinate vector of $\begin{bmatrix} -3 \\ -2 \end{bmatrix}$ relative to B is

$$\begin{bmatrix} -3 \\ -2 \end{bmatrix}_B = \begin{bmatrix} -\frac{3}{5} \\ -\frac{4}{5} \end{bmatrix}$$

We can now evaluate T, using matrix multiplication, so that

$$[T(\mathbf{v})]_{B'} = \begin{bmatrix} -1 & -3 \\ 4 & 2 \\ -1 & 2 \end{bmatrix} \begin{bmatrix} -\frac{3}{5} \\ -\frac{4}{5} \end{bmatrix} = \begin{bmatrix} 3 \\ -4 \\ -1 \end{bmatrix}$$

Hence,

$$T(\mathbf{v}) = 3 \begin{bmatrix} 1 \\ 0 \\ 0 \end{bmatrix} - 4 \begin{bmatrix} 1 \\ 1 \\ 0 \end{bmatrix} - \begin{bmatrix} 1 \\ 1 \\ 1 \end{bmatrix} = \begin{bmatrix} -2 \\ -5 \\ -1 \end{bmatrix}$$

which agrees with the direct computation.

EXAMPLE 3   Define a linear transformation $T: \mathcal{P}_2 \longrightarrow \mathcal{P}_3$ by

$$T(f(x)) = x^2 f''(x) - 2f'(x) + xf(x)$$

Find the matrix representation of T relative to the standard bases for $\mathcal{P}_2$ and $\mathcal{P}_3$.

Solution   Since the standard basis for $\mathcal{P}_2$ is $B = \{1, x, x^2\}$, we first compute

$$T(1) = x \qquad T(x) = x^2 - 2 \qquad T(x^2) = x^2(2) - 2(2x) + x(x^2) = x^3 + 2x^2 - 4x$$

Since the standard basis for $\mathcal{P}_3$ is $B' = \{1, x, x^2, x^3\}$, the coordinates relative to B' are

$$[T(1)]_{B'} = \begin{bmatrix} 0 \\ 1 \\ 0 \\ 0 \end{bmatrix} \qquad [T(x)]_{B'} = \begin{bmatrix} -2 \\ 0 \\ 1 \\ 0 \end{bmatrix} \qquad \text{and} \qquad [T(x^2)]_{B'} = \begin{bmatrix} 0 \\ -4 \\ 2 \\ 1 \end{bmatrix}$$

Hence, the matrix of the transformation is given by

$$[T]_B^{B'} = \begin{bmatrix} 0 & -2 & 0 \\ 1 & 0 & -4 \\ 0 & 1 & 2 \\ 0 & 0 & 1 \end{bmatrix}$$

As an example, let $f(x) = x^2 - 3x + 1$. Since $f'(x) = 2x - 3$ and $f''(x) = 2$, we have

$$T(f(x)) = x^2(2) - 2(2x - 3) + x(x^2 - 3x + 1)$$
$$= x^3 - x^2 - 3x + 6$$

Using the matrix representation of T to find the same image, we observe that

$$[f(x)]_B = \begin{bmatrix} 1 \\ -3 \\ 1 \end{bmatrix}$$

The coordinates of the image of $f(x)$ under the mapping T relative to B' are then given by

$$[T(f(x))]_{B'} = [T]_B^{B'}[f(x)]_B = \begin{bmatrix} 0 & -2 & 0 \\ 1 & 0 & -4 \\ 0 & 1 & 2 \\ 0 & 0 & 1 \end{bmatrix} \begin{bmatrix} 1 \\ -3 \\ 1 \end{bmatrix} = \begin{bmatrix} 6 \\ -3 \\ -1 \\ 1 \end{bmatrix}$$

The image $T(f(x))$ is the linear combination of the monomials in B' with coefficients the components of $[T(f(x))]_{B'}$, that is,

$$T(f(x)) = 6(1) - 3(x) - x^2 + x^3 = x^3 - x^2 - 3x + 6$$

This agrees with the direct calculation.

In Sec. 4.1 we discussed the addition, scalar multiplication, and composition of linear maps. The matrix representations for these combinations are given in a natural way, as described by Theorems 13 and 14. The proofs are omitted.

THEOREM 13   Let V and W be finite dimensional vector spaces with ordered bases B and B', respectively. If S and T are linear transformations from V to W, then

1. $[S + T]_B^{B'} = [S]_B^{B'} + [T]_B^{B'}$

2. $[kT]_B^{B'} = k[T]_B^{B'}$ for any scalar k

As before in the special case for which S and T are linear operators on a finite dimensional vector space V, and B is a fixed ordered basis for V, the notation becomes $[S + T]_B = [S]_B + [T]_B$ and $[kT]_B = k[T]_B$.

EXAMPLE 4   Let S and T be linear operators on $\mathbb{R}^2$ with

$$S\left(\begin{bmatrix} x \\ y \end{bmatrix}\right) = \begin{bmatrix} x + 2y \\ -y \end{bmatrix} \quad \text{and} \quad T\left(\begin{bmatrix} x \\ y \end{bmatrix}\right) = \begin{bmatrix} -x + y \\ 3x \end{bmatrix}$$

If B is the standard basis for $\mathbb{R}^2$, find $[S + T]_B$ and $[3S]_B$.

Solution   The matrix representations for the linear operators S and T are, respectively,

$$[S]_B = \begin{bmatrix} \begin{bmatrix} & \\ S(\mathbf{e}_1) & \\ & \end{bmatrix}_B \begin{bmatrix} & \\ S(\mathbf{e}_2) & \\ & \end{bmatrix}_B \end{bmatrix} = \begin{bmatrix} 1 & 2 \\ 0 & -1 \end{bmatrix}$$

and

$$[T]_B = \begin{bmatrix} \begin{bmatrix} & \\ T(\mathbf{e}_1) & \\ & \end{bmatrix}_B \begin{bmatrix} & \\ T(\mathbf{e}_2) & \\ & \end{bmatrix}_B \end{bmatrix} = \begin{bmatrix} -1 & 1 \\ 3 & 0 \end{bmatrix}$$

Then by Theorem 13,

$$[S + T]_B = \begin{bmatrix} 1 & 2 \\ 0 & -1 \end{bmatrix} + \begin{bmatrix} -1 & 1 \\ 3 & 0 \end{bmatrix} = \begin{bmatrix} 0 & 3 \\ 3 & -1 \end{bmatrix}$$

and

$$[3S]_B = 3 \begin{bmatrix} 1 & 2 \\ 0 & -1 \end{bmatrix} = \begin{bmatrix} 3 & 6 \\ 0 & -3 \end{bmatrix}$$

As we mentioned in Sec. 4.1, the matrix of the composition is the product of the matrices of the individual maps, as given in Theorem 14.

THEOREM 14   Let U, V, and W be finite dimensional vector spaces with ordered bases B, B', and B'', respectively. If $T: U \to V$ and $S: V \to W$ are linear transformations, then

$$[S \circ T]_B^{B''} = [S]_{B'}^{B''}[T]_B^{B'}$$

Again, if S and T are linear operators on a finite dimensional vector space V, and B is a fixed ordered basis for V, then

$$[S \circ T]_B = [S]_B [T]_B$$

Repeated application of Theorem 14 gives the following result.

COROLLARY 1   Let V be a finite dimensional vector space with ordered basis B. If T is a linear operator on V, then

$$[T^n]_B = ([T]_B)^n$$

EXAMPLE 5   Let $D: \mathcal{P}_3 \to \mathcal{P}_3$ be the linear operator defined by

$$D(p(x)) = p'(x)$$

a. Find the matrix of D relative to the standard basis $B = \{1, x, x^2, x^3\}$. Use the matrix to find the derivative of $p(x) = 1 - x + 2x^3$.

b. Find the matrix needed to compute the second derivative of a polynomial in $\mathcal{P}_3$. Use this matrix to find the second derivative of $p(x) = 1 - x + 2x^3$.

Solution   **a.** By Theorem 12, we have

$$[D]_B = \left[\left[D(1) \right]_B \quad \left[D(x) \right]_B \quad \left[D(x^2) \right]_B \quad \left[D(x^3) \right]_B \right]$$

$$= \begin{bmatrix} 0 & 1 & 0 & 0 \\ 0 & 0 & 2 & 0 \\ 0 & 0 & 0 & 3 \\ 0 & 0 & 0 & 0 \end{bmatrix}$$

Since the coordinate vector of $p(x) = 1 - x + 2x^3$, relative to B, is given by

$$[p(x)]_B = \begin{bmatrix} 1 \\ -1 \\ 0 \\ 2 \end{bmatrix}$$

then

$$[D(p(x))]_B = \begin{bmatrix} 0 & 1 & 0 & 0 \\ 0 & 0 & 2 & 0 \\ 0 & 0 & 0 & 3 \\ 0 & 0 & 0 & 0 \end{bmatrix} \begin{bmatrix} 1 \\ -1 \\ 0 \\ 2 \end{bmatrix} = \begin{bmatrix} -1 \\ 0 \\ 6 \\ 0 \end{bmatrix}$$

Therefore, as expected, $D(p(x)) = -1 + 6x^2$.

b. By Corollary 1, the matrix we need is given by

$$[D^2]_B = ([D]_B)^2 = \begin{bmatrix} 0 & 0 & 2 & 0 \\ 0 & 0 & 0 & 6 \\ 0 & 0 & 0 & 0 \\ 0 & 0 & 0 & 0 \end{bmatrix}$$

If $p(x) = 1 - x + 2x^3$, then

$$[D^2(p(x))]_B = \begin{bmatrix} 0 & 0 & 2 & 0 \\ 0 & 0 & 0 & 6 \\ 0 & 0 & 0 & 0 \\ 0 & 0 & 0 & 0 \end{bmatrix} \begin{bmatrix} 1 \\ -1 \\ 0 \\ 2 \end{bmatrix} = \begin{bmatrix} 0 \\ 12 \\ 0 \\ 0 \end{bmatrix}$$

so that $p''(x) = 12x$.

The final result of this section describes how to find the matrix representation of the inverse map of an invertible linear operator.

COROLLARY 2   Let T be an invertible linear operator on a finite dimensional vector space V and B an ordered basis for V. Then

$$[T^{-1}]_B = ([T]_B)^{-1}$$

Proof   Since $T^{-1} \circ T$ is the identity map, by Theorem 14, we have

$$[I]_B = [T^{-1} \circ T]_B = [T^{-1}]_B [T]_B$$

Since $[I]_B$ is the identity matrix, $[T^{-1}]_B = ([T]_B)^{-1}$.

Fact Summary

Let V and W be vector spaces, $B = \{v_1, \ldots, v_n\}$ and $B' = \{w_1, \ldots, w_m\}$ ordered bases of V and W, respectively, and T a linear transformation from V into W.

1. The matrix of T relative to B and B' is given by

$$[T]_B^{B'} = [\, [T(v_1)]_{B'} \ [T(v_2)]_{B'} \ \cdots \ [T(v_n)]_{B'}]$$

2. If v is a vector in V, the coordinates of $T(v)$ relative to the basis B' can be computed by

$$[T(v)]_{B'} = [T]_B^{B'} [v]_B$$

3. To find $T(\mathbf{v})$ multiply each basis vector in B' by the corresponding component of $[T(\mathbf{v})]_{B'}$. That is, if $[T(\mathbf{v})]_{B'} = [b'_1 \; b'_2 \; \ldots \; b'_m]^t$, then

$$T(\mathbf{v}) = b'_1 \mathbf{w}_1 + b'_2 \mathbf{w}_2 + \cdots + b'_m \mathbf{w}_m$$

4. If S is another linear transformation from V into W, then the matrix representation of $S + T$ relative to B and B' is the sum of the matrix representations for S and T. That is, $[S + T]_B^{B'} = [S]_B^{B'} + [T]_B^{B'}$.

5. If c is a scalar, then to find the matrix representation of cT relative to B and B', multiply the matrix representation for T by c. That is, $[cT]_B^{B'} = c[T]_B^{B'}$.

6. If S is a linear transformation from W into Z and B'' is an ordered basis for Z, then $[S \circ T]_B^{B''} = [S]_{B'}^{B''}[T]_B^{B'}$.

7. $[T^n]_B = ([T]_B)^n$

8. If T is invertible, then $[T^{-1}]_B = ([T]_B)^{-1}$.

Exercise Set 4.4

In Exercises 1–4, $T: \mathbb{R}^n \to \mathbb{R}^n$ is a linear operator.

a. Find the matrix representation for T relative to the standard basis for $\mathbb{R}^n$.

b. Find $T(\mathbf{v})$, using a direct computation and using the matrix representation.

1. $T: \mathbb{R}^2 \to \mathbb{R}^2$,

$$T\left(\begin{bmatrix} x \\ y \end{bmatrix}\right) = \begin{bmatrix} 5x - y \\ -x + y \end{bmatrix} \quad \mathbf{v} = \begin{bmatrix} 2 \\ 1 \end{bmatrix}$$

2. $T: \mathbb{R}^2 \to \mathbb{R}^2$,

$$T\left(\begin{bmatrix} x \\ y \end{bmatrix}\right) = \begin{bmatrix} -x \\ y \end{bmatrix}$$

$$\mathbf{v} = \begin{bmatrix} -1 \\ 3 \end{bmatrix}$$

3. $T: \mathbb{R}^3 \to \mathbb{R}^3$,

$$T\left(\begin{bmatrix} x \\ y \\ z \end{bmatrix}\right) = \begin{bmatrix} -x + y + 2z \\ 3y + z \\ x - z \end{bmatrix}$$

$$\mathbf{v} = \begin{bmatrix} 1 \\ -2 \\ 3 \end{bmatrix}$$

4. $T: \mathbb{R}^3 \to \mathbb{R}^3$,

$$T\left(\begin{bmatrix} x \\ y \\ z \end{bmatrix}\right) = \begin{bmatrix} x \\ y \\ -z \end{bmatrix}$$

$$\mathbf{v} = \begin{bmatrix} 2 \\ -5 \\ 1 \end{bmatrix}$$

In Exercises 5–12, $T: V \to V$ is a linear operator with B and B' ordered bases for V.

a. Find the matrix representation for T relative to the ordered bases B and B'.

b. Find $T(\mathbf{v})$, using a direct computation and using the matrix representation.

5. $T: \mathbb{R}^2 \to \mathbb{R}^2$,

$$T\left(\begin{bmatrix} x \\ y \end{bmatrix}\right) = \begin{bmatrix} -x + 2y \\ 3x \end{bmatrix}$$

$$B = \left\{ \begin{bmatrix} 1 \\ -1 \end{bmatrix}, \begin{bmatrix} 2 \\ 0 \end{bmatrix} \right\}$$

$$B' = \left\{ \begin{bmatrix} 1 \\ 0 \end{bmatrix}, \begin{bmatrix} 0 \\ 1 \end{bmatrix} \right\}$$

$$\mathbf{v} = \begin{bmatrix} -1 \\ -2 \end{bmatrix}$$

6. $T: \mathbb{R}^3 \to \mathbb{R}^3$,

$$T\left(\begin{bmatrix} x \\ y \\ z \end{bmatrix} \right) = \begin{bmatrix} 2x - z \\ -x + y + z \\ 2z \end{bmatrix}$$

$$B = \left\{ \begin{bmatrix} -1 \\ 0 \\ 1 \end{bmatrix}, \begin{bmatrix} 1 \\ 2 \\ 0 \end{bmatrix}, \begin{bmatrix} 1 \\ 2 \\ 1 \end{bmatrix} \right\}$$

$$B' = \left\{ \begin{bmatrix} 1 \\ 0 \\ 0 \end{bmatrix}, \begin{bmatrix} 0 \\ 1 \\ 0 \end{bmatrix}, \begin{bmatrix} 0 \\ 0 \\ 1 \end{bmatrix} \right\}$$

$$\mathbf{v} = \begin{bmatrix} 1 \\ -1 \\ 1 \end{bmatrix}$$

7. $T: \mathbb{R}^2 \to \mathbb{R}^2$,

$$T\left(\begin{bmatrix} x \\ y \end{bmatrix} \right) = \begin{bmatrix} 2x \\ x + y \end{bmatrix}$$

$$B = \left\{ \begin{bmatrix} -1 \\ -2 \end{bmatrix}, \begin{bmatrix} 1 \\ 1 \end{bmatrix} \right\}$$

$$B' = \left\{ \begin{bmatrix} 3 \\ -2 \end{bmatrix}, \begin{bmatrix} 0 \\ -2 \end{bmatrix} \right\}$$

$$\mathbf{v} = \begin{bmatrix} -1 \\ -3 \end{bmatrix}$$

8. $T: \mathbb{R}^3 \to \mathbb{R}^3$,

$$T\left(\begin{bmatrix} x \\ y \\ z \end{bmatrix} \right) = \begin{bmatrix} x + z \\ 2y - x \\ y + z \end{bmatrix}$$

$$B = \left\{ \begin{bmatrix} -1 \\ 1 \\ 1 \end{bmatrix}, \begin{bmatrix} -1 \\ -1 \\ 1 \end{bmatrix}, \begin{bmatrix} 0 \\ 1 \\ 1 \end{bmatrix} \right\}$$

$$B' = \left\{ \begin{bmatrix} 0 \\ 0 \\ 1 \end{bmatrix}, \begin{bmatrix} 1 \\ 0 \\ -1 \end{bmatrix}, \begin{bmatrix} -1 \\ -1 \\ 0 \end{bmatrix} \right\}$$

$$\mathbf{v} = \begin{bmatrix} -2 \\ 1 \\ 3 \end{bmatrix}$$

9. $T: \mathcal{P}_2 \to \mathcal{P}_2$,

$$T(ax^2 + bx + c) = ax^2 + bx + c$$

$$B = \{1, 1 - x, (1 - x)^2\}$$

$$B' = \{1, x, x^2\}$$

$$\mathbf{v} = x^2 - 3x + 3$$

10. $T: \mathcal{P}_2 \to \mathcal{P}_2$,

$$T(p(x)) = p'(x) + p(x)$$

$$B = \{1 - x - x^2, 1, 1 + x^2\}$$

$$B' = \{-1 + x, -1 + x + x^2, x\}$$

$$\mathbf{v} = 1 - x$$

11. Let

$$H = \begin{bmatrix} 1 & 0 \\ 0 & -1 \end{bmatrix}$$

and let T be the linear operator on all 2×2 matrices with trace 0, defined by

$$T(A) = AH - HA$$

$$B = \left\{ \begin{bmatrix} 1 & 0 \\ 0 & -1 \end{bmatrix}, \begin{bmatrix} 0 & 1 \\ 0 & 0 \end{bmatrix}, \begin{bmatrix} 0 & 0 \\ 1 & 0 \end{bmatrix} \right\}$$

$$B' = B$$

$$\mathbf{v} = \begin{bmatrix} 2 & 1 \\ 3 & -2 \end{bmatrix}$$

12. $T: M_{2\times 2} \to M_{2\times 2}$,

$$T(A) = 2A^t + A$$

B and B' the standard basis on $M_{2\times 2}$

$$\mathbf{v} = \begin{bmatrix} 1 & 3 \\ -1 & 2 \end{bmatrix}$$

13. Let $T: \mathbb{R}^2 \to \mathbb{R}^2$ be the linear operator defined by

$$T\left(\begin{bmatrix} x \\ y \end{bmatrix} \right) = \begin{bmatrix} x + 2y \\ x - y \end{bmatrix}$$

Let B be the standard ordered basis for $\mathbb{R}^2$ and B' the ordered basis for $\mathbb{R}^2$ defined by

$$B' = \left\{ \begin{bmatrix} 1 \\ 2 \end{bmatrix}, \begin{bmatrix} 4 \\ -1 \end{bmatrix} \right\}$$

a. Find $[T]_B$.

b. Find $[T]_{B'}$.

c. Find $[T]_B^{B'}$.

d. Find $[T]_{B'}^{B}$.

e. Let C be the ordered basis obtained by switching the order of the vectors in B. Find $[T]_C^{B'}$.

f. Let C' be the ordered basis obtained by switching the order of the vectors in B'. Find $[T]_{C'}^{B'}$.

14. Let $T: \mathbb{R}^2 \to \mathbb{R}^3$ be the linear transformation defined by

$$T\left(\begin{bmatrix} x \\ y \end{bmatrix}\right) = \begin{bmatrix} x - y \\ x \\ x + 2y \end{bmatrix}$$

Let B and B' be ordered bases for $\mathbb{R}^2$ and B'' the ordered basis for $\mathbb{R}^3$ defined by

$$B = \left\{ \begin{bmatrix} 1 \\ 1 \end{bmatrix}, \begin{bmatrix} 1 \\ 3 \end{bmatrix} \right\}$$

$$B' = \left\{ \begin{bmatrix} 1 \\ 0 \end{bmatrix}, \begin{bmatrix} 0 \\ 1 \end{bmatrix} \right\}$$

$$B'' = \left\{ \begin{bmatrix} 1 \\ 1 \\ 0 \end{bmatrix}, \begin{bmatrix} 0 \\ 1 \\ 1 \end{bmatrix}, \begin{bmatrix} 1 \\ 1 \\ 2 \end{bmatrix} \right\}$$

a. Find $[T]_B^{B''}$.

b. Find $[T]_{B'}^{B''}$.

c. Let C be the ordered basis obtained by switching the order of the vectors in B. Find $[T]_C^{B''}$.

d. Let C' be the ordered basis obtained by switching the order of the vectors in B'. Find $[T]_{C'}^{B''}$.

e. Let C'' be the ordered basis obtained by switching the order of the first and third vectors in B''. Find $[T]_B^{C''}$.

15. Let $T: \mathcal{P}_1 \to \mathcal{P}_2$ be the linear transformation defined by

$$T(a + bx) = ax + \frac{b}{2}x^2$$

Let B and B' be the standard ordered bases for $\mathcal{P}_1$ and $\mathcal{P}_2$, respectively.

a. Find $[T]_B^{B'}$.

b. Let C be the ordered basis obtained by switching the order of the vectors in B. Find $[T]_C^{B'}$.

c. Let C' be the ordered basis obtained by switching the first and second vectors in B'. Find $[T]_C^{C'}$.

d. Define $S: \mathcal{P}_2 \to \mathcal{P}_1$ by

$$S(a + bx + cx^2) = b + 2cx$$

Find $[S]_{B'}^B$.

e. Verify that $[S]_{B'}^B[T]_B^{B'} = I$, but that $[T]_B^{B'}[S]_{B'}^B \neq I$.

f. Interpret the statement

$$[S]_{B'}^B[T]_B^{B'} = I$$

in terms of the functions T and S.

16. Define a linear operator $T: M_{2 \times 2} \to M_{2 \times 2}$ by

$$T\left(\begin{bmatrix} a & b \\ c & d \end{bmatrix}\right) = \begin{bmatrix} 2a & c - b \\ -d & d - a \end{bmatrix}$$

Let B be the standard ordered basis for $M_{2 \times 2}$ and B' the ordered basis

$$B' = \left\{ \begin{bmatrix} 1 & 0 \\ 0 & 1 \end{bmatrix}, \begin{bmatrix} 0 & 1 \\ 1 & 0 \end{bmatrix}, \begin{bmatrix} -1 & 1 \\ -1 & 1 \end{bmatrix}, \begin{bmatrix} -1 & -1 \\ 1 & 1 \end{bmatrix} \right\}$$

a. Find $[T]_B^{B'}$.

b. Find $[T]_{B'}^B$.

c. Find $[T]_{B'}$.

d. Find $[I]_B^{B'}$ and $[I]_{B'}^B$.

e. Verify that

$$[T]_B^{B'} = [T]_{B'}[I]_B^{B'}$$
$$[T]_{B'}^B = [I]_{B'}^B[T]_{B'}$$

17. Define a linear operator $T: \mathbb{R}^2 \to \mathbb{R}^2$ by

$$T\left(\begin{bmatrix} x \\ y \end{bmatrix}\right) = \begin{bmatrix} x \\ -y \end{bmatrix}$$

Find the matrix for T relative to the standard basis for $\mathbb{R}^2$. Describe geometrically the action of T on a vector in $\mathbb{R}^2$.

18. Define a linear operator $T: \mathbb{R}^2 \to \mathbb{R}^2$ by

$$T\left(\begin{bmatrix} x \\ y \end{bmatrix}\right) = \begin{bmatrix} \cos\theta & -\sin\theta \\ \sin\theta & \cos\theta \end{bmatrix} \begin{bmatrix} x \\ y \end{bmatrix}$$

Describe geometrically the action of T on a vector in $\mathbb{R}^2$.

19. Let c be a fixed scalar and define $T: \mathbb{R}^n \to \mathbb{R}^n$ by

$$T\left(\begin{bmatrix} x_1 \\ x_2 \\ \vdots \\ x_n \end{bmatrix}\right) = c \begin{bmatrix} x_1 \\ x_2 \\ \vdots \\ x_n \end{bmatrix}$$

Find the matrix for T relative to the standard basis for $\mathbb{R}^n$.

20. Define $T: M_{2\times2} \to M_{2\times2}$ by

$$T(A) = A - A^t$$

Find the matrix for T relative to the standard basis for $M_{2\times2}$.

21. Define $T: M_{2\times2} \to \mathbb{R}$ by

$$T(A) = \mathbf{tr}(A)$$

Find the matrix $[T]_B^{B'}$, where B is the standard basis for $M_{2\times2}$ and $B' = \{1\}$.

In Exercises 22–25, let $S, T: \mathbb{R}^2 \to \mathbb{R}^2$ be defined by

$$T\left(\begin{bmatrix} x \\ y \end{bmatrix}\right) = \begin{bmatrix} 2x + y \\ -x + 3y \end{bmatrix}$$

and

$$S\left(\begin{bmatrix} x \\ y \end{bmatrix}\right) = \begin{bmatrix} x \\ x + y \end{bmatrix}$$

a. Find the matrix representation for the given linear operator relative to the standard basis.

b. Compute the image of $\mathbf{v} = \begin{bmatrix} -2 \\ 3 \end{bmatrix}$ directly and using the matrix found in part (a).

22. $-3S$

23. $2T + S$

24. $T \circ S$

25. $S \circ T$

In Exercises 26–29, let $S, T: \mathbb{R}^3 \to \mathbb{R}^3$ be defined by

$$T\left(\begin{bmatrix} x \\ y \\ z \end{bmatrix}\right) = \begin{bmatrix} x - y - z \\ 2y + 2z \\ -x + y + z \end{bmatrix}$$

and

$$S\left(\begin{bmatrix} x \\ y \\ z \end{bmatrix}\right) = \begin{bmatrix} 3x - z \\ x \\ z \end{bmatrix}$$

a. Find the matrix representation for the given linear operator relative to the standard basis.

b. Compute the image of

$$\mathbf{v} = \begin{bmatrix} -1 \\ 1 \\ 3 \end{bmatrix}$$

directly and using the matrix found in part (a).

26. $2T$

27. $-3T + 2S$

28. $T \circ S$

29. $S \circ T$

30. Let B be the basis for $\mathbb{R}^2$ defined by

$$B = \left\{ \begin{bmatrix} 1 \\ 1 \end{bmatrix}, \begin{bmatrix} 1 \\ 3 \end{bmatrix} \right\}$$

If $T: \mathbb{R}^2 \to \mathbb{R}^2$ is the linear operator defined by

$$T\left(\begin{bmatrix} x \\ y \end{bmatrix}\right) = \begin{bmatrix} 9x - 5y \\ 15x - 11y \end{bmatrix}$$

find the matrix for T^k, for $k \geq 1$, relative to the basis B.

31. Define $T: \mathcal{P}_4 \to \mathcal{P}_4$ by

$$T(p(x)) = p'''(x)$$

Find the matrix for T relative to the standard basis for $\mathcal{P}_4$. Use the matrix to find the third derivative of $p(x) = -2x^4 - 2x^3 + x^2 - 2x - 3$.

32. Let $T: \mathcal{P}_2 \to \mathcal{P}_2$ be defined by

$$T(p(x)) = p(x) + xp'(x)$$

Find the matrix $[T]_B$ where B is the standard basis for $\mathcal{P}_2$.

33. Let $S: \mathcal{P}_2 \rightarrow \mathcal{P}_3$ and $D: \mathcal{P}_3 \rightarrow \mathcal{P}_2$ be defined by

$$S(p(x)) = xp(x)$$

and

$$D(p(x)) = p'(x)$$

Find the matrices $[S]_B^{B'}$ and $[D]_{B'}^B$, where $B = \{1, x, x^2\}$ and $B' = \{1, x, x^2, x^3\}$. Observe that the operator T in Exercise 32 satisfies $T = D \circ S$. Verify Theorem 14 by showing that $[T]_B = [D]_{B'}^B [S]_B^{B'}$.

34. a. Define a basis for $\mathbb{R}^2$ by

$$B = \left\{ \begin{bmatrix} 1 \\ 1 \end{bmatrix}, \begin{bmatrix} 0 \\ 1 \end{bmatrix} \right\}$$

Find $[T]_B$ where $T: \mathbb{R}^2 \rightarrow \mathbb{R}^2$ is the linear operator that reflects a vector $\mathbf{v}$ through the line perpendicular to $\begin{bmatrix} 1 \\ 1 \end{bmatrix}$.

b. Let $B = \{\mathbf{v}_1, \mathbf{v}_2\}$ be a basis for $\mathbb{R}^2$. Find $[T]_B$ where $T: \mathbb{R}^2 \rightarrow \mathbb{R}^2$ is the linear operator that reflects a vector $\mathbf{v}$ through the line perpendicular to $\mathbf{v}_1$.

35. Let A be a fixed 2×2 matrix and define $T: M_{2\times 2} \rightarrow M_{2\times 2}$ by

$$T(B) = AB - BA$$

Find the matrix for T relative to the standard basis for $M_{2\times 2}$.

36. Let $B = \{\mathbf{v}_1, \mathbf{v}_2, \mathbf{v}_3\}$ and $B' = \{\mathbf{v}_2, \mathbf{v}_1, \mathbf{v}_3\}$ be ordered bases for the vector space V. If $T: V \rightarrow V$ is defined by $T(\mathbf{v}) = \mathbf{v}$, then find $[T]_B^{B'}$. Describe the relationship between $[\mathbf{v}]_B$ and $[\mathbf{v}]_{B'}$ and the relationship between the identity matrix I and $[T]_B^{B'}$.

37. Let V be a vector space and $B = \{\mathbf{v}_1, \mathbf{v}_2, \ldots, \mathbf{v}_n\}$ be an ordered basis for V. Define $\mathbf{v}_0 = \mathbf{0}$ and $T: V \rightarrow V$ by

$$T(\mathbf{v}_i) = \mathbf{v}_i + \mathbf{v}_{i-1} \qquad \text{for } i = 1, \ldots, n$$

Find $[T]_B$.

4.5 ▶ Similarity

We have just seen in Sec. 4.4 that if $T: V \longrightarrow V$ is a linear operator on the vector space V, and B is an ordered basis for V, then T has a matrix representation relative to B. The specific matrix for T depends on the particular basis; consequently, the matrix associated with a linear operator is not unique. However, the *action of the operator T on V is always the same regardless of the particular matrix representation*, as illustrated in Example 1.

EXAMPLE 1   Let $T: \mathbb{R}^2 \longrightarrow \mathbb{R}^2$ be the linear operator defined by

$$T\left(\begin{bmatrix} x \\ y \end{bmatrix} \right) = \begin{bmatrix} x + y \\ -2x + 4y \end{bmatrix}$$

Also let $B_1 = \{\mathbf{e}_1, \mathbf{e}_2\}$ be the standard basis for $\mathbb{R}^2$ and let $B_2 = \left\{ \begin{bmatrix} 1 \\ 1 \end{bmatrix}, \begin{bmatrix} 1 \\ 2 \end{bmatrix} \right\}$ be a second basis for $\mathbb{R}^2$. Verify that the action on the vector $\mathbf{v} = \begin{bmatrix} 2 \\ 3 \end{bmatrix}$ by the operator T is the same regardless of the matrix representation used for T.

Solution   The matrix representations for T relative to B_1 and B_2 are

$$[T]_{B_1} = \begin{bmatrix} 1 & 1 \\ -2 & 4 \end{bmatrix} \quad \text{and} \quad [T]_{B_2} = \begin{bmatrix} 2 & 0 \\ 0 & 3 \end{bmatrix}$$

respectively. Next, observe that

$$[\mathbf{v}]_{B_1} = \begin{bmatrix} 2 \\ 3 \end{bmatrix} \quad \text{and} \quad [\mathbf{v}]_{B_2} = \begin{bmatrix} 1 \\ 1 \end{bmatrix}$$

Applying the matrix representations of the operator T relative to B_1 and B_2, we obtain

$$[T(\mathbf{v})]_{B_1} = [T]_{B_1}[\mathbf{v}]_{B_1} = \begin{bmatrix} 1 & 1 \\ -2 & 4 \end{bmatrix}\begin{bmatrix} 2 \\ 3 \end{bmatrix} = \begin{bmatrix} 5 \\ 8 \end{bmatrix}$$

and

$$[T(\mathbf{v})]_{B_2} = [T]_{B_2}[\mathbf{v}]_{B_2} = \begin{bmatrix} 2 & 0 \\ 0 & 3 \end{bmatrix}\begin{bmatrix} 1 \\ 1 \end{bmatrix} = \begin{bmatrix} 2 \\ 3 \end{bmatrix}$$

To see that the result is the same, observe that

$$T(\mathbf{v}) = 5\begin{bmatrix} 1 \\ 0 \end{bmatrix} + 8\begin{bmatrix} 0 \\ 1 \end{bmatrix} = \begin{bmatrix} 5 \\ 8 \end{bmatrix} \quad \text{and} \quad T(\mathbf{v}) = 2\begin{bmatrix} 1 \\ 1 \end{bmatrix} + 3\begin{bmatrix} 1 \\ 2 \end{bmatrix} = \begin{bmatrix} 5 \\ 8 \end{bmatrix}$$

Theorem 15 gives the relationship between the matrices for a linear operator relative to two distinct bases.

THEOREM 15   Let V be a finite dimensional vector space, B_1 and B_2 two ordered bases for V, and $T: V \longrightarrow V$ a linear operator. Let $P = [I]_{B_2}^{B_1}$ be the transition matrix from B_2 to B_1. Then

$$[T]_{B_2} = P^{-1}[T]_{B_1} P$$

Proof   Let $\mathbf{v}$ be any vector in V. By Theorem 12 of Sec. 4.4, we have

$$[T(\mathbf{v})]_{B_2} = [T]_{B_2}[\mathbf{v}]_{B_2}$$

Alternatively, we can compute $[T(\mathbf{v})]_{B_2}$ as follows: First, since P is the transition matrix from B_2 to B_1,

$$[\mathbf{v}]_{B_1} = P[\mathbf{v}]_{B_2}$$

Thus, the coordinates of $T(\mathbf{v})$ relative to B_1 are given by

$$[T(\mathbf{v})]_{B_1} = [T]_{B_1}[\mathbf{v}]_{B_1} = [T]_{B_1} P[\mathbf{v}]_{B_2}$$

Now, to find the coordinates of $T(\mathbf{v})$ relative to B_2, we multiply on the left by P^{-1}, which is the transition matrix from B_1 to B_2, to obtain

$$[T(\mathbf{v})]_{B_2} = P^{-1}[T]_{B_1} P[\mathbf{v}]_{B_2}$$

Since both representations for $[T(\mathbf{v})]_{B_2}$ hold for all vectors $\mathbf{v}$ in V, then $[T]_{B_2} = P^{-1}[T]_{B_1} P$. See Fig. 1.

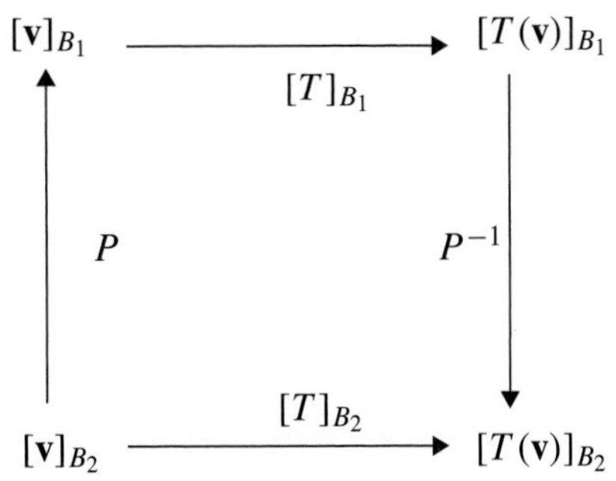

Figure 1

EXAMPLE 2   Let T, B_1, and B_2 be the linear operator and bases of Example 1. Then

$$[T]_{B_1} = \begin{bmatrix} 1 & 1 \\ -2 & 4 \end{bmatrix}$$

Use Theorem 15 to verify that

$$[T]_{B_2} = \begin{bmatrix} 2 & 0 \\ 0 & 3 \end{bmatrix}$$

Solution   Since B_1 is the standard basis for $\mathbb{R}^2$, by Theorem 14 of Sec. 3.4 the transition matrix from B_2 to B_1 is

$$P = [I]_{B_2}^{B_1} = \begin{bmatrix} \begin{bmatrix} 1 \\ 1 \end{bmatrix}_{B_1} & \begin{bmatrix} 1 \\ 2 \end{bmatrix}_{B_1} \end{bmatrix} = \begin{bmatrix} 1 & 1 \\ 1 & 2 \end{bmatrix}$$

and hence

$$P^{-1} = \begin{bmatrix} 2 & -1 \\ -1 & 1 \end{bmatrix}$$

Then

$$P^{-1}[T]_{B_1}P = \begin{bmatrix} 2 & -1 \\ -1 & 1 \end{bmatrix}\begin{bmatrix} 1 & 1 \\ -2 & 4 \end{bmatrix}\begin{bmatrix} 1 & 1 \\ 1 & 2 \end{bmatrix} = \begin{bmatrix} 2 & 0 \\ 0 & 3 \end{bmatrix} = [T]_{B_2}$$

EXAMPLE 3   Let $T: \mathbb{R}^2 \longrightarrow \mathbb{R}^2$ be the linear operator given by

$$T\left(\begin{bmatrix} x \\ y \end{bmatrix}\right) = \begin{bmatrix} -x + 2y \\ 3x + y \end{bmatrix}$$

and let

$$B_1 = \left\{ \begin{bmatrix} 2 \\ -1 \end{bmatrix}, \begin{bmatrix} 1 \\ 0 \end{bmatrix} \right\} \quad \text{and} \quad B_2 = \left\{ \begin{bmatrix} 1 \\ -1 \end{bmatrix}, \begin{bmatrix} 0 \\ 1 \end{bmatrix} \right\}$$

be ordered bases for $\mathbb{R}^2$. Find the matrix of T relative to B_1, and then use Theorem 15 to find the matrix of T relative to B_2.

Solution   Since

$$T\left(\begin{bmatrix} 2 \\ -1 \end{bmatrix}\right) = \begin{bmatrix} -4 \\ 5 \end{bmatrix} \quad \text{and} \quad T\left(\begin{bmatrix} 1 \\ 0 \end{bmatrix}\right) = \begin{bmatrix} -1 \\ 3 \end{bmatrix}$$

we have

$$[T]_{B_1} = \begin{bmatrix} \begin{bmatrix} -4 \\ 5 \end{bmatrix}_{B_1} & \begin{bmatrix} -1 \\ 3 \end{bmatrix}_{B_1} \end{bmatrix} = \begin{bmatrix} -5 & -3 \\ 6 & 5 \end{bmatrix}$$

The transition matrix from B_2 to B_1 is

$$P = [I]_{B_2}^{B_1} = \begin{bmatrix} \begin{bmatrix} 1 \\ -1 \end{bmatrix}_{B_1} & \begin{bmatrix} 0 \\ 1 \end{bmatrix}_{B_1} \end{bmatrix} = \begin{bmatrix} 1 & -1 \\ -1 & 2 \end{bmatrix}$$

Therefore, by Theorem 15

$$[T]_{B_2} = P^{-1}[T]_{B_1}P = \begin{bmatrix} 2 & 1 \\ 1 & 1 \end{bmatrix} \begin{bmatrix} -5 & -3 \\ 6 & 5 \end{bmatrix} \begin{bmatrix} 1 & -1 \\ -1 & 2 \end{bmatrix}$$

$$= \begin{bmatrix} -3 & 2 \\ -1 & 3 \end{bmatrix}$$

In general, if the square matrices A and B are matrix representations for the same linear operator, then the matrices are called **similar**. Using Theorem 15, we can define similarity for square matrices without reference to a linear operator.

DEFINITION 1   **Similar Matrices**   Let A and B be $n \times n$ matrices. We say that A is **similar** to B if there is an invertible matrix P such that $B = P^{-1}AP$.

The notion of similarity establishes a *relation* between matrices. This relation is *symmetric*; that is, if the matrix A is similar to the matrix B, then B is similar to A. To see this, let A be similar to B; that is, there is an invertible matrix P such that

$$B = P^{-1}AP$$

Now let $Q = P^{-1}$, so that B can be written as

$$B = QAQ^{-1}$$

Hence, $A = Q^{-1}BQ$, establishing that B is similar to A. For this reason we say that A and B are **similar** if either A is similar to B or B is similar to A. In addition, the relation is *reflexive* since any matrix is similar to itself with P being the identity matrix. This relation is also *transitive*; that is, if A is similar to B and B is similar to C, then A is similar to C. See Exercise 17. Any relation satisfying these three properties is called an **equivalence** relation.

Fact Summary

Let V be a finite dimensional vector space, B_1 and B_2 ordered bases of V, and T a linear operator on V.

1. The matrix representations $[T]_{B_1}$ and $[T]_{B_2}$ are similar. That is, there is an invertible matrix P such that $[T]_{B_2} = P^{-1}[T]_{B_1}P$. In addition, the matrix P is the transition matrix from B_2 to B_1.

2. A matrix is similar to itself. If A is similar to B, then B is similar to A. If A is similar to B and B is similar to C, then A is similar to C.

Exercise Set 4.5

In Exercises 1 and 2, $[T]_{B_1}$ is the matrix representation of a linear operator relative to the basis B_1, and $[T]_{B_2}$ is the matrix representation of the same operator relative to the basis B_2. Show that the action of the operator on the vector $\mathbf{v}$ is the same whether using $[T]_{B_1}$ or $[T]_{B_2}$.

1. $[T]_{B_1} = \begin{bmatrix} 1 & 2 \\ -1 & 3 \end{bmatrix}$, $[T]_{B_2} = \begin{bmatrix} 2 & 1 \\ -1 & 2 \end{bmatrix}$,

$B_1 = \left\{ \begin{bmatrix} 1 \\ 0 \end{bmatrix}, \begin{bmatrix} 0 \\ 1 \end{bmatrix} \right\}$

$B_2 = \left\{ \begin{bmatrix} 1 \\ 1 \end{bmatrix}, \begin{bmatrix} -1 \\ 0 \end{bmatrix} \right\}$

$\mathbf{v} = \begin{bmatrix} 4 \\ -1 \end{bmatrix}$

2. $[T]_{B_1} = \begin{bmatrix} 0 & 1 \\ 2 & -1 \end{bmatrix}$, $[T]_{B_2} = \begin{bmatrix} -2 & 0 \\ 4 & 1 \end{bmatrix}$,

$B_1 = \left\{ \begin{bmatrix} 1 \\ 0 \end{bmatrix}, \begin{bmatrix} 0 \\ 1 \end{bmatrix} \right\}$

$B_2 = \left\{ \begin{bmatrix} 1 \\ 2 \end{bmatrix}, \begin{bmatrix} 1 \\ 1 \end{bmatrix} \right\}$

$\mathbf{v} = \begin{bmatrix} 5 \\ 2 \end{bmatrix}$

In Exercises 3–6, a linear operator T and bases B_1 and B_2 are given.

  a. Find $[T]_{B_1}$ and $[T]_{B_2}$.

  b. Verify that the action on $\mathbf{v}$ of the linear operator T is the same when using the matrix representation of T relative to the bases B_1 and B_2.

3. $T\left(\begin{bmatrix} x \\ y \end{bmatrix} \right) = \begin{bmatrix} x + y \\ x + y \end{bmatrix}$,

$B_1 = \{\mathbf{e}_1, \mathbf{e}_2\}$

$B_2 = \left\{ \begin{bmatrix} 1 \\ 1 \end{bmatrix}, \begin{bmatrix} -1 \\ 1 \end{bmatrix} \right\}$

$\mathbf{v} = \begin{bmatrix} 3 \\ -2 \end{bmatrix}$

4. $T\left(\begin{bmatrix} x \\ y \end{bmatrix} \right) = \begin{bmatrix} -x \\ y \end{bmatrix}$

$B_1 = \{\mathbf{e}_1, \mathbf{e}_2\}$

$B_2 = \left\{ \begin{bmatrix} 2 \\ -1 \end{bmatrix}, \begin{bmatrix} -1 \\ 2 \end{bmatrix} \right\}$

$\mathbf{v} = \begin{bmatrix} 2 \\ -2 \end{bmatrix}$

5. $T\left(\begin{bmatrix} x \\ y \\ z \end{bmatrix} \right) = \begin{bmatrix} x \\ 0 \\ z \end{bmatrix}$

$B_1 = \{\mathbf{e}_1, \mathbf{e}_2, \mathbf{e}_3\}$

$B_2 = \left\{ \begin{bmatrix} 1 \\ 0 \\ 1 \end{bmatrix}, \begin{bmatrix} -1 \\ 1 \\ 0 \end{bmatrix}, \begin{bmatrix} 0 \\ 0 \\ 1 \end{bmatrix} \right\}$

$\mathbf{v} = \begin{bmatrix} 1 \\ 2 \\ -1 \end{bmatrix}$

6. $T\left(\begin{bmatrix} x \\ y \\ z \end{bmatrix}\right) = \begin{bmatrix} x + y \\ x - y + z \\ y - z \end{bmatrix}$

$B_1 = \{\mathbf{e}_1, \mathbf{e}_2, \mathbf{e}_3\}$

$B_2 = \left\{ \begin{bmatrix} -1 \\ 1 \\ 0 \end{bmatrix}, \begin{bmatrix} 0 \\ 0 \\ 1 \end{bmatrix}, \begin{bmatrix} 1 \\ 0 \\ 1 \end{bmatrix} \right\}$

$\mathbf{v} = \begin{bmatrix} 2 \\ -1 \\ -1 \end{bmatrix}$

In Exercises 7–10, $[T]_{B_1}$ and $[T]_{B_2}$ are, respectively, the matrix representations of a linear operator relative to the bases B_1 and B_2. Find the transition matrix $P = [I]_{B_2}^{B_1}$, and use Theorem 15 to show directly that the matrices are similar.

7. $[T]_{B_1} = \begin{bmatrix} 1 & 1 \\ 3 & 2 \end{bmatrix}$ $[T]_{B_2} = \begin{bmatrix} \frac{9}{2} & -\frac{1}{2} \\ \frac{23}{2} & -\frac{3}{2} \end{bmatrix}$

$B_1 = \{\mathbf{e}_1, \mathbf{e}_2\}$

$B_2 = \left\{ \begin{bmatrix} 3 \\ -1 \end{bmatrix}, \begin{bmatrix} -1 \\ 1 \end{bmatrix} \right\}$

8. $[T]_{B_1} = \begin{bmatrix} 0 & 2 \\ 2 & -3 \end{bmatrix}$ $[T]_{B_2} = \begin{bmatrix} -4 & 1 \\ 0 & 1 \end{bmatrix}$

$B_1 = \{\mathbf{e}_1, \mathbf{e}_2\}$

$B_2 = \left\{ \begin{bmatrix} -1 \\ 2 \end{bmatrix}, \begin{bmatrix} 1 \\ 0 \end{bmatrix} \right\}$

9. $[T]_{B_1} = \begin{bmatrix} 1 & 0 \\ 0 & -1 \end{bmatrix}$ $[T]_{B_2} = \begin{bmatrix} 0 & 3 \\ \frac{1}{3} & 0 \end{bmatrix}$

$B_1 = \left\{ \begin{bmatrix} 1 \\ 1 \end{bmatrix}, \begin{bmatrix} 2 \\ -1 \end{bmatrix} \right\}$

$B_2 = \left\{ \begin{bmatrix} 1 \\ 0 \end{bmatrix}, \begin{bmatrix} -1 \\ 2 \end{bmatrix} \right\}$

10. $[T]_{B_1} = \begin{bmatrix} -1 & 0 \\ 0 & 1 \end{bmatrix}$ $[T]_{B_2} = \begin{bmatrix} -2 & -1 \\ 3 & 2 \end{bmatrix}$

$B_1 = \left\{ \begin{bmatrix} 1 \\ -1 \end{bmatrix}, \begin{bmatrix} 1 \\ 1 \end{bmatrix} \right\}$

$B_2 = \left\{ \begin{bmatrix} -1 \\ 2 \end{bmatrix}, \begin{bmatrix} 0 \\ 1 \end{bmatrix} \right\}$

In Exercises 11–14, find the matrix representation of the linear operator T relative to B_1. Then use Theorem 15 to find $[T]_{B_2}$.

11. $T\left(\begin{bmatrix} x \\ y \end{bmatrix}\right) = \begin{bmatrix} 2x \\ 3y \end{bmatrix}$

$B_1 = \{\mathbf{e}_1, \mathbf{e}_2\}$

$B_2 = \left\{ \begin{bmatrix} 2 \\ 3 \end{bmatrix}, \begin{bmatrix} 1 \\ 2 \end{bmatrix} \right\}$

12. $T\left(\begin{bmatrix} x \\ y \end{bmatrix}\right) = \begin{bmatrix} x - y \\ x + 2y \end{bmatrix}$

$B_1 = \{\mathbf{e}_1, \mathbf{e}_2\}$

$B_2 = \left\{ \begin{bmatrix} 3 \\ 5 \end{bmatrix}, \begin{bmatrix} 1 \\ 2 \end{bmatrix} \right\}$

13. $T\left(\begin{bmatrix} x \\ y \end{bmatrix}\right) = \begin{bmatrix} y \\ -x \end{bmatrix}$

$B_1 = \left\{ \begin{bmatrix} 1 \\ -1 \end{bmatrix}, \begin{bmatrix} 1 \\ 0 \end{bmatrix} \right\}$

$B_2 = \left\{ \begin{bmatrix} 1 \\ 1 \end{bmatrix}, \begin{bmatrix} 0 \\ 1 \end{bmatrix} \right\}$

14. $T\left(\begin{bmatrix} x \\ y \end{bmatrix}\right) = \begin{bmatrix} -2x + y \\ 2y \end{bmatrix}$

$B_1 = \left\{ \begin{bmatrix} 2 \\ 0 \end{bmatrix}, \begin{bmatrix} -1 \\ 1 \end{bmatrix} \right\}$

$B_2 = \left\{ \begin{bmatrix} 2 \\ 2 \end{bmatrix}, \begin{bmatrix} 0 \\ -1 \end{bmatrix} \right\}$

15. Let $T: \mathcal{P}_2 \longrightarrow \mathcal{P}_2$ be the linear operator defined by $T(p(x)) = p'(x)$. Find the matrix representation $[T]_{B_1}$ relative to the basis $B_1 = \{1, x, x^2\}$ and the matrix representation $[T]_{B_2}$ relative to $B_2 = \{1, 2x, x^2 - 2\}$. Find the transition matrix $P = [I]_{B_2}^{B_1}$, and use Theorem 15 to show directly that the matrices $[T]_{B_1}$ and $[T]_{B_2}$ are similar.

16. Let $T: \mathcal{P}_2 \longrightarrow \mathcal{P}_2$ be the linear operator defined by $T(p(x)) = xp'(x) + p''(x)$. Find the matrix representation $[T]_{B_1}$ relative to the basis $B_1 = \{1, x, x^2\}$ and the matrix representation $[T]_{B_2}$ relative to $B_2 = \{1, x, 1 + x^2\}$. Find the transition matrix $P = [I]_{B_2}^{B_1}$, and use Theorem 15 to show directly that the matrices $[T]_{B_1}$ and $[T]_{B_2}$ are similar.

17. Show that if A and B are similar matrices and B and C are similar matrices, then A and C are similar matrices.

18. Show that if A and B are similar matrices, then $\det(A) = \det(B)$.

19. Show that if A and B are similar matrices, then $\mathbf{tr}(A) = \mathbf{tr}(B)$.

20. Show that if A and B are similar matrices, then A^t and B^t are similar matrices.

21. Show that if A and B are similar matrices, then A^n and B^n are similar matrices for each positive integer n.

22. Show that if A and B are similar matrices and λ is any scalar, then $\det(A - \lambda I) = \det(B - \lambda I)$.

4.6 ▶ **Application: Computer Graphics**

The rapid development of increasingly more powerful computers has led to the explosive growth of digital media. Computer-generated visual content is ubiquitous, found in almost every arena from advertising and entertainment to science and medicine. The branch of computer science known as *computer graphics* is devoted to the study of the generation and manipulation of digital images. Computer graphics are based on displaying two- or three-dimensional objects in two-dimensional space. Images displayed on a computer screen are stored in memory using data items called **pixels**, which is short for picture elements. A single picture can be comprised of millions of pixels, which collectively determine the image. Each pixel contains information on how to color the corresponding point on a computer screen, as shown in Fig. 1. If an image contains curves or lines, the pixels which describe the object may be connected by a mathematical formula. The saddle shown in Fig. 1 is an example.

Figure 1

Graphics Operations in $\mathbb{R}^2$

To manipulate images, computer programmers use linear transformations. Most of the examples we consider in this section use linear operators on $\mathbb{R}^2$. One of the properties of linear transformations that is especially useful to our work here is that *linear transformations map lines to lines, and hence polygons to polygons.* (See Exercise 10 of the Review Exercises for Chapter 4.) Therefore, to visualize the result of a linear transformation on a polygon, we only need to transform the vertices. Connecting the images of the vertices then gives the transformed polygon.

Scaling and Shearing

A transformation on an object that results in a horizontal contraction or dilation (stretching) is called a **horizontal scaling**. For example, let T be the triangle shown in Fig. 2 with vertices $(1, 1)$, $(2, 1)$, and $\left(\frac{3}{2}, 3\right)$. Suppose that we wish to perform a horizontal scaling of T by a factor of 3. The transformed triangle T' is obtained by multiplying the x coordinate of each vertex by 3. Joining the new vertices with straight lines produces the result shown in Fig. 3.

The linear operator $S: \mathbb{R}^2 \longrightarrow \mathbb{R}^2$ that accomplishes this is given by

$$S\left(\begin{bmatrix} x \\ y \end{bmatrix}\right) = \begin{bmatrix} 3x \\ y \end{bmatrix}$$

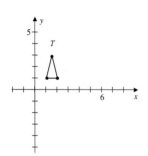

Figure 2

To find a matrix representation of S, let $B = \{\mathbf{e}_1, \mathbf{e}_2\}$ be the standard basis for $\mathbb{R}^2$. Then by Theorem 12 of Sec. 4.4, we have

$$[S]_B = [\, [S(\mathbf{e}_1)] \quad [S(\mathbf{e}_2)] \,] = \begin{bmatrix} 3 & 0 \\ 0 & 1 \end{bmatrix}$$

Let $\mathbf{v}_i$ and $\mathbf{v}_i'$, for $i = 1, 2$, and 3, be, respectively, the vertices (in vector form) of T and T'. Since the coordinates of the vertices of T are given relative to B, the vertices of T' can be found by matrix multiplication. Specifically,

$$\mathbf{v}_1' = \begin{bmatrix} 3 & 0 \\ 0 & 1 \end{bmatrix}\begin{bmatrix} 1 \\ 1 \end{bmatrix} = \begin{bmatrix} 3 \\ 1 \end{bmatrix} \qquad \mathbf{v}_2' = \begin{bmatrix} 3 & 0 \\ 0 & 1 \end{bmatrix}\begin{bmatrix} 2 \\ 1 \end{bmatrix} = \begin{bmatrix} 6 \\ 1 \end{bmatrix}$$

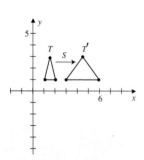

Figure 3

and

$$\mathbf{v}_3' = \begin{bmatrix} 3 & 0 \\ 0 & 1 \end{bmatrix}\begin{bmatrix} \frac{3}{2} \\ 3 \end{bmatrix} = \begin{bmatrix} \frac{9}{2} \\ 3 \end{bmatrix}$$

These results are consistent with the transformed triangle T', as shown in Fig. 3.

In general, a horizontal scaling by a factor k is given by the linear transformation S_h defined by

$$S_h\left(\begin{bmatrix} x \\ y \end{bmatrix}\right) = \begin{bmatrix} kx \\ y \end{bmatrix}$$

The matrix representation of S_h, relative to the standard basis for $\mathbb{R}^2$, is given by

$$[S_h]_B = \begin{bmatrix} k & 0 \\ 0 & 1 \end{bmatrix}$$

Similarly, a **vertical scaling** is given by the linear operator

$$S_v\left(\begin{bmatrix} x \\ y \end{bmatrix}\right) = \begin{bmatrix} x \\ ky \end{bmatrix}$$

The matrix representation of S_v, relative to the standard basis for $\mathbb{R}^2$, is given by

$$[S_v]_B = \begin{bmatrix} 1 & 0 \\ 0 & k \end{bmatrix}$$

If both components are multiplied by the same number k, then the result is called a **uniform scaling**. In all the above cases, if $k > 1$, then the transformation is called a **dilation**, or stretching; and if $0 < k < 1$, then the operator is a **contraction**.

EXAMPLE 1   Let T denote the triangle with vertices given by the vectors

$$\mathbf{v}_1 = \begin{bmatrix} 0 \\ 1 \end{bmatrix} \quad \mathbf{v}_2 = \begin{bmatrix} 2 \\ 1 \end{bmatrix} \quad \mathbf{v}_3 = \begin{bmatrix} 1 \\ 3 \end{bmatrix}$$

as shown in Fig. 4.

a. Stretch the triangle horizontally by a factor of 2.

b. Contract the triangle vertically by a factor of 3.

c. Stretch the triangle horizontally by a factor of 2, and contract the triangle vertically by a factor of 3.

Solution   a. To stretch the triangle horizontally by a factor of 2, we apply the matrix

$$\begin{bmatrix} 2 & 0 \\ 0 & 1 \end{bmatrix}$$

to each vertex to obtain

$$\mathbf{v}_1' = \begin{bmatrix} 0 \\ 1 \end{bmatrix} \quad \mathbf{v}_2' = \begin{bmatrix} 4 \\ 1 \end{bmatrix} \quad \mathbf{v}_3' = \begin{bmatrix} 2 \\ 3 \end{bmatrix}$$

Connecting the new vertices by straight-line segments gives the triangle T' shown in Fig. 5(a).

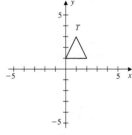

Figure 4

b. To contract the triangle vertically by a factor of 3, we apply the matrix

$$\begin{bmatrix} 1 & 0 \\ 0 & \frac{1}{3} \end{bmatrix}$$

to each vertex to obtain

$$\mathbf{v}_1'' = \begin{bmatrix} 0 \\ \frac{1}{3} \end{bmatrix} \quad \mathbf{v}_2'' = \begin{bmatrix} 2 \\ \frac{1}{3} \end{bmatrix} \quad \mathbf{v}_3'' = \begin{bmatrix} 1 \\ 1 \end{bmatrix}$$

The contracted triangle T'' is shown in Fig. 5(b).

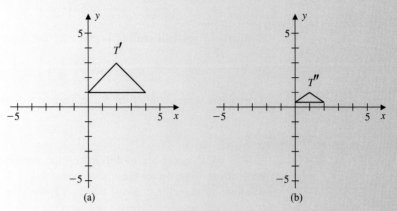

(a)   (b)

Figure 5

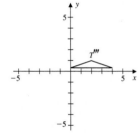

Figure 6

c. This operator is the composition of the linear operators of parts (a) and (b). By Theorem 14 of Sec. 4.4, the matrix of the operator, relative to the standard basis for $\mathbb{R}^2$, is given by the product

$$\begin{bmatrix} 1 & 0 \\ 0 & \frac{1}{3} \end{bmatrix} \begin{bmatrix} 2 & 0 \\ 0 & 1 \end{bmatrix} = \begin{bmatrix} 2 & 0 \\ 0 & \frac{1}{3} \end{bmatrix}$$

Applying this matrix to the vertices of the original triangle gives

$$\mathbf{v}_1''' = \begin{bmatrix} 0 \\ \frac{1}{3} \end{bmatrix} \qquad \mathbf{v}_2''' = \begin{bmatrix} 4 \\ \frac{1}{3} \end{bmatrix} \qquad \mathbf{v}_3''' = \begin{bmatrix} 2 \\ 1 \end{bmatrix}$$

as shown in Fig. 6.

Another type of transformation, called **shearing**, produces the visual effect of slanting. The linear operator $S: \mathbb{R}^2 \to \mathbb{R}^2$ used to produce a **horizontal shear** has the form

$$S\left(\begin{bmatrix} x \\ y \end{bmatrix} \right) = \begin{bmatrix} x + ky \\ y \end{bmatrix}$$

where k is a real number. Relative to the standard basis B, the matrix representation of S is given by

$$[S]_B = \begin{bmatrix} 1 & k \\ 0 & 1 \end{bmatrix}$$

As an illustration, let T be the triangle of Fig. 7(a) with vertices $\mathbf{v}_1 = \begin{bmatrix} 0 \\ 0 \end{bmatrix}$, $\mathbf{v}_2 = \begin{bmatrix} 2 \\ 0 \end{bmatrix}$, and $\mathbf{v}_3 = \begin{bmatrix} 1 \\ 1 \end{bmatrix}$, and let $k = 2$. After applying the matrix

$$[S]_B = \begin{bmatrix} 1 & 2 \\ 0 & 1 \end{bmatrix}$$

to each of the vertices of T, we obtain $\mathbf{v}'_1 = \begin{bmatrix} 0 \\ 0 \end{bmatrix}$, $\mathbf{v}'_2 = \begin{bmatrix} 2 \\ 0 \end{bmatrix}$, and $\mathbf{v}'_3 = \begin{bmatrix} 3 \\ 1 \end{bmatrix}$.
The resulting triangle T' is shown in Fig. 7(b).

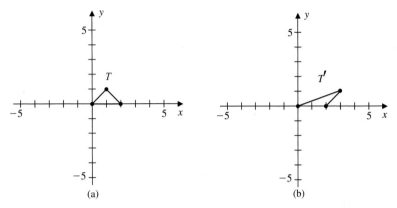

(a)   (b)

Figure 7

A **vertical shear** is defined similarly by

$$S\left(\begin{bmatrix} x \\ y \end{bmatrix}\right) = \begin{bmatrix} x \\ y + kx \end{bmatrix}$$

In this case the matrix for S, relative to the standard basis B, is given by

$$[S]_B = \begin{bmatrix} 1 & 0 \\ k & 1 \end{bmatrix}$$

EXAMPLE 2   Perform a vertical shear, with $k = 2$, on the triangle of Fig. 2.

Solution   The matrix of this operator, relative to the standard basis for $\mathbb{R}^2$, is given by

$$\begin{bmatrix} 1 & 0 \\ 2 & 1 \end{bmatrix}$$

Applying this matrix to the vertices

$$\mathbf{v}_1 = \begin{bmatrix} 1 \\ 1 \end{bmatrix} \qquad \mathbf{v}_2 = \begin{bmatrix} 2 \\ 1 \end{bmatrix} \qquad \mathbf{v}_3 = \begin{bmatrix} \frac{3}{2} \\ 3 \end{bmatrix}$$

we obtain

$$\mathbf{v}'_1 = \begin{bmatrix} 1 \\ 3 \end{bmatrix} \qquad \mathbf{v}'_2 = \begin{bmatrix} 2 \\ 5 \end{bmatrix} \qquad \mathbf{v}'_3 = \begin{bmatrix} \frac{3}{2} \\ 6 \end{bmatrix}$$

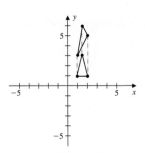

Figure 8

The images of the original triangle and the sheared triangle are shown in Fig. 8.

Reflection

The **reflection** of a geometric object through a line produces the *mirror image* of the object across the line. The linear operator that reflects a vector through the x axis is given by

$$R_x\left(\begin{bmatrix} x \\ y \end{bmatrix}\right) = \begin{bmatrix} x \\ -y \end{bmatrix}$$

A reflection through the y axis is given by

$$R_y\left(\begin{bmatrix} x \\ y \end{bmatrix}\right) = \begin{bmatrix} -x \\ y \end{bmatrix}$$

and a reflection through the line $y = x$ is given by

$$R_{y=x}\left(\begin{bmatrix} x \\ y \end{bmatrix}\right) = \begin{bmatrix} y \\ x \end{bmatrix}$$

The matrix representations, relative to the standard basis B, for each of these are given by

$$[R_x]_B = \begin{bmatrix} 1 & 0 \\ 0 & -1 \end{bmatrix} \qquad [R_y]_B = \begin{bmatrix} -1 & 0 \\ 0 & 1 \end{bmatrix} \qquad [R_{y=x}]_B = \begin{bmatrix} 0 & 1 \\ 1 & 0 \end{bmatrix}$$

EXAMPLE 3   Perform the following reflections on the triangle T of Fig. 4.

   a. Reflection through the x axis.
   b. Reflection through the y axis.
   c. Reflection through the line $y = x$.

Solution   **a.** The vertices of the triangle in Fig. 4 are given by

$$\mathbf{v}_1 = \begin{bmatrix} 0 \\ 1 \end{bmatrix} \qquad \mathbf{v}_2 = \begin{bmatrix} 2 \\ 1 \end{bmatrix} \qquad \mathbf{v}_3 = \begin{bmatrix} 1 \\ 3 \end{bmatrix}$$

Applying the matrix $[R_x]_B$ to the vertices of the original triangle, we obtain

$$\mathbf{v}'_1 = \begin{bmatrix} 0 \\ -1 \end{bmatrix} \qquad \mathbf{v}'_2 = \begin{bmatrix} 2 \\ -1 \end{bmatrix} \qquad \mathbf{v}'_3 = \begin{bmatrix} 1 \\ -3 \end{bmatrix}$$

The image of the triangle is shown in Fig. 9(a).

b. Applying the matrix $[R_y]_B$ to the vertices of the original triangle, we obtain

$$\mathbf{v}'_1 = \begin{bmatrix} 0 \\ 1 \end{bmatrix} \qquad \mathbf{v}'_2 = \begin{bmatrix} -2 \\ 1 \end{bmatrix} \qquad \mathbf{v}'_3 = \begin{bmatrix} -1 \\ 3 \end{bmatrix}$$

The image of the triangle with this reflection is shown in Fig. 9(b).

c. Finally, applying the matrix $[R_{x=y}]_B$ to the vertices of the original triangle, we obtain

$$\mathbf{v}'_1 = \begin{bmatrix} 1 \\ 0 \end{bmatrix} \qquad \mathbf{v}'_2 = \begin{bmatrix} 1 \\ 2 \end{bmatrix} \qquad \mathbf{v}'_3 = \begin{bmatrix} 3 \\ 1 \end{bmatrix}$$

The image of the triangle is shown in Fig. 9(c).

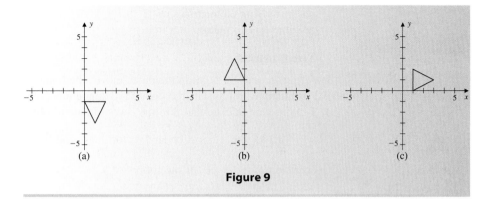

Figure 9

Reversing Graphics Operations

The operations of scaling, shearing, and reflection are all reversible, and hence the matrix representations for each of these operators are invertible. By Corollary 2 of Sec. 4.4, to reverse one of these operations, we apply the inverse matrix to the transformed image.

EXAMPLE 4

Let S be the linear operator that performs a reflection through the line $y = x$, followed by a horizontal stretching by a factor of 2.

  a. Find the matrix representation of S, relative to the standard basis B.

  b. Find the matrix representation of the reverse operator, again relative to B.

Solution

  a. Using the matrices given above for these operations, and by Theorem 14 of Sec. 4.4, the matrix of the transformation, relative to the standard basis for $\mathbb{R}^2$, is given by the product

$$[S]_B = \begin{bmatrix} 2 & 0 \\ 0 & 1 \end{bmatrix} \begin{bmatrix} 0 & 1 \\ 1 & 0 \end{bmatrix} = \begin{bmatrix} 0 & 2 \\ 1 & 0 \end{bmatrix}$$

  b. By Corollary 2 of Sec. 4.4, the matrix which reverses the operation of part (a) is given by

$$[S^{-1}]_B = ([S]_B)^{-1} = -\frac{1}{2} \begin{bmatrix} 0 & -2 \\ -1 & 0 \end{bmatrix} = \begin{bmatrix} 0 & 1 \\ \frac{1}{2} & 0 \end{bmatrix}$$

As we noted in Example 4(a), if a graphics operation S is given by a sequence of linear operators $S_1, S_2, \ldots, S_n$, then

$$S = S_n \circ S_{n-1} \circ \cdots \circ S_1$$

The matrix representation, relative to the basis B, is then given by the matrix product

$$[S]_B = [S_n]_B [S_{n-1}]_B \cdots [S_1]_B$$

The reverse process is given by

$$[S]_B^{-1} = [S_1]_B^{-1} [S_2]_B^{-1} \cdots [S_n]_B^{-1}$$

Thus, applying the matrices $[S_1]_B^{-1}, [S_2]_B^{-1}, \ldots, [S_n]_B^{-1}$ in succession reverses the process one transformation at a time.

Translation

A *translation* of a point in the plane moves the point vertically, horizontally, or both. For example, to translate the point $(1, 3)$ three units to the right and two units up, add 3 to the x coordinate and 2 to the y coordinate to obtain the point $(4, 5)$.

Now let $\mathbf{v} = \begin{bmatrix} v_1 \\ v_2 \end{bmatrix}$ be any vector in $\mathbb{R}^2$ and $\mathbf{b} = \begin{bmatrix} b_1 \\ b_2 \end{bmatrix}$ some fixed vector. An operation $S: \mathbb{R}^2 \to \mathbb{R}^2$ of the form

$$S(\mathbf{v}) = \mathbf{v} + \mathbf{b} = \begin{bmatrix} v_1 + b_1 \\ v_2 + b_2 \end{bmatrix}$$

is called a **translation** by the vector $\mathbf{b}$. This transformation is a linear operator if and only if $\mathbf{b} = \mathbf{0}$. Consequently, when $\mathbf{b} \neq \mathbf{0}$, then S cannot be accomplished by means of a 2×2 matrix. However, by using *homogeneous coordinates*, translation of a vector in $\mathbb{R}^2$ can be represented by a 3×3 matrix. The **homogeneous coordinates** of a vector in $\mathbb{R}^2$ are obtained by adding a third component whose value is 1. Thus, the homogeneous coordinates for the vector $\mathbf{v} = \begin{bmatrix} x \\ y \end{bmatrix}$ are given by

$$\mathbf{w} = \begin{bmatrix} x \\ y \\ 1 \end{bmatrix}$$

Now, to translate $\mathbf{w}$ by the vector $\mathbf{b} = \begin{bmatrix} b_1 \\ b_2 \end{bmatrix}$, we let

$$A = \begin{bmatrix} 1 & 0 & b_1 \\ 0 & 1 & b_2 \\ 0 & 0 & 1 \end{bmatrix}$$

so that

$$A\mathbf{w} = \begin{bmatrix} 1 & 0 & b_1 \\ 0 & 1 & b_2 \\ 0 & 0 & 1 \end{bmatrix} \begin{bmatrix} x \\ y \\ 1 \end{bmatrix} = \begin{bmatrix} x + b_1 \\ y + b_2 \\ 1 \end{bmatrix}$$

To return to $\mathbb{R}^2$, we select the first two components of $A\mathbf{w}$ so that

$$S(\mathbf{v}) = \begin{bmatrix} x + b_1 \\ y + b_2 \end{bmatrix}$$

as desired.

As an illustration of this, let $\mathbf{b} = \begin{bmatrix} 1 \\ -2 \end{bmatrix}$. Using homogeneous coordinates, the 3×3 matrix to perform the translation is

$$A = \begin{bmatrix} 1 & 0 & 1 \\ 0 & 1 & -2 \\ 0 & 0 & 1 \end{bmatrix}$$

Now let $\mathbf{v} = \begin{bmatrix} 3 \\ 2 \end{bmatrix}$. Then

$$\begin{bmatrix} 1 & 0 & 1 \\ 0 & 1 & -2 \\ 0 & 0 & 1 \end{bmatrix} \begin{bmatrix} 3 \\ 2 \\ 1 \end{bmatrix} = \begin{bmatrix} 4 \\ 0 \\ 1 \end{bmatrix}$$

The vector $S(\mathbf{v}) = \begin{bmatrix} 4 \\ 0 \end{bmatrix}$ is the translation of $\mathbf{v}$ by the vector $\mathbf{b}$.

In the previous illustration the translation can be accomplished with less work by simply adding the vector $\mathbf{b}$ to $\mathbf{v}$. The benefits of using a matrix representation are realized when we combine translation with other types of transformations. To do this, we note that all the previous linear operators can be represented by 3×3 matrices. For example, the 3×3 matrix for reflecting a point (in homogeneous coordinates) through the x axis is

$$\begin{bmatrix} 1 & 0 & 0 \\ 0 & -1 & 0 \\ 0 & 0 & 1 \end{bmatrix}$$

EXAMPLE 5   Find the image of the triangle T of Fig. 4 under a translation by the vector $\mathbf{b} = \begin{bmatrix} -5 \\ 3 \end{bmatrix}$, followed by a horizontal scaling by a factor of 1.5, followed by a reflection through the x axis.

Solution   The matrix for the composition of these operations is given by the product

$$\begin{bmatrix} 1 & 0 & 0 \\ 0 & -1 & 0 \\ 0 & 0 & 1 \end{bmatrix} \begin{bmatrix} 1.5 & 0 & 0 \\ 0 & 1 & 0 \\ 0 & 0 & 1 \end{bmatrix} \begin{bmatrix} 1 & 0 & -5 \\ 0 & 1 & 3 \\ 0 & 0 & 1 \end{bmatrix} = \begin{bmatrix} 1.5 & 0 & -7.5 \\ 0 & -1 & -3 \\ 0 & 0 & 1 \end{bmatrix}$$

The vertices of the original triangle in homogeneous coordinates are given by

$$\mathbf{v}_1 = \begin{bmatrix} 0 \\ 1 \\ 1 \end{bmatrix} \qquad \mathbf{v}_2 = \begin{bmatrix} 2 \\ 1 \\ 1 \end{bmatrix} \qquad \mathbf{v}_3 = \begin{bmatrix} 1 \\ 3 \\ 1 \end{bmatrix}$$

After applying the above matrix to each of these vectors, we obtain

$$\mathbf{v}_1' = \begin{bmatrix} -7.5 \\ -4 \\ 1 \end{bmatrix} \qquad \mathbf{v}_2' = \begin{bmatrix} -4.5 \\ -4 \\ 1 \end{bmatrix} \qquad \mathbf{v}_3' = \begin{bmatrix} -6 \\ -6 \\ 1 \end{bmatrix}$$

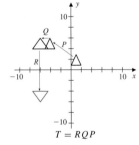

$T = RQP$

Figure 10

The resulting triangle, along with the intermediate steps, are shown in Fig. 10.

EXAMPLE 6   Find a 3×3 matrix that will transform the triangle shown in Fig. 11(a) to the triangle shown in Fig. 11(b).

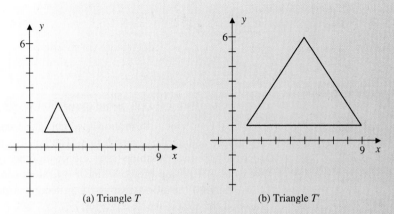

(a) Triangle T (b) Triangle T'

Figure 11

Solution   Triangle T' is obtained from triangle T through a horizontal scaling by a factor of 3, followed by a vertical scaling by a factor of 2, without changing the left vertex $(1, 1)$. The scalings alone will move the point $(1, 1)$ to $(3, 2)$. One way to correct this is to first translate the triangle so that the left vertex is located at the origin, perform the scaling, and then translate back. The matrix to perform all these operations is the product of the matrices for each transformation. The matrix is given by

$$\begin{bmatrix} 1 & 0 & 1 \\ 0 & 1 & 1 \\ 0 & 0 & 1 \end{bmatrix} \begin{bmatrix} 1 & 0 & 0 \\ 0 & 2 & 0 \\ 0 & 0 & 1 \end{bmatrix} \begin{bmatrix} 3 & 0 & 0 \\ 0 & 1 & 0 \\ 0 & 0 & 1 \end{bmatrix} \begin{bmatrix} 1 & 0 & -1 \\ 0 & 1 & -1 \\ 0 & 0 & 1 \end{bmatrix} = \begin{bmatrix} 3 & 0 & -2 \\ 0 & 2 & -1 \\ 0 & 0 & 1 \end{bmatrix}$$

Notice that

$$\begin{bmatrix} 1 & 0 & 1 \\ 0 & 1 & 1 \\ 0 & 0 & 1 \end{bmatrix} = \begin{bmatrix} 1 & 0 & -1 \\ 0 & 1 & -1 \\ 0 & 0 & 1 \end{bmatrix}^{-1}$$

that is, the matrix representation for translation by $\begin{bmatrix} 1 \\ 1 \end{bmatrix}$ is the inverse of the matrix representation for translation by $\begin{bmatrix} -1 \\ -1 \end{bmatrix}$.

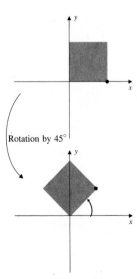

Rotation by 45°

Figure 12

Rotation

Another common graphics operation is a **rotation** through an angle θ. See Fig. 12. To describe how a point is rotated, let (x, y) be the coordinates of a point in $\mathbb{R}^2$ and θ a real number. From trigonometry, the new coordinates (x', y') of a point (x, y) rotated by θ rad about the origin are given by

$$x' = x \cos \theta - y \sin \theta$$
$$y' = x \sin \theta + y \cos \theta$$

If $\theta > 0$, then $\mathbf{v}$ is revolved about the origin in a counterclockwise direction. If $\theta < 0$, the direction is clockwise. These equations define a linear operator $S_\theta : \mathbb{R}^2 \to \mathbb{R}^2$

given by

$$S_\theta\left(\begin{bmatrix} x \\ y \end{bmatrix}\right) = \begin{bmatrix} x\cos\theta - y\sin\theta \\ x\sin\theta + y\cos\theta \end{bmatrix}$$

The matrix of S_θ relative to the standard basis $B = \{e_1, e_2\}$ for $\mathbb{R}^2$ is given by

$$[S_\theta]_B = \begin{bmatrix} \cos\theta & -\sin\theta \\ \sin\theta & \cos\theta \end{bmatrix}$$

When using homogeneous coordinates, we apply the matrix

$$\begin{bmatrix} \cos\theta & -\sin\theta & 0 \\ \sin\theta & \cos\theta & 0 \\ 0 & 0 & 1 \end{bmatrix}$$

EXAMPLE 7 Find the image of the triangle of Fig. 4 under a translation by the vector $\mathbf{b} = \begin{bmatrix} 1 \\ -1 \end{bmatrix}$, followed by a rotation of $30°$, or $\pi/6$ rad, in the counterclockwise direction.

Solution The matrix for the combined operations is given by

$$\begin{bmatrix} \cos\frac{\pi}{6} & -\sin\frac{\pi}{6} & 0 \\ \sin\frac{\pi}{6} & \cos\frac{\pi}{6} & 0 \\ 0 & 0 & 1 \end{bmatrix}\begin{bmatrix} 1 & 0 & 1 \\ 0 & 1 & -1 \\ 0 & 0 & 1 \end{bmatrix} = \begin{bmatrix} \frac{\sqrt{3}}{2} & -\frac{1}{2} & 0 \\ \frac{1}{2} & \frac{\sqrt{3}}{2} & 0 \\ 0 & 0 & 1 \end{bmatrix}\begin{bmatrix} 1 & 0 & 1 \\ 0 & 1 & -1 \\ 0 & 0 & 1 \end{bmatrix}$$

$$= \begin{bmatrix} \frac{\sqrt{3}}{2} & -\frac{1}{2} & \frac{\sqrt{3}}{2} + \frac{1}{2} \\ \frac{1}{2} & \frac{\sqrt{3}}{2} & \frac{1}{2} - \frac{\sqrt{3}}{2} \\ 0 & 0 & 1 \end{bmatrix}$$

The vertices of the triangle in homogeneous coordinates are given by

$$\mathbf{v}_1 = \begin{bmatrix} 0 \\ 1 \\ 1 \end{bmatrix} \quad \mathbf{v}_2 = \begin{bmatrix} 2 \\ 1 \\ 1 \end{bmatrix} \quad \text{and} \quad \mathbf{v}_3 = \begin{bmatrix} 1 \\ 3 \\ 1 \end{bmatrix}$$

After applying the above matrix to each of these vectors, we obtain

$$\mathbf{v}_1' = \begin{bmatrix} \frac{\sqrt{3}}{2} \\ \frac{1}{2} \\ 1 \end{bmatrix} \quad \mathbf{v}_2' = \begin{bmatrix} \frac{3\sqrt{3}}{2} \\ \frac{3}{2} \\ 1 \end{bmatrix} \quad \text{and} \quad \mathbf{v}_3' = \begin{bmatrix} \sqrt{3}-1 \\ \sqrt{3}+1 \\ 1 \end{bmatrix}$$

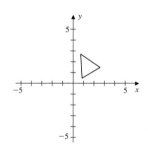

Figure 13 The resulting triangle is shown in Fig. 13.

Projection

Rendering a picture of a three-dimensional object on a flat computer screen requires projecting points in 3-space to points in 2-space. We discuss only one of many methods to project points in $\mathbb{R}^3$ to points in $\mathbb{R}^2$ that preserve the natural appearance of an object.

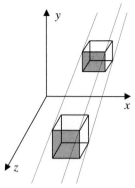

Figure 14

Parallel projection simulates the shadow that is cast onto a flat surface by a far away light source, such as the sun. Shown in Fig. 14 are rays intersecting an object in 3-space and the projection into 2-space. The orientation of the axes in Fig. 14 is such that the xy plane represents the computer screen.

To show how to find the xy coordinates of a projected point, let the vector

$$\mathbf{v}_d = \begin{bmatrix} x_d \\ y_d \\ z_d \end{bmatrix}$$

represent the direction of the rays. If (x_0, y_0, z_0) is a point in $\mathbb{R}^3$, then the *parametric equations* of the line going through the point and in the direction of $\mathbf{v}_d$ are given by

$$\begin{cases} x(t) &= x_0 + tx_d \\ y(t) &= y_0 + ty_d \\ z(t) &= z_0 + tz_d \end{cases}$$

for all $t \in \mathbb{R}$. The coordinates of the projection of (x_0, y_0, z_0) onto the xy plane are found by letting $z(t) = 0$. Solving for t, we obtain

$$t = -\frac{z_0}{z_d}$$

Now, substituting this value of t into the first two equations above, we find the coordinates of the projected point, which are given by

$$x_p = x_0 - \frac{z_0}{z_d}x_d \qquad y_p = y_0 - \frac{z_0}{z_d}y_d \qquad \text{and} \qquad z_p = 0$$

The components of $\mathbf{v}_d$ can also be used to find the angles that the rays make with the z axis and the xz plane. In particular, we have

$$\tan \psi = \frac{y_d}{x_d} \qquad \text{and} \qquad \tan \varphi = \frac{\sqrt{x_d^2 + y_d^2}}{z_d}$$

where ψ is the angle $\mathbf{v}_d$ makes with the xz plane and ϕ is the angle made with the z axis. On the other hand, if the angles ψ and ϕ are given, then these equations can be used to find the components of the projection vector $\mathbf{v}_d$.

EXAMPLE 8   Let $\psi = 30°$ and $\phi = 26.6°$.

a. Find the direction vector $\mathbf{v}_d$ and project the cube, shown in Fig. 15, into $\mathbb{R}^2$. The vertices of the cube are located at the points $(0, 0, 1)$, $(1, 0, 1)$, $(1, 0, 0)$, $(0, 0, 0)$, $(0, 1, 1)$, $(1, 1, 1)$, $(1, 1, 0)$, and $(0, 1, 0)$.

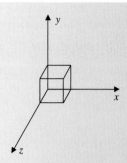

Figure 15

b. Find a 3×3 matrix that will rotate the (projected) vertices of the cube by $30°$ and another that will translate the cube by the vector $\begin{bmatrix} 2 \\ 1 \end{bmatrix}$.

Solution
a. We can arbitrarily set $z_d = -1$. Then

$$\tan \psi = \tan 30° \approx 0.577 = \frac{y_d}{x_d} \quad \text{and} \quad (\tan \phi)^2 = (\tan 26.6°)^2 \approx (0.5)^2 = x_d^2 + y_d^2$$

so that

$$y_d = 0.577 x_d \quad \text{and} \quad x_d^2 + y_d^2 = \tfrac{1}{4}$$

Solving the last two equations gives $x_d \approx 0.433$ and $y_d \approx 0.25$, so that the direction vector is

$$\mathbf{v}_d = \begin{bmatrix} 0.433 \\ 0.25 \\ -1 \end{bmatrix}$$

Using the formulas for a projected point given above, we can project each vertex of the cube into $\mathbb{R}^2$. Connecting the images by line segments gives the picture shown in Fig. 16. The projected points are given in Table 1.

Table 1

Vertex	Projected Point
(0,0,1)	(0.433, 0.25)
(1,0,1)	(1.433, 0.25)
(1,0,0)	(1, 0)
(0,0,0)	(0, 0)
(0,1,1)	(0.433, 1.25)
(1,1,1)	(1.433, 1.25)
(1,1,0)	(1, 1)
(0,1,0)	(0, 1)

b. Using homogeneous coordinates, we find the matrices to rotate the cube counterclockwise by 30° and translate the cube by the vector $\begin{bmatrix} 2 \\ 1 \end{bmatrix}$ are given by

$$\begin{bmatrix} \cos(\frac{\pi}{6}) & -\sin(\frac{\pi}{6}) & 0 \\ \sin(\frac{\pi}{6}) & \cos(\frac{\pi}{6}) & 0 \\ 0 & 0 & 1 \end{bmatrix} \quad \text{and} \quad \begin{bmatrix} 1 & 0 & 2 \\ 0 & 1 & 1 \\ 0 & 0 & 1 \end{bmatrix}$$

respectively. Depictions of the results when the original cube is rotated and
then the result is translated are shown in Figs. 17 and 18.

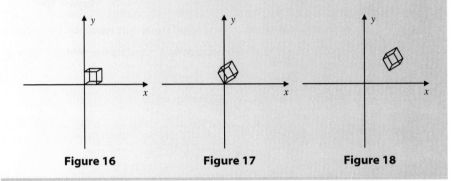

Figure 16   **Figure 17**   **Figure 18**

Exercise Set 4.6

1. Find the matrix representation relative to the
standard basis for the linear transformation
$T: \mathbb{R}^2 \longrightarrow \mathbb{R}^2$ that transforms the triangle with
vertices at the points $(0, 0)$, $(1, 1)$, and $(2, 0)$ to
the triangle shown in the figure.

a.

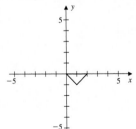

b.

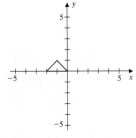

c.

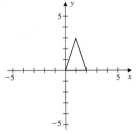

2. Find the matrix representation relative to the
standard basis for the linear transformation
$T: \mathbb{R}^2 \longrightarrow \mathbb{R}^2$ that transforms the square with
vertices at the points $(0, 0)$, $(1, 0)$, $(1, 1)$, and
$(0, 1)$ to the polygon shown in the figure.

a.

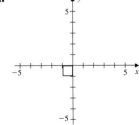

b.

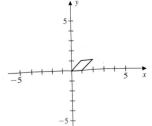

c.

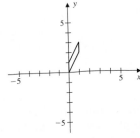

3. Let $T: \mathbb{R}^2 \longrightarrow \mathbb{R}^2$ be the transformation that performs a horizontal stretching by a factor of 3, followed by a vertical contraction by a factor of 2, followed by a reflection through the x axis.

   a. Find the matrix of T relative to the standard basis.

   b. Apply the transformation to the triangle with vertices $(1, 0)$, $(3, 0)$, and $(2, 2)$, and give a sketch of the result.

   c. Find the matrix relative to the standard basis that reverses T.

4. Let $T: \mathbb{R}^2 \longrightarrow \mathbb{R}^2$ be the transformation that performs a reflection through the y axis, followed by a horizontal shear by a factor of 3.

   a. Find the matrix of T relative to the standard basis.

   b. Apply the transformation to the rectangle with vertices $(1, 0)$, $(2, 0)$, $(2, 3)$, and $(1, 3)$, and give a sketch of the result.

   c. Find the matrix relative to the standard basis that reverses T.

5. Let $T: \mathbb{R}^2 \longrightarrow \mathbb{R}^2$ be the transformation that performs a rotation by $45°$, followed by a reflection through the origin.

   a. Find the matrix of T relative to the standard basis.

b. Apply the transformation to the square with vertices $(0, 0)$, $(1, 0)$, $(1, 1)$, and $(0, 1)$, and give a sketch of the result.

c. Find the matrix relative to the standard basis that reverses T.

6. Let $T: \mathbb{R}^2 \longrightarrow \mathbb{R}^2$ be the transformation that performs a reflection through the line $y = x$, followed by a rotation of $90°$.

   a. Find the matrix of T relative to the standard basis.

   b. Apply the transformation to the triangle with vertices $(0, 0)$, $(2, 0)$, and $(1, 3)$, and give a sketch of the result.

   c. Find the matrix relative to the standard basis that reverses T.

   d. Describe this transformation in another way. Verify your answer.

7. Let $T: \mathbb{R}^2 \longrightarrow \mathbb{R}^2$ be the (nonlinear) transformation that performs a translation by the vector $\begin{bmatrix} 1 \\ 1 \end{bmatrix}$, followed by a rotation of $30°$.

   a. Using homogeneous coordinates, find the 3×3 matrix that performs the translation and rotation.

   b. Apply the transformation to the parallelogram with vertices $(0, 0)$, $(2, 0)$, $(3, 1)$, and $(1, 1)$, and give a sketch of the result.

   c. Find the matrix that reverses T.

8. Let $T: \mathbb{R}^2 \longrightarrow \mathbb{R}^2$ be the (nonlinear) transformation that performs a translation by the vector $\begin{bmatrix} -4 \\ 2 \end{bmatrix}$, followed by a reflection through the y axis.

   a. Using homogeneous coordinates, find the 3×3 matrix that performs the translation and reflection.

   b. Apply the transformation to the trapezoid with vertices $(0, 0)$, $(3, 0)$, $(2, 1)$, and $(1, 1)$, and give a sketch of the result.

   c. Find the matrix that reverses T.

9. Let
$$B = \left\{ \begin{bmatrix} 1 \\ 1 \end{bmatrix}, \begin{bmatrix} -1 \\ 1 \end{bmatrix} \right\}$$

be a basis for $\mathbb{R}^2$, and let A be the triangle in the xy coordinate system with vertices $(0, 0)$, $(2, 2)$, and $(0, 2)$.

a. Find the coordinates of the vertices of A relative to B.

b. Let T be the transformation that performs a reflection through the line $y = x$. Find $[T]_B^S$, where S is the standard basis for $\mathbb{R}^2$.

c. Apply the matrix found in part (b) to the coordinates found in part (a). Sketch the result.

d. Show the same result is obtained by applying $\begin{bmatrix} 0 & 1 \\ 1 & 0 \end{bmatrix}$ to the original coordinates.

10. Let

$$B = \left\{ \begin{bmatrix} 1 \\ 0 \end{bmatrix}, \begin{bmatrix} 1 \\ 1 \end{bmatrix} \right\}$$

be a basis for $\mathbb{R}^2$, and let A be the parallelogram in the xy coordinate system with vertices $(0, 0)$, $(1, 1)$, $(1, 0)$, and $(2, 1)$.

a. Find the coordinates of the vertices of A relative to B.

b. Find the matrix representation relative to B of the transformation T that performs a reflection through the horizontal axis.

c. Apply the matrix found in part (b) to the coordinates found in part (a). Write the resulting vectors relative to the standard basis, and sketch the result.

d. Find the matrix representation relative to the standard basis for the transformation that performs the same operation on the parallelogram. Apply this matrix to the original coordinates, and verify the result agrees with part (c).

Review Exercises for Chapter 4

1. Let $T: \mathbb{R}^2 \longrightarrow \mathbb{R}^4$ be a linear transformation.

a. Verify that

$$S = \left\{ \begin{bmatrix} 1 \\ 1 \end{bmatrix}, \begin{bmatrix} 3 \\ -1 \end{bmatrix} \right\}$$

is a basis for $\mathbb{R}^2$.

b. If

$$T \begin{bmatrix} 1 \\ 1 \end{bmatrix} = \begin{bmatrix} 1 \\ 2 \\ 0 \\ 2 \end{bmatrix} \quad \text{and}$$

$$T \begin{bmatrix} 3 \\ -1 \end{bmatrix} = \begin{bmatrix} 3 \\ 2 \\ 4 \\ -2 \end{bmatrix}$$

determine

$$T \begin{bmatrix} x \\ y \end{bmatrix} \quad \text{for all} \quad \begin{bmatrix} x \\ y \end{bmatrix} \in \mathbb{R}^2$$

(*Hint*: Find the coordinates of $\begin{bmatrix} x \\ y \end{bmatrix}$ relative to S.)

c. Describe all vectors in $N(T)$.

d. Is the linear map T one-to-one? Explain.

e. Find a basis for $R(T)$.

f. Is T onto? Give two reasons.

g. Find a basis for $\mathbb{R}^4$ that contains the vectors

$$\begin{bmatrix} 1 \\ 0 \\ 1 \\ 1 \end{bmatrix} \quad \text{and} \quad \begin{bmatrix} -1 \\ 1 \\ 0 \\ 1 \end{bmatrix}$$

h. Use the basis $B = \left\{ \begin{bmatrix} 1 \\ 2 \end{bmatrix}, \begin{bmatrix} 1 \\ -1 \end{bmatrix} \right\}$ of $\mathbb{R}^2$ and the basis found in part (g) for $\mathbb{R}^4$, call it C, and find a matrix representation for T with respect to the bases B and C.

i. Apply the matrix A found in part (h) to an arbitrary vector $\begin{bmatrix} x \\ y \end{bmatrix}$.

2. Define linear transformations $S, T: \mathcal{P}_3 \to \mathcal{P}_4$ and $H: \mathcal{P}_4 \to \mathcal{P}_4$ by

$$S(p(x)) = p'(0)$$
$$T(p(x)) = (x+1)p(x)$$
$$H(p(x)) = p'(x) + p(0)$$

a. Compute $H \circ T$ and $S \circ (H \circ T)$.

b. Find the matrix for S, T, and H relative to the standard bases for $\mathcal{P}_3$ and $\mathcal{P}_4$.

c. Show that T is one-to-one.

d. Find $R(T)$.

3. Let $S, T: \mathbb{R}^2 \to \mathbb{R}^2$ be transformations so S reflects every vector through the x axis and T reflects every vector through the y axis.

a. Give definitions for S and T. Show the mappings are linear transformations.

b. Find the matrix for S and for T relative to the standard basis for $\mathbb{R}^2$.

c. Find the matrix for the linear transformations $T \circ S$ and $S \circ T$. Describe geometrically the action of $T \circ S$ and $S \circ T$.

4. a. Let $T: M_{2 \times 2} \to M_{2 \times 2}$ be defined by

$$T(A) = \begin{bmatrix} 1 & 3 \\ -1 & 1 \end{bmatrix} A$$

Is T a linear transformation? Is T one-to-one? Is T an isomorphism?

b. Let $T: M_{2 \times 2} \to M_{2 \times 2}$ be defined by

$$T(A) = \begin{bmatrix} 1 & 0 \\ 1 & 0 \end{bmatrix} A$$

Is T a linear transformation? Is T one-to-one? Show $R(T)$ is isomorphic to $\mathbb{R}^2$.

5. Let $\mathbf{v}_1$ and $\mathbf{v}_2$ be linearly independent vectors in $\mathbb{R}^2$ and $T: \mathbb{R}^2 \to \mathbb{R}^2$ a linear operator such that

$$T(\mathbf{v}_1) = \mathbf{v}_2 \quad \text{and} \quad T(\mathbf{v}_2) = \mathbf{v}_1$$

Let $B = \{\mathbf{v}_1, \mathbf{v}_2\}$ and $B' = \{\mathbf{v}_2, \mathbf{v}_1\}$.

a. Find $[T]_B$.

b. Find $[T]_B^{B'}$.

6. Let $T: \mathbb{R}^2 \to \mathbb{R}^2$ be the linear operator that projects a vector across the line $\mathbf{span}\left\{ \begin{bmatrix} 1 \\ -1 \end{bmatrix} \right\}$ and $S: \mathbb{R}^2 \to \mathbb{R}^2$ the linear operator that reflects a vector across the line $\mathbf{span}\left\{ \begin{bmatrix} 1 \\ 0 \end{bmatrix} \right\}$. Let B denote the standard basis for $\mathbb{R}^2$.

a. Find $[T]_B$ and $[S]_B$.

b. Find $T\left(\begin{bmatrix} -2 \\ 1 \end{bmatrix} \right)$ and $S\left(\begin{bmatrix} 2 \\ 3 \end{bmatrix} \right)$.

c. Find the matrix representation for the linear operator $H: \mathbb{R}^2 \to \mathbb{R}^2$ that reflects a vector across the subspace $\mathbf{span}\left\{ \begin{bmatrix} 1 \\ -1 \end{bmatrix} \right\}$ and across the subspace $\mathbf{span}\left\{ \begin{bmatrix} 1 \\ 0 \end{bmatrix} \right\}$.

d. Find $H\left(\begin{bmatrix} -2 \\ -1 \end{bmatrix} \right)$.

e. Find $N(T)$ and $N(S)$.

f. Find all vectors $\mathbf{v}$ such that $T(\mathbf{v}) = \mathbf{v}$ and all vectors $\mathbf{v}$ such that $S(\mathbf{v}) = \mathbf{v}$.

7. Let $T: \mathbb{R}^3 \to \mathbb{R}^3$ be the linear operator that reflects a vector across the plane

$$\mathbf{span}\left\{ \begin{bmatrix} 1 \\ 0 \\ 0 \end{bmatrix}, \begin{bmatrix} 0 \\ 1 \\ 1 \end{bmatrix} \right\}$$

The projection of a vector $\mathbf{u}$ onto a vector $\mathbf{v}$ is the vector

$$\text{proj}_{\mathbf{v}} \mathbf{u} = \frac{\mathbf{u} \cdot \mathbf{v}}{\mathbf{v} \cdot \mathbf{v}} \mathbf{v}$$

and the reflection of $\mathbf{v}$ across the plane with normal vector $\mathbf{n}$ is

$$\mathbf{v} - 2 \, \text{proj}_{\mathbf{n}} \mathbf{v}$$

Let B denote the standard basis for $\mathbb{R}^3$.

a. Find $[T]_B$.

b. Find $T\left(\begin{bmatrix} -1 \\ 2 \\ 1 \end{bmatrix} \right)$.

c. Find $N(T)$.

d. Find $R(T)$.

e. Find the matrix relative to B for T^n, $n \geq 2$.

8. Define a transformation $T: \mathcal{P}_2 \to \mathbb{R}$ by

$$T(p(x)) = \int_0^1 p(x)\,dx$$

a. Show that T is a linear transformation.

b. Compute $T(-x^2 - 3x + 2)$.

c. Describe $N(T)$. Is T one-to-one?

d. Find a basis for $N(T)$.

e. Show that T is onto.

f. Let B be the standard basis for $\mathcal{P}_2$ and $B' = \{1\}$, a basis for $\mathbb{R}$. Find $[T]_B^{B'}$.

g. Compute $T(-x^2 - 3x + 2)$, using the matrix found in part (f).

h. Define linear operators $T: C^{(1)}[0, 1] \to C^{(1)}[0, 1]$ and $S: C^{(1)}[0, 1] \to C^{(1)}[0, 1]$ by

$$T(f) = \frac{d}{dx} f(x)$$

and

$$S(f) = F, \quad \text{where} \quad F(x) = \int_0^x f(t)\,dt$$

Find $T(xe^x)$ and $S(xe^x)$. Describe $S \circ T$ and $T \circ S$.

9. Let $T: V \to V$ be a linear operator such that $T^2 - T + I = \mathbf{0}$, where I denotes the identity mapping. Show that T^{-1} exists and is equal to $I - T$.

10. Let $T: \mathbb{R}^2 \to \mathbb{R}^2$ be a linear operator.

a. Show that the line segment between two vectors $\mathbf{u}$ and $\mathbf{v}$ in $\mathbb{R}^2$ can be described by

$$t\mathbf{u} + (1 - t)\mathbf{v} \quad \text{for} \quad 0 \leq t \leq 1$$

b. Show that the image of a line segment under the map T is another line segment.

c. A set in $\mathbb{R}^2$ is called **convex** if for every pair of vectors in the set, the line segment between the vectors is in the set. See the figure.

Convex set   Not a convex set

Suppose $T: \mathbb{R}^2 \to \mathbb{R}^2$ is an isomorphism and S is a convex set in $\mathbb{R}^2$. Show that $T(S)$ is a convex set.

d. Define $T: \mathbb{R}^2 \to \mathbb{R}^2$ by

$$T\left(\begin{bmatrix} x \\ y \end{bmatrix}\right) = \begin{bmatrix} 2x \\ y \end{bmatrix}$$

Show that T is an isomorphism. Let

$$S = \left\{ \begin{bmatrix} x \\ y \end{bmatrix} \,\middle|\, x^2 + y^2 = 1 \right\}$$

Describe the image of the set S under the transformation T.

Chapter 4: Chapter Test

In Exercises 1–40, determine whether the statement is true or false.

1. The transformation $T: \mathbb{R}^2 \to \mathbb{R}^2$ defined by

$$T\left(\begin{bmatrix} x \\ y \end{bmatrix}\right) = \begin{bmatrix} 2x - 3y \\ x + y + 2 \end{bmatrix}$$

is a linear transformation.

2. The transformation $T: \mathbb{R} \to \mathbb{R}$ defined by $T(x) = 2x - 1$ is a linear transformation.

3. If $b = 0$, then the transformation $T: \mathbb{R} \to \mathbb{R}$ defined by $T(x) = mx + b$ is a linear transformation.

4. If A is an $m \times n$ matrix, then T defined by
$$T(\mathbf{v}) = A\mathbf{v}$$
is a linear transformation from $\mathbb{R}^n$ into $\mathbb{R}^m$.

5. Let A be a fixed matrix in $M_{n \times n}$. Define a transformation $T: M_{n \times n} \to M_{n \times n}$ by

$$T(B) = (B + A)^2 - (B + 2A)(B - 3A)$$

If $A^2 = \mathbf{0}$, then T is a linear transformation.

6. Let $\mathbf{u} = \begin{bmatrix} 1 \\ 0 \end{bmatrix}$ and $\mathbf{v} = \begin{bmatrix} 0 \\ 1 \end{bmatrix}$. If $T: \mathbb{R}^2 \longrightarrow \mathbb{R}^2$ is a linear operator and

$$T(\mathbf{u} + \mathbf{v}) = \mathbf{v} \quad \text{and} \quad T(2\mathbf{u} - \mathbf{v}) = \mathbf{u} + \mathbf{v}$$

then

$$T(\mathbf{u}) = \begin{bmatrix} \frac{2}{3} \\ \frac{1}{3} \end{bmatrix}$$

7. If $T: \mathbb{R}^2 \to \mathbb{R}^2$ is defined by

$$T\left(\begin{bmatrix} x \\ y \end{bmatrix} \right) = \begin{bmatrix} 2 & -4 \\ 1 & -2 \end{bmatrix} \begin{bmatrix} x \\ y \end{bmatrix}$$

then T is an isomorphism.

8. If $T: V \longrightarrow W$ is a linear transformation and $\{\mathbf{v}_1, \ldots, \mathbf{v}_n\}$ is a linearly independent set in V, then $\{T(\mathbf{v}_1), \ldots, T(\mathbf{v}_n)\}$ is a linearly independent subset of W.

9. The vector spaces $\mathcal{P}_8$ and $M_{3 \times 3}$ are isomorphic.

10. If a linear map $T: \mathcal{P}_4 \longrightarrow \mathcal{P}_3$ is defined by $T(p(x)) = p'(x)$, then T is a one-to-one map.

11. If A is an $n \times n$ invertible matrix, then as a mapping from $\mathbb{R}^n$ into $\mathbb{R}^n$ the null space of A consists of only the zero vector.

12. The linear operator $T: \mathbb{R}^2 \to \mathbb{R}^2$ defined by

$$T\left(\begin{bmatrix} x \\ y \end{bmatrix} \right) = \begin{bmatrix} x - y \\ 0 \end{bmatrix}$$

is one-to-one.

13. If $T: \mathbb{R}^2 \to \mathbb{R}^2$ is the transformation that reflects each vector through the origin, then the matrix for T relative to the standard basis for $\mathbb{R}^2$ is

$$\begin{bmatrix} -1 & 0 \\ 0 & -1 \end{bmatrix}$$

14. A linear transformation preserves the operations of vector addition and scalar multiplication.

15. Every linear transformation between finite dimensional vector spaces can be defined using a matrix product.

16. A transformation $T: V \to W$ is a linear transformation if and only if

$$T(c_1\mathbf{v}_1 + c_2\mathbf{v}_2) = c_1 T(\mathbf{v}_1) + c_2 T(\mathbf{v}_2)$$

for all vectors $\mathbf{v}_1$ and $\mathbf{v}_2$ in V and scalars c_1 and c_2.

17. If $f: \mathbb{R} \longrightarrow \mathbb{R}$ is a linear operator and $\phi: \mathbb{R}^2 \longrightarrow \mathbb{R}^2$ is defined by

$$\phi(x, y) = (x, y - f(x))$$

then the mapping ϕ is an isomorphism.

18. Let U, V, and W be finite dimensional vector spaces. If U is isomorphic to V and V is isomorphic to W, then U is isomorphic to W.

19. If $T: V \to V$ is a linear operator and $\mathbf{u} \in N(T)$, then

$$T(c\mathbf{u} + \mathbf{v}) = T(\mathbf{v})$$

for all $\mathbf{v} \in V$ and scalars c.

20. If $P: \mathbb{R}^3 \to \mathbb{R}^3$ is the projection defined by

$$P\left(\begin{bmatrix} x \\ y \\ z \end{bmatrix} \right) = \begin{bmatrix} x \\ y \\ 0 \end{bmatrix}$$

then $P^2 = P$.

21. If $T: V \to W$ is a linear transformation between vector spaces such that T assigns each element of a basis for V to the same element of W, then T is the identity mapping.

22. If $T: \mathbb{R}^4 \to \mathbb{R}^5$ and $\dim(N(T)) = 2$, then $\dim(R(T)) = 3$.

23. If $T: \mathbb{R}^4 \to \mathbb{R}^5$ and $\dim(R(T)) = 2$, then $\dim(N(T)) = 2$.

24. If $T: \mathbb{R}^3 \to \mathbb{R}^3$ is defined by

$$T\left(\begin{bmatrix} x \\ y \\ z \end{bmatrix} \right) = \begin{bmatrix} 2x - y + z \\ x \\ y - x \end{bmatrix}$$

then the matrix for T^{-1} relative to the standard basis for $\mathbb{R}^3$ is

$$\begin{bmatrix} 0 & 1 & 0 \\ 0 & 0 & 1 \\ -1 & -2 & 1 \end{bmatrix}$$

25. If $T: \mathbb{R}^2 \to \mathbb{R}^2$ is defined by

$$T\left(\begin{bmatrix} x \\ y \end{bmatrix}\right) = \begin{bmatrix} 2x + 3y \\ -x + y \end{bmatrix}$$

$B = \{e_1, e_2\}$, and $B' = \{e_2, e_1\}$, then

$$[T]_B^{B'} = \begin{bmatrix} -1 & 1 \\ 2 & 3 \end{bmatrix}$$

26. There exists a linear transformation T between vector spaces such that $T(\mathbf{0}) \neq \mathbf{0}$.

27. The linear transformation $T: \mathbb{R}^3 \to \mathbb{R}^3$ defined by

$$T\left(\begin{bmatrix} x \\ y \\ z \end{bmatrix}\right) = \begin{bmatrix} x \\ 0 \\ y \end{bmatrix}$$

projects each vector in $\mathbb{R}^3$ onto the xy plane.

28. The linear operator $T: \mathbb{R}^2 \to \mathbb{R}^2$ defined by

$$T\left(\begin{bmatrix} x \\ y \end{bmatrix}\right) = \begin{bmatrix} 0 & 1 \\ 1 & 0 \end{bmatrix}\begin{bmatrix} x \\ y \end{bmatrix}$$

reflects each vector in $\mathbb{R}^2$ across the line $y = x$.

29. Let $T: V \to W$ be a linear transformation and $B = \{\mathbf{v}_1, \ldots, \mathbf{v}_n\}$ a basis for V. If T is onto, then $\{T(\mathbf{v}_1), \ldots, T(\mathbf{v}_n)\}$ is a basis for W.

30. The vector space $\mathcal{P}_2$ is isomorphic to the subspace of $\mathbb{R}^5$

$$W = \left\{ \begin{bmatrix} a \\ b \\ 0 \\ c \\ 0 \end{bmatrix} \middle| a, b, c \in \mathbb{R} \right\}$$

31. If $T: V \to V$ is the identity transformation, then the matrix for T relative to any pair of bases B and B' for V is the identity matrix.

32. If $T: \mathbb{R}^3 \to \mathbb{R}^3$ is defined by

$$T\left(\begin{bmatrix} x \\ y \\ z \end{bmatrix}\right) = \begin{bmatrix} x + y + z \\ y - x \\ y \end{bmatrix}$$

then $\dim(N(T)) = 1$.

33. If $T: \mathcal{P}_2 \to \mathcal{P}_2$ is defined by

$$T(ax^2 + bx + c) = 2ax + b$$

then a basis for $N(T)$ is $\{-3\}$.

34. If $T: M_{2\times2} \to M_{2\times2}$ is defined by

$$T(A) = A^2 - A$$

then $N(T) = \{\mathbf{0}\}$.

35. If $T: \mathcal{P}_3 \to \mathcal{P}_3$ is defined by

$$T(p(x)) = p''(x) - xp'(x)$$

then T is onto.

36. If $T: \mathcal{P}_3 \to \mathcal{P}_3$ is defined by

$$T(p(x)) = p''(x) - xp'(x)$$

then $q(x) = x^2$ is in $R(T)$.

37. The linear operator $T: \mathbb{R}^3 \to \mathbb{R}^3$ defined by

$$T\left(\begin{bmatrix} x \\ y \\ z \end{bmatrix}\right) = \begin{bmatrix} 3 & -3 & 0 \\ 1 & 2 & 1 \\ 3 & -1 & 1 \end{bmatrix}\begin{bmatrix} x \\ y \\ z \end{bmatrix}$$

is an isomorphism.

38. If A is an $m \times n$ matrix and $T: \mathbb{R}^n \to \mathbb{R}^m$ is defined by

$$T(\mathbf{v}) = A\mathbf{v}$$

then the range of T is the set of all linear combinations of the column vectors of A.

39. If A is an $m \times n$ matrix with $m > n$ and $T: \mathbb{R}^n \to \mathbb{R}^m$ is defined by

$$T(\mathbf{v}) = A\mathbf{v}$$

then T cannot be one-to-one.

40. If A is an $m \times n$ matrix with $m > n$ and $T: \mathbb{R}^n \to \mathbb{R}^m$ is defined by

$$T(\mathbf{v}) = A\mathbf{v}$$

then T cannot be onto.

CHAPTER 5

Eigenvalues and Eigenvectors

A *Markov chain* is a mathematical model used to describe a random process that, at any given time $t = 1, 2, 3, \ldots$, is in one of a finite number of *states*. Between the times t and $t + 1$ the process moves from state j to state i with a probability p_{ij}. Markov processes are also *memoryless*; that is, the next state of the system depends only on the current state. As an example, consider a city C with surrounding residential areas N, S, E, and W. Residents can move between any two locations or stay in their current location, with fixed probabilities. In this case a state is the location of a resident at any given time. The *state diagram* shown in Fig. 1 describes the situation with the probabilities of moving from one location to another shown in the corresponding *transition matrix* $A = (p_{ij})$. For example, entry $p_{34} = 0.2$ is the probability that

U.S. Geological Survery/DAL

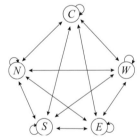

Figure 1

$$
\begin{array}{ccccc}
 & C & N & S & E & W \\
\begin{matrix} C \\ N \\ S \\ E \\ W \end{matrix} &
\left[\begin{matrix}
0.3 & 0.2 & 0.4 & 0.1 & 0.1 \\
0.2 & 0.4 & 0.1 & 0.2 & 0.1 \\
0.1 & 0.2 & 0.2 & 0.2 & 0.1 \\
0.2 & 0.1 & 0.2 & 0.3 & 0.2 \\
0.2 & 0.1 & 0.1 & 0.2 & 0.5
\end{matrix}\right]
\end{array}
$$

a resident in region E moves to region S. Since a resident is assumed to be living in one of the five regions, the probability of being in one of these regions is 1, and hence each column sum of A is equal to 1. A square matrix with each entry between 0 and 1 and column sums all equal to 1 is called a *stochastic matrix*. The initial distribution of the population is called the *initial probability vector*. Assume that the initial population distribution is given by the vector

$$
\mathbf{v} = \begin{bmatrix} 0.3 \\ 0.2 \\ 0.1 \\ 0.2 \\ 0.2 \end{bmatrix}
$$

Then the population distribution after one time step is $A\mathbf{v}$, after two time steps is $A(A\mathbf{v}) = A^2\mathbf{v}$, and so on. For example, after 10 time steps, the population distribution (rounded to two decimal places) is

$$A^{10}\mathbf{v} = \begin{bmatrix} 0.21 \\ 0.20 \\ 0.16 \\ 0.20 \\ 0.23 \end{bmatrix}$$

Notice that the sum of the entries of the population distribution vector is equal to 1. Starting with some initial distribution vector, the long-term behavior of the Markov chain, that is, $A^n\mathbf{v}$ as n tends to infinity, gives the limiting population distribution in the five regions into the future. When $A^n\mathbf{v}$ approaches a distribution vector $\mathbf{s}$ as n tends toward infinity, we say that $\mathbf{s}$ is the *steady-state vector*. If a transition matrix for a Markov chain is a stochastic matrix with positive terms, then for any initial probability vector $\mathbf{v}$, there is a unique steady-state vector $\mathbf{s}$. Moreover, if $\mathbf{s}$ is the steady-state vector, then $A\mathbf{s} = \mathbf{s}$. Finding the steady-state vector is equivalent to solving the matrix equation

$$A\mathbf{x} = \lambda\mathbf{x}$$

with $\lambda = 1$. In general, if there is a scalar λ and a nonzero vector $\mathbf{v}$ such that $A\mathbf{v} = \lambda\mathbf{v}$, then λ is called an *eigenvalue* for the matrix A and $\mathbf{v}$ is an *eigenvector* corresponding to the eigenvalue λ. In our Markov chain example, the steady-state vector corresponds to the eigenvalue $\lambda = 1$ for the transition matrix A.

In the last decade the growth in the power of modern computers has, quite miraculously, made it possible to compute the eigenvalues of a matrix with rows and columns in the billions. Google's page rank algorithm is essentially a Markov chain with transition matrix consisting of numerical weights for each site on the World Wide Web used as a *measure* of its relative importance within the set. The algorithm was developed by Larry Page and Sergey Brin, the founders of Google.

For any $n \times n$ matrix A, there exists at least one number-vector pair λ, $\mathbf{v}$ such that $A\mathbf{v} = \lambda\mathbf{v}$ (although λ may be a complex number). That is, the product of A and $\mathbf{v}$ is a scaling of the vector $\mathbf{v}$. Many applications require finding such number-vector pairs.

5.1 ▶ Eigenvalues and Eigenvectors

One of the most important problems in linear algebra is the *eigenvalue problem*. It can be stated thus: If A is an $n \times n$ matrix, does there exist a nonzero vector $\mathbf{v}$ such that $A\mathbf{v}$ is a scalar multiple of $\mathbf{v}$?

DEFINITION 1   **Eigenvalue and Eigenvector**   Let A be an $n \times n$ matrix. A number λ is called an **eigenvalue** of A provided that there exists a nonzero vector $\mathbf{v}$ in $\mathbb{R}^n$ such that

$$A\mathbf{v} = \lambda\mathbf{v}$$

Every nonzero vector satisfying this equation is called an **eigenvector of** A **corresponding to the eigenvalue** λ.

The zero vector is a trivial solution to the eigenvalue equation for any number λ and is not considered as an eigenvector.

As an illustration, let

$$A = \begin{bmatrix} 1 & 2 \\ 0 & -1 \end{bmatrix}$$

Observe that

$$\begin{bmatrix} 1 & 2 \\ 0 & -1 \end{bmatrix} \begin{bmatrix} 1 \\ 0 \end{bmatrix} = \begin{bmatrix} 1 \\ 0 \end{bmatrix}$$

so $\mathbf{v}_1 = \begin{bmatrix} 1 \\ 0 \end{bmatrix}$ is an eigenvector of A corresponding to the eigenvalue $\lambda_1 = 1$. We also have

$$\begin{bmatrix} 1 & 2 \\ 0 & -1 \end{bmatrix} \begin{bmatrix} 1 \\ -1 \end{bmatrix} = \begin{bmatrix} -1 \\ 1 \end{bmatrix} = -1 \begin{bmatrix} 1 \\ -1 \end{bmatrix}$$

so $\mathbf{v}_2 = \begin{bmatrix} 1 \\ -1 \end{bmatrix}$ is another eigenvector of A corresponding to the eigenvalue $\lambda_2 = -1$.

In Example 1 we show how to find eigenvalues and eigenvectors for a 2×2 matrix.

EXAMPLE 1   Let

$$A = \begin{bmatrix} 0 & 1 \\ 1 & 0 \end{bmatrix}$$

a. Find the eigenvalues of A.

b. Find the eigenvectors corresponding to each of the eigenvalues found in part (a).

Solution   **a.** The number λ is an eigenvalue of A if there is a nonzero vector $\mathbf{v} = \begin{bmatrix} x \\ y \end{bmatrix}$ such that

$$\begin{bmatrix} 0 & 1 \\ 1 & 0 \end{bmatrix} \begin{bmatrix} x \\ y \end{bmatrix} = \lambda \begin{bmatrix} x \\ y \end{bmatrix} \quad \text{which is equivalent to} \quad \begin{bmatrix} y \\ x \end{bmatrix} = \begin{bmatrix} \lambda x \\ \lambda y \end{bmatrix}$$

This matrix equation is equivalent to the homogeneous linear system

$$\begin{cases} -\lambda x + y = 0 \\ x - \lambda y = 0 \end{cases}$$

By Theorem 17 of Sec. 1.6, the linear system has a nontrivial solution if and only if

$$\begin{vmatrix} -\lambda & 1 \\ 1 & -\lambda \end{vmatrix} = 0$$

Consequently, λ is an eigenvalue of A if and only if

$$\lambda^2 - 1 = 0 \quad \text{so that} \quad \lambda_1 = 1 \quad \text{and} \quad \lambda_2 = -1$$

b. For $\lambda_1 = 1$, a vector $\mathbf{v}_1 = \begin{bmatrix} x \\ y \end{bmatrix}$ is an eigenvector if

$$\begin{bmatrix} 0 & 1 \\ 1 & 0 \end{bmatrix} \begin{bmatrix} x \\ y \end{bmatrix} = \begin{bmatrix} x \\ y \end{bmatrix}$$

This yields the linear system

$$\begin{cases} -x + y = 0 \\ x - y = 0 \end{cases} \quad \text{with solution set} \quad S = \left\{ \begin{bmatrix} t \\ t \end{bmatrix} \middle| t \in \mathbb{R} \right\}$$

Thus, any vector of the form $\mathbf{v}_1 = \begin{bmatrix} t \\ t \end{bmatrix}$, for $t \neq 0$, is an eigenvector corresponding to the eigenvalue $\lambda_1 = 1$. In a similar way, we find that any vector of the form $\mathbf{v}_2 = \begin{bmatrix} t \\ -t \end{bmatrix}$, for $t \neq 0$, is an eigenvector of A corresponding to the eigenvalue $\lambda_2 = -1$. Specific eigenvectors of A can be found by choosing any value for t so that neither $\mathbf{v}_1$ nor $\mathbf{v}_2$ is the zero vector. For example, letting $t = 1$, we know that

$$\mathbf{v}_1 = \begin{bmatrix} 1 \\ 1 \end{bmatrix}$$

is an eigenvector corresponding to $\lambda_1 = 1$ and

$$\mathbf{v}_2 = \begin{bmatrix} 1 \\ -1 \end{bmatrix}$$

is an eigenvector corresponding to $\lambda_2 = -1$.

Geometric Interpretation of Eigenvalues and Eigenvectors

A nonzero vector $\mathbf{v}$ is an eigenvector of a matrix A only when $A\mathbf{v}$ is a scaling of the vector $\mathbf{v}$. For example, let $A = \begin{bmatrix} 1 & -1 \\ 2 & 4 \end{bmatrix}$. Using the techniques just introduced, the eigenvalues of A are $\lambda_1 = 2$ and $\lambda_2 = 3$ with corresponding eigenvectors $\mathbf{v}_1 = \begin{bmatrix} 1 \\ -1 \end{bmatrix}$ and $\mathbf{v}_2 = \begin{bmatrix} 1 \\ -2 \end{bmatrix}$, respectively. Observe that

$$A\mathbf{v}_1 = \begin{bmatrix} 1 & -1 \\ 2 & 4 \end{bmatrix} \begin{bmatrix} 1 \\ -1 \end{bmatrix} = \begin{bmatrix} 2 \\ -2 \end{bmatrix} = 2\mathbf{v}_1$$

and

$$A\mathbf{v}_2 = \begin{bmatrix} 1 & -1 \\ 2 & 4 \end{bmatrix} \begin{bmatrix} 1 \\ -2 \end{bmatrix} = \begin{bmatrix} 3 \\ -6 \end{bmatrix} = 3\mathbf{v}_2$$

In Fig. 2, we provide sketches of the vectors $\mathbf{v}_1$, $\mathbf{v}_2$, $A\mathbf{v}_1$, and $A\mathbf{v}_2$ to underscore that the action of A on each of its eigenvectors is a scaling. Observe that this is not the case for an arbitrary vector. For example, if $\mathbf{v} = \begin{bmatrix} 1 \\ 1 \end{bmatrix}$, then

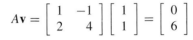

$$A\mathbf{v} = \begin{bmatrix} 1 & -1 \\ 2 & 4 \end{bmatrix} \begin{bmatrix} 1 \\ 1 \end{bmatrix} = \begin{bmatrix} 0 \\ 6 \end{bmatrix}$$

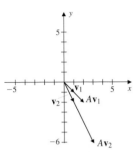

Figure 2

Eigenspaces

Notice that in Example 1, for each of the eigenvalues there are infinitely many eigenvectors. This is the case in general. To show this, let $\mathbf{v}$ be an eigenvector of the matrix A corresponding to the eigenvalue λ. If c is any nonzero real number, then

$$A(c\mathbf{v}) = cA(\mathbf{v}) = c(\lambda\mathbf{v}) = \lambda(c\mathbf{v})$$

so $c\mathbf{v}$ is another eigenvector associated with the eigenvalue λ. Notice that all eigenvectors corresponding to an eigenvalue are parallel but can have opposite directions.

Building on the procedure used in Example 1, we now describe a general method for finding eigenvalues and eigenvectors. If A is an $n \times n$ matrix, then

$$A\mathbf{v} = \lambda\mathbf{v}$$

for some number λ if and only if

$$A\mathbf{v} - \lambda\mathbf{v} = \mathbf{0} \qquad \text{that is} \qquad (A - \lambda I)\mathbf{v} = A\mathbf{v} - \lambda I\mathbf{v} = \mathbf{0}$$

Again by Theorem 17, of Sec. 1.6, this equation has a nontrivial solution if and only if

$$\det(A - \lambda I) = 0$$

We summarize this result in Theorem 1.

THEOREM 1   The number λ is an eigenvalue of the matrix A if and only if

$$\det(A - \lambda I) = 0$$

The equation $\det(A - \lambda I) = 0$ is called the **characteristic equation** of the matrix A, and the expression $\det(A - \lambda I)$ is called the **characteristic polynomial** of A. If A is an $n \times n$ matrix and λ is an eigenvalue of A, then the set

$$V_\lambda = \{\mathbf{v} \in \mathbb{R}^n \mid A\mathbf{v} = \lambda\mathbf{v}\}$$

is called the **eigenspace** of A corresponding to λ. Notice that V_λ is the union of the set of eigenvectors corresponding to λ and the zero vector.

We have already shown that V_λ is closed under scalar multiplication. Therefore, to show that V_λ is a subspace of $\mathbb{R}^n$, we need to show that it is also closed under addition. To see this, let $\mathbf{u}$ and $\mathbf{v}$ be vectors in V_λ; that is, $A\mathbf{u} = \lambda\mathbf{u}$ and $A\mathbf{v} = \lambda\mathbf{v}$ for a particular eigenvalue λ. Then

$$A(\mathbf{u} + \mathbf{v}) = A\mathbf{u} + A\mathbf{v} = \lambda\mathbf{u} + \lambda\mathbf{v} = \lambda(\mathbf{u} + \mathbf{v})$$

Alternatively, the set

$$V_\lambda = \{\mathbf{v} \in \mathbb{R}^n \mid A\mathbf{v} = \lambda\mathbf{v}\} = \{\mathbf{v} \in \mathbb{R}^n \mid (A - \lambda I)\mathbf{v} = 0\} = N(A - \lambda I)$$

Since V_λ is the null space of the matrix $A - \lambda I$, by Theorem 3 of Sec. 4.2 it is a subspace of $\mathbb{R}^n$.

EXAMPLE 2   Find the eigenvalues and corresponding eigenvectors of

$$A = \begin{bmatrix} 2 & -12 \\ 1 & -5 \end{bmatrix}$$

Give a description of the eigenspace corresponding to each eigenvalue.

Solution   By Theorem 1 to find the eigenvalues, we solve the characteristic equation

$$\begin{aligned}
\det(A - \lambda I) &= \begin{vmatrix} 2 - \lambda & -12 \\ 1 & -5 - \lambda \end{vmatrix} \\
&= (2 - \lambda)(-5 - \lambda) - (1)(-12) \\
&= \lambda^2 + 3\lambda + 2 \\
&= (\lambda + 1)(\lambda + 2) = 0
\end{aligned}$$

Thus, the eigenvalues are $\lambda_1 = -1$ and $\lambda_2 = -2$. To find the eigenvectors, we need to find all nonzero vectors in the null spaces of $A - \lambda_1 I$ and $A - \lambda_2 I$. First, for $\lambda_1 = -1$,

$$A - \lambda_1 I = A + I = \begin{bmatrix} 2 & -12 \\ 1 & -5 \end{bmatrix} + \begin{bmatrix} 1 & 0 \\ 0 & 1 \end{bmatrix} = \begin{bmatrix} 3 & -12 \\ 1 & -4 \end{bmatrix}$$

The null space of $A + I$ is found by row-reducing the augmented matrix

$$\begin{bmatrix} 3 & -12 & 0 \\ 1 & -4 & 0 \end{bmatrix} \quad \text{to} \quad \begin{bmatrix} 1 & -4 & 0 \\ 0 & 0 & 0 \end{bmatrix}$$

The solution set for this linear system is given by $S = \left\{ \begin{bmatrix} 4t \\ t \end{bmatrix} \middle| t \in \mathbb{R} \right\}$. Choosing $t = 1$, we obtain the eigenvector $\mathbf{v}_1 = \begin{bmatrix} 4 \\ 1 \end{bmatrix}$. Hence, the eigenspace corresponding to $\lambda_1 = -1$ is

$$V_{\lambda_1} = \left\{ t \begin{bmatrix} 4 \\ 1 \end{bmatrix} \middle| t \text{ is any real number} \right\}$$

For $\lambda_2 = -2$,

$$A - \lambda_2 I = \begin{bmatrix} 4 & -12 \\ 1 & -3 \end{bmatrix}$$

In a similar way we find that the vector $\mathbf{v}_2 = \begin{bmatrix} 3 \\ 1 \end{bmatrix}$ is an eigenvector corresponding to $\lambda_2 = -2$. The corresponding eigenspace is

$$V_{\lambda_2} = \left\{ t \begin{bmatrix} 3 \\ 1 \end{bmatrix} \middle| t \text{ is any real number} \right\}$$

The eigenspaces V_{λ_1} and V_{λ_2} are lines in the direction of the eigenvectors $\begin{bmatrix} 4 \\ 1 \end{bmatrix}$ and $\begin{bmatrix} 3 \\ 1 \end{bmatrix}$, respectively. The images of the eigenspaces, after multiplication by A, are the same lines, since the direction vectors $A \begin{bmatrix} 4 \\ 1 \end{bmatrix}$ and $A \begin{bmatrix} 3 \\ 1 \end{bmatrix}$ are scalar multiples of $\begin{bmatrix} 4 \\ 1 \end{bmatrix}$ and $\begin{bmatrix} 3 \\ 1 \end{bmatrix}$, respectively.

In Example 3 we illustrate how the eigenspace associated with a single eigenvalue can have dimension greater than 1.

EXAMPLE 3   Find the eigenvalues of

$$A = \begin{bmatrix} 1 & 0 & 0 & 0 \\ 0 & 1 & 5 & -10 \\ 1 & 0 & 2 & 0 \\ 1 & 0 & 0 & 3 \end{bmatrix}$$

and find a basis for each of the corresponding eigenspaces.

Solution   The characteristic equation of A is

$$\det(A - \lambda I) = \begin{vmatrix} 1-\lambda & 0 & 0 & 0 \\ 0 & 1-\lambda & 5 & -10 \\ 1 & 0 & 2-\lambda & 0 \\ 1 & 0 & 0 & 3-\lambda \end{vmatrix} = (\lambda - 1)^2(\lambda - 2)(\lambda - 3) = 0$$

Thus, the eigenvalues are

$$\lambda_1 = 1 \qquad \lambda_2 = 2 \qquad \text{and} \qquad \lambda_3 = 3$$

Since the exponent of the factor $\lambda - 1$ is 2, we say that the eigenvalue $\lambda_1 = 1$ has **algebraic multiplicity** 2. To find the eigenspace for $\lambda_1 = 1$, we reduce the matrix

$$A - (1)I = \begin{bmatrix} 0 & 0 & 0 & 0 \\ 0 & 0 & 5 & -10 \\ 1 & 0 & 1 & 0 \\ 1 & 0 & 0 & 2 \end{bmatrix} \qquad \text{to} \qquad \begin{bmatrix} 1 & 0 & 0 & 2 \\ 0 & 0 & 1 & -2 \\ 0 & 0 & 0 & 0 \\ 0 & 0 & 0 & 0 \end{bmatrix}$$

Hence, the eigenspace corresponding to $\lambda_1 = 1$ is

$$V_1 = \left\{ s \begin{bmatrix} 0 \\ 1 \\ 0 \\ 0 \end{bmatrix} + t \begin{bmatrix} -2 \\ 0 \\ 2 \\ 1 \end{bmatrix} \middle| s, t \in \mathbb{R} \right\}$$

Observe that the two vectors

$$\begin{bmatrix} 0 \\ 1 \\ 0 \\ 0 \end{bmatrix} \quad \text{and} \quad \begin{bmatrix} -2 \\ 0 \\ 2 \\ 1 \end{bmatrix}$$

are linearly independent and hence form a basis for V_{λ_1}. Since $\dim(V_{\lambda_1}) = 2$, we say that λ_1 has **geometric multiplicity** equal to 2. Alternatively, we can write

$$V_{\lambda_1} = \textbf{span} \left\{ \begin{bmatrix} 0 \\ 1 \\ 0 \\ 0 \end{bmatrix}, \begin{bmatrix} -2 \\ 0 \\ 2 \\ 1 \end{bmatrix} \right\}$$

Similarly, the eigenspaces corresponding to $\lambda_2 = 2$ and $\lambda_3 = 3$ are, respectively,

$$V_{\lambda_2} = \textbf{span} \left\{ \begin{bmatrix} 0 \\ 5 \\ 1 \\ 0 \end{bmatrix} \right\} \quad \text{and} \quad V_{\lambda_3} = \textbf{span} \left\{ \begin{bmatrix} 0 \\ -5 \\ 0 \\ 1 \end{bmatrix} \right\}$$

In Example 3 the algebraic and geometric multiplicities of each eigenvalue are equal. This is not the case in general. For example, if

$$A = \begin{bmatrix} 1 & 1 \\ 0 & 1 \end{bmatrix}$$

then the characteristic equation is $(\lambda - 1)^2 = 0$. Thus, $\lambda = 1$ has algebraic multiplicity 2. However,

$$V_\lambda = \left\{ \begin{bmatrix} t \\ 0 \end{bmatrix} \middle| t \in \mathbb{R} \right\}$$

so $\lambda = 1$ has geometric multiplicity 1.

Although eigenvectors are always nonzero, an eigenvalue can be zero. Also, as mentioned at the beginning of this section, eigenvalues can be complex numbers. These cases are illustrated in Example 4.

EXAMPLE 4 Find the eigenvalues of

$$A = \begin{bmatrix} 0 & 0 & 0 \\ 0 & 0 & -1 \\ 0 & 1 & 0 \end{bmatrix}$$

Solution   The characteristic equation is

$$\det(A - \lambda I) = \begin{vmatrix} -\lambda & 0 & 0 \\ 0 & -\lambda & -1 \\ 0 & 1 & -\lambda \end{vmatrix} = -\lambda^3 - \lambda = -\lambda(\lambda^2 + 1) = 0$$

Thus, the eigenvalues are $\lambda_1 = 0$, $\lambda_2 = i$, and $\lambda_3 = -i$. The corresponding eigenvectors are

$$\begin{bmatrix} 1 \\ 0 \\ 0 \end{bmatrix} \qquad \begin{bmatrix} 0 \\ 1 \\ -i \end{bmatrix} \quad \text{and} \quad \begin{bmatrix} 0 \\ 1 \\ i \end{bmatrix}$$

A fact that will be useful in the next section has to do with the eigenvalues of a square triangular matrix. For example, let

$$A = \begin{bmatrix} 2 & 4 \\ 0 & -3 \end{bmatrix}$$

Since $\det(A - \lambda I) = 0$ if and only if $(2 - \lambda)(-3 - \lambda) = 0$, we see that the eigenvalues of A are precisely the diagonal entries of A. In general, we have the following result.

PROPOSITION 1   The eigenvalues of an $n \times n$ triangular matrix are the numbers on the diagonal.

Proof   Let A be an $n \times n$ triangular matrix. By Theorem 13 of Sec. 1.6, the characteristic polynomial is given by

$$\det(A - \lambda I) = (a_{11} - \lambda)(a_{22} - \lambda) \cdots (a_{nn} - \lambda)$$

Hence, $\det(A - \lambda I) = 0$ if and only if $\lambda_1 = a_{11}, \lambda_2 = a_{22}, \ldots, \lambda_n = a_{nn}$.

Eigenvalues and Eigenvectors of Linear Operators

The definitions of eigenvalues and eigenvectors can be extended to linear operators.

DEFINITION 2   **Eigenvalue and Eigenvector of a Linear Operator**   Let V be a vector space and $T{:}V \longrightarrow V$ be a linear operator. A number λ is an **eigenvalue** of T provided that there is a nonzero vector $\mathbf{v}$ in V such that $T(\mathbf{v}) = \lambda\mathbf{v}$. Every nonzero vector that satisfies this equation is an **eigenvector of T corresponding to the eigenvalue** λ.

As an illustration define $T{:}\, \mathcal{P}_2 \to \mathcal{P}_2$ by

$$T(ax^2 + bx + c) = (-a + b + c)x^2 + (-b - 2c)x - 2b - c$$

Observe that

$$T(-x^2 + x + 1) = 3x^2 - 3x - 3 = -3(-x^2 + x + 1)$$

so $p(x) = -x^2 + x + 1$ is an eigenvector of T corresponding to the eigenvalue $\lambda = -3$.

Example 5 is from ordinary differential equations.

EXAMPLE 5 Interpret the solutions to the equation

$$f'(x) = kf(x)$$

as an eigenvalue problem of a linear operator.

Solution Let D denote the collection of all real-valued functions of one variable that have derivatives of all orders. Examples of such functions are polynomials, the trigonometric functions $\sin(x)$ and $\cos(x)$, and the natural exponential function e^x on $\mathbb{R}$. Define a linear operator $T: D \longrightarrow D$ by

$$T(f(x)) = f'(x)$$

Then λ is an eigenvalue of T if there is a function $f(x)$, not identically zero, such that $T(f(x)) = \lambda f(x)$. That is, $f(x)$ satisfies the differential equation

$$f'(x) = \lambda f(x)$$

Nonzero solutions to this differential equation are eigenvectors of the operator T, called **eigenfunctions**, corresponding to the eigenvalue λ. The general solution to this equation is given by

$$f(x) = ke^{\lambda x}$$

where k is an arbitrary constant. This class of functions is a model for exponential growth and decay with extensive applications.

Fact Summary

Let A be an $n \times n$ matrix.

1. The number λ is an eigenvalue of A if and only if $\det(A - \lambda I) = 0$.
2. The expression $\det(A - \lambda I)$ is a polynomial of degree n.
3. If λ is an eigenvalue of A and c is a nonzero scalar, then $c\lambda$ is another eigenvalue of A.
4. If λ is an eigenvalue of A, then the eigenspace

$$V_\lambda = \{\mathbf{v} \in \mathbb{R}^n \mid A\mathbf{v} = \lambda\mathbf{v}\}$$

is a subspace of $\mathbb{R}^n$.

5. The eigenspace corresponding to λ is the null space of the matrix $A - \lambda I$.
6. The eigenvalues of a square triangular matrix are the diagonal entries.

Exercise Set 5.1

In Exercises 1–6, a matrix A and an eigenvector $\mathbf{v}$ are given. Find the corresponding eigenvalue directly by solving $A\mathbf{v} = \lambda\mathbf{v}$.

1. $A = \begin{bmatrix} 3 & 0 \\ 1 & 3 \end{bmatrix}$ $\mathbf{v} = \begin{bmatrix} 0 \\ 1 \end{bmatrix}$

2. $A = \begin{bmatrix} -1 & 1 \\ 0 & -2 \end{bmatrix}$ $\mathbf{v} = \begin{bmatrix} -1 \\ 1 \end{bmatrix}$

3. $A = \begin{bmatrix} -3 & 2 & 3 \\ -1 & -2 & 1 \\ -3 & 2 & 3 \end{bmatrix}$ $\mathbf{v} = \begin{bmatrix} 1 \\ 0 \\ 1 \end{bmatrix}$

4. $A = \begin{bmatrix} 1 & 0 & 1 \\ 3 & 2 & 0 \\ 3 & 0 & -1 \end{bmatrix}$

$\mathbf{v} = \begin{bmatrix} -\frac{4}{3} \\ 1 \\ 4 \end{bmatrix}$

5. $A = \begin{bmatrix} 1 & 0 & 1 & 1 \\ 0 & 1 & 0 & 0 \\ 1 & 1 & 0 & 0 \\ 0 & 1 & 0 & 1 \end{bmatrix}$

$\mathbf{v} = \begin{bmatrix} -1 \\ 0 \\ -1 \\ 1 \end{bmatrix}$

6. $A = \begin{bmatrix} 1 & 1 & 1 & 0 \\ -1 & -1 & 0 & -1 \\ -1 & 1 & 0 & 1 \\ 0 & -1 & -1 & 0 \end{bmatrix}$

$\mathbf{v} = \begin{bmatrix} 0 \\ 1 \\ -1 \\ 0 \end{bmatrix}$

In Exercises 7–16, a matrix A is given.
  a. Find the characteristic equation for A.
  b. Find the eigenvalues of A.
  c. Find the eigenvectors corresponding to each eigenvalue.
  d. Verify the result of part (c) by showing that $A\mathbf{v}_i = \lambda_i\mathbf{v}_i$.

7. $A = \begin{bmatrix} -2 & 2 \\ 3 & -3 \end{bmatrix}$

8. $A = \begin{bmatrix} -2 & -1 \\ -1 & -2 \end{bmatrix}$

9. $A = \begin{bmatrix} 1 & -2 \\ 0 & 1 \end{bmatrix}$

10. $A = \begin{bmatrix} 0 & 2 \\ -1 & -3 \end{bmatrix}$

11. $A = \begin{bmatrix} -1 & 0 & 1 \\ 0 & 1 & 0 \\ 0 & 2 & -1 \end{bmatrix}$

12. $A = \begin{bmatrix} 0 & 2 & 0 \\ 0 & -1 & 1 \\ 0 & 0 & 1 \end{bmatrix}$

13. $A = \begin{bmatrix} 2 & 1 & 2 \\ 0 & 2 & -1 \\ 0 & 1 & 0 \end{bmatrix}$

14. $A = \begin{bmatrix} 1 & 1 & 1 \\ 0 & 1 & 0 \\ 0 & 0 & 1 \end{bmatrix}$

15. $A = \begin{bmatrix} -1 & 0 & 0 & 0 \\ 0 & 2 & 0 & 0 \\ 0 & 0 & -2 & 0 \\ 0 & 0 & 0 & 4 \end{bmatrix}$

16. $A = \begin{bmatrix} 3 & 2 & 3 & -1 \\ 0 & 1 & 2 & 1 \\ 0 & 0 & 2 & 0 \\ 0 & 0 & 0 & -1 \end{bmatrix}$

17. Show that if $\lambda^2 + b\lambda + c$ is the characteristic polynomial of the 2×2 matrix A, then $b = -\mathbf{tr}(A)$ and $c = \det(A)$.

18. Let A be an invertible matrix. Show that if λ is an eigenvalue of A, then $1/\lambda$ is an eigenvalue of A^{-1}.

19. Let A be an $n \times n$ matrix. Show that A is not invertible if and only if $\lambda = 0$ is an eigenvalue of A.

20. Let V be a vector space with $\dim(V) = n$ and $T: V \longrightarrow V$ a linear operator. If λ is an eigenvalue of T with geometric multiplicity n, then show that every nonzero vector of V is an eigenvector.

21. Let A be an idempotent matrix. Show that if λ is an eigenvalue of A, then $\lambda = 0$ or $\lambda = 1$.

22. Show that A and A^t have the same eigenvalues. Give an example to show A and A^t can have different eigenvectors.

23. Show that if there is a positive integer n such that $A^n = \mathbf{0}$, then $\lambda = 0$ is the only eigenvalue of A.

24. Let $A = \begin{bmatrix} 1 & 0 \\ 0 & -1 \end{bmatrix}$. Define an operator $T: M_{2\times 2} \to M_{2\times 2}$ by

$$T(B) = AB - BA$$

a. Show that $e = \begin{bmatrix} 0 & 1 \\ 0 & 0 \end{bmatrix}$ is an eigenvector corresponding to the eigenvalue $\lambda = 2$.

b. Show that $f = \begin{bmatrix} 0 & 0 \\ 1 & 0 \end{bmatrix}$ is an eigenvector corresponding to the eigenvalue $\lambda = -2$.

25. Let A and B be $n \times n$ matrices with A invertible. Show that AB and BA have the same eigenvalues.

26. Show that no such matrices A and B exist such that

$$AB - BA = I$$

27. Show that the eigenvalues of a square triangular matrix are the diagonal entries of the matrix.

28. Let λ be an eigenvalue of A. Use mathematical induction to show that for all n in the set of all natural numbers $\mathbb{N}$, if λ is an eigenvalue of A, then λ^n is an eigenvalue of A^n. What can be said about corresponding eigenvectors?

29. Let $C = B^{-1}AB$. Show that if $\mathbf{v}$ is an eigenvector of C corresponding to the eigenvalue λ, then $B\mathbf{v}$ is an eigenvector of A corresponding to λ.

30. Let A be an $n \times n$ matrix and suppose $\mathbf{v}_1, \ldots, \mathbf{v}_m$ are eigenvectors of A. If $S = \mathbf{span}\{\mathbf{v}_1, \ldots, \mathbf{v}_m\}$, show that if $\mathbf{v} \in S$, then $A\mathbf{v} \in S$.

31. Let $T: \mathbb{R}^2 \to \mathbb{R}^2$ be the linear operator that reflects a vector through the x axis. Find the eigenvalues and corresponding eigenvectors for T.

32. Define a linear operator $T: \mathbb{R}^2 \to \mathbb{R}^2$ by

$$T \begin{bmatrix} x \\ y \end{bmatrix} = \begin{bmatrix} y \\ x \end{bmatrix}$$

Show that the only eigenvalues of T are $\lambda = \pm 1$. Find the corresponding eigenvectors.

33. Define a linear operator $T: \mathbb{R}^2 \to \mathbb{R}^2$ by

$$T \begin{bmatrix} x \\ y \end{bmatrix} = \begin{bmatrix} \cos\theta & -\sin\theta \\ \sin\theta & \cos\theta \end{bmatrix} \begin{bmatrix} x \\ y \end{bmatrix}$$

That is, the action of T is a counterclockwise rotation of a vector by a nonnegative angle θ. Argue that if $\theta \neq 0, \pi$, then T has no real eigenvalues; if $\theta = 0$, then $\lambda = 1$ is an eigenvalue; and if $\theta = \pi$, then $\lambda = -1$ is an eigenvalue.

34. Let D denote the function space of all real-valued functions that have two derivatives, and define a linear operator T on D by

$$T(f) = f'' - 2f' - 3f$$

a. Show that for each k, the function $f(x) = e^{kx}$ is an eigenfunction for the operator T.

b. Find the corresponding eigenvalues for each eigenfunction $f(x) = e^{kx}$.

c. Find two nonzero functions f such that

$$f''(x) - 2f'(x) - 3f(x) = 0$$

35. Define a linear operator $T: \mathcal{P}_2 \to \mathcal{P}_2$ by

$$T(ax^2 + bx + c) = (a - b)x^2 + cx$$

Define two ordered bases for $\mathcal{P}_2$ by
$B = \{x - 1, x + 1, x^2\}$ and $B' = \{x + 1, 1, x^2\}$.

a. Find the matrix representation for T relative to the basis B.

b. Find the matrix representation for T relative to the basis B'.

c. Show that the eigenvalues for the matrices found in parts (a) and (b) are the same.

5.2 ▶ Diagonalization

Many applications of linear algebra involve factoring a matrix and writing it as the product of other matrices with special properties. For example, in Sec. 1.7, we saw how the LU factorization of a matrix can be used to develop efficient algorithms for solving a linear system with multiple input vectors. In this section, we determine if a matrix A has a factorization of the form

$$A = PDP^{-1}$$

where P is an invertible matrix and D is a diagonal matrix. The ideas presented here build on the concept of similarity of matrices, which we discussed in Sec. 4.5. Recall that if A and B are $n \times n$ matrices, then A is similar to B if there exists an invertible matrix P such that

$$B = P^{-1}AP$$

If B is a diagonal matrix, then the matrix A is called **diagonalizable**. Observe that if D is a diagonal matrix, then A is diagonalizable if either

$$D = P^{-1}AP \quad \text{or} \quad A = PDP^{-1}$$

for some invertible matrix P. One of the immediate benefits of diagonalizing a matrix A is realized when computing powers of A. This is often necessary when one is solving systems of differential equations. To see this, suppose that A is diagonalizable with

$$A = PDP^{-1}$$

Then

$$A^2 = (PDP^{-1})(PDP^{-1}) = PD(P^{-1}P)DP^{-1} = PD^2P^{-1}$$

Continuing in this way (see Exercise 27), we see that

$$A^k = PD^kP^{-1}$$

for any positive whole number k. Since D is a diagonal matrix, the entries of D^k are simply the diagonal entries of D raised to the k power.

As we shall soon see, diagonalization of a matrix A depends on the number of linearly independent eigenvectors, and fails when A is deficient in this way. We note that a connection does not exist between a matrix being diagonalizable and the matrix having an inverse. A square matrix has an inverse if and only if the matrix has only nonzero eigenvalues (see Exercise 19 of Sec. 5.1).

EXAMPLE 1   Let

$$A = \begin{bmatrix} 1 & 2 & 0 \\ 2 & 1 & 0 \\ 0 & 0 & -3 \end{bmatrix}$$

Show that A is diagonalizable with

$$P = \begin{bmatrix} 1 & 1 & 0 \\ -1 & 1 & 0 \\ 0 & 0 & 1 \end{bmatrix}$$

Solution   The inverse matrix is given by

$$P^{-1} = \begin{bmatrix} \frac{1}{2} & -\frac{1}{2} & 0 \\ \frac{1}{2} & \frac{1}{2} & 0 \\ 0 & 0 & 1 \end{bmatrix}$$

so that

$$P^{-1}AP = \begin{bmatrix} -1 & 0 & 0 \\ 0 & 3 & 0 \\ 0 & 0 & -3 \end{bmatrix}$$

Therefore, the matrix A is diagonalizable.

The diagonal entries of the matrix $P^{-1}AP$, in Example 1, are the eigenvalues of the matrix A, and the column vectors of P are the corresponding eigenvectors. For example, the product of A and the first column vector of P is given by

$$A \begin{bmatrix} 1 \\ -1 \\ 0 \end{bmatrix} = \begin{bmatrix} 1 & 2 & 0 \\ 2 & 1 & 0 \\ 0 & 0 & -3 \end{bmatrix} \begin{bmatrix} 1 \\ -1 \\ 0 \end{bmatrix} = -1 \begin{bmatrix} 1 \\ -1 \\ 0 \end{bmatrix}$$

Similarly, the second and third diagonal entries of $P^{-1}AP$ are the eigenvalues of A with corresponding eigenvectors the second and third column vectors of P, respectively. With Theorem 2 this idea is extended to $n \times n$ matrices.

THEOREM 2   An $n \times n$ matrix A is diagonalizable if and only if A has n linearly independent eigenvectors. Moreover, if $D = P^{-1}AP$, with D a diagonal matrix, then the diagonal entries of D are the eigenvalues of A and the column vectors of P are the corresponding eigenvectors.

Proof   First suppose that A has n linearly independent eigenvectors $\mathbf{v}_1, \mathbf{v}_2, \ldots, \mathbf{v}_n$, corresponding to the eigenvalues $\lambda_1, \lambda_2, \ldots, \lambda_n$. Note that the eigenvalues may not all be distinct. Let

$$\mathbf{v}_1 = \begin{bmatrix} p_{11} \\ p_{21} \\ \vdots \\ p_{n1} \end{bmatrix} \qquad \mathbf{v}_2 = \begin{bmatrix} p_{12} \\ p_{22} \\ \vdots \\ p_{n2} \end{bmatrix} \qquad \cdots \qquad \mathbf{v}_n = \begin{bmatrix} p_{1n} \\ p_{2n} \\ \vdots \\ p_{nn} \end{bmatrix}$$

and define the $n \times n$ matrix P so that the ith column vector is $\mathbf{v}_i$. Since the column vectors of P are linearly independent, by Theorem 9 of Sec. 2.3 the matrix P is invertible. Next, since the ith column vector of the product AP is

$$A\mathbf{P}_i = A\mathbf{v}_i = \lambda_i \mathbf{v}_i$$

we have

$$AP = \begin{bmatrix} \lambda_1 p_{11} & \lambda_2 p_{12} & \cdots & \lambda_n p_{1n} \\ \lambda_1 p_{21} & \lambda_2 p_{22} & \cdots & \lambda_n p_{2n} \\ \vdots & \vdots & \ddots & \vdots \\ \lambda_1 p_{n1} & \lambda_2 p_{n2} & \cdots & \lambda_n p_{nn} \end{bmatrix}$$

$$= \begin{bmatrix} p_{11} & p_{12} & \cdots & p_{1n} \\ p_{21} & p_{22} & \cdots & p_{2n} \\ \vdots & \vdots & \ddots & \vdots \\ p_{n1} & p_{n2} & \cdots & p_{nn} \end{bmatrix} \begin{bmatrix} \lambda_1 & 0 & \cdots & 0 \\ 0 & \lambda_2 & \cdots & 0 \\ \vdots & \vdots & \ddots & \vdots \\ 0 & 0 & \cdots & \lambda_n \end{bmatrix}$$

$$= PD$$

where D is a diagonal matrix with diagonal entries the eigenvalues of A. So $AP = PD$ and multiplying both sides on the left by P^{-1} gives

$$P^{-1}AP = D$$

The matrix A is similar to a diagonal matrix and is therefore diagonalizable.

Conversely, suppose that A is diagonalizable, that is, a diagonal matrix D and an invertible matrix P exist such that

$$D = P^{-1}AP$$

As above, denote the column vectors of the matrix P by $\mathbf{v}_1, \mathbf{v}_2, \ldots, \mathbf{v}_n$ and the diagonal entries of D by $\lambda_1, \lambda_2, \ldots, \lambda_n$. Since $AP = PD$, for each $i = 1, \ldots, n$, we have

$$A\mathbf{v}_i = \lambda_i \mathbf{v}_i$$

Hence, $\mathbf{v}_1, \mathbf{v}_2, \ldots, \mathbf{v}_n$ are eigenvectors of A. Since P is invertible, then by Theorem 9 of Sec. 2.3 the vectors $\mathbf{v}_1, \mathbf{v}_2, \ldots, \mathbf{v}_n$ are linearly independent.

EXAMPLE 2   Use Theorem 2 to diagonalize the matrix

$$A = \begin{bmatrix} 1 & 0 & 0 \\ 6 & -2 & 0 \\ 7 & -4 & 2 \end{bmatrix}$$

Solution   Since A is a triangular matrix, by Proposition 1 of Sec. 5.1, the eigenvalues of the matrix A are the diagonal entries

$$\lambda_1 = 1 \qquad \lambda_2 = -2 \qquad \text{and} \qquad \lambda_3 = 2$$

The corresponding eigenvectors, which are linearly independent, are given, respectively, by

$$\mathbf{v}_1 = \begin{bmatrix} 1 \\ 2 \\ 1 \end{bmatrix} \qquad \mathbf{v}_2 = \begin{bmatrix} 0 \\ 1 \\ 1 \end{bmatrix} \qquad \text{and} \qquad \mathbf{v}_3 = \begin{bmatrix} 0 \\ 0 \\ 1 \end{bmatrix}$$

Therefore, by Theorem 2, $D = P^{-1}AP$, where

$$D = \begin{bmatrix} 1 & 0 & 0 \\ 0 & -2 & 0 \\ 0 & 0 & 2 \end{bmatrix} \qquad \text{and} \qquad P = \begin{bmatrix} 1 & 0 & 0 \\ 2 & 1 & 0 \\ 1 & 1 & 1 \end{bmatrix}$$

To verify that $D = P^{-1}AP$, we can avoid finding P^{-1} by showing that

$$PD = AP$$

In this case,

$$PD = \begin{bmatrix} 1 & 0 & 0 \\ 2 & 1 & 0 \\ 1 & 1 & 1 \end{bmatrix} \begin{bmatrix} 1 & 0 & 0 \\ 0 & -2 & 0 \\ 0 & 0 & 2 \end{bmatrix} = \begin{bmatrix} 1 & 0 & 0 \\ 2 & -2 & 0 \\ 1 & -2 & 2 \end{bmatrix}$$

$$= \begin{bmatrix} 1 & 0 & 0 \\ 6 & -2 & 0 \\ 7 & -4 & 2 \end{bmatrix} \begin{bmatrix} 1 & 0 & 0 \\ 2 & 1 & 0 \\ 1 & 1 & 1 \end{bmatrix}$$

$$= AP$$

EXAMPLE 3   Let

$$A = \begin{bmatrix} 0 & 1 & 1 \\ 1 & 0 & 1 \\ 1 & 1 & 0 \end{bmatrix} \qquad \text{and} \qquad B = \begin{bmatrix} -1 & 1 & 0 \\ 0 & -1 & 1 \\ 0 & 0 & 2 \end{bmatrix}$$

Show that A is diagonalizable but that B is not diagonalizable.

Solution   To find the eigenvalues of A, we solve the characteristic equation

$$\det(A - \lambda I) = \det \begin{bmatrix} -\lambda & 1 & 1 \\ 1 & -\lambda & 1 \\ 1 & 1 & -\lambda \end{bmatrix}$$

$$= -(\lambda + 1)^2(\lambda - 2) = 0$$

Thus, the eigenvalues of A are $\lambda_1 = -1$, with algebraic multiplicity 2, and $\lambda_2 = 2$, with algebraic multiplicity 1. To find the eigenvectors, we find the null space of

$A - \lambda I$ for each eigenvalue. For $\lambda_1 = -1$ we reduce the matrix

$$\begin{bmatrix} 1 & 1 & 1 \\ 1 & 1 & 1 \\ 1 & 1 & 1 \end{bmatrix} \quad \text{to} \quad \begin{bmatrix} 1 & 1 & 1 \\ 0 & 0 & 0 \\ 0 & 0 & 0 \end{bmatrix}$$

Hence,

$$N(A + I) = \text{span} \left\{ \begin{bmatrix} -1 \\ 1 \\ 0 \end{bmatrix}, \begin{bmatrix} -1 \\ 0 \\ 1 \end{bmatrix} \right\}$$

In a similar manner we find

$$N(A - 2I) = \text{span} \left\{ \begin{bmatrix} 1 \\ 1 \\ 1 \end{bmatrix} \right\}$$

Since the three vectors $\begin{bmatrix} -1 \\ 1 \\ 0 \end{bmatrix}$, $\begin{bmatrix} -1 \\ 0 \\ 1 \end{bmatrix}$, and $\begin{bmatrix} 1 \\ 1 \\ 1 \end{bmatrix}$ are linearly independent, by Theorem 2 the matrix A is diagonalizable.

Using the same approach, we find that B has the same characteristic polynomial and hence the same eigenvalues. However, in this case

$$N(B + I) = \text{span} \left\{ \begin{bmatrix} 1 \\ 0 \\ 0 \end{bmatrix} \right\} \quad \text{and} \quad N(B - 2I) = \text{span} \left\{ \begin{bmatrix} 1 \\ 3 \\ 9 \end{bmatrix} \right\}$$

Since B does not have three linearly independent eigenvectors, by Theorem 2, B is not diagonalizable.

The matrix P that diagonalizes an $n \times n$ matrix A is not unique. For example, if the columns of P are permuted, then the resulting matrix also diagonalizes A. As an illustration, the matrix A of Example 3 is diagonalized by

$$P = \begin{bmatrix} -1 & -1 & 1 \\ 1 & 0 & 1 \\ 0 & 1 & 1 \end{bmatrix} \quad \text{with} \quad P^{-1}AP = \begin{bmatrix} -1 & 0 & 0 \\ 0 & -1 & 0 \\ 0 & 0 & 2 \end{bmatrix}$$

However, if Q is the matrix obtained from interchanging columns 2 and 3 of P, then Q also diagonalizes A, with

$$Q^{-1}AQ = \begin{bmatrix} -1 & 0 & 0 \\ 0 & 2 & 0 \\ 0 & 0 & -1 \end{bmatrix}$$

Notice, in this case, that the second and third diagonal entries are also interchanged.

Theorem 3 gives sufficient conditions for a matrix to be diagonalizable.

THEOREM 3   Let A be an $n \times n$ matrix, and let $\lambda_1, \lambda_2, \ldots, \lambda_n$ be distinct eigenvalues with corresponding eigenvectors $\mathbf{v}_1, \mathbf{v}_2, \ldots, \mathbf{v}_n$. Then the set $\{\mathbf{v}_1, \mathbf{v}_2, \ldots, \mathbf{v}_n\}$ is linearly independent.

Proof   The proof is by contradiction. Assume that $\lambda_1, \lambda_2, \ldots, \lambda_n$ are distinct eigenvalues of A with corresponding eigenvectors $\mathbf{v}_1, \mathbf{v}_2, \ldots, \mathbf{v}_n$, and assume that the set of eigenvectors is linearly dependent. Then by Theorem 5 of Sec. 2.3, at least one of the vectors can be written as a linear combination of the others. Moreover, the eigenvectors can be reordered so that $\mathbf{v}_1, \mathbf{v}_2, \ldots, \mathbf{v}_m$, with $m < n$, are linearly independent, but $\mathbf{v}_1, \mathbf{v}_2, \ldots, \mathbf{v}_{m+1}$ are linearly dependent with $\mathbf{v}_{m+1}$ a nontrivial linear combination of the first m vectors. Therefore, there are scalars $c_1, \ldots, c_m$, not all 0, such that

$$\mathbf{v}_{m+1} = c_1 \mathbf{v}_1 + \cdots + c_m \mathbf{v}_m$$

This is the statement that will result in a contradiction. We multiply the last equation by A to obtain

$$A\mathbf{v}_{m+1} = A(c_1 \mathbf{v}_1 + \cdots + c_m \mathbf{v}_m)$$
$$= c_1 A(\mathbf{v}_1) + \cdots + c_m A(\mathbf{v}_m)$$

Further, since $\mathbf{v}_i$ is an eigenvector corresponding to the eigenvalue λ_i, then $A\mathbf{v}_i = \lambda_i \mathbf{v}_i$, and after substitution in the previous equation, we have

$$\lambda_{m+1}\mathbf{v}_{m+1} = c_1 \lambda_1 \mathbf{v}_1 + \cdots + c_m \lambda_m \mathbf{v}_m$$

Now multiplying both sides of $\mathbf{v}_{m+1} = c_1 \mathbf{v}_1 + \cdots + c_m \mathbf{v}_m$ by λ_{m+1}, we also have

$$\lambda_{m+1}\mathbf{v}_{m+1} = c_1 \lambda_{m+1} \mathbf{v}_1 + \cdots + c_m \lambda_{m+1} \mathbf{v}_m$$

By equating the last two expressions for $\lambda_{m+1}\mathbf{v}_{m+1}$ we obtain

$$c_1 \lambda_1 \mathbf{v}_1 + \cdots + c_m \lambda_m \mathbf{v}_m = c_1 \lambda_{m+1} \mathbf{v}_1 + \cdots + c_m \lambda_{m+1} \mathbf{v}_m$$

or equivalently,

$$c_1 (\lambda_1 - \lambda_{m+1}) \mathbf{v}_1 + \cdots + c_m (\lambda_m - \lambda_{m+1}) \mathbf{v}_m = \mathbf{0}$$

Since the vectors $\mathbf{v}_1, \mathbf{v}_2, \ldots, \mathbf{v}_m$ are linearly independent, the only solution to the previous equation is the trivial solution, that is,

$$c_1 (\lambda_1 - \lambda_{m+1}) = 0 \qquad c_2 (\lambda_2 - \lambda_{m+1}) = 0 \qquad \ldots \qquad c_m (\lambda_m - \lambda_{m+1}) = 0$$

Since all the eigenvalues are distinct, we have

$$\lambda_1 - \lambda_{m+1} \neq 0 \qquad \lambda_2 - \lambda_{m+1} \neq 0 \qquad \ldots \qquad \lambda_m - \lambda_{m+1} \neq 0$$

and consequently

$$c_1 = 0 \qquad c_2 = 0 \qquad \ldots \qquad c_m = 0$$

This contradicts the assumption that the nonzero vector $\mathbf{v}_{m+1}$ is a nontrivial linear combination of $\mathbf{v}_1, \mathbf{v}_2, \ldots, \mathbf{v}_m$.

COROLLARY 1   If A is an $n \times n$ matrix with n distinct eigenvalues, then A is diagonalizable.

EXAMPLE 4   Show that every 2×2 real symmetric matrix is diagonalizable.

Solution   Recall that the matrix A is symmetric if and only if $A = A^t$. Every 2×2 symmetric matrix has the form

$$A = \begin{bmatrix} a & b \\ b & d \end{bmatrix}$$

See Example 5 of Sec. 1.3. The eigenvalues are found by solving the characteristic equation

$$\det(A - \lambda I) = \begin{vmatrix} a - \lambda & b \\ b & d - \lambda \end{vmatrix} = \lambda^2 - (a + d)\lambda + ad - b^2 = 0$$

By the quadratic formula, the eigenvalues are

$$\lambda = \frac{a + d \pm \sqrt{(a - d)^2 + 4b^2}}{2}$$

Since the discriminant $(a - d)^2 + 4b^2 \geq 0$, the characteristic equation has either one or two real roots. If $(a - d)^2 + 4b^2 = 0$, then $(a - d)^2 = 0$ and $b^2 = 0$, which holds if and only if $a = d$ and $b = 0$. Hence, the matrix A is diagonal. If $(a - d)^2 + 4b^2 > 0$, then A has two distinct eigenvalues; so by Corollary 1, the matrix A is diagonalizable.

By Theorem 2, if A is diagonalizable, then A is similar to a diagonal matrix whose eigenvalues are the same as the eigenvalues of A. In Theorem 4 we show that the same can be said about any two similar matrices.

THEOREM 4   Let A and B be similar $n \times n$ matrices. Then A and B have the same eigenvalues.

Proof   Since A and B are similar matrices, there is an invertible matrix P such that $B = P^{-1}AP$. Now

$$\begin{aligned} \det(B - \lambda I) &= \det(P^{-1}AP - \lambda I) \\ &= \det(P^{-1}(AP - P(\lambda I))) \\ &= \det(P^{-1}(AP - \lambda I P)) \\ &= \det(P^{-1}(A - \lambda I)P) \end{aligned}$$

Applying Theorem 15 and Corollary 1 of Sec. 1.6, we have

$$\det(B - \lambda I) = \det(P^{-1}) \det(A - \lambda I) \det(P)$$
$$= \det(P^{-1}) \det(P) \det(A - \lambda I)$$
$$= \det(A - \lambda I)$$

Since the characteristic polynomials of A and B are the same, their eigenvalues are equal.

EXAMPLE 5   Let

$$A = \begin{bmatrix} 1 & 2 \\ 0 & 3 \end{bmatrix} \quad \text{and} \quad P = \begin{bmatrix} 1 & 1 \\ 1 & 2 \end{bmatrix}$$

Verify that the matrices A and $B = P^{-1}AP$ have the same eigenvalues.

Solution   The characteristic equation for A is

$$\det(A - \lambda I) = (1 - \lambda)(3 - \lambda) = 0$$

so the eigenvalues of A are $\lambda_1 = 1$ and $\lambda_2 = 3$. Since

$$B = P^{-1}AP = \begin{bmatrix} 2 & -1 \\ -1 & 1 \end{bmatrix} \begin{bmatrix} 1 & 2 \\ 0 & 3 \end{bmatrix} \begin{bmatrix} 1 & 1 \\ 1 & 2 \end{bmatrix} = \begin{bmatrix} 3 & 4 \\ 0 & 1 \end{bmatrix}$$

the characteristic equation for B is

$$\det(B - \lambda I) = (1 - \lambda)(3 - \lambda) = 0$$

and hence, the eigenvalues of B are also $\lambda_1 = 1$ and $\lambda_2 = 3$.

In Sec. 4.5, we saw that a linear operator on a finite dimensional vector space can have different matrix representations depending on the basis used to construct the matrix. However, in every case the action of the linear operator on a vector remains the same. These matrix representations also have the same eigenvalues.

COROLLARY 2   Let V be a finite dimensional vector space, $T: V \longrightarrow V$ a linear operator, and B_1 and B_2 ordered bases for V. Then $[T]_{B_1}$ and $[T]_{B_2}$ have the same eigenvalues.

Proof   Let P be the transition matrix from B_2 to B_1. Then by Theorem 15 of Sec. 4.5, P is invertible and $[T]_{B_2} = P^{-1}[T]_{B_1} P$. Therefore, by Theorem 4, $[T]_{B_1}$ and $[T]_{B_2}$ have the same eigenvalues.

Recall that in Example 3, the characteristic polynomial for A and B is $-(\lambda + 1)^2(\lambda - 2)$. For the matrix A the eigenspaces corresponding to $\lambda_1 = -1$ and $\lambda_2 = 2$ are

$$V_{\lambda_1} = \mathbf{span}\left\{ \begin{bmatrix} -1 \\ 1 \\ 0 \end{bmatrix}, \begin{bmatrix} -1 \\ 0 \\ 1 \end{bmatrix} \right\} \quad \text{and} \quad V_{\lambda_2} = \mathbf{span}\left\{ \begin{bmatrix} 1 \\ 1 \\ 1 \end{bmatrix} \right\}$$

whereas the eigenspaces for B are

$$V_{\lambda_1} = \mathbf{span}\left\{ \begin{bmatrix} 1 \\ 0 \\ 0 \end{bmatrix} \right\} \quad \text{and} \quad V_{\lambda_2} = \mathbf{span}\left\{ \begin{bmatrix} 1 \\ 3 \\ 9 \end{bmatrix} \right\}$$

Notice that for the matrix A, we have $\dim(V_{\lambda_1}) = 2$ and $\dim(V_{\lambda_2}) = 1$, which, respectively, are equal to the corresponding algebraic multiplicities in the characteristic polynomial. This is not the case for the matrix B, since $\dim(V_{\lambda_1}) = 1$ and the corresponding algebraic multiplicity is 2. Moreover, for A, we have $\dim(V_{\lambda_1}) + \dim(V_{\lambda_2}) = 3 = n$.

The general result describing this situation is given, without proof, in Theorem 5.

THEOREM 5   Let A be an $n \times n$ matrix, and suppose that the characteristic polynomial is $c(x - \lambda_1)^{d_1}(x - \lambda_2)^{d_2} \cdots (x - \lambda_k)^{d_k}$. The matrix A is diagonalizable if and only if $d_i = \dim(V_{\lambda_i})$, for each $i = 1, \ldots, n$, and

$$d_1 + d_2 + \cdots + d_k = \dim(V_{\lambda_1}) + \dim(V_{\lambda_2}) + \cdots + \dim(V_{\lambda_k}) = n$$

To summarize Theorem 5, an $n \times n$ matrix A is diagonalizable if and only if the algebraic multiplicity for each eigenvalue is equal to the dimension of the corresponding eigenspace, which is the corresponding geometric multiplicity, and the common sum of these multiplicities is n.

Diagonalizable Linear Operators

In Theorem 12 of Sec. 4.4, we established that every linear operator on a finite dimensional vector space has a matrix representation. The particular matrix for the operator depends on the ordered basis used. From Corollary 2, we know that all matrix representations for a given linear operator are similar. This allows us to make the following definition.

DEFINITION 1   **Diagonalizable Linear Operator**   Let V be a finite dimensional vector space and $T: V \longrightarrow V$ a linear operator. The operator T is called **diagonalizable** if there is a basis B for V such that the matrix for T relative to B is a diagonal matrix.

Now suppose that V is a vector space of dimension n, $T: V \longrightarrow V$ a linear operator, and $B = \{\mathbf{v}_1, \mathbf{v}_2, \ldots, \mathbf{v}_n\}$ a basis for V consisting of n eigenvectors. Then

$$[T]_B = \left[\left[T(\mathbf{v}_1) \right]_B \ \left[T(\mathbf{v}_2) \right]_B \ \cdots \ \left[T(\mathbf{v}_n) \right]_B \right]$$

Since for each $i = 1, \ldots, n$ the vector $\mathbf{v}_i$ is an eigenvector, then $T(\mathbf{v}_i) = \lambda_i \mathbf{v}_i$, where λ_i is the corresponding eigenvalue. Since for each i, the basis vector $\mathbf{v}_i$ can be written uniquely as a linear combination of $\mathbf{v}_1, \ldots, \mathbf{v}_n$, we have

$$\mathbf{v}_i = 0\mathbf{v}_1 + \cdots + 0\mathbf{v}_{i-1} + \mathbf{v}_i + 0\mathbf{v}_{i+1} + \cdots + 0\mathbf{v}_n$$

Then the coordinate vector of $T(\mathbf{v}_i)$ relative to B is

$$[T(\mathbf{v}_i)]_B = \begin{bmatrix} 0 \\ \vdots \\ 0 \\ \lambda_i \\ 0 \\ \vdots \\ 0 \end{bmatrix}$$

Therefore, $[T]_B$ is a diagonal matrix. Alternatively, we can say that T is diagonalizable if there is a basis for V consisting of eigenvectors of T. As an illustration, define the linear operator $T: \mathbb{R}^2 \longrightarrow \mathbb{R}^2$ by

$$T\left(\begin{bmatrix} x \\ y \end{bmatrix} \right) = \begin{bmatrix} 2x \\ x + y \end{bmatrix}$$

Observe that

$$\mathbf{v}_1 = \begin{bmatrix} 1 \\ 1 \end{bmatrix} \quad \text{and} \quad \mathbf{v}_2 = \begin{bmatrix} 0 \\ 1 \end{bmatrix}$$

are eigenvectors of T with corresponding eigenvalues $\lambda_1 = 2$ and $\lambda_2 = 1$, respectively. Let $B = \{\mathbf{v}_1, \mathbf{v}_2\}$, so that

$$[T]_B = \left[\ [T(\mathbf{v}_1)]_B \ [T(\mathbf{v}_2)]_B \ \right] = \begin{bmatrix} 2 & 0 \\ 0 & 1 \end{bmatrix}$$

is a diagonal matrix.

In practice it is not always so easy to determine the eigenvalues and eigenvectors of T. However, if B is any basis for V such that $[T]_B$ is diagonalizable with diagonalizing matrix P, then T is diagonalizable. That is, if B' is the basis consisting of the column vectors of P, then $[T]_{B'} = P^{-1}[T]_B P$ is a diagonal matrix. This procedure is illustrated in Example 6.

EXAMPLE 6   Define the linear operator $T: \mathbb{R}^3 \longrightarrow \mathbb{R}^3$ by

$$T\left(\begin{bmatrix} x_1 \\ x_2 \\ x_3 \end{bmatrix}\right) = \begin{bmatrix} 3x_1 - x_2 + 2x_3 \\ 2x_1 + 2x_3 \\ x_1 + 3x_2 \end{bmatrix}$$

Show that T is diagonalizable.

Solution   Let $B = \{\mathbf{e}_1, \mathbf{e}_2, \mathbf{e}_3\}$ be the standard basis for $\mathbb{R}^3$. Then the matrix for T relative to B is

$$[T]_B = \begin{bmatrix} 3 & -1 & 2 \\ 2 & 0 & 2 \\ 1 & 3 & 0 \end{bmatrix}$$

Observe that the eigenvalues of $[T]_B$ are $\lambda_1 = -2$, $\lambda_2 = 4$, and $\lambda_3 = 1$ with corresponding eigenvectors, respectively,

$$\mathbf{v}_1 = \begin{bmatrix} 1 \\ 1 \\ -2 \end{bmatrix} \qquad \mathbf{v}_2 = \begin{bmatrix} 1 \\ 1 \\ 1 \end{bmatrix} \qquad \text{and} \qquad \mathbf{v}_3 = \begin{bmatrix} -5 \\ 4 \\ 7 \end{bmatrix}$$

Now let $B' = \{\mathbf{v}_1, \mathbf{v}_2, \mathbf{v}_3\}$ and

$$P = \begin{bmatrix} 1 & 1 & -5 \\ 1 & 1 & 4 \\ -2 & 1 & 7 \end{bmatrix}$$

Then

$$[T]_{B'} = \frac{1}{9}\begin{bmatrix} -1 & 4 & -3 \\ 5 & 1 & 3 \\ -1 & 1 & 0 \end{bmatrix}\begin{bmatrix} 3 & -1 & 2 \\ 2 & 0 & 2 \\ 1 & 3 & 0 \end{bmatrix}\begin{bmatrix} 1 & 1 & -5 \\ 1 & 1 & 4 \\ -2 & 1 & 7 \end{bmatrix} = \begin{bmatrix} -2 & 0 & 0 \\ 0 & 4 & 0 \\ 0 & 0 & 1 \end{bmatrix}$$

Fact Summary

Let A be an $n \times n$ matrix.

1. If A is diagonalizable, then $A = PDP^{-1}$ or equivalently $D = P^{-1}AP$. The matrix D is a diagonal matrix with diagonal entries the eigenvalues of A. The matrix P is invertible whose column vectors are the corresponding eigenvectors.

2. If A is diagonalizable, then the diagonalizing matrix P is not unique. If the columns of P are permuted, then the diagonal entries of D are permuted in the same way.

3. The matrix A is diagonalizable if and only if A has n linearly independent eigenvectors.

4. If A has n distinct eigenvalues, then A is diagonalizable.

5. Every 2×2 real symmetric matrix is diagonalizable and has real eigenvalues.

6. Similar matrices have the same eigenvalues.

7. If A is diagonalizable, then the algebraic multiplicity for each eigenvalue is equal to the dimension of the corresponding eigenspace (the geometric multiplicity). The common sum of these multiplicities is n.

8. Let $T: V \longrightarrow V$ be a linear operator on a finite dimensional vector space V. If V has an ordered basis B consisting of eigenvectors of T, then $[T]_B$ is a diagonal matrix.

9. Let $T: V \longrightarrow V$ be a linear operator and B_1 and B_2 ordered bases for V. Then $[T]_{B_1}$ and $[T]_{B_2}$ have the same eigenvalues.

Exercise Set 5.2

In Exercises 1–4, show that A is diagonalizable, using the matrix P.

1. $A = \begin{bmatrix} 1 & 0 \\ -2 & -3 \end{bmatrix}$ $P = \begin{bmatrix} -2 & 0 \\ 1 & 1 \end{bmatrix}$

2. $A = \begin{bmatrix} -1 & 1 \\ -3 & -5 \end{bmatrix}$

   $P = \begin{bmatrix} 1 & -1 \\ -3 & 1 \end{bmatrix}$

3. $A = \begin{bmatrix} 1 & 0 & 0 \\ 2 & -2 & 0 \\ 0 & 2 & 0 \end{bmatrix}$

   $P = \begin{bmatrix} 0 & 0 & \frac{3}{2} \\ 0 & -1 & 1 \\ 1 & 1 & 2 \end{bmatrix}$

4. $A = \begin{bmatrix} -1 & 2 & 2 \\ 0 & 2 & 0 \\ 2 & -1 & 2 \end{bmatrix}$

   $P = \begin{bmatrix} 1 & -2 & -2 \\ 0 & -4 & 0 \\ 2 & 1 & 1 \end{bmatrix}$

In Exercises 5–18, find the eigenvalues, and if necessary the corresponding eigenvectors, of A and determine whether A is diagonalizable.

5. $A = \begin{bmatrix} -1 & 1 \\ 0 & -2 \end{bmatrix}$

6. $A = \begin{bmatrix} -2 & -3 \\ -2 & -2 \end{bmatrix}$

7. $A = \begin{bmatrix} -1 & -1 \\ 0 & -1 \end{bmatrix}$

8. $A = \begin{bmatrix} -3 & -2 \\ 2 & 1 \end{bmatrix}$

9. $A = \begin{bmatrix} 0 & 1 \\ 0 & 1 \end{bmatrix}$

10. $A = \begin{bmatrix} 2 & 2 \\ -2 & -2 \end{bmatrix}$

11. $A = \begin{bmatrix} 2 & 2 & 0 \\ 2 & 2 & 2 \\ 0 & 0 & 3 \end{bmatrix}$

12. $A = \begin{bmatrix} -1 & 3 & 2 \\ -1 & 2 & 3 \\ -1 & 2 & 3 \end{bmatrix}$

13. $A = \begin{bmatrix} 2 & 1 & 1 \\ 2 & -1 & -1 \\ -1 & 1 & 2 \end{bmatrix}$

14. $A = \begin{bmatrix} -1 & 0 & 0 \\ -1 & 2 & -1 \\ 0 & -1 & 2 \end{bmatrix}$

15. $A = \begin{bmatrix} 1 & 0 & 0 \\ -1 & 0 & 0 \\ -1 & 0 & 0 \end{bmatrix}$

16. $A = \begin{bmatrix} 0 & 1 & 0 \\ 1 & 0 & -1 \\ 0 & -1 & 0 \end{bmatrix}$

17. $A = \begin{bmatrix} 0 & 0 & 0 & 0 \\ 1 & 0 & 1 & 0 \\ 0 & 1 & 0 & 1 \\ 1 & 1 & 1 & 1 \end{bmatrix}$

18. $A = \begin{bmatrix} 1 & 0 & 0 & 1 \\ 0 & 0 & 1 & 1 \\ 1 & 0 & 0 & 1 \\ 0 & 1 & 0 & 1 \end{bmatrix}$

In Exercises 19–26, diagonalize the matrix A.

19. $A = \begin{bmatrix} 2 & 0 \\ -1 & -1 \end{bmatrix}$

20. $A = \begin{bmatrix} -2 & 1 \\ 1 & 2 \end{bmatrix}$

21. $A = \begin{bmatrix} 1 & 0 & 0 \\ 0 & -2 & 1 \\ 1 & -2 & 1 \end{bmatrix}$

22. $A = \begin{bmatrix} 0 & -1 & 2 \\ 0 & 2 & 2 \\ 0 & 0 & 1 \end{bmatrix}$

23. $A = \begin{bmatrix} -1 & 0 & 0 \\ -1 & 1 & 0 \\ 0 & 0 & 1 \end{bmatrix}$

24. $A = \begin{bmatrix} 1 & 0 & 0 \\ 0 & 0 & 0 \\ 0 & -1 & 1 \end{bmatrix}$

25. $A = \begin{bmatrix} 1 & 0 & 1 & 0 \\ 0 & 1 & 0 & 0 \\ 1 & 0 & 1 & 1 \\ 0 & 0 & 0 & 1 \end{bmatrix}$

26. $A = \begin{bmatrix} 1 & 0 & 1 & 1 \\ 0 & 1 & 0 & 0 \\ 1 & 1 & 1 & 1 \\ 1 & 1 & 1 & 1 \end{bmatrix}$

27. Suppose A is diagonalizable with $D = P^{-1}AP$. Show that for any positive integer k,

$$A^k = PD^kP^{-1}$$

28. Let

$$A = \begin{bmatrix} 2 & 1 \\ 2 & 1 \end{bmatrix}$$

Factor A in the form $A = PDP^{-1}$, where D is a diagonal matrix. Then find A^6. See Exercise 27.

29. Let

$$A = \begin{bmatrix} 3 & -1 & -2 \\ 2 & 0 & -2 \\ 2 & -1 & -1 \end{bmatrix}$$

Factor A in the form $A = PDP^{-1}$, where D is a diagonal matrix. Then find A^k, for any positive integer k. See Exercise 27.

30. Suppose A is an $n \times n$ matrix that is diagonalized by P. Find a matrix that diagonalizes A^t.

31. Suppose A is an $n \times n$ matrix that is diagonalizable. Show that if B is a matrix similar to A, then B is diagonalizable.

32. Show that if A is invertible and diagonalizable, then A^{-1} is diagonalizable. Find a 2×2 matrix that is not a diagonal matrix, is not invertible, but is diagonalizable.

33. Suppose A is an $n \times n$ matrix and λ is an eigenvalue of multiplicity n. Show that A is diagonalizable if and only if $A = \lambda I$.

34. An $n \times n$ matrix A is called *nilpotent* if there is a positive integer k such that $A^k = \mathbf{0}$. Show that a nonzero nilpotent matrix is not diagonalizable.

35. Define a linear operator $T: \mathcal{P}_2 \to \mathcal{P}_2$ by

$$T(p(x)) = p'(x)$$

a. Find the matrix A for T relative to the standard basis $\{1, x, x^2\}$.

b. Find the matrix B for T relative to the basis $\{x, x - 1, x^2\}$.

c. Show the eigenvalues of A and B are the same.

d. Explain why T is not diagonalizable.

36. Define a vector space $V = \mathbf{span}\{\sin x, \cos x\}$ and a linear operator $T: V \to V$ by $T(f(x)) = f'(x)$. Show that T is diagonalizable.

37. Define a linear operator $T: \mathbb{R}^3 \to \mathbb{R}^3$ by

$$T\left(\begin{bmatrix} x_1 \\ x_2 \\ x_3 \end{bmatrix}\right) = \begin{bmatrix} 2x_1 + 2x_2 + 2x_3 \\ -x_1 + 2x_2 + x_3 \\ x_1 - x_2 \end{bmatrix}$$

Show that T is not diagonalizable.

38. Define a linear operator $T: \mathbb{R}^3 \to \mathbb{R}^3$ by

$$T\left(\begin{bmatrix} x_1 \\ x_2 \\ x_3 \end{bmatrix}\right) = \begin{bmatrix} 4x_1 + 2x_2 + 4x_3 \\ 4x_1 + 2x_2 + 4x_3 \\ 4x_3 \end{bmatrix}$$

Show that T is diagonalizable.

39. Let T be a linear operator on a finite dimensional vector space, A the matrix for T relative to a basis B_1, and B the matrix for T relative to a basis B_2. Show that A is diagonalizable if and only if B is diagonalizable.

5.3 ▶ Application: Systems of Linear Differential Equations

In Sec. 3.5 we considered only a single differential equation where the solution involved a single function. However, in many modeling applications, an equation that involves the derivatives of only one function is not sufficient. It is more likely that the rate of change of a variable quantity will be linked to other functions outside itself. This is the fundamental idea behind the notion of a *dynamical system*. One of the most familiar examples of this is the *predator-prey* model. For example, suppose we wish to create a model to predict the number of foxes and rabbits in some habitat. The growth rate of the foxes is dependent on not only the number of foxes but also the number of rabbits in their territory. Likewise, the growth rate of the rabbit population in part is dependent on their current number, but is obviously mitigated by the number of foxes in their midst. The mathematical model required to describe this relationship is a *system of differential equations* of the form

$$\begin{cases} y_1'(t) &= f(t, y_1, y_2) \\ y_2'(t) &= g(t, y_1, y_2) \end{cases}$$

In this section we consider systems of linear differential equations. Problems such as predator-prey problems involve systems of *nonlinear* differential equations.

Uncoupled Systems

At the beginning of Sec. 3.5 we saw that the differential equation given by

$$y' = ay$$

has the solution $y(t) = Ce^{at}$, where $C = y(0)$. An extension of this to two dimensions is the **system of differential equations**

$$\begin{cases} y_1' &= ay_1 \\ y_2' &= by_2 \end{cases}$$

where a and b are constants and y_1 and y_2 are functions of a common variable t. This system is called **uncoupled** since y_1' and y_2' depend only on y_1 and y_2, respectively. The general solution of the system is found by solving each equation separately and is given by

$$y_1(t) = C_1 e^{at} \qquad \text{and} \qquad y_2(t) = C_2 e^{bt}$$

where $C_1 = y_1(0)$ and $C_2 = y_2(0)$.

The previous system of two differential equations can also be written in matrix form. To do this, define

$$\mathbf{y}' = \begin{bmatrix} y_1' \\ y_2' \end{bmatrix} \qquad A = \begin{bmatrix} a & 0 \\ 0 & b \end{bmatrix} \qquad \text{and} \qquad \mathbf{y} = \begin{bmatrix} y_1 \\ y_2 \end{bmatrix}$$

Then the uncoupled system above is equivalent to the matrix equation

$$\mathbf{y}' = A\mathbf{y}$$

The matrix form of the solution is given by

$$\mathbf{y}(t) = \begin{bmatrix} e^{at} & 0 \\ 0 & e^{bt} \end{bmatrix} \mathbf{y}(0)$$

where $\mathbf{y}(0) = \begin{bmatrix} y_1(0) \\ y_2(0) \end{bmatrix}$

As an illustration, consider the system of differential equations

$$\begin{cases} y_1' & = -y_1 \\ y_2' & = 2y_2 \end{cases}$$

In matrix form the system is written as

$$\mathbf{y}' = A\mathbf{y} = \begin{bmatrix} -1 & 0 \\ 0 & 2 \end{bmatrix} \mathbf{y}$$

The solution to the system is

$$\mathbf{y} = \begin{bmatrix} e^{-t} & 0 \\ 0 & e^{2t} \end{bmatrix} \mathbf{y}(0)$$

that is,

$$y_1(t) = y_1(0)e^{-t} \qquad \text{and} \qquad y_2(t) = y_2(0)e^{2t}$$

The Phase Plane

In the case of a single differential equation, it is possible to sketch particular solutions in the plane to see explicitly how $y(t)$ depends on the independent variable t. However, for a system of two differential equations, the solutions are vectors which depend on a common parameter t, which is usually time. A particular solution can be viewed as a parameterized curve or **trajectory** in the plane, called the **phase plane**. Shown in Fig. 1 are trajectories for several particular solutions of the system

$$\begin{cases} y_1' & = -y_1 \\ y_2' & = 2y_2 \end{cases}$$

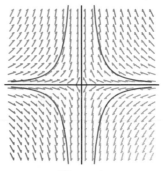

Figure 1

The vectors shown in Fig. 1 comprise the **direction field** for the system and describe the motion along a trajectory for increasing t. This sketch is called the **phase portrait** for the system. Phase portraits are usually drawn without the direction field. We have done so here to give a more complete picture of the system and its solutions.

Diagonalization

In the previous example, the matrix A is diagonal, as this is the case for any uncoupled system of differential equations. We now consider more general systems of the form

$$\mathbf{y}' = A\mathbf{y}$$

for which A is not a diagonal matrix, but is diagonalizable with real distinct eigenvalues. To solve problems of this type, our strategy is to reduce the system $\mathbf{y}' = A\mathbf{y}$ to one that is uncoupled.

To develop this idea, let A be a 2×2 diagonalizable matrix with distinct real eigenvalues. Consider the system of differential equations given by

$$\mathbf{y}' = A\mathbf{y}$$

Since A is diagonalizable, then by Theorem 2 of Sec. 5.2 there is a diagonal matrix D and an invertible matrix P such that

$$D = P^{-1}AP$$

The diagonal matrix D is given by

$$D = \begin{bmatrix} \lambda_1 & 0 \\ 0 & \lambda_2 \end{bmatrix}$$

where λ_1 and λ_2 are the eigenvalues of A. The column vectors of P are the corresponding eigenvectors. To uncouple the system $\mathbf{y}' = A\mathbf{y}$, let

$$\mathbf{w} = P^{-1}\mathbf{y}$$

Differentiating both sides of the last equation gives

$$\begin{aligned}
\mathbf{w}' = (P^{-1}\mathbf{y})' &= P^{-1}\mathbf{y}' \\
&= P^{-1}A\mathbf{y} \\
&= P^{-1}(PDP^{-1})\mathbf{y} = (P^{-1}P)(DP^{-1})\mathbf{y} \\
&= DP^{-1}\mathbf{y} \\
&= D\mathbf{w}
\end{aligned}$$

Since D is a diagonal matrix, the original linear system $\mathbf{y}' = A\mathbf{y}$ is transformed into the uncoupled linear system

$$\mathbf{w}' = P^{-1}AP\mathbf{w} = D\mathbf{w}$$

The general solution of this new system is given by

$$\mathbf{w}(t) = \begin{bmatrix} e^{\lambda_1 t} & 0 \\ 0 & e^{\lambda_2 t} \end{bmatrix} \mathbf{w}(0)$$

Now, to find the solution to the original system, we again use the substitution $\mathbf{w} = P^{-1}\mathbf{y}$ to obtain

$$P^{-1}\mathbf{y}(t) = \begin{bmatrix} e^{\lambda_1 t} & 0 \\ 0 & e^{\lambda_2 t} \end{bmatrix} P^{-1}\mathbf{y}(0)$$

Hence, the solution to the original system is

$$\mathbf{y}(t) = P \begin{bmatrix} e^{\lambda_1 t} & 0 \\ 0 & e^{\lambda_2 t} \end{bmatrix} P^{-1}\mathbf{y}(0)$$

EXAMPLE 1

Find the general solution to the system of differential equations

$$\begin{cases} y_1' &= -y_1 \\ y_2' &= 3y_1 + 2y_2 \end{cases}$$

Sketch several trajectories in the phase plane.

Solution

The differential equation is given in matrix form by

$$\mathbf{y}' = A\mathbf{y} = \begin{bmatrix} -1 & 0 \\ 3 & 2 \end{bmatrix} \mathbf{y}$$

After solving the characteristic equation $\det(A - \lambda I) = 0$, we know that the eigenvalues of A are $\lambda_1 = -1$ and $\lambda_2 = 2$ with corresponding eigenvectors

$$\mathbf{v}_1 = \begin{bmatrix} 1 \\ -1 \end{bmatrix} \quad \text{and} \quad \mathbf{v}_2 = \begin{bmatrix} 0 \\ 1 \end{bmatrix}$$

Hence, the matrix P which diagonalizes A (see Theorem 2 of Sec. 5.2) is

$$P = \begin{bmatrix} 1 & 0 \\ -1 & 1 \end{bmatrix} \quad \text{with} \quad P^{-1} = \begin{bmatrix} 1 & 0 \\ 1 & 1 \end{bmatrix}$$

The related uncoupled system is then given by

$$\mathbf{w}' = P^{-1}AP\mathbf{w}$$

$$= \begin{bmatrix} 1 & 0 \\ 1 & 1 \end{bmatrix} \begin{bmatrix} -1 & 0 \\ 3 & 2 \end{bmatrix} \begin{bmatrix} 1 & 0 \\ -1 & 1 \end{bmatrix} \mathbf{w}$$

$$= \begin{bmatrix} -1 & 0 \\ 0 & 2 \end{bmatrix} \mathbf{w}$$

whose general solution is

$$\mathbf{w}(t) = \begin{bmatrix} e^{-t} & 0 \\ 0 & e^{2t} \end{bmatrix} \mathbf{w}(0)$$

Hence, the solution to the original system is given by

$$\mathbf{y}(t) = \begin{bmatrix} 1 & 0 \\ -1 & 1 \end{bmatrix} \begin{bmatrix} e^{-t} & 0 \\ 0 & e^{2t} \end{bmatrix} \begin{bmatrix} 1 & 0 \\ 1 & 1 \end{bmatrix} \mathbf{y}(0)$$

$$= \begin{bmatrix} e^{-t} & 0 \\ -e^{-t} + e^{2t} & e^{2t} \end{bmatrix} \mathbf{y}(0)$$

The general solution can also be written in the form

$$y_1(t) = y_1(0)e^{-t} \quad \text{and} \quad y_2(t) = -y_1(0)e^{-t} + \left[y_1(0) + y_2(0) \right] e^{2t}$$

The phase portrait is shown in Fig. 2. The signs of the eigenvalues and the direction of the corresponding eigenvectors help to provide qualitative information about the trajectories in the phase portrait. In particular, notice in Fig. 2 that along the line spanned by the eigenvector $\mathbf{v}_1 = \begin{bmatrix} 1 \\ -1 \end{bmatrix}$ the flow is directed toward the origin. This is so because the sign of $\lambda_1 = -1$ is negative. On the other hand, flow along the line spanned by $\mathbf{v}_2 = \begin{bmatrix} 0 \\ 1 \end{bmatrix}$ is away from the origin, since in this case $\lambda_2 = 2$ is positive.

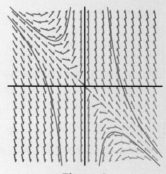

Figure 2

In Example 2 we describe the solution for a system when the eigenvalues have the same sign.

EXAMPLE 2   Find the general solution to the system of differential equations

$$\begin{cases} y_1' = y_1 + 3y_2 \\ y_2' = 2y_2 \end{cases}$$

Solution   The system of differential equations is given in matrix form by

$$\mathbf{y}' = A\mathbf{y} = \begin{bmatrix} 1 & 3 \\ 0 & 2 \end{bmatrix} \mathbf{y}$$

The eigenvalues of A are $\lambda_1 = 1$ and $\lambda_2 = 2$ with corresponding eigenvectors

$$\mathbf{v}_1 = \begin{bmatrix} 1 \\ 0 \end{bmatrix} \quad \text{and} \quad \mathbf{v}_2 = \begin{bmatrix} 3 \\ 1 \end{bmatrix}$$

The matrix that diagonalizes A is then

$$P = \begin{bmatrix} 1 & 3 \\ 0 & 1 \end{bmatrix} \quad \text{with} \quad P^{-1} = \begin{bmatrix} 1 & -3 \\ 0 & 1 \end{bmatrix}$$

The uncoupled system is given by

$$\mathbf{w}' = \begin{bmatrix} 1 & -3 \\ 0 & 1 \end{bmatrix} \begin{bmatrix} 1 & 3 \\ 0 & 2 \end{bmatrix} \begin{bmatrix} 1 & 3 \\ 0 & 1 \end{bmatrix} \mathbf{w}$$

$$= \begin{bmatrix} 1 & 0 \\ 0 & 2 \end{bmatrix} \mathbf{w}$$

with general solution

$$\mathbf{w}(t) = \begin{bmatrix} e^t & 0 \\ 0 & e^{2t} \end{bmatrix} \mathbf{w}(0)$$

Hence, the solution to the original system is given by

$$\mathbf{y}(t) = \begin{bmatrix} 1 & 3 \\ 0 & 1 \end{bmatrix} \begin{bmatrix} e^t & 0 \\ 0 & e^{2t} \end{bmatrix} \begin{bmatrix} 1 & -3 \\ 0 & 1 \end{bmatrix} \mathbf{y}(0)$$

$$= \begin{bmatrix} e^t & -3e^t + 3e^{2t} \\ 0 & e^{2t} \end{bmatrix} \mathbf{y}(0)$$

The general solution can also be written in the form

$$y_1(t) = \left[y_1(0) - 3y_2(0) \right] e^t + 3y_2(0)e^{2t} \quad \text{and} \quad y_2(t) = y_2(0)e^{2t}$$

The phase portrait is shown in Fig. 3. For this example, since λ_1 and λ_2 are both positive, the flow is oriented outward along the lines spanned by $\mathbf{v}_1$ and $\mathbf{v}_2$.

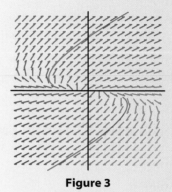

Figure 3

The process described for solving a system of two equations can be extended to higher dimensions provided that the matrix A of the system is diagonalizable.

EXAMPLE 3   Find the general solution to the system of differential equations

$$\begin{cases} y_1' & = -y_1 \\ y_2' & = 2y_1 + y_2 \\ y_3' & = 4y_1 + y_2 + 2y_3 \end{cases}$$

Solution   The system of differential equations in matrix form is

$$\mathbf{y}' = A\mathbf{y} = \begin{bmatrix} -1 & 0 & 0 \\ 2 & 1 & 0 \\ 4 & 1 & 2 \end{bmatrix} \mathbf{y}$$

Since A is triangular, the eigenvalues of A are the diagonal entries $\lambda_1 = -1, \lambda_2 = 1,$ and $\lambda_3 = 2$ with corresponding eigenvectors

$$\mathbf{v}_1 = \begin{bmatrix} -1 \\ 1 \\ 1 \end{bmatrix} \qquad \mathbf{v}_2 = \begin{bmatrix} 0 \\ 1 \\ -1 \end{bmatrix} \quad \text{and} \quad \mathbf{v}_3 = \begin{bmatrix} 0 \\ 0 \\ 2 \end{bmatrix}$$

respectively. Since A is a 3×3 matrix with three distinct eigenvalues, by Corollary 1 of Sec. 5.2, A is diagonalizable. Now, by Theorem 2 of Sec. 5.2, the diagonalizing matrix is given by

$$P = \begin{bmatrix} -1 & 0 & 0 \\ 1 & 1 & 0 \\ 1 & -1 & 2 \end{bmatrix} \quad \text{with} \quad P^{-1} = \begin{bmatrix} -1 & 0 & 0 \\ 1 & 1 & 0 \\ 1 & \frac{1}{2} & \frac{1}{2} \end{bmatrix}$$

The related uncoupled system then becomes

$$\mathbf{w'} = \begin{bmatrix} -1 & 0 & 0 \\ 1 & 1 & 0 \\ 1 & \frac{1}{2} & \frac{1}{2} \end{bmatrix} \begin{bmatrix} -1 & 0 & 0 \\ 2 & 1 & 0 \\ 4 & 1 & 2 \end{bmatrix} \begin{bmatrix} -1 & 0 & 0 \\ 1 & 1 & 0 \\ 1 & -1 & 2 \end{bmatrix} \mathbf{w}$$

$$= \begin{bmatrix} -1 & 0 & 0 \\ 0 & 1 & 0 \\ 0 & 0 & 2 \end{bmatrix} \mathbf{w}$$

with general solution

$$\mathbf{w}(t) = \begin{bmatrix} e^{-t} & 0 & 0 \\ 0 & e^t & 0 \\ 0 & 0 & e^{2t} \end{bmatrix} \mathbf{w}(0)$$

Hence, the solution to the original system is given by

$$\mathbf{y}(t) = \begin{bmatrix} -1 & 0 & 0 \\ 1 & 1 & 0 \\ 1 & -1 & 2 \end{bmatrix} \begin{bmatrix} e^{-t} & 0 & 0 \\ 0 & e^t & 0 \\ 0 & 0 & e^{2t} \end{bmatrix} \begin{bmatrix} -1 & 0 & 0 \\ 1 & 1 & 0 \\ 1 & \frac{1}{2} & \frac{1}{2} \end{bmatrix} \mathbf{y}(0)$$

$$= \begin{bmatrix} e^{-t} & 0 & 0 \\ -e^{-t}+e^t & e^t & 0 \\ -e^{-t}-e^t+2e^{2t} & -e^t+e^{2t} & e^{2t} \end{bmatrix} \mathbf{y}(0)$$

The general solution can also be written in the form

$$y_1(t) = y_1(0)e^{-t} \qquad y_2(t) = -y_1(0)e^{-t} + [y_1(0) + y_2(0)]e^t \qquad \text{and}$$
$$y_3(t) = -y_1(0)e^{-t} - [y_1(0) + y_2(0)]e^t + [2y_1(0) + y_2(0) + y_3(0)]e^{2t}.$$

Example 4 gives an illustration of how a linear system of differential equations can be used to model the concentration of salt in two interconnected tanks.

EXAMPLE 4

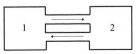

Figure 4

Suppose that two brine storage tanks are connected with two pipes used to exchange solutions between them. The first pipe allows water from tank 1 to enter tank 2 at a rate of 5 gal/min. The second pipe reverses the process allowing water to flow from tank 2 to tank 1, also at a rate of 5 gal/min. Initially, the first tank contains a well-mixed solution of 8 lb of salt in 50 gal of water, while the second tank contains 100 gal of pure water.

a. Find the linear system of differential equations to describe the amount of salt in each tank at time t.

b. Solve the system of equations by reducing it to an uncoupled system.

c. Determine the amount of salt in each tank as t increases to infinity and explain the result.

Solution   **a.** Let $y_1(t)$ and $y_2(t)$ be the amount of salt (in pounds) in each tank after t min. Thus, $y_1'(t)$ and $y_2'(t)$ are, respectively, the rates of change for the amount of salt in tank 1 and tank 2. To develop a system of equations, note that for each tank

$$\text{Rate of change of salt} = \text{rate in} - \text{rate out}$$

Since the volume of brine in each tank remains constant, for tank 1, the rate in is $\frac{5}{100} y_2(t)$ while the rate out is $\frac{5}{50} y_1(t)$. For tank 2, the rate in is $\frac{5}{50} y_1(t)$ while the rate out is $\frac{5}{100} y_2(t)$. The system of differential equations is then given by

$$\begin{cases} y_1'(t) = \frac{5}{100} y_2(t) - \frac{5}{50} y_1(t) \\ y_2'(t) = \frac{5}{50} y_1(t) - \frac{5}{100} y_2(t) \end{cases} \quad \text{that is,} \quad \begin{cases} y_1'(t) = -\frac{1}{10} y_1(t) + \frac{1}{20} y_2(t) \\ y_2'(t) = \frac{1}{10} y_1(t) - \frac{1}{20} y_2(t) \end{cases}$$

Since the initial amounts of salt in tank 1 and tank 2 are 8 and 0 lb, respectively, the initial conditions on the system are $y_1(0) = 8$ and $y_2(0) = 0$.

b. The system of equations in matrix form is given by

$$\mathbf{y}' = \begin{bmatrix} -\frac{1}{10} & \frac{1}{20} \\ \frac{1}{10} & -\frac{1}{20} \end{bmatrix} \mathbf{y} \quad \text{with} \quad \mathbf{y}(0) = \begin{bmatrix} 8 \\ 0 \end{bmatrix}$$

The eigenvalues of the matrix are $\lambda_1 = -\frac{3}{20}$ and $\lambda_2 = 0$ with corresponding eigenvectors $\begin{bmatrix} -1 \\ 1 \end{bmatrix}$ and $\begin{bmatrix} 1 \\ 2 \end{bmatrix}$. Thus, the matrix that uncouples the system is

$$P = \begin{bmatrix} -1 & 1 \\ 1 & 2 \end{bmatrix} \quad \text{with} \quad P^{-1} = \begin{bmatrix} -\frac{2}{3} & \frac{1}{3} \\ \frac{1}{3} & \frac{1}{3} \end{bmatrix}$$

The uncoupled system is then given by

$$\mathbf{w}' = \begin{bmatrix} -\frac{2}{3} & \frac{1}{3} \\ \frac{1}{3} & \frac{1}{3} \end{bmatrix} \begin{bmatrix} -\frac{1}{10} & \frac{1}{20} \\ \frac{1}{10} & -\frac{1}{20} \end{bmatrix} \begin{bmatrix} -1 & 1 \\ 1 & 2 \end{bmatrix} \mathbf{w}$$

$$= \begin{bmatrix} -\frac{3}{20} & 0 \\ 0 & 0 \end{bmatrix} \mathbf{w}$$

The solution to the uncoupled system is

$$\mathbf{w}(t) = \begin{bmatrix} e^{-\frac{3}{20}t} & 0 \\ 0 & 1 \end{bmatrix} \mathbf{w}(0)$$

Hence, the solution to the original system is given by

$$\mathbf{y}(t) = \begin{bmatrix} -1 & 1 \\ 1 & 2 \end{bmatrix} \begin{bmatrix} e^{-\frac{3}{20}t} & 0 \\ 0 & 1 \end{bmatrix} \begin{bmatrix} -\frac{2}{3} & \frac{1}{3} \\ \frac{1}{3} & \frac{1}{3} \end{bmatrix} \mathbf{y}(0)$$

$$= \frac{1}{3} \begin{bmatrix} 2e^{-\frac{3}{20}t} + 1 & -e^{-\frac{3}{20}t} + 1 \\ -2e^{-\frac{3}{20}t} + 2 & e^{-\frac{3}{20}t} + 2 \end{bmatrix} \begin{bmatrix} 8 \\ 0 \end{bmatrix}$$

$$= \frac{8}{3} \begin{bmatrix} 2e^{-\frac{3}{20}t} + 1 \\ -2e^{-\frac{3}{20}t} + 2 \end{bmatrix}$$

c. The solution to the system in equation form is given by

$$y_1(t) = \frac{8}{3}\left(2e^{-\frac{3}{20}t} + 1\right) \qquad \text{and} \qquad y_2(t) = \frac{8}{3}\left(-2e^{-\frac{3}{20}t} + 2\right)$$

To find the amount of salt in each tank as t goes to infinity, we compute the limits

$$\lim_{t\to\infty} \frac{8}{3}\left(2e^{-\frac{3}{20}t} + 1\right) = \frac{8}{3}(0+1) = \frac{8}{3}$$

and

$$\lim_{t\to\infty} \frac{8}{3}\left(-2e^{-\frac{3}{20}} + 2\right) = \frac{8}{3}(0+2) = \frac{16}{3}$$

These values make sense intuitively as we expect that the 8 lb of salt should eventually be thoroughly mixed, and divided proportionally between the two tanks in a ratio of $1:2$.

Exercise Set 5.3

In Exercises 1–6, find the general solution to the system of differential equations.

1. $\begin{cases} y_1' = -y_1 + y_2 \\ y_2' = \qquad - 2y_2 \end{cases}$

2. $\begin{cases} y_1' = -y_1 + 2y_2 \\ y_2' = \quad y_1 \end{cases}$

3. $\begin{cases} y_1' = \quad y_1 - 3y_2 \\ y_2' = -3y_1 + y_2 \end{cases}$

4. $\begin{cases} y_1' = \quad y_1 - y_2 \\ y_2' = -y_1 + y_2 \end{cases}$

5. $\begin{cases} y_1' = -4y_1 - 3y_2 - 3y_3 \\ y_2' = \quad 2y_1 + 3y_2 + 2y_3 \\ y_3' = \quad 4y_1 + 2y_2 + 3y_3 \end{cases}$

6. $\begin{cases} y_1' = -3y_1 - 4y_2 - 4y_3 \\ y_2' = \quad 7y_1 + 11y_2 + 13y_3 \\ y_3' = -5y_1 - 8y_2 - 10y_3 \end{cases}$

In Exercises 7 and 8, solve the initial-value problem.

7. $\begin{cases} y_1' = -y_1 \\ y_2' = 2y_1 + y_2 \end{cases} \qquad y_1(0) = 1 \qquad y_2(0) = -1$

8.
$$\begin{cases} y_1' = 5y_1 - 12y_2 + 20y_3 \\ y_2' = 4y_1 - 9y_2 + 16y_3 \\ y_3' = 2y_1 - 4y_2 + 7y_3 \end{cases}$$

$$y_1(0) = 2 \qquad y_2(0) = -1 \qquad y_3(0) = 0$$

9. Suppose that two brine storage tanks are connected with two pipes used to exchange solutions between them. The first pipe allows water from tank 1 to enter tank 2 at a rate of 1 gal/min. The second pipe reverses the process, allowing water to flow from tank 2 to tank 1, also at a rate of 1 gal/min. Initially, the first tank contains a well-mixed solution of 12 lb of salt in 60 gal of water, while the second tank contains 120 gal of pure water.

a. Find the linear system of differential equations to describe the amount of salt in each tank at time t.

b. Solve the system of equations by reducing it to an uncoupled system.

c. Determine the amount of salt in each tank as t increases to infinity and explain the result.

10. On a cold winter night when the outside temperature is 0 degrees Fahrenheit (0°F) at 9:00 p.m. the furnace in a two-story home fails.

Suppose the rates of heat flow between the upstairs, downstairs, and outside are as shown in the figure. Further suppose the temperature of the first floor is 70°F and that of the second floor is 60°F when the furnace fails.

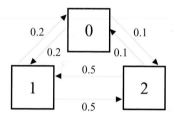

a. Use the *balance law*

$$\text{Net rate of change} = \text{rate in} - \text{rate out}$$

to set up an initial-value problem to model the heat flow.

b. Solve the initial-value problem found in part (a).

c. Compute how long it takes for each floor to reach 32°F.

5.4 ▶ Application: Markov Chains

In probability theory a *Markov process* refers to a type of mathematical model used to analyze a sequence of random events. A critical factor when computing the probabilities of a succession of events is whether the events are dependent on one another. For example, each toss of a fair coin is an *independent* event as the coin has no memory of a previous toss. A Markov process is useful in describing the tendencies of *conditionally dependent* random events, where the likelihood of each event depends on what happened previously.

As an illustration, we consider a simple weather model based on the two observations that

1. If today is sunny, then there is a 70 percent chance that tomorrow will be sunny.

2. If today is cloudy, then there is a 50 percent chance that tomorrow will be cloudy.

The conditional probabilities for the weather tomorrow, given the weather for today, are given in Table 1.

The column headings in Table 1 describe today's weather, and the row headings the weather for tomorrow. For example, the probability that a sunny day today is

Table 1

	Sunny	Cloudy
Sunny	0.7	0.5
Cloudy	0.3	0.5

followed by another sunny day tomorrow is 0.7, and the probability that a sunny day today is followed by a cloudy day tomorrow is 0.3. Notice that the column sums are both 1, since, for example, it is certain that a sunny day today is followed by either a sunny day or a cloudy day tomorrow.

In a Markov process, these observations are applied iteratively, giving us the ability to entertain questions such as, If today is sunny, what is the probability that it will be sunny one week from today?

State Vectors and Transition Matrices

To develop the Markov process required to make predictions about the weather using the observations above, we start with a vector $\mathbf{v} = \begin{bmatrix} v_1 \\ v_2 \end{bmatrix}$ whose components are the probabilities for the current weather conditions. In particular, let v_1 be the probability that today is sunny and v_2 the probability that today is cloudy. Each day the components of $\mathbf{v}$ change in accordance with the probabilities, listed in Table 1, giving us the current *state* of the weather. In a Markov process, the vector $\mathbf{v}$ is called a **state vector**, and a sequence of state vectors a **Markov chain**. Using Table 1, the state vector $\mathbf{v}' = \begin{bmatrix} v'_1 \\ v'_2 \end{bmatrix}$ for the weather tomorrow has components

$$v'_1 = 0.7v_1 + 0.5v_2 \qquad \text{and} \qquad v'_2 = 0.3v_1 + 0.5v_2$$

That is, the probability v'_1 of a sunny day tomorrow is 0.7 times the probability of a sunny day today plus 0.5 times the probability of a cloudy day today. Likewise, the probability v'_2 of a cloudy day tomorrow is 0.3 times the probability of a sunny day today plus 0.5 times the probability of a cloudy day today. For example, if today is sunny, then $v_1 = 1$ and $v_2 = 0$ so that

$$v'_1 = 0.7(1) + 0.5(0) = 0.7 \qquad \text{and} \qquad v'_2 = 0.3(1) + 0.5(0) = 0.3$$

which is in agreement with the observations above. Observe that if we let T be the matrix

$$T = \begin{bmatrix} 0.7 & 0.5 \\ 0.3 & 0.5 \end{bmatrix}$$

then the relationship above between $\mathbf{v}$ and $\mathbf{v}'$ can be written using matrix multiplication as

$$\begin{bmatrix} v'_1 \\ v'_2 \end{bmatrix} = \begin{bmatrix} 0.7 & 0.5 \\ 0.3 & 0.5 \end{bmatrix} \begin{bmatrix} v_1 \\ v_2 \end{bmatrix}$$

In a Markov chain, the matrix used to move from one state to the next is called the **transition matrix**. If n is the number of possible states, then the transition matrix T is an $n \times n$ matrix where the ij entry is the probability of moving from state j to state i. In the above example $t_{12} = 0.5$ gives the probability that a cloudy day is followed by one that is sunny. A vector with positive entries whose sum is 1 is called a **probability vector**. A matrix whose column vectors are probability vectors is called a **stochastic matrix**. The transition matrix T given above is an example of a stochastic matrix.

Returning to the weather example, to predict the weather 2 days forward, we apply the transition matrix T to the vector $\mathbf{v}'$ so that

$$\begin{bmatrix} v_1'' \\ v_2'' \end{bmatrix} = \begin{bmatrix} 0.7 & 0.5 \\ 0.3 & 0.5 \end{bmatrix} \begin{bmatrix} v_1' \\ v_2' \end{bmatrix}$$

$$= \begin{bmatrix} 0.7 & 0.5 \\ 0.3 & 0.5 \end{bmatrix}^2 \begin{bmatrix} v_1 \\ v_2 \end{bmatrix} = \begin{bmatrix} 0.64 & 0.60 \\ 0.36 & 0.40 \end{bmatrix} \begin{bmatrix} v_1 \\ v_2 \end{bmatrix}$$

Thus, for example, if today is sunny, the state vector for the weather 2 days from now is given by

$$\begin{bmatrix} v_1'' \\ v_2'' \end{bmatrix} = \begin{bmatrix} 0.64 & 0.60 \\ 0.36 & 0.40 \end{bmatrix} \begin{bmatrix} 1 \\ 0 \end{bmatrix} = \begin{bmatrix} 0.64 \\ 0.36 \end{bmatrix}$$

In general, after n days the state vector for the weather is given by

$$T^n \mathbf{v} = \begin{bmatrix} 0.7 & 0.5 \\ 0.3 & 0.5 \end{bmatrix}^n \begin{bmatrix} v_1 \\ v_2 \end{bmatrix}$$

To answer the question posed earlier about the weather one week after a sunny day, we compute

$$\begin{bmatrix} 0.7 & 0.5 \\ 0.3 & 0.5 \end{bmatrix}^7 \begin{bmatrix} 1 \\ 0 \end{bmatrix} = \begin{bmatrix} 0.625 & 0.625 \\ 0.375 & 0.375 \end{bmatrix} \begin{bmatrix} 1 \\ 0 \end{bmatrix} = \begin{bmatrix} 0.625 \\ 0.375 \end{bmatrix}$$

That is, if today is sunny, then the probability that it will be sunny one week after today is 0.625, and the probability it will be cloudy is 0.375.

Diagonalizing the Transition Matrix

As we have just seen, determining future states in a Markov process involves computing powers of the transition matrix. To facilitate the computations, we use the methods of Sec. 5.2 to diagonalize the transition matrix. To illustrate, we again consider the transition matrix

$$T = \begin{bmatrix} \frac{7}{10} & \frac{5}{10} \\ \frac{3}{10} & \frac{5}{10} \end{bmatrix}$$

of the weather example above. Observe that T has distinct eigenvalues given by

$$\lambda_1 = 1 \qquad \text{and} \qquad \lambda_2 = \frac{2}{10}$$

with corresponding eigenvectors

$$\mathbf{v}_1 = \begin{bmatrix} \frac{5}{3} \\ 1 \end{bmatrix} \qquad \text{and} \qquad \mathbf{v}_2 = \begin{bmatrix} -1 \\ 1 \end{bmatrix}$$

For reasons that will soon be clear, we scale $\mathbf{v}_1$ (by the reciprocal of the sum of its components) so that it becomes a probability vector. Observe that this new vector

$$\widehat{\mathbf{v}}_1 = \begin{bmatrix} \frac{5}{8} \\ \frac{3}{8} \end{bmatrix}$$

is also an eigenvector since it is in the eigenspace V_{λ_1}. Since the 2×2 transition matrix has two distinct eigenvalues, by Corollary 1 of Sec. 5.2, T is diagonalizable and, by Theorem 2 of Sec. 5.2, can be written as

$$T = PDP^{-1}$$

$$= \begin{bmatrix} \frac{5}{8} & -1 \\ \frac{3}{8} & 1 \end{bmatrix} \begin{bmatrix} 1 & 0 \\ 0 & \frac{2}{10} \end{bmatrix} \begin{bmatrix} 1 & 1 \\ -\frac{3}{8} & \frac{5}{8} \end{bmatrix}$$

By Exercise 27 of Sec. 5.2, the powers of T are given by

$$T^n = PD^nP^{-1} = P \begin{bmatrix} 1^n & 0 \\ 0 & \left(\frac{2}{10}\right)^n \end{bmatrix} P^{-1}$$

As mentioned above, this gives us an easier way to compute the state vector for large values of n. Another benefit from this representation is that the matrix D^n approaches

$$\begin{bmatrix} 1 & 0 \\ 0 & 0 \end{bmatrix}$$

as n gets large. This suggests that the eigenvector corresponding to $\lambda = 1$ is useful in determining the limiting proportion of sunny days to cloudy days far into the future.

Steady-State Vector

Given an initial state vector $\mathbf{v}$, of interest is the long-run behavior of this vector in a Markov chain, that is, the tendency of the vector $T^n\mathbf{v}$ for large n. If for any initial state vector $\mathbf{v}$ there is some vector $\mathbf{s}$ such that $T^n\mathbf{v}$ approaches $\mathbf{s}$, then $\mathbf{s}$ is called a **steady-state vector** for the Markov process.

In our weather model we saw that the transition matrix T has an eigenvalue $\lambda = 1$ and a corresponding probability eigenvector given by

$$\widehat{\mathbf{v}}_1 = \begin{bmatrix} \frac{5}{8} \\ \frac{3}{8} \end{bmatrix} = \begin{bmatrix} 0.625 \\ 0.375 \end{bmatrix}$$

We claim that this vector is a steady-state vector for the weather model. As verification, let $\mathbf{u}$ be an initial probability vector, say, $\mathbf{u} = \begin{bmatrix} 0.4 \\ 0.6 \end{bmatrix}$. We then compute

$$T^{10}\mathbf{u} = \begin{bmatrix} 0.6249999954 \\ 0.3750000046 \end{bmatrix} \quad \text{and} \quad T^{20}\mathbf{u} = \begin{bmatrix} 0.6250000002 \\ 0.3750000002 \end{bmatrix}$$

which suggests that $T^n\mathbf{u}$ approaches $\widehat{\mathbf{v}}_1$. That this is in fact the case is stated in Theorem 6. Before doing so, we note that a **regular** transition matrix T is a transition matrix such that for some n, all the entries of T^n are positive.

THEOREM 6

If a Markov chain has a regular stochastic transition matrix T, then there is a unique probability vector $\mathbf{s}$ with $T\mathbf{s} = \mathbf{s}$. Moreover, $\mathbf{s}$ is the steady-state vector for any initial probability vector.

EXAMPLE 1

A group insurance plan allows three different options for participants, plan A, B, or C. Suppose that the percentages of the total number of participants enrolled in each plan are 25 percent, 30 percent, and 45 percent, respectively. Also, from past experience assume that participants change plans as shown in the table.

	A	B	C
A	0.75	0.25	0.2
B	0.15	0.45	0.4
C	0.1	0.3	0.4

a. Find the percent of participants enrolled in each plan after 5 years.

b. Find the steady-state vector for the system.

Solution Let T be the matrix given by

$$T = \begin{bmatrix} 0.75 & 0.25 & 0.2 \\ 0.15 & 0.45 & 0.4 \\ 0.1 & 0.3 & 0.4 \end{bmatrix}$$

a. The number of participants enrolled in each plan after 5 years is approximated by the vector

$$T^5\mathbf{v} = \begin{bmatrix} 0.49776 & 0.46048 & 0.45608 \\ 0.28464 & 0.30432 & 0.30664 \\ 0.21760 & 0.23520 & 0.23728 \end{bmatrix} \begin{bmatrix} 0.25 \\ 0.30 \\ 0.45 \end{bmatrix} = \begin{bmatrix} 0.47 \\ 0.30 \\ 0.22 \end{bmatrix}$$

so approximately 47 percent will be enrolled in plan A, 30 percent in plan B, and 22 percent in plan C.

b. The steady-state vector for the system is the probability eigenvector corresponding to the eigenvalue $\lambda = 1$, that is,

$$\mathbf{s} = \begin{bmatrix} 0.48 \\ 0.30 \\ 0.22 \end{bmatrix}$$

Exercise Set 5.4

1. Each year it is estimated that 15 percent of the population in a city moves to the surrounding suburbs and 8 percent of people living in the suburbs move to the city. Currently, the total population of the city and surrounding suburbs is 2 million people with 1.4 million living in the city.

   a. Write the transition matrix for the Markov chain describing the migration pattern.

   b. Compute the expected population after 10 years.

   c. Find the steady-state probability vector.

2. After opening a new mass transit system, the transit authority studied the user patterns to try to determine the number of people who switched from using an automobile to the system. They estimated that each year 30 percent of those who tried the mass transit system decided to go back to driving and 20 percent switched from driving to using mass transit. Suppose that the population remains constant and that initially 35 percent of the commuters use mass transit.

   a. Write the transition matrix for the Markov chain describing the system.

   b. Compute the expected number of commuters who will be using the mass transit system in 2 years. In 5 years.

   c. Find the steady-state probability vector.

3. A plant blooms with red, pink, or white flowers. When a variety with red flowers is cross-bred with another variety, the probabilities of the new plant having red, pink, or white flowers are given in the table.

	R	P	W
R	0.5	0.4	0.1
P	0.4	0.4	0.2
W	0.1	0.2	0.7

   Suppose initially there are only plants with pink flowers which are bred with other varieties with the same likelihood. Find the probabilities of each variety occurring after three generations. After 10 generations.

4. A fleet of taxis picks up and delivers commuters between two nearby cities A and B and the surrounding suburbs S. The probability of a driver picking up a passenger in location X and delivering the passenger to location Y is given in the table. The taxi company is interested in knowing on average where the taxis are.

	A	B	S
A	0.6	0.3	0.4
B	0.1	0.4	0.3
S	0.3	0.3	0.3

   a. If a taxi is in city A, what is the probability it will be in location S after three fares?

   b. Suppose 30 percent of the taxis are in city A, 35 percent are in city B, and 35 percent are in the suburbs. Calculate the probability of a taxi being in location A, B, or S after five fares.

   c. Find the steady-state probability vector.

5. An endemic disease that has reached epidemic proportions takes the lives of one-quarter of those who are ill each month while one-half of those who are healthy become ill. Determine whether the epidemic will be eradicated. If so, estimate how long it will take.

6. A regional study of smokers revealed that from one year to the next 55 percent of smokers quit while 20 percent of nonsmokers either became new smokers or started smoking again. If 70 percent of the population are smokers, what fraction will be smoking in 5 years? In 10 years? In the long run?

7. A frog is confined to sitting on one of four lily pads. The pads are arranged in a square. Label the corners of the square A, B, C, and D clockwise. Each time the frog jumps, the probability of

jumping to an adjacent pad is 1/4, the probability of jumping to the diagonal pad is 1/6, and the probability of landing on the same pad is 1/3.

a. Write the transition matrix for the Markov process.

b. Find the probability state vector after the frog has made n jumps starting at pad A.

c. Find the steady-state vector.

8. Let the transition matrix for a Markov process be

$$T = \begin{bmatrix} 0 & 1 \\ 1 & 0 \end{bmatrix}$$

a. Find the eigenvalues of T.

b. Find T^n for $n \geq 1$. Use T^n to explain why the Markov process does have a steady-state vector.

c. Suppose T is the transition matrix describing the population distribution at any time for a constant population where residents can move between two locations. Describe the interaction in the population.

9. Show that for all p and q such that $0 < p < 1$ and $0 < q < 1$, the transition matrix

$$T = \begin{bmatrix} 1-p & q \\ p & 1-q \end{bmatrix}$$

has steady-state probability vector

$$\begin{bmatrix} \frac{q}{p+q} \\ \frac{p}{p+q} \end{bmatrix}$$

10. Suppose the transition matrix T for a Markov process is a 2×2 stochastic matrix that is also symmetric.

a. Find the eigenvalues for the matrix T.

b. Find the steady-state probability vector for the Markov process.

Review Exercises for Chapter 5

1. Let

$$A = \begin{bmatrix} a & b \\ b & a \end{bmatrix}$$

for some real numbers a and b.

a. Show that $\begin{bmatrix} 1 \\ 1 \end{bmatrix}$ is an eigenvector of A.

b. Find the eigenvalues of A.

c. Find the eigenvectors corresponding to each eigenvalue found in part (b).

d. Diagonalize the matrix A, using the eigenvectors found in part (b). That is, find the matrix P such that $P^{-1}AP$ is a diagonal matrix. Specify the diagonal matrix.

2. Let

$$A = \begin{bmatrix} 0 & 0 & 2 \\ 0 & 2 & 0 \\ 0 & 0 & -1 \end{bmatrix}$$

a. Find the eigenvalues of A.

b. From your result in part (a) can you conclude whether A is diagonalizable? Explain.

c. Find the eigenvectors corresponding to each eigenvalue.

d. Are the eigenvectors found in part (c) linearly independent? Explain.

e. From your result in part (c) can you conclude whether A is diagonalizable? Explain.

f. If your answer to part (e) is yes, find a matrix P that diagonalizes A. Specify the diagonal matrix D such that $D = P^{-1}AP$.

3. Repeat Exercise 2 with

$$A = \begin{bmatrix} 1 & 0 & 1 & 0 \\ 1 & 1 & 1 & 0 \\ 0 & 0 & 0 & 0 \\ 1 & 0 & 1 & 0 \end{bmatrix}$$

4. Let T be a linear operator on a finite dimensional vector space with a matrix representation

$$A = \begin{bmatrix} 1 & 0 & 0 \\ 6 & 3 & 2 \\ -3 & -1 & 0 \end{bmatrix}$$

a. Find the characteristic polynomial for A.

b. Find the eigenvalues of A.

c. Find the dimension of each eigenspace of A.

d. Using part (c), explain why the operator T is diagonalizable.

e. Find a matrix P and diagonal matrix D such that $D = P^{-1}AP$.

f. Find two other matrices P_1 and P_2 and corresponding diagonal matrices D_1 and D_2 such that $D_1 = P_1^{-1}AP_1$ and $D_2 = P_2^{-1}AP_2$.

5. Let

$$A = \begin{bmatrix} 0 & 1 & 0 \\ 0 & 0 & 1 \\ -k & 3 & 0 \end{bmatrix}$$

a. Show the characteristic equation of A is $\lambda^3 - 3\lambda + k = 0$.

b. Sketch the graph of $y(\lambda) = \lambda^3 - 3\lambda + k$ for $k < -2$, $k = 0$, and $k > 2$.

c. Determine the values of k for which the matrix A has three distinct real eigenvalues.

6. Suppose that $B = P^{-1}AP$ and $\mathbf{v}$ is an eigenvector of B corresponding to the eigenvalue λ. Show that $P\mathbf{v}$ is an eigenvector of A corresponding to the eigenvalue λ.

7. Suppose that A is an $n \times n$ matrix such that every row of A has the same sum λ.

a. Show that λ is an eigenvalue of A.

b. Does the same result hold if the sum of every column of A is equal to λ?

8. Let V be a vector space and $T: V \longrightarrow V$ a linear operator. A subspace W of V is **invariant under** T if for each vector $\mathbf{w}$ in W, the vector $T(\mathbf{w})$ is in W.

a. Explain why V and $\{\mathbf{0}\}$ are invariant subspaces of every linear operator on the vector space.

b. Show that if there is a one-dimensional subspace of V that is invariant under T, then T has a nonzero eigenvector.

c. Let T be a linear operator on $\mathbb{R}^2$ with matrix representation relative to the standard basis given by

$$A = \begin{bmatrix} 0 & -1 \\ 1 & 0 \end{bmatrix}$$

Show that the only invariant subspaces of T are $\mathbb{R}^2$ and $\{\mathbf{0}\}$.

9. a. Two linear operators S and T on a vector space V are said to *commute* if $S(T(\mathbf{v})) = T(S(\mathbf{v}))$ for every vector $\mathbf{v}$ in V. If S and T are commuting linear operators on V and λ_0 is an eigenvalue of T, show that V_{λ_0} is invariant under S, that is, $S(V_{\lambda_0}) \subseteq V_{\lambda_0}$.

b. Let S and T be commuting linear operators on an n-dimensional vector space V. Suppose that T has n distinct eigenvalues. Show that S and T have a common eigenvector.

c. A pair of linear operators T and S on a vector space V is called *simultaneously diagonalizable* if there is an ordered basis B for V such that $[T]_B$ and $[S]_B$ are both diagonal. Show that if S and T are simultaneously diagonalizable linear operators on an n-dimensional vector space V, then S and T commute.

d. Show directly that the matrices

$$A = \begin{bmatrix} 3 & 0 & 1 \\ 0 & 2 & 0 \\ 1 & 0 & 3 \end{bmatrix}$$

and

$$B = \begin{bmatrix} 1 & 0 & -2 \\ 0 & 1 & 0 \\ -2 & 0 & 1 \end{bmatrix}$$

are simultaneously diagonalizable.

10. The Taylor series expansion (about $x = 0$) for the natural exponential function is

$$e^x = 1 + x + \frac{1}{2!}x^2 + \frac{1}{3!}x^3 + \cdots = \sum_{k=0}^{\infty} \frac{1}{n!}x^k$$

If A is an $n \times n$ matrix, we can define the *matrix exponential* as

$$e^A = I + A + \frac{1}{2!}A^2 + \frac{1}{3!}A^3 + \cdots$$

$$= \lim_{m \to \infty} \left(I + A + \frac{1}{2!}A^2 + \frac{1}{3!}A^3 + \cdots + \frac{1}{m!}A^m \right)$$

a. Let D be the diagonal matrix

$$D = \begin{bmatrix} \lambda_1 & 0 & 0 & \cdots & 0 \\ 0 & \lambda_2 & 0 & \cdots & 0 \\ \vdots & \vdots & \ddots & & \vdots \\ 0 & \cdots & \cdots & \cdots & \lambda_n \end{bmatrix}$$

and find e^D.

b. Suppose A is diagonalizable and $D = P^{-1}AP$. Show that $e^A = Pe^D P^{-1}$.

c. Use parts (a) and (b) to compute e^A for the matrix

$$A = \begin{bmatrix} 6 & -1 \\ 3 & 2 \end{bmatrix}$$

Chapter 5: Chapter Test

In Exercises 1–40, determine whether the statement is true or false.

1. The matrix

$$P = \begin{bmatrix} 1 & 1 \\ 0 & 1 \end{bmatrix}$$

diagonalizes the matrix

$$A = \begin{bmatrix} -1 & 1 \\ 0 & -2 \end{bmatrix}$$

2. The matrix

$$A = \begin{bmatrix} -1 & 1 \\ 0 & -2 \end{bmatrix}$$

is similar to the matrix

$$D = \begin{bmatrix} -1 & 0 \\ 0 & -1 \end{bmatrix}$$

3. The matrix

$$A = \begin{bmatrix} -1 & 0 & 0 \\ 0 & 1 & 0 \\ -1 & -1 & 1 \end{bmatrix}$$

is diagonalizable.

4. The eigenvalues of

$$A = \begin{bmatrix} -1 & 0 \\ -4 & -3 \end{bmatrix}$$

are $\lambda_1 = -3$ and $\lambda_2 = -1$.

5. The characteristic polynomial of

$$A = \begin{bmatrix} -1 & -1 & -1 \\ 0 & 0 & -1 \\ 2 & -2 & -1 \end{bmatrix}$$

is $\lambda^3 + 2\lambda^2 + \lambda - 4$.

6. The eigenvectors of

$$A = \begin{bmatrix} -4 & 0 \\ 3 & -5 \end{bmatrix}$$

are $\begin{bmatrix} 0 \\ 1 \end{bmatrix}$ and $\begin{bmatrix} 1 \\ 3 \end{bmatrix}$.

7. The matrix

$$A = \begin{bmatrix} 3 & -2 \\ 2 & -1 \end{bmatrix}$$

has an eigenvalue $\lambda_1 = 1$ and V_{λ_1} has dimension 1.

8. If

$$A = \begin{bmatrix} -\frac{1}{2} & \frac{\sqrt{3}}{2} \\ \frac{\sqrt{3}}{2} & \frac{1}{2} \end{bmatrix}$$

then $AA^t = I$.

9. If A is a 2×2 matrix with $\det(A) < 0$, then A has two real eigenvalues.

10. If A is a 2×2 matrix that has two distinct eigenvalues λ_1 and λ_2, then $\text{tr}(A) = \lambda_1 + \lambda_2$.

11. If $A = \begin{bmatrix} a & b \\ b & a \end{bmatrix}$, then the eigenvalues of A are $\lambda_1 = a + b$ and $\lambda_2 = b - a$.

12. For all integers k the matrix $A = \begin{bmatrix} 1 & k \\ 1 & 1 \end{bmatrix}$ has only one eigenvalue.

13. If A is a 2×2 invertible matrix, then A and A^{-1} have the same eigenvalues.

14. If A is similar to B, then $\text{tr}(A) = \text{tr}(B)$.

15. The matrix $A = \begin{bmatrix} 1 & 1 \\ 0 & 1 \end{bmatrix}$ is diagonalizable.

16. If $A = \begin{bmatrix} a & b \\ c & d \end{bmatrix}$ and

$$a + c = b + d = \lambda$$

then λ is an eigenvalue of A.

In Exercises 17–19, let

$$A = \begin{bmatrix} 1 & 0 & 0 \\ 0 & 2 & 0 \\ 0 & 0 & -1 \end{bmatrix}$$

and

$$B = \begin{bmatrix} -1 & 0 & 0 \\ 0 & 1 & 0 \\ 0 & 0 & 2 \end{bmatrix}$$

17. The matrices A and B have the same eigenvalues.

18. The matrices A and B are similar.

19. If

$$P = \begin{bmatrix} 0 & 1 & 0 \\ 0 & 0 & 1 \\ 1 & 0 & 0 \end{bmatrix}$$

then $B = P^{-1}AP$.

20. If a 2×2 matrix has eigenvectors $\begin{bmatrix} -1 \\ 1 \end{bmatrix}$ and $\begin{bmatrix} 1 \\ -2 \end{bmatrix}$, then it has the form

$$\begin{bmatrix} 2\alpha - \beta & \alpha - \beta \\ \beta - 2\alpha & 2\beta - \alpha \end{bmatrix}$$

21. The only matrix similar to the identity matrix is the identity matrix.

22. If $\lambda = 0$ is an eigenvalue of A, then the matrix A is not invertible.

23. If A is diagonalizable, then A is similar to a unique diagonal matrix.

24. If an $n \times n$ matrix A has only m distinct eigenvalues with $m < n$, then A is not diagonalizable.

25. If an $n \times n$ matrix A has n distinct eigenvalues, then A is diagonalizable.

26. If an $n \times n$ matrix A has a set of eigenvectors that is a basis for $\mathbb{R}^n$, then A is diagonalizable.

27. If an $n \times n$ matrix A is diagonalizable, then A has n linearly independent eigenvectors.

28. If A and B are $n \times n$ matrices, then AB and BA have the same eigenvalues.

29. If D is a diagonal matrix and $A = PDP^{-1}$, then A is diagonalizable.

30. If A is invertible, then A is diagonalizable.

31. If A and B are $n \times n$ invertible matrices, then AB^{-1} and $B^{-1}A$ have the same eigenvalues.

32. A 3×3 matrix of the form

$$\begin{bmatrix} a & 1 & 0 \\ 0 & a & 1 \\ 0 & 0 & b \end{bmatrix}$$

always has fewer than three distinct eigenvalues.

33. If A and B are $n \times n$ diagonalizable matrices with the same diagonalizing matrix, then $AB = BA$.

34. If λ is an eigenvalue of the $n \times n$ matrix A, then the set of all eigenvectors corresponding to λ is a subspace of $\mathbb{R}^n$.

35. If each column sum of an $n \times n$ matrix A is a constant c, then c is an eigenvalue of A.

36. If A and B are similar, then they have the same characteristic equation.

37. If λ is an eigenvalue of A, then λ^2 is an eigenvalue of A^2.

38. If A is a 2×2 matrix with characteristic polynomial $\lambda^2 + \lambda - 6$, then the eigenvalues of A^2 are $\lambda_1 = 4$ and $\lambda_2 = 9$.

39. Define a linear operator $T: \mathcal{P}_1 \to \mathcal{P}_1$, by $T(a + bx) = a + (a + b)x$. Then the matrix representation for A relative to the standard basis is

$$A = \begin{bmatrix} 1 & 0 \\ 1 & 1 \end{bmatrix}$$

and so T is not diagonalizable.

40. If $V = \text{span}\{e^x, e^{-x}\}$ and $T: V \to V$ is defined by $T(f(x)) = f'(x)$, then T is diagonalizable.

Inner Product Spaces

*A*ccording to a growing number of scientists, a contributing factor in the rise in global temperatures is the emission of greenhouse gases such as carbon dioxide. The primary source of carbon dioxide in the atmosphere is from the burning of fossil fuels. Table 1* gives the global carbon emissions, in billions of tons, from burning fossil fuels during the period from 1950 through 2000. A *scatterplot* of the data, shown in Fig. 1, exhibits an increasing trend which can be approximated with a straight line, also shown in Fig. 1, which *best fits* the data even though

Figure 1

Table 1

Global Carbon Emissions 1950–2000			
1950	1.63	1980	5.32
1955	2.04	1985	5.43
1960	2.58	1990	6.14
1965	3.14	1995	6.40
1970	4.08	2000	6.64
1975	4.62		

*Worldwatch Institute, Vital Signs 2006–2007. *The trends that are shaping our future*, W. W. Norton and Company, New York London, 2006.

there is no one line that passes through all the points. To find this line, let (x_i, y_i), for $i = 1, 2, \ldots, 11$, denote the data points where x_i is the year, starting with $x_1 = 1950$, and y_i is the amount of greenhouse gas being released into the atmosphere. The linear equation $y = mx + b$ will best fit these data if we can find values for m and b such that the sum of the *square errors*

$$\sum_{i=1}^{11} \left[y_i - (mx_i + b)\right]^2 = [1.63 - (1950m - b)]^2 + \cdots + [6.64 - (2000m - b)]^2$$

is minimized. One method for finding the numbers m and b uses results from multi-variable calculus. An alternative approach, using linear algebra, is derived from the ideas developed in this chapter. To use this approach, we attempt to look for numbers m and b such that the linear system

$$\begin{cases} m(1950) + b & = 1.63 \\ m(1955) + b & = 2.04 \\ \quad\quad\vdots \\ m(2000) + b & = 6.64 \end{cases}$$

is satisfied. In matrix form, this system is given by $Ax = b$, where

$$A = \begin{bmatrix} 1950 & 1 \\ 1955 & 1 \\ 1960 & 1 \\ 1965 & 1 \\ 1970 & 1 \\ 1975 & 1 \\ 1980 & 1 \\ 1985 & 1 \\ 1990 & 1 \\ 1995 & 1 \\ 2000 & 1 \end{bmatrix} \quad x = \begin{bmatrix} m \\ b \end{bmatrix} \quad \text{and} \quad b = \begin{bmatrix} 1.63 \\ 2.04 \\ 2.58 \\ 3.14 \\ 4.08 \\ 4.62 \\ 5.32 \\ 5.43 \\ 6.14 \\ 6.40 \\ 6.64 \end{bmatrix}$$

Now, since there is no one line going through each of the data points, an exact solution to the previous linear system does not exist! However, as we will see, the best-fit line comes from finding a vector x so that Ax is as *close as possible* to b. In this case, the equation of the best-fit line, shown in Fig. 1, is given by

$$y = 0.107x - 207.462$$

In the last several chapters we have focused our attention on algebraic properties of abstract vector spaces derived from our knowledge of Euclidean space. For example, the observations made in Sec. 2.1 regarding the behavior of vectors in $\mathbb{R}^n$ provided us with a model for the axiomatic development of general vector spaces given in

Sec. 3.1. In this chapter we follow a similar approach as we describe the additional structures required to generalize the geometric notions of *length*, *distance*, and *angle* from $\mathbb{R}^2$ and $\mathbb{R}^3$ to abstract vector spaces. These geometric ideas are developed from a generalization of the dot product of two vectors in $\mathbb{R}^n$, called the *inner product*, which we define in Sec. 6.2. We begin with a description of the properties of the dot product on $\mathbb{R}^n$ and its relation to the geometry in Euclidean space.

6.1 ▶ The Dot Product on $\mathbb{R}^n$

In Definition 2 of Sec. 1.3, we defined the dot product of two vectors

$$\mathbf{u} = \begin{bmatrix} u_1 \\ u_2 \\ \vdots \\ u_n \end{bmatrix} \quad \text{and} \quad \mathbf{v} = \begin{bmatrix} v_1 \\ v_2 \\ \vdots \\ v_n \end{bmatrix}$$

in $\mathbb{R}^n$ as

$$\mathbf{u} \cdot \mathbf{v} = u_1 v_1 + u_2 v_2 + \cdots + u_n v_n$$

To forge a connection between the dot product and the geometry of Euclidean space, recall that in $\mathbb{R}^3$ the distance from a point (x_1, x_2, x_3) to the origin is given by

$$d = \sqrt{x_1^2 + x_2^2 + x_3^2}$$

Now let

$$\mathbf{v} = \begin{bmatrix} v_1 \\ v_2 \\ v_3 \end{bmatrix}$$

be a vector in $\mathbb{R}^3$ in standard position. Using the distance formula, the **length** (or **norm**) of $\mathbf{v}$, which we denote by $\| \mathbf{v} \|$, is defined as the distance from the terminal point of $\mathbf{v}$ to the origin and is given by

$$\| \mathbf{v} \| = \sqrt{v_1^2 + v_2^2 + v_3^2}$$

Observe that the quantity under the square root symbol can be written as the dot product of $\mathbf{v}$ with itself. So the length of $\mathbf{v}$ can be written equivalently as

$$\| \mathbf{v} \| = \sqrt{\mathbf{v} \cdot \mathbf{v}}$$

Generalizing this idea to $\mathbb{R}^n$, we have the following definition.

DEFINITION 1   **Length of a Vector in $\mathbb{R}^n$**   The **length** (or **norm**) of a vector

$$\mathbf{v} = \begin{bmatrix} v_1 \\ v_2 \\ \vdots \\ v_n \end{bmatrix}$$

in $\mathbb{R}^n$, denoted by $\| \mathbf{v} \|$, is defined as

$$\| \mathbf{v} \| = \sqrt{v_1^2 + v_2^2 + \cdots + v_n^2}$$
$$= \sqrt{\mathbf{v} \cdot \mathbf{v}}$$

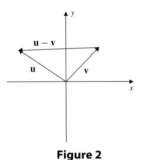

Figure 2

As an illustration, let $\mathbf{v} = \begin{bmatrix} 1 \\ 2 \\ -1 \end{bmatrix}$. Then

$$\| \mathbf{v} \| = \sqrt{\mathbf{v} \cdot \mathbf{v}} = \sqrt{1^2 + 2^2 + (-1)^2} = \sqrt{6}$$

In Sec. 2.1, it was shown that the difference $\mathbf{u} - \mathbf{v}$, of two vectors $\mathbf{u}$ and $\mathbf{v}$ in standard position, is a vector from the terminal point of $\mathbf{v}$ to the terminal point of $\mathbf{u}$, as shown in Fig. 2. This provides the rationale for the following definition.

DEFINITION 2   **Distance Between Vectors in $\mathbb{R}^n$**   Let

$$\mathbf{u} = \begin{bmatrix} u_1 \\ u_2 \\ \vdots \\ u_n \end{bmatrix} \qquad \text{and} \qquad \mathbf{v} = \begin{bmatrix} v_1 \\ v_2 \\ \vdots \\ v_n \end{bmatrix}$$

be vectors in $\mathbb{R}^n$. The **distance** between $\mathbf{u}$ and $\mathbf{v}$ is defined by

$$\| \mathbf{u} - \mathbf{v} \| = \sqrt{(\mathbf{u} - \mathbf{v}) \cdot (\mathbf{u} - \mathbf{v})}$$

Since the orientation of a vector does not affect its length, the distance from $\mathbf{u}$ to $\mathbf{v}$ is equal to the distance from $\mathbf{v}$ to $\mathbf{u}$, so that

$$\| \mathbf{u} - \mathbf{v} \| = \| \mathbf{v} - \mathbf{u} \|$$

EXAMPLE 1   Show that if $\mathbf{v}$ is a vector in $\mathbb{R}^n$ and c is a real number, then

$$\| c\mathbf{v} \| = | c | \| \mathbf{v} \|$$

Solution   Let

$$\mathbf{v} = \begin{bmatrix} v_1 \\ v_2 \\ \vdots \\ v_n \end{bmatrix}$$

Then

$$\| c\mathbf{v} \| = \sqrt{(c\mathbf{v}) \cdot (c\mathbf{v})} = \sqrt{(cv_1)^2 + (cv_2)^2 + \cdots + (cv_n)^2}$$

$$= \sqrt{c^2 v_1^2 + c^2 v_2^2 + \cdots + c^2 v_n^2} = \sqrt{c^2 (v_1^2 + v_2^2 + \cdots + v_n^2)}$$

$$= |c| \sqrt{v_1^2 + v_2^2 + \cdots + v_n^2} = |c| \, \| \mathbf{v} \|$$

The result of Example 1 provides verification of the remarks following Definition 2 of Sec. 2.1 on the effect of multiplying a vector $\mathbf{v}$ by a real number c. Indeed, as a consequence of Example 1, if $|c| > 1$, then $c\mathbf{v}$ is a stretching or *dilation* of $\mathbf{v}$; and is a shrinking or *contraction* of $\mathbf{v}$ if $|c| < 1$. If, in addition, $c < 0$, then the direction of $c\mathbf{v}$ is reversed. As an illustration, let $\mathbf{v}$ be a vector in $\mathbb{R}^n$ with $\| \mathbf{v} \| = 10$. Then $2\mathbf{v}$ has length 20. The vector $-3\mathbf{v}$ has length 30 and points in the opposite direction of $\mathbf{v}$.

If the length of a vector in $\mathbb{R}^n$ is 1, then $\mathbf{v}$ is called a **unit** vector.

PROPOSITION 1

Let $\mathbf{v}$ be a nonzero vector in $\mathbb{R}^n$. Then

$$\mathbf{u_v} = \frac{1}{\| \mathbf{v} \|} \mathbf{v}$$

is a unit vector in the direction of $\mathbf{v}$.

Proof   Using Definition 1 and the result of Example 1, we have

$$\left\| \frac{1}{\| \mathbf{v} \|} \mathbf{v} \right\| = \left| \frac{1}{\| \mathbf{v} \|} \right| \| \mathbf{v} \| = \frac{\| \mathbf{v} \|}{\| \mathbf{v} \|} = 1$$

Since $1 / \| \mathbf{v} \| > 0$, then the vector $\mathbf{u_v}$ has the same direction as $\mathbf{v}$.

EXAMPLE 2

Let

$$\mathbf{v} = \begin{bmatrix} 1 \\ 2 \\ -2 \end{bmatrix}$$

Find the unit vector $\mathbf{u_v}$ in the direction of $\mathbf{v}$.

Solution   Observe that $\| \mathbf{v} \| = \sqrt{1^2 + 2^2 + (-2)^2} = 3$. Then by Proposition 1, we have

$$\mathbf{u_v} = \frac{1}{3} \mathbf{v} = \frac{1}{3} \begin{bmatrix} 1 \\ 2 \\ -2 \end{bmatrix}$$

Theorem 1 gives useful properties of the dot product. The proofs are straightforward and are left to the reader.

THEOREM 1   Let $\mathbf{u}$, $\mathbf{v}$, and $\mathbf{w}$ be vectors in $\mathbb{R}^n$ and c a scalar.

1. $\mathbf{u} \cdot \mathbf{u} \geq 0$
2. $\mathbf{u} \cdot \mathbf{u} = 0$ if and only if $\mathbf{u} = \mathbf{0}$
3. $\mathbf{u} \cdot \mathbf{v} = \mathbf{v} \cdot \mathbf{u}$
4. $\mathbf{u} \cdot (\mathbf{v} + \mathbf{w}) = \mathbf{u} \cdot \mathbf{v} + \mathbf{u} \cdot \mathbf{w}$ and $(\mathbf{u} + \mathbf{v}) \cdot \mathbf{w} = \mathbf{u} \cdot \mathbf{w} + \mathbf{v} \cdot \mathbf{w}$
5. $(c\mathbf{u}) \cdot \mathbf{v} = c(\mathbf{u} \cdot \mathbf{v})$

EXAMPLE 3   Let $\mathbf{u}$ and $\mathbf{v}$ be vectors in $\mathbb{R}^n$. Use Theorem 1 to expand $(\mathbf{u} + \mathbf{v}) \cdot (\mathbf{u} + \mathbf{v})$.

Solution   By repeated use of part 4, we have

$$(\mathbf{u} + \mathbf{v}) \cdot (\mathbf{u} + \mathbf{v}) = (\mathbf{u} + \mathbf{v}) \cdot \mathbf{u} + (\mathbf{u} + \mathbf{v}) \cdot \mathbf{v}$$
$$= \mathbf{u} \cdot \mathbf{u} + \mathbf{v} \cdot \mathbf{u} + \mathbf{u} \cdot \mathbf{v} + \mathbf{v} \cdot \mathbf{v}$$

Now, by part 3, $\mathbf{v} \cdot \mathbf{u} = \mathbf{u} \cdot \mathbf{v}$, so that

$$(\mathbf{u} + \mathbf{v}) \cdot (\mathbf{u} + \mathbf{v}) = \mathbf{u} \cdot \mathbf{u} + 2\mathbf{u} \cdot \mathbf{v} + \mathbf{v} \cdot \mathbf{v}$$

or equivalently,

$$(\mathbf{u} + \mathbf{v}) \cdot (\mathbf{u} + \mathbf{v}) = \| \mathbf{u} \|^2 + 2\mathbf{u} \cdot \mathbf{v} + \| \mathbf{v} \|^2$$

The next result, know as the *Cauchy-Schwartz inequality*, is fundamental in developing a geometry on $\mathbb{R}^n$. In particular, this inequality makes it possible to define the angle between vectors.

THEOREM 2   **Cauchy-Schwartz Inequality**   If $\mathbf{u}$ and $\mathbf{v}$ are in vectors in $\mathbb{R}^n$, then

$$|\mathbf{u} \cdot \mathbf{v}| \leq \| \mathbf{u} \| \, \| \mathbf{v} \|$$

Proof   If $\mathbf{u} = \mathbf{0}$, then $\mathbf{u} \cdot \mathbf{v} = 0$. We also know, in this case, that $\| \mathbf{u} \| \, \| \mathbf{v} \| = 0 \| \mathbf{v} \| = 0$ so that equality holds. Now suppose that $\mathbf{u} \neq \mathbf{0}$ and k is a real number. Consider the dot product of the vector $k\mathbf{u} + \mathbf{v}$ with itself. By Theorem 1, part 1, we have

$$(k\mathbf{u} + \mathbf{v}) \cdot (k\mathbf{u} + \mathbf{v}) \geq 0$$

Now, by Theorem 1, part 4, the left-hand side can be expanded to obtain

$$k^2(\mathbf{u} \cdot \mathbf{u}) + 2k(\mathbf{u} \cdot \mathbf{v}) + \mathbf{v} \cdot \mathbf{v} \geq 0$$

Observe that the expression on the left-hand side is quadratic in the variable k with real coefficients. Letting $a = \mathbf{u} \cdot \mathbf{u}$, $b = \mathbf{u} \cdot \mathbf{v}$, and $c = \mathbf{v} \cdot \mathbf{v}$, we rewrite this inequality as

$$ak^2 + 2bk + c \geq 0$$

This inequality imposes conditions on the coefficients a, b, and c. Specifically, the equation $ak^2 + 2bk + c = 0$ must have at most one real zero. Thus, by the quadratic formula, the discriminant $(2b)^2 - 4ac \leq 0$, or equivalently,

$$(\mathbf{u} \cdot \mathbf{v})^2 \leq (\mathbf{u} \cdot \mathbf{u})(\mathbf{v} \cdot \mathbf{v})$$

After taking the square root of both sides, we obtain

$$|\mathbf{u} \cdot \mathbf{u}| \leq \|\mathbf{v}\| \|\mathbf{v}\|$$

as desired.

The Angle between Two Vectors

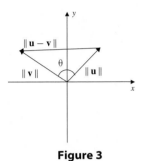

Figure 3

With the Cauchy-Schwartz inequality in hand, we are now in a position to define the angle between two vectors. To motivate this idea, let $\mathbf{u}$ and $\mathbf{v}$ be nonzero vectors in $\mathbb{R}^2$ with $\mathbf{u} - \mathbf{v}$ the vector connecting the terminal point of $\mathbf{v}$ to the terminal point of $\mathbf{u}$, as shown in Fig. 3. As these three vectors form a triangle in $\mathbb{R}^2$, we apply the law of cosines to obtain

$$\|\mathbf{u} - \mathbf{v}\|^2 = \|\mathbf{u}\|^2 + \|\mathbf{v}\|^2 - 2\|\mathbf{u}\| \|\mathbf{v}\| \cos \theta$$

Using Theorem 1, we rewrite this equation as

$$\mathbf{u} \cdot \mathbf{u} - 2\mathbf{u} \cdot \mathbf{v} + \mathbf{v} \cdot \mathbf{v} = \mathbf{u} \cdot \mathbf{u} + \mathbf{v} \cdot \mathbf{v} - 2\|\mathbf{u}\| \|\mathbf{v}\| \cos \theta$$

After simplifying and solving for $\cos \theta$, we obtain

$$\cos \theta = \frac{\mathbf{u} \cdot \mathbf{v}}{\|\mathbf{u}\| \|\mathbf{v}\|}$$

Our aim now is to extend this result and use it as the definition of the cosine of the angle between vectors in n-dimensional Euclidean space. To do so, we need $|\cos \theta| \leq 1$ for every angle θ, that is,

$$-1 \leq \frac{\mathbf{u} \cdot \mathbf{v}}{\|\mathbf{u}\| \|\mathbf{v}\|} \leq 1$$

for all vectors $\mathbf{u}$ and $\mathbf{v}$ in $\mathbb{R}^n$. But this fact follows immediately from the Cauchy-Schwartz inequality. Indeed, dividing both sides of

$$|\mathbf{u} \cdot \mathbf{v}| \leq \|\mathbf{u}\| \|\mathbf{v}\|$$

by $\|\mathbf{u}\| \|\mathbf{v}\|$, we obtain

$$\frac{|\mathbf{u} \cdot \mathbf{v}|}{\|\mathbf{u}\| \|\mathbf{v}\|} \leq 1$$

so that

$$-1 \leq \frac{\mathbf{u} \cdot \mathbf{v}}{\|\mathbf{u}\| \|\mathbf{v}\|} \leq 1$$

This permits us to make the following definition.

DEFINITION 3   **Angle Between Vectors in $\mathbb{R}^n$**   If $\mathbf{u}$ and $\mathbf{v}$ are vectors in $\mathbb{R}^n$, then the **cosine of the angle θ between the vectors** is defined by

$$\cos\theta = \frac{\mathbf{u} \cdot \mathbf{v}}{\|\mathbf{u}\|\,\|\mathbf{v}\|}$$

EXAMPLE 4   Find the angle between the two vectors

$$\mathbf{u} = \begin{bmatrix} 2 \\ -2 \\ 3 \end{bmatrix} \quad \text{and} \quad \mathbf{v} = \begin{bmatrix} -1 \\ 2 \\ 2 \end{bmatrix}$$

Solution   The lengths of the vectors are

$$\|\mathbf{u}\| = \sqrt{2^2 + (-2)^2 + 3^2} = \sqrt{17} \quad \text{and} \quad \|\mathbf{v}\| = \sqrt{(-1)^2 + 2^2 + 2^2} = 3$$

and the dot product of the vectors is

$$\mathbf{u} \cdot \mathbf{v} = 2(-1) + (-2)2 + 3(2) = 0$$

By Definition 3, the cosine of the angle between $\mathbf{u}$ and $\mathbf{v}$ is given by

$$\cos\theta = \frac{\mathbf{u} \cdot \mathbf{v}}{\|\mathbf{u}\|\,\|\mathbf{v}\|} = 0$$

Hence, $\theta = \pi/2$ and the vectors are perpendicular. Such vectors are also called *orthogonal*.

DEFINITION 4   **Orthogonal Vectors**   The vectors $\mathbf{u}$ and $\mathbf{v}$ are called **orthogonal** if the angle between them is $\pi/2$.

As a direct consequence of Definition 3, we see that if $\mathbf{u}$ and $\mathbf{v}$ are nonzero vectors in $\mathbb{R}^n$ with $\mathbf{u} \cdot \mathbf{v} = 0$, then $\cos\theta = 0$, so that $\theta = \pi/2$. On the other hand, if $\mathbf{u}$ and $\mathbf{v}$ are orthogonal, then $\cos\theta = 0$, so that

$$\frac{\mathbf{u} \cdot \mathbf{v}}{\|\mathbf{u}\|\,\|\mathbf{v}\|} = 0 \quad \text{therefore} \quad \mathbf{u} \cdot \mathbf{v} = 0$$

The zero vector is orthogonal to every vector in $\mathbb{R}^n$ since $\mathbf{0} \cdot \mathbf{v} = 0$, for every vector $\mathbf{v}$. These results are given in Proposition 2.

PROPOSITION 2

Two nonzero vectors $\mathbf{u}$ and $\mathbf{v}$ in $\mathbb{R}^n$ are orthogonal if and only if $\mathbf{u} \cdot \mathbf{v} = 0$. The zero vector is orthogonal to every vector in $\mathbb{R}^n$.

One consequence of Proposition 2 is that if $\mathbf{u}$ and $\mathbf{v}$ are orthogonal, then

$$\| \mathbf{u} + \mathbf{v} \|^2 = (\mathbf{u} + \mathbf{v}) \cdot (\mathbf{u} + \mathbf{v}) = \| \mathbf{u} \|^2 + 2\mathbf{u} \cdot \mathbf{v} + \| \mathbf{v} \|^2$$
$$= \| \mathbf{u} \|^2 + \| \mathbf{v} \|^2$$

This is a generalization of the Pythagorean theorem to $\mathbb{R}^n$.

Theorem 3 gives several useful properties of the norm in $\mathbb{R}^n$.

THEOREM 3

Properties of the Norm in $\mathbb{R}^n$ Let $\mathbf{v}$ be a vector in $\mathbb{R}^n$ and c a scalar.

1. $\| \mathbf{v} \| \geq 0$
2. $\| \mathbf{v} \| = 0$ if and only if $\mathbf{v} = \mathbf{0}$
3. $\| c\mathbf{v} \| = |c| \, \| \mathbf{v} \|$
4. (Triangle inequality) $\| \mathbf{u} + \mathbf{v} \| \leq \| \mathbf{u} \| + \| \mathbf{v} \|$

Proof   Parts 1 and 2 follow immediately from Definition 1 and Theorem 1. Part 3 is established in Example 1. To establish part 4, we have

$$\| \mathbf{u} + \mathbf{v} \|^2 = (\mathbf{u} + \mathbf{v}) \cdot (\mathbf{u} + \mathbf{v})$$
$$= (\mathbf{u} \cdot \mathbf{u}) + 2(\mathbf{u} \cdot \mathbf{v}) + (\mathbf{v} \cdot \mathbf{v})$$
$$= \| \mathbf{u} \|^2 + 2(\mathbf{u} \cdot \mathbf{v}) + \| \mathbf{v} \|^2$$
$$\leq \| \mathbf{u} \|^2 + 2|\mathbf{u} \cdot \mathbf{v}| + \| \mathbf{v} \|^2$$

Now, by the Cauchy-Schwartz inequality, $|\mathbf{u} \cdot \mathbf{v}| \leq \| \mathbf{u} \| \, \| \mathbf{v} \|$, so that

$$\| \mathbf{u} + \mathbf{v} \|^2 \leq \| \mathbf{u} \|^2 + 2 \| \mathbf{u} \| \, \| \mathbf{v} \| + \| \mathbf{v} \|^2$$
$$= (\| \mathbf{u} \| + \| \mathbf{v} \|)^2$$

After taking square roots of both sides of this equation, we obtain

$$\| \mathbf{u} + \mathbf{v} \| \leq \| \mathbf{u} \| + \| \mathbf{v} \|$$

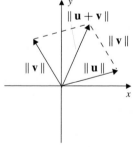

Figure 4

Geometrically, part 4 of Theorem 3 confirms our intuition that the shortest distance between two points is a straight line, as seen in Fig. 4.

PROPOSITION 3

Let $\mathbf{u}$ and $\mathbf{v}$ be vectors in $\mathbb{R}^n$. Then $\| \mathbf{u} + \mathbf{v} \| = \| \mathbf{u} \| + \| \mathbf{v} \|$ if and only if the vectors have the same direction.

Proof   First suppose that the vectors have the same direction. Then the angle between the vectors is 0, so that $\cos \theta = 1$ and $\mathbf{u} \cdot \mathbf{v} = \| \mathbf{u} \| \| \mathbf{v} \|$. Therefore,

$$\begin{aligned} \| \mathbf{u} + \mathbf{v} \|^2 &= (\mathbf{u} + \mathbf{v}) \cdot (\mathbf{u} + \mathbf{v}) \\ &= \| \mathbf{u} \|^2 + 2(\mathbf{u} \cdot \mathbf{v}) + \| \mathbf{v} \|^2 \\ &= \| \mathbf{u} \|^2 + 2 \| \mathbf{u} \| \| \mathbf{v} \| + \| \mathbf{v} \|^2 \\ &= (\| \mathbf{u} \| + \| \mathbf{v} \|)^2 \end{aligned}$$

Taking square roots of both sides of the previous equation gives $\| \mathbf{u} + \mathbf{v} \| = \| \mathbf{u} \| + \| \mathbf{v} \|$.

Conversely, suppose that $\| \mathbf{u} + \mathbf{v} \| = \| \mathbf{u} \| + \| \mathbf{v} \|$. After squaring both sides, we obtain

$$\| \mathbf{u} + \mathbf{v} \|^2 = \| \mathbf{u} \|^2 + 2 \| \mathbf{u} \| \| \mathbf{v} \| + \| \mathbf{v} \|^2$$

However, we also have

$$\| \mathbf{u} + \mathbf{v} \|^2 = (\mathbf{u} + \mathbf{v}) \cdot (\mathbf{u} + \mathbf{v}) = \| \mathbf{u} \|^2 + 2\mathbf{u} \cdot \mathbf{v} + \| \mathbf{v} \|^2$$

Equating both expressions for $\| \mathbf{u} + \mathbf{v} \|^2$ gives

$$\| \mathbf{u} \|^2 + 2 \| \mathbf{u} \| \| \mathbf{v} \| + \| \mathbf{v} \|^2 = \| \mathbf{u} \|^2 + 2\mathbf{u} \cdot \mathbf{v} + \| \mathbf{v} \|^2$$

Simplifying the last equation, we obtain $\mathbf{u} \cdot \mathbf{v} = \| \mathbf{u} \| \| \mathbf{v} \|$ and hence

$$\frac{\mathbf{u} \cdot \mathbf{v}}{\| \mathbf{u} \| \| \mathbf{v} \|} = 1$$

Therefore, $\cos \theta = 1$, so that $\theta = 0$ and the vectors have the same direction.

Fact Summary

All vectors are in $\mathbb{R}^n$.

1. The length of a vector and the distance between two vectors are natural extensions of the same geometric notions in $\mathbb{R}^2$ and $\mathbb{R}^3$.
2. The dot product of a vector with itself gives the square of its length and is 0 only when the vector is the zero vector. The dot product of two vectors is commutative and distributes through vector addition.
3. By using the Cauchy-Schwartz inequality $|\mathbf{u} \cdot \mathbf{v}| \leq \| \mathbf{u} \| \| \mathbf{v} \|$, the angle between vectors is defined by

$$\cos \theta = \frac{\mathbf{u} \cdot \mathbf{v}}{\| \mathbf{u} \| \| \mathbf{v} \|}$$

4. Two vectors are orthogonal if and only if the dot product of the vectors is 0.

5. The norm of a vector is nonnegative, is 0 only when the vector is the zero vector, and satisfies

$$\| c\mathbf{u} \| = |c| \| \mathbf{u} \| \qquad \text{and} \qquad \| \mathbf{u} + \mathbf{v} \| \leq \| \mathbf{u} \| + \| \mathbf{v} \|$$

Equality holds in the last inequality only when the vectors are in the same direction.

6. If $\mathbf{u}$ and $\mathbf{v}$ are orthogonal vectors, then the Pythagorean theorem

$$\| \mathbf{u} + \mathbf{v} \|^2 = \| \mathbf{u} \|^2 + \| \mathbf{v} \|^2$$

holds.

Exercise Set 6.1

In Exercises 1–4, let

$$\mathbf{u} = \begin{bmatrix} 0 \\ 1 \\ 3 \end{bmatrix} \qquad \mathbf{v} = \begin{bmatrix} 1 \\ -1 \\ 2 \end{bmatrix}$$

$$\mathbf{w} = \begin{bmatrix} 1 \\ 1 \\ -3 \end{bmatrix}$$

Compute the quantity.

1. $\mathbf{u} \cdot \mathbf{v}$

2. $\dfrac{\mathbf{u} \cdot \mathbf{v}}{\mathbf{v} \cdot \mathbf{v}}$

3. $\mathbf{u} \cdot (\mathbf{v} + 2\mathbf{w})$

4. $\dfrac{\mathbf{u} \cdot \mathbf{w}}{\mathbf{w} \cdot \mathbf{w}} \mathbf{w}$

In Exercises 5–10, let

$$\mathbf{u} = \begin{bmatrix} 1 \\ 5 \end{bmatrix} \qquad \mathbf{v} = \begin{bmatrix} 2 \\ 1 \end{bmatrix}$$

5. Find $\| \mathbf{u} \|$.

6. Find the distance between $\mathbf{u}$ and $\mathbf{v}$.

7. Find a unit vector in the direction of $\mathbf{u}$.

8. Find the cosine of the angle between the two vectors. Are the vectors orthogonal? Explain.

9. Find a vector in the direction of $\mathbf{v}$ with length 10.

10. Find a vector $\mathbf{w}$ that is orthogonal to both $\mathbf{u}$ and $\mathbf{v}$.

In Exercises 11–16, let

$$\mathbf{u} = \begin{bmatrix} -3 \\ -2 \\ 3 \end{bmatrix} \qquad \mathbf{v} = \begin{bmatrix} -1 \\ -1 \\ -3 \end{bmatrix}$$

11. Find $\| \mathbf{u} \|$.

12. Find the distance between $\mathbf{u}$ and $\mathbf{v}$.

13. Find a unit vector in the direction of $\mathbf{u}$.

14. Find the cosine of the angle between the two vectors. Are the vectors orthogonal? Explain.

15. Find a vector in the opposite direction of $\mathbf{v}$ with length 3.

16. Find a vector $\mathbf{w}$ that is orthogonal to both $\mathbf{u}$ and $\mathbf{v}$.

17. Find a scalar c, so that $\begin{bmatrix} c \\ 3 \end{bmatrix}$ is orthogonal to $\begin{bmatrix} -1 \\ 2 \end{bmatrix}$.

18. Find a scalar c, so that $\begin{bmatrix} -1 \\ c \\ 2 \end{bmatrix}$ is orthogonal to $\begin{bmatrix} 0 \\ 2 \\ -1 \end{bmatrix}$.

In Exercises 19–22, let

$$\mathbf{v}_1 = \begin{bmatrix} 1 \\ 2 \\ -1 \end{bmatrix} \qquad \mathbf{v}_2 = \begin{bmatrix} 6 \\ -2 \\ 2 \end{bmatrix}$$

$$\mathbf{v}_3 = \begin{bmatrix} -1 \\ -2 \\ 1 \end{bmatrix} \qquad \mathbf{v}_4 = \begin{bmatrix} -1/\sqrt{3} \\ 1/\sqrt{3} \\ 1/\sqrt{3} \end{bmatrix}$$

$$\mathbf{v}_5 = \begin{bmatrix} 3 \\ -1 \\ 1 \end{bmatrix}$$

19. Determine which of the vectors are orthogonal.

20. Determine which of the vectors are in the same direction.

21. Determine which of the vectors are in the opposite direction.

22. Determine which of the vectors are unit vectors.

In Exercises 23–28, find the projection of $\mathbf{u}$ onto $\mathbf{v}$ given by

$$\mathbf{w} = \frac{\mathbf{u} \cdot \mathbf{v}}{\mathbf{v} \cdot \mathbf{v}} \mathbf{v}$$

The vector $\mathbf{w}$ is called the *orthogonal projection* of $\mathbf{u}$ onto $\mathbf{v}$. Sketch the three vectors $\mathbf{u}$, $\mathbf{v}$, and $\mathbf{w}$.

23. $\mathbf{u} = \begin{bmatrix} 2 \\ 3 \end{bmatrix}$ $\mathbf{v} = \begin{bmatrix} 4 \\ 0 \end{bmatrix}$

24. $\mathbf{u} = \begin{bmatrix} -2 \\ 3 \end{bmatrix}$ $\mathbf{v} = \begin{bmatrix} 4 \\ 0 \end{bmatrix}$

25. $\mathbf{u} = \begin{bmatrix} 4 \\ 3 \end{bmatrix}$ $\mathbf{v} = \begin{bmatrix} 3 \\ 1 \end{bmatrix}$

26. $\mathbf{u} = \begin{bmatrix} 5 \\ 2 \\ 1 \end{bmatrix}$ $\mathbf{v} = \begin{bmatrix} 1 \\ 0 \\ 0 \end{bmatrix}$

27. $\mathbf{u} = \begin{bmatrix} 1 \\ 0 \\ 0 \end{bmatrix}$ $\mathbf{v} = \begin{bmatrix} 5 \\ 2 \\ 1 \end{bmatrix}$

28. $\mathbf{u} = \begin{bmatrix} 2 \\ 3 \\ -1 \end{bmatrix}$ $\mathbf{v} = \begin{bmatrix} 0 \\ 2 \\ 3 \end{bmatrix}$

29. Let $S = \{\mathbf{u}_1, \mathbf{u}_2, \ldots, \mathbf{u}_n\}$ and suppose $\mathbf{v} \cdot \mathbf{u}_i = 0$ for each $i = 1, \ldots, n$. Show that $\mathbf{v}$ is orthogonal to every vector in $\mathbf{span}(S)$.

30. Let $\mathbf{v}$ be a fixed vector in $\mathbb{R}^n$ and define $S = \{\mathbf{u} \mid \mathbf{u} \cdot \mathbf{v} = 0\}$. Show that S is a subspace of $\mathbb{R}^n$.

31. Let $S = \{\mathbf{v}_1, \mathbf{v}_2, \ldots, \mathbf{v}_n\}$ be a set of nonzero vectors which are pairwise orthogonal. That is, if $i \neq j$, then $\mathbf{v}_i \cdot \mathbf{v}_j = 0$. Show that S is linearly independent.

32. Let A be an $n \times n$ invertible matrix. Show that if $i \neq j$, then row vector i of A and column vector j of A^{-1} are orthogonal.

33. Show that for all vectors $\mathbf{u}$ and $\mathbf{v}$ in $\mathbb{R}^n$,

$$\| \mathbf{u} + \mathbf{v} \|^2 + \| \mathbf{u} - \mathbf{v} \|^2 = 2 \| \mathbf{u} \|^2 + 2 \| \mathbf{v} \|^2$$

34. a. Find a vector that is orthogonal to every vector in the plane P: $x + 2y - z = 0$.

   b. Find a matrix A such that the null space $N(A)$ is the plane $x + 2y - z = 0$.

35. Suppose that the column vectors of an $n \times n$ matrix A are pairwise orthogonal. Find $A^t A$.

36. Let A be an $n \times n$ matrix and $\mathbf{u}$ and $\mathbf{v}$ vectors in $\mathbb{R}^n$. Show that

$$\mathbf{u} \cdot (A\mathbf{v}) = (A^t\mathbf{u}) \cdot \mathbf{v}$$

37. Let A be an $n \times n$ matrix. Show that A is symmetric if and only if

$$(A\mathbf{u}) \cdot \mathbf{v} = \mathbf{u} \cdot (A\mathbf{v})$$

for all $\mathbf{u}$ and $\mathbf{v}$ in $\mathbb{R}^n$. *Hint*: See Exercise 36.

6.2 ▶ Inner Product Spaces

In Sec. 6.1 we introduced the concepts of the length of a vector and the angle between vectors in Euclidean space. Both of these notions are defined in terms of the dot product and provide a geometry on $\mathbb{R}^n$. Notice that the dot product on $\mathbb{R}^n$ defines a function from $\mathbb{R}^n \times \mathbb{R}^n$ into $\mathbb{R}$. That is, the dot product operates on two vectors in $\mathbb{R}^n$, producing a real number. To extend these ideas to an abstract vector space V, we require a function from $V \times V$ into $\mathbb{R}$ that generalizes the properties of the dot product given in Theorem 1 of Sec. 6.1.

DEFINITION 1

Inner Product   Let V be a vector space over $\mathbb{R}$. An **inner product** on V is a function that associates with each pair of vectors $\mathbf{u}$ and $\mathbf{v}$ in V a real number, denoted by $\langle \mathbf{u}, \mathbf{v} \rangle$, that satisfies the following axioms:

1. $\langle \mathbf{u}, \mathbf{u} \rangle \geq 0$ and $\langle \mathbf{u}, \mathbf{u} \rangle = 0$ if and only if $\mathbf{u} = \mathbf{0}$ (positive definite)
2. $\langle \mathbf{u}, \mathbf{v} \rangle = \langle \mathbf{v}, \mathbf{u} \rangle$ (symmetry)
3. $\langle \mathbf{u} + \mathbf{v}, \mathbf{w} \rangle = \langle \mathbf{u}, \mathbf{w} \rangle + \langle \mathbf{v}, \mathbf{w} \rangle$
4. $\langle c\mathbf{u}, \mathbf{v} \rangle = c \langle \mathbf{u}, \mathbf{v} \rangle$

The last two properties make the inner product *linear in the first variable*. Using the symmetry axiom, it can also be shown that the inner product is *linear in the second variable*, that is,

$3'$. $\langle \mathbf{u}, \mathbf{v} + \mathbf{w} \rangle = \langle \mathbf{u}, \mathbf{v} \rangle + \langle \mathbf{u}, \mathbf{w} \rangle$
$4'$. $\langle \mathbf{u}, c\mathbf{v} \rangle = c \langle \mathbf{u}, \mathbf{v} \rangle$

With these additional properties, the inner product is said to be *bilinear*.

A vector space V with an inner product is called an **inner product space**.

By Theorem 1 of Sec. 6.1, the dot product is an inner product on Euclidean n-space. Thus, $\mathbb{R}^n$ with the dot product is an inner product space.

EXAMPLE 1   Let $\mathbf{v} = \begin{bmatrix} 1 \\ 3 \end{bmatrix}$. Find all vectors $\mathbf{u}$ in $\mathbb{R}^2$ such that $\langle \mathbf{u}, \mathbf{v} \rangle = 0$, where the inner product is the dot product.

Solution   If $\mathbf{u} = \begin{bmatrix} x \\ y \end{bmatrix}$, then

$$\langle \mathbf{u}, \mathbf{v} \rangle = \mathbf{u} \cdot \mathbf{v} = x + 3y$$

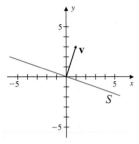

Figure 1

so that $\langle \mathbf{u}, \mathbf{v} \rangle = 0$ if and only if $y = -\frac{1}{3}x$. Therefore, the set of all vectors such that $\langle \mathbf{u}, \mathbf{v} \rangle = 0$ is given by $S = \mathbf{span}\left\{ \begin{bmatrix} 1 \\ -\frac{1}{3} \end{bmatrix} \right\}$. The vector $\mathbf{v}$ and the set S are shown in Fig. 1. Notice that each vector in S is perpendicular to $\mathbf{v}$.

For another example, consider the vector space of polynomials $\mathcal{P}_2$. To define an inner product on $\mathcal{P}_2$, let $p(x) = a_0 + a_1x + a_2x^2$ and $q(x) = b_0 + b_1x + b_2x^2$. Now let $\langle \cdot, \cdot \rangle \colon \mathcal{P}_2 \times \mathcal{P}_2 \to \mathbb{R}$ be the function defined by

$$\langle p, q \rangle = a_0b_0 + a_1b_1 + a_2b_2$$

Notice that this function is similar to the dot product on $\mathbb{R}^3$. The proof that $\mathcal{P}_2$ is an inner product space follows along the same lines as the proof of Theorem 1 of Sec. 6.1.

Another way to define an inner product on $\mathcal{P}_2$ is to use the definite integral. Specifically, let $p(x)$ and $q(x)$ be polynomials in $\mathcal{P}_2$, and let $\langle \cdot, \cdot \rangle$ be the function defined by

$$\langle p, q \rangle = \int_0^1 p(x)q(x)\, dx$$

This function is also an inner product on $\mathcal{P}_2$. The justification, in this case, is based on the fundamental properties of the Riemann integral which can be found in any text on real analysis.

EXAMPLE 2   Let $V = \mathcal{P}_2$ with inner product defined by

$$\langle p, q \rangle = \int_0^1 p(x)q(x)\, dx$$

a. Let $p(x) = 1 - x^2$ and $q(x) = 1 - x + 2x^2$. Find $\langle p, q \rangle$.
b. Let $p(x) = 1 - x^2$. Verify that $\langle p, p \rangle > 0$.

Solution   **a.** Using the definition given for the inner product, we have

$$\langle p, q \rangle = \int_0^1 (1 - x^2)(1 - x + 2x^2)\, dx$$

$$= \int_0^1 (1 - x + x^2 + x^3 - 2x^4)\, dx$$

$$= \left(x - \frac{1}{2}x^2 + \frac{1}{3}x^3 + \frac{1}{4}x^4 - \frac{2}{5}x^5 \right)\Big|_0^1$$

$$= \frac{41}{60}$$

b. The inner product of p with itself is given by

$$\langle p, p \rangle = \int_0^1 (1 - x^2)(1 - x^2)\, dx$$

$$= \int_0^1 (1 - 2x^2 + x^4)\, dx$$

$$= \left(x - \frac{2}{3}x^3 + \frac{1}{5}x^5 \right) \Big|_0^1$$

$$= \frac{8}{15} > 0$$

Example 3 gives an illustration of an inner product on $\mathbb{R}^n$ that is not the dot product.

EXAMPLE 3   Let $V = \mathbb{R}^2$ and

$$\mathbf{u} = \begin{bmatrix} u_1 \\ u_2 \end{bmatrix} \qquad \text{and} \qquad \mathbf{v} = \begin{bmatrix} v_1 \\ v_2 \end{bmatrix}$$

be vectors in V. Let k be a fixed positive real number, and define the function $\langle \cdot, \cdot \rangle : \mathbb{R}^2 \times \mathbb{R}^2 \to \mathbb{R}$ by

$$\langle \mathbf{u}, \mathbf{v} \rangle = u_1 v_1 + k u_2 v_2$$

Show that V is an inner product space.

Solution   First we show that $\langle \cdot, \cdot \rangle$ is nonnegative. From the definition above, we have

$$\langle \mathbf{u}, \mathbf{u} \rangle = u_1^2 + k u_2^2$$

Since $k > 0$, then $u_1^2 + k u_2^2 \geq 0$ for every vector $\mathbf{u}$. In addition,

$$u_1^2 + k u_2^2 = 0 \qquad \text{if and only if} \qquad u_1 = u_2 = 0$$

or equivalently if $\mathbf{u} = \mathbf{0}$. The property of symmetry also holds since

$$\langle \mathbf{u}, \mathbf{v} \rangle = u_1 v_1 + k u_2 v_2 = v_1 u_1 + k v_2 u_2 = \langle \mathbf{v}, \mathbf{u} \rangle$$

Next, let $\mathbf{w} = \begin{bmatrix} w_1 \\ w_2 \end{bmatrix}$ be another vector in $\mathbb{R}^2$. Then

$$\langle \mathbf{u} + \mathbf{v}, \mathbf{w} \rangle = (u_1 + v_1)w_1 + k(u_2 + v_2)w_2$$

$$= (u_1 w_1 + k u_2 w_2) + (v_1 w_1 + k v_2 w_2)$$

$$= \langle \mathbf{u}, \mathbf{w} \rangle + \langle \mathbf{v}, \mathbf{w} \rangle$$

Finally, if c is a scalar, then

$$\langle c\mathbf{u}, \mathbf{v} \rangle = (cu_1)v_1 + k(cu_2)v_2 = c(u_1v_1 + ku_2v_2) = c\,\langle \mathbf{u}, \mathbf{v} \rangle$$

Therefore, $\mathbb{R}^2$ with this definition for $\langle \cdot, \cdot \rangle$ is an inner product space.

Notice that in Example 3 the requirement that $k > 0$ is necessary. For example, if $k = -1$ and $\mathbf{u} = \begin{bmatrix} 1 \\ 2 \end{bmatrix}$, then the inner product of $\mathbf{u}$ with itself is given by $\langle \mathbf{u}, \mathbf{u} \rangle = (1)^2 + (-1)(2)^2 = -3$, which violates the first axiom of Definition 1.

Again using $\mathbb{R}^n$ as our model, we now define the **length** (or **norm**) of a vector $\mathbf{v}$ in an inner product space V as

$$\| \mathbf{v} \| = \sqrt{\langle \mathbf{v}, \mathbf{v} \rangle}$$

The **distance between** two vectors $\mathbf{u}$ and $\mathbf{v}$ in V is then defined by

$$\| \mathbf{u} - \mathbf{v} \| = \sqrt{\langle \mathbf{u} - \mathbf{v}, \mathbf{u} - \mathbf{v} \rangle}$$

The norm in an inner product space satisfies the same properties as the norm in $\mathbb{R}^n$. The results are summarized in Theorem 4.

THEOREM 4

Properties of the Norm in an Inner Product Space   Let $\mathbf{u}$ and $\mathbf{v}$ be vectors in an inner product space V and c a scalar.

1. $\| \mathbf{v} \| \geq 0$
2. $\| \mathbf{v} \| = 0$ if and only if $\mathbf{v} = \mathbf{0}$
3. $\| c\mathbf{v} \| = |c| \| \mathbf{v} \|$
4. $| \langle \mathbf{u}, \mathbf{v} \rangle | \leq \| \mathbf{u} \| \| \mathbf{v} \|$ (Cauchy-Schwartz inequality)
5. $\| \mathbf{u} + \mathbf{v} \| \leq \| \mathbf{u} \| + \| \mathbf{v} \|$ (Triangle inequality)

EXAMPLE 4

Let $V = \mathbb{R}^2$ with inner product defined by

$$\langle \mathbf{u}, \mathbf{v} \rangle = u_1v_1 + 3u_2v_2$$

Let

$$\mathbf{u} = \begin{bmatrix} 2 \\ -2 \end{bmatrix} \quad \text{and} \quad \mathbf{v} = \begin{bmatrix} 1 \\ 4 \end{bmatrix}$$

a. Verify that the Cauchy-Schwartz inequality is upheld.
b. Verify that the Triangle Inequality is upheld.

Solution **a.** Using the given definition for the inner product, we have

$$|\langle \mathbf{u}, \mathbf{v} \rangle| = |(2)(1) + 3(-2)(4)| = |-22| = 22$$

The norms of $\mathbf{u}$ and $\mathbf{v}$ are given, respectively, by

$$\|\mathbf{u}\| = \sqrt{\langle \mathbf{u}, \mathbf{u} \rangle} = \sqrt{(2)^2 + 3(-2)^2} = \sqrt{16} = 4$$

and

$$\|\mathbf{v}\| = \sqrt{\langle \mathbf{v}, \mathbf{v} \rangle} = \sqrt{(1)^2 + 3(4)^2} = \sqrt{49} = 7$$

Since

$$22 = |\langle \mathbf{u}, \mathbf{v} \rangle| < \|\mathbf{u}\| \, \|\mathbf{v}\| = 28$$

the Cauchy-Schwartz inequality is satisfied for the vectors $\mathbf{u}$ and $\mathbf{v}$.

b. To verify the triangle inequality, observe that

$$\mathbf{u} + \mathbf{v} = \begin{bmatrix} 2 \\ -2 \end{bmatrix} + \begin{bmatrix} 1 \\ 4 \end{bmatrix} = \begin{bmatrix} 3 \\ 2 \end{bmatrix}$$

so that

$$\|\mathbf{u} + \mathbf{v}\| = \sqrt{(3)^2 + 3(2)^2} = \sqrt{21}$$

Since

$$\sqrt{21} = \|\mathbf{u} + \mathbf{v}\| < \|\mathbf{u}\| + \|\mathbf{v}\| = 4 + 7 = 11$$

the triangle inequality holds for $\mathbf{u}$ and $\mathbf{v}$.

Orthogonal Sets

Taking the same approach as in Sec. 6.1, we define the **cosine of the angle between the vectors** $\mathbf{u}$ and $\mathbf{v}$ in an inner product space V by

$$\cos \theta = \frac{\langle \mathbf{u}, \mathbf{v} \rangle}{\|\mathbf{u}\| \, \|\mathbf{v}\|}$$

As before, the vectors $\mathbf{u}$ and $\mathbf{v}$ in V are **orthogonal** provided that $\langle \mathbf{u}, \mathbf{v} \rangle = 0$.

EXAMPLE 5 Let $V = \mathcal{P}_2$ with inner product defined by

$$\langle p, q \rangle = \int_{-1}^{1} p(x)q(x) \, dx$$

a. Show that the vectors in

$$S = \left\{ 1, x, \tfrac{1}{2}(3x^2 - 1) \right\}$$

are mutually orthogonal.

b. Find the length of each vector in S.

Solution   **a.** The inner product of each pair of vectors in S is

$$\langle 1, x \rangle = \int_{-1}^{1} x \, dx = \frac{1}{2}x^2 \Big|_{-1}^{1} = 0$$

$$\left\langle 1, \frac{1}{2}(3x^2 - 1) \right\rangle = \int_{-1}^{1} \frac{1}{2}(3x^2 - 1) \, dx = \frac{1}{2}(x^3 - x) \Big|_{-1}^{1} = 0$$

$$\left\langle x, \frac{1}{2}(3x^2 - 1) \right\rangle = \int_{-1}^{1} \frac{1}{2}(3x^3 - x) \, dx = \frac{1}{2}\left(\frac{3}{4}x^4 - \frac{1}{2}x^2 \right) \Big|_{-1}^{1} = 0$$

Since each pair of distinct vectors is orthogonal, the vectors in S are mutually orthogonal.

b. For the lengths of the vectors in S, we have

$$\| 1 \| = \sqrt{\langle 1, 1 \rangle} = \sqrt{\int_{-1}^{1} dx} = \sqrt{2}$$

$$\| x \| = \sqrt{\langle x, x \rangle} = \sqrt{\int_{-1}^{1} x^2 \, dx} = \sqrt{\frac{2}{3}}$$

$$\left\| \frac{1}{2}(3x^2 - 1) \right\| = \sqrt{\left\langle \frac{1}{2}(3x^2 - 1), \frac{1}{2}(3x^2 - 1) \right\rangle} = \sqrt{\frac{1}{4}\int_{-1}^{1}(3x^2 - 1)^2 dx} = \sqrt{\frac{2}{5}}$$

DEFINITION 2   **Orthogonal Set**   A set of vectors $\{v_1, v_2, \ldots, v_n\}$ in an inner product space is called **orthogonal** if the vectors are mutually orthogonal; that is, if $i \neq j$, then $\langle v_i, v_j \rangle = 0$. If in addition, $\| v_i \| = 1$, for all $i = 1, \ldots n$, then the set of vectors is called **orthonormal**.

Observe that the vectors of Example 5 form an orthogonal set. They do not, however, form an orthonormal set. Proposition 4 shows that the zero vector is orthogonal to every vector in an inner product space.

PROPOSITION 4   Let V be an inner product space. Then $\langle v, 0 \rangle = 0$ for every vector v in V.

Proof   Let v be a vector in V. Then

$$\langle v, 0 \rangle = \langle v, 0 + 0 \rangle = \langle v, 0 \rangle + \langle v, 0 \rangle$$

After subtracting $\langle v, 0 \rangle$ from both sides of the previous equation, we have $\langle v, 0 \rangle = 0$ as desired.

A useful property of orthogonal sets of nonzero vectors is that they are linearly independent. For example, the set of coordinate vectors $\{e_1, e_2, e_3\}$ in $\mathbb{R}^3$ is orthogonal and linearly independent. Theorem 5 relates the notions of orthogonality and linear independence in an inner product space.

THEOREM 5   If $S = \{\mathbf{v}_1, \mathbf{v}_2, \ldots, \mathbf{v}_n\}$ is an orthogonal set of nonzero vectors in an inner product space V, then S is linearly independent.

Proof   Since the set S is an orthogonal set of nonzero vectors,

$$\langle \mathbf{v}_i, \mathbf{v}_j \rangle = 0 \quad \text{for } i \neq j \quad \text{and} \quad \langle \mathbf{v}_i, \mathbf{v}_i \rangle = \|\mathbf{v}_i\|^2 \neq 0 \quad \text{for all } i$$

Now suppose that

$$c_1\mathbf{v}_1 + c_2\mathbf{v}_2 + \cdots + c_n\mathbf{v}_n = \mathbf{0}$$

The vectors are linearly independent if and only if the only solution to the previous equation is the trivial solution $c_1 = c_2 = \cdots = c_n = 0$. Now let $\mathbf{v}_j$ be an element of S. Take the inner product on both sides of the previous equation with $\mathbf{v}_j$ so that

$$\langle \mathbf{v}_j, (c_1\mathbf{v}_1 + c_2\mathbf{v}_2 + \cdots + c_{j-1}\mathbf{v}_{j-1} + c_j\mathbf{v}_j + c_{j+1}\mathbf{v}_{j+1} + \cdots + c_n\mathbf{v}_n) \rangle = \langle \mathbf{v}_j, \mathbf{0} \rangle$$

By the linearity of the inner product and the fact that S is orthogonal, this equation reduces to

$$c_j \langle \mathbf{v}_j, \mathbf{v}_j \rangle = \langle \mathbf{v}_j, \mathbf{0} \rangle$$

Now, by Proposition 4 and the fact that $\|\mathbf{v}_j\| \neq 0$, we have

$$c_j \|\mathbf{v}_j\|^2 = 0 \quad \text{so that} \quad c_j = 0$$

Since this holds for each $j = 1, \ldots, n$, then $c_1 = c_2 = \cdots = c_n = 0$ and therefore S is linearly independent.

COROLLARY 1   If V is an inner product space of dimension n, then any orthogonal set of n nonzero vectors is a basis for V.

The proof of this corollary is a direct result of Theorem 12 of Sec. 3.3. Theorem 6 provides us with an easy way to find the coordinates of a vector relative to an orthonormal basis. This property underscores the usefulness and desirability of orthonormal bases.

THEOREM 6   If $B = \{\mathbf{v}_1, \mathbf{v}_2, \ldots, \mathbf{v}_n\}$ is an ordered orthonormal basis for an inner product space V and $\mathbf{v} = c_1\mathbf{v}_1 + c_2\mathbf{v}_2 + \cdots + c_n\mathbf{v}_n$, then the coordinates of $\mathbf{v}$ relative to B are given by $c_i = \langle \mathbf{v}_i, \mathbf{v} \rangle$ for each $i = 1, 2, \ldots, n$.

Proof   Let $\mathbf{v}_i$ be a vector in B. Taking the inner product on both sides of

$$\mathbf{v} = c_1\mathbf{v}_1 + c_2\mathbf{v}_2 + \cdots + c_{i-1}\mathbf{v}_{i-1} + c_i\mathbf{v}_i + c_{i+1}\mathbf{v}_{i+1} + \cdots + c_n\mathbf{v}_n$$

with $\mathbf{v}_i$ on the right gives

$$\langle \mathbf{v}, \mathbf{v}_i \rangle = \langle (c_1\mathbf{v}_1 + c_2\mathbf{v}_2 + \cdots + c_{i-1}\mathbf{v}_{i-1} + c_i\mathbf{v}_i + c_{i+1}\mathbf{v}_{i+1} + \cdots + c_n\mathbf{v}_n), \mathbf{v}_i \rangle$$

$$= c_1 \langle \mathbf{v}_1, \mathbf{v}_i \rangle + \cdots + c_i \langle \mathbf{v}_i, \mathbf{v}_i \rangle + \cdots + c_n \langle \mathbf{v}_n, \mathbf{v}_i \rangle$$

Since B is an orthonormal set, this reduces to

$$\langle \mathbf{v}, \mathbf{v}_i \rangle = c_i \langle \mathbf{v}_i, \mathbf{v}_i \rangle = c_i$$

As this argument can be carried out for any vector in B, then $c_i = \langle \mathbf{v}, \mathbf{v}_i \rangle$ for all $i = 1, 2, \ldots, n$.

In Theorem 6, if the ordered basis B is orthogonal and $\mathbf{v}$ is any vector in V, then the coordinates relative to B are given by

$$c_i = \frac{\langle \mathbf{v}, \mathbf{v}_i \rangle}{\langle \mathbf{v}_i, \mathbf{v}_i \rangle} \qquad \text{for each } i = 1, \ldots, n$$

so that

$$\mathbf{v} = \frac{\langle \mathbf{v}, \mathbf{v}_1 \rangle}{\langle \mathbf{v}_1, \mathbf{v}_1 \rangle}\mathbf{v}_1 + \frac{\langle \mathbf{v}, \mathbf{v}_2 \rangle}{\langle \mathbf{v}_2, \mathbf{v}_2 \rangle}\mathbf{v}_2 + \cdots + \frac{\langle \mathbf{v}, \mathbf{v}_n \rangle}{\langle \mathbf{v}_n, \mathbf{v}_n \rangle}\mathbf{v}_n$$

Fact Summary

All vectors are in an inner product space.

1. An inner product on a vector space is a function that assigns to each pair of vectors a real number and generalizes the properties of the dot product on $\mathbb{R}^n$.

2. The norm of a vector is defined analogously to the definition in $\mathbb{R}^n$ by $\|\mathbf{v}\| = \sqrt{\langle \mathbf{v}, \mathbf{v} \rangle}$.

3. An orthogonal set of vectors is linearly independent. Thus, any set of n orthogonal vectors is a basis for an inner product space of dimension n.

4. When an arbitrary vector is written in terms of the vectors in an orthogonal basis, the coefficients are given explicitly by an expression in terms of the inner product. If $\{\mathbf{v}_1, \ldots, \mathbf{v}_n\}$ is the orthogonal basis and $\mathbf{v}$ is an arbitrary vector, then

$$\mathbf{v} = \frac{\langle \mathbf{v}, \mathbf{v}_1 \rangle}{\langle \mathbf{v}_1, \mathbf{v}_1 \rangle}\mathbf{v}_1 + \frac{\langle \mathbf{v}, \mathbf{v}_2 \rangle}{\langle \mathbf{v}_2, \mathbf{v}_2 \rangle}\mathbf{v}_2 + \cdots + \frac{\langle \mathbf{v}, \mathbf{v}_n \rangle}{\langle \mathbf{v}_n, \mathbf{v}_n \rangle}\mathbf{v}_n$$

If case B is an orthonormal basis, then

$$\mathbf{v} = \langle \mathbf{v}, \mathbf{v}_1 \rangle \mathbf{v}_1 + \cdots + \langle \mathbf{v}, \mathbf{v}_n \rangle \mathbf{v}_n$$

Exercise Set 6.2

In Exercises 1–10, determine whether V is an inner product space.

1. $V = \mathbb{R}^2$

   $\langle \mathbf{u}, \mathbf{v} \rangle = u_1 v_1 - 2u_1 v_2 - 2u_2 v_1 + 3u_2 v_2$

2. $V = \mathbb{R}^2$

   $\langle \mathbf{u}, \mathbf{v} \rangle = -u_1 v_1 + 2u_1 v_2$

3. $V = \mathbb{R}^2$

   $\langle \mathbf{u}, \mathbf{v} \rangle = u_1^2 v_1^2 + u_2^2 v_2^2$

4. $V = \mathbb{R}^3$

   $\langle \mathbf{u}, \mathbf{v} \rangle = u_1 v_1 + 2u_2 v_2 + 3u_3 v_3$

5. $V = \mathbb{R}^n$

   $\langle \mathbf{u}, \mathbf{v} \rangle = \mathbf{u} \cdot \mathbf{v}$

6. $V = M_{m \times n}$

   $\langle A, B \rangle = \mathbf{tr}(B^t A)$

7. $V = M_{m \times n}$

   $\langle A, B \rangle = \displaystyle\sum_{i=1}^{m} \sum_{j=1}^{n} a_{ij} b_{ij}$

8. $V = \mathcal{P}_n$

   $\langle p, q \rangle = \displaystyle\sum_{i=0}^{n} p_i q_i$

9. $V = C^{(0)}[-1, 1]$

   $\langle f, g \rangle = \int_{-1}^{1} f(x) g(x) e^{-x} \, dx$

10. $V = C^{(0)}[-1, 1]$

   $\langle f, g \rangle = \int_{-1}^{1} f(x) g(x) x \, dx$

In Exercises 11–14, let $V = C^{(0)}[a, b]$ with inner product

$$\langle f, g \rangle = \int_a^b f(x) g(x) \, dx$$

Verify that the set of vectors is orthogonal.

11. $\{1, \cos x, \sin x\}$; $a = -\pi, b = \pi$

12. $\left\{1, x, \frac{1}{2}(5x^3 - 3x)\right\}$; $a = -1, b = 1$

13. $\left\{1, 2x - 1, -x^2 + x - \frac{1}{6}\right\}$; $a = 0, b = 1$

14. $\{1, \cos x, \sin x, \cos 2x, \sin 2x\}$; $a = -\pi, b = \pi$

In Exercises 15–18, let $V = C^{(0)}[a, b]$ with inner product

$$\langle f, g \rangle = \int_a^b f(x) g(x) \, dx$$

a. Find the distance between the vectors f and g.

b. Find the cosine of the angle between the vectors f and g.

15. $f(x) = 3x - 2, g(x) = x^2 + 1$; $a = 0, b = 1$

16. $f(x) = \cos x, g(x) = \sin x$; $a = -\pi, b = \pi$

17. $f(x) = x, g(x) = e^x$; $a = 0, b = 1$

18. $f(x) = e^x, g(x) = e^{-x}$; $a = -1, b = 1$

In Exercises 19 and 20, let $V = \mathcal{P}_2$ with inner product

$$\langle p, q \rangle = \sum_{i=0}^{2} p_i q_i$$

a. Find the distance between the vectors p and q.

b. Find the cosine of the angle between the vectors p and q.

19. $p(x) = x^2 + x - 2, q(x) = -x^2 + x + 2$

20. $p(x) = x - 3, q(x) = 2x - 6$

In Exercises 21–24, let $V = M_{n \times n}$ with inner product

$$\langle A, B \rangle = \mathbf{tr}(B^t A)$$

a. Find the distance between the vectors A and B.

b. Find the cosine of the angle between the vectors A and B.

21. $A = \begin{bmatrix} 1 & 2 \\ 2 & -1 \end{bmatrix}$ $B = \begin{bmatrix} 2 & 1 \\ 1 & 3 \end{bmatrix}$

22. $A = \begin{bmatrix} 3 & 1 \\ 0 & -1 \end{bmatrix}$ $B = \begin{bmatrix} 0 & 2 \\ 1 & -2 \end{bmatrix}$

23. $A = \begin{bmatrix} 1 & 0 & -2 \\ -3 & 1 & 1 \\ -3 & -3 & -2 \end{bmatrix}$

   $B = \begin{bmatrix} 3 & -1 & -1 \\ -3 & 2 & 3 \\ -1 & -2 & 1 \end{bmatrix}$

24. $A = \begin{bmatrix} 2 & 1 & 2 \\ 3 & 1 & 0 \\ 3 & 2 & 1 \end{bmatrix}$

$B = \begin{bmatrix} 0 & 0 & 1 \\ 3 & 3 & 2 \\ 1 & 0 & 2 \end{bmatrix}$

25. Describe the set of all vectors in $\mathbb{R}^2$ that are orthogonal to $\begin{bmatrix} 2 \\ 3 \end{bmatrix}$.

26. Describe the set of all vectors in $\mathbb{R}^2$ that are orthogonal to $\begin{bmatrix} 1 \\ -b \end{bmatrix}$.

27. Describe the set of all vectors in $\mathbb{R}^3$ that are orthogonal to $\begin{bmatrix} 2 \\ -3 \\ 1 \end{bmatrix}$.

28. Describe the set of all vectors in $\mathbb{R}^3$ that are orthogonal to $\begin{bmatrix} 1 \\ 1 \\ 0 \end{bmatrix}$.

29. For f and g in $C^{(0)}[0, 1]$ define the inner product by

$$\langle f, g \rangle = \int_0^1 f(x)g(x)\,dx$$

a. Find $\langle x^2, x^3 \rangle$.

b. Find $\langle e^x, e^{-x} \rangle$.

c. Find $\| 1 \|$ and $\| x \|$.

d. Find the angle between $f(x) = 1$ and $g(x) = x$.

e. Find the distance between $f(x) = 1$ and $g(x) = x$.

30. Let A be a fixed 2×2 matrix, and define a function on $\mathbb{R}^2 \times \mathbb{R}^2$ by

$$\langle \mathbf{u}, \mathbf{v} \rangle = \mathbf{u}^t A \mathbf{v}$$

a. Verify that if $A = I$, then the function defines an inner product.

b. Show that if $A = \begin{bmatrix} 2 & -1 \\ -1 & 2 \end{bmatrix}$, then the function defines an inner product.

c. Show that if $A = \begin{bmatrix} 3 & 2 \\ 2 & 0 \end{bmatrix}$, then the function does not define an inner product.

31. Define an inner product on $C^{(0)}[-a, a]$ by

$$\langle f, g \rangle = \int_{-a}^a f(x)g(x)\,dx$$

Show that if f is an even function and g is an odd function, then f and g are orthogonal.

32. Define an inner product on $C^{(0)}[-\pi, \pi]$ by

$$\langle f, g \rangle = \int_{-\pi}^\pi f(x)g(x)\,dx$$

Show

$$\{1, \cos x, \sin x, \cos 2x, \sin 2x, \ldots\}$$

is an orthogonal set. (See Exercise 31.)

33. In an inner product space, show that if the set $\{\mathbf{u}_1, \mathbf{u}_2\}$ is orthogonal, then for scalars c_1 and c_2 the set $\{c_1\mathbf{u}_1, c_2\mathbf{u}_2\}$ is also orthogonal.

34. Show that if $\langle \mathbf{u}, \mathbf{v} \rangle$ and $\langle\langle \mathbf{u}, \mathbf{v} \rangle\rangle$ are two different inner products on V, then their sum

$$\langle\langle\langle \mathbf{u}, \mathbf{v} \rangle\rangle\rangle = \langle \mathbf{u}, \mathbf{v} \rangle + \langle\langle \mathbf{u}, \mathbf{v} \rangle\rangle$$

defines another inner product.

6.3 ▶ Orthonormal Bases

In Theorem 6 of Sec. 6.2 we saw that if $B = \{\mathbf{v}_1, \mathbf{v}_2, \ldots, \mathbf{v}_n\}$ is an ordered orthonormal basis of an inner product space V, then the coordinates of any vector $\mathbf{v}$ in V are given by an explicit formula using the inner product on the space. In particular, these coordinates relative to B are given by $c_i = \langle \mathbf{v}, \mathbf{v}_i \rangle$ for $i = 1, 2, \ldots n$. For this reason, an orthonormal basis for an inner product space is desirable. As we have already seen, the set of coordinate vectors $S = \{\mathbf{e}_1, \mathbf{e}_2, \ldots, \mathbf{e}_n\}$ is an orthonormal basis for $\mathbb{R}^n$. In this section we develop a method for constructing an orthonormal basis for any finite dimensional inner product space.

Orthogonal Projections

Of course, most of the bases we encounter are not orthonormal, or even orthogonal. We can, however, in a finite dimensional inner product space, transform any basis to an orthonormal basis. The method, called the *Gram-Schmidt process*, involves projections of vectors onto other vectors.

To motivate this topic, let $\mathbf{u}$ and $\mathbf{v}$ be vectors in $\mathbb{R}^2$, as shown in Fig. 1(a).

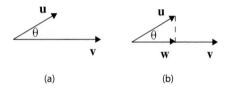

(a) (b)

Figure 1

Our aim is to find a vector $\mathbf{w}$ that results from an *orthogonal projection* of $\mathbf{u}$ onto $\mathbf{v}$, as shown in Fig. 1(b). To do this, recall from trigonometry that

$$\cos\theta = \frac{\|\mathbf{w}\|}{\|\mathbf{u}\|} \qquad \text{so that} \qquad \|\mathbf{w}\| = \|\mathbf{u}\|\cos\theta$$

Moreover, using the expression for $\cos\theta$, established at the beginning of Sec. 6.1, we have

$$\|\mathbf{w}\| = \|\mathbf{u}\|\cos\theta = \|\mathbf{u}\|\frac{\mathbf{u}\cdot\mathbf{v}}{\|\mathbf{u}\|\|\mathbf{v}\|} = \frac{\mathbf{u}\cdot\mathbf{v}}{\|\mathbf{v}\|}$$

This quantity is called the **scalar projection of u onto v**. Now, to find $\mathbf{w}$, we take the product of the scalar projection with a unit vector in the direction of $\mathbf{v}$, so that

$$\mathbf{w} = \left(\frac{\mathbf{u}\cdot\mathbf{v}}{\|\mathbf{v}\|}\right)\frac{\mathbf{v}}{\|\mathbf{v}\|} = \frac{\mathbf{u}\cdot\mathbf{v}}{\|\mathbf{v}\|^2}\mathbf{v}$$

Moreover, since $\|\mathbf{v}\|^2 = \mathbf{v}\cdot\mathbf{v}$, the vector $\mathbf{w}$ can be written in the form

$$\mathbf{w} = \left(\frac{\mathbf{u}\cdot\mathbf{v}}{\mathbf{v}\cdot\mathbf{v}}\right)\mathbf{v}$$

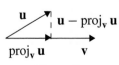

Figure 2

This vector is called the **orthogonal projection of u onto v** and is denoted by $\text{proj}_\mathbf{v}\,\mathbf{u}$, so that

$$\text{proj}_\mathbf{v}\,\mathbf{u} = \left(\frac{\mathbf{u}\cdot\mathbf{v}}{\mathbf{v}\cdot\mathbf{v}}\right)\mathbf{v}$$

Another useful vector, shown in Fig. 2, is the vector

$$\mathbf{u} - \text{proj}_\mathbf{v}\,\mathbf{u}$$

from **u** to $\text{proj}_v\,\mathbf{u}$. From the manner in which $\text{proj}_v\,\mathbf{u}$ is defined, the vector $\mathbf{u} - \text{proj}_v\,\mathbf{u}$ is orthogonal to $\text{proj}_v\,\mathbf{u}$, as shown in Fig. 2. To verify algebraically that $\text{proj}_v\,\mathbf{u}$ and $\mathbf{u} - \text{proj}_v\,\mathbf{u}$ are orthogonal, we show that the dot product of these two vectors is zero. That is,

$$\left(\text{proj}_v\,\mathbf{u}\right) \cdot \left(\mathbf{u} - \text{proj}_v\,\mathbf{u}\right) = \left(\frac{\mathbf{u} \cdot \mathbf{v}}{\mathbf{v} \cdot \mathbf{v}}\mathbf{v}\right) \cdot \left(\mathbf{u} - \frac{\mathbf{u} \cdot \mathbf{v}}{\mathbf{v} \cdot \mathbf{v}}\mathbf{v}\right)$$

$$= \frac{\mathbf{u} \cdot \mathbf{v}}{\mathbf{v} \cdot \mathbf{v}}\left(\mathbf{v} \cdot \mathbf{u} - \frac{\mathbf{u} \cdot \mathbf{v}}{\mathbf{v} \cdot \mathbf{v}}\mathbf{v} \cdot \mathbf{v}\right)$$

$$= \frac{\mathbf{u} \cdot \mathbf{v}}{\mathbf{v} \cdot \mathbf{v}}\left(\mathbf{v} \cdot \mathbf{u} - \mathbf{u} \cdot \mathbf{v}\right) = \frac{\mathbf{u} \cdot \mathbf{v}}{\mathbf{v} \cdot \mathbf{v}}\left(\mathbf{u} \cdot \mathbf{v} - \mathbf{u} \cdot \mathbf{v}\right) = 0$$

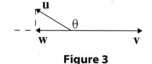

In Fig. 1, the angle θ shown is an acute angle. If θ is an obtuse angle, then $\text{proj}_v\,\mathbf{u}$ gives the orthogonal projection of **u** onto the negative of **v**, as shown in Fig. 3. If $\theta = 90°$, then $\text{proj}_v\,\mathbf{u} = \mathbf{0}$.

Figure 3

EXAMPLE 1   Let

$$\mathbf{u} = \begin{bmatrix} 1 \\ 3 \end{bmatrix} \quad \text{and} \quad \mathbf{v} = \begin{bmatrix} 1 \\ 1 \end{bmatrix}$$

a. Find $\text{proj}_v\,\mathbf{u}$.

b. Find $\mathbf{u} - \text{proj}_v\,\mathbf{u}$, and verify that $\text{proj}_v\,\mathbf{u}$ is orthogonal to $\mathbf{u} - \text{proj}_v\,\mathbf{u}$.

Solution   **a.** From the formula given above, we have

$$\text{proj}_v\,\mathbf{u} = \left(\frac{\mathbf{u} \cdot \mathbf{v}}{\mathbf{v} \cdot \mathbf{v}}\right)\mathbf{v} = \left(\frac{(1)(1) + (3)(1)}{(1)(1) + (1)(1)}\right)\begin{bmatrix} 1 \\ 1 \end{bmatrix} = 2\begin{bmatrix} 1 \\ 1 \end{bmatrix} = \begin{bmatrix} 2 \\ 2 \end{bmatrix}$$

b. Using the result of part (a), we have

$$\mathbf{u} - \text{proj}_v\,\mathbf{u} = \begin{bmatrix} 1 \\ 3 \end{bmatrix} - \begin{bmatrix} 2 \\ 2 \end{bmatrix} = \begin{bmatrix} -1 \\ 1 \end{bmatrix}$$

To show that $\text{proj}_v\,\mathbf{u}$ is orthogonal to $\mathbf{u} - \text{proj}_v\mathbf{u}$, we compute the dot product. Here we have

$$\text{proj}_v\,\mathbf{u} \cdot \left(\mathbf{u} - \text{proj}_v\,\mathbf{u}\right) = \begin{bmatrix} 2 \\ 2 \end{bmatrix} \cdot \begin{bmatrix} -1 \\ 1 \end{bmatrix} = (2)(-1) + (2)(1) = 0$$

See Fig. 4.

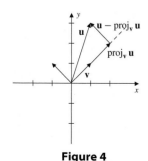

Figure 4

Definition 1 provides an extension of this idea to general inner product spaces.

DEFINITION 1   **Orthogonal Projection**   Let $\mathbf{u}$ and $\mathbf{v}$ be vectors in an inner product space. The **orthogonal projection** of $\mathbf{u}$ onto $\mathbf{v}$, denoted by $\text{proj}_{\mathbf{v}}\,\mathbf{u}$, is defined by

$$\text{proj}_{\mathbf{v}}\,\mathbf{u} = \frac{\langle \mathbf{u}, \mathbf{v} \rangle}{\langle \mathbf{v}, \mathbf{v} \rangle}\mathbf{v}$$

The vector $\mathbf{u} - \text{proj}_{\mathbf{v}}\,\mathbf{u}$ is orthogonal to $\text{proj}_{\mathbf{v}}\,\mathbf{u}$.

EXAMPLE 2   Define an inner product on $\mathcal{P}_3$ by

$$\langle p, q \rangle = \int_0^1 p(x)q(x)\,dx$$

Let $p(x) = x$ and $q(x) = x^2$.

a. Find $\text{proj}_q\,p$.

b. Find $p - \text{proj}_q\,p$ and verify that $\text{proj}_q\,p$ and $p - \text{proj}_q\,p$ are orthogonal.

Solution   **a.** In this case

$$\langle p, q \rangle = \int_0^1 x^3\,dx = \frac{1}{4} \quad \text{and} \quad \langle q, q \rangle = \int_0^1 x^4\,dx = \frac{1}{5}$$

Now the projection of p onto q is given by

$$\text{proj}_q\,p = \frac{\langle p, q \rangle}{\langle q, q \rangle}q = \frac{5}{4}x^2$$

b. From part (a), we have

$$p - \text{proj}_q\,p = x - \frac{5}{4}x^2$$

To show that the vectors p and $p - \text{proj}_q\,p$ are orthogonal, we show that the inner product is zero. Here we have

$$\int_0^1 \frac{5}{4}x^2\left(x - \frac{5}{4}x^2\right)dx = \int_0^1 \left(\frac{5}{4}x^3 - \frac{25}{16}x^4\right)dx = \left(\frac{5}{16}x^4 - \frac{5}{16}x^5\right)\Big|_0^1 = 0$$

We now turn our attention to the construction of an orthonormal basis for an inner product space. The key to this construction is the projection of one vector onto another. As a preliminary step, let $V = \mathbb{R}^2$ and let $B = \{\mathbf{v}_1, \mathbf{v}_2\}$ be a basis, as shown in Fig. 5. Now, define the vectors $\mathbf{w}_1$ and $\mathbf{w}_2$ by

$$\mathbf{w}_1 = \mathbf{v}_1 \quad \text{and} \quad \mathbf{w}_2 = \mathbf{v}_2 - \text{proj}_{\mathbf{v}_1}\,\mathbf{v}_2$$

Recall from Example 1 that $\mathbf{w}_2$ defined in this way is orthogonal to $\mathbf{w}_1$ as shown in Fig. 6. We *normalize* these vectors by dividing each by its length, so that

$$B' = \left\{ \frac{\mathbf{w}_1}{\|\mathbf{w}_1\|}, \frac{\mathbf{w}_2}{\|\mathbf{w}_2\|} \right\}$$

is an orthonormal basis for $\mathbb{R}^2$.

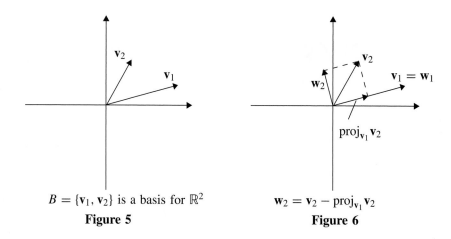

$B = \{\mathbf{v}_1, \mathbf{v}_2\}$ is a basis for $\mathbb{R}^2$

Figure 5

$\mathbf{w}_2 = \mathbf{v}_2 - \text{proj}_{\mathbf{v}_1} \mathbf{v}_2$

Figure 6

To construct an orthonormal basis for $\mathbb{R}^n$, we first need to extend this idea to general inner product spaces.

THEOREM 7   Every finite dimensional inner product space has an orthogonal basis.

Proof   The proof is by induction on the dimension n of the inner product space. First if $n = 1$, then any basis $\{\mathbf{v}_1\}$ is orthogonal. Now assume that every inner product space of dimension n has an orthogonal basis. Let V be an inner product space with $\dim(V) = n + 1$, and suppose that $\{\mathbf{v}_1, \mathbf{v}_2, \ldots, \mathbf{v}_n, \mathbf{v}_{n+1}\}$ is a basis. Let $W = \mathbf{span}\{\mathbf{v}_1, \mathbf{v}_2, \ldots, \mathbf{v}_n\}$. Observe that $\dim(W) = n$. By the inductive hypothesis, W has an orthogonal basis B. Let $B = \{\mathbf{w}_1, \mathbf{w}_2, \ldots, \mathbf{w}_n\}$. Notice that $B' = \{\mathbf{w}_1, \mathbf{w}_2, \ldots, \mathbf{w}_n, \mathbf{v}_{n+1}\}$ is another basis for V. By Theorem 5 of Sec. 6.2, it suffices to find a nonzero vector $\mathbf{w}$ that is orthogonal to each vector in B. (Here is where we extend the idea presented just prior to the theorem.) Let

$$\mathbf{w} = \mathbf{v}_{n+1} - \text{proj}_{\mathbf{w}_1} \mathbf{v}_{n+1} - \text{proj}_{\mathbf{w}_2} \mathbf{v}_{n+1} - \cdots - \text{proj}_{\mathbf{w}_n} \mathbf{v}_{n+1}$$

$$= \mathbf{v}_{n+1} - \frac{\langle \mathbf{v}_{n+1}, \mathbf{w}_1 \rangle}{\langle \mathbf{w}_1, \mathbf{w}_1 \rangle} \mathbf{w}_1 - \frac{\langle \mathbf{v}_{n+1}, \mathbf{w}_2 \rangle}{\langle \mathbf{w}_2, \mathbf{w}_2 \rangle} \mathbf{w}_2 - \cdots - \frac{\langle \mathbf{v}_{n+1}, \mathbf{w}_n \rangle}{\langle \mathbf{w}_n, \mathbf{w}_n \rangle} \mathbf{w}_n$$

Observe that $\mathbf{w} \neq \mathbf{0}$ since if $\mathbf{w} = \mathbf{0}$, then B' will be linearly dependent and therefore not a basis for V. To complete the proof, we must show that $\mathbf{w}$ is orthogonal to each vector in B. To see this, let $\mathbf{w}_i$ be a vector in B. Then

$$\langle \mathbf{w}, \mathbf{w}_i \rangle = \left\langle \mathbf{v}_{n+1} - \text{proj}_{\mathbf{w}_1}\mathbf{v}_{n+1} - \text{proj}_{\mathbf{w}_2}\mathbf{v}_{n+1} - \cdots - \text{proj}_{\mathbf{w}_n}\mathbf{v}_{n+1}, \mathbf{w}_i \right\rangle$$

$$= \langle \mathbf{v}_{n+1}, \mathbf{w}_i \rangle - \frac{\langle \mathbf{v}_{n+1}, \mathbf{w}_1 \rangle}{\langle \mathbf{w}_1, \mathbf{w}_1 \rangle}\langle \mathbf{w}_1, \mathbf{w}_i \rangle - \frac{\langle \mathbf{v}_{n+1}, \mathbf{w}_2 \rangle}{\langle \mathbf{w}_2, \mathbf{w}_2 \rangle}\langle \mathbf{w}_2, \mathbf{w}_i \rangle$$

$$- \cdots - \frac{\langle \mathbf{v}_{n+1}, \mathbf{w}_i \rangle}{\langle \mathbf{w}_i, \mathbf{w}_i \rangle}\langle \mathbf{w}_i, \mathbf{w}_i \rangle - \cdots - \frac{\langle \mathbf{v}_{n+1}, \mathbf{w}_n \rangle}{\langle \mathbf{w}_n, \mathbf{w}_n \rangle}\langle \mathbf{w}_n, \mathbf{w}_i \rangle$$

Now, as each vector in $B = \{\mathbf{w}_1, \mathbf{w}_2, \ldots, \mathbf{w}_n\}$ is mutually orthogonal, the previous equation reduces to

$$\langle \mathbf{w}, \mathbf{w}_i \rangle = \langle \mathbf{v}_{n+1}, \mathbf{w}_i \rangle - 0 - 0 - \cdots - \frac{\langle \mathbf{v}_{n+1}, \mathbf{w}_i \rangle}{\langle \mathbf{w}_i, \mathbf{w}_i \rangle}\langle \mathbf{w}_i, \mathbf{w}_i \rangle - 0 - \cdots - 0$$

$$= \langle \mathbf{v}_{n+1}, \mathbf{w}_i \rangle - \langle \mathbf{v}_{n+1}, \mathbf{w}_i \rangle = 0$$

Therefore $B' = \{\mathbf{w}_1, \mathbf{w}_2, \ldots, \mathbf{w}_n, \mathbf{w}\}$ is an orthogonal set of $n+1$ vectors in V. That B' is a basis for V is due to Corollary 1 of Sec. 6.2.

From Theorem 7, we also know that every finite dimensional vector space has an orthonormal basis. That is, if $B = \{\mathbf{w}_1, \mathbf{w}_2, \ldots, \mathbf{w}_n\}$ is an orthogonal basis, then dividing each vector by its length gives the orthonormal basis

$$B' = \left\{ \frac{\mathbf{w}_1}{\|\mathbf{w}_1\|}, \frac{\mathbf{w}_2}{\|\mathbf{w}_2\|}, \ldots, \frac{\mathbf{w}_n}{\|\mathbf{w}_n\|} \right\}$$

Gram-Schmidt Process

Theorem 7 guarantees the existence of an orthogonal basis in a finite dimensional inner product space. The proof of Theorem 7 also provides a procedure for constructing an orthogonal basis from any basis of the vector space. The algorithm, called the **Gram-Schmidt process**, is summarized here.

1. Let $B = \{\mathbf{v}_1, \mathbf{v}_2, \ldots, \mathbf{v}_n\}$ be any basis for the inner product space V.

2. Use B to define a set of n vectors as follows:

$$\mathbf{w}_1 = \mathbf{v}_1$$

$$\mathbf{w}_2 = \mathbf{v}_2 - \text{proj}_{\mathbf{w}_1}\mathbf{v}_2 = \mathbf{v}_2 - \frac{\langle \mathbf{v}_2, \mathbf{w}_1 \rangle}{\langle \mathbf{w}_1, \mathbf{w}_1 \rangle}\mathbf{w}_1$$

$$\mathbf{w}_3 = \mathbf{v}_3 - \text{proj}_{\mathbf{w}_1}\mathbf{v}_3 - \text{proj}_{\mathbf{w}_2}\mathbf{v}_3$$
$$= \mathbf{v}_3 - \frac{\langle \mathbf{v}_3, \mathbf{w}_1 \rangle}{\langle \mathbf{w}_1, \mathbf{w}_1 \rangle}\mathbf{w}_1 - \frac{\langle \mathbf{v}_3, \mathbf{w}_2 \rangle}{\langle \mathbf{w}_2, \mathbf{w}_2 \rangle}\mathbf{w}_2$$

$$\vdots$$

$$\mathbf{w}_n = \mathbf{v}_n - \text{proj}_{\mathbf{w}_1}\mathbf{v}_n - \text{proj}_{\mathbf{w}_2}\mathbf{v}_n - \cdots - \text{proj}_{\mathbf{w}_{n-1}}\mathbf{v}_n$$
$$= \mathbf{v}_n - \frac{\langle \mathbf{v}_n, \mathbf{w}_1 \rangle}{\langle \mathbf{w}_1, \mathbf{w}_1 \rangle}\mathbf{w}_1 - \frac{\langle \mathbf{v}_n, \mathbf{w}_2 \rangle}{\langle \mathbf{w}_2, \mathbf{w}_2 \rangle}\mathbf{w}_2 - \cdots - \frac{\langle \mathbf{v}_n, \mathbf{w}_{n-1} \rangle}{\langle \mathbf{w}_{n-1}, \mathbf{w}_{n-1} \rangle}\mathbf{w}_{n-1}$$

3. The set $B' = \{\mathbf{w}_1, \mathbf{w}_2, \ldots, \mathbf{w}_n\}$ is an orthogonal basis for V.
4. Dividing each of the vectors in B' by its length gives an orthonormal basis for the vector space V

$$B'' = \left\{ \frac{\mathbf{w}_1}{\|\mathbf{w}_1\|}, \frac{\mathbf{w}_2}{\|\mathbf{w}_2\|}, \ldots, \frac{\mathbf{w}_n}{\|\mathbf{w}_n\|} \right\}$$

A Geometric Interpretation of the Gram-Schmidt Process

The orthogonal projection $\text{proj}_{\mathbf{v}}\mathbf{u}$ of the vector $\mathbf{u}$ onto the vector $\mathbf{v}$, in $\mathbb{R}^n$, is the projection of $\mathbf{u}$ onto the one-dimensional subspace $W = \mathbf{span}\{\mathbf{v}\}$. See Figs. 2 and 3. As seen above, to construct an orthogonal basis from the basis $B = \{\mathbf{v}_1, \mathbf{v}_2, \mathbf{v}_3\}$, the first step in the Gram-Schmidt process is to let $\mathbf{w}_1 = \mathbf{v}_1$, and then perform an orthogonal projection of $\mathbf{v}_2$ onto $\mathbf{span}\{\mathbf{v}_1\}$. As a result, $\mathbf{w}_1$ is orthogonal to $\mathbf{w}_2 = \mathbf{v}_2 - \text{proj}_{\mathbf{w}_1}\mathbf{v}_2$. Our aim in the next step is to find a vector $\mathbf{w}_3$ that is orthogonal to the two-dimensional subspace $\mathbf{span}\{\mathbf{w}_1, \mathbf{w}_2\}$. This is accomplished by projecting $\mathbf{v}_3$ separately onto the one-dimensional subspaces $\mathbf{span}\{\mathbf{w}_1\}$ and $\mathbf{span}\{\mathbf{w}_2\}$, as shown in Fig. 7. The orthogonal projections are

$$\text{proj}_{\mathbf{w}_1}\mathbf{v}_3 = \frac{\langle \mathbf{v}_3, \mathbf{w}_1 \rangle}{\langle \mathbf{w}_1, \mathbf{w}_1 \rangle}\mathbf{w}_1 \qquad \text{and} \qquad \text{proj}_{\mathbf{w}_2}\mathbf{v}_3 = \frac{\langle \mathbf{v}_3, \mathbf{w}_2 \rangle}{\langle \mathbf{w}_2, \mathbf{w}_2 \rangle}\mathbf{w}_2$$

Hence, the orthogonal projection of $\mathbf{v}_3$ onto $\mathbf{span}\{\mathbf{w}_1, \mathbf{w}_2\}$ is the sum of the projections

$$\text{proj}_{\mathbf{w}_1}\mathbf{v}_3 + \text{proj}_{\mathbf{w}_2}\mathbf{v}_3$$

also shown in Fig. 7. Finally, the required vector is

$$\mathbf{w}_3 = \mathbf{v}_3 - \left(\text{proj}_{\mathbf{w}_1}\mathbf{v}_3 + \text{proj}_{\mathbf{w}_2}\mathbf{v}_3\right) = \mathbf{v}_3 - \text{proj}_{\mathbf{w}_1}\mathbf{v}_3 - \text{proj}_{\mathbf{w}_2}\mathbf{v}_3$$

which is orthogonal to both $\mathbf{w}_1$ and $\mathbf{w}_2$, as shown in Fig. 7.

In general, when $\dim(W) = n > 1$, then the Gram-Schmidt process describes projecting the vector $\mathbf{v}_{n+1}$ onto n one-dimensional subspaces $\mathbf{span}\{\mathbf{w}_1\}$, $\mathbf{span}\{\mathbf{w}_2\}, \ldots, \mathbf{span}\{\mathbf{w}_n\}$. Then the vector $\mathbf{w}_{n+1}$ that is orthogonal to each of the vectors $\mathbf{w}_1, \mathbf{w}_2, \ldots, \mathbf{w}_n$ is obtained by subtracting each projection from the vector $\mathbf{v}_n$.

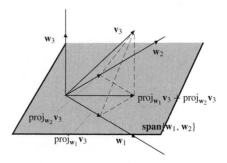

Figure 7

EXAMPLE 3   Let B be the basis for $\mathbb{R}^3$ given by

$$B = \{\mathbf{v}_1, \mathbf{v}_2, \mathbf{v}_3\} = \left\{ \begin{bmatrix} 1 \\ 1 \\ 1 \end{bmatrix}, \begin{bmatrix} -1 \\ 1 \\ 0 \end{bmatrix}, \begin{bmatrix} -1 \\ 0 \\ 1 \end{bmatrix} \right\}$$

Apply the Gram-Schmidt process to B to find an orthonormal basis for $\mathbb{R}^3$.

Solution   In this case the inner product on $\mathbb{R}^3$ is the dot product. Notice that $\mathbf{v}_1 \cdot \mathbf{v}_2 = 0$, so that the vectors $\mathbf{v}_1$ and $\mathbf{v}_2$ are already orthogonal. Applying the Gram-Schmidt process results in $\mathbf{w}_1 = \mathbf{v}_1$ and $\mathbf{w}_2 = \mathbf{v}_2$. Following the steps outlined above, we have

$$\mathbf{w}_1 = \mathbf{v}_1 = \begin{bmatrix} 1 \\ 1 \\ 1 \end{bmatrix}$$

$$\mathbf{w}_2 = \mathbf{v}_2 - \left(\frac{\mathbf{v}_2 \cdot \mathbf{w}_1}{\mathbf{w}_1 \cdot \mathbf{w}_1} \right) \mathbf{w}_1 = \mathbf{v}_2 - \tfrac{0}{3}\mathbf{w}_1 = \mathbf{v}_2 = \begin{bmatrix} -1 \\ 1 \\ 0 \end{bmatrix}$$

Next note that $\mathbf{v}_1$ and $\mathbf{v}_3$ are also orthogonal, so that in this case only one projection is required. That is,

$$\mathbf{w}_3 = \mathbf{v}_3 - \left(\frac{\mathbf{v}_3 \cdot \mathbf{w}_1}{\mathbf{w}_1 \cdot \mathbf{w}_1} \right) \mathbf{w}_1 - \left(\frac{\mathbf{v}_3 \cdot \mathbf{w}_2}{\mathbf{w}_2 \cdot \mathbf{w}_2} \right) \mathbf{w}_2$$

$$= \begin{bmatrix} -1 \\ 0 \\ 1 \end{bmatrix} - 0\mathbf{w}_1 - \tfrac{1}{2} \begin{bmatrix} -1 \\ 1 \\ 0 \end{bmatrix} = \begin{bmatrix} -\tfrac{1}{2} \\ -\tfrac{1}{2} \\ 1 \end{bmatrix}$$

Then

$$B' = \{\mathbf{w}_1, \mathbf{w}_2, \mathbf{w}_3\} = \left\{ \begin{bmatrix} 1 \\ 1 \\ 1 \end{bmatrix}, \begin{bmatrix} -1 \\ 1 \\ 0 \end{bmatrix}, \begin{bmatrix} -\tfrac{1}{2} \\ -\tfrac{1}{2} \\ 1 \end{bmatrix} \right\}$$

is an orthogonal basis for $\mathbb{R}^3$. See Fig. 8. An orthonormal basis is then given by

$$B'' = \left\{ \frac{\mathbf{w}_1}{\| \mathbf{w}_1 \|}, \frac{\mathbf{w}_2}{\| \mathbf{w}_2 \|}, \frac{\mathbf{w}_3}{\| \mathbf{w}_3 \|} \right\} = \left\{ \frac{1}{\sqrt{3}} \begin{bmatrix} 1 \\ 1 \\ 1 \end{bmatrix}, \frac{1}{\sqrt{2}} \begin{bmatrix} -1 \\ 1 \\ 0 \end{bmatrix}, \frac{1}{\sqrt{6}} \begin{bmatrix} -1 \\ -1 \\ 2 \end{bmatrix} \right\}$$

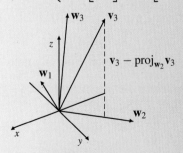

Figure 8

Example 4 illustrates the use of the Gram-Schmidt process on a space of polynomials.

EXAMPLE 4   Define an inner product on $\mathcal{P}_3$ by

$$\langle p, q \rangle = \int_{-1}^{1} p(x) q(x) \, dx$$

Use the standard basis $B = \{\mathbf{v}_1, \mathbf{v}_2, \mathbf{v}_3, \mathbf{v}_4\} = \{1, x, x^2, x^3\}$ to construct an orthogonal basis for $\mathcal{P}_3$.

Solution   First note that B is not orthogonal, since

$$\langle x, x^3 \rangle = \int_{-1}^{1} x^4 \, dx = \frac{2}{5}$$

We can simplify some of the work by noting that since the interval $[-1, 1]$ is a symmetric interval,

When p is an odd function, then $\quad \displaystyle\int_{-1}^{1} p(x) \, dx = 0$

When p is an even function, then $\quad \displaystyle\int_{-1}^{1} p(x) \, dx = 2 \int_{0}^{1} p(x) \, dx$

Now, since $f(x) = x$, $g(x) = x^3$, and $h(x) = x^5$ are all odd functions,

$$\langle \mathbf{v}_1, \mathbf{v}_2 \rangle = \int_{-1}^{1} x \, dx = 0 \qquad \langle \mathbf{v}_2, \mathbf{v}_3 \rangle = \int_{-1}^{1} x^3 \, dx = 0$$

$$\langle \mathbf{v}_4, \mathbf{v}_1 \rangle = \int_{-1}^{1} x^3 \, dx = 0 \qquad \langle \mathbf{v}_4, \mathbf{v}_3 \rangle = \int_{-1}^{1} x^5 \, dx = 0$$

Since $\mathbf{v}_1$ and $\mathbf{v}_2$ are orthogonal, proceeding with the Gram-Schmidt process, we have

$$\mathbf{w}_1 = \mathbf{v}_1 \qquad \text{and} \qquad \mathbf{w}_2 = \mathbf{v}_2$$

Next, to find $\mathbf{w}_3$, the required computation is

$$\mathbf{w}_3 = \mathbf{v}_3 - \frac{\langle \mathbf{v}_3, \mathbf{w}_1 \rangle}{\langle \mathbf{w}_1, \mathbf{w}_1 \rangle} \mathbf{w}_1 - \frac{\langle \mathbf{v}_3, \mathbf{w}_2 \rangle}{\langle \mathbf{w}_2, \mathbf{w}_2 \rangle} \mathbf{w}_2$$

$$= \mathbf{v}_3 - \frac{\langle \mathbf{v}_3, \mathbf{v}_1 \rangle}{\langle \mathbf{v}_1, \mathbf{v}_1 \rangle} \mathbf{v}_1 - \frac{\langle \mathbf{v}_3, \mathbf{v}_2 \rangle}{\langle \mathbf{v}_2, \mathbf{v}_2 \rangle} \mathbf{v}_2$$

$$= \mathbf{v}_3 - \frac{\langle \mathbf{v}_3, \mathbf{v}_1 \rangle}{\langle \mathbf{v}_1, \mathbf{v}_1 \rangle} \mathbf{v}_1 - \frac{\langle \mathbf{v}_2, \mathbf{v}_3 \rangle}{\langle \mathbf{v}_2, \mathbf{v}_2 \rangle} \mathbf{v}_2$$

But we have already noted above that $0 = \langle \mathbf{v}_2, \mathbf{v}_3 \rangle$ and since

$$\langle \mathbf{v}_3, \mathbf{v}_1 \rangle = \int_{-1}^{1} x^2 \, dx = 2 \int_{0}^{1} x^2 \, dx = \frac{2}{3} \qquad \text{and} \qquad \langle \mathbf{v}_1, \mathbf{v}_1 \rangle = \int_{-1}^{1} dx = 2$$

then

$$\mathbf{w}_3 = x^2 - \frac{1}{3}$$

To find $\mathbf{w}_4$, we first note that since $\langle \mathbf{v}_4, \mathbf{v}_1 \rangle = 0$, $\mathbf{w}_1 = \mathbf{v}_1$, and $\mathbf{w}_2 = \mathbf{v}_2$, then

$$\mathbf{w}_4 = \mathbf{v}_4 - \frac{\langle \mathbf{v}_4, \mathbf{w}_1 \rangle}{\langle \mathbf{w}_1, \mathbf{w}_1 \rangle} \mathbf{w}_1 - \frac{\langle \mathbf{v}_4, \mathbf{w}_2 \rangle}{\langle \mathbf{w}_2, \mathbf{w}_2 \rangle} \mathbf{w}_2 - \frac{\langle \mathbf{v}_4, \mathbf{w}_3 \rangle}{\langle \mathbf{w}_3, \mathbf{w}_3 \rangle} \mathbf{w}_3$$

$$= \mathbf{v}_4 - \frac{\langle \mathbf{v}_4, \mathbf{v}_2 \rangle}{\langle \mathbf{v}_2, \mathbf{v}_2 \rangle} \mathbf{v}_2 - \frac{\langle \mathbf{v}_4, \mathbf{w}_3 \rangle}{\langle \mathbf{w}_3, \mathbf{w}_3 \rangle} \mathbf{w}_3$$

Next, observe that $p(x) = x^5 - \frac{1}{3}x^3$ is an odd function. Hence,

$$\langle \mathbf{v}_4, \mathbf{w}_3 \rangle = \int_{-1}^{1} \left(x^5 - \tfrac{1}{3}x^3 \right) dx = 0$$

Consequently,

$$\mathbf{w}_4 = \mathbf{v}_4 - \frac{\langle \mathbf{v}_4, \mathbf{v}_2 \rangle}{\langle \mathbf{v}_2, \mathbf{v}_2 \rangle} \mathbf{v}_2 = x^3 - \tfrac{3}{5}x$$

An orthogonal basis for $\mathcal{P}_3$ is therefore given by

$$B' = \left\{ 1, x, x^2 - \tfrac{1}{3}, x^3 - \tfrac{3}{5}x \right\}$$

By normalizing each of these vectors, we obtain the orthonormal basis

$$B'' = \left\{ \frac{\sqrt{2}}{2}, \frac{\sqrt{6}}{2}x, \frac{3\sqrt{10}}{4}\left(x^2 - \tfrac{1}{3} \right), \frac{5\sqrt{14}}{4}\left(x^3 - \tfrac{3}{5}x \right) \right\}$$

EXAMPLE 5   Let U be the subspace of $\mathbb{R}^4$ with basis

$$B = \{\mathbf{u}_1, \mathbf{u}_2, \mathbf{u}_3\} = \left\{ \begin{bmatrix} -1 \\ 1 \\ 1 \\ 0 \end{bmatrix}, \begin{bmatrix} -1 \\ 0 \\ 1 \\ 0 \end{bmatrix}, \begin{bmatrix} 1 \\ 0 \\ 0 \\ 1 \end{bmatrix} \right\}$$

where the inner product is the dot product. Find an orthonormal basis for U.

Solution   Following the Gram-Schmidt process, we let $\mathbf{w}_1 = \mathbf{u}_1$. Next we have

$$\mathbf{w}_2 = \mathbf{u}_2 - \frac{\mathbf{u}_2 \cdot \mathbf{w}_1}{\mathbf{w}_1 \cdot \mathbf{w}_1} \mathbf{w}_1 = \begin{bmatrix} -1 \\ 0 \\ 1 \\ 0 \end{bmatrix} - \frac{2}{3} \begin{bmatrix} -1 \\ 1 \\ 1 \\ 0 \end{bmatrix} = \begin{bmatrix} -\frac{1}{3} \\ -\frac{2}{3} \\ \frac{1}{3} \\ 0 \end{bmatrix} = -\frac{1}{3} \begin{bmatrix} 1 \\ 2 \\ -1 \\ 0 \end{bmatrix}$$

To facilitate the computations, we replace $\mathbf{w}_2$ with

$$\mathbf{w}_2 = \begin{bmatrix} 1 \\ 2 \\ -1 \\ 0 \end{bmatrix}$$

To justify this substitution, note that $\mathbf{w}_1 \cdot \mathbf{w}_2 = 0$; that is, multiplying $\mathbf{w}_2$ by a scalar does not change the fact that it is orthogonal to $\mathbf{w}_1$. To find $\mathbf{w}_3$, we use the computation

$$\mathbf{w}_3 = \mathbf{u}_3 - \frac{\mathbf{u}_3 \cdot \mathbf{w}_1}{\mathbf{w}_1 \cdot \mathbf{w}_1}\mathbf{w}_1 - \frac{\mathbf{u}_3 \cdot \mathbf{w}_2}{\mathbf{w}_2 \cdot \mathbf{w}_2}\mathbf{w}_2$$

$$= \begin{bmatrix} 1 \\ 0 \\ 0 \\ 1 \end{bmatrix} - \left(-\frac{1}{3}\right)\begin{bmatrix} -1 \\ 1 \\ 1 \\ 0 \end{bmatrix} - \frac{1}{6}\begin{bmatrix} 1 \\ 2 \\ -1 \\ 0 \end{bmatrix} = \frac{1}{2}\begin{bmatrix} 1 \\ 0 \\ 1 \\ 2 \end{bmatrix}$$

As before we replace $\mathbf{w}_3$ with

$$\mathbf{w}_3 = \begin{bmatrix} 1 \\ 0 \\ 1 \\ 2 \end{bmatrix}$$

An orthogonal basis for U is then given by

$$B' = \left\{ \begin{bmatrix} -1 \\ 1 \\ 1 \\ 0 \end{bmatrix}, \begin{bmatrix} 1 \\ 2 \\ -1 \\ 0 \end{bmatrix}, \begin{bmatrix} 1 \\ 0 \\ 1 \\ 2 \end{bmatrix} \right\}$$

Normalizing each of the vectors of B' produces the orthonormal basis

$$B'' = \left\{ \frac{1}{\sqrt{3}}\begin{bmatrix} -1 \\ 1 \\ 1 \\ 0 \end{bmatrix}, \frac{1}{\sqrt{6}}\begin{bmatrix} 1 \\ 2 \\ -1 \\ 0 \end{bmatrix}, \frac{1}{\sqrt{6}}\begin{bmatrix} 1 \\ 0 \\ 1 \\ 2 \end{bmatrix} \right\}$$

Fact Summary

1. Every finite dimensional inner product space has an orthonormal basis.
2. The Gram-Schmidt process is an algorithm to construct an orthonormal basis from any basis of the vector space.

Exercise Set 6.3

In Exercises 1–8, use the standard inner product on $\mathbb{R}^n$.

    a. Find $\text{proj}_{\mathbf{v}}\,\mathbf{u}$.

    b. Find the vector $\mathbf{u} - \text{proj}_{\mathbf{v}}\mathbf{u}$ and verify this vector is orthogonal to $\mathbf{v}$.

1. $\mathbf{u} = \begin{bmatrix} -1 \\ 2 \end{bmatrix}$ $\mathbf{v} = \begin{bmatrix} -1 \\ 1 \end{bmatrix}$

2. $\mathbf{u} = \begin{bmatrix} 3 \\ -2 \end{bmatrix}$ $\mathbf{v} = \begin{bmatrix} 1 \\ -2 \end{bmatrix}$

3. $\mathbf{u} = \begin{bmatrix} 1 \\ -2 \end{bmatrix}$ $\mathbf{v} = \begin{bmatrix} 1 \\ 2 \end{bmatrix}$

4. $\mathbf{u} = \begin{bmatrix} 1 \\ -1 \end{bmatrix}$ $\mathbf{v} = \begin{bmatrix} -2 \\ -2 \end{bmatrix}$

5. $\mathbf{u} = \begin{bmatrix} -1 \\ 3 \\ 0 \end{bmatrix}$ $\mathbf{v} = \begin{bmatrix} 1 \\ -1 \\ -1 \end{bmatrix}$

6. $\mathbf{u} = \begin{bmatrix} 1 \\ 0 \\ 1 \end{bmatrix}$ $\mathbf{v} = \begin{bmatrix} 3 \\ 2 \\ -1 \end{bmatrix}$

7. $\mathbf{u} = \begin{bmatrix} 1 \\ -1 \\ -1 \end{bmatrix}$ $\mathbf{v} = \begin{bmatrix} 0 \\ 0 \\ 1 \end{bmatrix}$

8. $\mathbf{u} = \begin{bmatrix} 3 \\ 2 \\ 0 \end{bmatrix}$ $\mathbf{v} = \begin{bmatrix} 1 \\ 0 \\ -1 \end{bmatrix}$

In Exercises 9–12, use the inner product on $\mathcal{P}_2$ defined by

$$\langle p, q \rangle = \int_0^1 p(x)q(x)\,dx$$

a. Find $\text{proj}_q p$.

b. Find the vector $p - \text{proj}_q p$ and verify that this vector is orthogonal to q.

9. $p(x) = x^2 - x + 1, q(x) = 3x - 1$

10. $p(x) = x^2 - x + 1, q(x) = 2x - 1$

11. $p(x) = 2x^2 + 1, q(x) = x^2 - 1$

12. $p(x) = -4x + 1, q(x) = x$

In Exercises 13–16, use the standard inner product on $\mathbb{R}^n$. Use the basis B and the Gram-Schmidt process to find an orthonormal basis for $\mathbb{R}^n$.

13. $B = \left\{ \begin{bmatrix} 1 \\ -1 \end{bmatrix}, \begin{bmatrix} 1 \\ -2 \end{bmatrix} \right\}$

14. $B = \left\{ \begin{bmatrix} 2 \\ -1 \end{bmatrix}, \begin{bmatrix} 3 \\ -2 \end{bmatrix} \right\}$

15. $B = \left\{ \begin{bmatrix} 1 \\ 0 \\ 1 \end{bmatrix}, \begin{bmatrix} 0 \\ -1 \\ 1 \end{bmatrix}, \begin{bmatrix} 0 \\ -1 \\ -1 \end{bmatrix} \right\}$

16. $B = \left\{ \begin{bmatrix} 1 \\ 0 \\ -1 \end{bmatrix}, \begin{bmatrix} 0 \\ 1 \\ 1 \end{bmatrix}, \begin{bmatrix} 1 \\ 1 \\ 1 \end{bmatrix} \right\}$

In Exercises 17 and 18, use the inner product on $\mathcal{P}_2$ defined by

$$\langle p, q \rangle = \int_0^1 p(x)q(x)\,dx$$

Use the given basis B and the Gram-Schmidt process to find an orthonormal basis for $\mathcal{P}_2$.

17. $B = \{x - 1, x + 2, x^2\}$

18. $B = \{x^2 - x, x, 2x + 1\}$

In Exercises 19–22, use the standard inner product on $\mathbb{R}^n$ to find an orthonormal basis for the subspace **span**(W).

19. $W = \left\{ \begin{bmatrix} 1 \\ 1 \\ 1 \end{bmatrix}, \begin{bmatrix} 1 \\ -1 \\ -1 \end{bmatrix} \right\}$

20. $W = \left\{ \begin{bmatrix} 0 \\ 1 \\ 1 \end{bmatrix}, \begin{bmatrix} -1 \\ -1 \\ 1 \end{bmatrix} \right\}$

21. $W = \left\{ \begin{bmatrix} -1 \\ -2 \\ 0 \\ 1 \end{bmatrix}, \begin{bmatrix} -1 \\ 3 \\ -1 \\ -1 \end{bmatrix}, \begin{bmatrix} 1 \\ -2 \\ 0 \\ 1 \end{bmatrix} \right\}$

22. $W = \left\{ \begin{bmatrix} 1 \\ -2 \\ 0 \\ 0 \end{bmatrix}, \begin{bmatrix} -1 \\ 3 \\ 1 \\ -1 \end{bmatrix}, \begin{bmatrix} 0 \\ -1 \\ 0 \\ -1 \end{bmatrix} \right\}$

In Exercises 23 and 24, use the inner product on $\mathcal{P}_3$ defined by

$$\langle p, q \rangle = \int_0^1 p(x)q(x)\,dx$$

to find an orthonormal basis for the subspace **span**(W).

23. $W = \{x, 2x + 1\}$

24. $W = \{1, x + 2, x^3 - 1\}$

25. In $\mathbb{R}^4$ with the standard inner product find an orthonormal basis for

$$\text{span} \left\{ \begin{bmatrix} 1 \\ 0 \\ 1 \\ 1 \end{bmatrix}, \begin{bmatrix} 0 \\ 1 \\ -1 \\ 1 \end{bmatrix}, \begin{bmatrix} 2 \\ -3 \\ 5 \\ -1 \end{bmatrix}, \begin{bmatrix} -1 \\ 2 \\ -3 \\ 1 \end{bmatrix} \right\}$$

26. In $\mathbb{R}^3$ with the standard inner product find an orthonormal basis for

$$\text{span} \left\{ \begin{bmatrix} 2 \\ 0 \\ 1 \end{bmatrix}, \begin{bmatrix} 3 \\ 1 \\ 1 \end{bmatrix}, \begin{bmatrix} 3 \\ -1 \\ 2 \end{bmatrix}, \begin{bmatrix} 1 \\ 1 \\ 0 \end{bmatrix} \right\}$$

27. Let $\{\mathbf{u}_1, \mathbf{u}_2, \dots, \mathbf{u}_n\}$ be an orthonormal basis for $\mathbb{R}^n$. Show that

$$\| \mathbf{v} \|^2 = |\mathbf{v} \cdot \mathbf{u}_1|^2 + \cdots + |\mathbf{v} \cdot \mathbf{u}_n|^2$$

for every vector $\mathbf{v}$ in $\mathbb{R}^n$.

28. Let A be an $n \times n$ matrix. Show that the following conditions are equivalent.

a. $A^{-1} = A^t$

b. The row vectors of A form an orthonormal basis for $\mathbb{R}^n$.

c. The column vectors of A form an orthonormal basis for $\mathbb{R}^n$.

29. Show that an $n \times n$ matrix A has orthonormal column vectors if and only if $A^t A = I$.

30. Let A be an $m \times n$ matrix, $\mathbf{x}$ a vector in $\mathbb{R}^m$, and $\mathbf{y}$ a vector in $\mathbb{R}^n$. Show that $\mathbf{x} \cdot (A\mathbf{y}) = A^t \mathbf{x} \cdot \mathbf{y}$.

31. Show that if A is an $m \times n$ matrix with orthonormal column vectors, then $\| A\mathbf{x} \| = \| \mathbf{x} \|$.

32. Show that if A is an $m \times n$ matrix with orthonormal column vectors and $\mathbf{x}$ and $\mathbf{y}$ are in $\mathbb{R}^n$, then $(A\mathbf{x}) \cdot (A\mathbf{y}) = \mathbf{x} \cdot \mathbf{y}$.

33. Show that if A is an $m \times n$ matrix with orthonormal column vectors and $\mathbf{x}$ and $\mathbf{y}$ are in $\mathbb{R}^n$, then $(A\mathbf{x}) \cdot (A\mathbf{y}) = 0$ if and only if $\mathbf{x} \cdot \mathbf{y} = 0$.

34. In $\mathbb{R}^4$ with the standard inner product show that the set of all vectors orthogonal to both $\begin{bmatrix} 1 \\ 0 \\ -1 \\ 1 \end{bmatrix}$

and $\begin{bmatrix} 2 \\ 3 \\ -1 \\ 2 \end{bmatrix}$ is a subspace. Find a basis for the subspace.

35. Let $S = \{\mathbf{u}_1, \dots, \mathbf{u}_m\}$ be a set of vectors in $\mathbb{R}^n$. Show that the set of all vectors orthogonal to every $\mathbf{u}_i$ is a subspace of $\mathbb{R}^n$.

In Exercises 36–41, a (real) $n \times n$ matrix A is called *positive semidefinite* if A is symmetric and $\mathbf{u}^t A \mathbf{u} \geq 0$ for every nonzero vector $\mathbf{u}$ in $\mathbb{R}^n$. If the inequality is strict, then A is *positive definite*.

36. Let A be a positive definite matrix. Show that the function $\langle \mathbf{u}, \mathbf{v} \rangle = \mathbf{u}^t A \mathbf{v}$ defines an inner product on $\mathbb{R}^n$. (Note that when $A = I$ this function corresponds to the dot product.)

37. Let $A = \begin{bmatrix} 3 & 1 \\ 1 & 3 \end{bmatrix}$. Show that A is positive definite.

38. Show that if A is positive definite, then the diagonal entries are positive.

39. Let A be an $m \times n$ matrix. Show that $A^t A$ is positive semidefinite.

40. Show that a positive definite matrix is invertible.

41. Show that the eigenvalues of a positive definite matrix are positive.

42. Let

$$\mathbf{v}_1 = \begin{bmatrix} -2 \\ -1 \end{bmatrix} \qquad \mathbf{v}_2 = \begin{bmatrix} 2 \\ -4 \end{bmatrix}$$

a. Are the vectors $\mathbf{v}_1 = \begin{bmatrix} -2 \\ -1 \end{bmatrix}$ and $\mathbf{v}_2 = \begin{bmatrix} 2 \\ -4 \end{bmatrix}$ orthogonal? Let $A = \begin{bmatrix} -2 & -1 \\ 2 & -4 \end{bmatrix}$.

b. Find $\det(A^t A)$.

c. Show that the area of the rectangle spanned by $\mathbf{v}_1 = \begin{bmatrix} -2 \\ -1 \end{bmatrix}$ and $\mathbf{v}_2 = \begin{bmatrix} 2 \\ -4 \end{bmatrix}$ is $\sqrt{\det(A^t A)}$.

d. Show that the area of the rectangle is $|\det(A)|$.

e. If $\mathbf{v}_1$ and $\mathbf{v}_2$ are any two orthogonal vectors in $\mathbb{R}^2$, show that the area of the rectangle spanned by the vectors is $|\det(A)|$, where A is the matrix with row vectors $\mathbf{v}_1$ and $\mathbf{v}_2$.

f. Let $\mathbf{v}_1$ and $\mathbf{v}_2$ be two vectors in $\mathbb{R}^2$ that span a parallelogram, as shown in the figure. Show that the area of the parallelogram is $|\det(A)|$, where A is the matrix with row vectors $\mathbf{v}_1$ and $\mathbf{v}_2$.

g. If $\mathbf{v}_1$, $\mathbf{v}_2$, and $\mathbf{v}_3$ are mutually orthogonal vectors in $\mathbb{R}^3$, show that the volume of the box spanned by the three vectors is $|\det(A)|$, where A is the matrix with row vectors $\mathbf{v}_1$, $\mathbf{v}_2$, and $\mathbf{v}_3$.

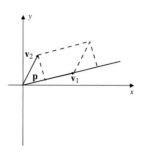

6.4 ▶ Orthogonal Complements

Throughout this chapter we have seen the importance of orthogonal vectors and bases in inner product spaces. Recall that two vectors $\mathbf{u}$ and $\mathbf{v}$ in an inner product space V are orthogonal if and only if

$$\langle \mathbf{u}, \mathbf{v} \rangle = 0$$

A collection of vectors forms an orthogonal basis if the vectors are a basis and are pairwise orthogonal. In this section we extend the notion of orthogonality to subspaces of inner product spaces. As a first step, let $\mathbf{v}$ be a vector in an inner product space V and W a subspace of V. We say that $\mathbf{v}$ is **orthogonal to** W if and only if

$$\langle \mathbf{v}, \mathbf{w} \rangle = 0 \qquad \text{for each vector} \quad \mathbf{w} \in W$$

As an illustration, let W be the yz plane in the Euclidean space $\mathbb{R}^3$. Observe that W is closed under addition and scalar multiplication, so that by Theorem 3 of Sec. 3.2 it is a subspace. Using the dot product as the inner product on $\mathbb{R}^3$, the coordinate vector

$$\mathbf{e}_1 = \begin{bmatrix} 1 \\ 0 \\ 0 \end{bmatrix}$$

is orthogonal to W since

$$\begin{bmatrix} 1 \\ 0 \\ 0 \end{bmatrix} \cdot \begin{bmatrix} 0 \\ y \\ z \end{bmatrix} = 0$$

for every $y, z \in \mathbb{R}$. Note that any scalar multiple of $\mathbf{e}_1$ is also orthogonal to W.

Example 1 gives an illustration of how to find vectors orthogonal to a subspace.

EXAMPLE 1   Let $V = \mathbb{R}^3$, with the dot product as the inner product, and let W be the subspace defined by

$$W = \mathbf{span} \left\{ \begin{bmatrix} 1 \\ -2 \\ 3 \end{bmatrix} \right\}$$

Describe all vectors in $\mathbb{R}^3$ that are orthogonal to W.

Solution   Let

$$\mathbf{w} = \begin{bmatrix} 1 \\ -2 \\ 3 \end{bmatrix}$$

Thus, any vector in W has the form $c\mathbf{w}$, for some real number c. Consequently, a vector

$$\mathbf{v} = \begin{bmatrix} x \\ y \\ z \end{bmatrix}$$

in $\mathbb{R}^3$ is orthogonal to W if and only if $\mathbf{v} \cdot \mathbf{w} = 0$. This last equation is equivalent to the equation

$$x - 2y + 3z = 0$$

whose solution set is given by

$$S = \{(2s - 3t, s, t) \mid s, t \in \mathbb{R}\}$$

Therefore, the set of vectors orthogonal to W is given by

$$S' = \left\{ \begin{bmatrix} 2s - 3t \\ s \\ t \end{bmatrix} \,\middle|\, s, t \in \mathbb{R} \right\} = \left\{ s \begin{bmatrix} 2 \\ 1 \\ 0 \end{bmatrix} + t \begin{bmatrix} -3 \\ 0 \\ 1 \end{bmatrix} \,\middle|\, s, t \in \mathbb{R} \right\}$$

Letting $s = t = 1$ gives the particular vector

$$\mathbf{v} = \begin{bmatrix} -1 \\ 1 \\ 1 \end{bmatrix}$$

which is orthogonal to W since

$$\mathbf{v} \cdot \mathbf{w} = \begin{bmatrix} -1 \\ 1 \\ 1 \end{bmatrix} \cdot \begin{bmatrix} 1 \\ -2 \\ 3 \end{bmatrix} = (-1)(1) + (1)(-2) + (1)(3) = 0$$

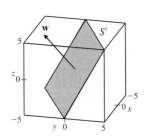

Figure 1

If the vectors in S' are placed in standard position, then the solution set describes a plane in $\mathbb{R}^3$, as shown in Fig. 1. This is in support of our intuition as the set of vectors orthogonal to a single vector in $\mathbb{R}^3$ should all lie in a plane perpendicular to that vector, which is called the *normal* vector.

The set of vectors found in Example 1, orthogonal to the subspace W, is called the *orthogonal complement* of W. The following definition generalizes this idea to inner product spaces.

DEFINITION 1

Orthogonal Complement   Let W be a subspace of an inner product space V. The **orthogonal complement** of W, denoted by $W^\perp$, is the set of all vectors in V that are orthogonal to W. That is,

$$W^\perp = \{\mathbf{v} \in V \mid \langle \mathbf{v}, \mathbf{w} \rangle = 0 \text{ for all } \mathbf{w} \in W\}$$

EXAMPLE 2

Let $V = \mathcal{P}_3$ and define an inner product on V by

$$\langle p, q \rangle = \int_0^1 p(x)q(x)\, dx$$

Find $W^\perp$ if W is the subspace of constant polynomials.

Solution   Let $f(x) = a + bx + cx^2 + dx^3$ be an arbitrary polynomial in $\mathcal{P}_3$ and $p(x) = k$ be an arbitrary constant polynomial. Then f is in $W^\perp$ if and only if

$$\langle f, p \rangle = \int_0^1 k(a + bx + cx^2 + dx^3)\, dx = k\left(a + \frac{b}{2} + \frac{c}{3} + \frac{d}{4} \right) = 0$$

Since this equation must hold for all $k \in \mathbb{R}$,

$$W^\perp = \left\{ a + bx + cx^2 + dx^3 \;\middle|\; a + \frac{b}{2} + \frac{c}{3} + \frac{d}{4} = 0 \right\}$$

Notice in Examples 1 and 2 that the zero vector is an element of the orthogonal complement $W^\perp$. It can also be shown for these examples that $W^\perp$ is closed under vector space addition and scalar multiplication. This leads to Theorem 8.

THEOREM 8

Let W be a subspace of an inner product space V.

1. The orthogonal complement $W^\perp$ is a subspace of V.
2. The only vector in W and $W^\perp$ is the zero vector; that is, $W \cap W^\perp = \{\mathbf{0}\}$.

Proof   (1) Let $\mathbf{u}$ and $\mathbf{v}$ be vectors in $W^\perp$, and $\mathbf{w}$ a vector in W, so that

$$\langle \mathbf{u}, \mathbf{w} \rangle = 0 \qquad \text{and} \qquad \langle \mathbf{v}, \mathbf{w} \rangle = 0$$

Now for any scalar c, we have

$$\begin{aligned}
\langle \mathbf{u} + c\mathbf{v}, \mathbf{w} \rangle &= \langle \mathbf{u}, \mathbf{w} \rangle + \langle c\mathbf{v}, \mathbf{w} \rangle \\
&= \langle \mathbf{u}, \mathbf{w} \rangle + c \langle \mathbf{v}, \mathbf{w} \rangle \\
&= 0 + 0 = 0
\end{aligned}$$

Thus, $\mathbf{u} + c\mathbf{v}$ is in $W^\perp$, and therefore by Theorem 4 of Sec. 3.2, $W^\perp$ is a subspace of V.

(2) Let $\mathbf{w}$ be any vector in $W \cap W^\perp$. Then

$$\langle \mathbf{w}, \mathbf{w} \rangle = 0$$

and hence $\mathbf{w} = \mathbf{0}$ (see Definition 1 of Sec. 6.2). Thus, $W \cap W^\perp = \{\mathbf{0}\}$.

To determine whether a vector $\mathbf{v}$ is in the orthogonal complement of a subspace, it suffices to show that $\mathbf{v}$ is orthogonal to each one of the vectors in a basis for the subspace.

PROPOSITION 5   Let W be a subspace of an inner product space V and $B = \{\mathbf{w}_1, \ldots, \mathbf{w}_m\}$ a basis for W. The vector $\mathbf{v}$ is in $W^\perp$ if and only if $\mathbf{v}$ is orthogonal to each vector in B.

Proof   First suppose that $\mathbf{v}$ is orthogonal to each vector in B. Let $\mathbf{w}$ be a vector in W. Then there are scalars $c_1, \ldots, c_m$ such that

$$\mathbf{w} = c_1 \mathbf{w}_1 + \cdots + c_m \mathbf{w}_m$$

To show that $\mathbf{v}$ is in $W^\perp$, take the inner product of both sides of the previous equation with $\mathbf{v}$, so that

$$\langle \mathbf{v}, \mathbf{w} \rangle = c_1 \langle \mathbf{v}, \mathbf{w}_1 \rangle + c_2 \langle \mathbf{v}, \mathbf{w}_2 \rangle + \cdots + c_m \langle \mathbf{v}, \mathbf{w}_m \rangle$$

Since $\langle \mathbf{v}, \mathbf{w}_j \rangle = 0$ for all $j = 1, 2, \ldots, m$, we have $\langle \mathbf{v}, \mathbf{w} \rangle = 0$ and hence $\mathbf{v} \in W^\perp$.

On the other hand, if $\mathbf{v} \in W^\perp$, then $\mathbf{v}$ is orthogonal to each vector in W. In particular, $\mathbf{v}$ is orthogonal to $\mathbf{w}_j$, for all $j = 1, 2, \ldots, m$.

EXAMPLE 3   Let $V = \mathbb{R}^4$ with the dot product as the inner product, and let

$$W = \mathbf{span} \left\{ \begin{bmatrix} 1 \\ 0 \\ -1 \\ -1 \end{bmatrix}, \begin{bmatrix} 0 \\ 1 \\ -1 \\ 1 \end{bmatrix} \right\}$$

a. Find a basis for W.
b. Find a basis for $W^\perp$.
c. Find an orthonormal basis for $\mathbb{R}^4$.
d. Let

$$\mathbf{v}_0 = \begin{bmatrix} 1 \\ 0 \\ 0 \\ 0 \end{bmatrix}$$

Show that $\mathbf{v}_0$ can be written as the sum of a vector from W and a vector from $W^\perp$.

Solution a. Let

$$\mathbf{w}_1 = \begin{bmatrix} 1 \\ 0 \\ -1 \\ -1 \end{bmatrix} \quad \text{and} \quad \mathbf{w}_2 = \begin{bmatrix} 0 \\ 1 \\ -1 \\ 1 \end{bmatrix}$$

Notice that $\mathbf{w}_1$ and $\mathbf{w}_2$ are orthogonal and hence by Theorem 5 of Sec. 6.2 are linearly independent. Thus, $\{\mathbf{w}_1, \mathbf{w}_2\}$ is a basis for W.

b. Now by Proposition 5, the vector

$$\mathbf{v} = \begin{bmatrix} x \\ y \\ z \\ w \end{bmatrix}$$

is in $W^\perp$ if and only if $\mathbf{v} \cdot \mathbf{w}_1 = 0$ and $\mathbf{v} \cdot \mathbf{w}_2 = 0$. This requirement leads to the linear system

$$\begin{cases} x & -z - w = 0 \\ & y - z + w = 0 \end{cases}$$

The two-parameter solution set for this linear system is

$$S = \left\{ \begin{bmatrix} s+t \\ s-t \\ s \\ t \end{bmatrix} \Bigg| \, s, t \in \mathbb{R} \right\}$$

The solution to this system, in vector form, provides a description of the orthogonal complement of W and is given by

$$W^\perp = \mathbf{span} \left\{ \begin{bmatrix} 1 \\ 1 \\ 1 \\ 0 \end{bmatrix}, \begin{bmatrix} 1 \\ -1 \\ 0 \\ 1 \end{bmatrix} \right\}$$

Let

$$\mathbf{v}_1 = \begin{bmatrix} 1 \\ 1 \\ 1 \\ 0 \end{bmatrix} \quad \text{and} \quad \mathbf{v}_2 = \begin{bmatrix} 1 \\ -1 \\ 0 \\ 1 \end{bmatrix}$$

Since $\mathbf{v}_1$ and $\mathbf{v}_2$ are orthogonal, by Theorem 5 of Sec. 6.2 they are linearly independent and hence a basis for $W^\perp$.

c. Let B be the set of vectors $B = \{\mathbf{w}_1, \mathbf{w}_2, \mathbf{v}_1, \mathbf{v}_2\}$. Since B is an orthogonal set of four vectors in $\mathbb{R}^4$, then by Corollary 1 of Sec. 6.2, B is a basis for $\mathbb{R}^4$. Dividing each of these vectors by its length, we obtain the (ordered) orthonormal basis for $\mathbb{R}^4$ given by

$$B' = \{\mathbf{b}_1, \mathbf{b}_2, \mathbf{b}_3, \mathbf{b}_4\} = \left\{ \frac{1}{\sqrt{3}} \begin{bmatrix} 1 \\ 0 \\ -1 \\ -1 \end{bmatrix}, \frac{1}{\sqrt{3}} \begin{bmatrix} 0 \\ 1 \\ -1 \\ 1 \end{bmatrix}, \frac{1}{\sqrt{3}} \begin{bmatrix} 1 \\ 1 \\ 1 \\ 0 \end{bmatrix}, \frac{1}{\sqrt{3}} \begin{bmatrix} 1 \\ -1 \\ 0 \\ 1 \end{bmatrix} \right\}$$

d. By Theorem 6 of Sec. 6.2 the coordinates of $\mathbf{v}_0$ relative to B' are given by $c_i = \mathbf{v}_0 \cdot \mathbf{b}_i$, for $1 \le i \le 4$. So

$$c_1 = \frac{1}{\sqrt{3}} \qquad c_2 = 0 \qquad c_3 = \frac{1}{\sqrt{3}} \qquad c_4 = \frac{1}{\sqrt{3}}$$

Now, observe that the first two vectors of B' are an orthonormal basis for W while the second two vectors are an orthonormal basis for $W^\perp$. Let $\mathbf{w}$ be the vector in W given by

$$\mathbf{w} = c_1 \mathbf{b}_1 + c_2 \mathbf{b}_2 = \frac{1}{3} \begin{bmatrix} 1 \\ 0 \\ -1 \\ -1 \end{bmatrix}$$

and $\mathbf{u}$ be the vector in $W^\perp$ given by

$$\mathbf{u} = c_3 \mathbf{b}_3 + c_4 \mathbf{b}_4 = \frac{1}{3} \begin{bmatrix} 2 \\ 0 \\ 1 \\ 1 \end{bmatrix}$$

Then

$$\mathbf{w} + \mathbf{u} = \frac{1}{3} \begin{bmatrix} 1 \\ 0 \\ -1 \\ -1 \end{bmatrix} + \frac{1}{3} \begin{bmatrix} 2 \\ 0 \\ 1 \\ 1 \end{bmatrix} = \begin{bmatrix} 1 \\ 0 \\ 0 \\ 0 \end{bmatrix} = \mathbf{v}_0$$

The vector $\mathbf{w}$ in Example 3 is called the *orthogonal projection* of $\mathbf{v}$ onto the subspace W, and the vector $\mathbf{u}$ is called the *component of $\mathbf{v}$ orthogonal to W*. The situation, in general, is the content of Definition 2 and Theorem 9.

DEFINITION 2   **Direct Sum**   Let W_1 and W_2 be subspaces of a vector space V. If each vector in V can be written uniquely as the sum of a vector from W_1 and a vector from W_2, then V is called the **direct sum** of W_1 and W_2. In this case we write, $V = W_1 \oplus W_2$.

PROPOSITION 6
Let W_1 and W_2 be subspaces of a vector space V with $V = W_1 \oplus W_2$. Then $W_1 \cap W_2 = \{\mathbf{0}\}$.

Proof   Let $\mathbf{v} \in W_1 \cap W_2$. Then

$$\mathbf{v} = \mathbf{w}_1 + \mathbf{0} \qquad \text{and} \qquad \mathbf{v} = \mathbf{0} + \mathbf{w}_2$$

with $\mathbf{w}_1 \in W_1$ and $\mathbf{w}_2 \in W_2$. Hence, by the uniqueness of direct sum representations, we have $\mathbf{w}_1 = \mathbf{w}_2 = \mathbf{0}$.

THEOREM 9
Projection Theorem   If W is a finite dimensional subspace of an inner product space V, then

$$V = W \oplus W^\perp$$

Proof   The proof of this theorem has two parts. First we must show that for any vector $\mathbf{v} \in V$ there exist vectors $\mathbf{w} \in W$ and $\mathbf{u} \in W^\perp$ such that $\mathbf{w} + \mathbf{u} = \mathbf{v}$. Then we must show that this representation is unique.

For the first part, let $B = \{\mathbf{w}_1, \ldots, \mathbf{w}_n\}$ be a basis for W. By Theorem 7 of Sec. 6.3, we can take B to be an orthonormal basis for W. Now, let $\mathbf{v}$ be a vector in V, and let the vectors $\mathbf{w}$ and $\mathbf{u}$ be defined by

$$\mathbf{w} = \langle \mathbf{v}, \mathbf{w}_1 \rangle \mathbf{w}_1 + \langle \mathbf{v}, \mathbf{w}_2 \rangle \mathbf{w}_2 + \cdots + \langle \mathbf{v}, \mathbf{w}_n \rangle \mathbf{w}_n \qquad \text{and} \qquad \mathbf{u} = \mathbf{v} - \mathbf{w}$$

Since $\mathbf{w}$ is a linear combination of the vectors in B, then $\mathbf{w} \in W$. To show that $\mathbf{u}$ is in $W^\perp$, we show that $\langle \mathbf{u}, \mathbf{w}_i \rangle = 0$ for each $i = 1, 2, \ldots, n$ and invoke Proposition 5. To this end, let $\mathbf{w}_i$ be a vector in B. Then

$$\langle \mathbf{u}, \mathbf{w}_i \rangle = \langle \mathbf{v} - \mathbf{w}, \mathbf{w}_i \rangle$$
$$= \langle \mathbf{v}, \mathbf{w}_i \rangle - \langle \mathbf{w}, \mathbf{w}_i \rangle$$
$$= \langle \mathbf{v}, \mathbf{w}_i \rangle - \sum_{j=1}^{n} \langle \mathbf{v}, \mathbf{w}_j \rangle \langle \mathbf{w}_j, \mathbf{w}_i \rangle$$

Since B is an orthonormal basis,

$$\langle \mathbf{w}_i, \mathbf{w}_i \rangle = 1 \qquad \text{and} \qquad \langle \mathbf{w}_j, \mathbf{w}_i \rangle = 0 \qquad \text{for } i \neq j$$

Hence,

$$\langle \mathbf{u}, \mathbf{w}_i \rangle = \langle \mathbf{v}, \mathbf{w}_i \rangle - \langle \mathbf{v}, \mathbf{w}_i \rangle \langle \mathbf{w}_i, \mathbf{w}_i \rangle = 0$$

Since this holds for each $i = 1, \ldots, n$, then by Proposition 5 the vector $\mathbf{u} \in W^\perp$ as claimed.

For the second part of the proof, let

$$\mathbf{v} = \mathbf{w} + \mathbf{u} \qquad \text{and} \qquad \mathbf{v} = \mathbf{w}' + \mathbf{u}'$$

with $\mathbf{w}$ and $\mathbf{w}'$ in W and $\mathbf{u}$ and $\mathbf{u}'$ in $W^\perp$. Subtracting the previous equations gives

$$(\mathbf{w} - \mathbf{w}') + (\mathbf{u} - \mathbf{u}') = \mathbf{0}$$

or equivalently,

$$\mathbf{w}' - \mathbf{w} = \mathbf{u} - \mathbf{u}'$$

Now, from this last equation we know that the vector $\mathbf{u} - \mathbf{u}'$ is in W, as it is a linear combination of the vectors $\mathbf{w}$ and $\mathbf{w}'$, which are in W. However, $\mathbf{u} - \mathbf{u}'$ is also in $W^\perp$ since it is the difference of two vectors in $W^\perp$. Therefore, by Theorem 8, part 2, $\mathbf{u} - \mathbf{u}' = \mathbf{0}$ and hence $\mathbf{u} = \mathbf{u}'$. This being the case, we now have $\mathbf{w}' - \mathbf{w} = \mathbf{0}$, so that $\mathbf{w}' = \mathbf{w}$. Thus, we have shown that for every $\mathbf{v}$ in V, there are unique vectors $\mathbf{w}$ in W and $\mathbf{u}$ in $W^\perp$ such that $\mathbf{v} = \mathbf{w} + \mathbf{u}$ and hence $V = W \oplus W^\perp$.

Motivated by the terminology of Example 3, we call the vector $\mathbf{w}$, of Theorem 9, the **orthogonal projection of v onto** W, which we denote by $\text{proj}_W \mathbf{v}$, and call $\mathbf{u}$ the **component of v orthogonal to** W.

Matrices

In Definition 4 of Sec. 3.2 we defined the null space of an $m \times n$ matrix A, denoted by $N(A)$, as the set of all vectors $\mathbf{x}$ in $\mathbb{R}^n$ such that $A\mathbf{x} = \mathbf{0}$. The column space of A, denoted by $\mathbf{col}(A)$, is the subspace of $\mathbb{R}^m$ spanned by the column vectors of A. In a similar way, the **left null space of** A, denoted by $N(A^t)$, is the set of vectors $\mathbf{y}$ in $\mathbb{R}^m$ such that $A^t\mathbf{y} = \mathbf{0}$. Finally, the row space of A, which we discussed in Sec. 4.2, denoted by $\mathbf{row}(A)$, is the subspace of $\mathbb{R}^n$ spanned by the row vectors of A. Since the rows of A are the columns of A^t, then $\mathbf{row}(A) = \mathbf{col}(A^t)$. These four subspaces

$$N(A) \qquad N(A^t) \qquad \mathbf{col}(A) \qquad \text{and} \qquad \mathbf{col}(A^t)$$

are referred to as the **four fundamental subspaces associated with the matrix** A. Theorem 10 gives relationships among them.

THEOREM 10 Let A be an $m \times n$ matrix.

1. $N(A) = \mathbf{col}(A^t)^\perp$
2. $N(A^t) = \mathbf{col}(A)^\perp$

Proof (1) Let $\mathbf{v}_1, \ldots, \mathbf{v}_m$ denote the row vectors of A. So that

$$A\mathbf{x} = \begin{bmatrix} \mathbf{v}_1 \cdot \mathbf{x} \\ \mathbf{v}_2 \cdot \mathbf{x} \\ \vdots \\ \mathbf{v}_m \cdot \mathbf{x} \end{bmatrix}$$

First let $\mathbf{x}$ be a vector in $N(A)$ so that $A\mathbf{x} = \mathbf{0}$. Then $\mathbf{v}_i \cdot \mathbf{x} = 0$ for $i = 1, 2, \ldots, m$. Thus, $\mathbf{x} \in \mathbf{row}(A)^\perp = \mathbf{col}(A^t)^\perp$ and $N(A) \subseteq \mathbf{col}(A^t)^\perp$. On the other hand, let $\mathbf{x}$ be a vector in $\mathbf{col}(A^t)^\perp = \mathbf{row}(A)^\perp$. Then $\mathbf{x} \cdot \mathbf{v}_i = 0$, for $i = 1, \ldots, m$, so that $A\mathbf{x} = \mathbf{0}$. Therefore, $\mathbf{col}(A^t)^\perp \subseteq N(A)$. Hence, $N(A) = \mathbf{col}(A^t)^\perp$.

For part 2, substitute A^t for A in part 1.

Linear Systems

Let A be an $m \times n$ matrix. In light of Theorem 10, we are now in a position to provide an analysis of the linear system $A\mathbf{x} = \mathbf{b}$ in terms of the geometric structure of Euclidean space and the fundamental subspaces of A. As a first step, we describe the action of A on a vector $\mathbf{x}$ in $\mathbb{R}^n$. Since $\mathbf{row}(A) = \mathbf{col}(A^t)$, by Theorem 10, $N(A)$ is the orthogonal complement of $\mathbf{row}(A)$. Thus, by Theorem 9, a vector $\mathbf{x}$ in $\mathbb{R}^n$ can be uniquely written as

$$\mathbf{x} = \mathbf{x}_{\text{row}} + \mathbf{x}_{\text{null}}$$

where $\mathbf{x}_{\text{row}}$ is in the row space of A and $\mathbf{x}_{\text{null}}$ is in the null space of A. Now, multiplying $\mathbf{x}$ by A, we have

$$A\mathbf{x} = A(\mathbf{x}_{\text{row}} + \mathbf{x}_{\text{null}}) = A\mathbf{x}_{\text{row}} + A\mathbf{x}_{\text{null}}$$

Since $A\mathbf{x}_{\text{null}} = \mathbf{0}$, the mapping $T: \mathbb{R}^n \longrightarrow \mathbb{R}^m$ defined by $T(\mathbf{x}) = A\mathbf{x}$ maps the row space of A to the column space of A. Observe that no vector in $\mathbb{R}^n$ is mapped to a nonzero vector in $N(A^t)$, which by Theorem 10 is the orthogonal complement of the column space of A. See Fig. 2.

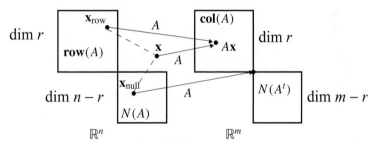

Figure 2

We now consider, again from a geometric point of view, the consistency of the linear system $A\mathbf{x} = \mathbf{b}$ for an $m \times n$ matrix A and a given vector $\mathbf{b}$ in $\mathbb{R}^m$. We have already observed in Sec. 3.2 that $A\mathbf{x} = \mathbf{b}$ is consistent if and only if $\mathbf{b}$ is in the column space of A. By Theorem 10, this system is consistent if and only if $\mathbf{b}$ is perpendicular to the left null space of A, or equivalently, if and only if $\mathbf{b}$ is orthogonal to every vector in $\mathbb{R}^m$, which is orthogonal to the column vectors of A. This sounds a bit awkward. However, in cases where a basis for the null space of A^t consists of only a few vectors, we can perform an easy check to see if $A\mathbf{x} = \mathbf{b}$ is consistent. As an illustration, let

$$A = \begin{bmatrix} 1 & 0 & 0 \\ 0 & 1 & 1 \\ -1 & -1 & -1 \end{bmatrix} \quad \text{and} \quad \mathbf{b} = \begin{bmatrix} 2 \\ 1 \\ -3 \end{bmatrix}$$

Since

$$A^t = \begin{bmatrix} 1 & 0 & -1 \\ 0 & 1 & -1 \\ 0 & 1 & -1 \end{bmatrix} \quad \text{then} \quad N(A^t) = \mathbf{span} \left\{ \begin{bmatrix} 1 \\ 1 \\ 1 \end{bmatrix} \right\}$$

Now as **b** is orthogonal to $\begin{bmatrix} 1 \\ 1 \\ 1 \end{bmatrix}$, by Proposition 5, **b** is orthogonal to $N(A^t)$ and hence in **col**(A). Therefore, the linear system $A\mathbf{x} = \mathbf{b}$ is consistent.

Fact Summary

Let W be a subspace of an inner product space V.

1. The orthogonal complement of the span of a single nonzero vector **v** in 3-space is the plane with normal vector **v**.
2. The orthogonal complement of W is a subspace of V.
3. The only vector common to both W and its orthogonal complement is the zero vector.
4. If W is finite dimensional, then the vector space is the direct sum of W and its orthogonal complement. That is, $V = W \oplus W^{\perp}$.
5. If B is a basis for W, then **v** is in $W^{\perp}$ if and only if **v** is orthogonal to each vector in B.
6. If A is an $m \times n$ matrix, then $N(A) = \mathbf{col}(A^t)^{\perp}$ and $N(A^t) = \mathbf{col}(A)^{\perp}$.

Exercise Set 6.4

In Exercises 1–8, find the orthogonal complement of W in $\mathbb{R}^n$ with the standard inner product.

1. $W = \text{span}\left\{ \begin{bmatrix} 1 \\ -2 \end{bmatrix} \right\}$

2. $W = \text{span}\left\{ \begin{bmatrix} 1 \\ 0 \end{bmatrix} \right\}$

3. $W = \text{span}\left\{ \begin{bmatrix} 2 \\ 1 \\ -1 \end{bmatrix} \right\}$

4. $W = \text{span}\left\{ \begin{bmatrix} 1 \\ 0 \\ 2 \end{bmatrix} \right\}$

5. $W = \text{span}\left\{ \begin{bmatrix} 2 \\ 1 \\ -1 \end{bmatrix}, \begin{bmatrix} 1 \\ 2 \\ 0 \end{bmatrix} \right\}$

6. $W = \text{span}\left\{ \begin{bmatrix} -3 \\ 1 \\ -1 \end{bmatrix}, \begin{bmatrix} 0 \\ 1 \\ 1 \end{bmatrix} \right\}$

7. $W = \text{span}\left\{ \begin{bmatrix} 3 \\ 1 \\ 1 \\ -1 \end{bmatrix}, \begin{bmatrix} 0 \\ 2 \\ 1 \\ 2 \end{bmatrix} \right\}$

8. $W = \text{span}\left\{ \begin{bmatrix} 1 \\ 1 \\ 0 \\ 1 \end{bmatrix}, \begin{bmatrix} 1 \\ 0 \\ 1 \\ 1 \end{bmatrix}, \begin{bmatrix} 0 \\ 1 \\ 1 \\ 1 \end{bmatrix} \right\}$

In Exercises 9–12, find a basis for the orthogonal complement of W in $\mathbb{R}^n$ with the standard inner product.

9. $W = \text{span}\left\{ \begin{bmatrix} 2 \\ 1 \\ 1 \end{bmatrix}, \begin{bmatrix} -1 \\ 1 \\ 0 \end{bmatrix} \right\}$

10. $W = \text{span}\left\{ \begin{bmatrix} 1 \\ -1 \\ 1 \end{bmatrix}, \begin{bmatrix} -2 \\ 2 \\ -2 \end{bmatrix} \right\}$

11. $W = \text{span}\left\{\begin{bmatrix} 3 \\ 1 \\ -1 \\ 2 \end{bmatrix}, \begin{bmatrix} 1 \\ 1 \\ 4 \\ 0 \end{bmatrix}\right\}$

12. $W = \text{span}\left\{\begin{bmatrix} 1 \\ 1 \\ 1 \\ 1 \end{bmatrix}, \begin{bmatrix} 2 \\ 0 \\ -1 \\ 1 \end{bmatrix}, \begin{bmatrix} 0 \\ 2 \\ 3 \\ 1 \end{bmatrix}\right\}$

In Exercises 13 and 14, find a basis for the orthogonal complement of W in $\mathcal{P}_2$ with the inner product

$$\langle p, q \rangle = \int_0^1 p(x)q(x)\,dx$$

13. $W = \text{span}\{x - 1, x^2\}$

14. $W = \text{span}\{1, x^2\}$

15. Let W be the subspace of $\mathbb{R}^4$, with the standard inner product, consisting of all vectors $\mathbf{w}$ such that $w_1 + w_2 + w_3 + w_4 = 0$. Find a basis for $W^\perp$.

In Exercises 16–21, W is a subspace of $\mathbb{R}^n$ with the standard inner product. If $\mathbf{v}$ is in $\mathbb{R}^n$ and $\{\mathbf{w}_1, \ldots, \mathbf{w}_m\}$ is an orthogonal basis for W, then the *orthogonal projection* of $\mathbf{v}$ onto W is given by

$$\text{proj}_W \mathbf{v} = \sum_{i=1}^m \frac{\langle \mathbf{v}, \mathbf{w}_i \rangle}{\langle \mathbf{w}_i, \mathbf{w}_i \rangle} \mathbf{w}_i$$

Find the orthogonal projection of $\mathbf{v}$ onto W. If necessary, first find an orthogonal basis for W.

16. $W = \text{span}\left\{\begin{bmatrix} 1 \\ 0 \\ -1 \end{bmatrix}, \begin{bmatrix} 2 \\ 1 \\ 1 \end{bmatrix}\right\}$

$\mathbf{v} = \begin{bmatrix} 1 \\ -2 \\ 2 \end{bmatrix}$

17. $W = \text{span}\left\{\begin{bmatrix} 2 \\ 0 \\ 0 \end{bmatrix}, \begin{bmatrix} 0 \\ -1 \\ 0 \end{bmatrix}\right\}$

$\mathbf{v} = \begin{bmatrix} 1 \\ 2 \\ -3 \end{bmatrix}$

18. $W = \text{span}\left\{\begin{bmatrix} 3 \\ -1 \\ 1 \end{bmatrix}, \begin{bmatrix} -2 \\ 2 \\ 0 \end{bmatrix}\right\}$

$\mathbf{v} = \begin{bmatrix} 5 \\ -3 \\ 1 \end{bmatrix}$

19. $W = \text{span}\left\{\begin{bmatrix} 1 \\ 2 \\ 1 \end{bmatrix}, \begin{bmatrix} -1 \\ 3 \\ 2 \end{bmatrix}\right\}$

$\mathbf{v} = \begin{bmatrix} 1 \\ -3 \\ 5 \end{bmatrix}$

20. $W = \text{span}\left\{\begin{bmatrix} 1 \\ 2 \\ -1 \\ 1 \end{bmatrix}, \begin{bmatrix} 1 \\ 3 \\ -1 \\ 0 \end{bmatrix}, \begin{bmatrix} 3 \\ 0 \\ 1 \\ -1 \end{bmatrix}\right\}$

$\mathbf{v} = \begin{bmatrix} 0 \\ 0 \\ 1 \\ 0 \end{bmatrix}$

21. $W = \text{span}\left\{\begin{bmatrix} 3 \\ 0 \\ -1 \\ 2 \end{bmatrix}, \begin{bmatrix} -6 \\ 0 \\ 2 \\ 4 \end{bmatrix}\right\}$

$\mathbf{v} = \begin{bmatrix} 1 \\ 2 \\ 1 \\ -1 \end{bmatrix}$

In Exercises 22–25, W is a subspace of $\mathbb{R}^n$ with the standard inner product.

    a. Find $W^\perp$.

    b. Find the orthogonal projection of $\mathbf{v}$ onto W. (See Exercises 16–21.)

    c. Compute $\mathbf{u} = \mathbf{v} - \text{proj}_W \mathbf{v}$.

    d. Show $\mathbf{u}$ is in $W^\perp$ so $\mathbf{v}$ is a sum of a vector in W and one in $W^\perp$.

    e. Make a sketch of W, $W^\perp$, $\mathbf{v}$, $\text{proj}_W \mathbf{v}$, and $\mathbf{u}$.

22. $W = \text{span}\left\{\begin{bmatrix} 1 \\ 2 \end{bmatrix}\right\}$ $\mathbf{v} = \begin{bmatrix} 1 \\ 1 \end{bmatrix}$

23. $W = \text{span}\left\{\begin{bmatrix} 3 \\ -1 \end{bmatrix}\right\}$ $\mathbf{v} = \begin{bmatrix} 0 \\ 1 \end{bmatrix}$

24. $W = \text{span}\left\{\begin{bmatrix} 1 \\ 1 \\ 0 \end{bmatrix}\right\}$

$\mathbf{v} = \begin{bmatrix} 1 \\ 1 \\ 1 \end{bmatrix}$

25. $W = \text{span}\left\{\begin{bmatrix} 1 \\ 1 \\ -1 \end{bmatrix}, \begin{bmatrix} -1 \\ 2 \\ 4 \end{bmatrix}\right\}$

$\mathbf{v} = \begin{bmatrix} 2 \\ 1 \\ 1 \end{bmatrix}$

26. Show that if V is an inner product space, then $V^{\perp} = \{\mathbf{0}\}$ and $\{\mathbf{0}\}^{\perp} = V$.

27. Show that if W_1 and W_2 are finite dimensional subspaces of an inner product space and $W_1 \subset W_2$, then $W_2^{\perp} \subset W_1^{\perp}$.

28. Let $V = C^{(0)}[-1, 1]$ with the inner product

$$\langle f, g \rangle = \int_{-1}^{1} f(x)g(x)\, dx$$

and $W = \{f \in V \mid f(-x) = f(x)\}$.

a. Show that W is a subspace of V.

b. Show $W^{\perp} = \{f \in V \mid f(-x) = -f(x)\}$.

c. Verify that $W \cap W^{\perp} = \{\mathbf{0}\}$.

d. Let $g(x) = \frac{1}{2}[f(x) + f(-x)]$ and $h(x) = \frac{1}{2}[f(x) - f(-x)]$. Verify $g(-x) = g(x)$ and $h(-x) = -h(x)$, so every f can be written as the sum of a function in W and a function in $W^{\perp}$.

29. Let $V = M_{2 \times 2}$ with the inner product

$$\langle A, B \rangle = \mathbf{tr}(B^t A)$$

Let $W = \{A \in V \mid A \text{ is symmetric}\}$.

a. Show that

$$W^{\perp} = \{A \in V \mid A \text{ is skew symmetric}\}$$

b. Show that every A in V can be written as the sum of matrices from W and $W^{\perp}$.

30. In $\mathbb{R}^2$ with the standard inner product, the transformation that sends a vector to the orthogonal projection onto a subspace W is a linear transformation. Let $W = \text{span}\left\{\begin{bmatrix} 2 \\ 1 \end{bmatrix}\right\}$.

a. Find the matrix representation P relative to the standard basis for the orthogonal projection of $\mathbb{R}^2$ onto W.

b. Let $\mathbf{v} = \begin{bmatrix} 1 \\ 1 \end{bmatrix}$. Find $\text{proj}_W \mathbf{v}$ and verify the result is the same by applying the matrix P found in part (a).

c. Show $P^2 = P$.

31. If W is a finite dimensional subspace of an inner product space, show that $(W^{\perp})^{\perp} = W$.

6.5 ▶ Application: Least Squares Approximation

There are many applications in mathematics and science where an exact solution to a problem cannot be found, but an approximate solution exists that is sufficient to satisfy the demands of the application. Consider the problem of finding the equation of a line going through the points $(1, 2)$, $(2, 1)$, and $(3, 3)$. Observe from Fig. 1 that this problem has no solution as the three points are noncollinear.

This leads to the problem of finding the line that is the *best fit* for these three points based on some criteria for measuring *goodness of fit*. There are different ways of solving this new problem. One way, which uses calculus, is based on the idea

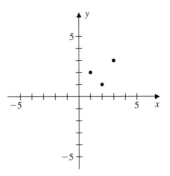

Figure 1

of finding the line that minimizes the sum of the square distances between itself and each of the points. Another quite elegant method uses the concepts of linear algebra to produce the same result. To illustrate the technique, we consider the original problem of finding an equation of the form $y = mx + b$ that is satisfied by the points $(1, 2)$, $(2, 1)$, and $(3, 3)$. Substitution of these points into the equation $y = mx + b$ yields the linear system

$$\begin{cases} m + b = 2 \\ 2m + b = 1 \\ 3m + b = 3 \end{cases}$$

As noted above, this system is inconsistent. As a first step toward finding an optimal approximate solution, we let

$$A = \begin{bmatrix} 1 & 1 \\ 2 & 1 \\ 3 & 1 \end{bmatrix} \qquad \mathbf{x} = \begin{bmatrix} m \\ b \end{bmatrix} \qquad \text{and} \qquad \mathbf{b} = \begin{bmatrix} 2 \\ 1 \\ 3 \end{bmatrix}$$

and we write the linear system as $A\mathbf{x} = \mathbf{b}$. From this perspective, we see that the linear system is inconsistent as $\mathbf{b}$ is not in $\mathbf{col}(A)$. Thus, the best we can do is to look for a vector $\widehat{\mathbf{w}}$ in $\mathbf{col}(A)$ that is as *close as possible* to $\mathbf{b}$, as shown in Fig. 2.

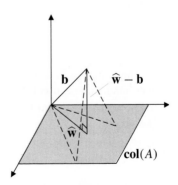

Figure 2

We will soon see that the optimal choice is to let $\widehat{\mathbf{w}}$ be the orthogonal projection of $\mathbf{b}$ onto $\mathbf{col}(A)$. In this case, to find $\widehat{\mathbf{w}}$ we let

$$W = \mathbf{col}(A) = \mathbf{span} \left\{ \begin{bmatrix} 1 \\ 2 \\ 3 \end{bmatrix}, \begin{bmatrix} 1 \\ 1 \\ 1 \end{bmatrix} \right\}$$

By Theorem 9 of Sec. 6.4, the vector $\mathbf{b}$ can be written uniquely as

$$\mathbf{b} = \widehat{\mathbf{w}} + \mathbf{y}$$

where $\mathbf{y}$ is in $W^\perp$. By Theorem 10 of Sec. 6.4, we have $W^\perp = N(A^t)$. By row reducing A^t, the orthogonal complement of W is

$$W^\perp = \mathbf{span} \left\{ \begin{bmatrix} 1 \\ -2 \\ 1 \end{bmatrix} \right\}$$

As this space is one-dimensional, the computations are simplified by finding $\mathbf{y}$ first, which in the terminology of Sec. 6.4 is the component of $\mathbf{b}$ orthogonal to W. To find $\mathbf{y}$, we use Definition 1 of Sec. 6.3 and compute the orthogonal projection of $\mathbf{b}$ onto $W^\perp$. Let $\mathbf{v} = \begin{bmatrix} 1 \\ -2 \\ 1 \end{bmatrix}$, so that

$$\mathbf{y} = \frac{\mathbf{b} \cdot \mathbf{v}}{\mathbf{v} \cdot \mathbf{v}} \mathbf{v} = \frac{1}{2} \begin{bmatrix} 1 \\ -2 \\ 1 \end{bmatrix}$$

Hence,

$$\widehat{\mathbf{w}} = \mathbf{b} - \mathbf{y} = \begin{bmatrix} 2 \\ 1 \\ 3 \end{bmatrix} - \frac{1}{2} \begin{bmatrix} 1 \\ -2 \\ 1 \end{bmatrix} = \frac{1}{2} \begin{bmatrix} 3 \\ 4 \\ 5 \end{bmatrix}$$

Finally, to find values for m and b, we solve the linear system $A\mathbf{x} = \widehat{\mathbf{w}}$, that is, the system

$$\begin{bmatrix} 1 & 1 \\ 2 & 1 \\ 3 & 1 \end{bmatrix} \begin{bmatrix} m \\ b \end{bmatrix} = \frac{1}{2} \begin{bmatrix} 3 \\ 4 \\ 5 \end{bmatrix}$$

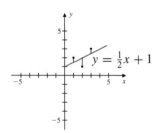

By Theorem 6 of Sec. 3.2, this last linear system is consistent since the vector on the right-hand side is in $\mathbf{col}(A)$. Solving the linear system, we obtain $m = \frac{1}{2}$ and $b = 1$, giving us the slope and the y intercept, respectively, for the best-fit line $y = \frac{1}{2}x + 1$. The vector

$$\begin{bmatrix} m \\ b \end{bmatrix} = \begin{bmatrix} \frac{1}{2} \\ 1 \end{bmatrix}$$

Figure 3

is the *least squares solution* to the system $A\mathbf{x} = \mathbf{b}$ since it produces a line whose total squared distances from the given points are minimal, as shown in Fig. 3. Finding the line that best fits a set of data points is called **linear regression**.

Least Squares Solutions

We now consider the general problem of finding a least squares solution to the $m \times n$ linear system $A\mathbf{x} = \mathbf{b}$. An exact solution exists if $\mathbf{b}$ is in $\mathbf{col}(A)$; moreover, the solution is unique if the columns of A are linearly independent. In the case where $\mathbf{b}$ is not in $\mathbf{col}(A)$, we look for a vector $\mathbf{x}$ in $\mathbb{R}^n$ that makes the *error term* $\| \mathbf{b} - A\mathbf{x} \|$ as small as possible. Using the standard inner product on $\mathbb{R}^m$ to define the length of a vector, we have

$$\| \mathbf{b} - A\mathbf{x} \|^2 = (\mathbf{b} - A\mathbf{x}) \cdot (\mathbf{b} - A\mathbf{x})$$
$$= [b_1 - (A\mathbf{x})_1]^2 + [b_2 - (A\mathbf{x})_2]^2 + \cdots + [b_m - (A\mathbf{x})_m]^2$$

This equation gives the rationale for the term *least squares solution*.

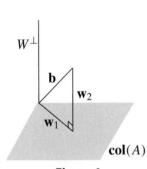

To find the least squares solution $\widehat{\mathbf{x}}$ to the linear system $A\mathbf{x} = \mathbf{b}$, we let $W = \mathbf{col}(A)$. As W is a finite dimensional subspace of $\mathbb{R}^m$, by Theorem 9 of Sec. 6.4, the vector $\mathbf{b}$ can be written uniquely as

$$\mathbf{b} = \mathbf{w}_1 + \mathbf{w}_2$$

where $\mathbf{w}_1$ is the orthogonal projection of $\mathbf{b}$ onto W and $\mathbf{w}_2$ is the component of $\mathbf{b}$ orthogonal to W, as shown in Fig. 4.

Figure 4

We now show that the orthogonal projection minimizes the error term $\| \mathbf{b} - A\mathbf{x} \|$, for all $\mathbf{x}$ in $\mathbf{col}(A)$. First, we have

$$\| \mathbf{b} - A\mathbf{x} \|^2 = \| \mathbf{w}_1 + \mathbf{w}_2 - A\mathbf{x} \|^2$$
$$= \langle \mathbf{w}_2 + (\mathbf{w}_1 - A\mathbf{x}), \mathbf{w}_2 + (\mathbf{w}_1 - A\mathbf{x}) \rangle$$
$$= \langle \mathbf{w}_2, \mathbf{w}_2 \rangle + 2 \langle \mathbf{w}_2, \mathbf{w}_1 - A\mathbf{x} \rangle + \langle \mathbf{w}_1 - A\mathbf{x}, \mathbf{w}_1 - A\mathbf{x} \rangle$$

Since $\mathbf{w}_1$ and $A\mathbf{x}$ are in W and $\mathbf{w}_2$ is in $W^\perp$, the middle term vanishes, giving

$$\| \mathbf{b} - A\mathbf{x} \|^2 = \langle \mathbf{w}_2, \mathbf{w}_2 \rangle + \langle \mathbf{w}_1 - A\mathbf{x}, \mathbf{w}_1 - A\mathbf{x} \rangle$$
$$= \| \mathbf{w}_2 \|^2 + \| \mathbf{w}_1 - A\mathbf{x} \|^2$$

The quantity on the right-hand side is minimized if $\mathbf{x}$ is any solution to

$$A\mathbf{x} = \mathbf{w}_1$$

Since $\mathbf{w}_1$ is in $\mathbf{col}(A)$, this linear system is consistent. We call any vector $\widehat{\mathbf{x}}$ in $\mathbb{R}^n$ such that $A\mathbf{x} = \mathbf{w}_1$ a **least squares solution** of $A\mathbf{x} = \mathbf{b}$. Moreover, the solution is unique if the columns of A are linearly independent.

Occasionally, as was the case for the example at the beginning of this section, it is possible to find $\mathbf{w}_1$ directly. The least squares solution can then be found by solving $A\mathbf{x} = \mathbf{w}_1$. In most cases, however, the vector $\mathbf{w}_1$ is hard to obtain. Solving the **normal equation**

$$A^t A\mathbf{x} = A^t \mathbf{b}$$

circumvents this difficulty.

THEOREM 11   Let A be an $m \times n$ matrix and $\mathbf{b}$ a vector in $\mathbb{R}^m$. A vector $\widehat{\mathbf{x}}$ in $\mathbb{R}^n$ is a solution to the normal equation

$$A^t A\mathbf{x} = A^t \mathbf{b}$$

if and only if it is a least squares solution to $A\mathbf{x} = \mathbf{b}$.

Proof   From the discussion just before the theorem, we know that a least squares solution $\widehat{\mathbf{x}}$ to $A\mathbf{x} = \mathbf{b}$ exists. By Theorem 9 of Sec. 6.4, there are unique vectors $\mathbf{w}_1$ in $W = \mathbf{col}(A)$ and $\mathbf{w}_2$ in $W^\perp$ such that $\mathbf{b} = \mathbf{w}_1 + \mathbf{w}_2$.

First assume that $\widehat{\mathbf{x}}$ is a least squares solution. Since $\mathbf{w}_2$ is orthogonal to the columns of A, then $A^t \mathbf{w}_2 = \mathbf{0}$. Moreover, since $\widehat{\mathbf{x}}$ is a least squares solution, $A\widehat{\mathbf{x}} = \mathbf{w}_1$. Therefore,

$$A^t A\widehat{\mathbf{x}} = A^t \mathbf{w}_1 = A^t(\mathbf{b} - \mathbf{w}_2) = A^t \mathbf{b}$$

so that $\widehat{\mathbf{x}}$ is a solution to the normal equation.

Conversely, we now show that if $\widehat{\mathbf{x}}$ is a solution to $A^t A\mathbf{x} = A^t \mathbf{b}$, then it is also a least squares solution to $A\mathbf{x} = \mathbf{b}$. Suppose that $A^t A\widehat{\mathbf{x}} = A^t \mathbf{b}$, or equivalently,

$$A^t(\mathbf{b} - A\widehat{\mathbf{x}}) = \mathbf{0}$$

Consequently, the vector $\mathbf{b} - A\widehat{\mathbf{x}}$ is orthogonal to each row of A^t and hence to each column of A. Since the columns of A span W, the vector $\mathbf{b} - A\widehat{\mathbf{x}}$ is in $W^\perp$. Hence, $\mathbf{b}$ can be written as

$$\mathbf{b} = A\widehat{\mathbf{x}} + (\mathbf{b} - A\widehat{\mathbf{x}})$$

where $A\widehat{\mathbf{x}}$ is in $W = \mathbf{col}(A)$ and $\mathbf{b} - A\widehat{\mathbf{x}}$ is in $W^\perp$. Again, by Theorem 9 of Sec. 6.4, this decomposition of the vector $\mathbf{b}$ is unique and hence $A\widehat{\mathbf{x}} = \mathbf{w}_1$. Therefore, $\widehat{\mathbf{x}}$ is a least squares solution.

EXAMPLE 1   Let

$$A = \begin{bmatrix} -2 & 3 \\ 1 & -2 \\ 1 & -1 \end{bmatrix} \quad \text{and} \quad \mathbf{b} = \begin{bmatrix} 1 \\ -1 \\ 2 \end{bmatrix}$$

a. Find the least squares solution to $A\mathbf{x} = \mathbf{b}$.

b. Find the orthogonal projection of $\mathbf{b}$ onto $W = \mathbf{col}(A)$ and the decomposition $\mathbf{b} = \mathbf{w}_1 + \mathbf{w}_2$, where $\mathbf{w}_1$ is in W and $\mathbf{w}_2$ is in $W^\perp$.

Solution   **a.** Since the linear system $A\mathbf{x} = \mathbf{b}$ is inconsistent, the least squares solution is the best approximation we can find. By Theorem 11, the least squares solution can be found by solving the normal equation

$$A^t A\mathbf{x} = A^t \mathbf{b}$$

In this case the normal equation becomes

$$\begin{bmatrix} -2 & 1 & 1 \\ 3 & -2 & -1 \end{bmatrix} \begin{bmatrix} -2 & 3 \\ 1 & -2 \\ 1 & -1 \end{bmatrix} \begin{bmatrix} x \\ y \end{bmatrix} = \begin{bmatrix} -2 & 1 & 1 \\ 3 & -2 & -1 \end{bmatrix} \begin{bmatrix} 1 \\ -1 \\ 2 \end{bmatrix}$$

which simplifies to

$$\begin{bmatrix} 6 & -9 \\ -9 & 14 \end{bmatrix} \begin{bmatrix} x \\ y \end{bmatrix} = \begin{bmatrix} -1 \\ 3 \end{bmatrix}$$

The matrix on the left-hand side is invertible, so that

$$\begin{bmatrix} x \\ y \end{bmatrix} = \frac{1}{3} \begin{bmatrix} 14 & 9 \\ 9 & 6 \end{bmatrix} \begin{bmatrix} -1 \\ 3 \end{bmatrix}$$

The least squares solution is then given by

$$\widehat{\mathbf{x}} = \begin{bmatrix} x \\ y \end{bmatrix} = \begin{bmatrix} \frac{13}{3} \\ 3 \end{bmatrix}$$

b. To find the orthogonal projection $\mathbf{w}_1$ of $\mathbf{b}$ onto $\mathbf{col}(A)$, we use the fact that $\mathbf{w}_1 = A\widehat{\mathbf{x}}$. So

$$\mathbf{w}_1 = \begin{bmatrix} -2 & 3 \\ 1 & -2 \\ 1 & -1 \end{bmatrix} \begin{bmatrix} \frac{13}{3} \\ 3 \end{bmatrix} = \frac{1}{3} \begin{bmatrix} 1 \\ -5 \\ 4 \end{bmatrix}$$

We now find $\mathbf{w}_2$ from the equation $\mathbf{w}_2 = \mathbf{b} - \mathbf{w}_1$, so that

$$\mathbf{w}_2 = \mathbf{b} - \mathbf{w}_1 = \begin{bmatrix} 1 \\ -1 \\ 2 \end{bmatrix} - \frac{1}{3} \begin{bmatrix} 1 \\ -5 \\ 4 \end{bmatrix} = \frac{2}{3} \begin{bmatrix} 1 \\ 1 \\ 1 \end{bmatrix}$$

The decomposition of $\mathbf{b}$ is then given by

$$\mathbf{b} = \mathbf{w}_1 + \mathbf{w}_2 = \frac{1}{3} \begin{bmatrix} 1 \\ -5 \\ 4 \end{bmatrix} + \frac{2}{3} \begin{bmatrix} 1 \\ 1 \\ 1 \end{bmatrix}$$

Note that $\mathbf{w}_2$ is orthogonal to each of the columns of A.

Linear Regression

Example 2 illustrates the use of least squares approximation to find trends in data sets.

EXAMPLE 2   The data in Table 1, which are also shown in the scatter plot in Fig. 5, give the average temperature, in degree celsius ($^\circ C$), of the earth's surface from 1975 through 2002.* Find the equation of the line that best fits these data points.

---

*Worldwatch Institute, Vital Signs 2006–2007. *The trends that are shaping our future*, W. W. Norton and Company, New York London, 2006.

Table 1

Average Global Temperatures 1975–2002					
1975	13.94	1985	14.03	1994	14.25
1976	13.86	1986	14.12	1995	14.37
1977	14.11	1987	14.27	1996	14.23
1978	14.02	1988	14.29	1997	14.40
1979	14.09	1989	14.19	1998	14.56
1980	14.16	1990	14.37	1999	14.32
1981	14.22	1991	14.32	2000	14.31
1982	14.04	1992	14.14	2001	14.46
1983	14.25	1993	14.14	2002	14.52
1984	14.07				

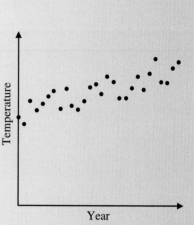

Figure 5

Solution   Denote the data points by (x_i, y_i), for $i = 1, 2, \ldots, 28$, where x_i is the year starting with $x_1 = 1975$ and y_i is the average global temperature for that year. A line with equation $y = mx + b$ will pass through all the data points if the linear system

$$\begin{cases} m(1975) + b = 13.94 \\ m(1976) + b = 13.86 \\ \vdots \\ m(2002) + b = 14.52 \end{cases}$$

has a solution. In matrix form, this linear system becomes

$$\begin{bmatrix} 1975 & 1 \\ 1976 & 1 \\ \vdots & \\ 2002 & 1 \end{bmatrix} \begin{bmatrix} m \\ b \end{bmatrix} = \begin{bmatrix} 13.94 \\ 13.86 \\ \vdots \\ 14.52 \end{bmatrix}$$

Since the linear system is inconsistent, to obtain the best fit of the data we seek values for m and b such that $\mathbf{x} = \begin{bmatrix} m \\ b \end{bmatrix}$ is a least squares solution. The normal equation for this system is given by

$$\begin{bmatrix} 1975 & \cdots & 2002 \\ 1 & \cdots & 1 \end{bmatrix} \begin{bmatrix} 1975 & 1 \\ \vdots & \\ 2002 & 1 \end{bmatrix} \begin{bmatrix} m \\ b \end{bmatrix} = \begin{bmatrix} 1975 & \cdots & 2002 \\ 1 & \cdots & 1 \end{bmatrix} \begin{bmatrix} 13.94 \\ \vdots \\ 14.52 \end{bmatrix}$$

which simplifies to

$$\begin{bmatrix} 110{,}717{,}530 & 55{,}678 \\ 55{,}678 & 28 \end{bmatrix} \begin{bmatrix} m \\ b \end{bmatrix} = \begin{bmatrix} 791{,}553.23 \\ 398.05 \end{bmatrix}$$

The least squares solution is

$$\begin{bmatrix} m \\ b \end{bmatrix} = \begin{bmatrix} 0.0168609742 \\ -19.31197592 \end{bmatrix}$$

The line that best fits the data is then given by $y = 0.0168609742x - 19.31197592$, as shown in Fig. 6.

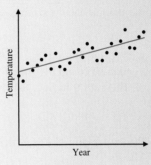

Figure 6

The procedure used in Example 2 can be extended to fit data with a polynomial of any degree $n \geq 1$. For example, if $n = 2$, to find the best-fit parabola of the form $y = ax^2 + bx + c$ for a set of data points requires finding the least squares solution to an $n \times 3$ linear system. See Exercise 6.

Fourier Polynomials

A **trigonometric polynomial** of degree n is an expression in cosines and sines of the form

$$a_0 + a_1 \cos x + b_1 \sin x + a_2 \cos 2x + b_2 \sin 2x + \cdots + a_n \cos nx + b_n \sin nx$$

where the coefficients $a_0, a_1, b_1, a_2, b_2, \ldots, a_n, b_n$ are real numbers. Let $\mathcal{PC}[-\pi, \pi]$ denote the vector space of piecewise continuous functions on the interval $[-\pi, \pi]$. The vector space $\mathcal{PC}[-\pi, \pi]$ is an inner product space with inner product defined by

$$\langle f, g \rangle = \int_{-\pi}^{\pi} f(x)g(x)\,dx$$

Suppose now that given a piecewise continuous function f defined on $[-\pi, \pi]$, which may or may not be a trigonometric polynomial, we wish to find the trigonometric polynomial of degree n that *best* approximates the function.

To solve this problem using linear algebra, let W be the subspace of $\mathcal{PC}[-\pi, \pi]$ of trigonometric polynomials. Let $f_0(x) = 1/\sqrt{2\pi}$, and for $k \geq 1$, let

$$f_k(x) = \frac{1}{\sqrt{\pi}} \cos kx \qquad \text{and} \qquad g_k(x) = \frac{1}{\sqrt{\pi}} \sin kx$$

Define the set B by

$$B = \{f_0, f_1, f_2, \ldots, f_n, g_1, g_2, \ldots, g_n\}$$

$$= \left\{ \frac{1}{\sqrt{2\pi}}, \frac{1}{\sqrt{\pi}} \cos x, \frac{1}{\sqrt{\pi}} \cos 2x, \ldots, \frac{1}{\sqrt{\pi}} \cos nx, \frac{1}{\sqrt{\pi}} \sin x, \frac{1}{\sqrt{\pi}} \sin 2x, \ldots, \frac{1}{\sqrt{\pi}} \sin nx \right\}$$

It can be verified that relative to the inner product above, B is an orthonormal basis for W. Now let f be a function in $\mathcal{PC}[-\pi, \pi]$. Since W is finite dimensional, f has the unique decomposition

$$f = f_W + f_{W^\perp}$$

with f_W in W and $f_{W^\perp}$ in $W^\perp$. Since B is already an orthonormal basis for W, then f_W can be found by using the formula for the orthogonal projection given in the proof of Theorem 9 in Sec. 6.4. In this case we have

$$f_W = \langle f, f_0 \rangle f_0 + \langle f, f_1 \rangle f_1 + \cdots + \langle f, f_n \rangle f_n + \langle f, g_1 \rangle g_1 + \cdots + \langle f, g_n \rangle g_n$$

We now claim that f_W, defined in this way, is the best approximation for f in W. That is,

$$\| f - f_W \| \leq \| f - \mathbf{w} \| \qquad \text{for all } \mathbf{w} \in W$$

To establish the claim, observe that

$$\begin{aligned}
\| f - \mathbf{w} \|^2 &= \left\| f_W + f_{W^\perp} - \mathbf{w} \right\|^2 \\
&= \left\| f_{W^\perp} + (f_W - \mathbf{w}) \right\|^2 \\
&= \left\langle f_{W^\perp} + (f_W - \mathbf{w}), f_{W^\perp} + (f_W - \mathbf{w}) \right\rangle \\
&= \left\langle f_{W^\perp}, f_{W^\perp} \right\rangle + 2 \left\langle f_{W^\perp}, f_W - \mathbf{w} \right\rangle + \left\langle f_W - \mathbf{w}, f_W - \mathbf{w} \right\rangle
\end{aligned}$$

The middle term of the last equation is zero since $f_{W^\perp}$ and $f_W - \mathbf{w}$ are orthogonal. So

$$\| f - \mathbf{w} \|^2 = \left\| f_{W^\perp} \right\|^2 + \| f_W - \mathbf{w} \|^2$$

Observe that the right-hand side is minimized if $\mathbf{w} = f_W$, that is, if we choose $\mathbf{w}$ to be the orthogonal projection of f onto W. The function f_W is called the **Fourier polynomial** of degree n for f.

EXAMPLE 3   Let

$$f(x) = \begin{cases} -1 & -\pi \leq x < 0 \\ 1 & 0 < x \leq \pi \end{cases}$$

Find the Fourier polynomial for f of degree $n = 5$.

Solution   The graph of $y = f(x)$ is shown in Fig. 7. Since $f(x)$ is an odd function and $f_k(x)$ is an even function for $k \geq 0$, the product $f(x)f_k(x)$ is also an odd function. Hence, the integral on any symmetric interval about the origin is 0, and we have

$$\langle f, f_k \rangle = 0 \qquad \text{for } k \geq 0$$

Now for $k \geq 1$, we have

$$\langle f, g_k \rangle = \int_{-\pi}^{\pi} f(x)g_k(x)\, dx$$

$$= -\frac{1}{\sqrt{\pi}} \int_{-\pi}^{0} \sin kx \, dx + \frac{1}{\sqrt{\pi}} \int_{0}^{\pi} \sin kx \, dx$$

$$= \frac{1}{k\sqrt{\pi}}[2 - 2\cos k\pi]$$

$$= \begin{cases} 0 & \text{if } k \text{ is even} \\ \frac{4}{k\sqrt{\pi}} & \text{if } k \text{ is odd} \end{cases}$$

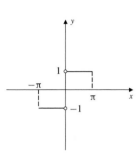

Figure 7

Therefore, the Fourier polynomial of degree 5 that best approximates the function f on the interval $[-\pi, \pi]$ is

$$p(x) = \frac{4}{\pi}\sin x + \frac{4}{3\pi}\sin 3x + \frac{4}{5\pi}\sin 5x$$

In Fig. 8 we see the function and its Fourier approximations for $n = 1, 3,$ and 5.

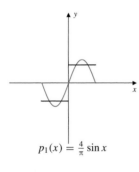

$p_1(x) = \frac{4}{\pi}\sin x$

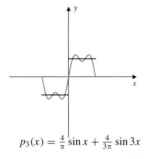

$p_3(x) = \frac{4}{\pi}\sin x + \frac{4}{3\pi}\sin 3x$

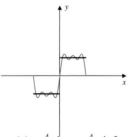

$p_3(x) = \frac{4}{\pi}\sin x + \frac{4}{3\pi}\sin 3x + \frac{4}{5\pi}\sin 5x$

Figure 8

Exercise Set 6.5

1. Let

$$A = \begin{bmatrix} 1 & 3 \\ 1 & 3 \\ 2 & 3 \end{bmatrix} \quad \text{and} \quad \mathbf{b} = \begin{bmatrix} 4 \\ 1 \\ 5 \end{bmatrix}$$

a. Find the least squares solution to $A\mathbf{x} = \mathbf{b}$.

b. Find the orthogonal projection of $\mathbf{b}$ onto $W = \mathbf{col}(A)$ and the decomposition of the vector $\mathbf{b} = \mathbf{w}_1 + \mathbf{w}_2$, where $\mathbf{w}_1$ is in W and $\mathbf{w}_2$ is in $W^{\perp}$.

2. Let

$$A = \begin{bmatrix} 2 & 2 \\ 1 & 2 \\ 1 & 1 \end{bmatrix} \quad \text{and} \quad \mathbf{b} = \begin{bmatrix} -2 \\ 0 \\ 1 \end{bmatrix}$$

a. Find the least squares solution to $A\mathbf{x} = \mathbf{b}$.

b. Find the orthogonal projection of $\mathbf{b}$ onto $W = \text{col}(A)$ and the decomposition of the vector $\mathbf{b} = \mathbf{w}_1 + \mathbf{w}_2$, where $\mathbf{w}_1$ is in W and $\mathbf{w}_2$ is in $W^{\perp}$.

3. The table gives world hydroelectricity use in thousands of terawatthours.

1965	927	1990	2185
1970	1187	1995	2513
1975	1449	2000	2713
1980	1710	2004	2803
1985	2004		

a. Sketch a scatter plot of the data.

b. Find the linear function that is the best fit to the data.

4. The table gives world infant mortality rates in deaths per 1000 live births.

1955	157	1985	78
1960	141	1990	70
1965	119	1995	66
1970	104	2000	62
1975	93	2005	57
1980	87		

a. Sketch a scatter plot of the data.

b. Find the linear function that is the best fit to the data.

5. The table gives world population in billions.

1950	2.56
1960	3.04
1970	3.71
1980	4.46
1990	5.28
2000	6.08

a. Sketch a scatter plot of the data.

b. Find the linear function that is the best fit to the data.

6. The table gives the worldwide cumulative HIV infections in millions.

1980	0.1	1995	29.8
1982	0.7	1997	40.9
1985	2.4	2000	57.9
1987	4.5	2002	67.9
1990	10	2005	82.7
1992	16.1		

a. Sketch a scatter plot of the data.

b. Find a curve of the form $y = ax^2 + bx + c$ that best fits the data.

7. Let $f(x) = x$ on the interval $-\pi \leq x \leq \pi$.

a. Find the Fourier polynomials for f of degrees $n = 2, 3, 4,$ and 5.

b. Sketch $y = f(x)$ along with the polynomials found in part (a).

8. Let

$$f(x) = \begin{cases} x & \text{if } 0 \leq x \leq \pi \\ x + \pi & \text{if } -\pi \leq x < 0 \end{cases}$$

a. Find the Fourier polynomials for f of degrees $n = 2, 3, 4,$ and 5.

b. Sketch $y = f(x)$ along with the polynomials found in part (a).

9. Let $f(x) = x^2$ on the interval $-\pi \le x \le \pi$.

a. Find the Fourier polynomials for f of degrees $n = 2, 3, 4,$ and 5.

b. Sketch $y = f(x)$ along with the polynomials found in part (a).

10. Let A be an $m \times n$ matrix with $\mathbf{rank}(A) = n$, and suppose $A = QR$ is a QR factorization of A. (See Exercise 9, Review Exercises for Chapter 6.) Show that the best least squares solution to the linear system $A\mathbf{x} = \mathbf{b}$ can be found by back substitution on the upper triangular system $R\mathbf{x} = Q^t\mathbf{b}$.

6.6 ▶ Diagonalization of Symmetric Matrices

In Sec. 5.2 a methodology was given for diagonalizing a square matrix. A characterization was also provided to determine which $n \times n$ matrices were diagonalizable. Recall, specifically, from Theorem 2 of Sec. 5.2 that an $n \times n$ matrix is diagonalizable if and only if it has n linearly independent eigenvectors. As we have seen, the application of this theorem requires finding all eigenvectors of a matrix. In certain cases, however, we can tell by inspection if a matrix is diagonalizable. An example of such a case was given in Example 4 of Sec. 5.2, where it was shown that any 2×2 real symmetric matrix is diagonalizable with real eigenvalues. That this is the case in general is the subject of this section.

In the remarks preceding Example 8 of Sec. 3.1, we defined the set of complex numbers $\mathbb{C}$. The proof of our main result requires that the reader be familiar with some of the terminology and notation from complex variables. In particular, if $z = a + bi$ is a complex number, then the **conjugate** of z, denoted by $\bar{z}$, is given by $\bar{z} = a - bi$.

Two complex numbers are equal if and only if their real and imaginary parts are equal. From this we know that a complex number $z = \bar{z}$ if and only if z is a real number. To see this, first suppose that $z = \bar{z}$. Then $bi = -bi$ or $2bi = 0$ and hence $b = 0$. We therefore have $z = a + 0i$ and z is a real number. Conversely, if z is a real number, then $z = a + 0i = a$ and $\bar{z} = a - 0i = a$ so that $z = \bar{z}$.

We can also define this *bar* notation for vectors and matrices. So if $\mathbf{v}$ is a vector with complex components and M is a matrix with complex entries, then

$$\bar{\mathbf{v}} = \begin{bmatrix} \overline{v_1} \\ \overline{v_2} \\ \vdots \\ \overline{v_n} \end{bmatrix} \quad \text{and} \quad \overline{M} = \begin{bmatrix} \overline{a_{11}} & \overline{a_{12}} & \cdots & \overline{a_{1n}} \\ \overline{a_{21}} & \overline{a_{22}} & \cdots & \overline{a_{2n}} \\ \vdots & \vdots & \ddots & \vdots \\ \overline{a_{m1}} & \overline{a_{12}} & \cdots & \overline{a_{mn}} \end{bmatrix}$$

We are now ready to state our main result.

THEOREM 12   The eigenvalues of an $n \times n$ real symmetric matrix A are all real numbers.

Proof   Let $\mathbf{v}$ be an eigenvector of A corresponding to the eigenvalue λ. To show that λ is a real number, we will show that $\overline{\lambda} = \lambda$. We first consider the matrix product $(\overline{\mathbf{v}}^t A\mathbf{v})^t$, which by Theorem 6 of Sec. 1.3 can be written as

$$(\overline{\mathbf{v}}^t A\mathbf{v})^t = \mathbf{v}^t A^t \overline{\mathbf{v}}$$

Since A is symmetric, $A^t = A$. Also since A has real entries, then $\overline{A} = A$. Therefore,

$$\mathbf{v}^t A^t \overline{\mathbf{v}} = \mathbf{v}^t A\overline{\mathbf{v}} = \mathbf{v}^t \overline{A}\overline{\mathbf{v}} = \mathbf{v}^t \overline{A\mathbf{v}}$$

Now as $\mathbf{v}$ is an eigenvector of A corresponding to the eigenvalue λ, then $A\mathbf{v} = \lambda\mathbf{v}$, so that

$$\mathbf{v}^t \overline{A\mathbf{v}} = \mathbf{v}^t \overline{\lambda\mathbf{v}} = \mathbf{v}^t \overline{\lambda}\overline{\mathbf{v}} = \overline{\lambda}\mathbf{v}^t \overline{\mathbf{v}}$$

Alternatively, the original expression can be evaluated by

$$(\overline{\mathbf{v}}^t A\mathbf{v})^t = (\overline{\mathbf{v}}^t \lambda\mathbf{v})^t = \mathbf{v}^t \lambda\overline{\mathbf{v}} = \lambda\mathbf{v}^t \overline{\mathbf{v}}$$

Equating these results gives

$$\lambda\mathbf{v}^t \overline{\mathbf{v}} = \overline{\lambda}\mathbf{v}^t \overline{\mathbf{v}} \qquad \text{that is} \qquad (\lambda - \overline{\lambda})\mathbf{v}^t \overline{\mathbf{v}} = \mathbf{0}$$

Since $\mathbf{v}$ is an eigenvector of A, and therefore nonzero, so is $\mathbf{v}^t \overline{\mathbf{v}}$. By an extension to the complex numbers of Theorem 2 (part 4) of Sec. 3.1, we have $\lambda - \overline{\lambda} = 0$; hence $\lambda = \overline{\lambda}$, establishing that λ is a real number.

One consequence of Theorem 12 is that the eigenvectors of a real symmetric matrix have real components. To see this, let A be a symmetric matrix with real entries and $\mathbf{v}$ an eigenvector corresponding to the real eigenvalue $\lambda = a$. Observe that $\mathbf{v}$ is a vector in the null space of the real $n \times n$ matrix $A - aI$. By Theorem 7 of Sec. 3.2, $N(A - aI)$ is a subspace of $\mathbb{R}^n$. Thus, $\mathbf{v}$ being a vector in $\mathbb{R}^n$, has real components as claimed.

EXAMPLE 1   Let A be the symmetric matrix defined by

$$A = \begin{bmatrix} 2 & 0 & 2 \\ 0 & 0 & -2 \\ 2 & -2 & 1 \end{bmatrix}$$

Verify that the eigenvalues and corresponding eigenvectors of A are real.

Solution The characteristic equation of A is

$$\det(A - \lambda I) = -\lambda^3 + 3\lambda^2 + 6\lambda - 8 = 0$$

After factoring the characteristic polynomial we obtain

$$(\lambda - 1)(\lambda + 2)(\lambda - 4) = 0$$

Thus, the eigenvalues of A are $\lambda_1 = 1$, $\lambda_2 = -2$, and $\lambda_3 = 4$.

To find eigenvectors corresponding to $\lambda = 1$, we find the null space of $A - I$. To do this, we see that

$$A - I = \begin{bmatrix} 1 & 0 & 2 \\ 0 & -1 & -2 \\ 2 & -2 & 0 \end{bmatrix} \quad \text{reduces to} \quad \begin{bmatrix} 1 & 0 & 2 \\ 0 & 1 & 2 \\ 0 & 0 & 0 \end{bmatrix}$$

Thus, an eigenvector corresponding to $\lambda_1 = 1$ is $\mathbf{v}_1 = \begin{bmatrix} -2 \\ -2 \\ 1 \end{bmatrix}$. In a similar way

we have that eigenvectors corresponding to $\lambda_2 = -2$ and $\lambda_3 = 4$ are, respectively,

$$\mathbf{v}_2 = \begin{bmatrix} 1 \\ -2 \\ -2 \end{bmatrix} \quad \text{and} \quad \mathbf{v}_3 = \begin{bmatrix} -2 \\ 1 \\ -2 \end{bmatrix}.$$

Orthogonal Diagonalization

In Sec. 6.1 we showed that two vectors $\mathbf{u}$ and $\mathbf{v}$ in $\mathbb{R}^n$ are orthogonal if and only if their dot product $\mathbf{u} \cdot \mathbf{v} = 0$. An equivalent formulation of this condition can be developed by using matrix multiplication. To do this, observe that if $\mathbf{u}$ and $\mathbf{v}$ are vectors in $\mathbb{R}^n$, then $\mathbf{v}^t \mathbf{u}$ is a matrix with a single entry equal to $\mathbf{u} \cdot \mathbf{v}$. Hence, we know that $\mathbf{u}$ and $\mathbf{v}$ are orthogonal if and only if $\mathbf{v}^t \mathbf{u} = \mathbf{0}$.

Theorem 13 shows that eigenvectors which correspond to distinct eigenvalues of a real symmetric matrix are orthogonal.

THEOREM 13 Let A be a real symmetric matrix and $\mathbf{v}_1$ and $\mathbf{v}_2$ be eigenvectors corresponding, respectively, to the distinct eigenvalues λ_1 and λ_2. Then $\mathbf{v}_1$ and $\mathbf{v}_2$ are orthogonal.

Proof We have already shown that $\mathbf{v}_1$ and $\mathbf{v}_2$ are vectors in $\mathbb{R}^n$. To show that they are orthogonal, we show that $\mathbf{v}_1^t \mathbf{v}_2 = \mathbf{0}$. Now, since λ_2 is an eigenvalue of A, then $A\mathbf{v}_2 = \lambda_2 \mathbf{v}_2$, so that

$$\mathbf{v}_1^t A \mathbf{v}_2 = \mathbf{v}_1^t \lambda_2 \mathbf{v}_2 = \lambda_2 \mathbf{v}_1^t \mathbf{v}_2$$

Also since $A^t = A$,

$$\mathbf{v}_1^t A \mathbf{v}_2 = \mathbf{v}_1^t A^t \mathbf{v}_2 = (A\mathbf{v}_1)^t \mathbf{v}_2 = \lambda_1 \mathbf{v}_1^t \mathbf{v}_2$$

Equating the two expressions for $\mathbf{v}_1^t A \mathbf{v}_2$, we obtain

$$(\lambda_1 - \lambda_2)\mathbf{v}_1^t \mathbf{v}_2 = \mathbf{0}$$

Since $\lambda_1 \neq \lambda_2$ then $\lambda_1 - \lambda_2 \neq 0$. Hence, by Theorem 2, part 4, of Sec. 3.1, we have $\mathbf{v}_1^t \mathbf{v}_2 = \mathbf{0}$, which, by the remarks preceding this theorem, gives that $\mathbf{v}_1$ is orthogonal to $\mathbf{v}_2$.

EXAMPLE 2

Let A be the real symmetric matrix given by

$$A = \begin{bmatrix} 1 & 0 & 0 \\ 0 & 0 & 1 \\ 0 & 1 & 0 \end{bmatrix}$$

Show that the eigenvectors corresponding to distinct eigenvalues of A are orthogonal.

Solution

The characteristic equation of A is

$$\det(A - \lambda I) = -(\lambda - 1)^2(\lambda + 1) = 0$$

so the eigenvalues are $\lambda_1 = 1$ and $\lambda_2 = -1$. Then the eigenspaces (see Sec. 5.1) are given by

$$V_{\lambda_1} = \mathbf{span}\left\{ \begin{bmatrix} 1 \\ 0 \\ 0 \end{bmatrix}, \begin{bmatrix} 0 \\ 1 \\ 1 \end{bmatrix} \right\} \quad \text{and} \quad V_{\lambda_2} = \mathbf{span}\left\{ \begin{bmatrix} 0 \\ -1 \\ 1 \end{bmatrix} \right\}$$

Since every vector in V_{λ_1} is a linear combination of $\mathbf{u} = \begin{bmatrix} 1 \\ 0 \\ 0 \end{bmatrix}$ and $\mathbf{v} = \begin{bmatrix} 0 \\ 1 \\ 1 \end{bmatrix}$,

and $\mathbf{w} = \begin{bmatrix} 0 \\ -1 \\ 1 \end{bmatrix}$ is orthogonal to both $\mathbf{u}$ and $\mathbf{v}$, then by Proposition 5 of Sec. 6.4, $\mathbf{w}$ is orthogonal to every eigenvector in V_{λ_1}. Hence, every eigenvector in V_{λ_2} is orthogonal to every eigenvector in V_{λ_1}.

In Example 2, we showed that every vector in the eigenspace V_{λ_1} is orthogonal to every vector in the eigenspace V_{λ_2}. Notice, moreover, that the vectors within V_{λ_1} are orthogonal to one another. In this case, the matrix has a special factorization. We normalize the spanning vectors of the eigenspaces to obtain

$$\begin{bmatrix} 1 \\ 0 \\ 0 \end{bmatrix} \qquad \begin{bmatrix} 0 \\ \frac{1}{\sqrt{2}} \\ \frac{1}{\sqrt{2}} \end{bmatrix} \qquad \text{and} \qquad \begin{bmatrix} 0 \\ -\frac{1}{\sqrt{2}} \\ \frac{1}{\sqrt{2}} \end{bmatrix}$$

Using these vectors, we construct the matrix

$$P = \begin{bmatrix} 1 & 0 & 0 \\ 0 & \frac{1}{\sqrt{2}} & -\frac{1}{\sqrt{2}} \\ 0 & \frac{1}{\sqrt{2}} & \frac{1}{\sqrt{2}} \end{bmatrix}$$

which is then used to diagonalize A. That is,

$$P^{-1}AP = \begin{bmatrix} 1 & 0 & 0 \\ 0 & \frac{1}{\sqrt{2}} & \frac{1}{\sqrt{2}} \\ 0 & -\frac{1}{\sqrt{2}} & \frac{1}{\sqrt{2}} \end{bmatrix} \begin{bmatrix} 1 & 0 & 0 \\ 0 & 0 & 1 \\ 0 & 1 & 0 \end{bmatrix} \begin{bmatrix} 1 & 0 & 0 \\ 0 & \frac{1}{\sqrt{2}} & -\frac{1}{\sqrt{2}} \\ 0 & \frac{1}{\sqrt{2}} & \frac{1}{\sqrt{2}} \end{bmatrix} = \begin{bmatrix} 1 & 0 & 0 \\ 0 & 1 & 0 \\ 0 & 0 & -1 \end{bmatrix}$$

Observe in this case that the diagonalizing matrix P has the special property that $PP^t = I$, so that $P^{-1} = P^t$. This leads to Definition 1.

DEFINITION 1   **Orthogonal Matrix**   A square matrix P is called an **orthogonal matrix** if it is invertible and $P^{-1} = P^t$.

One important property of orthogonal matrices is that the column (and row) vectors of an $n \times n$ orthogonal matrix are an *orthonormal* basis for $\mathbb{R}^n$. That is, the vectors of this basis are all mutually orthogonal and have unit length.

As we mentioned at the beginning of this section, one particularly nice fact about symmetric matrices is that they are diagonalizable. So by Theorem 2 of Sec. 5.2 a real symmetric matrix has n linearly independent eigenvectors. For the matrix A of Example 2, the eigenvectors are all mutually orthogonal. Producing an orthogonal matrix P to diagonalize A required only that we normalize the eigenvectors. In many cases there is more to do. Specifically, by Theorem 2, eigenvectors corresponding to distinct eigenvalues are orthogonal. However, if the geometric multiplicity of an eigenvalue λ is greater than 1, then the vectors within V_λ (while linearly independent), might not be mutually orthogonal. In this case we can use the Gram-Schmidt process, given in Sec. 6.3, to find an orthonormal basis from the linearly independent eigenvectors. The previous discussion is summarized in Theorem 14.

THEOREM 14   Let A be an $n \times n$ real symmetric matrix. Then there is an orthogonal matrix P and a diagonal matrix D such that $P^{-1}AP = P^tAP = D$. The eigenvalues are the diagonal entries of D.

The following steps can be used to diagonalize an $n \times n$ real symmetric matrix A.

1. Find the eigenvalues and corresponding eigenvectors of A.
2. Since A is diagonalizable, there are n linearly independent eigenvectors. If necessary, use the Gram-Schmidt process to find an orthonormal set of eigenvectors.
3. Form the orthogonal matrix P with column vectors determined in Step 2.
4. The matrix $P^{-1}AP = P^t AP = D$ is a diagonal matrix.

EXAMPLE 3   Let

$$A = \begin{bmatrix} 0 & 1 & 1 \\ 1 & 0 & 1 \\ 1 & 1 & 0 \end{bmatrix}$$

Find an orthogonal matrix P such that $P^{-1}AP$ is a diagonal matrix.

Solution   The characteristic equation for A is given by

$$\det(A - \lambda I) = -\lambda^3 + 3\lambda + 2 = -(\lambda - 2)(\lambda + 1)^2 = 0$$

Thus, the eigenvalues are $\lambda_1 = -1$ and $\lambda_2 = 2$. The corresponding eigenspaces are

$$V_{\lambda_1} = \mathbf{span} \left\{ \begin{bmatrix} -1 \\ 1 \\ 0 \end{bmatrix}, \begin{bmatrix} -1 \\ 0 \\ 1 \end{bmatrix} \right\} \quad \text{and} \quad V_{\lambda_2} = \mathbf{span} \left\{ \begin{bmatrix} 1 \\ 1 \\ 1 \end{bmatrix} \right\}$$

Let B be the set of vectors

$$B = \{\mathbf{v}_1, \mathbf{v}_2, \mathbf{v}_3\} = \left\{ \begin{bmatrix} 1 \\ 1 \\ 1 \end{bmatrix}, \begin{bmatrix} -1 \\ 1 \\ 0 \end{bmatrix}, \begin{bmatrix} -1 \\ 0 \\ 1 \end{bmatrix} \right\}$$

Since B is a linearly independent set of three vectors, by Theorem 2 of Sec. 5.2, A is diagonalizable. To find an orthogonal matrix P which diagonalizes A, we use the Gram-Schmidt process on B. This was done in Example 3 of Sec. 6.3, yielding the orthonormal basis

$$B' = \left\{ \frac{1}{\sqrt{3}} \begin{bmatrix} 1 \\ 1 \\ 1 \end{bmatrix}, \frac{1}{\sqrt{2}} \begin{bmatrix} -1 \\ 1 \\ 0 \end{bmatrix}, \frac{1}{\sqrt{6}} \begin{bmatrix} -1 \\ -1 \\ 2 \end{bmatrix} \right\}$$

Now let P be the matrix given by

$$P = \begin{bmatrix} \frac{\sqrt{3}}{3} & -\frac{\sqrt{2}}{2} & -\frac{\sqrt{6}}{6} \\ \frac{\sqrt{3}}{3} & \frac{\sqrt{2}}{2} & -\frac{\sqrt{6}}{6} \\ \frac{\sqrt{3}}{3} & 0 & \frac{\sqrt{6}}{3} \end{bmatrix}$$

Observe that P is an orthogonal matrix with $P^{-1} = P^t$. Moreover,

$$P^t AP = \begin{bmatrix} 2 & 0 & 0 \\ 0 & -1 & 0 \\ 0 & 0 & -1 \end{bmatrix}$$

Fact Summary

Let A be an $n \times n$ real symmetric matrix.

1. The eigenvalues of A are all real numbers.
2. The eigenvectors corresponding to distinct eigenvalues of A are orthogonal.
3. The matrix A is diagonalizable.
4. There is an orthogonal matrix P such that $D = P^{-1}AP = P^t AP$, where D is a diagonal matrix with diagonal entries the eigenvalues of A.

Exercise Set 6.6

In Exercises 1–4, verify that the eigenvalues of the symmetric matrix are all real numbers.

1. $A = \begin{bmatrix} 1 & 2 \\ 2 & 1 \end{bmatrix}$

2. $A = \begin{bmatrix} -1 & 3 \\ 3 & -1 \end{bmatrix}$

3. $A = \begin{bmatrix} 1 & 2 & 0 \\ 2 & -1 & 2 \\ 0 & 2 & 1 \end{bmatrix}$

4. $A = \begin{bmatrix} 1 & 1 & -2 \\ 1 & -1 & 2 \\ -2 & 2 & 1 \end{bmatrix}$

In Exercises 5–8, verify that the eigenvectors of the symmetric matrix corresponding to distinct eigenvalues are orthogonal.

5. $A = \begin{bmatrix} 1 & 2 \\ 2 & -2 \end{bmatrix}$

6. $A = \begin{bmatrix} -3 & 2 \\ 2 & -3 \end{bmatrix}$

7. $A = \begin{bmatrix} 1 & 2 & 0 \\ 2 & -1 & -2 \\ 0 & -2 & 1 \end{bmatrix}$

8. $A = \begin{bmatrix} 1 & 0 & -2 \\ 0 & -1 & 0 \\ -2 & 0 & 1 \end{bmatrix}$

In Exercises 9–12, find the eigenspaces of the $n \times n$ symmetric matrix, and verify that the sum of the dimensions of the eigenspaces is n.

9. $A = \begin{bmatrix} 1 & 0 & 2 \\ 0 & -1 & 0 \\ 2 & 0 & 1 \end{bmatrix}$

10. $A = \begin{bmatrix} 1 & 0 & 1 \\ 0 & -1 & 0 \\ 1 & 0 & 1 \end{bmatrix}$

11. $A = \begin{bmatrix} 2 & 1 & 1 & 1 \\ 1 & -2 & 1 & 1 \\ 1 & 1 & -1 & 0 \\ 1 & 1 & 0 & -1 \end{bmatrix}$

12. $A = \begin{bmatrix} 1 & 0 & 0 & 0 \\ 0 & -2 & 0 & 0 \\ 0 & 0 & -1 & 0 \\ 0 & 0 & 0 & 1 \end{bmatrix}$

In Exercises 13–16, determine whether the matrix is orthogonal.

13. $A = \begin{bmatrix} \frac{\sqrt{3}}{2} & \frac{1}{2} \\ -\frac{1}{2} & \frac{\sqrt{3}}{2} \end{bmatrix}$

14. $A = \begin{bmatrix} -1 & -\frac{\sqrt{5}}{5} \\ 0 & -\frac{2\sqrt{5}}{5} \end{bmatrix}$

15. $A = \begin{bmatrix} \frac{\sqrt{2}}{2} & \frac{\sqrt{2}}{2} & 0 \\ -\frac{\sqrt{2}}{2} & \frac{\sqrt{2}}{2} & 0 \\ 0 & 0 & 1 \end{bmatrix}$

16. $A = \begin{bmatrix} \frac{2}{3} & \frac{2}{3} & \frac{1}{3} \\ -\frac{2}{3} & \frac{2}{3} & \frac{1}{3} \\ \frac{1}{3} & 0 & 1 \end{bmatrix}$

In Exercises 17–22, for the given matrix A find an orthogonal matrix P and a diagonal matrix D such that $D = P^{-1}AP$.

17. $A = \begin{bmatrix} 3 & 4 \\ 4 & 3 \end{bmatrix}$

18. $A = \begin{bmatrix} 5 & 2 \\ 2 & 5 \end{bmatrix}$

19. $A = \begin{bmatrix} -1 & 3 \\ 3 & -1 \end{bmatrix}$

20. $A = \begin{bmatrix} 1 & 2 \\ 2 & -2 \end{bmatrix}$

21. $A = \begin{bmatrix} 1 & -1 & 1 \\ -1 & -1 & 1 \\ 1 & 1 & 1 \end{bmatrix}$

22. $A = \begin{bmatrix} 1 & 0 & -1 \\ 0 & -1 & 0 \\ -1 & 0 & 1 \end{bmatrix}$

23. Show that if A and B are orthogonal matrices, then AB and BA are orthogonal matrices.

24. Show that if A is an orthogonal matrix, then $\det(A) = \pm 1$.

25. Show that if A is an orthogonal matrix, then A^t is an orthogonal matrix.

26. Show that if A is an orthogonal matrix, then A^{-1} is an orthogonal matrix.

27. a. Show that the matrix
$$A = \begin{bmatrix} \cos\theta & -\sin\theta \\ \sin\theta & \cos\theta \end{bmatrix}$$
is orthogonal.

b. Suppose that A is a 2×2 orthogonal matrix. Show that there is a real number θ such that
$$A = \begin{bmatrix} \cos\theta & -\sin\theta \\ \sin\theta & \cos\theta \end{bmatrix}$$
or
$$A = \begin{bmatrix} \cos\theta & \sin\theta \\ \sin\theta & -\cos\theta \end{bmatrix}$$
(*Hint*: Consider the equation $A^t A = I$.)

c. Suppose that A is an orthogonal 2×2 matrix and $T: \mathbb{R}^2 \to \mathbb{R}^2$ is a linear operator defined by $T(\mathbf{v}) = A\mathbf{v}$. Show that if $\det(A) = 1$, then T is a rotation and if $\det(A) = -1$, then T is a reflection about the x axis followed by a rotation.

28. Matrices A are B are *orthogonally similar* if there is an orthogonal matrix P such that $B = P^t AP$. Suppose that A and B are orthogonally similar.
a. Show that A is symmetric if and only if B is symmetric.
b. Show that A is orthogonal if and only if B is orthogonal.

29. Suppose that A is an $n \times n$ matrix such that there exists a diagonal matrix D and an orthogonal matrix P such that $D = P^t AP$. (Matrix A is called **orthogonally diagonalizable**.) Show that A is symmetric.

30. Suppose A is invertible and orthogonally diagonalizable. Show that A^{-1} is orthogonally diagonalizable. (See Exercise 29.)

31. Let A be an $n \times n$ skew-symmetric matrix.
a. If $\mathbf{v}$ is in $\mathbb{R}^n$, expand $\mathbf{v}^t \mathbf{v}$ in terms of the components of the vector.
b. Show that the only possible real eigenvalue of A is $\lambda = 0$. [*Hint*: Consider the quantity $\mathbf{v}^t(\lambda\mathbf{v})$.]

6.7 ▶ **Application: Quadratic Forms**

When working with complicated algebraic expressions, mathematicians will often attempt to simplify problems by applying transformations designed to make these expressions easier to interpret, or at least better suited to the task at hand. In this section we show how certain transformations of the coordinate axes in $\mathbb{R}^2$ can be used to simplify equations that describe *conic sections*, that is, equations in x and y whose graphs are parabolas, hyperbolas, circles, and ellipses. As an illustration, consider the equation

$$x^2 - 4x + y^2 - 6y - 3 = 0$$

To simplify this equation, we complete the square on $x^2 - 4x$ and $y^2 - 6y$ to obtain

$$(x^2 - 4x + 4) + (y^2 - 6y + 9) = 3 + 4 + 9$$

that is,

$$(x - 2)^2 + (y - 3)^2 = 16$$

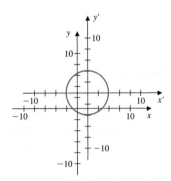

Figure 1

This last equation describes a circle of radius 4 centered at the point $(2, 3)$. The graph is shown in Fig. 1. To further simplify this equation, we can translate the coordinate axes by means of the equations

$$x' = x - 2 \quad \text{and} \quad y' = y - 3$$

The equation of the circle then becomes

$$\left(x'\right)^2 + \left(y'\right)^2 = 16$$

This is the equation of the circle in *standard position* in the $x'y'$ plane with center at the origin.

Rotation of Axes

The most general **quadratic equation** in two variables has the form

$$ax^2 + bxy + cy^2 + dx + ey + f = 0$$

where a, b, c, d, e, and f are real numbers such that at least one of a, b, or c is not zero. The graph of a quadratic equation in x and y is a conic section (including possible degenerate cases), the particular one being dependent on the values of the coefficients. When $b \neq 0$, the conic section is rotated from standard position. The expression

$$ax^2 + bxy + cy^2$$

is called the **associated quadratic form**. For example, the quadratic equation

$$2x^2 + 5xy - 7y^2 + 2x - 4y + 1 = 0$$

has an associated quadratic form given by

$$2x^2 + 5xy - 7y^2$$

The quadratic equation

$$ax^2 + bxy + cy^2 + dx + ey + f = 0$$

is also given in matrix form by setting

$$\mathbf{x} = \begin{bmatrix} x \\ y \end{bmatrix} \qquad A = \begin{bmatrix} a & \frac{b}{2} \\ \frac{b}{2} & c \end{bmatrix} \qquad \text{and} \qquad \mathbf{b} = \begin{bmatrix} d \\ e \end{bmatrix}$$

Then the quadratic equation above is equivalent to

$$\mathbf{x}^t A \mathbf{x} + \mathbf{b}^t \mathbf{x} + f = 0$$

The quadratic form (in matrix form) is then given by

$$\mathbf{x}^t A \mathbf{x}$$

As an illustration, the quadratic equation $2x^2 + 5xy + y^2 + 3x - y + 1 = 0$ in matrix form is given by

$$[x \; y] \begin{bmatrix} 2 & \frac{5}{2} \\ \frac{5}{2} & 1 \end{bmatrix} \begin{bmatrix} x \\ y \end{bmatrix} + [3 \;\; -1] \begin{bmatrix} x \\ y \end{bmatrix} + 1 = 0$$

The associated quadratic form is

$$[x \; y] \begin{bmatrix} 2 & \frac{5}{2} \\ \frac{5}{2} & 1 \end{bmatrix} \begin{bmatrix} x \\ y \end{bmatrix}$$

Observe that the matrix A, for any quadratic equation in two variables, is symmetric; that is, $A^t = A$. This fact enables us to develop a transformation that we can use to simplify the equation. Specifically, the map we desire will rotate the coordinate axes by the precise angle needed to situate the conic section in standard position with respect to a new coordinate system.

To produce such a mapping, first recall from Theorem 14 of Sec. 6.6 that if A is a real symmetric matrix, then there exists an orthogonal matrix P and a diagonal matrix D such that $A = PDP^{-1} = PDP^t$. Next, we need to examine which orthogonal 2×2 matrices are rotations.

Now by Exercise 27(b) of Sec. 6.6, a real orthogonal 2×2 matrix has the form

$$B = \begin{bmatrix} \cos\theta & -\sin\theta \\ \sin\theta & \cos\theta \end{bmatrix} \quad \text{or} \quad B' = \begin{bmatrix} \cos\theta & \sin\theta \\ \sin\theta & -\cos\theta \end{bmatrix}$$

Next, recall from Sec. 4.6 that B is the matrix representation, relative to the standard basis for $\mathbb{R}^2$, of a linear operator which rotates a vector in the plane by θ rad. The matrix B' is not a rotation (relative to any basis). To see this, let $Q = \{\mathbf{v}_1, \mathbf{v}_2\}$ be a basis for $\mathbb{R}^2$ and $\theta = 0$. Then

$$B' = \begin{bmatrix} 1 & 0 \\ 0 & -1 \end{bmatrix}$$

Relative to the basis Q, this matrix produces a reflection through the line spanned by $\mathbf{v}_1$. For example, if Q is the standard basis for $\mathbb{R}^2$, then B' is a reflection through the x axis. These results are summarized in Theorem 15.

THEOREM 15   Let B be a real orthogonal 2×2 matrix. The change of coordinates given by

$$\begin{bmatrix} x' \\ y' \end{bmatrix} = B \begin{bmatrix} x \\ y \end{bmatrix}$$

is a rotation if and only if $\det(B) = 1$.

We are now in a position to analyze quadratic equations in two variables. Start with C a conic section with equation

$$\mathbf{x}^t A \mathbf{x} + \mathbf{b}^t \mathbf{x} + f = 0$$

Let P be the orthogonal matrix that diagonalizes A, so that

$$A = PDP^t \quad \text{where} \quad D = \begin{bmatrix} \lambda_1 & 0 \\ 0 & \lambda_2 \end{bmatrix}$$

with λ_1 and λ_2 being the eigenvalues of A. As P is orthogonal, by the above remarks on the form of P, its determinant is either $+1$ or -1. If $\det(P) = -1$, then interchange the column vectors of P, along with the diagonal entries of D. Since

$$\begin{bmatrix} \sin\theta & \cos\theta \\ -\cos\theta & \sin\theta \end{bmatrix} = \begin{bmatrix} \cos\left(\frac{\pi}{2} - \theta\right) & \sin\left(\frac{\pi}{2} - \theta\right) \\ -\sin\left(\frac{\pi}{2} - \theta\right) & \cos\left(\frac{\pi}{2} - \theta\right) \end{bmatrix}$$

a rearrangement of the column vectors of P is a rotation. To obtain the equation for C in the $x'y'$ coordinate system, substitute $\mathbf{x} = P\mathbf{x}'$ into $\mathbf{x}^t A \mathbf{x} + \mathbf{b}^t \mathbf{x} + f = 0$ to obtain

$$\left(P\mathbf{x}'\right)^t A \left(P\mathbf{x}'\right) + \mathbf{b}^t P\mathbf{x}' + f = 0$$

By Theorem 6 of Sec. 1.3, if the product of A and B is defined, then $(AB)^t = B^t A^t$, and since matrix multiplication is associative, we have

$$(\mathbf{x}')^t P^t A P \mathbf{x}' + \mathbf{b}^t P\mathbf{x}' + f = 0 \quad \text{that is,} \quad (\mathbf{x}')^t D\mathbf{x}' + \mathbf{b}^t P\mathbf{x}' + f = 0$$

Let $\mathbf{b}^t P = \begin{bmatrix} d' \\ e' \end{bmatrix}$. The last equation can now be written as

$$\lambda_1(x')^2 + \lambda_2(y')^2 + d'x' + e'y' + f = 0$$

This equation gives the conic section C in standard position in the $x'y'$ coordinate system. The type of conic section depends on the eigenvalues. Specifically, C is

1. An ellipse if λ_1 and λ_2 have the same sign
2. A hyperbola if λ_1 and λ_2 have opposite signs
3. A parabola if either λ_1 or λ_2 is zero

EXAMPLE 1   Let C be the conic section whose equation is $x^2 - xy + y^2 - 8 = 0$.

    a. Transform the equation to $x'y'$ coordinates so that C is in standard position with no $x'y'$ term.

    b. Find the angle of rotation between the standard coordinate axes and the $x'y'$ coordinate system.

Solution   **a.** The matrix form of this equation is given by

$$\mathbf{x}^t A \mathbf{x} - 8 = 0 \qquad \text{with} \quad A = \begin{bmatrix} 1 & -\frac{1}{2} \\ -\frac{1}{2} & 1 \end{bmatrix}$$

The eigenvalues of A are $\lambda_1 = \frac{1}{2}$ and $\lambda_2 = \frac{3}{2}$, with corresponding (unit) eigenvectors

$$\mathbf{v}_1 = \frac{1}{\sqrt{2}} \begin{bmatrix} 1 \\ 1 \end{bmatrix} \qquad \text{and} \qquad \mathbf{v}_2 = \frac{1}{\sqrt{2}} \begin{bmatrix} -1 \\ 1 \end{bmatrix}$$

Then the orthogonal matrix

$$P = \frac{1}{\sqrt{2}} \begin{bmatrix} 1 & -1 \\ 1 & 1 \end{bmatrix}$$

diagonalizes A. Moreover, since $\det(P) = 1$, then by Theorem 15, the coordinate transformation is a rotation. Making the substitution $\mathbf{x} = P\mathbf{x}'$ in the matrix equation above gives

$$(\mathbf{x}')^t P^t A P \mathbf{x}' - 8 = 0$$

that is,

$$(\mathbf{x}')^t D \mathbf{x}' - 8 = 0 \qquad \text{where} \quad D = \begin{bmatrix} \frac{1}{2} & 0 \\ 0 & \frac{3}{2} \end{bmatrix}$$

This last equation can now be written as

$$[x' \ y'] \begin{bmatrix} \frac{1}{2} & 0 \\ 0 & \frac{3}{2} \end{bmatrix} \begin{bmatrix} x' \\ y' \end{bmatrix} - 8 = 0$$

so that the standard form for the equation of the ellipse in the $x'y'$ coordinate system is

$$\frac{(x')^2}{16} + \frac{3(y')^2}{16} = 1$$

This is the equation of an ellipse with x' as the major axis and y' as the minor axis, as shown in Fig. 2.

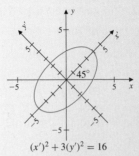

$$(x')^2 + 3(y')^2 = 16$$

Figure 2

b. To find the angle between the original axes and the $x'y'$ coordinate system, observe that the eigenvector $\mathbf{v}_1$ points in the direction of the x' axis. Now using Definition 3 of Sec. 6.1, the cosine of the angle between $\mathbf{e}_1$ and $\mathbf{v}_1$ is given by

$$\cos\theta = \frac{\mathbf{e}_1 \cdot \mathbf{v}_1}{\|\mathbf{e}_1\|\,\|\mathbf{v}_1\|} = \frac{1}{\sqrt{2}} \qquad \text{so that} \qquad \theta = \frac{\pi}{4}$$

An alternative way to find the angle between the axes is to note that the matrix P, which is the transition matrix from $x'y'$ coordinates to xy coordinates, can be written as

$$P = \begin{bmatrix} \frac{1}{\sqrt{2}} & -\frac{1}{\sqrt{2}} \\ \frac{1}{\sqrt{2}} & \frac{1}{\sqrt{2}} \end{bmatrix} = \begin{bmatrix} \cos\theta & -\sin\theta \\ \sin\theta & \cos\theta \end{bmatrix} \qquad \text{with } \theta = \frac{\pi}{4}$$

Example 2 involves a rotation and a translation.

EXAMPLE 2

Describe the conic section C whose equation is

$$2x^2 - 4xy - y^2 - 4x - 8y + 14 = 0$$

Solution

The equation for C has the form $\mathbf{x}^t A\mathbf{x} + \mathbf{b}^t\mathbf{x} + f = 0$ given by

$$[x \; y]\begin{bmatrix} 2 & -2 \\ -2 & -1 \end{bmatrix}\begin{bmatrix} x \\ y \end{bmatrix} + [-4 \; -8]\begin{bmatrix} x \\ y \end{bmatrix} + 14 = 0$$

The eigenvalues of $A = \begin{bmatrix} 2 & -2 \\ -2 & -1 \end{bmatrix}$ are $\lambda_1 = -2$ and $\lambda_2 = 3$, with corresponding (unit) eigenvectors

$$\mathbf{v}_1 = \frac{1}{\sqrt{5}}\begin{bmatrix} 1 \\ 2 \end{bmatrix} \qquad \text{and} \qquad \mathbf{v}_2 = \frac{1}{\sqrt{5}}\begin{bmatrix} -2 \\ 1 \end{bmatrix}$$

Since the eigenvalues have opposite sign, the conic section C is a hyperbola. To describe the hyperbola, we first diagonalize A. Using the unit eigenvectors, the orthogonal matrix that diagonalizes A is

$$P = \frac{1}{\sqrt{5}} \begin{bmatrix} 1 & -2 \\ 2 & 1 \end{bmatrix} \qquad \text{with} \qquad \begin{bmatrix} -2 & 0 \\ 0 & 3 \end{bmatrix} = P^{t}AP$$

Making the substitution $\mathbf{x} = P\mathbf{x}'$ in the equation $\mathbf{x}^{t}A\mathbf{x} + \mathbf{b}^{t}\mathbf{x} + f = 0$ gives

$$[x'\ y'] \begin{bmatrix} -2 & 0 \\ 0 & 3 \end{bmatrix} \begin{bmatrix} x' \\ y' \end{bmatrix} + [-4\ -8] \begin{bmatrix} \frac{1}{\sqrt{5}} & \frac{-2}{\sqrt{5}} \\ \frac{2}{\sqrt{5}} & \frac{1}{\sqrt{5}} \end{bmatrix} \begin{bmatrix} x' \\ y' \end{bmatrix} + 14 = 0$$

After simplification of this equation we obtain

$$-2(x')^2 - 4\sqrt{5}x' + 3(y')^2 + 14 = 0$$

that is,

$$-2[(x')^2 + 2\sqrt{5}(x')] + 3(y')^2 + 14 = 0$$

After completing the square on x', we obtain

$$-2[(x')^2 + 2\sqrt{5}(x') + 5] + 3(y')^2 = -14 - 10$$

that is,

$$\frac{(x' + \sqrt{5})^2}{12} - \frac{(y')^2}{8} = 1$$

This last equation describes a hyperbola with x' as the major axis. An additional transformation translating the x' axis allows us to simplify the result even further. If we let

$$x'' = x' + \sqrt{5} \qquad \text{and} \qquad y'' = y'$$

then the equation now becomes

$$\frac{(x'')^2}{12} - \frac{(y'')^2}{8} = 1$$

The graph is shown in Fig. 3.

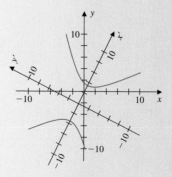

Figure 3

Quadric Surfaces

The graph of a quadratic equation in three variables of the form

$$ax^2 + bxy + cxz + dy^2 + eyz + fz^2 + gx + hy + iz + j = 0$$

is an ellipsoid, a hyperboloid, a paraboloid, or a cone. As in the two-dimensional case, the terms gx, hy, and iz produce translations from standard form, while the *mixed terms xy, xz,* and *yz* produce rotations. The quadratic form

$$ax^2 + bxy + cxz + dy^2 + eyz + fz^2$$

can be written in matrix form as

$$\mathbf{x}^t A \mathbf{x} = [x \ y \ z] \begin{bmatrix} a & \frac{b}{2} & \frac{c}{2} \\ \frac{b}{2} & d & \frac{e}{2} \\ \frac{c}{2} & \frac{e}{2} & f \end{bmatrix} \begin{bmatrix} x \\ y \\ z \end{bmatrix}$$

As before, a rotation developed from the eigenvectors of A can be used to transform the quadric surface to one in standard form

$$\lambda_1 (x')^2 + \lambda_2 (y')^2 + \lambda_3 (z')^2 + j = 0$$

where λ_1, λ_2, and λ_3 are the eigenvalues of A. We omit the details.

EXAMPLE 3 Write the quadratic equation

$$5x^2 + 4y^2 - 5z^2 + 8xz = 36$$

in standard form by eliminating the xz term.

Solution Let

$$A = \begin{bmatrix} 5 & 0 & 4 \\ 0 & 4 & 0 \\ 4 & 0 & -5 \end{bmatrix}$$

Then the quadratic equation can be written as

$$[x \ y \ z] \begin{bmatrix} 5 & 0 & 4 \\ 0 & 4 & 0 \\ 4 & 0 & -5 \end{bmatrix} \begin{bmatrix} x \\ y \\ z \end{bmatrix} = 36$$

The eigenvalues of the matrix A are

$$\lambda_1 = \sqrt{41} \qquad \lambda_2 = -\sqrt{41} \qquad \lambda_3 = 4$$

Hence, the quadric surface, in standard position, has the equation

$$\sqrt{41}(x')^2 - \sqrt{41}(y')^2 + 4(z')^2 = 36$$

The graph of the surface, which is a hyperboloid of one sheet, is shown in Fig. 4.

Figure 4

Exercise Set 6.7

In Exercises 1–6, let C denote the conic section given by the equation. Transform the equation to $x'y'$ coordinates so that C is in standard position with no $x'y'$ term.

1. $27x^2 - 18xy + 3y^2 + x + 3y = 0$

2. $2x^2 - 8xy + 8y^2 + 2x + y = 0$

3. $12x^2 + 8xy + 12y^2 - 8 = 0$

4. $11x^2 - 6xy + 19y^2 + 2x + 4y - 12 = 0$

5. $-x^2 - 6xy - y^2 + 8 = 0$

6. $xy = 1$

7. Let C denote the conic section in standard position given by the equation $4x^2 + 16y^2 = 16$.
   a. Write the quadratic equation in matrix form.
   b. Find the quadratic equation that describes the conic C rotated by $45°$.

8. Let C denote the conic section in standard position given by the equation $x^2 - y^2 = 1$.

a. Write the quadratic equation in matrix form.
b. Find the quadratic equation that describes the conic C rotated by $-30°$.

9. Let C denote the conic section in standard position given by the equation $16x^2 + 4y^2 = 16$.
   a. Find the quadratic equation for the conic section obtained by rotating C by $60°$.
   b. Find the quadratic equation that describes the conic found in part (a) after a translation 3 units to the right and 2 units upward.

10. Let C denote the conic section in standard position given by the equation $x^2 - y = 0$.
    a. Find the quadratic equation for the conic section obtained by rotating C by $30°$.
    b. Find the quadratic equation that describes the conic found in part (a) after a translation 2 units to the right and 1 unit downward.

6.8 ▶ Application: Singular Value Decomposition

In earlier sections we have examined various ways to write a given matrix as a product of other matrices with special properties. For example, with the LU factorization of Sec. 1.7, we saw that an $m \times n$ matrix A could be written as $A = LU$ with L being an invertible lower triangular matrix and U an upper triangular matrix. Also in Sec. 1.7, we showed that if A is invertible, then it could be written as the product of elementary matrices. In Sec. 5.2 it was shown that an $n \times n$ matrix A with n linearly independent eigenvectors can be written as

$$A = PDP^{-1}$$

where D is a diagonal matrix of eigenvalues of A. As a special case, if A is symmetric, then A has the factorization

$$A = QDQ^t$$

where Q is an orthogonal matrix.

   In this section we consider a generalization of this last result for $m \times n$ matrices. Specifically, we introduce the *singular value decomposition*, abbreviated as SVD, which enables us to write any $m \times n$ matrix as

$$A = U\Sigma V^t$$

where U is an $m \times m$ orthogonal matrix, V is an $n \times n$ orthogonal matrix, and Σ is an $m \times n$ matrix with numbers, called *singular values*, on its diagonal.

Singular Values of an $m \times n$ Matrix

To define the singular values of an $m \times n$ matrix A, we consider the matrix $A^t A$. Observe that since A is an $m \times n$ matrix, A^t is an $n \times m$ matrix, so the product $A^t A$ is a square $n \times n$ matrix. This new matrix is symmetric since $(A^t A)^t = A^t A^{tt} = A^t A$. Hence, by Theorem 14 of Sec. 6.6, there is an orthogonal matrix P such that

$$P^t (A^t A) P = D$$

where D is a diagonal matrix of the eigenvalues of $A^t A$ given by

$$D = \begin{bmatrix} \lambda_1 & 0 & \cdots & 0 \\ 0 & \lambda_2 & \cdots & 0 \\ \vdots & \vdots & \ddots & \vdots \\ 0 & \cdots & \cdots & \lambda_n \end{bmatrix}$$

Since by Exercise 39 of Sec. 6.3 the matrix $A^t A$ is positive semidefinite, we also have, by Exercise 41 of Sec. 6.3, that $\lambda_i \geq 0$ for $1 \leq i \leq n$. This permits us to make the following definition.

DEFINITION 1

Singular Values   Let A be an $m \times n$ matrix. The **singular values** of A, denoted by σ_i for $1 \leq i \leq n$, are the positive square roots of the eigenvalues $\lambda_1, \ldots, \lambda_n$ of $A^t A$. That is,

$$\sigma_i = \sqrt{\lambda_i} \qquad \text{for} \quad 1 \leq i \leq n$$

It is customary to write the singular values of A in decreasing order

$$\sigma_1 \geq \sigma_2 \geq \cdots \geq \sigma_n$$

As mentioned in Sec. 5.2, this can be accomplished by permuting the columns of the diagonalizing matrix P.

EXAMPLE 1

Let A be the matrix given by

$$A = \begin{bmatrix} 1 & 1 \\ 0 & 1 \\ 1 & 0 \end{bmatrix}$$

Find the singular values of A.

Solution   The singular values of A are found by first computing the eigenvalues of the square matrix

$$A^t A = \begin{bmatrix} 1 & 0 & 1 \\ 1 & 1 & 0 \end{bmatrix} \begin{bmatrix} 1 & 1 \\ 0 & 1 \\ 1 & 0 \end{bmatrix} = \begin{bmatrix} 2 & 1 \\ 1 & 2 \end{bmatrix}$$

The characteristic equation, in this case, is given by

$$\det(A^t A - \lambda I) = (\lambda - 3)(\lambda - 1) = 0$$

The eigenvalues of $A^t A$ are then $\lambda_1 = 3$ and $\lambda_2 = 1$, so that the singular values are $\sigma_1 = \sqrt{3}$ and $\sigma_2 = 1$.

We have already seen that orthogonal bases are desirable and the Gram-Schmidt process can be used to construct an orthogonal basis from any basis. If A is an $m \times n$ matrix and $\mathbf{v}_1, \ldots, \mathbf{v}_r$ are the eigenvectors of $A^t A$, then we will see that $\{A\mathbf{v}_1, \ldots, A\mathbf{v}_r\}$ is an orthogonal basis for $\mathbf{col}(A)$. We begin with the connection between the singular values of A and the vectors $A\mathbf{v}_1, \ldots, A\mathbf{v}_r$.

THEOREM 16   Let A be an $m \times n$ matrix and let $B = \{\mathbf{v}_1, \mathbf{v}_2, \ldots, \mathbf{v}_n\}$ be an orthonormal basis of $\mathbb{R}^n$ consisting of eigenvectors of $A^t A$, with corresponding eigenvectors λ_1, $\lambda_2, \ldots, \lambda_n$. Then

1. $\| A\mathbf{v}_i \| = \sigma_i$ for each $i = 1, 2, \ldots, n$.
2. $A\mathbf{v}_i$ is orthogonal to $A\mathbf{v}_j$ for $i \neq j$.

Proof   For the first statement recall from Sec. 6.6 that the length of a vector $\mathbf{v}$ in Euclidean space can be given by the matrix product $\| \mathbf{v} \| = \sqrt{\mathbf{v}^t \mathbf{v}}$. Therefore,

$$\| A\mathbf{v}_i \|^2 = (A\mathbf{v}_i)^t (A\mathbf{v}_i) = \mathbf{v}_i^t (A^t A)\mathbf{v}_i = \mathbf{v}_i^t \lambda_i \mathbf{v}_i = \lambda_i \| \mathbf{v}_i \| = \lambda_i$$

The last equality is due to the fact that $\mathbf{v}_i$ is a unit vector. Part 1 is established by noting that $\sigma_i = \sqrt{\lambda_i} = \| A\mathbf{v}_i \|$. For part 2 of the theorem, we know that (as in Sec. 6.6) the dot product of two vectors $\mathbf{u}$ and $\mathbf{v}$ in Euclidean space can be given by the matrix product $\mathbf{u} \cdot \mathbf{v} = \mathbf{u}^t \mathbf{v}$. Thus, since B is an orthonormal basis of $\mathbb{R}^n$, if $i \neq j$, then

$$(A\mathbf{v}_i) \cdot (A\mathbf{v}_j) = (A\mathbf{v}_i)^t (A\mathbf{v}_j) = \mathbf{v}_i^t (A^t A)\mathbf{v}_j = \mathbf{v}_i^t \lambda_j \mathbf{v}_j = \lambda_j \mathbf{v}_i^t \mathbf{v}_j = 0$$

In Theorem 16, the set of vectors $\{A\mathbf{v}_1, A\mathbf{v}_2, \ldots, A\mathbf{v}_n\}$ is shown to be orthogonal. In Theorem 17 we establish that the eigenvectors of $A^t A$, after multiplication by A, are an orthogonal basis for $\mathbf{col}(A)$.

THEOREM 17   Let A be an $m \times n$ matrix and $B = \{\mathbf{v}_1, \mathbf{v}_2, \ldots, \mathbf{v}_n\}$ an orthonormal basis of $\mathbb{R}^n$ consisting of eigenvectors of $A^t A$. Suppose that the corresponding eigenvalues satisfy $\lambda_1 \geq \lambda_2 \geq \cdots \geq \lambda_r > \lambda_{r+1} = \cdots \lambda_n = 0$, that is, $A^t A$ has r nonzero eigenvalues. Then $B' = \{A\mathbf{v}_1, A\mathbf{v}_2, \ldots, A\mathbf{v}_r\}$ is an orthogonal basis for the column space of A and $\mathbf{rank}(A) = r$.

Proof   First observe that since $\sigma_i = \sqrt{\lambda_i}$ are all nonzero for $1 \le i \le r$; then by Theorem 16, part 1, we have $A\mathbf{v}_1, A\mathbf{v}_2, \ldots, A\mathbf{v}_r$ are all nonzero vectors in $\mathbf{col}(A)$. By part 2 of Theorem 16, we have $B' = \{A\mathbf{v}_1, A\mathbf{v}_2, \ldots, A\mathbf{v}_r\}$ is an orthogonal set of vectors in $\mathbb{R}^m$. Hence, by Theorem 5 of Sec. 6.2, B' is linearly independent. Now to show that these vectors span the column space of A, let $\mathbf{w}$ be a vector in $\mathbf{col}(A)$. Thus, there exists a vector $\mathbf{v}$ in $\mathbb{R}^n$ such that $A\mathbf{v} = \mathbf{w}$. Since $B = \{\mathbf{v}_1, \mathbf{v}_2, \ldots, \mathbf{v}_n\}$ is a basis for $\mathbb{R}^n$, there are scalars $c_1, c_2, \ldots, c_n$ such that

$$\mathbf{v} = c_1\mathbf{v}_1 + c_2\mathbf{v}_2 + \cdots + c_n\mathbf{v}_n$$

Multiplying both sides of the last equation by A, we obtain

$$A\mathbf{v} = c_1 A\mathbf{v}_1 + c_2 A\mathbf{v}_2 + \cdots + c_n A\mathbf{v}_n$$

Now, using the fact that $A\mathbf{v}_{r+1} = A\mathbf{v}_{r+2} = \cdots = A\mathbf{v}_n = \mathbf{0}$, then

$$A\mathbf{v} = c_1 A\mathbf{v}_1 + c_2 A\mathbf{v}_2 + \cdots + c_r A\mathbf{v}_r$$

so that $\mathbf{w} = A\mathbf{v}$ is in $\mathbf{span}\{A\mathbf{v}_1, A\mathbf{v}_2, \ldots, A\mathbf{v}_r\}$. Consequently, $B' = \{A\mathbf{v}_1, A\mathbf{v}_2, \ldots, A\mathbf{v}_r\}$ is an orthogonal basis for the column space of A, and the rank of A is equal to the number of its nonzero singular values.

EXAMPLE 2

Let A be the matrix given by

$$A = \begin{bmatrix} 1 & 1 \\ 0 & 1 \\ 1 & 0 \end{bmatrix}$$

Find the image of the unit circle under the linear transformation $T: \mathbb{R}^2 \to \mathbb{R}^3$ defined by $T(\mathbf{v}) = A\mathbf{v}$.

Solution   From Example 1, the eigenvalues of $A^t A$ are $\lambda_1 = 3$ and $\lambda_2 = 1$, with eigenvectors

$$\mathbf{v}_1 = \begin{bmatrix} 1/\sqrt{2} \\ 1/\sqrt{2} \end{bmatrix} \quad \text{and} \quad \mathbf{v}_2 = \begin{bmatrix} -1/\sqrt{2} \\ 1/\sqrt{2} \end{bmatrix}$$

respectively. The singular values of A are then $\sigma_1 = \sqrt{3}$ and $\sigma_2 = 1$. Let $C(t)$ be the unit circle given by $\cos(t)\mathbf{v}_1 + \sin(t)\mathbf{v}_2$ for $0 \le t \le 2\pi$. The image of $C(t)$ under T is given by

$$T(C(t)) = \cos(t)A\mathbf{v}_1 + \sin(t)A\mathbf{v}_2$$

By Theorem 17, $B' = \left\{ \frac{1}{\sigma_1} A\mathbf{v}_1, \frac{1}{\sigma_2} A\mathbf{v}_2 \right\}$ is a basis for the range of T. Hence, the coordinates of $T(C(t))$ relative to B' are $x' = \sigma_1 \cos t = \sqrt{3} \cos t$ and $y' = \sigma_2 \sin t = \sin t$. Observe that

$$\left(\frac{x'}{\sqrt{3}} \right)^2 + (y')^2 = \frac{(x')^2}{3} + (y')^2 = \cos^2 t + \sin^2 t = 1$$

which is an ellipse with the length of the semimajor axis equal to σ_1 and length of the semiminor axis equal to σ_2, as shown in Fig. 1.

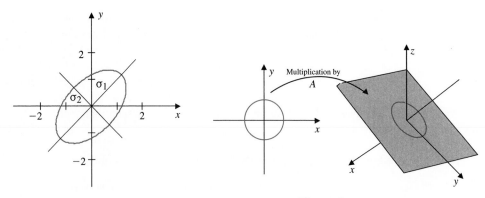

Figure 1

For certain matrices, some of the singular values may be zero. As an illustration, consider the matrix $A = \begin{bmatrix} 1 & 2 \\ 3 & 6 \end{bmatrix}$. For this matrix, we have $\textbf{col}(A) = \textbf{span} \begin{bmatrix} 1 \\ 3 \end{bmatrix}$. The reduced row echelon form for A is the matrix $\begin{bmatrix} 1 & 2 \\ 0 & 0 \end{bmatrix}$, which has only one pivot column. Hence, the rank of A is equal to 1. The eigenvalues of $A^t A$ are $\lambda_1 = 50$ and $\lambda_2 = 0$ with corresponding unit eigenvectors

$$\mathbf{v}_1 = \begin{bmatrix} 1/\sqrt{5} \\ 2/\sqrt{5} \end{bmatrix} \qquad \text{and} \qquad \mathbf{v}_2 = \begin{bmatrix} -2/\sqrt{5} \\ 1/\sqrt{5} \end{bmatrix}$$

The singular values of A are given by $\sigma_1 = 5\sqrt{2}$ and $\sigma_2 = 0$. Now, multiplying $\mathbf{v}_1$ and $\mathbf{v}_2$ by A gives

$$A\mathbf{v}_1 = \begin{bmatrix} \sqrt{5} \\ 3\sqrt{5} \end{bmatrix} \qquad \text{and} \qquad A\mathbf{v}_2 = \begin{bmatrix} 0 \\ 0 \end{bmatrix}$$

Observe that $A\mathbf{v}_1$ spans the one dimensional column space of A. In this case, the linear transformation $T: \mathbb{R}^2 \longrightarrow \mathbb{R}^2$ defined by $T(\mathbf{x}) = A\mathbf{x}$ maps the unit circle to the line segment

$$\left\{ t \begin{bmatrix} \sqrt{5} \\ 3\sqrt{5} \end{bmatrix} \middle| -1 \le t \le 1 \right\}$$

as shown in Fig. 2.

Singular Value Decomposition (SVD)

We now turn our attention to the problem of finding a singular value decomposition of an $m \times n$ matrix A.

THEOREM 18 **SVD** Let A be an $m \times n$ matrix of rank r, with r nonzero singular values $\sigma_1, \sigma_2, \ldots, \sigma_r$. Then there exists an $m \times n$ matrix Σ, an $m \times m$ orthogonal matrix

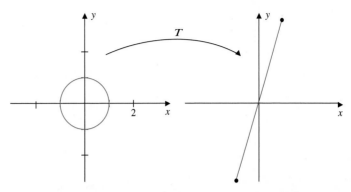

Figure 2

U, and an $n \times n$ orthogonal matrix V such that

$$A = U \Sigma V^t$$

Proof Since $A^t A$ is an $n \times n$ symmetric matrix, by Theorem 14 of Sec. 6.6 there is an orthonormal basis $\{\mathbf{v}_1, \ldots, \mathbf{v}_n\}$ of $\mathbb{R}^n$, consisting of eigenvectors of $A^t A$. Now by Theorem 17, $\{A\mathbf{v}_1, \ldots, A\mathbf{v}_r\}$ is an orthogonal basis for $\mathbf{col}(A)$. Let $\{\mathbf{u}_1, \ldots, \mathbf{u}_r\}$ be the orthonormal basis for $\mathbf{col}(A)$, given by

$$\mathbf{u}_i = \frac{1}{\|A\mathbf{v}_i\|} A\mathbf{v}_i = \frac{1}{\sigma_i} A\mathbf{v}_i \qquad \text{for} \qquad i = 1, \ldots, r$$

Next, extend $\{\mathbf{u}_1, \ldots, \mathbf{u}_r\}$ to the orthonormal basis $\{\mathbf{u}_1, \ldots, \mathbf{u}_m\}$ of $\mathbb{R}^m$. We can now define the orthogonal matrices V and U, using the vectors $\{\mathbf{v}_1, \ldots, \mathbf{v}_n\}$ and $\{\mathbf{u}_1, \ldots, \mathbf{u}_m\}$, respectively, as column vectors, so that

$$V = [\ \mathbf{v}_1 \quad \mathbf{v}_2 \quad \cdots \quad \mathbf{v}_n\ ] \qquad \text{and} \qquad U = [\ \mathbf{u}_1 \quad \mathbf{u}_2 \quad \cdots \quad \mathbf{u}_m\ ]$$

Moreover, since $A\mathbf{v}_i = \sigma_i \mathbf{u}_i$, for $i = 1, \ldots, r$, then

$$AV = \begin{bmatrix} A\mathbf{v}_1 & \cdots & A\mathbf{v}_r & \mathbf{0} & \cdots & \mathbf{0} \end{bmatrix} = \begin{bmatrix} \sigma_1\mathbf{u}_1 & \cdots & \sigma_r\mathbf{u}_r & \mathbf{0} & \cdots & \mathbf{0} \end{bmatrix}$$

Now let Σ be the $m \times n$ matrix given by

$$\Sigma = \left[\begin{array}{cccc|ccc} \sigma_1 & 0 & \cdots & 0 & 0 & \cdots & 0 \\ 0 & \sigma_2 & \cdots & 0 & 0 & \cdots & 0 \\ \vdots & \vdots & \ddots & \vdots & \vdots & \vdots & \vdots \\ 0 & \cdots & \cdots & \sigma_r & 0 & \cdots & 0 \\ \hline 0 & \cdots & \cdots & 0 & 0 & \cdots & 0 \\ \vdots & & & \vdots & \vdots & & \vdots \\ 0 & \cdots & \cdots & 0 & 0 & \cdots & 0 \end{array} \right]$$

Then

$$U\Sigma = \begin{bmatrix} \mathbf{u}_1 & \mathbf{u}_2 & \cdots & \mathbf{u}_m \end{bmatrix} \Sigma$$

$$= \begin{bmatrix} \sigma_1\mathbf{u}_1 & \cdots & \sigma_r\mathbf{u}_r & \mathbf{0} & \cdots & \mathbf{0} \end{bmatrix}$$

$$= AV$$

Since V is orthogonal, then $V^t = V^{-1}$, and hence, $A = U\Sigma V^t$.

EXAMPLE 3   Find a singular value decomposition of the matrix

$$A = \begin{bmatrix} -1 & 1 \\ -1 & 1 \\ 2 & -2 \end{bmatrix}$$

Solution   A procedure for finding an SVD of A is included in the proof of Theorem 18. We present the solution as a sequence of steps.

Step 1. *Find the eigenvalues and corresponding orthonormal eigenvectors of A^tA and define the matrix V.*

The eigenvalues of the matrix

$$A^tA = \begin{bmatrix} 6 & -6 \\ -6 & 6 \end{bmatrix}$$

in decreasing order are given by $\lambda_1 = 12$ and $\lambda_2 = 0$. The corresponding orthonormal eigenvectors are

$$\mathbf{v}_1 = \begin{bmatrix} -1/\sqrt{2} \\ 1/\sqrt{2} \end{bmatrix} \quad \text{and} \quad \mathbf{v}_2 = \begin{bmatrix} 1/\sqrt{2} \\ 1/\sqrt{2} \end{bmatrix}$$

Since the column vectors of V are given by the orthonormal eigenvectors of A^tA, the matrix V is given by

$$V = \begin{bmatrix} -1/\sqrt{2} & 1/\sqrt{2} \\ 1/\sqrt{2} & 1/\sqrt{2} \end{bmatrix}$$

Step 2. *Find the singular values of A and define the matrix Σ.*

The singular values of A are the square roots of the eigenvalues of A^tA, so that

$$\sigma_1 = \sqrt{\lambda_1} = 2\sqrt{3} \quad \text{and} \quad \sigma_2 = \sqrt{\lambda_2} = 0$$

Since Σ has the same dimensions as A, then Σ is 3×2. In this case,

$$\Sigma = \begin{bmatrix} 2\sqrt{3} & 0 \\ 0 & 0 \\ 0 & 0 \end{bmatrix}$$

Step 3. *Define the matrix U.*

The matrix A has one nonzero singular value, so by Theorem 17 the rank of A is 1. Therefore, the first column of U is

$$\mathbf{u}_1 = \frac{1}{\sigma_1} A\mathbf{v}_1 = \begin{bmatrix} 1/\sqrt{6} \\ 1/\sqrt{6} \\ -2/\sqrt{6} \end{bmatrix}$$

Next we extend the set $\{\mathbf{u}_1\}$ to an orthonormal basis for $\mathbb{R}^3$ by adding to it the vectors

$$\mathbf{u}_2 = \begin{bmatrix} 2/\sqrt{5} \\ 0 \\ 1/\sqrt{5} \end{bmatrix} \quad \text{and} \quad \mathbf{u}_3 = \begin{bmatrix} -1/\sqrt{2} \\ 1/\sqrt{2} \\ 0 \end{bmatrix}$$

so that

$$U = \begin{bmatrix} 1/\sqrt{6} & 2/\sqrt{5} & -1/\sqrt{2} \\ 1/\sqrt{6} & 0 & 1/\sqrt{2} \\ -2/\sqrt{6} & 1/\sqrt{5} & 0 \end{bmatrix}$$

The singular value decomposition of A is then given by

$$A = U \Sigma V^t = \begin{bmatrix} 1/\sqrt{6} & 2/\sqrt{5} & -1/\sqrt{2} \\ 1/\sqrt{6} & 0 & 1/\sqrt{2} \\ -2/\sqrt{6} & 1/\sqrt{5} & 0 \end{bmatrix} \begin{bmatrix} 2\sqrt{3} & 0 \\ 0 & 0 \\ 0 & 0 \end{bmatrix} \begin{bmatrix} -1/\sqrt{2} & 1/\sqrt{2} \\ 1/\sqrt{2} & 1/\sqrt{2} \end{bmatrix}$$

$$= \begin{bmatrix} -1 & 1 \\ -1 & 1 \\ 2 & -2 \end{bmatrix}$$

In Example 3, the process of finding a singular value decomposition of A was complicated by the task of extending the set $\{\mathbf{u}_1, \ldots, \mathbf{u}_r\}$ to an orthogonal basis for $\mathbb{R}^m$. Alternatively, we can use A^tA to find V and AA^t to find U. To see this, note that if $A = U\Sigma V^t$ is an SVD of A, then $A^t = V\Sigma^t U^t$. After multiplying A on the left by its transpose, we obtain

$$A^t A = V \Sigma^t U^t U \Sigma V^t = V D_1 V^t$$

where D_1 is an $n \times n$ diagonal matrix with diagonal entries the eigenvalues of $A^t A$. Hence, V is an orthogonal matrix that diagonalizes $A^t A$. On the other hand,

$$AA^t = U \Sigma V^t V \Sigma^t U^t = U D_2 U^t$$

where D_2 is an $m \times m$ diagonal matrix with diagonal entries the eigenvalues of AA^t and U is an orthogonal matrix that diagonalizes AA^t. Note that the matrices $A^t A$ and AA^t have the same eigenvalues. (See Exercise 22 of Sec. 5.1.) Therefore, the nonzero diagonal entries of D_1 and D_2 are the same. The matrices U and V found using this procedure are not unique. We also note that changing the signs of the column vectors in U and V also produces orthogonal matrices that diagonalize AA^t and A^tA. As a result, finding an SVD of A may require changing the signs of certain columns of U or V.

In Example 4 we use this idea to find an SVD for a matrix.

EXAMPLE 4   Find a singular value decomposition of the matrix

$$A = \begin{bmatrix} 1 & 1 \\ 3 & -3 \end{bmatrix}$$

Solution   First observe that

$$A^tA = \begin{bmatrix} 1 & 3 \\ 1 & -3 \end{bmatrix}\begin{bmatrix} 1 & 1 \\ 3 & -3 \end{bmatrix} = \begin{bmatrix} 10 & -8 \\ -8 & 10 \end{bmatrix}$$

By inspection we see that $\mathbf{v}_1 = \frac{1}{\sqrt{2}}\begin{bmatrix} 1 \\ -1 \end{bmatrix}$ is a unit eigenvector of A^tA with corresponding eigenvalue $\lambda_1 = 18$, and $\mathbf{v}_2 = \frac{1}{\sqrt{2}}\begin{bmatrix} 1 \\ 1 \end{bmatrix}$ is a unit eigenvector of A^tA with corresponding eigenvalue $\lambda_2 = 2$. Hence,

$$V = \frac{1}{\sqrt{2}}\begin{bmatrix} 1 & 1 \\ -1 & 1 \end{bmatrix}$$

The singular values of A are $\sigma_1 = 3\sqrt{2}$ and $\sigma_2 = \sqrt{2}$ so that

$$\Sigma = \begin{bmatrix} 3\sqrt{2} & 0 \\ 0 & \sqrt{2} \end{bmatrix}$$

To find U, we compute

$$AA^t = \begin{bmatrix} 1 & 1 \\ 3 & -3 \end{bmatrix}\begin{bmatrix} 1 & 3 \\ 1 & -3 \end{bmatrix} = \begin{bmatrix} 2 & 0 \\ 0 & 18 \end{bmatrix}$$

Observe that a unit eigenvector corresponding to $\lambda_1 = 18$ is $\mathbf{u}_1 = \begin{bmatrix} 0 \\ 1 \end{bmatrix}$ and a unit eigenvector corresponding to $\lambda_2 = 2$ is $\mathbf{u}_2 = \begin{bmatrix} 1 \\ 0 \end{bmatrix}$. Thus,

$$U = \begin{bmatrix} 0 & 1 \\ 1 & 0 \end{bmatrix}$$

A singular value decomposition of A is then given by

$$A = U\Sigma V^t = \begin{bmatrix} 0 & 1 \\ 1 & 0 \end{bmatrix}\begin{bmatrix} 3\sqrt{2} & 0 \\ 0 & \sqrt{2} \end{bmatrix}\begin{bmatrix} \frac{1}{\sqrt{2}} & \frac{-1}{\sqrt{2}} \\ \frac{1}{\sqrt{2}} & \frac{1}{\sqrt{2}} \end{bmatrix} = \begin{bmatrix} 1 & 1 \\ 3 & -3 \end{bmatrix}$$

The Four Fundamental Subspaces

In this subsection we show how the matrices U and V, which give the singular value decomposition of A, provide orthonormal bases for the four fundamental subspaces of A, introduced in Sec. 6.4. To develop this idea, let A be an $m \times n$ matrix of rank $r \leq n$ and $B = \{\mathbf{v}_1, \ldots, \mathbf{v}_n\}$ be an orthonormal basis of eigenvectors of $A^t A$ with corresponding eigenvalues $\lambda_1 \geq \lambda_2 \geq \cdots \geq \lambda_r > \lambda_{r+1} = \cdots = \lambda_n = 0$. First, from the proof of Theorem 17 if $\sigma_1, \cdots, \sigma_r$ are the nonzero singular values of A, then

$$C' = \left\{ \frac{1}{\sigma_1} A\mathbf{v}_1, \ldots, \frac{1}{\sigma_r} A\mathbf{v}_r \right\} = \{\mathbf{u}_1, \ldots, \mathbf{u}_r\}$$

is a basis for $\mathbf{col}(A)$. Next, the remaining columns of U are defined by extending C' to an orthonormal basis $C = \{\mathbf{u}_1, \ldots, \mathbf{u}_r, \mathbf{u}_{r+1}, \ldots, \mathbf{u}_m\}$ for $\mathbb{R}^m$. We claim that $C'' = \{\mathbf{u}_{r+1}, \ldots, \mathbf{u}_m\}$ is an orthonormal basis for $N(A^t)$. To see this, observe that each vector of C' is orthogonal to each vector of C''. Hence, by Proposition 5 of Sec. 6.4 and the fact that $\dim(\mathbb{R}^m) = m$, we have $\mathbf{span}\{\mathbf{u}_{r+1}, \ldots, \mathbf{u}_m\} = \mathbf{col}(A)^\perp$. By Theorem 10, part 2, of Sec. 6.4, $\mathbf{span}\{\mathbf{u}_{r+1}, \ldots, \mathbf{u}_m\} = N(A^t)$, so that $C'' = \{\mathbf{u}_{r+1}, \ldots, \mathbf{u}_m\}$ is a basis for $N(A^t)$ as claimed. We now turn our attention to the matrix V. From the proof of Theorem 16, we have $A\mathbf{v}_{r+1} = \cdots = A\mathbf{v}_n = \mathbf{0}$. Consequently, $\mathbf{span}\{\mathbf{v}_{r+1}, \ldots, \mathbf{v}_n\}$ is contained in $N(A)$. Now by Theorem 5 of Sec. 4.2,

$$\dim(N(A)) + \dim(\mathbf{col}(A)) = n$$

so that $\dim(N(A)) = n - r$. Since $B'' = \{\mathbf{v}_{r+1}, \ldots, \mathbf{v}_n\}$ is an orthogonal, and hence linearly independent, set of $n - r$ vectors in $N(A)$, by Theorem 12, part (1), of Sec. 3.3, B'' is a basis for $N(A)$. Finally, since $B = \{\mathbf{v}_1, \ldots, \mathbf{v}_n\}$ is an orthonormal basis for $\mathbb{R}^n$, each vector of B'' is orthogonal to every vector in $B' = \{\mathbf{v}_1, \ldots, \mathbf{v}_r\}$. Hence,

$$\mathbf{span}\{\mathbf{v}_1, \ldots, \mathbf{v}_r\} = N(A)^\perp = \mathbf{col}(A^t) = \mathbf{row}(A)$$

so that B' is a basis for $\mathbf{row}(A)$.

To illustrate the ideas of this discussion, consider the matrix A of Example 3 and its SVD. By the above discussion, we have

$$\mathbf{row}(A) = \mathbf{span}\left\{ \begin{bmatrix} -1 \\ 1 \end{bmatrix} \right\} \qquad \mathbf{col}(A) = \mathbf{span}\left\{ \begin{bmatrix} 1 \\ 1 \\ -2 \end{bmatrix} \right\}$$

$$N(A) = \mathbf{span}\left\{ \begin{bmatrix} 1 \\ 1 \end{bmatrix} \right\} \qquad N(A^t) = \mathbf{span}\left\{ \begin{bmatrix} 2 \\ 0 \\ 1 \end{bmatrix}, \begin{bmatrix} -1 \\ 1 \\ 0 \end{bmatrix} \right\}$$

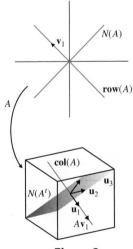

Figure 3

The four fundamental subspaces are shown in Fig. 3.

Data Compression

An important application that involves the singular value decomposition is data compression. As a preliminary step, suppose that a matrix A of rank r (with r nonzero singular values) has the SVD $A = U\Sigma V^t$. That is,

$$A = U\Sigma V^t = \begin{bmatrix} \sigma_1\mathbf{u}_1 & \cdots & \sigma_r\mathbf{u}_r & \mathbf{0} & \cdots & \mathbf{0} \end{bmatrix} V^t$$

$$= \sigma_1\mathbf{u}_1\mathbf{v}_1^t + \sigma_2\mathbf{u}_2\mathbf{v}_2^t + \cdots + \sigma_r\mathbf{u}_r\mathbf{v}_r^t$$

$$= \sigma_1\left(\frac{1}{\sigma_1}A\mathbf{v}_1\right)\mathbf{v}_1^t + \sigma_2\left(\frac{1}{\sigma_2}A\mathbf{v}_2\right)\mathbf{v}_2^t + \cdots + \sigma_r\left(\frac{1}{\sigma_r}A\mathbf{v}_r\right)\mathbf{v}_r^t$$

$$= (A\mathbf{v}_1)\mathbf{v}_1^t + (A\mathbf{v}_2)\mathbf{v}_2^t + \cdots + (A\mathbf{v}_r)\mathbf{v}_r^t$$

Observe that each of the terms $A\mathbf{v}_i\mathbf{v}_i^t$ is a matrix of rank 1. Consequently, the sum of the first k terms of the last equation is a matrix of rank $k \leq r$, which gives an approximation to the matrix A. This factorization of a matrix has application in many areas.

Figure 4

As an illustration of the utility of such an approximation, suppose that A is the 356×500 matrix, where each entry is a numeric value for a pixel, of the gray scale image of the surface of Mars shown in Fig. 4. A simple algorithm using the method above for approximating the image stored in the matrix A is given by the following:

1. Find the eigenvectors of the $n \times n$ symmetric matrix $A^t A$.
2. Compute $A\mathbf{v}_i$, for $i = 1, \ldots, k$, with $k \leq r = \mathbf{rank}(A)$.
3. The matrix $(A\mathbf{v}_1)\mathbf{v}_1^t + (A\mathbf{v}_2)\mathbf{v}_2^t + \cdots + (A\mathbf{v}_k)\mathbf{v}_k^t$ is an approximation of the original image.

To transmit the kth approximation of the image and reproduce it back on earth requires the eigenvectors $\mathbf{v}_1, \ldots, \mathbf{v}_k$ of $A^t A$ and the vectors $A\mathbf{v}_1, \ldots, A\mathbf{v}_k$.

The images in Fig. 5 are produced using matrices of ranks $1, 4, 10, 40, 80,$ and 100, respectively.

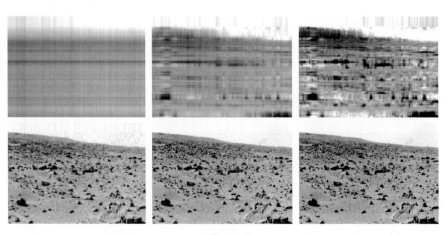

Figure 5

The storage requirements for each of the images are given in Table 1.

Table 1

Image	Storage Requirement	Percent of Original	Rank
Original	$356 \times 500 = 178,000$	100%	
Approximation 1	$2 \times 500 = 1,000$	0.6	1
Approximation 2	$8 \times 500 = 4,000$	1	4
Approximation 3	$20 \times 500 = 10,000$	1	10
Approximation 4	$80 \times 500 = 40,000$	22	40
Approximation 5	$160 \times 500 = 80,000$	45	80
Approximation 6	$200 \times 500 = 100,000$	56	100

Exercise Set 6.8

In Exercises 1–4, find the singular values for the matrix.

1. $A = \begin{bmatrix} -2 & -2 \\ 1 & 1 \end{bmatrix}$

2. $A = \begin{bmatrix} -1 & -2 \\ 1 & -2 \end{bmatrix}$

3. $A = \begin{bmatrix} 1 & 0 & 2 \\ 2 & -1 & -1 \\ -2 & 1 & 1 \end{bmatrix}$

4. $A = \begin{bmatrix} 1 & -1 & 0 \\ 0 & 0 & 1 \\ -1 & 1 & 0 \end{bmatrix}$

In Exercises 5–8, find a singular value decomposition for the matrix.

5. $A = \begin{bmatrix} 5 & 3 \\ 3 & 5 \end{bmatrix}$

6. $A = \begin{bmatrix} 2 & 2 \\ 4 & -1 \end{bmatrix}$

7. $A = \begin{bmatrix} 1 & 0 & 0 \\ 0 & 1 & 1 \end{bmatrix}$

8. $A = \begin{bmatrix} -2 & 1 & -1 \\ 0 & 1 & 1 \end{bmatrix}$

In Exercises 9 and 10, the **condition number** of a matrix A is the ratio σ_1/σ_r, of the largest to the smallest singular value. The condition number provides a measure of the sensitivity of the linear system $A\mathbf{x} = \mathbf{b}$ to perturbations to A or $\mathbf{b}$. A linear system is **ill-conditioned** when the condition number is *too large* and called **singular** when the condition number is infinite (the matrix is not invertible).

9. Let $A = \begin{bmatrix} 1 & 1 \\ 1 & 1.000000001 \end{bmatrix}$.

   a. Solve the linear system
   $$A\mathbf{x} = \begin{bmatrix} 2 \\ 2 \end{bmatrix}$$

   b. Solve the linear system
   $$A\mathbf{x} = \begin{bmatrix} 2 \\ 2.000000001 \end{bmatrix}$$

   c. Find the condition number for A.

10. Let $\mathbf{b} = \begin{bmatrix} 1 \\ 3 \\ -4 \end{bmatrix}$.

a. Let

$$A = \begin{bmatrix} -2 & -1 & 0 \\ -2 & -1 & -2 \\ 0 & -2 & 1 \end{bmatrix}$$

Solve the linear system $A\mathbf{x} = \mathbf{b}$.

b. Let

$$B = \begin{bmatrix} -2.00001 & -1.001 & 0 \\ -2.01 & -0.87 & -2 \\ 0 & -2 & 1 \end{bmatrix}$$

Solve the linear system $B\mathbf{x} = \mathbf{b}$.

c. Find the condition number for A.

Review Exercises for Chapter 6

1. Let V be the inner product space $\mathbb{R}^3$ with the standard inner product and let

$$B = \left\{ \begin{bmatrix} 1 \\ 0 \\ 1 \end{bmatrix}, \begin{bmatrix} 1 \\ 0 \\ 0 \end{bmatrix}, \begin{bmatrix} 2 \\ 1 \\ 0 \end{bmatrix} \right\}$$

a. Verify that B is a basis for $\mathbb{R}^3$.

b. Use B to find an orthonormal basis for $\mathbb{R}^3$.

c. Let $W = \mathbf{span} \left\{ \begin{bmatrix} 1 \\ 0 \\ 1 \end{bmatrix}, \begin{bmatrix} 1 \\ 0 \\ 0 \end{bmatrix} \right\}$ and

$\mathbf{v} = \begin{bmatrix} -2 \\ 1 \\ -1 \end{bmatrix}$. Find $\mathrm{proj}_W \mathbf{v}$. (*Hint*: First use

the Gram-Schmidt process to find an orthogonal basis for W; then refer to Exercise 16 of Sec. 6.4.)

2. Let

$$W = \mathbf{span} \left\{ \begin{bmatrix} -1 \\ 2 \\ 2 \\ -2 \end{bmatrix}, \begin{bmatrix} -3 \\ 0 \\ 0 \\ 0 \end{bmatrix}, \begin{bmatrix} 3 \\ -2 \\ -1 \\ 1 \end{bmatrix}, \begin{bmatrix} 0 \\ 0 \\ -1 \\ 1 \end{bmatrix} \right\}$$

be a subspace of $\mathbb{R}^4$ with the standard inner product.

a. Find a basis for W.

b. Find $W^\perp$.

c. Find an orthonormal basis for W.

d. Find an orthonormal basis for $W^\perp$.

e. Verify that $\dim(\mathbb{R}^4) = \dim(W) + \dim(W^\perp)$.

f. Find the orthogonal projection of $\mathbf{v} = \begin{bmatrix} -2 \\ 0 \\ 3 \\ -1 \end{bmatrix}$

onto W.

3. Let a, b, and c be real numbers and

$$W = \left\{ \begin{bmatrix} x \\ y \\ z \end{bmatrix} \in \mathbb{R}^3 \;\middle|\; ax + by + cz = 0 \right\}$$

where $\mathbb{R}^3$ is given the standard inner product.

a. Show that $\begin{bmatrix} a \\ b \\ c \end{bmatrix}$ is in $W^\perp$.

b. Describe $W^\perp$.

c. Let $\mathbf{v} = \begin{bmatrix} x_1 \\ x_2 \\ x_3 \end{bmatrix}$. Find $\mathrm{proj}_{W^\perp} \mathbf{v}$.

d. Find $\| \mathrm{proj}_{W^\perp} \mathbf{v} \|$.

4. Define on $\mathcal{P}_2$ an inner product by

$$\langle p, q \rangle = \int_{-1}^{1} p(x)q(x)\,dx$$

Let $p(x) = x$ and $q(x) = x^2 - x + 1$.

a. Find $\langle p, q \rangle$.

b. Find the distance between p and q.

c. Are p and q orthogonal? Explain.

d. Find the cosine of the angle between p and q.

e. Find $\mathrm{proj}_q \, p$.

f. Let $W = \mathbf{span}\{p\}$. Find $W^\perp$.

5. Let V be the inner product space $C^{(0)}[-\pi, \pi]$ with inner product defined by

$$\langle f, g \rangle = \int_{-\pi}^{\pi} f(x)g(x)\,dx$$

Let $W = \mathbf{span}\{1, \cos x, \sin x\}$.

a. Verify that the set $\{1, \cos x, \sin x\}$ is orthogonal.

b. Find an orthonormal basis for W.

c. Find $\mathrm{proj}_W x^2$.

d. Find $\| \mathrm{proj}_W x^2 \|$.

6. Let $B = \{\mathbf{v}_1, \ldots, \mathbf{v}_n\}$ be an orthonormal basis for an inner product space V, and let $\mathbf{v}$ be a vector in V.

a. Find the coordinate $\begin{bmatrix} c_1 \\ c_2 \\ \vdots \\ c_n \end{bmatrix}$ of $\mathbf{v}$ relative to B.

b. Show that $c_i \mathbf{v}_i = \mathrm{proj}_{\mathbf{v}_i} \mathbf{v}$ for each $i = 1, 2, \ldots, n$.

c. Let

$$B = \left\{ \frac{1}{\sqrt{2}} \begin{bmatrix} 1 \\ 1 \\ 0 \end{bmatrix}, \frac{1}{\sqrt{2}} \begin{bmatrix} 1 \\ -1 \\ 0 \end{bmatrix}, \frac{1}{\sqrt{6}} \begin{bmatrix} 1 \\ 1 \\ -2 \end{bmatrix} \right\}$$

be an orthonormal basis for $\mathbb{R}^3$, with the standard inner product, and let

$$\mathbf{v} = \begin{bmatrix} \frac{1}{\sqrt{2}} + \frac{1}{\sqrt{3}} \\ \frac{1}{\sqrt{2}} - \frac{1}{\sqrt{3}} \\ \frac{1}{\sqrt{3}} \end{bmatrix}. \text{ Find the coordinate}$$

$\begin{bmatrix} c_1 \\ c_2 \\ \vdots \\ c_n \end{bmatrix}$ of $\mathbf{v}$ relative to B.

7. Show that if B is an orthonormal basis for $\mathbb{R}^n$, with the standard inner product, and

$$[\mathbf{v}]_B = \begin{bmatrix} c_1 \\ c_2 \\ \vdots \\ c_n \end{bmatrix}, \text{ then}$$

$$\| \mathbf{v} \| = \sqrt{c_1^2 + c_2^2 + \cdots + c_n^2}$$

Give a similar formula for $\| \mathbf{v} \|$ if B is an orthogonal basis, not necessarily orthonormal.

8. Let $\{\mathbf{v}_1, \ldots, \mathbf{v}_m\}$ be an orthonormal subset of $\mathbb{R}^n$, with the standard inner product, and let $\mathbf{v}$ be any vector in $\mathbb{R}^n$. Show that

$$\| \mathbf{v} \|^2 \geq \sum_{i=1}^{m} (\mathbf{v} \cdot \mathbf{v}_i)^2$$

(*Hint*: Expand $\| \mathbf{v} - \sum_{i=1}^{m} \langle \mathbf{v}, \mathbf{v}_i \rangle \mathbf{v}_i \|^2$.)

9. (*QR* factorization) Let A be an $m \times n$ matrix with linearly independent column vectors. In this exercise we will describe a process to write $A = QR$, where Q is an $m \times n$ matrix whose column vectors form an orthonormal basis for $\mathbf{col}(A)$ and R is an $n \times n$ upper triangular matrix that is invertible. Let

$$A = \begin{bmatrix} 1 & 0 & -1 \\ 1 & -1 & 2 \\ 1 & 0 & 1 \\ 1 & -1 & 2 \end{bmatrix}$$

a. Let $B = \{\mathbf{v}_1, \mathbf{v}_2, \mathbf{v}_3\}$ be the set of column vectors of the matrix A. Verify that B is linearly independent and hence forms a basis for $\mathbf{col}(A)$.

b. Use the Gram-Schmidt process on B to find an orthogonal basis $B_1 = \{\mathbf{w}_1, \mathbf{w}_2, \mathbf{w}_3\}$.

c. Use B_1 to find an orthonormal basis $B_2 = \{\mathbf{q}_1, \mathbf{q}_2, \mathbf{q}_3\}$.

d. Define the matrix $Q = [\mathbf{q}_1 \ \mathbf{q}_2 \ \mathbf{q}_3]$. Define the upper triangle matrix R for $i = 1, 2, 3$ by

$$r_{ij} = \begin{cases} 0 & \text{if } i > j \\ \mathbf{v}_j \cdot \mathbf{q}_i & \text{if } i \leq j, \ j = i, \ldots, 3 \end{cases}$$

e. Verify that $A = QR$.

10. Let $B = \{\mathbf{v}_1, \mathbf{v}_2, \ldots, \mathbf{v}_n\}$ be an orthogonal basis for an inner product space V and $c_1, c_2, \ldots, c_n$ arbitrary nonzero scalars. Show that

$$B_1 = \{c_1 \mathbf{v}_1, c_2 \mathbf{v}_2, \ldots, c_n \mathbf{v}_n\}$$

is an orthogonal basis for V. How can the scalars be chosen so that B_1 is an orthonormal basis?

Chapter 6: Chapter Test

In Exercises 1–40, determine whether the statement is true or false.

1. If $\mathbf{u}$ is orthogonal to both $\mathbf{v}_1$ and $\mathbf{v}_2$, then $\mathbf{u}$ is orthogonal $\mathbf{span}\{\mathbf{v}_1, \mathbf{v}_2\}$.

2. If W is a subspace of an inner product space V and $\mathbf{v} \in V$, then $\mathbf{v} - \text{proj}_W \mathbf{v} \in W^{\perp}$.

3. If W is a subspace of an inner product space V, then $W \cap W^{\perp}$ contains a nonzero vector.

4. Not every orthogonal set in an inner product space is linearly independent.

In Exercises 5–10, let

$$\mathbf{v}_1 = \begin{bmatrix} 2 \\ 1 \\ -4 \\ 3 \end{bmatrix} \qquad \mathbf{v}_2 = \begin{bmatrix} -2 \\ 1 \\ 2 \\ 1 \end{bmatrix}$$

be vectors in $\mathbb{R}^4$ with inner product the standard dot product.

5. $\| \mathbf{v}_1 \| = 30$

6. The distance between the vectors $\mathbf{v}_1$ and $\mathbf{v}_2$ is $2\sqrt{14}$.

7. The vector $\mathbf{u} = \frac{1}{\sqrt{30}} \mathbf{v}_1$ is a unit vector.

8. The vectors $\mathbf{v}_1$ and $\mathbf{v}_2$ are orthogonal.

9. The cosine of the angle between the vectors $\mathbf{v}_1$ and $\mathbf{v}_2$ is $-\frac{4}{15}\sqrt{10}$.

10. $\text{proj}_{\mathbf{v}_1} \mathbf{v}_2 = \begin{bmatrix} -8/15 \\ -4/15 \\ -16/15 \\ -12/15 \end{bmatrix}$

In Exercises 11–16, let

$$\mathbf{v}_1 = \begin{bmatrix} 1 \\ 0 \\ 1 \end{bmatrix} \qquad \mathbf{v}_2 = \begin{bmatrix} -1 \\ 1 \\ 1 \end{bmatrix}$$

$$\mathbf{v}_3 = \begin{bmatrix} 2 \\ 4 \\ -2 \end{bmatrix}$$

be vectors in $\mathbb{R}^3$ with inner product the standard dot product.

11. The set $\{\mathbf{v}_1, \mathbf{v}_2, \mathbf{v}_3\}$ is orthogonal.

12. The set $\{\mathbf{v}_1, \mathbf{v}_2, \mathbf{v}_3\}$ is a basis for $\mathbb{R}^3$.

13. If $W = \mathbf{span}\{\mathbf{v}_1, \mathbf{v}_2\}$ and $\mathbf{u} = \begin{bmatrix} 1 \\ 1 \\ 1 \end{bmatrix}$, then

$\text{proj}_W \mathbf{u} = \mathbf{v}_1 + \frac{1}{3}\mathbf{v}_2$.

14. If $W = \mathbf{span}\{\mathbf{v}_1, \mathbf{v}_2, \mathbf{v}_3\}$, then $W^{\perp} = \{\mathbf{0}\}$.

15. If $W = \mathbf{span}\{\mathbf{v}_1, \mathbf{v}_2\}$, then $W^{\perp} = \{\mathbf{0}\}$.

16. $W = \mathbf{span}\{\mathbf{v}_1, \mathbf{v}_2, \mathbf{v}_3\}$, then $\text{proj}_W \mathbf{v} = \mathbf{v}$ for any vector $\mathbf{v} \in \mathbb{R}^3$.

In Exercises 17–23, use the inner product defined on P_2 defined by

$$\langle p, q \rangle = \int_{-1}^{1} p(x)q(x)\,dx$$

17. $\| x^2 - x \| = \frac{4}{\sqrt{15}}$

18. The polynomials $p(x) = x$ and $q(x) = x^2 - 1$ are orthogonal.

19. The polynomials $p(x) = 1$ and $q(x) = x^2 - 1$ are orthogonal.

20. The set $\{1, x, x^2 - \frac{1}{3}\}$ is orthogonal.

21. The vector $p(x) = \frac{1}{2}$ is a unit vector.

22. If $W = \mathbf{span}\{1, x\}$, then $\dim(W^{\perp}) = 1$.

23. If $W = \mathbf{span}\{1, x^2\}$, then a basis for $W^{\perp}$ is $\{1, x\}$.

24. An $n \times n$ symmetric matrix has n distinct real eigenvalues.

25. If $\mathbf{u}$ and $\mathbf{v}$ are vectors in $\mathbb{R}^2$, then $\langle \mathbf{u}, \mathbf{v} \rangle = 3u_1 v_2 - u_2 v_1$ defines an inner product.

26. For any inner product

$$\langle 2\mathbf{u}, 2\mathbf{v} + 2\mathbf{w} \rangle = 2\langle \mathbf{u}, \mathbf{v} \rangle + 2\langle \mathbf{u}, \mathbf{w} \rangle$$

27. If $W = \mathbf{span}\{1, x^2\}$ is a subspace of $\mathcal{P}_2$ with inner product

$$\langle p, q \rangle = \int_0^1 p(x)q(x)\,dx$$

then a basis for $W^\perp$ is $\{x\}$.

28. If $\{\mathbf{u}_1, \ldots, \mathbf{u}_k\}$ is a basis for a subspace W of an inner product space V and $\{\mathbf{v}_1, \ldots, \mathbf{v}_m\}$ is a basis for $W^\perp$, then $\{\mathbf{u}_1, \ldots, \mathbf{u}_k, \mathbf{v}_1, \ldots, \mathbf{v}_m\}$ is a basis for V.

29. If A is an $n \times n$ matrix whose column vectors form an orthogonal set in $\mathbb{R}^n$ with the standard inner product, then $\mathbf{col}(A) = \mathbb{R}^n$.

30. In $\mathbb{R}^2$ with the standard inner product, the orthogonal complement of $y = 2x$ is $y = \frac{1}{2}x$.

31. In $\mathbb{R}^3$ with the standard inner product, the orthogonal complement of $-3x + 3z = 0$ is

$$\mathbf{span}\left\{ \begin{bmatrix} -3 \\ 0 \\ 3 \end{bmatrix} \right\}.$$

32. Every finite dimensional inner product space has an orthonormal basis.

33. If

$$W = \mathbf{span}\left\{ \begin{bmatrix} 1 \\ 2 \\ 1 \end{bmatrix}, \begin{bmatrix} 0 \\ 1 \\ -1 \end{bmatrix} \right\}$$

then a basis for $W^\perp$ is also a basis for the null space of

$$A = \begin{bmatrix} 1 & 0 \\ 2 & 1 \\ 1 & -1 \end{bmatrix}.$$

34. If

$$W = \mathbf{span}\left\{ \begin{bmatrix} 1 \\ 0 \\ 1 \end{bmatrix}, \begin{bmatrix} -1 \\ 1 \\ 0 \end{bmatrix} \right\}$$

then $\dim(W^\perp) = 2$.

35. If

$$W = \mathbf{span}\left\{ \begin{bmatrix} 0 \\ 1 \\ 1 \end{bmatrix}, \begin{bmatrix} 1 \\ 0 \\ 1 \end{bmatrix} \right\}$$

then

$$W^\perp = \mathbf{span}\left\{ \begin{bmatrix} -1 \\ -1 \\ 1 \end{bmatrix} \right\}$$

36. In $\mathbb{R}^5$ with the standard inner product there exists a subspace W such that $\dim(W) = \dim(W^\perp)$.

37. If A is an $n \times n$ matrix whose column vectors are orthonormal, then $AA^t\mathbf{v}$ is the orthogonal projection of $\mathbf{v}$ onto $\mathbf{col}(A)$.

38. If $\mathbf{u}$ and $\mathbf{v}$ are orthogonal, $\mathbf{w}_1$ is a unit vector in the direction of $\mathbf{u}$, and $\mathbf{w}_2$ is a unit vector in the opposite direction of $\mathbf{v}$, then $\mathbf{w}_1$ and $\mathbf{w}_2$ are orthogonal.

39. If $\mathbf{u}$ and $\mathbf{v}$ are vectors in $\mathbb{R}^n$ and the vector projection of $\mathbf{u}$ onto $\mathbf{v}$ is equal to the vector projection of $\mathbf{v}$ onto $\mathbf{u}$, then $\mathbf{u}$ and $\mathbf{v}$ are linearly independent.

40. If A is an $m \times n$ matrix, then AA^t and A^tA have the same rank.

APPENDIX A

Preliminaries

A.1 ▶ Algebra of Sets

The notion of a *set* is a fundamental concept in mathematics allowing for the grouping and analysis of objects with common attributes. For example, we can consider the collection of all even numbers, or the collection of all polynomials of degree 3. A **set** is any *well-defined* collection of objects. By this we mean that a clear process exists for deciding whether an object is contained in the set. The colors of the rainbow—red, yellow, green, blue, and purple—can be grouped in the set

$$C = \{\text{red, yellow, green, blue, purple}\}$$

The objects contained in a set are called **members**, or **elements**, of the set. To indicate that x is an element of a set S, we write $x \in S$. Since green is one of the colors of the rainbow, we have that green $\in C$. The color orange, however, is not one of the colors of the rainbow and therefore is not an element of C. In this case we write orange $\notin C$.

There are several ways to write a set. If the number of elements is finite and small, then all the elements can be listed, as we did with the set C, separated by commas and enclosed in braces. Another example is

$$S = \{-3, -2, 0, 1, 4, 7\}$$

If a pattern exists among its elements, a set can be described by specifying only a few of them. For example,

$$S = \{2, 4, 6, \ldots, 36\}$$

is the set of all even numbers between 2 and 36, inclusive. The set of all even whole numbers can be written

$$T = \{2, 4, 6, \ldots\}$$

Special sets of numbers are often given special symbols. Several common ones are described here. The set of **natural numbers**, denoted by $\mathbb{N}$, is the set

$$\mathbb{N} = \{1, 2, 3, \ldots\}$$

The set of **integers**, denoted by $\mathbb{Z}$, is given by

$$\mathbb{Z} = \{\ldots, -3, -2, -1, 0, 1, 2, 3, \ldots\}$$

We use the symbol $\mathbb{Q}$ to denote the set of **rational numbers**, which can be described as

$$\mathbb{Q} = \left\{ \frac{p}{q} \,\middle|\, p, q \in \mathbb{Z}, q \neq 0 \right\}$$

Finally, the set of **real numbers**, denoted by $\mathbb{R}$, consists of all rational and irrational numbers. Examples of irrational numbers are $\sqrt{2}$ and π.

In many cases, the set we wish to consider is taken from a larger one. For example,

$$S = \{x \in \mathbb{R} \mid -1 \leq x < 4\}$$

is the set of all real numbers greater than or equal to -1 and less than 4. In general, the notation

$$\{x \in L \mid \text{restriction on } x\}$$

translates to "the set of all x in L such that x satisfies the restriction." In some cases L is omitted if a *universal set* is implied or understood.

Sets can be compared using the notion of containment. Denote two sets by A and B. The set A is **contained** in B if each element of A is also in B. When this happens, we say that A is a **subset** of B and write $A \subseteq B$. For example, let

$$A = \{1, 2\} \qquad B = \{1, 2, 3\} \qquad \text{and} \qquad C = \{2, 3, 4\}$$

Since every element of A is also in B, we have $A \subseteq B$. However, A is not a subset of C since $1 \in A$ but $1 \notin C$. In this case we write $A \nsubseteq C$. For the sets of natural numbers, integers, rational numbers, and real numbers, we have

$$\mathbb{N} \subseteq \mathbb{Z} \subseteq \mathbb{Q} \subseteq \mathbb{R}$$

The set with no elements is called the **empty set**, or **null set**, and is denoted by ϕ. One special property of the empty set ϕ is that it is a subset of every set.

Two sets A and B are equal if they have the same elements. Alternatively, A and B are equal if $A \subseteq B$ and $B \subseteq A$. In this case we write $A = B$.

Operations on Sets

Elements can be extracted from several sets and placed in one set by using the operations of *intersection* and *union*. The **intersection** of two sets A and B, denoted by $A \cap B$, is the set of all elements that are in both A and B, that is,

$$A \cap B = \{x \mid x \in A \text{ and } x \in B\}$$

The **union** of two sets A and B, denoted by $A \cup B$, is the set of all elements that are in A or B, that is,

$$A \cup B = \{x \mid x \in A \text{ or } x \in B\}$$

As an illustration, let $A = \{1, 3, 5\}$ and $B = \{1, 2, 4\}$. Then

$$A \cap B = \{1\} \quad \text{and} \quad A \cup B = \{1, 2, 3, 4, 5\}$$

A graphical device, called a **Venn diagram**, is helpful for visualizing set operations. The Venn diagrams for the intersection and union of two sets are shown in Fig. 1.

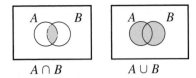

$$A \cap B \qquad\qquad A \cup B$$

Figure 1

EXAMPLE 1

Define two intervals of real numbers by $A = [-3, 2)$ and $B = [-7, 1)$. Find $A \cap B$ and $A \cup B$.

Solution

Since the intervals overlap with $-7 < -3 < 1 < 2$, the intersection is the interval

$$A \cap B = [-3, 1)$$

and the union is the interval

$$A \cup B = [-7, 2)$$

Notice that $x \notin A \cap B$ if and only if $x \notin A$ or $x \notin B$, and $x \notin A \cup B$ if and only if $x \notin A$ and $x \notin B$.

The **complement** of the set A relative to the set B, denoted by $B \backslash A$, consists of all elements of B that are not elements of A. In set notation this complement is given by

$$B \backslash A = \{x \in B \mid x \notin A\}$$

For example, let A and B be the intervals given by $A = [1, 2]$ and $B = [0, 5]$. Then

$$B \backslash A = [0, 1) \cup (2, 5]$$

If A is taken from a known universal set, then the complement of A is denoted by A^c. To illustrate, let $A = [1, 2]$ as before. Then the complement of A relative to the set of real numbers is

$$\mathbb{R} \backslash A = A^c = (-\infty, 1) \cup (2, \infty)$$

Another operation on sets is the *Cartesian product*. Specifically, the **Cartesian product** of two sets A and B, denoted by $A \times B$, is the set of all ordered pairs whose first component comes from A and whose second component comes from B. So

$$A \times B = \{(x, y) \mid x \in A \text{ and } y \in B\}$$

For example, if $A = \{1, 2\}$ and $B = \{10, 20\}$, then

$$A \times B = \{(1, 10), (1, 20), (2, 10), (2, 20)\}$$

This last set is a subset of the Euclidean plane, which can be written as the Cartesian product of $\mathbb{R}$ with itself, so that

$$\mathbb{R}^2 = \mathbb{R} \times \mathbb{R} = \{(x, y) \mid x, y \in \mathbb{R}\}$$

EXAMPLE 2   Let $A = [-3, 2)$ and $B = (-2, 1]$. Describe the set $A \times B$.

Solution   Since $A \times B$ consists of all ordered pairs whose first component comes from A and second from B, we have

$$-3 \le x < 2 \quad \text{and} \quad -2 < y \le 1$$

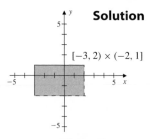

The points that satisfy these two conditions lie in the rectangular region shown in Fig. 2.

Figure 2

Example 3 shows that operations on sets can be combined to produce results similar to the arithmetic properties of real numbers.

EXAMPLE 3   Verify that if A, B, and C are sets, then $A \cap (B \cup C) = (A \cap B) \cup (A \cap C)$.

Solution   The Venn diagrams in Fig. 3 show that although the two sets are computed in different ways, the result is the same. The quantities inside the parentheses are carried out first. Of course, the picture alone does not constitute a proof. To establish the fact, we must show that the set on the left-hand side of the equation above is a subset of the set on the right, and vice versa.

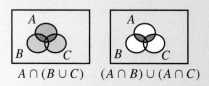

$A \cap (B \cup C) \qquad (A \cap B) \cup (A \cap C)$

Figure 3

Indeed, if $x \in A \cap (B \cup C)$, then $x \in A$ and $x \in B \cup C$. This is equivalent to the statement $x \in A$ and $(x \in B$ or $x \in C)$, which in turn is also equivalent to

$$(x \in A \text{ and } x \in B) \qquad \text{or} \qquad (x \in A \text{ and } x \in C)$$

Hence, $x \in (A \cap B) \cup (A \cap C)$, and we have shown that

$$A \cap (B \cup C) \subseteq (A \cap B) \cup (A \cap C)$$

On the other hand, let $x \in (A \cap B) \cup (A \cap C)$, which can also be written as

$$x \in (A \cap B) \qquad \text{or} \qquad x \in (A \cap C)$$

This gives

$$x \in A \quad \text{and} \quad x \in B \quad \text{or} \quad x \in A \quad \text{and} \quad x \in C$$

In either case, $x \in A$ and, in addition, $x \in B$ or $x \in C$, so that

$$x \in A \cap (B \cup C)$$

Therefore,

$$(A \cap B) \cup (A \cap C) \subseteq A \cap (B \cup C)$$

Since each set is a subset of the other, we have

$$A \cap (B \cup C) = (A \cap B) \cup (A \cap C)$$

Theorem 1 includes the result given in Example 3 along with other properties of set operations. The verifications of the remaining properties are left as exercises.

THEOREM 1   Let A, B, and C be sets contained in a universal set U.

1. $A \cap A = A$, $A \cup A = A$
2. $(A^c)^c = A$
3. $A \cap A^c = \phi$, $A \cup A^c = U$
4. $A \cap B = B \cap A$, $A \cup B = B \cup A$
5. $(A \cap B) \cap C = A \cap (B \cap C)$, $(A \cup B) \cup C = A \cup (B \cup C)$
6. $A \cap (B \cup C) = (A \cap B) \cup (A \cap C)$
   $A \cup (B \cap C) = (A \cup B) \cap (A \cup C)$

THEOREM 2   **DeMorgan's Laws**   Let A, B, and C be sets. Then

1. $A \backslash (B \cup C) = (A \backslash B) \cap (A \backslash C)$
2. $A \backslash (B \cap C) = (A \backslash B) \cup (A \backslash C)$

Proof   (1) We need to verify that the set on the left-hand side of the equation is a subset of the set on the right, and vice versa. We begin by letting $x \in A \backslash (B \cup C)$. This means that $x \in A$ and $x \notin B \cup C$. This is equivalent to the statement

$$x \in A \qquad \text{and} \qquad (x \notin B \text{ and } x \notin C)$$

which is then equivalent to

$$x \in A \quad \text{and} \quad x \notin B \quad \text{and} \quad x \in A \quad \text{and} \quad x \notin C$$

This last pair of statements gives

$$x \in (A \backslash B) \cap (A \backslash C) \qquad \text{so that} \qquad A \backslash (B \cup C) \subseteq (A \backslash B) \cap (A \backslash C)$$

To show containment in the other direction, we let $x \in (A \backslash B) \cap (A \backslash C)$. Rewriting this in equivalent forms, we have

$$
\begin{array}{llll}
 & x \in (A \backslash B) & \text{and} & x \in (A \backslash C) \\
x \in A \quad \text{and} & x \notin B & \text{and} & x \in A \quad \text{and} \quad x \notin C \\
 & x \in A & \text{and} & x \notin B \quad \text{and} \quad x \notin C \\
 & x \in A & \text{and} & x \notin (B \cup C)
\end{array}
$$

Therefore,

$$(A \backslash B) \cap (A \backslash C) \subseteq A \backslash (B \cup C)$$

(2) The proof is similar to the one given in part 1 and is left as an exercise.

Exercise Set A.1

In Exercises 1–6, let the universal set be $\mathbb{Z}$ and let

$$A = \{-4, -2, 0, 1, 2, 3, 5, 7, 9\}$$
$$B = \{-3, -2, -1, 2, 4, 6, 8, 9, 10\}$$

Compute the set.

1. $A \cap B$

2. $A \cup B$

3. $A \times B$

4. $(A \cup B)^c$

5. $A \backslash B$

6. $B \backslash A$

In Exercises 7–14, use the sets

$$A = (-11, 3] \qquad B = [0, 8] \qquad C = [-9, \infty)$$

Compute the set.

7. $A \cap B$

8. $(A \cup B)^c$

9. $A \backslash B$

10. $C \backslash A$

11. $A \backslash C$

12. $(A \cup B)^c \cap C$

13. $(A \cup B) \backslash C$

14. $B \backslash (A \cap C)$

In Exercises 15–20, use the sets

$$A = (-2, 3] \qquad B = [1, 4] \qquad C = [0, 2]$$

to sketch the specified set in the plane.

15. $A \times B$

16. $B \times C$

17. $C \times B$

18. $(A \times B) \backslash [C \times (B \cap C)]$

19. $A \times (B \cap C)$

20. $(A \times B) \cap (A \times C)$

In Exercises 21–26, let

$$A = \{1, 2, 3, 5, 7, 9, 11\}$$
$$B = \{2, 5, 10, 14, 20\}$$
$$C = \{1, 5, 7, 14, 30, 37\}$$

Verify that the statement holds.

21. $(A \cap B) \cap C = A \cap (B \cap C)$

22. $(A \cup B) \cup C = A \cup (B \cup C)$

23. $A \cap (B \cup C) = (A \cap B) \cup (A \cap C)$

24. $A \cup (B \cap C) = (A \cup B) \cap (A \cup C)$

25. $A \backslash (B \cup C) = (A \backslash B) \cap (A \backslash C)$

26. $A \backslash (B \cap C) = (A \backslash B) \cup (A \backslash C)$

In Exercises 27–34, show that the statement holds for all sets A, B, and C.

27. $(A^c)^c = A$

28. The set $A \cup A^c$ is the universal set.

29. $A \cap B = B \cap A$

30. $A \cup B = B \cup A$

31. $(A \cap B) \cap C = A \cap (B \cap C)$

32. $(A \cup B) \cup C = A \cup (B \cup C)$

33. $A \cup (B \cap C) = (A \cup B) \cap (A \cup C)$

34. $A \backslash (B \cap C) = (A \backslash B) \cup (A \backslash C)$

35. If A and B are sets, show that

$$A \backslash B = A \cap B^c$$

36. If A and B are sets, show that

$$(A \cup B) \cap A^c = B \setminus A$$

37. If A and B are sets, show that

$$(A \cup B) \setminus (A \cap B) = (A \setminus B) \cup (B \setminus A)$$

38. If A and B are sets, show that

$$(A \cap B) = A \setminus (A \setminus B)$$

39. If A, B, and C are sets, show that

$$A \times (B \cap C) = (A \times B) \cap (A \times C)$$

40. The *symmetric difference* operation Δ on two sets A and B is defined by

$$A \Delta B = (A \setminus B) \cup (B \setminus A)$$

Show that

$$A \Delta B = (A \cup B) \setminus (A \cap B)$$

A.2 ▶ Functions

The sets we described in Sec. A.1 along with *functions* are two of the fundamental objects of modern mathematics. Sets act as nouns defining objects and functions as verbs describing actions to be performed on the elements of a set. Functions connect each element of one set to a unique element of another set. The functions that are studied in calculus are defined on sets of real numbers. Other branches of mathematics require functions that are defined on other types of sets. The following definition is general enough for a wide variety of abstract settings.

DEFINITION 1

Function　A **function** f **from a set** $\mathbb{X}$ **to a set** $\mathbb{Y}$ is a rule of correspondence that associates with each element of $\mathbb{X}$ exactly one element of $\mathbb{Y}$.

Before continuing with a description of functions, we note that there are other ways of associating the elements of two sets. A **relation** is a rule of correspondence that does not (necessarily) assign a unique element of $\mathbb{Y}$ for each element of $\mathbb{X}$. A function, then, is a relation that is well defined with a clear procedure that associates a unique element of $\mathbb{Y}$ with each element of $\mathbb{X}$. A common metaphor for a function is a machine that produces a unique *output* for each *input*.

A function f is also called a **mapping** from $\mathbb{X}$ to $\mathbb{Y}$ and is written $f: \mathbb{X} \longrightarrow \mathbb{Y}$. If $x \in \mathbb{X}$ is associated with $y \in \mathbb{Y}$ via the function f, then we call y the **image** of x under f and write $y = f(x)$. The set $\mathbb{X}$ is called the **domain** of f and is denoted by $\text{dom}(f)$. The **range** of f, denoted by $\text{range}(f)$, is the set of all images of f. That is,

$$\text{range}(f) = \{f(x) \mid x \in \text{dom}(f)\}$$

If A is a subset of the domain, then the **image of** A is defined by

$$f(A) = \{f(x) \mid x \in A\}$$

Using this notation, we have $\text{range}(f) = f(\mathbb{X})$.

There are many ways of describing functions. The pictures shown in Fig. 1, give us one way while providing an illustration of the key idea distinguishing relations from functions.

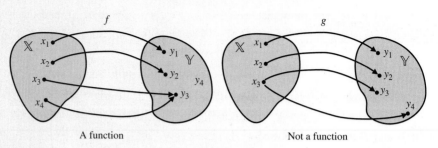

A function   Not a function

Figure 1

The relation f, shown in Fig. 1, is a function since it is well defined with each element of the set $\mathbb{X}$ corresponding to a unique element of the set $\mathbb{Y}$. Notice that more than one element in the domain of the function f can be associated with the same element in the range. In this case x_3 and x_4 both map to y_3. However, the relation g, also shown in Fig. 1, is not a function since x_3 corresponds to both y_3 and y_4. Notice in this example that $f(\mathbb{X})$ is not equal to $\mathbb{Y}$, since y_4 is not in the range of f. In general, for the mapping $f\colon \mathbb{X} \longrightarrow \mathbb{Y}$, the set $\mathbb{X}$ is always the domain, but range$(f) \subseteq \mathbb{Y}$.

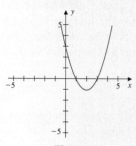

Figure 2

The **graph** of a function $f\colon \mathbb{X} \longrightarrow \mathbb{Y}$ is a subset of the Cartesian product $\mathbb{X} \times \mathbb{Y}$ and is defined by

$$\text{graph}(f) = \{(x, y) \mid x \in \mathbb{X} \text{ and } y = f(x) \in \text{range}(f)\}$$

For a function $f\colon \mathbb{R} \longrightarrow \mathbb{R}$ the graph is a subset of $\mathbb{R}^2$, the Cartesian plane.

A familiar function is $f\colon \mathbb{R} \longrightarrow \mathbb{R}$ defined by the rule

$$f(x) = x^2 - 4x + 3 = (x - 2)^2 - 1$$

Since the rule describing the function is defined for all real numbers, we have dom$(f) = \mathbb{R}$. For the range, since the vertex of the parabola is $(2, -1)$, then range$(f) = [-1, \infty)$. These sets are also evident from the graph of the function, as shown in Fig. 2. Also the image of $x = 0$ is $f(0) = 3$. Notice that in this example it is also the case that $f(4) = 3$, so $\{0, 4\}$ is the set of all real numbers with image equal to 3. The set $\{0, 4\}$ is called the *inverse image* of the set $\{3\}$. This motivates the next concept.

If $f\colon \mathbb{X} \longrightarrow \mathbb{Y}$ is a function and $B \subseteq \mathbb{Y}$, then the **inverse image** of B, denoted by $f^{-1}(B)$, is the set of all elements of the domain that are mapped to B. That is,

$$f^{-1}(B) = \{x \in \mathbb{X} \mid f(x) \in B\}$$

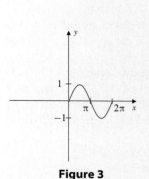

Figure 3

The set $f^{-1}(B)$ is also called the set of **preimages** of the set B. As another illustration let $f\colon [0, 2\pi] \longrightarrow [-1, 1]$ be defined by $f(x) = \sin x$. The graph is shown in Fig. 3. We see from the graph that

$$f^{-1}([0, 1]) = [0, \pi] \qquad \text{and} \qquad f^{-1}([-1, 0]) = [\pi, 2\pi]$$

EXAMPLE 1 Let $f: \mathbb{R} \longrightarrow \mathbb{R}$ be the function defined by $f(x) = x^2 - 4x + 3$. Define the sets

$$A = [0, 3] \qquad B = [1, 4] \qquad C = [-1, 3] \qquad D = [0, 3]$$

a. Compare the sets $f(A \cap B)$ and $f(A) \cap f(B)$.
b. Compare the sets $f(A \cup B)$ and $f(A) \cup f(B)$.
c. Compare the sets $f^{-1}(C \cap D)$ and $f^{-1}(C) \cap f^{-1}(D)$.
d. Compare the sets $f^{-1}(C \cup D)$ and $f^{-1}(C) \cup f^{-1}(D)$.

Solution **a.** Since $A \cap B = [1, 3]$, we see from Fig. 2 that $f(A \cap B) = [-1, 0]$. Again using the graph of f, we have $f(A) = f([0, 3]) = [-1, 3]$ and $f(B) = f([1, 4]) = [-1, 3]$, so that $f(A) \cap f(B) = [-1, 3]$. Hence, we have shown that

$$f(A \cap B) \subseteq f(A) \cap f(B) \quad \text{with} \quad f(A \cap B) \neq f(A) \cap f(B)$$

b. Since $A \cup B = [0, 4]$, we have $f(A \cup B) = [-1, 3]$. Also $f(A) = [-1, 3] = f(B)$, so that $f(A) \cup f(B) = [-1, 3]$. Therefore,

$$f(A \cup B) = f(A) \cup f(B)$$

c. Since $C \cap D = [0, 3]$, we have

$$\begin{aligned}
f^{-1}(C \cap D) &= \{x \in \mathbb{R} \mid f(x) \in [0, 3]\} \\
&= \{x \in \mathbb{R} \mid 0 \le f(x) \le 3\} \\
&= \{x \in \mathbb{R} \mid 0 \le (x - 2)^2 - 1 \le 3\}
\end{aligned}$$

We see from Fig. 2

$$f^{-1}(C \cap D) = [0, 1] \cup [3, 4]$$

On the other hand,

$$f^{-1}(C) = [0, 4]$$

The inverse image of the set D is

$$f^{-1}(D) = [0, 1] \cup [3, 4]$$

Finally,

$$f^{-1}(C) \cap f^{-1}(D) = f^{-1}(C \cap D)$$

d. Since $C \cup D = [-1, 3]$, we have from the results in part (c)

$$f^{-1}(C \cup D) = f^{-1}(C) \cup f^{-1}(D)$$

Theorem 3 summarizes several results about images of sets and inverse images of sets including the observations made in Example 1.

THEOREM 3   Let $f: \mathbb{X} \longrightarrow \mathbb{Y}$ be a function, and suppose A and B are subsets of $\mathbb{X}$ and C and D are subsets of $\mathbb{Y}$. Then

1. $f(A \cap B) \subseteq f(A) \cap f(B)$
2. $f(A \cup B) = f(A) \cup f(B)$
3. $f^{-1}(C \cap D) = f^{-1}(C) \cap f^{-1}(D)$
4. $f^{-1}(C \cup D) = f^{-1}(C) \cup f^{-1}(D)$
5. $A \subseteq f^{-1}(f(A))$
6. $f(f^{-1}(C)) \subseteq C$

Proof   (1) Let $y \in f(A \cap B)$. Then there is some $x \in A \cap B$ such that $y = f(x)$. This means that $y \in f(A)$ and $y \in f(B)$, and hence $y \in f(A) \cap f(B)$. Therefore, $f(A \cap B) \subseteq f(A) \cap f(B)$.

(3) To show that the sets are equal, we show that each set is a subset of the other. Let $x \in f^{-1}(C \cap D)$, so that $f(x) \in C \cap D$, which is equivalent to the statement $f(x) \in C$ and $f(x) \in D$. Therefore, $x \in f^{-1}(C)$ and $x \in f^{-1}(D)$, and we have $f^{-1}(C \cap D) \subseteq f^{-1}(C) \cap f^{-1}(D)$.

Now let $x \in f^{-1}(C) \cap f^{-1}(D)$, which is equivalent to the statement $x \in f^{-1}(C)$ and $x \in f^{-1}(D)$. Then $f(x) \in C$ and $f(x) \in D$, so that $f(x) \in C \cap D$. Therefore, $x \in f^{-1}(C \cap D)$ and hence $f^{-1}(C) \cap f^{-1}(D) \subseteq f^{-1}(C \cap D)$.

(5) If $x \in A$, then $f(x) \in f(A)$, and hence $x \in f^{-1}(f(A))$. This gives $A \subseteq f^{-1}(f(A))$.

The proofs of parts 2, 4, and 6 are left as exercises.

Example 1(a) provides a counterexample to show that the result in Theorem 3, part 1, cannot be replaced with equality.

Inverse Functions

An *inverse function* of a function f, when it exists, is a function that reverses the action of f. Observe that if g is an inverse function of f and $f(a) = b$, then $g(b) = a$. For example, if $f(x) = 3x - 1$ and $g(x) = (x + 1)/3$, then $f(2) = 5$ and $g(5) = 2$. One of the most important function-inverse pairs in mathematics and science is $f(x) = e^x$ and $g(x) = \ln x$.

For a function to have an inverse function, the inverse image for each element of the range of the function must be well defined. This is often not the case. For example, the function $f: \mathbb{R} \longrightarrow \mathbb{R}$ defined by $f(x) = x^2$ cannot be reversed as a function since the inverse image of the set $\{4\}$ is the set $\{-2, 2\}$. Notice that the inverse image of a set in the range of a function is always defined, but the function may not have an inverse function. A function that has an inverse is called **invertible**. Later in this section we show that if a function is invertible, then it has a unique inverse. This will justify the use of the definite article and the symbol f^{-1} when referring to the inverse of the function f. Functions that have inverses are characterized by the property called *one-to-one*. The function described in Fig. 1 is not one-to-one, since both x_3 and x_4 are sent to the same element of $\mathbb{Y}$. This cannot occur for a one-to-one function.

DEFINITION 2

One-to-One Function   Let $f: \mathbb{X} \longrightarrow \mathbb{Y}$ be a function. Then f is called **one-to-one**, or **injective**, if for all x_1 and x_2 with $x_1 \neq x_2$, then $f(x_1) \neq f(x_2)$.

Alternatively, f is one-to-one if whenever $f(x_1) = f(x_2)$, then $x_1 = x_2$. For a function $f: \mathbb{R} \longrightarrow \mathbb{R}$, this condition is met if every horizontal line intersects the graph of f in at most one point. When this happens, f passes the *horizontal line test* and is thus invertible. This test is similar to the *vertical line test* used to determine if f is a function. The inverse of f is denoted by f^{-1} with $f^{-1}: \text{range}(f) \longrightarrow \mathbb{X}$. Theorem 4 gives a characterization of functions that are invertible. We omit the proof.

THEOREM 4

Let $\mathbb{X}$ and $\mathbb{Y}$ be nonempty sets and $f: \mathbb{X} \longrightarrow \mathbb{Y}$ be a function. The function f has an inverse function if and only if f is one-to-one.

As an illustration, let $f: \mathbb{R} \longrightarrow \mathbb{R}$ be defined by $y = f(x) = 3x + 1$. Since the graph, which is a straight line, satisfies the horizontal line test, the function is one-to-one and hence has an inverse function. To find the inverse in this case is an easy matter. We can solve for x in terms of y to obtain

$$x = \frac{y - 1}{3}$$

The inverse function is then written using the same independent variable, so that

$$f^{-1}(x) = \frac{x - 1}{3}$$

It is also possible to show that a function has an inverse even when it is difficult to find the inverse.

EXAMPLE 2

Show the function $f: \mathbb{R} \longrightarrow \mathbb{R}$ defined by $f(x) = x^3 + x$ is invertible.

Solution

By Theorem 4, to show that f is invertible, we show that f is one-to-one. Suppose that $x_1 \neq x_2$ with $x_1 < x_2$. We wish to show that $f(x_1) \neq f(x_2)$. Since the cubing function is strictly increasing for all x, we have

$$x_1 < x_2 \quad \text{and} \quad x_1^3 < x_2^3$$

Therefore,

$$f(x_1) = x_1^3 + x_1 < x_2^3 + x_2 = f(x_2) \quad \text{so that} \quad f(x_1) \neq f(x_2)$$

The graph of an invertible function can be used to describe the graph of the inverse function. To see how, suppose that (a, b) is a point on the graph of $y = f(x)$. Then $b = f(a)$ and $a = f^{-1}(b)$. Consequently, the point (b, a) is on the graph of $y = f^{-1}(x)$. Since the point (b, a) is the reflection of (a, b) through the line $y = x$, the graphs of f and f^{-1} are also reflections through $y = x$. The graph of the function and its inverse in Example 2 are shown in Fig. 4.

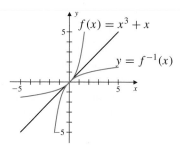

Figure 4

When $f: \mathbb{X} \longrightarrow \mathbb{Y}$ is a function such that the set of images is all of $\mathbb{Y}$, that is, $f(\mathbb{X}) = \mathbb{Y}$, we call the function *onto*.

DEFINITION 3   **Onto Function**   The function $f: \mathbb{X} \longrightarrow \mathbb{Y}$ is called **onto**, or **surjective**, if range$(f) = \mathbb{Y}$.

For example, the function of Example 2 is onto since the range of f is all of $\mathbb{R}$. See Fig. 4. A function is called **bijective** if it is both one-to-one and onto.

Notice that the function $f: \mathbb{R} \longrightarrow \mathbb{R}$ with $f(x) = x^2$ is not onto since range$(f) = [0, \infty)$. Of course, every function is a mapping onto its range. So the function $f: \mathbb{R} \longrightarrow [0, \infty)$ defined by $f(x) = x^2$ is onto. This new version of the original function is not one-to-one, but by restricting the domain to $[0, \infty)$, we can define a version that is one-to-one and onto. That is, the function $f: [0, \infty) \longrightarrow [0, \infty)$ defined by $f(x) = x^2$ is a bijection. The function defined in Example 2 is also bijective. Notice also that a function has an inverse if and only if it is bijective.

Composition of Functions

Functions can be combined in a variety of ways to create new functions. For example, if $f: \mathbb{X}_1 \longrightarrow \mathbb{Y}_1$ and $g: \mathbb{X}_2 \longrightarrow \mathbb{Y}_2$ are real-valued functions of a real variable, then the standard arithmetic operations on functions are defined by

$$(f + g)(x) = f(x) + g(x)$$
$$(f - g)(x) = f(x) - g(x)$$
$$(fg)(x) = f(x)g(x)$$
$$\left(\frac{f}{g}\right)(x) = \frac{f(x)}{g(x)}$$

The domains of these functions are given by $\text{dom}(f+g) = \text{dom}(f-g) = \text{dom}(fg) = \mathbb{X}_1 \cap \mathbb{X}_2$ and $\text{dom}(f/g) = (\mathbb{X}_1 \cap \mathbb{X}_2)\setminus\{x \mid g(x) = 0\}$. Another method of combining functions is through the *composition* of two functions. In the composition of two functions f and g, the output of one function is used as the input to the other. For example, if $f(x) = \sqrt{x}$ and $g(x) = x^2 - x - 2$, then $f(g(3)) = f(4) = 2$ is the composition of f with g evaluated at the number 3 and is denoted by $(f \circ g)(3)$.

DEFINITION 4

Composition   Let A, B, and C be nonempty sets and $f: B \longrightarrow C$ and $g: A \longrightarrow B$ be functions. The **composition** $f \circ g: A \longrightarrow C$ is defined by

$$(f \circ g)(x) = f(g(x))$$

The domain of the composition is $\text{dom}(f \circ g) = \{x \in \text{dom}(g) \mid g(x) \in \text{dom}(f)\}$.

A function and its inverse undo each other relative to composition. For example, let $f(x) = 2x - 1$. Since f is one-to-one, it is invertible with $f^{-1}(x) = (x+1)/2$. Notice that

$$(f^{-1} \circ f)(x) = f^{-1}(f(x)) = \frac{f(x)+1}{2} = \frac{2x-1+1}{2} = x$$

and

$$(f \circ f^{-1})(x) = f(f^{-1}(x)) = 2\left(\frac{x+1}{2}\right) - 1 = x + 1 - 1 = x$$

THEOREM 5

Suppose that $f: \mathbb{X} \longrightarrow \mathbb{Y}$ is a bijection. Then

1. $(f^{-1} \circ f)(x) = x$ for all $x \in \mathbb{X}$
2. $(f \circ f^{-1})(x) = x$ for all $x \in \mathbb{Y}$

As mentioned earlier, when an inverse function exists, it is unique. To see this, let $f: \mathbb{X} \longrightarrow \mathbb{Y}$ be an invertible function and f^{-1} an inverse function. Suppose that $g: \mathbb{Y} \longrightarrow \mathbb{X}$ is another inverse function for f. Let $I_{\mathbb{X}}$ be the identity function on $\mathbb{X}$ and $I_{\mathbb{Y}}$ the identity function on $\mathbb{Y}$. That is, $I_{\mathbb{X}}(x) = x$ for all $x \in \mathbb{X}$ and $I_{\mathbb{Y}}(y) = y$ for all $y \in \mathbb{Y}$. If y is in $\mathbb{Y}$, then

$$\begin{aligned} g(y) &= g \circ I_{\mathbb{Y}}(y) = g \circ (f \circ f^{-1})(y) \\ &= g(f(f^{-1}(y))) = (g \circ f) \circ f^{-1}(y) \\ &= I_{\mathbb{X}}(f^{-1}(y)) = f^{-1}(y) \end{aligned}$$

Since this holds for all y in $\mathbb{Y}$, then $g = f^{-1}$. Consequently, when it exists, the inverse function is unique. This justifies the use of the symbol f^{-1} for the inverse of f, when it exists.

THEOREM 6   Let $f: \mathbb{X} \longrightarrow \mathbb{Y}$ be a bijection. Then

1. $f^{-1}: \mathbb{Y} \longrightarrow \mathbb{X}$ is also a bijection,
2. $(f^{-1})^{-1} = f$

THEOREM 7   Let A, B, and C be nonempty sets and $f: B \longrightarrow C$ and $g: A \longrightarrow B$ be functions.

1. If f and g are injections, then $f \circ g$ is an injection.
2. If f and g are surjections, then $f \circ g$ is a surjection.
3. If f and g are bijections, then $f \circ g$ is a bijection.
4. If $f \circ g$ is an injection, then g is an injection.
5. If $f \circ g$ is a surjection, then f is a surjection.

Proof   (1) Suppose that x_1 and x_2 are in A and $(f \circ g)(x_1) = (f \circ g)(x_2)$. Then by the definition of composition, we have

$$f(g(x_1)) = f(g(x_2))$$

Since f is an injection, $g(x_1) = g(x_2)$. But since g is also an injection, we have $x_1 = x_2$. Therefore, $f \circ g$ is an injection.

(5) Let $c \in C$. Since $f \circ g: A \longrightarrow C$ is a surjection, there is some $a \in A$ such that $(f \circ g)(a) = c$. That is, $f(g(a)) = c$. But $g(a) \in B$, so there is an element of B with image under f equal to c. Since c was chosen arbitrarily, we know that f is a surjection.

The proofs of parts 2, 3, and 4 are left as exercises.

THEOREM 8   Let A, B, and C be nonempty sets and $f: B \longrightarrow C$ and $g: A \longrightarrow B$ be functions. If f and g are bijections, then the function $f \circ g$ has an inverse function and $(f \circ g)^{-1} = g^{-1} \circ f^{-1}$.

Proof   By Theorem 7, the composition $f \circ g: A \longrightarrow C$ is a bijection; hence by Theorem 4, the inverse function $(f \circ g)^{-1}: C \longrightarrow A$ exists. Moreover, the function $g^{-1} \circ f^{-1}$ also maps C to A. For each $c \in C$ we will show that $(f \circ g)^{-1}(c) = (g^{-1} \circ f^{-1})(c)$. Let $c \in C$. Since f is onto, there is $b \in B$ such that $f(b) = c$, so that $b = f^{-1}(c)$. Next, since g is onto, there is an $a \in A$ such that $g(a) = b$, which is equivalent to $a = g^{-1}(b)$. Taking compositions gives $(f \circ g)(a) = f(g(a)) = c$, and hence $(f \circ g)^{-1}(c) = a$. We also have $g^{-1}(f^{-1}(c)) = (g^{-1} \circ f^{-1})(c) = a$. Since this holds for all $c \in C$, the functions $(f \circ g)^{-1}$ and $g^{-1} \circ f^{-1}$ are identical, that is, $(f \circ g)^{-1} = g^{-1} \circ f^{-1}$.

Exercise Set A.2

In Exercises 1–10, let

$$\mathbb{X} = \{1, 2, 3, 4, 5, 6\}$$
$$\mathbb{Y} = \{-2, -1, 3, 5, 9, 11, 14\}$$

and define $f: \mathbb{X} \longrightarrow \mathbb{Y}$ by the set of ordered pairs

$$\{(1, -2), (2, 3), (3, 9), (4, -2), (5, 11), (6, -1)\}$$

1. Explain why f is a function.

2. Is f a one-to-one function? Explain.

3. Is f an onto function? Specify range(f).

4. Let $A = \{1, 2, 4\}$. Find $f(A)$.

5. Find $f^{-1}(\{-2\})$.

6. Find $f^{-1}(f(\{1\}))$.

7. Does f have an inverse function? Explain.

8. Is it possible to define a function with domain $\mathbb{X}$ that is onto $\mathbb{Y}$? Explain.

9. Define a function $g: \mathbb{X} \longrightarrow \mathbb{Y}$ that is one-to-one.

10. Is it possible to define a function $g: \mathbb{Y} \longrightarrow \mathbb{X}$ that is onto? Explain.

In Exercises 11–14, use the function $f: \mathbb{R} \to \mathbb{R}$ defined by

$$f(x) = x^2$$

11. Let $A = (-3, 5)$ and $B = [0, 7)$. Verify that

$$f(A \cup B) = f(A) \cup f(B)$$

12. Let $C = [1, \infty)$ and $D = [3, 5]$. Verify that

$$f^{-1}(C \cup D) = f^{-1}(C) \cup f^{-1}(D)$$

13. Let $A = [-2, 0]$ and $B = [0, 2]$. Verify that

$$f(A \cap B) \subset f(A) \cap f(B)$$

and the sets are not equal.

14. Define a function g by the rule $g(x) = x^2$, but with the domain restricted to the interval $[0, \infty)$. If $A = [0, 5)$ and $B = [2, 7)$, verify that

$$g(A \cap B) = g(A) \cap g(B)$$

What property does g have that f does not?

15. Define a function $f: \mathbb{R} \to \mathbb{R}$ by $f(x) = ax + b$, where a and b are real numbers with $a \neq 0$. Find the inverse function of f.

16. Define a function $f: \mathbb{R} \to \mathbb{R}$ by $f(x) = x^5 + 2x$. Show that the inverse function of f exists.

17. Given a function f, define for each positive integer n

$$f^n(x) = (f \circ f \circ \cdots \circ f)(x)$$

where the composition is taken $n - 1$ times. If c is a fixed real number and $f(x) = -x + c$, find $f^n(c)$ for all n.

18. Define a function $f: \mathbb{R} \to \mathbb{R}$ by

$$f(x) = \begin{cases} 2x & \text{if } 0 \leq x \leq \frac{1}{2} \\ 2 - 2x & \text{if } \frac{1}{2} < x \leq 1 \end{cases}$$

Sketch the graphs of $y = f(x)$ and $y = (f \circ f)(x)$.

19. Define a function $f: \mathbb{R} \to \mathbb{R}$ by

$$f(x) = e^{2x-1}$$

a. Show that f is one-to-one.

b. Is f onto? Justify your answer.

c. Define a function g with the same rule and domain as f but that is onto.

d. Find the inverse function for the g defined in part (c).

20. Define a function $f: \mathbb{R} \to \mathbb{R}$ by

$$f(x) = e^{-x^2}$$

Show the function is not one-to-one.

21. Define a function $f: \mathbb{N} \to \mathbb{N}$ by

$$f(n) = 2n$$

a. Show that f is one-to-one.

b. Is f onto? Explain.

c. If E denotes the set of even positive integers and O the odd positive integers, find $f^{-1}(E)$ and $f^{-1}(O)$.

22. Define a function $f: \mathbb{Z} \to \mathbb{Z}$ by

$$f(n) = \begin{cases} n+1 & \text{if } n \text{ is even} \\ n-3 & \text{if } n \text{ is odd} \end{cases}$$

Let E denote the set of even integers and O the set of odd integers. Find $f(E)$ and $f(O)$.

23. Define a function $f: \mathbb{Z} \times \mathbb{Z} \to \mathbb{Z}$ by

$$f((m, n)) = 2m + n$$

a. Let $A = \{(p, q) \mid p \text{ and } q \text{ are odd}\}$. Find $f(A)$.

b. Let $B = \{(p, q) \mid q \text{ is odd}\}$. Find $f(B)$.

c. Find $f^{-1}(\{0\})$.

d. Let E denote the set of even integers. Find $f^{-1}(E)$.

e. Let O denote the set of odd integers. Find $f^{-1}(O)$.

f. Show that f is not one-to-one.

g. Show that f is onto.

24. Define a function $f: \mathbb{R}^2 \to \mathbb{R}^2$ by

$$f((x, y)) = (2x, 2x + 3y)$$

a. Show that f is one-to-one.

b. Is f onto? Justify your answer.

c. Let A be the set of all points that lie on the line $y = x + 1$. Find $f(A)$.

In Exercises 25–27, $f: \mathbb{X} \to \mathbb{Y}$ is a function, A and B are subsets of $\mathbb{X}$, and C and D are subsets of $\mathbb{Y}$. Prove the statements.

25. $f(A \cup B) = f(A) \cup f(B)$

26. $f^{-1}(C \cup D) = f^{-1}(C) \cup f^{-1}(D)$

27. $f(f^{-1}(C)) \subseteq C$

In Exercises 28–30, $f: B \to C$ and $g: A \to B$ are functions. Prove the statements.

28. If f and g are surjections, then $f \circ g$ is a surjection.

29. If f and g are bijections, then $f \circ g$ is a bijection.

30. If $f \circ g$ is an injection, then g is an injection.

31. If $f: \mathbb{X} \to \mathbb{Y}$ is a function and A and B are subsets of $\mathbb{X}$, show that

$$f(A) \setminus f(B) \subseteq f(A \setminus B)$$

32. If $f: \mathbb{X} \to \mathbb{Y}$ is a function and C and D are subsets of $\mathbb{Y}$, show that

$$f^{-1}(C \setminus D) = f^{-1}(C) \setminus f^{-1}(D)$$

A.3 ▶ Techniques of Proof

Mathematics is built on facts. A few of these, called **axioms**, are accepted as self-evident and do not require justification. Every other statement of fact requires *proof*. A **proof** is the process of establishing the validity of a statement. Results in mathematics that require proof are called **theorems** and are made up of two parts. The first part, called the **hypothesis**, is a set of assumptions. The second part, called the **conclusion**, is the statement that requires proof. It is customary to use the letter P to denote the hypotheses (or hypothesis if there is only one) and the letter Q to denote the conclusion. A theorem is symbolized by

$$P \implies Q$$

which we read as "if P, then Q" or "P implies Q" or "P is sufficient for Q." The **converse** of a theorem is symbolized by

$$Q \implies P$$

read as "Q implies P" or "P is necessary for Q." For example, let P be the statement *Mary lives in Iowa* and Q the statement that *Mary lives in the United States.* Then certainly $P \implies Q$ is a theorem since every resident of Iowa is a resident of the United States. But $Q \implies P$ is not a theorem since, for example, if Mary is a resident of California, then she is a resident of the United States but not a resident of Iowa. So the statement $Q \implies P$ is not always true given that Q is true. In terms of sets, if A is the set of residents of Iowa and B is the set of residents of the United States, then the statement P is *Mary is in A* and Q is *Mary is in B.* Then *Mary is in A* implies *Mary is in B.* It is also clear that if *Mary is in $B \backslash A$,* then *Mary is in B* does not imply that *Mary is in A.*

A statement that is equivalent to the theorem $P \implies Q$ is the **contrapositive** statement $\sim Q \implies \sim P$, that is, *not Q implies not P.* In the example above, if Mary is not a resident of the United States, then Mary is not a resident of Iowa. An equivalent formulation of the statement, in the terminology of sets, is that if Mary $\notin B$, then it implies Mary $\notin A$.

There are other statements in mathematics that require proof. **Lemmas** are preliminary results used to prove theorems, **propositions** are results not as important as theorems, and **corollaries** are special cases of a theorem. A statement that is not yet proven is called a **conjecture**. One of the most famous conjectures is the celebrated Riemann hypothesis. A single **counterexample** is enough to refute a false conjecture. For example, the statement *All lions have green eyes* is rendered invalid by the discovery of a single blue-eyed lion.

In this section we briefly introduce three main types of proof. A fourth type, called mathematical induction, is discussed in Sec. A.4.

Direct Argument

In a **direct argument**, a sequence of logical steps links the hypotheses P to the conclusion Q. Example 1 provides an illustration of this technique.

EXAMPLE 1   Prove that if p and q are odd integers, then $p + q$ is an even integer.

Solution   To prove this statement with a direct argument, we assume that p and q are odd integers. Then there are integers m and n such that

$$p = 2m + 1 \quad \text{and} \quad q = 2n + 1$$

Adding p and q gives

$$p + q = 2m + 1 + 2n + 1$$
$$= 2(m + n) + 2$$
$$= 2(m + n + 1)$$

Since $p + q$ is a multiple of 2, it is an even integer.

Contrapositive Argument

The **contrapositive statement** of the statement $P \implies Q$ is the statement $\sim Q \implies \sim P$. The notation $\sim Q$ denotes the negation of the statement Q. A statement and the contrapositive statement are equivalent, so that if one holds, then the other also holds. In a **contrapositive argument** the hypothesis is $\sim Q$, and we proceed with a direct argument to show that $\sim P$ holds.

EXAMPLE 2   If p^2 is an even integer, then p is an even integer.

Solution   In a direct argument we assume that p^2 is even, so that we can write $p^2 = 2k$ for some integer k. Then

$$p = \sqrt{2k} = \sqrt{2}\sqrt{k}$$

which does not allow us to conclude that p is even.

To use a contrapositive argument, we assume that p is *not* an even integer. That is, we assume that p is an odd integer. Then there is an integer k such that $p = 2k + 1$. Squaring both sides of equation $p = 2k + 1$ gives

$$
\begin{aligned}
p^2 &= (2k + 1)^2 \\
&= 4k^2 + 4k + 1 \\
&= 2(2k^2 + 2k) + 1
\end{aligned}
$$

and hence p^2 is an odd integer. Therefore, the original statement holds.

Contradiction Argument

In a **contradiction argument** to show that a statement holds, we assume the contrary and use this assumption to arrive at some contradiction. For example, to prove that *the set of natural numbers $\mathbb{N}$ is infinite*, we would assume the set of natural numbers is finite and argue that this leads to a contradiction. A contrapositive argument is a form of contradiction where to prove $P \implies Q$, we assume that P holds and $\sim Q$ holds and arrive at the conclusion that $\sim P$ holds. Since both P and $\sim P$ cannot be true, we have a contradiction. In certain cases the contradiction may be hard to recognize.

EXAMPLE 3   Prove that $\sqrt{2}$ is an irrational number.

Solution   To use a contradiction argument, we assume that $\sqrt{2}$ is *not* irrational. That is, we assume that there are integers p and q such that

$$\sqrt{2} = \frac{p}{q}$$

where p and q have no common factors. We will arrive at a contradiction by showing that if $\sqrt{2} = p/q$, then p and q do have a common factor. Squaring both sides of the last equation gives

$$2 = \frac{p^2}{q^2} \qquad \text{so that} \qquad p^2 = 2q^2$$

Hence, p^2 is even. Since p^2 is an even integer, then by Example 2 so is p. Thus, there is an integer k such that $p = 2k$. Substituting $2k$ for p in the equation $2q^2 = p^2$ gives

$$2q^2 = p^2 = (2k)^2 = 4k^2 \qquad \text{so that} \qquad q^2 = 2k^2$$

Hence, q is also an even integer. Since p and q are both even, they have a common factor of 2, which contradicts the assumption that p and q are chosen to have no common factors.

Quantifiers

Often statements in mathematics are quantified using the **universal** quantifier *for all*, denoted by the symbol $\forall$, or by the **existential** quantifier *there exists*, denoted by the symbol $\exists$. If $P(x)$ is a statement that depends on the parameter x, then the symbols

$$\forall x, \, P(x)$$

are read *for all x, $P(x)$*. To prove that the statement is true, we have to verify that the statement $P(x)$ holds for every choice of x. To prove that the statement is false, we need to find only one x such that $P(x)$ is false, that is, we need to find a counterexample. To prove that a statement of the form

$$\exists x, \, P(x)$$

holds requires finding at least one x such that $P(x)$ holds. The statement is false if the statement

$$\sim(\exists x, \, P(x))$$

holds. When we negate a statement involving quantifiers, $\sim\exists$ becomes $\forall$ and $\sim\forall$ becomes $\exists$. So the statement

$$\sim(\exists x, \, P(x)) \qquad \text{is equivalent to} \qquad \forall x, \sim P(x)$$

and the statement

$$\sim(\forall x, \, P(x)) \qquad \text{is equivalent to} \qquad \exists x, \sim P(x)$$

Exercise Set A.3

1. Prove that in an isosceles right triangle, the hypotenuse is $\sqrt{2}$ times the length of one of the equal sides.

2. Prove that if ABC is an isosceles right triangle with C the vertex of the right angle and sides opposite the vertices a, b, and c, respectively, then the area of the triangle is $c^2/4$.

3. Prove that in an equilateral triangle the area of the triangle is $\sqrt{3}/4$ times the square of the length of a side.

4. Prove that if s and t are rational numbers with $t \neq 0$, then s/t is a rational number.

5. Prove that if a, b, and c are integers such that a divides b and b divides c, then a divides c.

6. Prove that if m and n are even integers, then $m + n$ is an even integer.

7. Prove that if n is an odd integer, then n^2 is an odd integer.

8. Prove that if n is in $\mathbb{N}$, then $n^2 + n + 3$ is odd.

9. Prove that if a and b are consecutive integers, then $(a + b)^2$ is an odd integer.

10. Prove that if m and n are odd integers, then mn is an odd integer.

11. Show that the statement *if m and n are two consecutive integers, then 4 divides $m^2 + n^2$* is false.

12. Let $f(x) = (x - 1)^2$ and $g(x) = x + 1$. Prove that if x is in the set $S = \{x \in \mathbb{R} \mid 0 \leq x \leq 3\}$, then $f(x) \leq g(x)$.

13. Prove that if n is an integer and n^2 is odd, then n is odd.

14. Prove that if n is an integer and n^3 is even, then n is even.

15. Prove that if p and q are positive real numbers such that $\sqrt{pq} \neq (p + q)/2$, then $p \neq q$.

16. Prove that if c is an odd integer, then the equation $n^2 + n - c = 0$ has no integer solution for n.

17. Prove that if x is a nonnegative real number such that $x < \epsilon$, for every real number $\epsilon > 0$, then $x = 0$.

18. Prove that if x is a rational number and $x + y$ is an irrational number, then y is an irrational number.

19. Prove that $\sqrt[3]{2}$ is irrational.

20. Prove that if n in $\mathbb{N}$, then
$$\frac{n}{n + 1} > \frac{n}{n + 2}$$

21. Suppose that x and y are real numbers with $x < 2y$. Prove that if $7xy \leq 3x^2 + 2y^2$, then $3x \leq y$.

22. Define a function $f: \mathbb{X} \to \mathbb{Y}$ and sets A and B in $\mathbb{X}$ that is a counterexample to show the statement
$$\text{If } f(A) \subseteq f(B), \text{ then } A \subset B$$
is false.

23. Define a function $f: \mathbb{X} \to \mathbb{Y}$ and sets C and D in $\mathbb{Y}$ that is a counterexample to show the statement
$$\text{If } f^{-1}(C) \subseteq f^{-1}(D), \text{ then } C \subset D$$
is false.

In Exercises 24–30, $f: \mathbb{X} \to \mathbb{Y}$ is a function, A and B are subsets of $\mathbb{X}$, and C and D are subsets of $\mathbb{Y}$. Prove the statements.

24. If $A \subseteq B$, then $f(A) \subseteq f(B)$.

25. If $C \subseteq D$, then $f^{-1}(C) \subseteq f^{-1}(D)$.

26. If f is an injection, then for all A and B
$$f(A \cap B) = f(A) \cap f(B)$$

27. If f is an injection, then for all A and B
$$f(A \backslash B) = f(A) \backslash f(B)$$

28. If f is an injection, then for all A
$$f^{-1}(f(A)) = A$$

29. If f is a surjection, then for all C
$$f(f^{-1}(C)) = C$$

A.4 ▶ **Mathematical Induction**

Throughout mathematics there are statements that depend on natural numbers and where the aim is to determine whether the statement is true or false for all natural numbers. Some simple examples are the following three statements, the third being a well-known puzzle, called the Tower of Hanoi puzzle.

1. For every natural number n, the sum of the first n natural numbers is given by

$$1 + 2 + 3 + \cdots + n = \frac{n(n + 1)}{2}$$

2. The expression $6n + 1$ is a prime number for every natural number n.
3. Given three pegs, labeled 1, 2, and 3, and a stack of n disks of decreasing diameters on peg–1, the disks can be moved to peg–3 in $2^n - 1$ moves. This is under the restriction that a disk can be placed on top of another disk only when it has smaller diameter.

When we are considering a statement involving natural numbers to provide insight, a useful first step is to substitute specific numbers for n and determine whether the statement is true. If the statement is false, often a counterexample is found quickly, allowing us to reject the statement. For example, in the second statement above, for $n = 1, 2$, and 3 the expression $6n + 1$ has values 7, 13, and 19, respectively, all of which are prime numbers. However, if $n = 4$, then $6(4) + 1 = 25$, which is not a prime number, and the statement is not true for all natural numbers n.

In the case of the first statement, the data in Table 1 provide more convincing evidence that the formula may indeed hold for all natural numbers. Of course, to establish the fact for all n requires a proof, which we postpone until Example 1.

For the Tower of Hanoi puzzle, when $n = 1$, the number of steps required is 1, and when $n = 2$, it is also easy to see a solution requiring 3 steps. A solution for $n = 3$ is given by the moves

$$D3 \longrightarrow P3, D2 \longrightarrow P2, D3 \longrightarrow P2, D1 \longrightarrow P3, D3 \longrightarrow P1,$$
$$D2 \longrightarrow P3, D1 \longrightarrow P3$$

Table 1

$1 + 2 + 3 + \cdots + n$	$\frac{n(n+1)}{2}$
1	$\frac{(1)(2)}{2} = 1$
$1 + 2 = 3$	$\frac{(2)(3)}{2} = 3$
$1 + 2 + 3 = 6$	$\frac{(3)(4)}{2} = 6$
$1 + 2 + 3 + 4 = 10$	$\frac{(4)(5)}{2} = 10$
$1 + 2 + 3 + 4 + 5 = 15$	$\frac{(5)(6)}{2} = 15$
$1 + 2 + 3 + 4 + 5 + 6 = 21$	$\frac{(6)(7)}{2} = 21$
$1 + 2 + 3 + 4 + 5 + 6 + 7 = 28$	$\frac{(7)(8)}{2} = 28$

where $D1$, $D2$, and $D3$ represent the three disks of decreasing diameters and $P1$, $P2$, and $P3$ represent the three pegs. So for $n = 3$, we have a solution with $7 = 2^3 - 1$ moves. Again, the evidence is leading toward the result being true, but we have not given a satisfactory proof. Let's push this example a bit further. *How can we use the result for three disks to argue that this result holds for four disks?* The same sequence of steps we gave for the solution of the three-disk problem can be used to move the stack from $P1$ to either $P2$ or $P3$. Now, suppose that there are four disks on $P1$. Since the bottom disk is the largest, $P1$ can be used as before to move the top three disks. So as a first step, move the top three disks to $P2$, which requires $2^3 - 1 = 7$ moves. Next, move the remaining (largest) disk on $P1$ to $P3$, which requires 1 move. Now, using the same procedure as before, move the three-disk stack on $P2$ over to $P3$, requiring another $2^3 - 1 = 7$ moves. The total number of moves is now

$$2(2^3 - 1) + 1 = 2^4 - 2 + 1 = 2^4 - 1 = 15$$

This approach contains the essentials of *mathematical induction*. We start with an initial case, called the **base case**, that we can argue holds. The next step, called the **inductive hypothesis**, provides a mechanism for advancing from one natural number to the next. In the Tower of Hanoi example, the base case is the case for $n = 1$, and one disk on $P1$ requires only $1 = 2^1 - 1$ move to transfer the disk to $P3$ or $P2$. The inductive hypothesis is to assume that the result holds when there are n disks on $P1$. We are required to argue the result holds for $n + 1$ disks on $P1$. We did this for $n = 3$.

Theorem 9 provides a formal statement of the *principle of mathematical induction*. The proof of this statement, which we omit, is based on the axiomatic foundations of the natural numbers. Specifically, the proof uses the *well-ordering* principle, which states that every nonempty subset of $\mathbb{N}$ has a smallest element.

THEOREM 9   **The Principle of Mathematical Induction**

Let P be a statement that depends on the natural number n. Suppose that

1. P is true for $n = 1$ and
2. When P is true for a natural number n, then P is true for the successor $n + 1$

Then the statement P is true for every natural number n.

The principle of mathematical induction is also referred to as *mathematical induction*, or simply *induction*.

An analogy to describe the process of mathematical induction is an infinite row of dominoes that are toppled one domino at a time, starting with the first domino. If the dominoes are set up so that whenever a domino falls its successor will fall (the inductive hypothesis), then the entire row of dominoes will fall once the first domino is toppled (base case).

The principle of mathematical induction is used to prove a statement holds for all natural numbers, or for all natural numbers beyond a fixed natural number. This is illustrated in the following examples.

EXAMPLE 1   Prove that for every natural number n,

$$\sum_{k=1}^{n} k = 1 + 2 + 3 + \cdots + n = \frac{n(n+1)}{2}$$

Solution   To establish the base case when $n = 1$, notice that

$$1 = \frac{(1)(2)}{2}$$

The inductive hypothesis is to assume that the statement is true for some fixed natural number n. That is, we assume

$$1 + 2 + 3 + \cdots + n = \frac{n(n+1)}{2}$$

Next, add $n + 1$ to both sides of the last equation to obtain

$$1 + 2 + 3 + \cdots + n + (n+1) = (1 + 2 + 3 + \cdots + n) + (n+1)$$

and we apply the inductive hypothesis to conclude

$$1 + 2 + 3 + \cdots + n + (n+1) = (1 + 2 + 3 + \cdots + n) + (n+1)$$

$$= \frac{n(n+1)}{2} + (n+1)$$

$$= \frac{(n+1)(n+2)}{2}$$

The last equality agrees with the stated formula for the successor of n, that is, for $n + 1$. Therefore, by induction the statement holds for all natural numbers.

EXAMPLE 2   Prove that for every natural number n, the number $3^n - 1$ is divisible by 2.

Solution   In Table 2 we have verified that for $n = 1, 2, 3, 4$, and 5 the number $3^n - 1$ is divisible by 2.

In particular, if $n = 1$, then $3^n - 1 = 2$, which is divisible by 2. Next, we assume that the statement $3^n - 1$ *is divisible by* 2 holds. To complete the proof, we must verify that the number $3^{n+1} - 1$ is also divisible by 2. Since $3^n - 1$ is

Table 2

n	$3^n - 1$
1	2
2	8
3	26
4	80
5	242

divisible by 2, then there is a natural number q such that

$$3^n - 1 = 2q \qquad \text{which gives} \qquad 3^n = 2q + 1$$

Next, we rewrite the expression $3^{n+1} - 1$ to include 3^n in order to use the inductive hypothesis. This gives

$$\begin{aligned}
3^{n+1} - 1 &= 3(3^n) - 1 \\
&= 3(2q + 1) - 1 \\
&= 6q + 2 \\
&= 2(3q + 1)
\end{aligned}$$

Therefore, the expression $3^{n+1} - 1$ is also divisible by 2.

Recall that *factorial notation* is used to express the product of consecutive natural numbers. Several examples are

$$\begin{aligned}
1! &= 1 \\
2! &= 1 \cdot 2 = 2 \\
3! &= 1 \cdot 2 \cdot 3 = 6 \\
4! &= 1 \cdot 2 \cdot 3 \cdot 4 = 24 \\
&\vdots \\
20! &= 2,432,902,008,176,640,000
\end{aligned}$$

For a natural number n, the definition of n **factorial** is the positive integer

$$n! = n(n - 1)(n - 2) \cdots 3 \cdot 2 \cdot 1$$

We also define $0! = 1$.

EXAMPLE 3   Verify that for every natural number n,

$$n! \geq 2^{n-1}$$

Solution   For $n = 1$ the statement is true, since $n! = 1! = 1$ and $2^{n-1} = 2^0 = 1$. Now assume that the statement $n! \geq 2^{n-1}$ holds. Next, we consider

$$(n + 1)! = (n + 1)n!$$

which we need to show is greater than or equal to 2^n. Applying the inductive hypothesis to $n!$ gives the inequality

$$(n + 1)! \geq (n + 1)2^{n-1}$$

Since for every natural number $n \geq 1$ it is also the case that $n + 1 \geq 2$, we have

$$(n + 1)! \geq (n + 1)2^{n-1} \geq 2 \cdot 2^{n-1} = 2^n$$

Consequently, the statement $n! \geq 2^{n-1}$ is true for every natural number n.

EXAMPLE 4

For any natural number n, find the sum of the odd natural numbers from 1 to $2n - 1$.

Solution

The first five cases are given in Table 3.

Table 3

n	$2n - 1$	$1 + 3 + \cdots + (2n - 1)$
1	1	1
2	3	$1 + 3 = 4$
3	5	$1 + 3 + 5 = 9$
4	7	$1 + 3 + 5 + 7 = 16$
5	9	$1 + 3 + 5 + 7 + 9 = 25$

The data in Table 3 suggest that for each $n \geq 1$,

$$1 + 3 + 5 + 7 + \cdots + (2n - 1) = n^2$$

Starting with the case for $n = 1$, we see that the left-hand side is 1 and the expression on the right is $1^2 = 1$. Hence, the statement holds when $n = 1$. Next, we assume that $1 + 3 + 5 + \cdots + (2n - 1) = n^2$. For the next case when the index is $n + 1$, we consider the sum

$$1 + 3 + 5 + \cdots + (2n - 1) + [2(n + 1) - 1] = 1 + 3 + 5 + \cdots + (2n - 1) + (2n + 1)$$

Using the inductive hypothesis, we get

$$\underbrace{1 + 3 + 5 + \cdots + (2n - 1)}_{n^2} + [2(n + 1) - 1] = n^2 + (2n + 1)$$

$$= n^2 + 2n + 1$$

$$= (n + 1)^2$$

Therefore, by induction the statement holds for all natural numbers.

EXAMPLE 5   Let $P_1, P_2, \ldots, P_n$ be n points in a coordinate plane with no three points collinear (in a line). Verify that the number of line segments joining all pairs of points is

$$\frac{n^2 - n}{2}$$

Solution   In Fig. 1 is a picture for the case with five points. The number of line segments connecting pairs of points is $10 = (5^2 - 5)/2$.

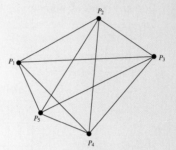

Figure 1

If one additional point is added to the graph in Fig. 1, the result is the graph shown in Fig. 2. Moreover, adding the one additional point requires adding five additional line segments, one to connect the new point to each of the five original points. In general, an additional n line segments are required to move from a graph with n points to one with $n + 1$ points.

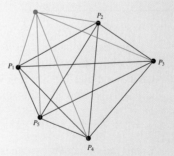

Figure 2

These observations lead to the following proof by induction.

If there is only one point, then the graph contains no line segments. Also since $(1^2 - 1)/2 = 0$, the statement holds for $n = 1$. Next, assume the number of line segments needed to join n points in a coordinate plane is $(n^2 - n)/2$. If there is one additional point, that is, $n + 1$ points, then n additional line segments are required. Hence, by the inductive hypothesis, the total number of line segments required for

$n + 1$ points is

$$\frac{n^2 - n}{2} + n = \frac{n^2 - n + 2n}{2}$$

$$= \frac{n^2 + 2n + 1 - 1 - n}{2}$$

$$= \frac{(n + 1)^2 - (n + 1)}{2}$$

Therefore, by induction the statement holds for all natural numbers.

Binomial Coefficients and the Binomial Theorem

In Fig. 3 are the first eight rows of *Pascal's triangle*. Notice that each element can be obtained from the sum of the two elements to the immediate left and right in the row above.

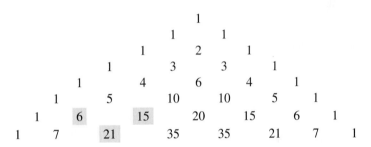

Figure 3

In Fig. 4 are the expansions for $(a + b)^n$ for $n = 0, 1, 2, 3, \ldots, 7$. The coefficients of the expansions are exactly the numbers in Pascal's triangle.

$(a + b)^0$	1
$(a + b)^1$	$a + b$
$(a + b)^2$	$a^2 + 2ab + b^2$
$(a + b)^3$	$a^3 + 3a^2b + 3ab^2 + b^3$
$(a + b)^4$	$a^4 + 4a^3b + 6a^2b^2 + 4ab^3 + b^4$
$(a + b)^5$	$a^5 + 5a^4b + 10a^3b^2 + 10a^2b^3 + 5ab^4 + b^5$
$(a + b)^6$	$a^6 + 6a^5b + 15a^4b^2 + 20a^3b^3 + 15a^2b^4 + 6ab^5 + b^6$
$(a + b)^7$	$a^7 + 7a^6b + 21a^5b^2 + 35a^4b^3 + 35a^3b^4 + 21a^2b^5 + 7ab^6 + b^7$

Figure 4

The numbers in Pascal's triangle or the coefficients of an expansion of the form $(a + b)^n$ are called the **binomial coefficients**. Notice that the number 20, in Fig. 3, is located in row 6 (starting with a row 0) and column 3 (starting with a column 0). In addition, using factorials, we have the formula

$$\frac{6!}{3!(6-3)!} = \frac{1 \cdot 2 \cdot 3 \cdot 4 \cdot 5 \cdot 6}{(1 \cdot 2 \cdot 3)(1 \cdot 2 \cdot 3)}$$
$$= \frac{4 \cdot 5 \cdot 6}{1 \cdot 2 \cdot 3}$$
$$= 20$$

DEFINITION 1

Binomial Coefficient   For $n \geq 0$ and $0 \leq r \leq n$, the **binomial coefficient** $\begin{pmatrix} n \\ r \end{pmatrix}$ is defined by

$$\begin{pmatrix} n \\ r \end{pmatrix} = \frac{n!}{r!(n-r)!}$$

We observed above that entries in Pascal's triangle can be obtained from the sum of the two elements to the immediate left and right in the row above. The next identity is the equivalent statement about binomial coefficients.

PROPOSITION 1

If k and r are natural numbers such that $0 \leq r \leq k$, then

$$\begin{pmatrix} k \\ r \end{pmatrix} = \begin{pmatrix} k-1 \\ r-1 \end{pmatrix} + \begin{pmatrix} k-1 \\ r \end{pmatrix}$$

Proof   First observe that

$$r! = r(r-1)! \qquad \text{and} \qquad (k-r)! = (k-r)(k-r-1)!$$

Expanding the binomial coefficients using factorials gives

$$\begin{pmatrix} k-1 \\ r-1 \end{pmatrix} + \begin{pmatrix} k-1 \\ r \end{pmatrix} = \frac{(k-1)!}{(r-1)![(k-1)-(r-1)]!} + \frac{(k-1)!}{r!(k-1-r)!}$$

$$= (k-1)! \left[\frac{1}{(r-1)!(k-r)!} + \frac{1}{r!(k-r-1)!} \right]$$

$$= \frac{(k-1)!}{(r-1)!(k-r-1)!} \left(\frac{1}{k-r} + \frac{1}{r} \right)$$

$$= \frac{(k-1)!}{(r-1)!(k-r-1)!} \left[\frac{r+(k-r)}{r(k-r)} \right]$$

$$= \frac{(k-1)!}{(r-1)!(k-r-1)!} \left[\frac{k}{r(k-r)} \right]$$

$$= \frac{k!}{r!(k-r)!}$$

$$= \begin{pmatrix} k \\ r \end{pmatrix}$$

THEOREM 10   **Binomial Theorem**   If a and b are any numbers and n is a nonnegative integer, then

$$(a+b)^n = \binom{n}{0} a^n + \binom{n}{1} a^{n-1}b + \binom{n}{2} a^{n-2}b^2$$

$$+ \cdots + \binom{n}{r} a^{n-r}b^r + \cdots + \binom{n}{n-1} ab^{n-1} + \binom{n}{n} b^n$$

Proof   The proof is by induction on the exponent n. If $n = 1$, then $(a+b)^n = a + b$, and

$$\binom{n}{0} a^1 + \binom{n}{1} b^1 = \binom{1}{0} a + \binom{1}{1} b = a + b$$

Therefore, the statement holds for the case $n = 1$. Next assume that the statement

$$(a+b)^n = \binom{n}{0} a^n + \binom{n}{1} a^{n-1}b + \cdots + \binom{n}{n-1} ab^{n-1} + \binom{n}{n} b^n$$

holds. For the next case, we consider $(a+b)^{n+1} = (a+b)(a+b)^n$ and apply the inductive hypothesis. This gives

$$(a+b)^{n+1} = (a+b)(a+b)^n$$

$$= (a+b) \left[\binom{n}{0} a^n + \binom{n}{1} a^{n-1}b + \cdots + \binom{n}{n-1} ab^{n-1} + \binom{n}{n} b^n \right]$$

$$= a \left[\binom{n}{0} a^n + \binom{n}{1} a^{n-1}b + \cdots + \binom{n}{n-1} ab^{n-1} + \binom{n}{n} b^n \right]$$

$$+ b \left[\binom{n}{0} a^n + \binom{n}{1} a^{n-1}b + \cdots + \binom{n}{n-1} ab^{n-1} + \binom{n}{n} b^n \right]$$

$$= \binom{n}{0} a^{n+1} + \binom{n}{1} a^n b + \cdots + \binom{n}{n-1} a^2 b^{n-1} + \binom{n}{n} ab^n$$

$$+ \binom{n}{0} a^n b + \binom{n}{1} a^{n-1}b^2 + \cdots + \binom{n}{n-1} ab^n + \binom{n}{n} b^{n+1}$$

Now, combine the terms with the same exponents on a and b to obtain

$$(a+b)^{n+1} = \binom{n}{0} a^{n+1} + \left[\binom{n}{0} + \binom{n}{1} \right] a^n b + \left[\binom{n}{1} + \binom{n}{2} \right] a^{n-1}b^2$$

$$+ \cdots + \left[\binom{n}{n-1} + \binom{n}{n} \right] ab^n + \binom{n}{n} b^{n+1}$$

Finally by repeated use of Proposition 1, we have

$$(a + b)^{n+1} = \binom{n+1}{0} a^{n+1} + \binom{n+1}{1} a^n b + \binom{n+1}{2} a^{n-1} b^2$$

$$+ \cdots + \binom{n+1}{n} ab^n + \binom{n+1}{n+1} b^{n+1}$$

Therefore, by induction the statement holds for all natural numbers.

Exercise Set A.4

In Exercises 1–10, use mathematical induction to show that the summation formula holds for all natural numbers.

1. $1^2 + 2^2 + 3^2 + \cdots + n^2 = \frac{n(n+1)(2n+1)}{6}$

2. $1^3 + 2^3 + 3^3 + \cdots + n^3 = \frac{n^2(n+1)^2}{4}$

3. $1 + 4 + 7 + \cdots + (3n - 2) = \frac{n(3n-1)}{2}$

4. $3 + 11 + 19 + \cdots + (8n - 5) = 4n^2 - n$

5. $2 + 5 + 8 + \cdots + (3n - 1) = \frac{n(3n+1)}{2}$

6. $3 + 7 + 11 + \cdots + (4n - 1) = n(2n + 1)$

7. $3 + 6 + 9 + \cdots + 3n = \frac{3n(n+1)}{2}$

8. $1 \cdot 2 + 2 \cdot 3 + 3 \cdot 4$

$$+ \cdots + n(n + 1) = \frac{n(n+1)(n+2)}{3}$$

9. $\sum_{k=1}^{n} 2^k = 2^{n+1} - 2$

10. $\sum_{k=1}^{n} k \cdot k! = (n + 1)! - 1$

11. Find a formula for all natural numbers n for the sum

$$2 + 4 + 6 + 8 + \cdots + 2n$$

Verify your answer, using mathematical induction.

12. Find a formula for all natural numbers n for the sum

$$\sum_{k=1}^{n} (4k - 3)$$

13. Show that for all natural numbers $n \geq 5$, the inequality $2^n > n^2$ holds. First show the inequality holds for $n = 5$, and then proceed to the second step when using mathematical induction.

14. Show that for all natural numbers $n \geq 3$, the inequality $n^2 > 2n + 1$ holds. First show that the inequality holds for $n = 3$, and then proceed to the second step when using mathematical induction.

15. Show that for all natural numbers n the expression $n^2 + n$ is divisible by 2.

16. Show that for all natural numbers n the expression $x^n - y^n$ is divisible by $x - y$. Note that $x^2 - y^2$ is divisible by $x - y$ since $x^2 - y^2 = (x + y)(x - y)$.

17. Use mathematical induction to show that for a real number r and all natural numbers n,

$$1 + r + r^2 + r^3 + \cdots + r^{n-1} = \frac{r^n - 1}{r - 1}$$

18. Let f_n denote the nth Fibonacci number.

a. Determine the sum of the first n Fibonacci numbers for $n = 2, 3, 4,$ and 5. That is, determine $f_1 + f_2, f_1 + f_2 + f_3, f_1 + f_2 + f_3 + f_4,$ and $f_1 + f_2 + f_3 + f_4 + f_5$.

b. Find a formula for the sum of the first n Fibonacci numbers.

c. Show that the formula found in part (b) holds for all natural numbers.

19. Let $A, B_1, B_2, \ldots$ be sets. Prove that for every natural number n,

$$A \cap (B_1 \cup B_2 \cup \cdots \cup B_n)$$
$$= (A \cap B_1) \cup \cdots \cup (A \cap B_n)$$

20. Show that for every natural number n, a $2^n \times 2^n$ grid of squares with one square removed can be covered with copies of the shape

as shown in the figure.

21. Verify that if $0 \le r \le n$, then

$$\binom{n}{r} = \binom{n}{n-r}$$

22. Verify that

$$\binom{n}{r-1} + \binom{n}{r} = \binom{n+1}{r}$$

23. Show that

$$\sum_{k=0}^{n} \binom{n}{k} = 2^n$$

24. Show that

$$\sum_{k=0}^{n} (-1)^k \binom{n}{k} = 0$$

Answers to Odd-Numbered Exercises

Chapter 1

Section 1.1

1. $x_1 = 3, x_2 = 8, x_3 = -4$

3. $x_1 = 2 - 3x_4, x_2 = 1 - x_4, x_3 = -1 - 2x_4, x_4 \in \mathbb{R}$

5. $x = 0, y = -\frac{2}{3}$

7. $x = 1, y = 0$

9. $S = \left\{ \left(\frac{2t+4}{3}, t \right) \Big| \ t \in \mathbb{R} \right\}$

11. $x = 0, y = 1, z = 0$

13. $S = \left\{ \left(-1 - 5t, 6t + \frac{1}{2}, t \right) \Big| t \in \mathbb{R} \right\}$

15. $S = \left\{ \left(-t + \frac{2}{3}, -\frac{1}{2}, t \right) \Big| t \in \mathbb{R} \right\}$

17. $S = \left\{ \left(3 - \frac{5}{3}t, -s - \frac{4}{3}t + 3, s, t \right) \Big| s, t \in \mathbb{R} \right\}$

19. $x = -2a + b, y = -3a + 2b$

21. $x = 2a + 6b - c, y = a + 3b, z = -2a - 7b + c$

23. Consistent if $a = -1$

25. Consistent if $b = -a$

27. Consistent for all a, b, and c such that $c - a - b = 0$

29. Inconsistent if $a = 2$

31. Inconsistent for $a \neq 6$

33. $y = \left(x - \frac{3}{2} \right)^2 - 2$; vertex: $\left(\frac{3}{2}, -2 \right)$

35. $y = -(x - 2)^2 + 3$; vertex: $(2, 3)$

37. a. $(2, 3)$

b.

39. a. $\begin{cases} x + y = 2 \\ x - y = 0 \end{cases}$

b. $\begin{cases} x + y = 1 \\ 2x + 2y = 2 \end{cases}$

c. $\begin{cases} x + y = 2 \\ 3x + 3y = -6 \end{cases}$

41. a. $S = \{(3 - 2s - t, 2 + s - 2t, s, t) \mid s, t \in \mathbb{R}\}$

b. $S = \{(7 - 2s - 5t, s, -2 + s + 2t, t) \mid s, t \in \mathbb{R}\}$

43. a. $k = 3$

b. $k = -3$

c. $k \neq \pm 3$

Section 1.2

1. $\left[\begin{array}{cc|c} 2 & -3 & 5 \\ -1 & 1 & -3 \end{array} \right]$

3. $\left[\begin{array}{ccc|c} 2 & 0 & -1 & 4 \\ 1 & 4 & 1 & 2 \\ 4 & 1 & -1 & 1 \end{array} \right]$

5. $\left[\begin{array}{ccc|c} 2 & 0 & -1 & 4 \\ 1 & 4 & 1 & 2 \end{array} \right]$

7. $\left[\begin{array}{cccc|c} 2 & 4 & 2 & 2 & -2 \\ 4 & -2 & -3 & -2 & 2 \\ 1 & 3 & 3 & -3 & -4 \end{array} \right]$

9. $x = -1, y = \frac{1}{2}, z = 0$

11. $x = -3 - 2z, y = 2 + z, z \in \mathbb{R}$

13. $x = -3 + 2y, z = 2, y \in \mathbb{R}$

15. Inconsistent

17. $x = 3 + 2z - 5w, y = 2 + z - 2w, z \in \mathbb{R}, w \in \mathbb{R}$

19. $x = 1 + 3w, y = 7 + w, z = -1 - 2w, w \in \mathbb{R}$

21. In reduced row echelon form

23. Not in reduced row echelon form

25. In reduced row echelon form

27. Not in reduced row echelon form

29. $\begin{bmatrix} 1 & 0 \\ 0 & 1 \end{bmatrix}$

31. $\begin{bmatrix} 1 & 0 & 0 \\ 0 & 1 & 0 \\ 0 & 0 & 1 \end{bmatrix}$

33. $\begin{bmatrix} 1 & 0 & -1 \\ 0 & 1 & 0 \end{bmatrix}$

35. $\begin{bmatrix} 1 & 0 & 0 & -2 \\ 0 & 1 & 0 & -1 \\ 0 & 0 & 1 & 0 \end{bmatrix}$

37. $x = -1, y = 2$

39. $x = 1, y = 0, z = \frac{1}{3}$

41. Inconsistent

43. $x_1 = -\frac{1}{2} - 2x_3, x_2 = -\frac{3}{4} + \frac{3}{2}x_3, x_3 \in \mathbb{R}$

45. $x_1 = 1 - \frac{1}{2}x_4, x_2 = 1 - \frac{1}{2}x_4, x_3 = 1 - \frac{1}{2}x_4, x_4 \in \mathbb{R}$

47. $x_1 = 1 + \frac{1}{3}x_3 + \frac{1}{2}x_4, x_2 = 2 + \frac{2}{3}x_3 + \frac{3}{2}x_4, x_3 \in \mathbb{R},$
$x_4 \in \mathbb{R}$

49. a. $c - a + b = 0$
   b. $c - a + b \neq 0$
   c. Infinitely many solutions.
   d. $a = 1, b = 0, c = 1; x = -2, y = 2, z = 1$

51. a. $a + 2b - c = 0$
   b. $a + 2b - c \neq 0$
   c. Infinitely many solutions
   d. $a = 0, b = 0, c = 0; x = \frac{4}{5}, y = \frac{1}{5}, z = 1$

Section 1.3

1. $A + B = \begin{bmatrix} 1 & 0 \\ 2 & 6 \end{bmatrix} = B + A$

3. $(A + B) + C = \begin{bmatrix} 2 & 1 \\ 7 & 4 \end{bmatrix} = A + (B + C)$

5. $(A - B) + C = \begin{bmatrix} -7 & -3 & 9 \\ 0 & 5 & 6 \\ 1 & -2 & 10 \end{bmatrix}$

$2A + B = \begin{bmatrix} -7 & 3 & 9 \\ -3 & 10 & 6 \\ 2 & 2 & 11 \end{bmatrix}$

7. $AB = \begin{bmatrix} 7 & -2 \\ 0 & -8 \end{bmatrix}; BA = \begin{bmatrix} 6 & 2 \\ 7 & -7 \end{bmatrix}$

9. $AB = \begin{bmatrix} -9 & 4 \\ -13 & 7 \end{bmatrix}$

11. $AB = \begin{bmatrix} 5 & -6 & 4 \\ 3 & 6 & -18 \\ 5 & -7 & 6 \end{bmatrix}$

13. $A(B + C) = \begin{bmatrix} 1 & 3 \\ 12 & 0 \end{bmatrix}$

15. $2A(B - 3C) = \begin{bmatrix} 10 & -18 \\ -24 & 0 \end{bmatrix}$

17. $2A^t - B^t = \begin{bmatrix} 7 & 5 \\ -1 & 3 \\ -3 & -2 \end{bmatrix}$

19. $AB^t = \begin{bmatrix} -7 & -4 \\ -5 & 1 \end{bmatrix}$

21. $(A^t + B^t)C = \begin{bmatrix} -1 & 7 \\ 6 & 8 \\ 4 & 12 \end{bmatrix}$

23. $(A^tC)B = \begin{bmatrix} 0 & 20 & 15 \\ 0 & 0 & 0 \\ -18 & -22 & -15 \end{bmatrix}$

25. $AB = AC = \begin{bmatrix} -5 & -1 \\ 5 & 1 \end{bmatrix}$

27. A has the form $\begin{bmatrix} 1 & 0 \\ 0 & 1 \end{bmatrix}, \begin{bmatrix} 1 & b \\ 0 & -1 \end{bmatrix}, \begin{bmatrix} -1 & b \\ 0 & 1 \end{bmatrix},$
or $\begin{bmatrix} -1 & 0 \\ 0 & -1 \end{bmatrix}$

29. $A = \begin{bmatrix} 1 & 1 \\ 0 & 0 \end{bmatrix}, B = \begin{bmatrix} -1 & -1 \\ 1 & 1 \end{bmatrix}$

31. $a = b = 4$

33. $A^{20} = \begin{bmatrix} 1 & 0 & 0 \\ 0 & 1 & 0 \\ 0 & 0 & 1 \end{bmatrix}$

35. If $AB = BA$, then $A^2B = AAB = ABA = BAA = BA^2$.

37. If $\mathbf{x} = \begin{bmatrix} 1 \\ 0 \\ \vdots \\ 0 \end{bmatrix}$, then $A\mathbf{x} = \mathbf{0}$ implies the first column of

A has all 0 entries. Then let $\mathbf{x} = \begin{bmatrix} 0 \\ 1 \\ \vdots \\ 0 \end{bmatrix}$ and so on, to

show that each column of A has all 0 entries.

39. The only matrix is the 2×2 zero matrix.

41. Since $(AA^t)^t = (A^t)^tA^t = AA^t$, the matrix AA^t is
symmetric. Similarly, $(A^tA)^t = A^t(A^t)^t = A^tA$.

43. If $A^t = -A$, then the diagonal entries satisfy $a_{ii} = -a_{ii}$
and hence $a_{ii} = 0$ for each i.

Section 1.4

1. $A^{-1} = \frac{1}{5}\begin{bmatrix} -1 & 2 \\ -3 & 1 \end{bmatrix}$

3. The matrix is not invertible.

5. $A^{-1} = \begin{bmatrix} 3 & 1 & -2 \\ -4 & -1 & 3 \\ -5 & -1 & 3 \end{bmatrix}$

7. The matrix is not invertible.

9. $A^{-1} = \begin{bmatrix} \frac{1}{3} & -1 & -2 & \frac{1}{2} \\ 0 & 1 & 2 & -1 \\ 0 & 0 & -1 & \frac{1}{2} \\ 0 & 0 & 0 & -\frac{1}{2} \end{bmatrix}$

11. $A^{-1} = \frac{1}{3}\begin{bmatrix} 3 & 0 & 0 & 0 \\ -6 & 3 & 0 & 0 \\ 1 & -2 & -1 & 0 \\ 1 & 1 & 1 & 1 \end{bmatrix}$

13. The matrix is not invertible.

15. $A^{-1} = \begin{bmatrix} 0 & 0 & -1 & 0 \\ 1 & -1 & -2 & 1 \\ 1 & -2 & -1 & 1 \\ 0 & -1 & -1 & 1 \end{bmatrix}$

17. $AB + A = \begin{bmatrix} 3 & 8 \\ 10 & -10 \end{bmatrix} = A(B+I)$

$AB + B = \begin{bmatrix} 2 & 9 \\ 6 & -3 \end{bmatrix} = (A+I)B$

19. a. Since $A^2 = \begin{bmatrix} -3 & 4 \\ -4 & -3 \end{bmatrix}$ and

$-2A = \begin{bmatrix} -2 & -4 \\ 4 & -2 \end{bmatrix}$, then $A^2 - 2A + 5I = 0$.

  b. $A^{-1} = \frac{1}{5}\begin{bmatrix} 1 & -2 \\ 2 & 1 \end{bmatrix} = \frac{1}{5}(2I - A)$

  c. If $A^2 - 2A + 5I = 0$, then $A^2 - 2A = -5I$, so that $A\left[\frac{1}{5}(2I - A)\right] = \frac{2}{5}A - \frac{1}{5}A^2 = -\frac{1}{5}(A^2 - 2A) = -\frac{1}{5}(-5I) = I$.

21. If $\lambda = -2$, then the matrix is not invertible.

23. a. If $\lambda \neq 1$, then the matrix is invertible.

  b. $\begin{bmatrix} -\frac{1}{\lambda-1} & \frac{\lambda}{\lambda-1} & -\frac{\lambda}{\lambda-1} \\ \frac{1}{\lambda-1} & -\frac{1}{\lambda-1} & \frac{1}{\lambda-1} \\ 0 & 0 & 1 \end{bmatrix}$

25. The matrices

$A = \begin{bmatrix} 1 & 0 \\ 0 & 0 \end{bmatrix}$ and $B = \begin{bmatrix} 0 & 0 \\ 0 & 1 \end{bmatrix}$

are not invertible, but $A + B = \begin{bmatrix} 1 & 0 \\ 0 & 1 \end{bmatrix}$ is invertible.

27. $(A+B)A^{-1}(A-B) = (AA^{-1} + BA^{-1})(A-B)$
$= (I + BA^{-1})(A-B)$
$= A - B + B - BA^{-1}B$
$= A - BA^{-1}B$

Similarly, $(A-B)A^{-1}(A+B) = A - BA^{-1}B$.

29. a. If A is invertible and $AB = 0$, then $A^{-1}(AB) = A^{-1}0$, so that $B = 0$.

  b. If A is not invertible, then $A\mathbf{x} = 0$ has infinitely many solutions. Let $\mathbf{x}_1, \ldots, \mathbf{x}_n$ be solutions of $A\mathbf{x} = 0$ and B be the matrix with nth column vector $\mathbf{x}_n$. Then $AB = 0$.

31. $(AB)^t = B^t A^t = BA = AB$

33. If $AB = BA$, then $B^{-1}AB = A$, so $B^{-1}A = AB^{-1}$. Now $(AB^{-1})^t = (B^{-1})^t A^t = (B^t)^{-1}A^t = B^{-1}A = AB^{-1}$.

35. If $A^t = A^{-1}$ and $B^t = B^{-1}$, then $(AB)^t = B^t A^t = B^{-1}A^{-1} = (AB)^{-1}$.

37. a. $(ABC)(C^{-1}B^{-1}A^{-1}) = (AB)CC^{-1}(B^{-1}A^{-1})$
$= ABB^{-1}A^{-1}$
$= AA^{-1} = I$

  b. Case 1, $k = 2$: $(A_1A_2)^{-1} = A_2^{-1}A_1^{-1}$
  Case 2: Suppose that
$$(A_1A_2 \cdots A_k)^{-1} = A_k^{-1}A_{k-1}^{-1} \cdots A_1^{-1}$$
  Then
$$(A_1A_2 \cdots A_kA_{k+1})^{-1} = ([A_1A_2 \cdots A_k]A_{k+1})^{-1}$$
$$= A_{k+1}^{-1}[A_1A_2 \cdots A_k]^{-1}$$
$$= A_{k+1}^{-1}A_k^{-1}A_{k-1}^{-1} \cdots A_1^{-1}$$

39. If A is invertible, then the augmented matrix $[A|I]$ can be row-reduced to $[I|A^{-1}]$. If A is upper triangular, then only terms on or above the main diagonal can be affected by the reduction process, and hence the inverse is upper triangular. Similarly, the inverse for an invertible lower triangle matrix is also lower triangular.

41. a. $\begin{bmatrix} ax_1 + bx_3 & ax_2 + bx_4 \\ cx_1 + dx_3 & cx_2 + dx_4 \end{bmatrix} = \begin{bmatrix} 1 & 0 \\ 0 & 1 \end{bmatrix}$

  b. From part (a), we have the two linear systems
$$\begin{cases} ax_1 + bx_3 = 1 \\ cx_1 + dx_3 = 0 \end{cases} \text{ and } \begin{cases} ax_2 + bx_4 = 0 \\ cx_2 + dx_4 = 1 \end{cases}$$
  so
$$(ad - bc)x_3 = d \quad \text{and} \quad (ad - bc)x_4 = -b$$
  If $ad - bc = 0$, then $b = d = 0$.

  c. From part (b), both $b = 0$ and $d = 0$. Notice that if in addition either $a = 0$ or $c = 0$, then the matrix is not invertible. Also from part(b), we have that $ax_1 = 1, ax_2 = 0, cx_1 = 0$, and $cx_2 = 1$. If a and c are not zero, then these equations are inconsistent and the matrix is not invertible.

Section 1.5

1. $A = \begin{bmatrix} 2 & 3 \\ -1 & 2 \end{bmatrix}$, $\mathbf{x} = \begin{bmatrix} x \\ y \end{bmatrix}$, and $\mathbf{b} = \begin{bmatrix} -1 \\ 4 \end{bmatrix}$

3. $A = \begin{bmatrix} 2 & -3 & 1 \\ -1 & -1 & 2 \\ 3 & -2 & -2 \end{bmatrix}$, $\mathbf{x} = \begin{bmatrix} x \\ y \\ z \end{bmatrix}$, and

$\mathbf{b} = \begin{bmatrix} -1 \\ -1 \\ 3 \end{bmatrix}$

5. $A = \begin{bmatrix} 4 & 3 & -2 & -3 \\ -3 & -3 & 1 & 0 \\ 2 & -3 & 4 & -4 \end{bmatrix}$, $\mathbf{x} = \begin{bmatrix} x_1 \\ x_2 \\ x_3 \\ x_4 \end{bmatrix}$, and

$\mathbf{b} = \begin{bmatrix} -1 \\ 4 \\ 3 \end{bmatrix}$

7. $\begin{cases} 2x & - & 5y & = & 3 \\ 2x & + & y & = & 2 \end{cases}$

9. $\begin{cases} & - & 2y & & & = & 3 \\ 2x & - & y & - & z & = & 1 \\ 3x & - & y & + & 2z & = & -1 \end{cases}$

11. $\begin{cases} 2x_1 & + & 5x_2 & - & 5x_3 & + & 3x_4 & = & 2 \\ 3x_1 & + & x_2 & - & 2x_3 & - & 4x_4 & = & 0 \end{cases}$

13. $\mathbf{x} = \begin{bmatrix} 1 \\ 4 \\ -3 \end{bmatrix}$

15. $\mathbf{x} = \begin{bmatrix} 9 \\ -3 \\ -8 \\ 7 \end{bmatrix}$

17. $\mathbf{x} = \dfrac{1}{10} \begin{bmatrix} -16 \\ 9 \end{bmatrix}$

19. $\mathbf{x} = \begin{bmatrix} -11 \\ 4 \\ 12 \end{bmatrix}$

21. $\mathbf{x} = \dfrac{1}{3} \begin{bmatrix} 0 \\ 0 \\ 1 \\ -1 \end{bmatrix}$

23. a. $\mathbf{x} = \dfrac{1}{5} \begin{bmatrix} 7 \\ -3 \end{bmatrix}$

     b. $\mathbf{x} = \dfrac{1}{5} \begin{bmatrix} -7 \\ 8 \end{bmatrix}$

25. The general solution is

$$S = \left\{ \begin{bmatrix} -4t \\ t \end{bmatrix} \,\middle|\, t \in \mathbb{R} \right\}$$

with a particular nontrivial solution of $x = -4$ and $y = 1$.

27. $A = \begin{bmatrix} 1 & 2 & 1 \\ 1 & 2 & 1 \\ 1 & 2 & 1 \end{bmatrix}$

29. From the fact that $A\mathbf{u} = A\mathbf{v}$, we have $A(\mathbf{u} - \mathbf{v}) = \mathbf{0}$. If A is invertible, then $\mathbf{u} - \mathbf{v} = \mathbf{0}$, that is, $\mathbf{u} = \mathbf{v}$, which contradicts the statement that $\mathbf{u} \neq \mathbf{v}$.

31. a. $\mathbf{x} = \begin{bmatrix} 1 \\ -1 \end{bmatrix}$

    b. $C = \dfrac{1}{3} \begin{bmatrix} 1 & -1 & 0 \\ 1 & 2 & 0 \end{bmatrix}$

    c. $C\mathbf{b} = \dfrac{1}{3} \begin{bmatrix} 1 & -1 & 0 \\ 1 & 2 & 0 \end{bmatrix} \begin{bmatrix} 1 \\ -2 \\ -1 \end{bmatrix} = \begin{bmatrix} 1 \\ -1 \end{bmatrix}$

Section 1.6

1. The determinant is the product of the terms on the diagonal and equals 24.

3. The determinant is the product of the terms on the diagonal and equals -10.

5. Since the determinant is 2, the matrix is invertible.

7. Since the determinant is -6, the matrix is invertible.

9. a–c. $\det(A) = -5$

    d. $\det \left(\begin{bmatrix} -4 & 1 & -2 \\ 3 & -1 & 4 \\ 2 & 0 & 1 \end{bmatrix} \right) = 5$

    e. Let B denote the matrix in part (d) and B' denote the new matrix. Then $\det(B') = -2 \det(B) = -10$. Then $\det(A) = \frac{1}{2} \det(B')$.

    f. Let B'' denote the new matrix. The row operation does not change the determinant, so $\det(B'') = \det(B') = -10$.

    g. Since $\det(A) \neq 0$, the matrix A does have an inverse.

11. Determinant: 13; invertible

13. Determinant: -16; invertible

15. Determinant: 0; not invertible

17. Determinant: 30; invertible

19. Determinant: -90; invertible

21. Determinant: 0; not invertible

23. Determinant: -32; invertible

25. Determinant: 0; not invertible

27. $\det(3A) = 3^3 \det(A) = 270$

29. $\det((2A)^{-1}) = \dfrac{1}{\det(2A)} = \dfrac{1}{2^3 \det(A)} = \dfrac{1}{80}$

31. Since the determinant of the matrix is $-5x^2 + 10x = -5x(x-2)$, the determinant is 0 if and only if $x = 0$ or $x = 2$.

33. $y = \frac{b_2 - a_2}{b_1 - a_1}x + \frac{b_1 a_2 - a_1 b_2}{b_1 - a_1}$

35. a. $A = \begin{bmatrix} 1 & -1 & -2 \\ -1 & 2 & 3 \\ 2 & -2 & -2 \end{bmatrix}$

  b. $\det(A) = 2$

  c. Since the coefficient matrix is invertible, the linear system has a unique solution.

  d. $\mathbf{x} = \begin{bmatrix} 3 \\ 8 \\ -4 \end{bmatrix}$,

37. a. $A = \begin{bmatrix} -1 & 0 & -1 \\ 2 & 0 & 2 \\ 1 & -3 & -3 \end{bmatrix}$

  b. $\det(A) = 0$

  c. Since the determinant of the coefficient matrix is 0, A is not invertible. Therefore, the linear system has either no solutions or infinitely many solutions.

  d. No solutions

39. a. $\begin{vmatrix} y^2 & x & y & 1 \\ 4 & -2 & -2 & 1 \\ 4 & 3 & 2 & 1 \\ 9 & 4 & -3 & 1 \end{vmatrix}$
$= -29y^2 + 20x - 25y + 106 = 0$

  b.

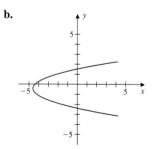

41. a. $\begin{vmatrix} x^2 & y^2 & x & y & 1 \\ 0 & 16 & 0 & -4 & 1 \\ 0 & 16 & 0 & 4 & 1 \\ 1 & 4 & 1 & -2 & 1 \\ 4 & 9 & 2 & 3 & 1 \end{vmatrix}$
$= 136x^2 - 16y^2 - 328x + 256 = 0$

b.

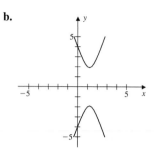

43. a. $\begin{vmatrix} x^2 & xy & y^2 & x & y & 1 \\ 1 & 0 & 0 & -1 & 0 & 1 \\ 0 & 0 & 1 & 0 & 1 & 1 \\ 1 & 0 & 0 & 1 & 0 & 1 \\ 4 & 4 & 4 & 2 & 2 & 1 \\ 9 & 3 & 1 & 3 & 1 & 1 \end{vmatrix}$
$= -12 + 12x^2 - 36xy + 42y^2 - 30y = 0$

  b.

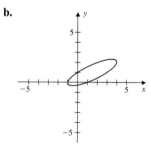

45. $x = \dfrac{\begin{vmatrix} 7 & -5 \\ 6 & -3 \\ \hline 5 & -5 \\ 2 & -3 \end{vmatrix}}{} = -\dfrac{9}{5}$, $y = \dfrac{\begin{vmatrix} 5 & 7 \\ 2 & 6 \\ \hline 5 & -5 \\ 2 & -3 \end{vmatrix}}{} = -\dfrac{16}{5}$

47. $x = \dfrac{\begin{vmatrix} 3 & -4 \\ -10 & 5 \end{vmatrix}}{\begin{vmatrix} -9 & -4 \\ -7 & 5 \end{vmatrix}} = \dfrac{25}{73}$, $y = \dfrac{\begin{vmatrix} -9 & 3 \\ -7 & -10 \end{vmatrix}}{\begin{vmatrix} -9 & -4 \\ -7 & 5 \end{vmatrix}}$
$= -\dfrac{111}{73}$

49. $x = \dfrac{\begin{vmatrix} 4 & -3 \\ 3 & 4 \end{vmatrix}}{\begin{vmatrix} -1 & -3 \\ -8 & 4 \end{vmatrix}} = -\dfrac{25}{28}$, $y = \dfrac{\begin{vmatrix} -1 & 4 \\ -8 & 3 \end{vmatrix}}{\begin{vmatrix} -1 & -3 \\ -8 & 4 \end{vmatrix}} = -\dfrac{29}{28}$

51. $x = -\dfrac{160}{103}, y = \dfrac{10}{103}, z = \dfrac{42}{103}$

53. Expansion of the determinant of A across row one equals the expansion down column one of A^t, so $\det(A) = \det(A^t)$.

Section 1.7

1. a. $E = \begin{bmatrix} 1 & 0 & 0 \\ 2 & 1 & 0 \\ 0 & 0 & 1 \end{bmatrix}$

b. $EA = \begin{bmatrix} 1 & 2 & 1 \\ 5 & 5 & 4 \\ 1 & 1 & -4 \end{bmatrix}$

3. a. $E = \begin{bmatrix} 1 & 0 & 0 \\ 0 & 1 & 0 \\ 0 & -3 & 1 \end{bmatrix}$

b. $EA = \begin{bmatrix} 1 & 2 & 1 \\ 3 & 1 & 2 \\ -8 & -2 & -10 \end{bmatrix}$

5. a. $I = E_3 E_2 E_1 A$

$$= \begin{bmatrix} 1 & -3 \\ 0 & 1 \end{bmatrix} \begin{bmatrix} 1 & 0 \\ 0 & \frac{1}{10} \end{bmatrix} \begin{bmatrix} 1 & 0 \\ 2 & 1 \end{bmatrix} A$$

b. $A = E_1^{-1} E_2^{-1} E_3^{-1}$

$$= \begin{bmatrix} 1 & 0 \\ -2 & 1 \end{bmatrix} \begin{bmatrix} 1 & 0 \\ 0 & 10 \end{bmatrix} \begin{bmatrix} 1 & 3 \\ 0 & 1 \end{bmatrix}$$

7. a. $I = E_5 E_4 E_3 E_2 E_1 A$

$$E_1 = \begin{bmatrix} 1 & 0 & 0 \\ -2 & 1 & 0 \\ 0 & 0 & 1 \end{bmatrix} \qquad E_2 = \begin{bmatrix} 1 & 0 & 0 \\ 0 & 1 & 0 \\ -1 & 0 & 1 \end{bmatrix}$$

$$E_3 = \begin{bmatrix} 1 & -2 & 0 \\ 0 & 1 & 0 \\ 0 & 0 & 1 \end{bmatrix} \qquad E_4 = \begin{bmatrix} 1 & 0 & 11 \\ 0 & 1 & 0 \\ 0 & 0 & 1 \end{bmatrix}$$

$$E_5 = \begin{bmatrix} 1 & 0 & 0 \\ 0 & 1 & -5 \\ 0 & 0 & 1 \end{bmatrix} .$$

b. $A = E_1^{-1} E_2^{-1} E_3^{-1} E_4^{-1} E_5^{-1}$

9. a. $I = E_6 \cdots E_1 A$

$$E_1 = \begin{bmatrix} 0 & 1 & 0 \\ 1 & 0 & 0 \\ 0 & 0 & 1 \end{bmatrix} \qquad E_2 = \begin{bmatrix} 1 & -2 & 0 \\ 0 & 1 & 0 \\ 0 & 0 & 1 \end{bmatrix}$$

$$E_3 = \begin{bmatrix} 1 & 0 & 0 \\ 0 & 1 & 0 \\ 0 & -1 & 1 \end{bmatrix} \qquad E_4 = \begin{bmatrix} 1 & 0 & 0 \\ 0 & 1 & 1 \\ 0 & 0 & 1 \end{bmatrix}$$

$$E_5 = \begin{bmatrix} 1 & 0 & 1 \\ 0 & 1 & 0 \\ 0 & 0 & 1 \end{bmatrix} \qquad E_6 = \begin{bmatrix} 1 & 0 & 0 \\ 0 & 1 & 0 \\ 0 & 0 & -1 \end{bmatrix}$$

b. $A = E_1^{-1} E_2^{-1} \cdots E_6^{-1}$

11. $A = LU = \begin{bmatrix} 1 & 0 \\ -3 & 1 \end{bmatrix} \begin{bmatrix} 1 & -2 \\ 0 & 1 \end{bmatrix}$

13. $A = LU = \begin{bmatrix} 1 & 0 & 0 \\ 2 & 1 & 0 \\ -3 & 0 & 1 \end{bmatrix} \begin{bmatrix} 1 & 2 & 1 \\ 0 & 1 & 3 \\ 0 & 0 & 1 \end{bmatrix}$

15. $A = LU = \begin{bmatrix} 1 & 0 & 0 \\ 1 & 1 & 0 \\ -1 & -\frac{1}{2} & 1 \end{bmatrix} \begin{bmatrix} 1 & \frac{1}{2} & -3 \\ 0 & 1 & 4 \\ 0 & 0 & 3 \end{bmatrix}$

17. • LU factorization:

$$L = \begin{bmatrix} 1 & 0 \\ -2 & 1 \end{bmatrix} \qquad U = \begin{bmatrix} -2 & 1 \\ 0 & 1 \end{bmatrix}$$

• $\mathbf{y} = U\mathbf{x} = \begin{bmatrix} -2x_1 + x_2 \\ x_2 \end{bmatrix}$

• Solve $L\mathbf{y} = \begin{bmatrix} -1 \\ 5 \end{bmatrix}$: $y_1 = -1, y_2 = 3$

• Solve $U\mathbf{x} = \mathbf{y}$: $x_1 = 2, x_2 = 3$

19. • LU factorization:

$$L = \begin{bmatrix} 1 & 0 & 0 \\ -1 & 1 & 0 \\ 2 & 0 & 1 \end{bmatrix} \qquad U = \begin{bmatrix} 1 & 4 & -3 \\ 0 & 1 & 2 \\ 0 & 0 & 1 \end{bmatrix}$$

• $\mathbf{y} = U\mathbf{x} = \begin{bmatrix} x_1 + 4x_2 - 3x_3 \\ x_2 + 2x_3 \\ x_3 \end{bmatrix}$

• Solve $L\mathbf{y} = \begin{bmatrix} 0 \\ -3 \\ 1 \end{bmatrix}$: $y_1 = 0, y_2 = -3, y_3 = 1$

• Solve $U\mathbf{x} = \mathbf{y}$: $x_1 = 23, x_2 = -5, x_3 = 1$

21. • LU factorization:

$$L = \begin{bmatrix} 1 & 0 & 0 & 0 \\ 1 & 1 & 0 & 0 \\ 2 & 0 & 1 & 0 \\ -1 & -1 & 0 & 1 \end{bmatrix}$$

$$U = \begin{bmatrix} 1 & -2 & 3 & 1 \\ 0 & 1 & 2 & 2 \\ 0 & 0 & 1 & 1 \\ 0 & 0 & 0 & 1 \end{bmatrix}$$

- $\mathbf{y} = U\mathbf{x} = \begin{bmatrix} x_1 - 2x_2 + 3x_3 + x_4 \\ x_2 + 2x_3 + 2x_4 \\ x_3 + x_4 \\ x_4 \end{bmatrix}$

- Solve $L\mathbf{y} = \begin{bmatrix} 5 \\ 6 \\ 14 \\ -8 \end{bmatrix}$:

  $y_1 = 5, y_2 = 1, y_3 = 4, y_4 = -2$

- Solve $U\mathbf{x} = \mathbf{y}$: $x_1 = -25, x_2 = -7, x_3 = 6, x_4 = -2$

23. $A = PLU$

$= \begin{bmatrix} 0 & 1 & 0 \\ 1 & 0 & 0 \\ 0 & 0 & 0 \end{bmatrix} \begin{bmatrix} 1 & 0 & 0 \\ 2 & 5 & 0 \\ 0 & 1 & -\frac{1}{5} \end{bmatrix} \begin{bmatrix} 1 & -3 & 2 \\ 0 & 1 & -\frac{4}{5} \\ 0 & 0 & 1 \end{bmatrix}$

25. $A = LU = \begin{bmatrix} 1 & 0 \\ -3 & 1 \end{bmatrix} \begin{bmatrix} 1 & 4 \\ 0 & 1 \end{bmatrix}$

$A^{-1} = U^{-1}L^{-1} = \begin{bmatrix} 1 & -4 \\ 0 & 1 \end{bmatrix} \begin{bmatrix} 1 & 0 \\ 3 & 1 \end{bmatrix}$

$= \begin{bmatrix} -11 & -4 \\ 3 & 1 \end{bmatrix}$

27. $A = LU = \begin{bmatrix} 1 & 0 & 0 \\ 1 & 1 & 0 \\ 1 & 1 & 1 \end{bmatrix} \begin{bmatrix} 2 & 1 & -1 \\ 0 & 1 & -1 \\ 0 & 0 & 3 \end{bmatrix}$

$A^{-1} = U^{-1}L^{-1}$

$= \begin{bmatrix} \frac{1}{2} & -\frac{1}{2} & 0 \\ 0 & 1 & \frac{1}{3} \\ 0 & 0 & \frac{1}{3} \end{bmatrix} \begin{bmatrix} 1 & 0 & 0 \\ -1 & 1 & 0 \\ 0 & -1 & 1 \end{bmatrix}$

$= \begin{bmatrix} 1 & -\frac{1}{2} & 0 \\ -1 & \frac{2}{3} & \frac{1}{3} \\ 0 & -\frac{1}{3} & \frac{1}{3} \end{bmatrix}$

29. Suppose

$\begin{bmatrix} a & 0 \\ b & c \end{bmatrix} \begin{bmatrix} d & e \\ 0 & f \end{bmatrix} = \begin{bmatrix} 0 & 1 \\ 1 & 0 \end{bmatrix}$

This gives the system of equations $ad = 0, ae = 1$, $bd = 1, be + cf = 0$. The first two equations are satisfied only when $a \neq 0$ and $d = 0$. But this is incompatible with the third equation.

31. If A is invertible, there are elementary matrices $E_1, \ldots, E_k$ such that $I = E_k \cdots E_1 A$. Similarly, there are elementary matrices $D_1, \ldots, D_\ell$ such that $I = D_\ell \cdots D_1 B$. Then $A = E_k^{-1} \cdots E_1^{-1} D_\ell \cdots D_1 B$, so A is row equivalent to B.

Section 1.8

1. $x_1 = 2, x_2 = 9, x_3 = 3, x_4 = 9$

3. Let $x_5 = 3$. Then $x_1 = x_5 = 3, x_2 = \frac{1}{3}x_5 = 1, x_3 = \frac{1}{3}x_5 = 1, x_4 = x_5 = 3$.

5. Let $x_1, x_2, \ldots, x_7$ be defined as in the figure.

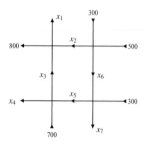

Then $x_1 = 1000 - x_4 - x_7, x_2 = 800 - x_6, x_3 = 1000 - x_4 + x_6 - x_7, x_5 = 300 + x_6 - x_7$
Since the network consists of one-way streets, the individual flows are nonnegative. As a sample solution let $x_4 = 200, x_6 = 300, x_7 = 100$; then $x_1 = 700$, $x_2 = 500, x_3 = 1000, x_5 = 500$.

7. $x_1 = 150 - x_4, x_2 = 50 - x_4 - x_5, x_3 = 50 + x_4 + x_5$. As a sample solution let $x_4 = x_5 = 20$; then $x_1 = 130, x_2 = 10, x_3 = 90$

9. $x_1 = 1.4, x_2 = 3.2, x_3 = 1.6, x_4 = 6.2$

11. a. $A = \begin{bmatrix} 0.02 & 0.04 & 0.05 \\ 0.03 & 0.02 & 0.04 \\ 0.03 & 0.3 & 0.1 \end{bmatrix}$

b. The internal demand vector is

$A \begin{bmatrix} 300 \\ 150 \\ 200 \end{bmatrix} = \begin{bmatrix} 22 \\ 20 \\ 74 \end{bmatrix}$. The total external demand

for the three sectors is $300 - 22 = 278, 150 - 20 = 130$, and $200 - 74 = 126$, respectively.

c. $(I - A)^{-1} \approx \begin{bmatrix} 1.02 & 0.06 & 0.06 \\ 0.03 & 1.04 & 0.05 \\ 0.05 & 0.35 & 1.13 \end{bmatrix}$

d. $X = (I - A)^{-1}D$

$= \begin{bmatrix} 1.02 & 0.06 & 0.06 \\ 0.03 & 1.04 & 0.05 \\ 0.05 & 0.35 & 1.13 \end{bmatrix} \begin{bmatrix} 350 \\ 400 \\ 600 \end{bmatrix}$

$= \begin{bmatrix} 418.2 \\ 454.9 \\ 832.3 \end{bmatrix}$

13. a.

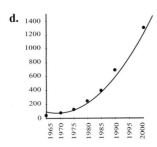

b. $\begin{cases} 3,880,900a + 1970b + c = 80 \\ 3,920,400a + 1980b + c = 250 \\ 3,960,100a + 1990b + c = 690 \end{cases}$

c. $a = \frac{27}{20}, b = -\frac{10,631}{2}, c = 5,232,400$

d.

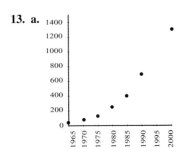

e. The model gives an estimate, in billions of dollars, for health care costs in 2010 at

$$\frac{27}{20}(2010)^2 - \frac{10,631}{2}(2010) + 5,232,400 = 2380$$

15. a. $A = \begin{bmatrix} 0.9 & 0.08 \\ 0.1 & 0.92 \end{bmatrix}$

b. $A \begin{bmatrix} 1,500,000 \\ 600,000 \end{bmatrix} = \begin{bmatrix} 1,398,000 \\ 702,000 \end{bmatrix}$

c. $A^2 \begin{bmatrix} 1,500,000 \\ 600,000 \end{bmatrix} = \begin{bmatrix} 1,314,360 \\ 785,640 \end{bmatrix}$

d. $A^n \begin{bmatrix} 1,500,000 \\ 600,000 \end{bmatrix}$

17. The transition matrix is

$$A = \begin{bmatrix} 0.9 & 0.2 & 0.1 \\ 0.1 & 0.5 & 0.3 \\ 0 & 0.3 & 0.6 \end{bmatrix}$$

so the numbers of people in each category after 1 month are given by

$$A \begin{bmatrix} 20,000 \\ 20,000 \\ 10,000 \end{bmatrix} = \begin{bmatrix} 23,000 \\ 15,000 \\ 12,000 \end{bmatrix}$$

after 2 months by

$$A^2 \begin{bmatrix} 20,000 \\ 20,000 \\ 10,000 \end{bmatrix} = \begin{bmatrix} 24,900 \\ 13,400 \\ 11,700 \end{bmatrix}$$

and after 1 year by

$$A^{12} \begin{bmatrix} 20,000 \\ 20,000 \\ 10,000 \end{bmatrix} \approx \begin{bmatrix} 30,530 \\ 11,120 \\ 8,350 \end{bmatrix}$$

19. a. $I_1 + I_3 = I_2$

b. $\begin{cases} 4I_1 + 3I_2 = 8 \\ 3I_2 + 5I_3 = 10 \end{cases}$

c. $\begin{cases} I_1 - I_2 + I_3 = 0 \\ 4I_1 + 3I_2 = 8 \\ 3I_2 + 5I_3 = 10 \end{cases}$

Solution: $I_1 \approx 0.72, I_2 \approx 1.7, I_3 \approx 0.98$

21. Denote the average temperatures of the four points by $a, b, c,$ and d clockwise, starting with the upper left point. The resulting linear system is

$$\begin{cases} 4a - b - d = 50 \\ -a + 4b - c = 55 \\ - b - d = 45 \\ -a - c + 4d = 40 \end{cases}$$

The solution is $a \approx 24.4, b \approx 25.6, c \approx 23.1, d \approx 21.9$.

Review Exercises Chapter 1

1. a. $A = \begin{bmatrix} 1 & 1 & 2 & 1 \\ -1 & 0 & 1 & 2 \\ 2 & 2 & 0 & 1 \\ 1 & 1 & 2 & 3 \end{bmatrix}$

b. $\det(A) = -8$

c. Since the determinant of the coefficient matrix is not 0, the matrix is invertible and the linear system is consistent and has a unique solution.

d. The only solution is the trivial solution.

e. From part (b), since the determinant is not zero, the inverse exists.

$$A^{-1} = \frac{1}{8} \begin{bmatrix} -3 & -8 & -2 & 7 \\ 5 & 8 & 6 & -9 \\ 5 & 0 & -2 & -1 \\ -4 & 0 & 0 & 4 \end{bmatrix}$$

f. $x = A^{-1} \begin{bmatrix} 3 \\ 1 \\ -2 \\ 5 \end{bmatrix} = \frac{1}{4} \begin{bmatrix} 11 \\ -17 \\ 7 \\ 4 \end{bmatrix}$

3. $a = 0, c = 0, b = 0; \ a = 0, c = 1, b \in \mathbb{R};$
$a = 1, c = 0, b \in \mathbb{R}; \ a = 1, b = 0, c = 1$

5. a. If

$$A = \begin{bmatrix} a_1 & b_1 \\ c_1 & d_1 \end{bmatrix} \qquad B = \begin{bmatrix} a_2 & b_2 \\ c_2 & d_2 \end{bmatrix}$$

then the sum of the diagonal entries is

$$(a_1 a_2 + b_1 c_2) - (a_1 a_2 + b_2 c_1)$$
$$+ (b_2 c_1 + d_1 d_2) - (b_1 c_2 + d_1 d_2) = 0$$

b.

$$\begin{bmatrix} a & b \\ c & -a \end{bmatrix} \begin{bmatrix} a & b \\ c & -a \end{bmatrix} = \begin{bmatrix} a^2 + bc & 0 \\ 0 & a^2 + bc \end{bmatrix}$$
$$= (a^2 + bc)I$$

c. Let $M = AB - BA$. By part (a), $M^2 = kI$ for some k. Then

$$(AB - BA)^2 C = M^2 C$$
$$= (kI)C = C(kI)$$
$$= CM^2 = C(AB - BA)^2$$

7. a. Since $\det(A) = 1$, then A is invertible.

b. Six 1s can be added, making 21 the maximum number of entries that can be 1 and the matrix is invertible.

9. a. $B^t = (A + A^t)^t = A^t + (A^t)^t = A^t + A = B;$
$C^t = (A - A^t)^t = A^t - (A^t)^t = A^t - A = -C$

b. $A = \frac{1}{2}(A + A^t) + \frac{1}{2}(A - A^t)$

Chapter Test: Chapter 1

1. T	**2.** F
3. F	**4.** T
5. F	**6.** T
7. F	**8.** T
9. T	**10.** T
11. T	**12.** T
13. T	**14.** T
15. F	**16.** T
17. F	**18.** T
19. F	**20.** T
21. F	**22.** T

23. T	**24.** T
25. T	**26.** T
27. F	**28.** T
29. F	**30.** T
31. T	**32.** T
33. F	**34.** T
35. F	**36.** T
37. T	**38.** F
39. T	**40.** T
41. T	**42.** F
43. T	**44.** F
45. T	

Chapter 2

Section 2.1

1. $\mathbf{u} + \mathbf{v} = \begin{bmatrix} -1 \\ 2 \\ 3 \end{bmatrix} = \mathbf{v} + \mathbf{u}$

3. $\mathbf{u} - 2\mathbf{v} + 3\mathbf{w} = \begin{bmatrix} 11 \\ -7 \\ 0 \end{bmatrix}$

5. $-3(\mathbf{u} + \mathbf{v}) - \mathbf{w} = \begin{bmatrix} 1 \\ -7 \\ -8 \end{bmatrix}$

7. $\begin{bmatrix} -17 \\ -14 \\ 9 \\ -6 \end{bmatrix}$

9.

$$(x_1 + x_2)\mathbf{u} = (x_1 + x_2)\begin{bmatrix} 1 \\ -2 \\ 3 \\ 0 \end{bmatrix}$$

$$= \begin{bmatrix} x_1 + x_2 \\ -2x_1 - 2x_2 \\ 3x_1 + 3x_2 \\ 0 \end{bmatrix}$$

$$= \begin{bmatrix} x_1 \\ -2x_1 \\ 3x_1 \\ 0 \end{bmatrix} + \begin{bmatrix} x_2 \\ -2x_2 \\ 3x_2 \\ 0 \end{bmatrix}$$

$$= x_1\mathbf{u} + x_2\mathbf{v}$$

11. $\mathbf{v} = 2\mathbf{e}_1 + 4\mathbf{e}_2 + \mathbf{e}_3$

13. $\mathbf{v} = 3\mathbf{e}_2 - 2\mathbf{e}_3$

15. $\mathbf{w} = \begin{bmatrix} \frac{7}{2} \\ 1 \\ -1 \end{bmatrix}$

17. $\begin{cases} c_1 + 3c_2 = -2 \\ -2c_1 - 2c_2 = -1 \end{cases}$

Solution: $c_1 = \frac{7}{4}$, $c_2 = -\frac{5}{4}$

The vector $\begin{bmatrix} -2 \\ -1 \end{bmatrix}$ is a combination of $\begin{bmatrix} 1 \\ -2 \end{bmatrix}$ and $\begin{bmatrix} 3 \\ -2 \end{bmatrix}$.

19. $\begin{cases} c_1 - c_2 = 3 \\ 2c_1 - 2c_2 = 1 \end{cases}$

Solution: The linear system is inconsistent.

The vector $\begin{bmatrix} 3 \\ 1 \end{bmatrix}$ cannot be written as a combination

of $\begin{bmatrix} 1 \\ 2 \end{bmatrix}$ and $\begin{bmatrix} -1 \\ -2 \end{bmatrix}$.

21. $\begin{cases} -4c_1 - 5c_3 = -3 \\ 4c_1 + 3c_2 + c_3 = -3 \\ 3c_1 - c_2 - 5c_3 = 4 \end{cases}$

Solution: $c_1 = \frac{87}{121}, c_2 = -\frac{238}{121}, c_3 = \frac{3}{121}$

The vector $\begin{bmatrix} -3 \\ -3 \\ 4 \end{bmatrix}$ is a combination of the three

vectors.

23. $\begin{cases} -c_1 - c_2 + c_3 = -1 \\ c_2 - c_3 = 0 \\ c_1 + c_2 - c_3 = 2 \end{cases}$

Solution: The linear system is inconsistent

The vector $\begin{bmatrix} -1 \\ 0 \\ 2 \end{bmatrix}$ cannot be written as a

combination of the other vectors.

25. All 2×2 vectors. Moreover, $c_1 = \frac{1}{3}a - \frac{2}{3}b$, $c_2 = \frac{1}{3}a + \frac{1}{3}b$

27. All vectors of the form $\begin{bmatrix} a \\ -a \end{bmatrix}$ such that $a \in \mathbb{R}$.

29. All 3×3 vectors. Moreover, $c_1 = \frac{1}{3}a - \frac{2}{3}b + \frac{2}{3}c$,
$c_2 = -\frac{1}{3}a + \frac{2}{3}b + \frac{1}{3}c, c_3 = \frac{1}{3}a + \frac{1}{3}b - \frac{1}{3}c$

31. All vectors of the form $\begin{bmatrix} a \\ b \\ 2a - 3b \end{bmatrix}$ such that

$a, b \in \mathbb{R}$.

Section 2.2

1. $\begin{bmatrix} 1 & -2 & -4 \\ 1 & 3 & 11 \end{bmatrix} \longrightarrow \begin{bmatrix} 1 & 0 & 2 \\ 0 & 1 & 3 \end{bmatrix}$; yes

3. $\begin{bmatrix} -2 & 3 & 1 \\ 4 & -6 & 1 \end{bmatrix} \longrightarrow \begin{bmatrix} 1 & -\frac{3}{2} & 0 \\ 0 & 0 & 1 \end{bmatrix}$; no

5. Yes

$\begin{bmatrix} -2 & 1 & -3 \\ 3 & 4 & 10 \\ 4 & 2 & 10 \end{bmatrix} \longrightarrow \begin{bmatrix} 1 & 0 & 2 \\ 0 & 1 & 1 \\ 0 & 0 & 0 \end{bmatrix}$

7. Yes

$\begin{bmatrix} 2 & 3 & -2 & 2 \\ -2 & 0 & 0 & 8 \\ 0 & -3 & -1 & 2 \end{bmatrix} \longrightarrow \begin{bmatrix} 1 & 0 & 0 & -4 \\ 0 & 1 & 0 & \frac{2}{3} \\ 0 & 0 & 1 & -4 \end{bmatrix}$

9. No

$\begin{bmatrix} 1 & -1 & 0 & -1 \\ 2 & -1 & 1 & 1 \\ -1 & 3 & 2 & 5 \end{bmatrix} \longrightarrow \begin{bmatrix} 1 & 0 & 0 & 0 \\ 0 & 1 & 1 & 0 \\ 0 & 0 & 0 & 1 \end{bmatrix}$

11. Yes

$\begin{bmatrix} 2 & 1 & -1 & 3 \\ -3 & 6 & -1 & -17 \\ 4 & -1 & 2 & 17 \\ 1 & 2 & 3 & 7 \end{bmatrix} \longrightarrow \begin{bmatrix} 1 & 0 & 0 & 3 \\ 0 & 1 & 0 & -1 \\ 0 & 0 & 1 & 2 \\ 0 & 0 & 0 & 0 \end{bmatrix}$

13. Infinitely many ways

$c_1 = 1 + \frac{1}{3}c_3, c_2 = 1 + \frac{7}{3}c_3, c_3 \in \mathbb{R}$

15. Infinitely many ways

$c_1 = 3 + 6c_4, c_2 = -2 - c_4, c_3 = 2 + 2c_4, c_4 \in \mathbb{R}$

17. Yes

$\begin{bmatrix} 1 & -2 & -1 & -2 \\ 2 & 3 & 3 & 4 \\ 1 & 1 & 2 & 4 \\ -1 & 4 & 1 & 0 \end{bmatrix} \longrightarrow \begin{bmatrix} 1 & 0 & 0 & -1 \\ 0 & 1 & 0 & -1 \\ 0 & 0 & 1 & 3 \\ 0 & 0 & 0 & 0 \end{bmatrix}$

19. No

$\begin{bmatrix} 2 & 3 & 3 & 2 \\ 2 & -1 & -1 & 1 \\ -1 & 2 & 2 & -1 \\ 3 & -2 & 2 & 2 \end{bmatrix} \longrightarrow \begin{bmatrix} 1 & 0 & 0 & 0 \\ 0 & 1 & 0 & 0 \\ 0 & 0 & 1 & 0 \\ 0 & 0 & 0 & 1 \end{bmatrix}$

21. $A\mathbf{x} = 2\begin{bmatrix} 1 \\ -2 \end{bmatrix} - \begin{bmatrix} 3 \\ 1 \end{bmatrix}$

23. $(\mathbf{AB})_1 = 3\begin{bmatrix} -1 \\ 3 \end{bmatrix} + 2\begin{bmatrix} -2 \\ 4 \end{bmatrix},$

$(\mathbf{AB})_2 = 2\begin{bmatrix} -1 \\ 3 \end{bmatrix} + 5\begin{bmatrix} -2 \\ 4 \end{bmatrix}$

25. Not possible.

27. $x^3 - 2x + 1 = \frac{1}{2}(1+x) + 2(-x) + 0(x^2 + 1) + \frac{1}{2}(2x^3 - x + 1)$

29. All vectors $\begin{bmatrix} a \\ b \\ c \end{bmatrix}$ such that $3a - b + c = 0$.

31. $\mathbf{v} = 2\mathbf{v}_1 - \mathbf{v}_2 + 4\mathbf{v}_3$

33. Since $c_1 \neq 0, \mathbf{v}_1 = -\frac{c_2}{c_1}\mathbf{v}_2 - \cdots - \frac{c_n}{c_1}\mathbf{v}_n.$

35. Let $\mathbf{v} \in S_1$. Since $c \neq 0$, then $\mathbf{v} = c_1\mathbf{v}_1 + \cdots + \frac{c_k}{c}(c\mathbf{v}_k)$, so $\mathbf{v} \in S_2$. If $\mathbf{v} \in S_2$, then $\mathbf{v} = c_1\mathbf{v}_1 + \cdots + (cc_k)\mathbf{v}_k$, so $\mathbf{v} \in S_1$. Therefore, $S_1 = S_2$.

37. If $\mathbf{A}_3 = c\mathbf{A}_1$, then $\det(A) = 0$. Since the linear system is assumed to be consistent, it must have infinitely many solutions.

Section 2.3

1. Since $\begin{vmatrix} -1 & 2 \\ 1 & -3 \end{vmatrix} = 1$, the vectors are linearly independent.

3. Since $\begin{vmatrix} 1 & -2 \\ -4 & 8 \end{vmatrix} = 0$, the vectors are linearly dependent.

5. Since $\begin{bmatrix} -1 & 2 \\ 2 & 2 \\ 1 & 3 \end{bmatrix} \longrightarrow \begin{bmatrix} -1 & 2 \\ 0 & 6 \\ 0 & 0 \end{bmatrix}$, the vectors are linearly independent.

7. Since $\begin{vmatrix} -4 & -5 & 3 \\ 4 & 3 & -5 \\ -1 & 3 & 5 \end{vmatrix} = 0$, the vectors are linearly dependent.

9. Since $\begin{bmatrix} 3 & 1 & 3 \\ -1 & 0 & -1 \\ -1 & 2 & 0 \\ 2 & 1 & 1 \end{bmatrix} \longrightarrow \begin{bmatrix} 3 & 1 & 3 \\ 0 & \frac{1}{3} & 0 \\ 0 & 0 & 1 \\ 0 & 0 & 0 \end{bmatrix}$, the vectors are linearly independent.

11. Since

$$\begin{bmatrix} 3 & 0 & 1 \\ 3 & 1 & -1 \\ 2 & 0 & -1 \\ 1 & 0 & -2 \end{bmatrix} \longrightarrow \begin{bmatrix} 3 & 0 & 1 \\ 0 & 1 & -2 \\ 0 & 0 & -\frac{5}{3} \\ 0 & 0 & 0 \end{bmatrix}$$

the matrices are linearly independent.

13. Since

$$\begin{bmatrix} 1 & 0 & -1 & 1 \\ -2 & -1 & 1 & 1 \\ -2 & 2 & -2 & -1 \\ -2 & 2 & 2 & -2 \end{bmatrix}$$

$$\longrightarrow \begin{bmatrix} 1 & 0 & -1 & 1 \\ 0 & -1 & -1 & 3 \\ 0 & 0 & -6 & 7 \\ 0 & 0 & 0 & \frac{11}{3} \end{bmatrix}$$

the matrices are linearly independent.

15. $\mathbf{v}_2 = -\frac{1}{2}\mathbf{v}_1$

17. Any set of vectors containing the zero vector is linearly dependent.

19. **a.** $\mathbf{A}_2 = -2\mathbf{A}_1$

  b. $\mathbf{A}_3 = \mathbf{A}_1 + \mathbf{A}_2$

21. $a \neq 6$

23. **a.** Since $\begin{vmatrix} 1 & 1 & 1 \\ 1 & 2 & 1 \\ 1 & 3 & 2 \end{vmatrix} = 1$, the vectors are linearly independent.

  b. $c_1 = 0, c_2 = -1, c_3 = 3$

25. Since $\begin{vmatrix} 1 & 2 & 0 \\ -1 & 0 & 3 \\ 2 & 1 & 2 \end{vmatrix} = 13$, the matrix is invertible so $A\mathbf{x} = \mathbf{b}$ has a unique solution for every vector $\mathbf{b}$.

27. Linear independent

29. Linearly dependent

31. If $x = 0$, then $c_1 = 0$, and if $x = \frac{1}{2}$, then $c_2 = 0$.

33. Let $x = 0$, then $c_3 = 0$. Now letting $x = 1$ and $x = -1$, $c_1 = c_2 = c_3 = 0$.

35. If $\mathbf{u}$ and $\mathbf{v}$ are linearly dependent, then there are scalars a and b, not both 0, such that $a\mathbf{u} + b\mathbf{v} = \mathbf{0}$. If $a \neq 0$, then $\mathbf{u} = -(b/a)\mathbf{v}$. On the other hand, if there is a scalar c such that $\mathbf{u} = c\mathbf{v}$, then $\mathbf{u} - c\mathbf{v} = \mathbf{0}$.

37. Setting a linear combination of $\mathbf{w}_1, \mathbf{w}_2, \mathbf{w}_3$ to $\mathbf{0}$, we have

$$\mathbf{0} = c_1\mathbf{w}_1 + c_2\mathbf{w}_2 + c_3\mathbf{w}_3$$
$$= c_1\mathbf{v}_1 + (c_1 + c_2 + c_3)\mathbf{v}_2 + (-c_2 + c_3)\mathbf{v}_3$$

if and only if $c_1 = 0, c_1 + c_2 + c_3 = 0$, and $-c_2 + c_3 = 0$ if and only if $c_1 = c_2 = c_3 = 0$.

39. Consider $c_1\mathbf{v}_1 + c_2\mathbf{v}_2 + c_3\mathbf{v}_3 = \mathbf{0}$, which is true if and only if $c_3\mathbf{v}_3 = -c_1\mathbf{v}_1 - c_2\mathbf{v}_2$. If $c_3 \neq 0$, then $\mathbf{v}_3$ would be a linear combination of $\mathbf{v}_1$ and $\mathbf{v}_2$ contradicting the hypothesis that it is not the case. Therefore, $c_3 = 0$. Now since $\mathbf{v}_1$ and $\mathbf{v}_2$ are linearly independent $c_1 = c_2 = 0$.

41. Since $\mathbf{A}_1, \mathbf{A}_2, \ldots, \mathbf{A}_n$ are linearly independent, if

$$A\mathbf{x} = x_1\mathbf{A}_1 + \cdots + x_n\mathbf{A}_n = \mathbf{0}$$

then $x_1 = x_2 = \cdots = x_n = 0$.

Review Exercises Chapter 2

1. Since $\begin{vmatrix} a & b \\ c & d \end{vmatrix} = ad - bc \neq 0$, the column vectors are linearly independent. If $ad - bc = 0$, then the column vectors are linearly dependent.

3. The determinant $\begin{vmatrix} a^2 & 0 & 1 \\ 0 & a & 0 \\ 1 & 2 & 1 \end{vmatrix} = a^3 - a \neq 0$ if and only if $a \neq \pm 1$, and $a \neq 0$. So the vectors are linearly independent if and only if $a \neq \pm 1$, and $a \neq 0$.

5. a. Since the vectors are not scalar multiples of each other, S is linearly independent.

b. Since

$$\left[\begin{array}{cc|c} 1 & 1 & a \\ 0 & 1 & b \\ 2 & 1 & c \end{array}\right] \rightarrow \left[\begin{array}{cc|c} 1 & 1 & a \\ 0 & 1 & b \\ 0 & 0 & -2a+b+c \end{array}\right]$$

the linear system is inconsistent for $-2a + b + c \neq 0$. If $a = 1, b = 1, c = 3$, then the system is inconsistent and $\mathbf{v} = \begin{bmatrix} 1 \\ 1 \\ 3 \end{bmatrix}$ is not a linear combination of the vectors.

c. All vectors $\begin{bmatrix} a \\ b \\ c \end{bmatrix}$ such that $-2a + b + c = 0$

d. Linearly independent

e. All vectors in $\mathbb{R}^3$

7. a. Let $A = \begin{bmatrix} 1 & 1 & 2 & 1 \\ -1 & 0 & 1 & 2 \\ 2 & 2 & 0 & 1 \\ 1 & 1 & 2 & 3 \end{bmatrix}, \mathbf{x} = \begin{bmatrix} x \\ y \\ z \\ w \end{bmatrix}$, and

$\mathbf{b} = \begin{bmatrix} 3 \\ 1 \\ -2 \\ 5 \end{bmatrix}$.

b. $\det(A) = -8$

c. Yes, since the determinant of A is nonzero.

d. Since the determinant of the coefficient matrix is nonzero, the matrix A is invertible, so the linear system has a unique solution.

e. $x = \frac{11}{4}, y = -\frac{17}{4}, z = \frac{7}{4}, w = 1$

9. a.

$$x_1\begin{bmatrix} 1 \\ 2 \\ 1 \end{bmatrix} + x_2\begin{bmatrix} 3 \\ -1 \\ 1 \end{bmatrix} + x_3\begin{bmatrix} 2 \\ 3 \\ -1 \end{bmatrix} = \begin{bmatrix} b_1 \\ b_2 \\ b_3 \end{bmatrix}$$

b. Since $\det(A) = 19$, the linear system has a unique solution equal to $\mathbf{x} = A^{-1}\mathbf{b}$.

c. Yes

d. Yes, since the determinant of A is nonzero, A^{-1} exists and the linear system has a unique solution.

Chapter Test: Chapter 2

1. T		**2.** F	
3. T		**4.** T	
5. F		**6.** F	
7. F		**8.** T	
9. F		**10.** F	
11. T		**12.** T	
13. F		**14.** T	
15. F		**16.** T	
17. T		**18.** F	
19. T		**20.** T	
21. F		**22.** F	
23. F		**24.** F	
25. T		**26.** F	
27. T		**28.** F	
29. T		**30.** F	
31. T		**32.** F	
33. T			

Chapter 3

Section 3.1

1. Since

$$\begin{bmatrix} x_1 \\ y_1 \\ z_1 \end{bmatrix} \oplus \begin{bmatrix} x_2 \\ y_2 \\ z_2 \end{bmatrix} = \begin{bmatrix} x_1 - x_2 \\ y_1 - y_2 \\ z_1 - z_2 \end{bmatrix}$$

and

$$\begin{bmatrix} x_2 \\ y_2 \\ z_2 \end{bmatrix} \oplus \begin{bmatrix} x_1 \\ y_1 \\ z_1 \end{bmatrix} = \begin{bmatrix} x_2 - x_1 \\ y_2 - y_1 \\ z_2 - z_1 \end{bmatrix}$$

do not agree for all pairs of vectors, the operation $\oplus$ is not commutative, so V is not a vector space.

3. The operation $\oplus$ is not associative, so V is not a vector space.

7. Since

$$(c+d) \odot \begin{bmatrix} x \\ y \end{bmatrix} = \begin{bmatrix} x+c+d \\ y \end{bmatrix}$$

does not equal

$$c \odot \begin{bmatrix} x \\ y \end{bmatrix} + d \odot \begin{bmatrix} x \\ y \end{bmatrix}$$

$$= \begin{bmatrix} x+c \\ y \end{bmatrix} + \begin{bmatrix} x+d \\ y \end{bmatrix}$$

$$= \begin{bmatrix} 2x+c+d \\ 2y \end{bmatrix}$$

for all vectors $\begin{bmatrix} x \\ y \end{bmatrix}$, then V is not a vector space.

9. Since the operation $\oplus$ is not commutative, V is not a vector space.

11. The zero vector is given by $\mathbf{0} = \begin{bmatrix} 0 \\ 0 \end{bmatrix}$. Since this vector is not in V, then V is not a vector space.

13. a. Since V is not closed under vector addition, V is not a vector space.

  b. Each of the 10 vector space axioms are satisfied with vector addition and scalar multiplication defined in this way.

15. Yes, V is a vector space.

17. No, V is not a vector space. Let $A = I$ and $B = -I$. Then $A + B$ is not invertible and hence not in V.

19. Yes, V is a vector space.

21. a. The additive identity is $\mathbf{0} = \begin{bmatrix} 1 & 0 \\ 0 & 1 \end{bmatrix}$, and the additive inverse of A is A^{-1}.

  b. If $c = 0$, then cA is not in V.

23. a. The additive identity is $\mathbf{0} = \begin{bmatrix} 1 \\ 2 \\ 3 \end{bmatrix}$. Let

$$\mathbf{u} = \begin{bmatrix} 1+a \\ 2-a \\ 3+2a \end{bmatrix}.$$ Then the additive inverse is

$$-\mathbf{u} = \begin{bmatrix} 1-a \\ 2+a \\ 3-2a \end{bmatrix}.$$

b. Each of the 10 vector space axioms is satisfied.

$$\textbf{c. } 0 \odot \begin{bmatrix} 1+t \\ 2-t \\ 3+2t \end{bmatrix} = \begin{bmatrix} 1+0t \\ 2-0t \\ 3+2(0)t \end{bmatrix} = \begin{bmatrix} 1 \\ 2 \\ 3 \end{bmatrix}$$

25. Each of the 10 vector space axioms is satisfied.

27. Each of the 10 vector space axioms is satisfied.

29. Since $(f + g)(0) = f(0) + g(0) = 1 + 1 = 2$, then V is not closed under addition and hence is not a vector space.

31. a. The zero vector is given by $f(x + 0) = x^3$ and $-f(x + t) = f(x - t)$.

  b. Each of the 10 vector space axioms is satisfied.

Section 3.2

1. The set S is a subspace of $\mathbb{R}^2$.

3. The set S is not a subspace of $\mathbb{R}^2$. If $\mathbf{u} = \begin{bmatrix} 2 \\ -1 \end{bmatrix}$ and $\mathbf{v} = \begin{bmatrix} -1 \\ 3 \end{bmatrix}$, then $\mathbf{u} + \mathbf{v} = \begin{bmatrix} 1 \\ 2 \end{bmatrix} \notin S$.

5. The set S is not a subspace of $\mathbb{R}^2$. If $\mathbf{u} = \begin{bmatrix} 0 \\ -1 \end{bmatrix}$ and $c = 0$, then $c\mathbf{v} = \begin{bmatrix} 0 \\ 0 \end{bmatrix} \notin S$.

7. Since

$$\begin{bmatrix} x_1 \\ x_2 \\ x_3 \end{bmatrix} + c \begin{bmatrix} y_1 \\ y_2 \\ y_3 \end{bmatrix} = \begin{bmatrix} x_1 + cy_1 \\ x_2 + cy_2 \\ x_3 + cy_3 \end{bmatrix}$$

and $(x_1 + cy_1) + (x_3 + cy_3) = -2(c + 1) = 2$ if and only if $c = -2$, so S is not a subspace of $\mathbb{R}^3$.

9. Since

$$\begin{bmatrix} s - 2t \\ s \\ t + s \end{bmatrix} + c \begin{bmatrix} x - 2y \\ x \\ y + x \end{bmatrix}$$

$$= \begin{bmatrix} (s + cx) - 2(t + cy) \\ s + cx \\ (t + cy) + (s + cx) \end{bmatrix}$$

is in S, then S is a subspace.

11. Yes, S is a subspace.

13. No, S is not a subspace.

15. Yes, S is a subspace.

17. Yes, S is a subspace.

19. No, S is not a subspace since $x^3 - x^3 = 0$, which is not a polynomial of degree 3.

21. Yes, S is a subspace.

23. No, S is not a subspace.

25. Since

$$\left[\begin{array}{rrr|r} 1 & -1 & -1 & 1 \\ 1 & -1 & 2 & -1 \\ 0 & 1 & 0 & 1 \end{array}\right] \rightarrow \left[\begin{array}{rrr|r} 1 & -1 & -1 & 1 \\ 0 & 1 & 0 & 1 \\ 0 & 0 & 3 & -2 \end{array}\right]$$

the vector $\mathbf{v}$ is in the span.

27. Since

$$\left[\begin{array}{rrr|r} 1 & 0 & 1 & -2 \\ 1 & 1 & -1 & 1 \\ 0 & 2 & -4 & 6 \\ -1 & 1 & -3 & 5 \end{array}\right] \rightarrow \left[\begin{array}{rrr|r} 1 & 0 & 1 & -2 \\ 0 & 1 & -2 & 3 \\ 0 & 0 & 0 & 0 \\ 0 & 0 & 0 & 0 \end{array}\right]$$

the vector $\mathbf{v}$ is in the span.

29. Since

$$c_1(1+x) + c_2(x^2 - 2) + c_3(3x) = 2x^2 - 6x - 11$$

implies $c_1 = -7, c_2 = 2, c_3 = \frac{1}{3}$, the polynomial is in the span.

31. $\mathbf{span}(S) = \left\{ \left[\begin{array}{c} a \\ b \\ c \end{array}\right] \middle| a + c = 0 \right\}$

33. $\mathbf{span}(S) = \left\{ \left[\begin{array}{cc} a & b \\ \frac{a+b}{3} & \frac{2a-b}{3} \end{array}\right] \middle| a, b \in \mathbb{R} \right\}$

35. $\mathbf{span}(S) = \left\{ ax^2 + bx + c \middle| a - c = 0 \right\}$

37. a. $\mathbf{span}(S) = \left\{ \left[\begin{array}{c} a \\ b \\ \frac{b-2a}{3} \end{array}\right] \middle| a, b \in \mathbb{R} \right\}$

   b. Yes, S is linearly independent.

39. a. $\mathbf{span}(S) = \mathbb{R}^3$

   b. Yes, S is linearly independent.

41. a. $\mathbf{span}(S) = \mathbb{R}^3$

   b. No, S is linearly dependent.

   c. $\mathbf{span}(T) = \mathbb{R}^3$; T is linearly dependent.

   d. $\mathbf{span}(H) = \mathbb{R}^3$; H is linearly independent.

43. a. $\mathbf{span}(S) = \mathcal{P}_2$

   b. No, S is linearly dependent.

   c. $2x^2 + 3x + 5 = 2(1) - (x - 3) + 2(x^2 + 2x)$

   d. T is linearly independent; $\mathbf{span}(T) = \mathcal{P}_3$

45. a–b Since

$$\left[\begin{array}{c} -s \\ s - 5t \\ 2s + 3t \end{array}\right] = s\left[\begin{array}{c} -1 \\ 1 \\ 2 \end{array}\right] + t\left[\begin{array}{c} 0 \\ -5 \\ 3 \end{array}\right]$$

then $S = \mathbf{span}\left\{ \left[\begin{array}{c} -1 \\ 1 \\ 2 \end{array}\right], \left[\begin{array}{c} 0 \\ -5 \\ 3 \end{array}\right] \right\}$.

Therefore, S is a subspace.

   c. Yes, the vectors are linearly independent.

   d. $S \neq \mathbb{R}^3$

47. Since $A(\mathbf{x} + c\mathbf{y}) = \left[\begin{array}{c} 1 \\ 2 \end{array}\right] + c\left[\begin{array}{c} 1 \\ 2 \end{array}\right] = \left[\begin{array}{c} 1 \\ 2 \end{array}\right]$ if and only if $c = 0$, then S is not a subspace.

49. Let $B_1, B_2 \in S$. Since

$$\begin{aligned} A(B_1 + cB_2) &= AB_1 + cAB_2 \\ &= B_1A + c(B_2A) \\ &= (B_1 + cB_2)A \end{aligned}$$

then $B_1 + cB_2 \in S$ and S is a subspace.

Section 3.3

1. The set S has only two vectors, while $\dim(\mathbb{R}^3) = 3$.

3. Since the third vector can be written as the sum of the first two, the set S is not linearly independent.

5. Since the third polynomial is a linear combination of the first two, the set S is not linearly independent.

7. The set S is a linearly independent set of two vectors in $\mathbb{R}^2$.

9. The set S is a linearly independent set of three vectors in $\mathbb{R}^3$.

11. The set S is a linearly independent set of four vectors in $M_{2\times2}$. Since $\dim(M_{2\times2}) = 4$, then S is a basis.

13. The set S is a linearly independent set of three vectors in $\mathbb{R}^3$ and so is a basis.

15. The set S is a linearly dependent and is therefore not a basis for $\mathbb{R}^4$.

17. The set S is a linearly independent set of three vectors in $\mathcal{P}_2$ so S is a basis.

19. A basis for S is $B = \left\{ \left[\begin{array}{c} 1 \\ -1 \\ 0 \end{array}\right], \left[\begin{array}{c} 2 \\ 1 \\ 1 \end{array}\right] \right\}$ and $\dim(S) = 2$.

21. A basis for S is

$$B = \left\{ \left[\begin{array}{cc} 1 & 0 \\ 0 & 0 \end{array}\right], \left[\begin{array}{cc} 0 & 1 \\ 1 & 0 \end{array}\right], \left[\begin{array}{cc} 0 & 0 \\ 0 & 1 \end{array}\right] \right\}$$

and $\dim(S) = 3$.

23. A basis for S is $B = \{x, x^2\}$ and $\dim(S) = 2$.

25. The set S is already a basis for $\mathbb{R}^3$ since it is a linearly independent set of three vectors in $\mathbb{R}^3$.

27. A basis for the span of S is given by

$$B = \left\{ \begin{bmatrix} 2 \\ -3 \\ 0 \end{bmatrix}, \begin{bmatrix} 0 \\ 2 \\ 2 \end{bmatrix}, \begin{bmatrix} -1 \\ -1 \\ 0 \end{bmatrix} \right\}. \text{ Observe that}$$

$\mathbf{span}(S) = \mathbb{R}^3$.

29. A basis for the span of S is given by

$$B = \left\{ \begin{bmatrix} 2 \\ -3 \\ 0 \end{bmatrix}, \begin{bmatrix} 0 \\ 2 \\ 2 \end{bmatrix}, \begin{bmatrix} 4 \\ 0 \\ 4 \end{bmatrix} \right\}. \text{ Observe that}$$

$\mathbf{span}(S) = \mathbb{R}^3$.

31. A basis for $\mathbb{R}^3$ containing S is

$$B = \left\{ \begin{bmatrix} 2 \\ -1 \\ 3 \end{bmatrix}, \begin{bmatrix} 1 \\ 0 \\ 2 \end{bmatrix}, \begin{bmatrix} 1 \\ 0 \\ 0 \end{bmatrix} \right\}$$

33. A basis for $\mathbb{R}^4$ containing S is

$$B = \left\{ \begin{bmatrix} 1 \\ -1 \\ 2 \\ 4 \end{bmatrix}, \begin{bmatrix} 3 \\ 1 \\ 1 \\ 2 \end{bmatrix}, \begin{bmatrix} 1 \\ 0 \\ 0 \\ 0 \end{bmatrix}, \begin{bmatrix} 0 \\ 0 \\ 1 \\ 0 \end{bmatrix} \right\}$$

35. A basis for $\mathbb{R}^3$ containing S is

$$B = \left\{ \begin{bmatrix} -1 \\ 1 \\ 3 \end{bmatrix}, \begin{bmatrix} 1 \\ 1 \\ 1 \end{bmatrix}, \begin{bmatrix} 1 \\ 0 \\ 0 \end{bmatrix} \right\}$$

37. $B = \{e_{ii} \mid 1 \le i \le n\}$

43. $\dim(W) = 2$

Section 3.4

1. $[\mathbf{v}]_B = \begin{bmatrix} 2 \\ -1 \end{bmatrix}$

3. $[\mathbf{v}]_B = \begin{bmatrix} 2 \\ -1 \\ 3 \end{bmatrix}$

5. $[\mathbf{v}]_B = \begin{bmatrix} 5 \\ 2 \\ -2 \end{bmatrix}$

7. $[\mathbf{v}]_B = \begin{bmatrix} -1 \\ 2 \\ -2 \\ 4 \end{bmatrix}$

9. $[\mathbf{v}]_{B_1} = \begin{bmatrix} -\frac{1}{4} \\ \frac{1}{8} \end{bmatrix}$; $[\mathbf{v}]_{B_2} = \begin{bmatrix} \frac{1}{2} \\ -\frac{1}{2} \end{bmatrix}$

11. $[\mathbf{v}]_{B_1} = \begin{bmatrix} 1 \\ 2 \\ -1 \end{bmatrix}$; $[\mathbf{v}]_{B_2} = \begin{bmatrix} 1 \\ 1 \\ 0 \end{bmatrix}$

13. $[I]_{B_1}^{B_2} = \begin{bmatrix} 1 & -1 \\ 1 & 1 \end{bmatrix}$

$[\mathbf{v}]_{B_2} = [I]_{B_1}^{B_2}[\mathbf{v}]_{B_1} = \begin{bmatrix} -1 \\ 5 \end{bmatrix}$

15. $[I]_{B_1}^{B_2} = \begin{bmatrix} 3 & 2 & 1 \\ -1 & -\frac{2}{3} & 0 \\ 0 & -\frac{1}{3} & 0 \end{bmatrix}$

$[\mathbf{v}]_{B_2} = [I]_{B_1}^{B_2}[\mathbf{v}]_{B_1} = \begin{bmatrix} -1 \\ 1 \\ 0 \end{bmatrix}$

17. $[I]_{B_1}^{B_2} = \begin{bmatrix} 0 & 0 & 1 \\ 1 & 0 & 0 \\ 0 & 1 & 0 \end{bmatrix}$

$[\mathbf{v}]_{B_2} = [I]_{B_1}^{B_2}[\mathbf{v}]_{B_1} = \begin{bmatrix} 5 \\ 2 \\ 3 \end{bmatrix}$

19. $\begin{bmatrix} a \\ b \\ c \end{bmatrix}_B = \begin{bmatrix} -a - b + c \\ a + b \\ a + 2b - c \end{bmatrix}$

21. a. $[I]_{B_1}^{B_2} = \begin{bmatrix} 0 & 1 & 0 \\ 1 & 0 & 0 \\ 0 & 0 & 1 \end{bmatrix}$

b. $[\mathbf{v}]_{B_2} = [I]_{B_1}^{B_2} \begin{bmatrix} 1 \\ 2 \\ 3 \end{bmatrix} = \begin{bmatrix} 2 \\ 1 \\ 3 \end{bmatrix}$

23. a. $[I]_S^B = \begin{bmatrix} 1 & 1 \\ 0 & 2 \end{bmatrix}$

b. $\begin{bmatrix} 1 \\ 2 \end{bmatrix}_B = \begin{bmatrix} 3 \\ 4 \end{bmatrix}$ $\begin{bmatrix} 1 \\ 4 \end{bmatrix}_B = \begin{bmatrix} 5 \\ 8 \end{bmatrix}$

$\begin{bmatrix} 4 \\ 2 \end{bmatrix}_B = \begin{bmatrix} 6 \\ 4 \end{bmatrix}$ $\begin{bmatrix} 4 \\ 4 \end{bmatrix}_B = \begin{bmatrix} 8 \\ 8 \end{bmatrix}$

c.

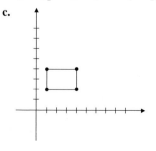

d.

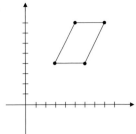

25. a. $[I]_{B_1}^{B_2} = \begin{bmatrix} -1 & -1 & 0 \\ 2 & 2 & -1 \\ 0 & -1 & 1 \end{bmatrix}$

b. $[2\mathbf{u}_1 - 2\mathbf{u}_2 + \mathbf{u}_3]_{B_2} = [I]_{B_1}^{B_2} \begin{bmatrix} 2 \\ -3 \\ 1 \end{bmatrix} = \begin{bmatrix} 1 \\ -3 \\ 4 \end{bmatrix}$

Section 3.5

1. a. $y_1 = e^{2x}, y_2 = e^{3x}$

b. $W[y_1, y_2](x) = \begin{vmatrix} e^{2x} & e^{3x} \\ 2e^{2x} & 3e^{3x} \end{vmatrix} = e^{5x} > 0$ for all x.

c. $y(x) = C_1 e^{2x} + C_2 e^{3x}$

3. a. $y_1 = e^{-2x}, y_2 = xe^{-2x}$

b. $W[y_1, y_2](x) = \begin{vmatrix} e^{-2x} & xe^{-2x} \\ -2e^{-2x} & e^{-2x} - 2xe^{-2x} \end{vmatrix} =$

$e^{-4x} > 0$ for all x.

c. $y(x) = C_1 e^{-2x} + C_2 xe^{-2x}$

5. $y(x) = e^x + 2xe^x$

7. a. $y_c(x) = C_1 e^{3x} + C_2 e^x$

b. $a = 1, b = 3, c = 4$

9. $y(x) = \frac{1}{4}\cos(8x)$

Review Exercises Chapter 3

1. $k \neq 69$

3. a. Since S is closed under vector addition and scalar multiplication, S is a subspace of $M_{2\times 2}$.

b. Yes, let $a = 3, b = -2, c = 0$.

c. $B = \left\{ \begin{bmatrix} 1 & 1 \\ 0 & 1 \end{bmatrix}, \begin{bmatrix} -1 & 0 \\ 1 & 0 \end{bmatrix}, \begin{bmatrix} 0 & 0 \\ 1 & -1 \end{bmatrix} \right\}$

d. The matrix $\begin{bmatrix} 0 & 1 \\ 2 & 1 \end{bmatrix}$ is not in S.

5. a. The set T is a basis since it is a linearly independent set of three vectors in the three-dimensional vector space V.

b. The set W is not a basis for V since it is not linearly independent.

7. Since $\mathbf{v}_1$ can be written as

$$\mathbf{v}_1 = \left(\frac{-c_2}{c_1}\right)\mathbf{v}_2 + \left(\frac{-c_3}{c_1}\right)\mathbf{v}_3 + \cdots + \left(\frac{-c_n}{c_1}\right)\mathbf{v}_n$$

then

$$V = \mathbf{span}\{\mathbf{v}_2, \mathbf{v}_3, \ldots, \mathbf{v}_n\}$$

9. a. The set $B = \{\mathbf{u}, \mathbf{v}\}$ is a basis for $\mathbb{R}^2$ since it is linearly independent. To see this, consider

$$a\mathbf{u} + b\mathbf{v} = \mathbf{0}$$

Now take the dot product of both sides with first $\mathbf{u}$, then $\mathbf{v}$, to show that $a = b = 0$.

b. If $[\mathbf{w}]_B = \begin{bmatrix} \alpha \\ \beta \end{bmatrix}$, then

$$\alpha = \frac{\begin{vmatrix} x & v_1 \\ y & v_2 \end{vmatrix}}{\begin{vmatrix} u_1 & v_1 \\ u_2 & v_2 \end{vmatrix}} = \frac{xv_2 - yv_1}{u_1v_2 - v_1u_2}$$

and

$$\beta = \frac{\begin{vmatrix} u_1 & x \\ u_2 & y \end{vmatrix}}{\begin{vmatrix} u_1 & v_1 \\ u_2 & v_2 \end{vmatrix}} = \frac{yu_1 - xu_2}{u_1v_2 - v_1u_2}.$$

Chapter Test: Chapter 3

1. F		**2.** T	
3. F		**4.** F	
5. T		**6.** F	
7. F		**8.** F	
9. T		**10.** T	
11. T		**12.** T	
13. T		**14.** T	
15. T		**16.** F	
17. F		**18.** F	
19. T		**20.** T	
21. T		**22.** T	
23. T		**24.** T	
25. T		**26.** F	
27. T		**28.** T	
29. T		**30.** F	
31. T		**32.** F	
33. T		**34.** F	
35. T			

Chapter 4

Section 4.1

1. T is linear.

3. T is not linear.

5. T is linear.

7. Since $T(x + y) \neq T(x) + T(y)$ for all real numbers x and y, T is not linear.

9. Since $T(c\mathbf{u}) \neq cT(\mathbf{u})$, T is not linear.

11. Since $T(\mathbf{0}) \neq \mathbf{0}$, T is not linear.

13. T is linear.

15. Since $T(cA) = c^2 T(A) \neq cT(A)$ for all scalars c, T is not linear.

17. a. $T(\mathbf{u}) = \begin{bmatrix} 2 \\ 3 \end{bmatrix}$; $T(\mathbf{v}) = \begin{bmatrix} -2 \\ -2 \end{bmatrix}$

   b. Yes

   c. Yes

19. a. $T(\mathbf{u}) = \begin{bmatrix} 0 \\ 0 \end{bmatrix}$; $T(\mathbf{v}) = \begin{bmatrix} 0 \\ -1 \end{bmatrix}$

   b. No. $T(\mathbf{u} + \mathbf{v}) = \begin{bmatrix} -1 \\ -1 \end{bmatrix}$; $T(\mathbf{u}) + T(\mathbf{v}) = \begin{bmatrix} 0 \\ -1 \end{bmatrix}$

   c. No, by part (b).

21. $T\left(\begin{bmatrix} 1 \\ -3 \end{bmatrix} \right) = \begin{bmatrix} 5 \\ -9 \end{bmatrix}$

23. $T(-3 + x - x^2) = -1 - 4x + 4x^2$

25. $T\left(\begin{bmatrix} 3 \\ 7 \end{bmatrix} \right) = \begin{bmatrix} 22 \\ -11 \end{bmatrix}$

27. a. No. The polynomial $2x^2 - 3x + 2$ cannot be written as a linear combination of x^2, $-3x$, and $-x^2 + 3x$.

   b. Yes. $T(3x^2 - 4x) = \frac{4}{3}x^2 + 6x - \frac{13}{3}$

29. a. $A = \begin{bmatrix} -1 & 0 \\ 0 & -1 \end{bmatrix}$

   b. $T(\mathbf{e}_1) = \begin{bmatrix} -1 \\ 0 \end{bmatrix}$ and $T(\mathbf{e}_2) = \begin{bmatrix} 0 \\ -1 \end{bmatrix}$. Observe that these are the column vectors of A.

31. $T\left(\begin{bmatrix} 0 \\ 0 \\ z \end{bmatrix} \right) = \begin{bmatrix} 0 \\ 0 \end{bmatrix}$, for all $z \in \mathbb{R}$

33. a. The zero vector is the only vector in $\mathbb{R}^3$ such that
$$T\left(\begin{bmatrix} x \\ y \\ z \end{bmatrix} \right) = \begin{bmatrix} 0 \\ 0 \end{bmatrix}$$

   b. $T\left(\begin{bmatrix} 1 \\ -2 \\ 2 \end{bmatrix} \right) = \begin{bmatrix} 7 \\ -6 \\ -9 \end{bmatrix}$

35. $T(c\mathbf{v} + \mathbf{w}) = \begin{bmatrix} cT_1(\mathbf{v}) + T_1(\mathbf{w}) \\ cT_2(\mathbf{v}) + T_2(\mathbf{w}) \end{bmatrix}$

$= c\begin{bmatrix} T_1(\mathbf{v}) \\ T_2(\mathbf{v}) \end{bmatrix} + \begin{bmatrix} T_1(\mathbf{w}) \\ T_2(\mathbf{w}) \end{bmatrix}$

$= cT(\mathbf{v}) + T(\mathbf{w})$

37. $T(kA + C) = (kA + C)B - B(kA + C)$

$= kAB - kBA + CB - BC$

$= kT(A) + T(C)$

39. a. $T(cf + g) = \int_0^1 \left[cf(x) + g(x) \right] dx$

$= \int_0^1 cf(x)\, dx + \int_0^1 g(x)\, dx$

$= c\int_0^1 f(x)\, dx + \int_0^1 g(x)\, dx$

$= cT(f) + T(g)$

   b. $T(2x^2 - x + 3) = \frac{19}{6}$

41. Since neither $\mathbf{v}$ nor $\mathbf{w}$ is the zero vector, if either $T(\mathbf{v}) = 0$ or $T(\mathbf{w}) = 0$, then the conclusion holds. Now assume that $T(\mathbf{v})$ and $T(\mathbf{w})$ are linearly dependent and not zero; then there exist scalars a_0 and b_0, not both 0, such that $a_0 T(\mathbf{v}) + b_0 T(\mathbf{w}) = \mathbf{0}$. Since $\mathbf{v}$ and $\mathbf{w}$ are linearly independent, then $a_0 \mathbf{v} + b_0 \mathbf{w} \neq \mathbf{0}$ and since T is linear, then $T(a_0 \mathbf{v} + b_0 \mathbf{w}) = \mathbf{0}$.

43. Let $T(\mathbf{v}) = \mathbf{0}$ for all $\mathbf{v}$ in $\mathbb{R}^3$.

Section 4.2

1. Since $T(\mathbf{v}) = \begin{bmatrix} 0 \\ 0 \end{bmatrix}$, $\mathbf{v}$ is in $N(T)$.

3. Since $T(\mathbf{v}) = \begin{bmatrix} -5 \\ 10 \end{bmatrix}$, $\mathbf{v}$ is not in $N(T)$.

5. Since $T(p(x)) = 2x$, $p(x)$ is not in $N(T)$.

7. Since $T(p(x)) = -2x$, $p(x)$ is not in $N(T)$.

9. Since $T\left(\begin{bmatrix} -1 \\ 2 \\ 1 \end{bmatrix} \right) = \mathbf{v}$, $\mathbf{v}$ is in $R(T)$.

11. The vector $\mathbf{v}$ is not in $R(T)$.

13. The matrix A is in $R(T)$ with $a = 1, b = 0, c = -2$, $d = -1$.

15. The matrix A is not in $R(T)$.

17. $\left\{ \begin{bmatrix} 0 \\ 0 \end{bmatrix} \right\}$

19. $\left\{ \begin{bmatrix} -2 \\ 1 \\ 1 \end{bmatrix} \right\}$

21. $\left\{ \begin{bmatrix} 2 \\ 1 \\ 0 \end{bmatrix}, \begin{bmatrix} 1 \\ 0 \\ 1 \end{bmatrix} \right\}$

23. $\{x, x^2\}$

25. $\left\{ \begin{bmatrix} 1 \\ 0 \\ 2 \end{bmatrix}, \begin{bmatrix} 1 \\ 1 \\ 0 \end{bmatrix}, \begin{bmatrix} 2 \\ -1 \\ 1 \end{bmatrix} \right\}$

27. $\left\{ \begin{bmatrix} 1 \\ 0 \\ 0 \end{bmatrix}, \begin{bmatrix} 0 \\ 1 \\ 0 \end{bmatrix} \right\}$

29. $\{1, x, x^2\}$

31. a. No, $\begin{bmatrix} -6 \\ 5 \\ 0 \end{bmatrix}$ is not in $R(T)$.

    b. $\left\{ \begin{bmatrix} -2 \\ 1 \\ 1 \end{bmatrix}, \begin{bmatrix} 0 \\ 1 \\ -1 \end{bmatrix} \right\}$

    c. Since $\dim(N(T)) + \dim(R(T)) = \dim(\mathbb{R}^3) = 3$ and $\dim(R(T)) = 2$, then $\dim(N(T)) = 1$.

33. a. The polynomial $2x^2 - 4x + 6$ is not in $R(T)$.

    b. $\{-2x + 1, x^2 + x\} = \{T(x), T(x^2)\}$

35. $T\left(\begin{bmatrix} x \\ y \\ z \end{bmatrix} \right) = \begin{bmatrix} x \\ y \end{bmatrix}$

37. a. The range $R(T)$ is the subspace of $\mathcal{P}_n$ consisting of all polynomials of degree $n - 1$ or less.

    b. $\dim(R(T)) = n$

    c. $\dim(N(T)) = 1$

39. a. $\dim(R(T)) = 2$

    b. $\dim(N(T)) = 1$

41. $\left\{ \begin{bmatrix} 1 & 0 \\ 0 & 0 \end{bmatrix}, \begin{bmatrix} 0 & 0 \\ 0 & 1 \end{bmatrix} \right\}$

43. a. The range of T is the set of symmetric matrices.

    b. The null space of T is the set of skew-symmetric matrices.

45. If the matrix A is invertible, then $R(T) = M_{n \times n}$.

Section 4.3

1. T is one-to-one.

3. T is one-to-one.

5. T is one-to-one.

7. T is onto $\mathbb{R}^2$.

9. T is onto $\mathbb{R}^3$.

11. Is a basis

13. Is a basis

15. Is a basis

17. Is a basis

19. Is a basis

21. a. Since $\det(A) = \det\left(\begin{bmatrix} 1 & 0 \\ -2 & -3 \end{bmatrix} \right) = -3 \neq 0$, then T is an isomorphism.

    b. $A^{-1} = -\frac{1}{3} \begin{bmatrix} -3 & 0 \\ 2 & 1 \end{bmatrix}$

    c. $A^{-1} T\left(\begin{bmatrix} x \\ y \end{bmatrix} \right)$
$$= \begin{bmatrix} 1 & 0 \\ -\frac{2}{3} & -\frac{1}{3} \end{bmatrix} \begin{bmatrix} x \\ -2x - 3y \end{bmatrix}$$
$$= \begin{bmatrix} x \\ y \end{bmatrix}$$

23. a. Since
$$\det(A) = \det\left(\begin{bmatrix} -2 & 0 & 1 \\ 1 & -1 & -1 \\ 0 & 1 & 0 \end{bmatrix} \right)$$
$$= -1 \neq 0$$
then T is an isomorphism.

    b. $A^{-1} = \begin{bmatrix} -1 & -1 & -1 \\ 0 & 0 & 1 \\ -1 & -2 & -2 \end{bmatrix}$

    c. $A^{-1} T\left(\begin{bmatrix} x \\ y \\ z \end{bmatrix} \right)$
$$= \begin{bmatrix} -1 & -1 & -1 \\ 0 & 0 & 1 \\ -1 & -2 & -2 \end{bmatrix} \begin{bmatrix} -2x + z \\ x - y - z \\ y \end{bmatrix}$$
$$= \begin{bmatrix} x \\ y \\ z \end{bmatrix}$$

25. T is an isomorphism.

27. T is an isomorphism.

29. Since $T(cA + B) = (cA + B)^t = cA^t + B^t = cT(A) + T(B)$, T is linear. Since $T(A) = \mathbf{0}$ implies that $A = \mathbf{0}$, T is one-to-one. If B is a matrix in $M_{n \times n}$ and $A = B^t$, then $T(A) = T(B^t) = (B^t)^t = B$, so T is onto. Hence, T is an isomorphism.

31. Since $T(kB + C) = A(kB + C)A^{-1} = kABA^{-1} + ACA^{-1} = kT(B) + T(C)$, T is linear. Since $T(B) = ABA^{-1} = \mathbf{0}$ implies that $B = \mathbf{0}$, T is one-to-one. If C is

a matrix in $M_{n \times n}$ and $B = A^{-1}CA$, then
$T(B) = T(A^{-1}CA) = A(A^{-1}CA)A^{-1} = C$, so T is onto.
Hence, T is an isomorphism.

33.
$$T\left(\begin{bmatrix} a \\ b \\ c \\ d \end{bmatrix}\right) = ax^3 + bx^2 + cx + d$$

35. Since
$$V = \left\{ \begin{bmatrix} x \\ y \\ x+2y \end{bmatrix} \middle| x, y \in \mathbb{R} \right\}$$

define $T: V \to \mathbb{R}^2$ by
$$T\left(\begin{bmatrix} x \\ y \\ x+2y \end{bmatrix}\right) = \begin{bmatrix} x \\ y \end{bmatrix}$$

37. Let $\mathbf{v}$ be a nonzero vector in $\mathbb{R}^3$. Then a line L through the origin can be given by
$$L = \{t\mathbf{v} \mid t \in \mathbb{R}\}$$
Now, let $T: \mathbb{R}^3 \longrightarrow \mathbb{R}^3$ be an isomorphism. Since T is linear, $T(t\mathbf{v}) = tT(\mathbf{v})$. Also, by Theorem 8, $T(\mathbf{v})$ is nonzero. Hence, the set
$$L' = \{tT(\mathbf{v}) \mid t \in \mathbb{R}\}$$
is also a line in $\mathbb{R}^3$ through the origin. The proof for a plane is similar with the plane being given by
$$P = \{s\mathbf{u} + t\mathbf{v} \mid s, t \in \mathbb{R}\}$$
for two linearly independent vectors $\mathbf{u}$ and $\mathbf{v}$ in $\mathbb{R}^3$.

Section 4.4

1. a. $[T]_B = \begin{bmatrix} 5 & -1 \\ -1 & 1 \end{bmatrix}$

b. $T\begin{bmatrix} 2 \\ 1 \end{bmatrix} = \begin{bmatrix} 9 \\ -1 \end{bmatrix}$;

$T\begin{bmatrix} 2 \\ 1 \end{bmatrix} = \begin{bmatrix} 5 & -1 \\ -1 & 1 \end{bmatrix}\begin{bmatrix} 2 \\ 1 \end{bmatrix} = \begin{bmatrix} 9 \\ -1 \end{bmatrix}$

3. a. $[T]_B = \begin{bmatrix} -1 & 1 & 2 \\ 0 & 3 & 1 \\ 1 & 0 & -1 \end{bmatrix}$

b. $T\begin{bmatrix} 1 \\ -2 \\ 3 \end{bmatrix} = \begin{bmatrix} 3 \\ -3 \\ -2 \end{bmatrix} = [T]_B \begin{bmatrix} 1 \\ -2 \\ 3 \end{bmatrix}$

5. a. $[T]_B^{B'} = \begin{bmatrix} -3 & -2 \\ 3 & 6 \end{bmatrix}$

b. $T\begin{bmatrix} -1 \\ -2 \end{bmatrix} = \begin{bmatrix} -3 \\ -3 \end{bmatrix}$;

$T\begin{bmatrix} -1 \\ -2 \end{bmatrix} = \begin{bmatrix} -3 & -2 \\ 3 & 6 \end{bmatrix}\begin{bmatrix} -1 \\ -2 \end{bmatrix}_B$

$\qquad = \begin{bmatrix} -3 & -2 \\ 3 & 6 \end{bmatrix}\begin{bmatrix} 2 \\ -\frac{3}{2} \end{bmatrix} = \begin{bmatrix} -3 \\ -3 \end{bmatrix}$

7. a. $[T]_B^{B'} = \begin{bmatrix} -\frac{2}{3} & \frac{2}{3} \\ \frac{13}{6} & -\frac{5}{3} \end{bmatrix}$

b. $T\begin{bmatrix} -1 \\ -3 \end{bmatrix} = \begin{bmatrix} -2 \\ -4 \end{bmatrix}$;

$\left[T\begin{bmatrix} -1 \\ -3 \end{bmatrix}\right]_{B'} = [T]_B^{B'}\begin{bmatrix} -1 \\ -3 \end{bmatrix}_B = [T]_B^{B'}\begin{bmatrix} 2 \\ 1 \end{bmatrix}$

$\qquad = \begin{bmatrix} -\frac{2}{3} \\ \frac{8}{3} \end{bmatrix}$

$T\begin{bmatrix} -1 \\ -3 \end{bmatrix} = -\frac{2}{3}\begin{bmatrix} 3 \\ -2 \end{bmatrix} + \frac{8}{3}\begin{bmatrix} 0 \\ -2 \end{bmatrix} = \begin{bmatrix} -2 \\ -4 \end{bmatrix}$

9. a. $[T]_B^{B'} = \begin{bmatrix} 1 & 1 & 1 \\ 0 & -1 & -2 \\ 0 & 0 & 1 \end{bmatrix}$

b. $T(x^2 - 3x + 3) = x^2 - 3x + 3$;

$\left[T(x^2 - 3x + 3)\right]_{B'} = [T]_B^{B'}[x^2 - 3x + 3]_B$

$\qquad = [T]_B^{B'}\begin{bmatrix} 1 \\ 1 \\ 1 \end{bmatrix} = \begin{bmatrix} 3 \\ -3 \\ 1 \end{bmatrix}$

$T(x^2 - 3x + 3) = 3 - 3x + x^2$

11. If $A = \begin{bmatrix} a & b \\ c & -a \end{bmatrix}$, then $T(A) = \begin{bmatrix} 0 & -2b \\ 2c & 0 \end{bmatrix}$.

a. $[T]_B = \begin{bmatrix} 0 & 0 & 0 \\ 0 & -2 & 0 \\ 0 & 0 & 2 \end{bmatrix}$

b. $T\left(\begin{bmatrix} 2 & 1 \\ 3 & -2 \end{bmatrix}\right) = \begin{bmatrix} 0 & -2 \\ 6 & 0 \end{bmatrix}$;

$\left[T\left(\begin{bmatrix} 2 & 1 \\ 3 & -2 \end{bmatrix}\right)\right]_B = [T]_B\begin{bmatrix} 2 \\ 1 \\ 3 \end{bmatrix} = \begin{bmatrix} 0 \\ -2 \\ 6 \end{bmatrix}$

$$T\left(\begin{bmatrix} 2 & 1 \\ 3 & -2 \end{bmatrix}\right) = 0 \begin{bmatrix} 1 & 0 \\ 0 & -1 \end{bmatrix} - 2 \begin{bmatrix} 0 & 1 \\ 0 & 0 \end{bmatrix}$$

$$+ 6 \begin{bmatrix} 0 & 0 \\ 1 & 0 \end{bmatrix}$$

$$= \begin{bmatrix} 0 & -2 \\ 6 & 0 \end{bmatrix}$$

13. a. $[T]_B = \begin{bmatrix} 1 & 2 \\ 1 & -1 \end{bmatrix}$

  b. $[T]_{B'} = \dfrac{1}{9} \begin{bmatrix} 1 & 22 \\ 11 & -1 \end{bmatrix}$

  c. $[T]_B^{B'} = \dfrac{1}{9} \begin{bmatrix} 5 & -2 \\ 1 & 5 \end{bmatrix}$

  d. $[T]_{B'}^{B} = \dfrac{1}{3} \begin{bmatrix} 5 & 2 \\ -1 & 5 \end{bmatrix}$

  e. $[T]_C^{B'} = \dfrac{1}{9} \begin{bmatrix} -2 & 5 \\ 5 & 1 \end{bmatrix}$

  f. $[T]_{C'}^{B'} = \dfrac{1}{9} \begin{bmatrix} 22 & 1 \\ -1 & 11 \end{bmatrix}$

15. a. $[T]_B^{B'} = \begin{bmatrix} 0 & 0 \\ 1 & 0 \\ 0 & \frac{1}{2} \end{bmatrix}$

  b. $[T]_C^{B'} = \begin{bmatrix} 0 & 0 \\ 0 & 1 \\ \frac{1}{2} & 0 \end{bmatrix}$

  c. $[T]_C^{C'} = \begin{bmatrix} 0 & 1 \\ 0 & 0 \\ \frac{1}{2} & 0 \end{bmatrix}$

  d. $[S]_{B'}^{B} = \begin{bmatrix} 0 & 1 & 0 \\ 0 & 0 & 2 \end{bmatrix}$

  e. $[S]_{B'}^{B}[T]_B^{B'} = \begin{bmatrix} 1 & 0 \\ 0 & 1 \end{bmatrix}$

  $[T]_B^{B'}[S]_{B'}^{B} = \begin{bmatrix} 0 & 0 & 0 \\ 0 & 1 & 0 \\ 0 & 0 & 1 \end{bmatrix}$

  f. The function $S \circ T$ is the identity map; that is, $(S \circ T)(ax + b) = ax + b$ so S reverses the action of T.

17. $[T]_B = \begin{bmatrix} 1 & 0 \\ 0 & -1 \end{bmatrix}$

  The transformation T reflects a vector across the x-axis.

19. $[T]_B = cI$

21. $[T]_B^{B'} = [1\ 0\ 0\ 1]$

23. a. $[2T + S]_B = 2[T]_B + [S]_B = \begin{bmatrix} 5 & 2 \\ -1 & 7 \end{bmatrix}$

  b. $\begin{bmatrix} -4 \\ 23 \end{bmatrix}$

25. a. $[S \circ T]_B = [S]_B[T]_B = \begin{bmatrix} 2 & 1 \\ 1 & 4 \end{bmatrix}$

  b. $\begin{bmatrix} -1 \\ 10 \end{bmatrix}$

27. a. $[-3T + 2S]_B = \begin{bmatrix} 3 & 3 & 1 \\ 2 & -6 & -6 \\ 3 & -3 & -1 \end{bmatrix}$

  b. $\begin{bmatrix} 3 \\ -26 \\ -9 \end{bmatrix}$

29. a. $[S \circ T]_B = \begin{bmatrix} 4 & -4 & -4 \\ 1 & -1 & -1 \\ -1 & 1 & 1 \end{bmatrix}$

  b. $\begin{bmatrix} -20 \\ -5 \\ 5 \end{bmatrix}$

31. $[T]_B = \begin{bmatrix} 0 & 0 & 0 & 6 & 0 \\ 0 & 0 & 0 & 0 & 24 \\ 0 & 0 & 0 & 0 & 0 \\ 0 & 0 & 0 & 0 & 0 \\ 0 & 0 & 0 & 0 & 0 \end{bmatrix}$

  $[T(p(x)]_B = \begin{bmatrix} -12 \\ -48 \\ 0 \\ 0 \\ 0 \end{bmatrix}$

  $T(p(x)) = p'''(x) = -12 - 48x$

33. $[S]_B^{B'} = \begin{bmatrix} 0 & 0 & 0 \\ 1 & 0 & 0 \\ 0 & 1 & 0 \\ 0 & 0 & 1 \end{bmatrix}$

  $[D]_B^{B'} = \begin{bmatrix} 0 & 1 & 0 & 0 \\ 0 & 0 & 2 & 0 \\ 0 & 0 & 0 & 3 \end{bmatrix}$

  $[D]_B^{B'}[S]_B^{B'} = \begin{bmatrix} 1 & 0 & 0 \\ 0 & 2 & 0 \\ 0 & 0 & 3 \end{bmatrix} = [T]_B$

35. If $A = \begin{bmatrix} a & b \\ c & d \end{bmatrix}$, then the matrix representation for T is

$$[T]_S = \begin{bmatrix} 0 & -c & b & 0 \\ -b & a-d & 0 & b \\ c & 0 & d-a & -c \\ 0 & c & -b & 0 \end{bmatrix}$$

37. $[T]_B = \begin{bmatrix} 1 & 1 & 0 & 0 & \cdots & \cdots & 0 & 0 \\ 0 & 1 & 1 & 0 & \cdots & \cdots & 0 & 0 \\ 0 & 0 & 1 & 1 & \cdots & \cdots & 0 & 0 \\ \vdots & \vdots & \vdots & \vdots & \vdots & \vdots & \vdots & \vdots \\ 0 & 0 & 0 & 0 & 0 & \cdots & 1 & 0 \\ 0 & 0 & 0 & 0 & 0 & \cdots & 1 & 1 \\ 0 & 0 & 0 & 0 & 0 & \cdots & 0 & 1 \end{bmatrix}$

Section 4.5

1.

$$[T]_{B_1}[\mathbf{v}]_{B_1} = \begin{bmatrix} 1 & 2 \\ -1 & 3 \end{bmatrix}\begin{bmatrix} 4 \\ -1 \end{bmatrix} = \begin{bmatrix} 2 \\ -7 \end{bmatrix}$$

$$[T]_{B_2}[\mathbf{v}]_{B_2} = \begin{bmatrix} 2 & 1 \\ -1 & 2 \end{bmatrix}\begin{bmatrix} -1 \\ -5 \end{bmatrix} = \begin{bmatrix} -7 \\ 9 \end{bmatrix}$$

To show the results are the same, observe that

$$-7\begin{bmatrix} 1 \\ 1 \end{bmatrix} + (-9)\begin{bmatrix} -1 \\ 0 \end{bmatrix} = \begin{bmatrix} 2 \\ -7 \end{bmatrix}.$$

3. a. $[T]_{B_1} = \begin{bmatrix} 1 & 1 \\ 1 & 1 \end{bmatrix}$; $[T]_{B_2} = \begin{bmatrix} 2 & 0 \\ 0 & 0 \end{bmatrix}$

b.

$$[T]_{B_1}[\mathbf{v}]_{B_1} = \begin{bmatrix} 1 & 1 \\ 1 & 1 \end{bmatrix}\begin{bmatrix} 3 \\ -2 \end{bmatrix} = \begin{bmatrix} 1 \\ 1 \end{bmatrix}$$

$$[T]_{B_2}[\mathbf{v}]_{B_2} = \begin{bmatrix} 2 & 0 \\ 0 & 0 \end{bmatrix}\begin{bmatrix} \frac{1}{2} \\ -\frac{5}{2} \end{bmatrix} = \begin{bmatrix} 1 \\ 0 \end{bmatrix}$$

To show the results are the same, observe that

$$1\begin{bmatrix} 1 \\ 1 \end{bmatrix} + 0\begin{bmatrix} -1 \\ 1 \end{bmatrix} = \begin{bmatrix} 1 \\ 1 \end{bmatrix}$$

5. a. $[T]_{B_1} = \begin{bmatrix} 1 & 0 & 0 \\ 0 & 0 & 0 \\ 0 & 0 & 1 \end{bmatrix}$

$$[T]_{B_2} = \begin{bmatrix} 1 & -1 & 0 \\ 0 & 0 & 0 \\ 0 & 1 & 1 \end{bmatrix}$$

b.

$$[T]_{B_1}[\mathbf{v}]_{B_1} = \begin{bmatrix} 1 & 0 & 0 \\ 0 & 0 & 0 \\ 0 & 0 & 1 \end{bmatrix}\begin{bmatrix} 1 \\ 2 \\ -1 \end{bmatrix} = \begin{bmatrix} 1 \\ 0 \\ -1 \end{bmatrix}$$

$$[T]_{B_2}[\mathbf{v}]_{B_2} = \begin{bmatrix} 1 & -1 & 0 \\ 0 & 0 & 0 \\ 0 & 1 & 1 \end{bmatrix}\begin{bmatrix} 3 \\ 2 \\ -4 \end{bmatrix} = \begin{bmatrix} 1 \\ 0 \\ -2 \end{bmatrix}$$

To show that the results are the same, observe that

$$1\begin{bmatrix} 1 \\ 0 \\ 1 \end{bmatrix} + 0\begin{bmatrix} -1 \\ 1 \\ 0 \end{bmatrix} + (-2)\begin{bmatrix} 0 \\ 0 \\ 1 \end{bmatrix} = \begin{bmatrix} 1 \\ 0 \\ -1 \end{bmatrix}$$

7. $P = [I]_{B_2}^{B_1} = \begin{bmatrix} 3 & -1 \\ -1 & 1 \end{bmatrix}$

$$[T]_{B_2} = P^{-1}[T]_{B_1}P$$

$$= \frac{1}{2}\begin{bmatrix} 1 & 1 \\ 1 & 3 \end{bmatrix}\begin{bmatrix} 1 & 1 \\ 3 & 2 \end{bmatrix}\begin{bmatrix} 3 & -1 \\ -1 & 1 \end{bmatrix}$$

$$= \begin{bmatrix} \frac{9}{2} & -\frac{1}{2} \\ \frac{23}{2} & -\frac{3}{2} \end{bmatrix}$$

9. $P = [I]_{B_2}^{B_1} = \begin{bmatrix} \frac{1}{3} & 1 \\ \frac{1}{3} & -1 \end{bmatrix}$

$$[T]_{B_2} = P^{-1}[T]_{B_1}P$$

$$= \begin{bmatrix} \frac{3}{2} & \frac{3}{2} \\ \frac{1}{2} & -\frac{1}{2} \end{bmatrix}\begin{bmatrix} 1 & 0 \\ 0 & -1 \end{bmatrix}\begin{bmatrix} \frac{1}{3} & 1 \\ \frac{1}{3} & -1 \end{bmatrix}$$

$$= \begin{bmatrix} 0 & 3 \\ \frac{1}{3} & 0 \end{bmatrix}$$

11. $P = [I]_{B_2}^{B_1} = \begin{bmatrix} 2 & 1 \\ 3 & 2 \end{bmatrix}$

$$[T]_{B_2} = P^{-1}[T]_{B_1}P$$

$$= \begin{bmatrix} 2 & -1 \\ -3 & 2 \end{bmatrix}\begin{bmatrix} 2 & 0 \\ 0 & 3 \end{bmatrix}\begin{bmatrix} 2 & 1 \\ 3 & 2 \end{bmatrix}$$

$$= \begin{bmatrix} -1 & -2 \\ 6 & 6 \end{bmatrix}$$

13. $[T]_{B_1} = \begin{bmatrix} 1 & -1 \\ -2 & 1 \end{bmatrix}$

$$P = [I]_{B_2}^{B_1} = \begin{bmatrix} -1 & -1 \\ 2 & 1 \end{bmatrix}$$

$$[T]_{B_2} = P^{-1}[T]_{B_1}P$$

$$= \begin{bmatrix} 1 & 1 \\ -2 & -1 \end{bmatrix}\begin{bmatrix} 1 & -1 \\ -2 & 1 \end{bmatrix}\begin{bmatrix} -1 & -1 \\ 2 & 1 \end{bmatrix}$$

$$= \begin{bmatrix} 1 & 1 \\ 2 & 1 \end{bmatrix}$$

At top right:

$$[T]_{B_2}[\mathbf{v}]_{B_2} = \begin{bmatrix} 1 & -1 & 0 \\ 0 & 0 & 0 \\ 0 & 1 & 1 \end{bmatrix}\begin{bmatrix} 3 \\ 2 \\ -4 \end{bmatrix} = \begin{bmatrix} 1 \\ 0 \\ -2 \end{bmatrix}$$

To show that the results are the same, observe that

$$1\begin{bmatrix} 1 \\ 0 \\ 1 \end{bmatrix} + 0\begin{bmatrix} -1 \\ 1 \\ 0 \end{bmatrix} + (-2)\begin{bmatrix} 0 \\ 0 \\ 1 \end{bmatrix} = \begin{bmatrix} 1 \\ 0 \\ -1 \end{bmatrix}$$

15. $[T]_{B_1} = \begin{bmatrix} 0 & 1 & 0 \\ 0 & 0 & 2 \\ 0 & 0 & 0 \end{bmatrix}$

$[T]_{B_2} = \begin{bmatrix} 0 & 2 & 0 \\ 0 & 0 & 1 \\ 0 & 0 & 0 \end{bmatrix}$

If

$$P = [I]_{B_2}^{B_1} = \begin{bmatrix} 1 & 0 & -2 \\ 0 & 2 & 0 \\ 0 & 0 & 1 \end{bmatrix}$$

then $[T]_{B_2} = P^{-1}[T]_{B_1}P$.

17. Since A and B are similar, there is an invertible matrix P such that $B = P^{-1}AP$. Also since B and C are similar, there is an invertible matrix Q such that $C = Q^{-1}BQ$. Therefore, $C = Q^{-1}P^{-1}APQ = (PQ)^{-1}A(PQ)$ so that A and C are also similar.

19. For any square matrices A and B the trace function satisfies the property $\mathbf{tr}(AB) = \mathbf{tr}(BA)$. Now, since A and B are similar matrices, there exists an invertible matrix P such that $B = P^{-1}AP$. Hence,

$$\mathbf{tr}(B) = \mathbf{tr}(P^{-1}AP) = \mathbf{tr}(APP^{-1}) = \mathbf{tr}(A)$$

21. Since A and B are similar matrices, there exists an invertible matrix P such that $B = P^{-1}AP$. Hence,

$$B^n = (P^{-1}AP)^n = P^{-1}A^nP$$

Thus, A^n and B^n are similar.

Section 4.6

1. a. $\begin{bmatrix} 1 & 0 \\ 0 & -1 \end{bmatrix}$

b. $\begin{bmatrix} -1 & 0 \\ 0 & 1 \end{bmatrix}$

c. $\begin{bmatrix} 1 & 0 \\ 0 & 3 \end{bmatrix}$

3. a. $[T]_S = \begin{bmatrix} 3 & 0 \\ 0 & -\frac{1}{2} \end{bmatrix}$

b.

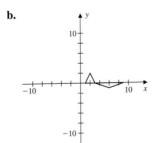

c. $[T]_S^{-1} = \begin{bmatrix} \frac{1}{3} & 0 \\ 0 & -2 \end{bmatrix}$

5. a. $[T]_S = \begin{bmatrix} -\sqrt{2}/2 & \sqrt{2}/2 \\ -\sqrt{2}/2 & -\sqrt{2}/2 \end{bmatrix}$

b.

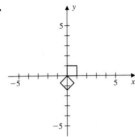

c. $[T]_S^{-1} = \begin{bmatrix} -\sqrt{2}/2 & -\sqrt{2}/2 \\ \sqrt{2}/2 & -\sqrt{2}/2 \end{bmatrix}$

7. a. $\begin{bmatrix} \sqrt{3}/2 & -1/2 & \sqrt{3}/2 - 1/2 \\ 1/2 & \sqrt{3}/2 & \sqrt{3}/2 + 1/2 \\ 0 & 0 & 1 \end{bmatrix}$

b.

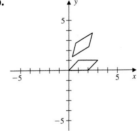

c. $\begin{bmatrix} \sqrt{3}/2 & 1/2 & -1 \\ -1/2 & \sqrt{3}/2 & -1 \\ 0 & 0 & 1 \end{bmatrix}$

9. a. $\left[\begin{bmatrix} 0 \\ 0 \end{bmatrix}\right]_B = \begin{bmatrix} 0 \\ 0 \end{bmatrix}$

$\left[\begin{bmatrix} 2 \\ 2 \end{bmatrix}\right]_B = \begin{bmatrix} 2 \\ 0 \end{bmatrix}$

$\left[\begin{bmatrix} 0 \\ 2 \end{bmatrix}\right]_B = \begin{bmatrix} 1 \\ 1 \end{bmatrix}$

b. $[T]_B^S = \begin{bmatrix} 1 & 1 \\ 1 & -1 \end{bmatrix}$

c. $\begin{bmatrix} 1 & 1 \\ 1 & -1 \end{bmatrix}\begin{bmatrix} 0 \\ 0 \end{bmatrix} = \begin{bmatrix} 0 \\ 0 \end{bmatrix}$

$\begin{bmatrix} 1 & 1 \\ 1 & -1 \end{bmatrix}\begin{bmatrix} 2 \\ 0 \end{bmatrix} = \begin{bmatrix} 2 \\ 2 \end{bmatrix}$

$$\begin{bmatrix} 1 & 1 \\ 1 & -1 \end{bmatrix} \begin{bmatrix} 1 \\ 1 \end{bmatrix} = \begin{bmatrix} 2 \\ 0 \end{bmatrix}$$

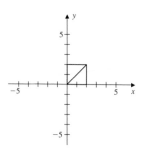

The original triangle is reflected across the line $y = x$.

d. $\begin{bmatrix} 0 & 1 \\ 1 & 0 \end{bmatrix} \begin{bmatrix} 0 \\ 0 \end{bmatrix} = \begin{bmatrix} 0 \\ 0 \end{bmatrix}$

$\begin{bmatrix} 0 & 1 \\ 1 & 0 \end{bmatrix} \begin{bmatrix} 2 \\ 2 \end{bmatrix} = \begin{bmatrix} 2 \\ 2 \end{bmatrix}$

$\begin{bmatrix} 0 & 1 \\ 1 & 0 \end{bmatrix} \begin{bmatrix} 0 \\ 2 \end{bmatrix} = \begin{bmatrix} 2 \\ 0 \end{bmatrix}$

Review Exercises Chapter 4

1. a. The vectors are not scalar multiples, so S is a basis.

b. $T \begin{bmatrix} x \\ y \end{bmatrix} = \begin{bmatrix} x \\ x+y \\ x-y \\ 2y \end{bmatrix}$

c. $N(T) = \{\mathbf{0}\}$

d. Since $N(T) = \{\mathbf{0}\}$, T is one-to-one.

e. $\left\{ \begin{bmatrix} 1 \\ 1 \\ 1 \\ 2 \end{bmatrix}, \begin{bmatrix} 0 \\ 1 \\ -1 \\ 2 \end{bmatrix} \right\}$

f. No, T is not onto since $\dim(R(T)) = 2$ and $\dim(\mathbb{R}^4) = 4$. Also $\begin{bmatrix} a \\ b \\ c \\ d \end{bmatrix}$ is in $R(T)$ if and only if $c + b - 2a = 0$.

g. $\left\{ \begin{bmatrix} 1 \\ 0 \\ 1 \\ 1 \end{bmatrix}, \begin{bmatrix} -1 \\ 1 \\ 0 \\ 1 \end{bmatrix}, \begin{bmatrix} 1 \\ 0 \\ 0 \\ 0 \end{bmatrix}, \begin{bmatrix} 0 \\ 1 \\ 0 \\ 0 \end{bmatrix} \right\}$

h. $[T]_B^C = \begin{bmatrix} -1 & 2 \\ 5 & -4 \\ 7 & -5 \\ -2 & 4 \end{bmatrix}$

i. $\left[A \begin{bmatrix} x \\ y \end{bmatrix} \right]_C = \begin{bmatrix} -1 & 2 \\ 5 & -4 \\ 7 & -5 \\ -2 & 4 \end{bmatrix} \left[\begin{bmatrix} x \\ y \end{bmatrix} \right]_B$

$= \begin{bmatrix} -1 & 2 \\ 5 & -4 \\ 7 & -5 \\ -2 & 4 \end{bmatrix} \begin{bmatrix} \frac{1}{3}x + \frac{1}{3}y \\ \frac{2}{3}x - \frac{1}{3}y \end{bmatrix}$

$= \begin{bmatrix} x - y \\ -x + 3y \\ -x + 4y \\ 2x - 2y \end{bmatrix}$

This implies that $A \begin{bmatrix} x \\ y \end{bmatrix} = \begin{bmatrix} x \\ x+y \\ x-y \\ 2y \end{bmatrix}$.

3. a. $S \begin{bmatrix} x \\ y \end{bmatrix} = \begin{bmatrix} x \\ -y \end{bmatrix}; T \begin{bmatrix} x \\ y \end{bmatrix} = \begin{bmatrix} -x \\ y \end{bmatrix}$

b. $[S]_B = \begin{bmatrix} 1 & 0 \\ 0 & -1 \end{bmatrix}; [T]_B = \begin{bmatrix} -1 & 0 \\ 0 & 1 \end{bmatrix}$

c. $[T \circ S]_B = \begin{bmatrix} -1 & 0 \\ 0 & -1 \end{bmatrix} = [S \circ T]_B$

The linear operators $S \circ T$ and $T \circ S$ reflect a vector through the origin.

5. a. $[T]_B = \begin{bmatrix} 0 & 1 \\ 1 & 0 \end{bmatrix}$

b. $[T]_B^{B'} = \begin{bmatrix} 1 & 0 \\ 0 & 1 \end{bmatrix}$

7. a. $[T]_B = \begin{bmatrix} 1 & 0 & 0 \\ 0 & 0 & 1 \\ 0 & 1 & 0 \end{bmatrix}$

b. $\left[T \begin{bmatrix} -1 \\ 2 \\ 1 \end{bmatrix} \right]_B = \begin{bmatrix} -1 \\ 1 \\ 2 \end{bmatrix} = T \begin{bmatrix} -1 \\ 2 \\ 1 \end{bmatrix}$

c. $N(T) = \left\{ \begin{bmatrix} 0 \\ 0 \\ 0 \end{bmatrix} \right\}$

d. $R(T) = \mathbb{R}^3$

e. $[T^n]_B = \begin{bmatrix} 1 & 0 & 0 \\ 0 & 0 & 1 \\ 0 & 1 & 0 \end{bmatrix}^n$

9. Since $T^2 - T + I = 0, T - T^2 = I$. Then

$$(T \circ (I - T))(\mathbf{v}) = T((I - T)(\mathbf{v})) = T(\mathbf{v} - T(\mathbf{v}))$$
$$= T(\mathbf{v}) - T^2(\mathbf{v}) = I(\mathbf{v}) = \mathbf{v}$$

Chapter Test: Chapter 4

1. F		**2.** F	
3. T		**4.** T	
5. T		**6.** F	
7. F		**8.** F	
9. T		**10.** F	
11. T		**12.** F	
13. T		**14.** T	
15. T		**16.** T	
17. T		**18.** T	
19. T		**20.** T	
21. F		**22.** F	
23. T		**24.** F	
25. T		**26.** F	
27. F		**28.** T	
29. F		**30.** T	
31. F		**32.** F	
33. T		**34.** F	
35. F		**36.** T	
37. T		**38.** T	
39. F		**40.** T	

Chapter 5

Section 5.1

1. $\lambda = 3$

3. $\lambda = 0$

5. $\lambda = 1$

7. a. $\lambda^2 + 5\lambda = 0$

b. $\lambda_1 = 0, \lambda_2 = -5$

c. $\mathbf{v}_1 = \begin{bmatrix} 1 \\ 1 \end{bmatrix}, \mathbf{v}_2 = \begin{bmatrix} -2 \\ 3 \end{bmatrix}$

d. $\begin{bmatrix} -2 & 2 \\ 3 & -3 \end{bmatrix} \begin{bmatrix} 1 \\ 1 \end{bmatrix} = \begin{bmatrix} 0 \\ 0 \end{bmatrix} = 0 \begin{bmatrix} 1 \\ 1 \end{bmatrix};$

$\begin{bmatrix} -2 & 2 \\ 3 & -3 \end{bmatrix} \begin{bmatrix} -2 \\ 3 \end{bmatrix} = \begin{bmatrix} 10 \\ -15 \end{bmatrix} = -5 \begin{bmatrix} -2 \\ 3 \end{bmatrix}$

9. a. $(\lambda - 1)^2 = 0$

b. $\lambda_1 = 1$

c. $\mathbf{v}_1 = \begin{bmatrix} 1 \\ 0 \end{bmatrix}$

d. $\begin{bmatrix} 1 & -2 \\ 0 & 1 \end{bmatrix} \begin{bmatrix} 1 \\ 0 \end{bmatrix} = \begin{bmatrix} 1 \\ 0 \end{bmatrix} = 1 \begin{bmatrix} 1 \\ 0 \end{bmatrix}$

11. a. $(\lambda + 1)^2(\lambda - 1) = 0$

b. $\lambda_1 = -1, \lambda_2 = 1$

c. $\mathbf{v}_1 = \begin{bmatrix} 1 \\ 0 \\ 0 \end{bmatrix}, \mathbf{v}_2 = \begin{bmatrix} 1 \\ 2 \\ 2 \end{bmatrix}$

d. $\begin{bmatrix} -1 & 0 & 1 \\ 0 & 1 & 0 \\ 0 & 2 & -1 \end{bmatrix} \begin{bmatrix} 1 \\ 0 \\ 0 \end{bmatrix} = \begin{bmatrix} -1 \\ 0 \\ 0 \end{bmatrix} = -1 \begin{bmatrix} 1 \\ 0 \\ 0 \end{bmatrix}$

$\begin{bmatrix} -1 & 0 & 1 \\ 0 & 1 & 0 \\ 0 & 2 & -1 \end{bmatrix} \begin{bmatrix} 1 \\ 2 \\ 2 \end{bmatrix} = \begin{bmatrix} 1 \\ 2 \\ 2 \end{bmatrix} = 1 \begin{bmatrix} 1 \\ 2 \\ 2 \end{bmatrix}$

13. a. $(\lambda - 2)(\lambda - 1)^2 = 0$

b. $\lambda_1 = 2, \lambda_2 = 1$

c. $\mathbf{v}_1 = \begin{bmatrix} 1 \\ 0 \\ 0 \end{bmatrix}, \mathbf{v}_2 = \begin{bmatrix} -3 \\ 1 \\ 1 \end{bmatrix}$

d. $\begin{bmatrix} 2 & 1 & 2 \\ 0 & 2 & -1 \\ 0 & 1 & 0 \end{bmatrix} \begin{bmatrix} 1 \\ 0 \\ 0 \end{bmatrix} = \begin{bmatrix} 2 \\ 0 \\ 0 \end{bmatrix} = 2 \begin{bmatrix} 1 \\ 0 \\ 0 \end{bmatrix}$

$\begin{bmatrix} 2 & 1 & 2 \\ 0 & 2 & -1 \\ 0 & 1 & 0 \end{bmatrix} \begin{bmatrix} -3 \\ 1 \\ 1 \end{bmatrix} = \begin{bmatrix} -3 \\ 1 \\ 1 \end{bmatrix} = 1 \begin{bmatrix} -3 \\ 1 \\ 1 \end{bmatrix}$

15. a. $(\lambda + 1)(\lambda - 2)(\lambda + 2)(\lambda - 4) = 0$

b. $\lambda_1 = -1, \lambda_2 = 2, \lambda_3 = -2, \lambda_4 = 4$

c. $\mathbf{v}_1 = \begin{bmatrix} 1 \\ 0 \\ 0 \\ 0 \end{bmatrix}, \mathbf{v}_2 = \begin{bmatrix} 0 \\ 1 \\ 0 \\ 0 \end{bmatrix}, \mathbf{v}_3 = \begin{bmatrix} 0 \\ 0 \\ 1 \\ 0 \end{bmatrix},$

$\mathbf{v}_4 = \begin{bmatrix} 0 \\ 0 \\ 0 \\ 1 \end{bmatrix}$

d.

$$\begin{bmatrix} -1 & 0 & 0 & 0 \\ 0 & 2 & 0 & 0 \\ 0 & 0 & -2 & 0 \\ 0 & 0 & 0 & 4 \end{bmatrix} \begin{bmatrix} 1 \\ 0 \\ 0 \\ 0 \end{bmatrix} = \begin{bmatrix} -1 \\ 0 \\ 0 \\ 0 \end{bmatrix}$$

$$= -1 \begin{bmatrix} 1 \\ 0 \\ 0 \\ 0 \end{bmatrix}$$

The other cases are similar.

17. Let $A = \begin{bmatrix} a & b \\ c & d \end{bmatrix}$. The characteristic equation is

$(a - \lambda)(d - \lambda) - bc = 0$, which simplifies to $\lambda^2 - (a + d)\lambda + (ad - bc) = 0$. Observe that the coefficient of λ is $-(a + d)$, which is equal to $-\mathbf{tr}(A)$. Also, the constant term $ad - bc$ is equal to $\det(A)$.

19. Suppose A is not invertible. Then the homogeneous equation $A\mathbf{x} = \mathbf{0}$ has a nontrivial solution $\mathbf{x}_0$. Observe that $\mathbf{x}_0$ is an eigenvector of A corresponding to the eigenvalue $\lambda = 0$ since $A\mathbf{x}_0 = \mathbf{0} = 0\mathbf{x}_0$. On the other hand, suppose that $\lambda = 0$ is an eigenvalue of A. Then there exists a nonzero vector $\mathbf{x}_0$ such that $A\mathbf{x}_0 = \mathbf{0}$, so A is not invertible.

21. Let A be such that $A^2 = A$, and let λ be an eigenvalue of A with corresponding eigenvector $\mathbf{v}$ so that $A\mathbf{v} = \lambda\mathbf{v}$. Then $A^2\mathbf{v} = \lambda A\mathbf{v}$, so $A\mathbf{v} = \lambda^2\mathbf{v}$. The two equations

$$A\mathbf{v} = \lambda\mathbf{v} \quad \text{and} \quad A\mathbf{v} = \lambda^2\mathbf{v}$$

imply that $\lambda^2\mathbf{v} = \lambda\mathbf{v}$, so that $(\lambda^2 - \lambda)\mathbf{v} = \mathbf{0}$. Since $\mathbf{v} \neq \mathbf{0}$, then $\lambda(\lambda - 1) = 0$, so that either $\lambda = 0$ or $\lambda = 1$.

23. Let A be such that $A^n = \mathbf{0}$ for some n, and let λ be an eigenvalue of A with corresponding eigenvector $\mathbf{v}$, so that $A\mathbf{v} = \lambda\mathbf{v}$. Then $A^2\mathbf{v} = \lambda A\mathbf{v} = \lambda^2\mathbf{v}$. Continuing in this way, we see that $A^n\mathbf{v} = \lambda^n\mathbf{v}$. Since $A^n = \mathbf{0}$, then $\lambda^n\mathbf{v} = \mathbf{0}$. Since $\mathbf{v} \neq \mathbf{0}$, then $\lambda^n = 0$, so that $\lambda = 0$.

25. If A is invertible, then

$$\det(AB - \lambda I) = \det(A^{-1}(AB - \lambda I)A)$$
$$= \det(BA - \lambda I)$$

27. Since

$$\det(A - \lambda I) = (\lambda - a_{11})(\lambda - a_{22}) \cdots (\lambda - a_{nn})$$

the eigenvalues are the diagonal entries.

29. Let λ be an eigenvalue of C with corresponding eigenvector $\mathbf{v}$. Let $C = B^{-1}AB$. Since $C\mathbf{v} = \lambda\mathbf{v}$, then $B^{-1}AB\mathbf{v} = \lambda\mathbf{v}$. Then $A(B\mathbf{v}) = \lambda(B\mathbf{v})$. Therefore, $B\mathbf{v}$ is an eigenvector of A corresponding to λ.

31. Let $T\begin{bmatrix} x \\ y \end{bmatrix} = \begin{bmatrix} x \\ -y \end{bmatrix}$. The eigenvalues are $\lambda = 1$ and $\lambda = -1$ with corresponding eigenvectors $\begin{bmatrix} 1 \\ 0 \end{bmatrix}$ and $\begin{bmatrix} 0 \\ 1 \end{bmatrix}$, respectively.

33. If $\theta \neq 0$ or $\theta \neq \pi$, then T can only be described as a rotation. Hence, $T\begin{bmatrix} x \\ y \end{bmatrix}$ cannot be expressed by scalar multiplication as this only performs a contraction or a dilation. When $\theta = 0$, then T is the identity map $T\begin{bmatrix} x \\ y \end{bmatrix} = \begin{bmatrix} x \\ y \end{bmatrix}$. In this case every vector in $\mathbb{R}^2$ is an eigenvector with corresponding eigenvalue equal to 1. Also, if $\theta = \pi$, then $T\begin{bmatrix} x \\ y \end{bmatrix} = \begin{bmatrix} -1 & 0 \\ 0 & -1 \end{bmatrix}\begin{bmatrix} x \\ y \end{bmatrix}$. In this case every vector in $\mathbb{R}^2$ is an eigenvector with eigenvalue equal to -1.

35. a. $[T]_B = \begin{bmatrix} -\frac{1}{2} & \frac{1}{2} & 0 \\ -\frac{1}{2} & \frac{1}{2} & 0 \\ -1 & -1 & 1 \end{bmatrix}$

b. $[T]_{B'} = \begin{bmatrix} 1 & 1 & 0 \\ -1 & -1 & 0 \\ -1 & 0 & 1 \end{bmatrix}$

c. The characteristic polynomial for the matrices in parts (a) and (b) is given by $p(x) = x^3 - x^2$. Hence, the eigenvalues are the same.

Section 5.2

1. $P^{-1}AP = \begin{bmatrix} 1 & 0 \\ 0 & -3 \end{bmatrix}$

3. $P^{-1}AP = \begin{bmatrix} 0 & 0 & 0 \\ 0 & -2 & 0 \\ 0 & 0 & 1 \end{bmatrix}$

5. Eigenvalues: $-2, -1$; A is diagonalizable since there are two distinct eigenvalues.

7. Eigenvalues: -1 with multiplicity 2; eigenvectors: $\begin{bmatrix} 1 \\ 0 \end{bmatrix}$; A is not diagonalizable.

9. Eigenvalues: $1, 0$; A is diagonalizable since there are two distinct eigenvalues.

11. Eigenvalues: $3, 4, 0$; A is diagonalizable since there are three distinct eigenvalues.

13. Eigenvalues: -1 and 2 with multiplicity 2;

eigenvectors: $\begin{bmatrix} 1 \\ -5 \\ 2 \end{bmatrix}, \begin{bmatrix} -1 \\ -1 \\ 1 \end{bmatrix}$; A is not

diagonalizable since there are only two linearly independent eigenvectors.

15. Eigenvalues: 1 and 0 with multiplicity 2; eigenvectors:

$\begin{bmatrix} -1 \\ 1 \\ 1 \end{bmatrix}, \begin{bmatrix} 0 \\ 1 \\ 0 \end{bmatrix}, \begin{bmatrix} 0 \\ 0 \\ 1 \end{bmatrix}$; A is diagonalizable since

there are three linearly independent eigenvectors.

17. Eigenvalues: $-1, 2, 0$ with multiplicity 2;

eigenvectors: $\begin{bmatrix} 0 \\ -1 \\ 1 \\ 0 \end{bmatrix}, \begin{bmatrix} 0 \\ 1 \\ 2 \\ 3 \end{bmatrix}, \begin{bmatrix} 0 \\ -1 \\ 0 \\ 1 \end{bmatrix}, \begin{bmatrix} -1 \\ 0 \\ 1 \\ 0 \end{bmatrix}$;

A is diagonalizable since there are four linearly independent eigenvectors.

19. $P = \begin{bmatrix} -3 & 0 \\ 1 & 1 \end{bmatrix}$; $P^{-1}AP = \begin{bmatrix} 2 & 0 \\ 0 & -1 \end{bmatrix}$

21. $P = \begin{bmatrix} 0 & 2 & 0 \\ 1 & 1 & 1 \\ 1 & 3 & 2 \end{bmatrix}$

$P^{-1}AP = \begin{bmatrix} -1 & 0 & 0 \\ 0 & 1 & 0 \\ 0 & 0 & 0 \end{bmatrix}$

23. $P = \begin{bmatrix} 2 & 0 & 0 \\ 1 & 1 & 0 \\ 0 & 0 & 1 \end{bmatrix}$

$P^{-1}AP = \begin{bmatrix} -1 & 0 & 0 \\ 0 & 1 & 0 \\ 0 & 0 & 1 \end{bmatrix}$

25. $P = \begin{bmatrix} -1 & 0 & -1 & 1 \\ 0 & 1 & 0 & 0 \\ 0 & 0 & 1 & 1 \\ 1 & 0 & 0 & 0 \end{bmatrix}$

$P^{-1}AP = \begin{bmatrix} 1 & 0 & 0 & 0 \\ 0 & 1 & 0 & 0 \\ 0 & 0 & 0 & 0 \\ 0 & 0 & 0 & 2 \end{bmatrix}$

27. By induction. If $k = 1$, then $A^k = A = PDP^{-1} = PD^kP^{-1}$. Suppose the result holds for a natural number k. Then

$$
\begin{aligned}
A^{k+1} &= (PDP^{-1})^{k+1} \\
&= (PDP^{-1})^k(PDP^{-1}) \\
&= (PD^kP^{-1})(PDP^{-1}) \\
&= (PD^k)(P^{-1}P)(DP^{-1}) \\
&= PD^{k+1}P^{-1}
\end{aligned}
$$

29. $P = \begin{bmatrix} 1 & 0 & 1 \\ 1 & -2 & 2 \\ 1 & 1 & 0 \end{bmatrix}$; $D = \begin{bmatrix} 0 & 0 & 0 \\ 0 & 1 & 0 \\ 0 & 0 & 1 \end{bmatrix}$;

$A^k = PD^kP^{-1} = \begin{bmatrix} 3 & -1 & -2 \\ 2 & 0 & -2 \\ 2 & -1 & -1 \end{bmatrix}$

31. Since A is diagonalizable, there is an invertible P and diagonal D such that $A = PDP^{-1}$. Since B is similar to A, there is an invertible Q such that $B = Q^{-1}AQ$. Then

$$D = P^{-1}QBQ^{-1}P = (Q^{-1}P)^{-1}B(Q^{-1}P)$$

33. If A is diagonalizable with an eigenvalue of multiplicity n, then $A = P(\lambda I)P^{-1} = (\lambda I)PP^{-1} = \lambda I$. On the other hand, if $A = \lambda I$, then A is a diagonal matrix.

35. a. $[T]_{B_1} = \begin{bmatrix} 0 & 1 & 0 \\ 0 & 0 & 2 \\ 0 & 0 & 0 \end{bmatrix}$

b. $[T]_{B_2} = \begin{bmatrix} 1 & 1 & 2 \\ -1 & -1 & 0 \\ 0 & 0 & 0 \end{bmatrix}$

c. The only eigenvalue of A and B is $\lambda = 0$, of multiplicity 3.

d. The only eigenvector corresponding to $\lambda = 0$ is

$\begin{bmatrix} -1 \\ 1 \\ 0 \end{bmatrix}$, so T is not diagonalizable.

37. If B is the standard basis for $\mathbb{R}^3$, then

$$[T]_B = \begin{bmatrix} 2 & 2 & 2 \\ -1 & 2 & 1 \\ 1 & -1 & 0 \end{bmatrix}$$

The eigenvalues are $\lambda_1 = 1$, multiplicity 2, and $\lambda_2 = 2$

with corresponding eigenvectors $\begin{bmatrix} 0 \\ -1 \\ 1 \end{bmatrix}$ and

$\begin{bmatrix} 1 \\ -1 \\ 1 \end{bmatrix}$, respectively. Since there are only two

linearly independent eigenvectors, T is not diagonalizable.

39. Since A and B are matrix representations for the same linear operator, they are similar. Let $A = Q^{-1}BQ$. The matrix A is diagonalizable if and only if $D = P^{-1}AP$ for some invertible matrix P and diagonal matrix D. Then

$$D = P^{-1}(Q^{-1}BQ)P = (QP)^{-1}B(QP)$$

so B is diagonalizable. The proof of the converse is identical.

Section 5.3

1.
$$y_1(t) = [y_1(0) + y_2(0)]e^{-t} - y_2(0)e^{-2t}$$
$$y_2(t) = y_2(0)e^{-2t}$$

3.
$$y_1(t) = \frac{1}{2}[y_1(0) - y_2(0)]e^{4t}$$
$$+ \frac{1}{2}[y_1(0) + y_2(0)]e^{-2t}$$
$$y_2(t) = \frac{1}{2}[-y_1(0) + y_2(0)]e^{4t}$$
$$+ \frac{1}{2}[y_1(0) + y_2(0)]e^{-2t}$$

5.
$$y_1(t) = [2y_1(0) + y_2(0) + y_3(0)]e^{-t}$$
$$+ [-y_1(0) - y_2(0) - y_3(0)]e^{2t}$$
$$y_2(t) = [-2y_1(0) - y_2(0) - 2y_3(0)]e^{t}$$
$$+ 2[y_1(0) + y_2(0) + y_3(0)]e^{2t}$$
$$y_3(t) = [-2y_1(0) - y_2(0) - y_3(0)]e^{-t}$$
$$+ [2y_1(0) + y_2(0) + 2y_3(0)]e^{t}$$

7. $y_1(t) = e^{-t}, y_2(t) = -e^{-t}$

9. a. $y_1'(t) = -\frac{1}{60}y_1 + \frac{1}{120}y_2$,
$$y_2'(t) = \frac{1}{60}y_1 - \frac{1}{120}y_2$$
$$y_1(0) = 12, \ y_2(0) = 0$$

b. $y_1(t) = 4 + 8e^{-\frac{1}{40}t}, y_2(t) = 8 - 8e^{-\frac{1}{40}t}$

c. $\lim_{t \to \infty} y_1(t) = 4, \lim_{t \to \infty} y_2(t) = 8$
The 12 lb of salt will be evenly distributed in a ratio of 1:2 between the two tanks.

Section 5.4

1. a. $T = \begin{bmatrix} 0.85 & 0.08 \\ 0.15 & 0.92 \end{bmatrix}$

b. $T^{10} \begin{bmatrix} 0.7 \\ 0.3 \end{bmatrix} \approx \begin{bmatrix} 0.37 \\ 0.63 \end{bmatrix}$

c. $\begin{bmatrix} 0.35 \\ 0.65 \end{bmatrix}$

3. $T = \begin{bmatrix} 0.5 & 0.4 & 0.1 \\ 0.4 & 0.4 & 0.2 \\ 0.1 & 0.2 & 0.7 \end{bmatrix}$

$$T^3 \begin{bmatrix} 0 \\ 1 \\ 0 \end{bmatrix} \approx \begin{bmatrix} 0.36 \\ 0.35 \\ 0.29 \end{bmatrix}$$

$$T^{10} \begin{bmatrix} 0 \\ 1 \\ 0 \end{bmatrix} \approx \begin{bmatrix} 0.33 \\ 0.33 \\ 0.33 \end{bmatrix}$$

5. $T = \begin{bmatrix} 0.5 & 0 & 0 \\ 0.5 & 0.75 & 0 \\ 0 & 0.25 & 1 \end{bmatrix}$

The steady-state probability vector is $\begin{bmatrix} 0 \\ 0 \\ 1 \end{bmatrix}$, and

hence the disease will not be eradicated.

7. a. $T = \begin{bmatrix} 0.33 & 0.25 & 0.17 & 0.25 \\ 0.25 & 0.33 & 0.25 & 0.17 \\ 0.17 & 0.25 & 0.33 & 0.25 \\ 0.25 & 0.17 & 0.25 & 0.33 \end{bmatrix}$

b. $T \begin{bmatrix} 1 \\ 0 \\ 0 \\ 0 \end{bmatrix} = \begin{bmatrix} 0.5(0.16)^n + 0.25 \\ 0.25 \\ -0.5(0.16)^n + 0.25 \\ 0.25 \end{bmatrix}$

c. $\begin{bmatrix} 0.25 \\ 0.25 \\ 0.25 \\ 0.25 \end{bmatrix}$

9. Eigenvalues of T: $\lambda_1 = -q + p + 1, \lambda_2 = 1$, with corresponding eigenvectors $\begin{bmatrix} -1 \\ 1 \end{bmatrix}$ and $\begin{bmatrix} q/p \\ 1 \end{bmatrix}$.
The steady-state probability vector is

$$\frac{1}{1 + q/p} \begin{bmatrix} q/p \\ 1 \end{bmatrix} = \begin{bmatrix} \frac{q}{p+q} \\ \frac{p}{p+q} \end{bmatrix}.$$

Review Exercises Chapter 5

1. a. $\begin{bmatrix} a & b \\ b & a \end{bmatrix} \begin{bmatrix} 1 \\ 1 \end{bmatrix} = \begin{bmatrix} a+b \\ a+b \end{bmatrix}$
$$= (a+b) \begin{bmatrix} 1 \\ 1 \end{bmatrix}$$

b. $\lambda_1 = a + b, \lambda_2 = a - b$

c. $\mathbf{v}_1 = \begin{bmatrix} 1 \\ 1 \end{bmatrix}, \mathbf{v}_2 = \begin{bmatrix} -1 \\ 1 \end{bmatrix}$

d. $P = \begin{bmatrix} 1 & -1 \\ 1 & 1 \end{bmatrix}, D = \begin{bmatrix} a+b & 0 \\ 0 & a-b \end{bmatrix}$

3. a. $\lambda_1 = 0, \lambda_2 = 1$

b. No conclusion can be drawn from part (a) about the diagonalizability of A.

c. $\lambda_1 = 0$: $\mathbf{v}_1 = \begin{bmatrix} 0 \\ 0 \\ 0 \\ 1 \end{bmatrix}$ $\mathbf{v}_2 = \begin{bmatrix} -1 \\ 0 \\ 1 \\ 0 \end{bmatrix}$

$\lambda_2 = 1$: $\mathbf{v}_3 = \begin{bmatrix} 0 \\ 1 \\ 0 \\ 0 \end{bmatrix}$

d. The eigenvectors $\{\mathbf{v}_1, \mathbf{v}_2, \mathbf{v}_3\}$ are linearly independent.

e. A is not diagonalizable as it is a 4×4 matrix with only three linearly independent eigenvectors.

5. a. $\det(A - \lambda I) = \begin{vmatrix} -\lambda & 1 & 0 \\ 0 & -\lambda & 1 \\ -k & 3 & -\lambda \end{vmatrix}$

$$= \lambda^3 - 3\lambda + k = 0$$

b.

c. $-2 < k < 2$

7. a. Let $\mathbf{v} = \begin{bmatrix} 1 \\ 1 \\ \vdots \\ 1 \end{bmatrix}$. Then

$$A\mathbf{v} = \begin{bmatrix} \lambda \\ \lambda \\ \vdots \\ \lambda \end{bmatrix} = \lambda \begin{bmatrix} 1 \\ 1 \\ \vdots \\ 1 \end{bmatrix}$$

so λ is an eigenvalue of A corresponding to the eigenvector $\mathbf{v}$.

b. Yes, since A and A^t have the same eigenvalues.

Chapter Test: Chapter 5

1. F		**2.** F
3. F		**4.** T
5. T		**6.** T
7. T		**8.** T
9. T		**10.** T
11. F		**12.** F
13. F		**14.** T
15. F		**16.** T
17. T		**18.** T
19. T		**20.** F
21. T		**22.** T
23. F		**24.** F
25. T		**26.** T
27. T		**28.** T
29. T		**30.** F
31. T		**32.** T
33. T		**34.** F
35. T		**36.** T
37. T		**38.** T
39. T		**40.** T

Chapter 6

Section 6.1

1. 5

3. -11

5. $\sqrt{26}$

7. $\dfrac{1}{\sqrt{26}} \begin{bmatrix} 1 \\ 5 \end{bmatrix}$

9. $\dfrac{10}{\sqrt{5}} \begin{bmatrix} 2 \\ 1 \end{bmatrix}$

11. $\sqrt{22}$

13. $\dfrac{1}{\sqrt{22}} \begin{bmatrix} -3 \\ -2 \\ 3 \end{bmatrix}$

15. $\dfrac{3}{\sqrt{11}} \begin{bmatrix} 1 \\ 1 \\ 3 \end{bmatrix}$

17. $c = 6$

19. $\mathbf{v}_1 \perp \mathbf{v}_2$; $\mathbf{v}_1 \perp \mathbf{v}_4$; $\mathbf{v}_1 \perp \mathbf{v}_5$; $\mathbf{v}_2 \perp \mathbf{v}_3$; $\mathbf{v}_3 \perp \mathbf{v}_4$; $\mathbf{v}_3 \perp \mathbf{v}_5$

21. Since $\mathbf{v}_3 = -\mathbf{v}_1$, the vectors $\mathbf{v}_1$ and $\mathbf{v}_3$ are in opposite directions.

23. $\mathbf{w} = \begin{bmatrix} 2 \\ 0 \end{bmatrix}$

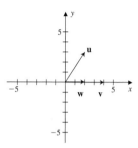

25. $\mathbf{w} = \dfrac{3}{2} \begin{bmatrix} 3 \\ 1 \end{bmatrix}$

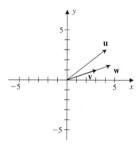

27. $\mathbf{w} = \dfrac{1}{6} \begin{bmatrix} 5 \\ 2 \\ 1 \end{bmatrix}$

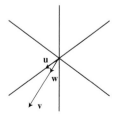

29. Let $\mathbf{u}$ be a vector in $\mathbf{span}\{\mathbf{u}_1, \mathbf{u}_2, \cdots, \mathbf{u}_n\}$. Then there exist scalars $c_1, c_2, \cdots, c_n$ such that

$$\mathbf{u} = c_1 \mathbf{u}_1 + c_2 \mathbf{u}_2 + \cdots + c_n \mathbf{u}_n$$

Then

$$
\begin{aligned}
\mathbf{v} \cdot \mathbf{u} &= \mathbf{v} \cdot (c_1 \mathbf{u}_1 + c_2 \mathbf{u}_2 + \cdots + c_n \mathbf{u}_n) \\
&= c_1 \mathbf{v} \cdot \mathbf{u}_1 + c_2 \mathbf{v} \cdot \mathbf{u}_2 + \cdots + c_n \mathbf{v} \cdot \mathbf{u}_n \\
&= c_1 (0) + c_2 (0) + \cdots + c_n (0) = 0
\end{aligned}
$$

31. Consider the equation

$$c_1 \mathbf{v}_1 + c_2 \mathbf{v}_2 + \cdots + c_n \mathbf{v}_n = \mathbf{0}$$

Since

$$\mathbf{v}_1 \cdot (c_1 \mathbf{v}_1 + c_2 \mathbf{v}_2 + \cdots + c_n \mathbf{v}_n) = \mathbf{v}_1 \cdot \mathbf{0}$$

so

$$c_1 \mathbf{v}_1 \cdot \mathbf{v}_1 + c_2 \mathbf{v}_1 \cdot \mathbf{v}_2 + \cdots + c_n \mathbf{v}_1 \cdot \mathbf{v}_n = 0$$

Since S is an orthogonal set of vectors, this equation reduces to

$$c_1 ||\mathbf{v}_1||^2 = 0$$

and since $||\mathbf{v}_1|| \neq 0$, then $c_1 = 0$. In a similar way we have $c_2 = c_3 = \cdots = c_n = 0$. Hence, S is linearly independent.

33. Since $||\mathbf{u}||^2 = \mathbf{u} \cdot \mathbf{u}$,

$$
\begin{aligned}
||\mathbf{u} + \mathbf{v}||^2 + ||\mathbf{u} - \mathbf{v}||^2 &= (\mathbf{u} + \mathbf{v}) \cdot (\mathbf{u} + \mathbf{v}) \\
&\quad + (\mathbf{u} - \mathbf{v}) \cdot (\mathbf{u} - \mathbf{v}) \\
&= \mathbf{u} \cdot \mathbf{u} + 2 \mathbf{u} \cdot \mathbf{v} + \mathbf{v} \cdot \mathbf{v} \\
&\quad + \mathbf{u} \cdot \mathbf{u} - 2 \mathbf{u} \cdot \mathbf{v} + \mathbf{v} \cdot \mathbf{v} \\
&= 2||\mathbf{u}||^2 + 2||\mathbf{v}||^2
\end{aligned}
$$

35. If the column vectors of A form an orthogonal set, then the row vectors of A^t are orthogonal to the column vectors of A. Consequently,

$$(A^t A)_{ij} = 0 \qquad \text{if } i \neq j$$

On the other hand, if $i = j$, then $(A^t A)_{ii} = ||\mathbf{A}_i||^2$. Thus,

$$A^t A = \begin{bmatrix} ||\mathbf{A}_1||^2 & 0 & \cdots & 0 \\ 0 & ||\mathbf{A}_2||^2 & 0 & \vdots \\ \vdots & 0 & \ddots & 0 \\ 0 & \cdots & 0 & ||\mathbf{A}_n||^2 \end{bmatrix}$$

37. Suppose that $(A\mathbf{u}) \cdot \mathbf{v} = \mathbf{u} \cdot (A\mathbf{v})$ for all $\mathbf{u}$ and $\mathbf{v}$ in $\mathbb{R}^n$. By Exercise 36,

$$\mathbf{u} \cdot (A\mathbf{v}) = (A^t \mathbf{u}) \cdot \mathbf{v}$$

and by hypothesis

$$\mathbf{u} \cdot (A\mathbf{v}) = (A\mathbf{u}) \cdot \mathbf{v}$$

for all $\mathbf{u}$ and $\mathbf{v}$ in $\mathbb{R}^n$. Thus,

$$(A^t\mathbf{u})\cdot\mathbf{v} = (A\mathbf{u})\cdot\mathbf{v}$$

for all $\mathbf{u}$ and $\mathbf{v}$ in $\mathbb{R}^n$. Let $\mathbf{u} = \mathbf{e}_i$ and $\mathbf{v} = \mathbf{e}_j$, so $(A^t)_{ij} = A_{ij}$. Hence $A^t = A$, so A is symmetric. For the converse, suppose that $A = A^t$. Then by Exercise 36,

$$\mathbf{u}\cdot(A\mathbf{v}) = (A^t\mathbf{u})\cdot\mathbf{v} = (A\mathbf{u})\cdot\mathbf{v}$$

Section 6.2

1. Since $\langle\mathbf{u},\mathbf{u}\rangle = 0$ when $u_1 = 3u_2$ or $u_1 = u_2$, V is not an inner product space.

3. Since $\langle\mathbf{u}+\mathbf{v},\mathbf{w}\rangle$ and $\langle\mathbf{u},\mathbf{v}\rangle + \langle\mathbf{v},\mathbf{w}\rangle$ are not equal for all $\mathbf{u},\mathbf{v}$, and $\mathbf{w}$, V is not an inner product space.

5. Yes, V is an inner product space.

7. Yes, V is an inner product space.

9. Yes, V is an inner product space.

11.
$$\int_{-\pi}^{\pi} \sin x \, dx = \int_{-\pi}^{\pi} \cos x \, dx$$
$$= \int_{-\pi}^{\pi} \cos x \sin x \, dx = 0$$

13. $\int_0^1 (2x-1)\,dx = 0$

$\int_0^1 \left(-x^2 + x - \frac{1}{6}\right)dx = 0$

$\int_0^1 \left(-2x^3 + 3x^2 - \frac{4}{3}x + \frac{1}{6}\right)dx = 0$

15. **a.** $\| -3 + 3x - x^2 \| = \sqrt{\frac{370}{10}}$

b. $\cos\theta = -\frac{5}{168}\sqrt{105}$

17. **a.** $\| x - e^x \| = \sqrt{\frac{1}{2}e^2 - \frac{13}{6}}$

b. $\cos\theta = \frac{2\sqrt{3}}{\sqrt{2e^2 - 2}}$

19. **a.** $\| 2x^2 - 4 \| = 2\sqrt{5}$

b. $\cos\theta = -\frac{2}{3}$

21. **a.** $\| A - B \| = \sqrt{\text{tr}\begin{bmatrix} 2 & -5 \\ -5 & 17 \end{bmatrix}} = \sqrt{19}$

b. $\cos\theta = \frac{3}{5\sqrt{6}}$

23. **a.** $\| A - B \| = \sqrt{\text{tr}\begin{bmatrix} 8 & 0 & 8 \\ 0 & 3 & 4 \\ 8 & 4 & 14 \end{bmatrix}} = \sqrt{25} = 5$

b. $\cos\theta = \frac{26}{\sqrt{38}\sqrt{39}}$

25. $\left\{ \begin{bmatrix} x \\ y \end{bmatrix} \middle| \ 2x + 3y = 0 \right\}$

27. $\left\{ \begin{bmatrix} x \\ y \\ z \end{bmatrix} \middle| \ 2x - 3y + z = 0 \right\}$

29. **a.** $\langle x^2, x^3\rangle = \int_0^1 x^5\,dx = \frac{1}{6}$

b. $\langle e^x, e^{-x}\rangle = \int_0^1 dx = 1$

c. $\| 1 \| = \sqrt{\int_0^1 dx} = 1$

$\| x \| = \sqrt{\int_0^1 x^2\,dx} = \frac{\sqrt{3}}{3}$

d. $\cos\theta = \frac{3}{2\sqrt{3}}$

e. $\| 1 - x \| = \frac{\sqrt{3}}{3}$

31. If f is an even function and g is an odd function, then fg is an odd function. Then

$$\int_{-a}^{a} f(x)g(x)\,dx = 0$$

so f and g are orthogonal.

33. $\langle c_1\mathbf{u}_1, c_2\mathbf{u}_2\rangle = c_1\langle\mathbf{u}_1, c_2\mathbf{u}_2\rangle$
$= c_1 c_2\langle\mathbf{u}_1, \mathbf{u}_2\rangle$
$= 0$

Section 6.3

1. **a.** $\text{proj}_\mathbf{v}\,\mathbf{u} = \begin{bmatrix} -\frac{3}{2} \\ \frac{3}{2} \end{bmatrix}$

b. $\mathbf{u} - \text{proj}_\mathbf{v}\,\mathbf{u} = \begin{bmatrix} \frac{1}{2} \\ \frac{1}{2} \end{bmatrix}$

$\mathbf{v}\cdot(\mathbf{u} - \text{proj}_\mathbf{v}\mathbf{u}) = \begin{bmatrix} -1 \\ 1 \end{bmatrix}\cdot\begin{bmatrix} \frac{1}{2} \\ \frac{1}{2} \end{bmatrix} = 0$

3. **a.** $\text{proj}_\mathbf{v}\,\mathbf{u} = \begin{bmatrix} -\frac{3}{5} \\ -\frac{6}{5} \end{bmatrix}$

b. $\mathbf{u} - \text{proj}_\mathbf{v}\,\mathbf{u} = \begin{bmatrix} \frac{8}{5} \\ -\frac{4}{5} \end{bmatrix}$

$\mathbf{v}\cdot(\mathbf{u} - \text{proj}_\mathbf{v}\,\mathbf{u}) = \begin{bmatrix} 1 \\ 2 \end{bmatrix}\cdot\begin{bmatrix} \frac{8}{5} \\ -\frac{4}{5} \end{bmatrix} = 0$

5. **a.** $\text{proj}_\mathbf{v}\,\mathbf{u} = \begin{bmatrix} -\frac{4}{3} \\ \frac{4}{3} \\ \frac{4}{3} \end{bmatrix}$

b. $\mathbf{u} - \text{proj}_{\mathbf{v}}\, \mathbf{u} = \begin{bmatrix} \frac{1}{3} \\ \frac{5}{3} \\ -\frac{4}{3} \end{bmatrix}$

$$\mathbf{v} \cdot (\mathbf{u} - \text{proj}_{\mathbf{v}}\, \mathbf{u}) = \begin{bmatrix} 1 \\ -1 \\ -1 \end{bmatrix} \cdot \begin{bmatrix} \frac{1}{3} \\ \frac{5}{3} \\ -\frac{4}{3} \end{bmatrix} = 0$$

7. a. $\text{proj}_{\mathbf{v}}\, \mathbf{u} = \begin{bmatrix} 0 \\ 0 \\ -1 \end{bmatrix}$

b. $\mathbf{u} - \text{proj}_{\mathbf{v}}\, \mathbf{u} = \begin{bmatrix} 1 \\ -1 \\ 0 \end{bmatrix}$

$$\mathbf{v} \cdot (\mathbf{u} - \text{proj}_{\mathbf{v}}\, \mathbf{u}) = \begin{bmatrix} 0 \\ 0 \\ 1 \end{bmatrix} \cdot \begin{bmatrix} 1 \\ -1 \\ 0 \end{bmatrix} = 0$$

9. a. $\text{proj}_{q}\, p = \frac{5}{4}x - \frac{5}{12}$

b. $p - \text{proj}_{q}\, p = x^2 - \frac{9}{4}x + \frac{17}{12}$

$$\langle q, p - \text{proj}_{q}\, p \rangle$$
$$= \int_0^1 (3x - 1)\left(x^2 - \frac{9}{4}x + \frac{17}{12}\right) dx = 0$$

11. a. $\text{proj}_{q}\, p = -\frac{7}{4}x^2 + \frac{7}{4}$

b. $p - \text{proj}_{q}\, p = \frac{15}{4}x^2 - \frac{3}{4}$

$$\langle q, p - \text{proj}_{q}\, p \rangle$$
$$= \int_0^1 (x^2 - 1)\left(\frac{15}{4}x^2 - \frac{3}{4}\right) dx = 0$$

13. $\left\{ \dfrac{1}{\sqrt{2}}\begin{bmatrix} 1 \\ -1 \end{bmatrix}, \dfrac{1}{\sqrt{2}}\begin{bmatrix} -1 \\ -1 \end{bmatrix} \right\}$

15. $\left\{ \dfrac{1}{\sqrt{2}}\begin{bmatrix} 1 \\ 0 \\ 1 \end{bmatrix}, \dfrac{1}{\sqrt{6}}\begin{bmatrix} 1 \\ 2 \\ -1 \end{bmatrix}, \dfrac{1}{\sqrt{3}}\begin{bmatrix} 1 \\ -1 \\ -1 \end{bmatrix} \right\}$

17. $\left\{ \sqrt{3}(x - 1),\ 3x - 1,\ 6\sqrt{5}(x^2 - x + \frac{1}{6}) \right\}$

19. $\left\{ \dfrac{1}{\sqrt{3}}\begin{bmatrix} 1 \\ 1 \\ 1 \end{bmatrix}, \dfrac{1}{\sqrt{6}}\begin{bmatrix} 2 \\ -1 \\ -1 \end{bmatrix} \right\}$

21. $\left\{ \dfrac{1}{\sqrt{6}}\begin{bmatrix} -1 \\ -2 \\ 0 \\ 1 \end{bmatrix}, \dfrac{1}{\sqrt{6}}\begin{bmatrix} -2 \\ 1 \\ -1 \\ 0 \end{bmatrix}, \dfrac{1}{\sqrt{6}}\begin{bmatrix} 1 \\ 0 \\ -2 \\ 1 \end{bmatrix} \right\}$

23. $\left\{ \sqrt{3}x,\ -3x + 2 \right\}$

25. $\left\{ \dfrac{1}{\sqrt{3}}\begin{bmatrix} 1 \\ 0 \\ 1 \\ 1 \end{bmatrix}, \dfrac{1}{\sqrt{3}}\begin{bmatrix} 0 \\ 1 \\ -1 \\ 1 \end{bmatrix} \right\}$

27. Let
$$\mathbf{v} = c_1\mathbf{u}_1 + c_2\mathbf{u}_2 + \cdots + c_n\mathbf{u}_n$$
Then
$$\|\mathbf{v}\|^2 = \mathbf{v} \cdot \mathbf{v}$$
$$= c_1^2(\mathbf{u}_1 \cdot \mathbf{u}_1) + c_2^2(\mathbf{u}_2 \cdot \mathbf{u}_2) + \cdots + c_n^2(\mathbf{u}_n \cdot \mathbf{u}_n)$$
$$= c_1^2 + c_2^2 + \cdots + c_n^2$$
$$= |\mathbf{v} \cdot \mathbf{u}_1|^2 + \cdots + |\mathbf{v} \cdot \mathbf{u}_n|^2$$

29. Since
$$\sum_{k=1}^{n} a_{ki} a_{kj} = \begin{cases} 0 & \text{if } i \neq j \\ 1 & \text{if } i = j \end{cases} = (A^t A)_{ij}$$
then $A^t A = I$.

31. Since $\|A\mathbf{x}\| = \sqrt{A\mathbf{x} \cdot A\mathbf{x}}$ and
$$A\mathbf{x} \cdot A\mathbf{x} = \mathbf{x}^t \cdot (A^t A\mathbf{x}) = \mathbf{x} \cdot \mathbf{x}$$
then $\|A\mathbf{x}\|^2 = \mathbf{x} \cdot \mathbf{x} = \|\mathbf{x}\|^2$ so $\|A\mathbf{x}\| = \|\mathbf{x}\|$.

33. By Exercise 32, $A\mathbf{x} \cdot A\mathbf{y} = \mathbf{x} \cdot \mathbf{y}$. Then $A\mathbf{x} \cdot A\mathbf{y} = 0$ if and only if $\mathbf{x} \cdot \mathbf{y} = 0$

35. Let
$$W = \{\mathbf{v} \mid \mathbf{v} \cdot \mathbf{u}_i = 0 \text{ for all } i = 1, 2, \ldots, m\}$$
If c is a real number and $\mathbf{x}$ and $\mathbf{y}$ are vectors in W, then
$$(\mathbf{x} + c\mathbf{y}) \cdot \mathbf{u}_i = \mathbf{x} \cdot \mathbf{u}_i + c\mathbf{y} \cdot \mathbf{u}_i = 0 + c(0) = 0$$
for all $i = 1, 2, \ldots, n$.

37.
$$\mathbf{v}^t A\mathbf{v} = [x\ y] \begin{bmatrix} 3 & 1 \\ 1 & 3 \end{bmatrix} \begin{bmatrix} x \\ y \end{bmatrix}$$
$$= 3x^2 + 2xy + 3y^2$$
$$\geq (x + y)^2 \geq 0$$

39.
$$\mathbf{x}^t A^t A\mathbf{x} = (A\mathbf{x})^t A\mathbf{x} = (A\mathbf{x}) \cdot (A\mathbf{x})$$
$$= \|A\mathbf{x}\|^2 \geq 0$$

41. Since $A\mathbf{x} = \lambda\mathbf{x}$, then $\mathbf{x}^t A\mathbf{x} = \lambda\|\mathbf{x}\|^2$. Since A is positive definite and $\mathbf{x}$ is not the zero vector, then $\mathbf{x}^t A\mathbf{x} > 0$, so $\lambda > 0$.

Section 6.4

1. $W^\perp = \text{span}\left\{\begin{bmatrix} 1 \\ \frac{1}{2} \end{bmatrix}\right\}$

3. $W^\perp = \text{span}\left\{\begin{bmatrix} 1 \\ -2 \\ 0 \end{bmatrix}, \begin{bmatrix} 0 \\ 1 \\ 1 \end{bmatrix}\right\}$

5. $W^\perp = \text{span}\left\{\begin{bmatrix} \frac{2}{3} \\ -\frac{1}{3} \\ 1 \end{bmatrix}\right\}$

7. $W^\perp = \text{span}\left\{\begin{bmatrix} -\frac{1}{6} \\ -\frac{1}{2} \\ 1 \\ 0 \end{bmatrix}, \begin{bmatrix} \frac{2}{3} \\ -1 \\ 0 \\ 1 \end{bmatrix}\right\}$

9. $\left\{\begin{bmatrix} -\frac{1}{3} \\ -\frac{1}{3} \\ 1 \end{bmatrix}\right\}$

11. $\left\{\begin{bmatrix} \frac{1}{2} \\ -\frac{3}{2} \\ 1 \\ 0 \end{bmatrix}, \begin{bmatrix} -\frac{1}{2} \\ \frac{1}{2} \\ 0 \\ 1 \end{bmatrix}\right\}$

13. $\left\{\frac{50}{9}x^2 - \frac{52}{9}x + 1\right\}$

15. $\left\{\begin{bmatrix} 1 \\ 1 \\ 1 \\ 1 \end{bmatrix}\right\}$

17. An orthogonal basis is
$$B = \left\{\begin{bmatrix} 2 \\ 0 \\ 0 \end{bmatrix}, \begin{bmatrix} 0 \\ -1 \\ 0 \end{bmatrix}\right\}$$

$$\text{proj}_W \mathbf{v} = \begin{bmatrix} 1 \\ 2 \\ 0 \end{bmatrix}$$

19. $\text{proj}_W \mathbf{v} = \begin{bmatrix} 0 \\ 0 \\ 0 \end{bmatrix}$

21. An orthogonal basis is
$$B = \left\{\begin{bmatrix} 3 \\ 0 \\ -1 \\ 2 \end{bmatrix}, \begin{bmatrix} -5 \\ 21 \\ -3 \\ 6 \end{bmatrix}\right\}$$

$$\text{proj}_W \mathbf{v} = \frac{4}{73}\begin{bmatrix} -5 \\ 21 \\ -3 \\ 6 \end{bmatrix}$$

23. a. $W^\perp = \text{span}\left\{\begin{bmatrix} 1 \\ 3 \end{bmatrix}\right\}$

b. $\text{proj}_W \mathbf{v} = \frac{1}{10}\begin{bmatrix} -3 \\ 1 \end{bmatrix}$

c. $\mathbf{u} = \mathbf{v} - \text{proj}_W \mathbf{v} = \frac{1}{10}\begin{bmatrix} 3 \\ 9 \end{bmatrix}$

d. $\frac{1}{10}\begin{bmatrix} 3 \\ 9 \end{bmatrix} \cdot \begin{bmatrix} -3 \\ 1 \end{bmatrix} = 0$

e.

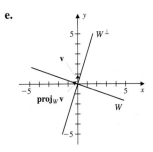

25. Notice that the vectors $\mathbf{v}_1$ and $\mathbf{v}_2$ are not orthogonal. Using the Gram-Schmidt process orthogonal vectors with the same span are
$$\begin{bmatrix} 1 \\ 1 \\ -1 \end{bmatrix} \qquad \begin{bmatrix} 0 \\ 3 \\ 3 \end{bmatrix}$$

a. $W^\perp = \text{span}\left\{\begin{bmatrix} 2 \\ -1 \\ 1 \end{bmatrix}\right\}$

b. $\text{proj}_W \mathbf{v} = \frac{1}{3}\begin{bmatrix} 2 \\ 5 \\ 1 \end{bmatrix}$

c. $\mathbf{u} = \mathbf{v} - \text{proj}_W \mathbf{v} = \frac{1}{3}\begin{bmatrix} 4 \\ -2 \\ 2 \end{bmatrix}$

d. Since **u** is a scalar multiple of $\begin{bmatrix} 2 \\ -1 \\ 1 \end{bmatrix}$, then **u** is

in $W^\perp$.

e.

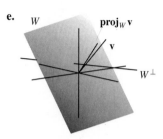

27. Let $\mathbf{w} \in W_2^\perp$, so $\langle \mathbf{w}, \mathbf{u} \rangle = 0$ for all $\mathbf{u} \in W_2$. Since $W_1 \subseteq W_2$, then $\langle \mathbf{w}, \mathbf{u} \rangle = 0$ for all $\mathbf{u} \in W_1$. Hence $\mathbf{w} \in W_1^\perp$, so $W_2^\perp \subseteq W_1^\perp$.

29. a. Let $A = \begin{bmatrix} d & e \\ f & g \end{bmatrix}$ and $B = \begin{bmatrix} a & b \\ b & c \end{bmatrix}$. Then

$$\langle A, B \rangle = \mathbf{tr}\left(\begin{bmatrix} a & b \\ b & c \end{bmatrix} \begin{bmatrix} d & e \\ f & g \end{bmatrix} \right)$$

$$= \mathbf{tr} \begin{bmatrix} ad + bf & ae + bg \\ bd + cf & be + cg \end{bmatrix}$$

So $A \in W^\perp$ if and only if $ad + bf + be + cg = 0$ for all real numbers $a, b,$ and c. This implies $A = \begin{bmatrix} 0 & e \\ -e & 0 \end{bmatrix}$. That is, A is skew-symmetric.

b. $\begin{bmatrix} a & b \\ c & d \end{bmatrix} = \begin{bmatrix} a & \frac{b+c}{2} \\ \frac{b+c}{2} & d \end{bmatrix} + \begin{bmatrix} 0 & \frac{b-c}{2} \\ -\frac{b-c}{2} & 0 \end{bmatrix}$

Section 6.5

1. a. $\widehat{\mathbf{x}} = \begin{bmatrix} \frac{5}{2} \\ 0 \end{bmatrix}$

b. $\mathbf{w}_1 = A\widehat{\mathbf{x}} = \begin{bmatrix} \frac{5}{2} \\ \frac{5}{2} \\ 5 \end{bmatrix}$

$\mathbf{w}_2 = \mathbf{b} - \mathbf{w}_1 = \begin{bmatrix} \frac{3}{2} \\ -\frac{3}{2} \\ 0 \end{bmatrix}$

3. a.

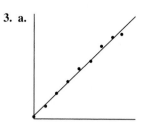

b. $y = \frac{653,089}{13,148} x - \frac{317,689,173}{3287}$

5. a.

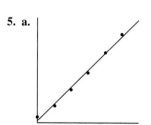

b. $y = 0.07162857143x - 137.2780952$

7. a.
$$p_2(x) = 2 \sin x - \sin 2x$$
$$p_3(x) = 2 \sin x - \sin 2x + \frac{2}{3} \sin 3x$$
$$p_4(x) = 2 \sin x - \sin 2x + \frac{2}{3} \sin 3x$$
$$- \frac{1}{2} \sin 4x$$
$$p_5(x) = 2 \sin x - \sin 2x + \frac{2}{3} \sin 3x$$
$$- \frac{1}{2} \sin 4x + \frac{2}{5} \sin 5x$$

b.

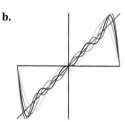

9. a.
$$p_2(x) = \frac{1}{3}\pi^2 - 4 \cos x + \cos 2x$$
$$p_3(x) = \frac{1}{3}\pi^2 - 4 \cos x + \cos 2x - \frac{4}{9} \cos 3x$$
$$p_4(x) = \frac{1}{3}\pi^2 - 4 \cos x + \cos 2x - \frac{4}{9} \cos 3x$$
$$+ \frac{1}{4} \cos 4x$$

$$p_5(x) = \frac{1}{3}\pi^2 - 4\cos x + \cos 2x - \frac{4}{9}\cos 3x$$
$$+ \frac{1}{4}\cos 4x - \frac{4}{25}\cos 5x$$

b.

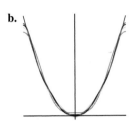

Section 6.6

1. $\lambda_1 = 3, \lambda_2 = -1$

3. $\lambda_1 = 1, \lambda_2 = -3, \lambda_3 = 3$

5. $\lambda_1 = -3$ with eigenvector $\mathbf{v}_1 = \begin{bmatrix} -1 \\ 2 \end{bmatrix}$; $\lambda_2 = 2$ with

eigenvector $\mathbf{v}_2 = \begin{bmatrix} 2 \\ 1 \end{bmatrix}$. Observe that $\mathbf{v}_1 \cdot \mathbf{v}_2 = 0$.

7. $\lambda_1 = 1$ with eigenvector $\mathbf{v}_1 = \begin{bmatrix} 1 \\ 0 \\ 1 \end{bmatrix}$; $\lambda_2 = -3$ with

eigenvector $\mathbf{v}_2 = \begin{bmatrix} -1 \\ 2 \\ 1 \end{bmatrix}$; $\lambda_3 = 3$ with eigenvector

$\mathbf{v}_3 = \begin{bmatrix} -1 \\ -1 \\ 1 \end{bmatrix}$. Observe that $\mathbf{v}_1 \cdot \mathbf{v}_2 = \mathbf{v}_1 \cdot \mathbf{v}_3 =$

$\mathbf{v}_2 \cdot \mathbf{v}_3 = 0$.

9. $V_3 = \mathbf{span}\left\{ \begin{bmatrix} 1 \\ 0 \\ 1 \end{bmatrix} \right\}$

$V_{-1} = \mathbf{span}\left\{ \begin{bmatrix} -1 \\ 0 \\ 1 \end{bmatrix}, \begin{bmatrix} 0 \\ 1 \\ 0 \end{bmatrix} \right\}$

$\dim(V_3) + \dim(V_{-1}) = 1 + 2 = 3$

11. $V_3 = \mathrm{span}\left\{ \begin{bmatrix} 3 \\ 1 \\ 1 \\ 1 \end{bmatrix} \right\}$

$V_{-3} = \mathrm{span}\left\{ \begin{bmatrix} 0 \\ -2 \\ 1 \\ 1 \end{bmatrix} \right\}$

$V_{-1} = \mathrm{span}\left\{ \begin{bmatrix} -1 \\ 1 \\ 2 \\ 0 \end{bmatrix}, \begin{bmatrix} 0 \\ 0 \\ -1 \\ 1 \end{bmatrix} \right\}$

$$\dim(V_3) + \dim(V_{-3}) + \dim(V_{-1})$$
$$= 1 + 1 + 2 = 4$$

13. Yes.

$$\begin{bmatrix} \sqrt{3}/2 & 1/2 \\ -1/2 & \sqrt{3}/2 \end{bmatrix} \begin{bmatrix} \sqrt{3}/2 & -1/2 \\ 1/2 & \sqrt{3}/2 \end{bmatrix} = \begin{bmatrix} 1 & 0 \\ 0 & 1 \end{bmatrix}$$

15. Yes.

$$\begin{bmatrix} \sqrt{2}/2 & \sqrt{2}/2 & 0 \\ -\sqrt{2}/2 & \sqrt{2}/2 & 0 \\ 0 & 0 & 1 \end{bmatrix} \begin{bmatrix} \sqrt{2}/2 & -\sqrt{2}/2 & 0 \\ \sqrt{2}/2 & \sqrt{2}/2 & 0 \\ 0 & 0 & 1 \end{bmatrix}$$

$$= \begin{bmatrix} 1 & 0 & 0 \\ 0 & 1 & 0 \\ 0 & 0 & 1 \end{bmatrix}$$

17. $P = \begin{bmatrix} -1/\sqrt{2} & 1/\sqrt{2} \\ 1/\sqrt{2} & 1/\sqrt{2} \end{bmatrix}; D = \begin{bmatrix} -1 & 0 \\ 0 & 7 \end{bmatrix}$

19. $P = \begin{bmatrix} -1/\sqrt{2} & 1/\sqrt{2} \\ 1/\sqrt{2} & 1/\sqrt{2} \end{bmatrix}; D = \begin{bmatrix} -4 & 0 \\ 0 & 2 \end{bmatrix}$

21. $P = \begin{bmatrix} -1/\sqrt{3} & 1/\sqrt{2} & -1/\sqrt{6} \\ 1/\sqrt{3} & 0 & -2/\sqrt{6} \\ 1/\sqrt{3} & 1/\sqrt{2} & 1/\sqrt{6} \end{bmatrix}$

$D = \begin{bmatrix} 1 & 0 & 0 \\ 0 & 2 & 0 \\ 0 & 0 & -2 \end{bmatrix}$

23. Since $AA^t = BB^t = I$, then

$$(AB)(AB)^t = AB(B^t A^t)$$
$$= A(BB^t)A^t = AIA^t$$
$$= AA^t = I$$

Similarly, $(BA)(BA)^t = I$.

25. Since $AA^t = I$, A^t is the inverse of A so $A^t A = I$ and hence A^t is also orthogonal.

27. a. Since $\cos^2\theta + \sin^2\theta = 1$, then

$$\begin{bmatrix} \cos\theta & -\sin\theta \\ \sin\theta & \cos\theta \end{bmatrix}\begin{bmatrix} \cos\theta & \sin\theta \\ -\sin\theta & \cos\theta \end{bmatrix} = \begin{bmatrix} 1 & 0 \\ 0 & 1 \end{bmatrix}$$

29. If $D = P^t A P$, then

$$D^t = (P^t A P)^t = P^t A^t P$$

Since D is a diagonal matrix, then $D^t = D$, so
$D = P^t A P$ and hence $P^t A P = P^t A^t P$. Then
$P(P^t A P)P^t = P(P^t A^t P)P^t$, so $A = A^t$.

31. a. $\mathbf{v}^t\mathbf{v} = v_1^2 + \cdots + v_n^2$

b. The transpose of both sides of the equation $A\mathbf{v} = \lambda\mathbf{v}$
gives $\mathbf{v}^t A^t = \lambda\mathbf{v}^t$. Since A is skew-symmetric,
$\mathbf{v}^t(-A) = \lambda\mathbf{v}^t$. Now, right multiplication of both
sides by $\mathbf{v}$ gives $\mathbf{v}^t(-A\mathbf{v}) = \lambda\mathbf{v}^t\mathbf{v}$, so $\mathbf{v}^t(-\lambda\mathbf{v}) =$
$\lambda\mathbf{v}^t\mathbf{v}$. Then $2\lambda\mathbf{v}^t\mathbf{v} = 0$ so $2\lambda(v_1^2 + \cdots + v_n^2) = 0$ and
this gives $\lambda = 0$.

Section 6.7

1. $30(y')^2 + \sqrt{10}x' = 0$

3. $2(x')^2 + (y')^2 = 1$

5. $\frac{(x')^2}{2} - \frac{(y')^2}{4} = 1$

7. a. $\begin{bmatrix} x & y \end{bmatrix}\begin{bmatrix} 4 & 0 \\ 0 & 16 \end{bmatrix}\begin{bmatrix} x \\ y \end{bmatrix} - 16 = 0$

b. $10x^2 - 12xy + 10y^2 - 16 = 0$

9. a. $7x^2 + 6\sqrt{3}xy + 13y^2 - 16 = 0$

b. $7(x-3)^2 + 6\sqrt{3}(x-3)(y-2) + 13(y-2)^2 - 16 = 0$

Section 6.8

1. $\sigma_1 = \sqrt{10}, \sigma_2 = 0$

3. $\sigma_1 = 2\sqrt{3}, \sigma_2 = \sqrt{5}, \sigma_3 = 0$

5. $A = \begin{bmatrix} \frac{1}{\sqrt{2}} & \frac{1}{\sqrt{2}} \\ \frac{1}{\sqrt{2}} & -\frac{1}{\sqrt{2}} \end{bmatrix}\begin{bmatrix} 8 & 0 \\ 0 & 2 \end{bmatrix}\begin{bmatrix} \frac{1}{\sqrt{2}} & \frac{1}{\sqrt{2}} \\ \frac{1}{\sqrt{2}} & -\frac{1}{\sqrt{2}} \end{bmatrix}$

7. $A = \begin{bmatrix} 0 & 1 \\ 1 & 0 \end{bmatrix}\begin{bmatrix} \sqrt{2} & 0 & 0 \\ 0 & 1 & 0 \end{bmatrix}\begin{bmatrix} 0 & \frac{1}{\sqrt{2}} & \frac{1}{\sqrt{2}} \\ 1 & 0 & 0 \\ 0 & -\frac{1}{\sqrt{2}} & \frac{1}{\sqrt{2}} \end{bmatrix}$

9. a. $x_1 = 2, x_2 = 0$ **b.** $x_1 = 1, x_2 = 1$
c. $\sigma_1/\sigma_2 \approx 6,324,555$

Review Exercises Chapter 6

1. a. $\begin{bmatrix} 1 & 1 & 2 \\ 0 & 0 & 1 \\ 1 & 0 & 0 \end{bmatrix} \longrightarrow \begin{bmatrix} 1 & 0 & 0 \\ 0 & 1 & 0 \\ 0 & 0 & 1 \end{bmatrix}$

b. $\left\{\begin{bmatrix} 0 \\ 1 \\ 0 \end{bmatrix}, \begin{bmatrix} \sqrt{2}/2 \\ 0 \\ -\sqrt{2}/2 \end{bmatrix}, \begin{bmatrix} \sqrt{2}/2 \\ 0 \\ \sqrt{2}/2 \end{bmatrix}\right\}$

c. $\mathbf{proj}_W \mathbf{v} = \begin{bmatrix} -2 \\ 0 \\ -1 \end{bmatrix}$

3. a. If $\begin{bmatrix} x \\ y \\ z \end{bmatrix} \in W$, then

$$\begin{bmatrix} x \\ y \\ z \end{bmatrix}\cdot\begin{bmatrix} a \\ b \\ c \end{bmatrix} = ax + by + cz = 0$$

so $\begin{bmatrix} a \\ b \\ c \end{bmatrix}$ is in $W^\perp$.

b. $W^\perp = \mathbf{span}\left\{\begin{bmatrix} a \\ b \\ c \end{bmatrix}\right\}$

That is, $W^\perp$ is the line in the direction of $\begin{bmatrix} a \\ b \\ c \end{bmatrix}$
and which is perpendicular (the normal vector) to
the plane $ax + by + cz = 0$.

c. $\mathbf{proj}_{W^\perp}\mathbf{v} = \frac{ax_1 + bx_2 + cx_3}{a^2 + b^2 + c^2}\begin{bmatrix} a \\ b \\ c \end{bmatrix}$

d. $\|\mathbf{proj}_{W^\perp}\mathbf{v}\| = \frac{|ax_1 + bx_2 + cx_3|}{\sqrt{a^2 + b^2 + c^2}}$

Note: This gives the distance from the point (x_1, x_2, x_3)
to the plane.

5. a. $\langle 1, \cos x\rangle = \int_{-\pi}^{\pi} \cos x\, dx = 0$

$\langle 1, \sin x\rangle = \int_{-\pi}^{\pi} \sin x\, dx = 0$

$\langle \cos x, \sin x\rangle = \int_{-\pi}^{\pi} \cos x \sin x\, dx = 0$

b. $\left\{\frac{1}{\sqrt{2\pi}}, \frac{1}{\sqrt{\pi}}\cos x, \frac{1}{\sqrt{\pi}}\sin x\right\}$

c. $\mathbf{proj}_W x^2 = \frac{1}{3}\pi^2 - 4\cos x$

d. $\|\mathbf{proj}_W x^2\| = \frac{1}{3}\sqrt{2\pi^5 + 144\pi}$

7. Using the properties of an inner product and the fact that the vectors are orthonormal,

$$\| \mathbf{v} \| = \sqrt{\mathbf{v} \cdot \mathbf{v}}$$

$$= \sqrt{c_1^2 \langle \mathbf{v}_1, \mathbf{v}_1 \rangle + \cdots + c_n^2 \langle \mathbf{v}_n, \mathbf{v}_n \rangle}$$

$$= \sqrt{c_1^2 + \cdots + c_n^2}$$

If the basis is orthogonal, then

$$\| \mathbf{v} \| = \sqrt{c_1^2 \langle \mathbf{v}_1, \mathbf{v}_1 \rangle + \cdots + c_n^2 \langle \mathbf{v}_n, \mathbf{v}_n \rangle}$$

9. a.
$$\begin{bmatrix} 1 & 0 & -1 \\ 1 & -1 & 2 \\ 1 & 0 & 1 \\ 1 & -1 & 2 \end{bmatrix} \longrightarrow \begin{bmatrix} 1 & 0 & 0 \\ 0 & 1 & 0 \\ 0 & 0 & 1 \\ 0 & 0 & 0 \end{bmatrix}$$

b. $B_1 = \left\{ \begin{bmatrix} 1 \\ 1 \\ 1 \\ 1 \end{bmatrix}, \begin{bmatrix} \frac{1}{2} \\ -\frac{1}{2} \\ \frac{1}{2} \\ -\frac{1}{2} \end{bmatrix}, \begin{bmatrix} -1 \\ 0 \\ 1 \\ 0 \end{bmatrix} \right\}$

c. $B_2 = \left\{ \begin{bmatrix} \frac{1}{2} \\ \frac{1}{2} \\ \frac{1}{2} \\ \frac{1}{2} \end{bmatrix}, \begin{bmatrix} \frac{1}{2} \\ -\frac{1}{2} \\ \frac{1}{2} \\ -\frac{1}{2} \end{bmatrix}, \begin{bmatrix} -\frac{\sqrt{2}}{2} \\ 0 \\ \frac{\sqrt{2}}{2} \\ 0 \end{bmatrix} \right\}$

d. $Q = \begin{bmatrix} \frac{1}{2} & \frac{1}{2} & -\frac{\sqrt{2}}{2} \\ \frac{1}{2} & -\frac{1}{2} & 0 \\ \frac{1}{2} & \frac{1}{2} & \frac{\sqrt{2}}{2} \\ \frac{1}{2} & -\frac{1}{2} & 0 \end{bmatrix}$

$R = \begin{bmatrix} 2 & -1 & 2 \\ 0 & 1 & -2 \\ 0 & 0 & \sqrt{2} \end{bmatrix}$

e. $A = QR$

Chapter Test: Chapter 6

1. T   **2.** T

3. F   **4.** F

5. F   **6.** T

7. T   **8.** F

9. F   **10.** F

11. T   **12.** T

13. T   **14.** T

15. F   **16.** T

17. T   **18.** T

19. F   **20.** T

21. F   **22.** T

23. F   **24.** F

25. F   **26.** F

27. T   **28.** T

29. T   **30.** F

31. T   **32.** T

33. F   **34.** F

35. T   **36.** F

37. T   **38.** T

39. F   **40.** T

Appendix A

Section A.1

1. $A \cap B = \{-2, 2, 9\}$

3. $A \times B = \{(a, b) \mid a \in A, b \in B\}$

There are $9 \times 9 = 81$ ordered pairs in $A \times B$.

5. $A \backslash B = \{-4, 0, 1, 3, 5, 7\}$

7. $A \cap B = [0, 3]$

9. $A \backslash B = (-11, 0)$

11. $A \backslash C = (-11, -9)$

13. $(A \cup B) \backslash C = (-11, -9)$

15.
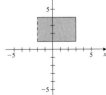

17.

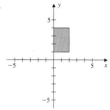

19.

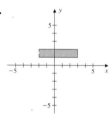

21. $(A \cap B) \cap C = \{5\} = A \cap (B \cap C)$

23. $A \cap (B \cup C) = \{1, 2, 5, 7\} = (A \cap B) \cup (A \cap C)$

25. $A \backslash (B \cup C) = \{3, 9, 11\} = (A \backslash B) \cap (A \backslash C)$

Section A.2

1. Since for each first coordinate there is a unique second coordinate, f is a function.

3. Since there is no x such that $f(x) = 14$, the function is not onto. The range of f is the set $\{-2, -1, 3, 9, 11\}$.

5. $f^{-1}(\{-2\}) = \{1, 4\}$

7. Since f is not one-to-one, f does not have an inverse.

9. $\{(1, -2), (2, -1), (3, 3), (4, 5), (5, 9), (6, 11)\}$

11. $f(A \cup B) = f((-3, 7)) = [0, 49)$

$f(A) \cup f(B) = [0, 25] \cup [0, 49) = [0, 49)$

13. $f(A \cap B) = f(\{0\}) = \{0\}$

$f(A) \cap f(B) = [0, 4] \cap [0, 4] = [0, 4]$

Therefore, $f(A \cap B) \subset f(A) \cap f(B)$, but $f(A \cap B) \neq f(A) \cap f(B)$.

15. $f^{-1}(x) = \frac{x-b}{a}$

17. If n is odd, then $f^{(n)}(x) = -x + c$. If n is even, then $f^{(n)}(x) = x$.

19. a. To show that f is one-to-one, we have

$$e^{2x_1 - 1} = e^{2x_2 - 1}$$
$$\Leftrightarrow 2x_1 - 1 = 2x_2 - 1$$
$$\Leftrightarrow x_1 = x_2$$

b. Since the exponential function is always positive, f is not onto $\mathbb{R}$.

c. Define $g : \mathbb{R} \to (0, \infty)$ by $g(x) = e^{2x-1}$.

d. $g^{-1}(x) = \frac{1}{2}(1 + \ln x)$.

21. a. To show that f is one-to-one, we have $2n_1 = 2n_2$ if and only if $n_1 = n_2$.

b. Since every image is an even number, the range of f is a proper subset of $\mathbb{N}$.

c. $f^{-1}(E) = \mathbb{N}$; $f^{-1}(O) = \phi$

23. a. $f(A) = \{2k + 1 \mid k \in \mathbb{Z}\}$

b. $f(B) = \{2k + 1 \mid k \in \mathbb{Z}\}$

c. $f^{-1}(\{0\}) = \{(m, n) \mid n = -2m\}$

d. $f^{-1}(E) = \{(m, n) \mid n \text{ is even}\}$

e. $f^{-1}(O) = \{(m, n) \mid n \text{ is odd}\}$

f. Since $f((1, -2)) = 0 = f((0, 0))$, then f is not one-to-one.

g. If $z \in \mathbb{Z}$, let $m = 0$ and $n = z$, so that $f(m, n) = z$.

Section A.3

1. If the side is x, then $h^2 = x^2 + x^2 = 2x^2$, so

$$h = \sqrt{2}x.$$

3. If the side is x, then the height is $h = \frac{\sqrt{3}}{2}x$, so the area is $A = \frac{1}{2}x \frac{\sqrt{3}}{2}x = \frac{\sqrt{3}}{4}x^2$.

5. If a divides b, there is some k such that $ak = b$; and if b divides c, there is some ℓ such that $b\ell = c$. Then $c = b\ell = (ak)\ell = (k\ell)a$, so a divides c.

7. If n is odd, there is some k such that $n = 2k + 1$. Then $n^2 = (2k + 1)^2 = 2(2k^2 + k) + 1$, so n^2 is odd.

9. If $b = a + 1$, then $(a + b)^2 = (2a + 1)^2 = 2(2a^2 + 2a) + 1$, so $(a + b)^2$ is odd.

11. Let $m = 2$ and $n = 3$. Then $m^2 + n^2 = 13$, which is not divisible by 4.

13. Contrapositive: Suppose n is even, so there is some k such that $n = 2k$. Then $n^2 = 4k^2$, so n^2 is even.

15. Contrapositive: Suppose $p = q$. Then $\sqrt{pq} = \sqrt{p^2} = p = (p + q)/2$.

17. Contrapositive: Suppose $x > 0$. If $\epsilon = x/2 > 0$, then $x > \epsilon$.

19. Contradiction: Suppose $\sqrt[3]{2} = p/q$ such that p and q have no common factors. Then $2q^3 = p^3$, so p^3 is even and hence p is even. This gives that q is also even, which contradicts the assumption that p and q have no common factors.

21. If $7xy \leq 3x^2 + 2y^2$, then $3x^2 - 7xy + 2y^2 = (3x - y)(x - 2y) \geq 0$. There are two cases: either both factors are greater than or equal to 0, or both are less than or equal to 0. The first case is not possible since the assumption is that $x < 2y$. Therefore, $3x \leq y$.

23. Define $f : \mathbb{R} \to \mathbb{R}$ by $f(x) = x^2$. Let $C = [-4, 4]$, $D = [0, 4]$. Then $f^{-1}(C) = [-2, 2] = f^{-1}(D)$ but $C \nsubseteq D$.

25. If $x \in f^{-1}(C)$, then $f(x) \in C$. Since $C \subset D$, then $f(x) \in D$. Hence, $x \in f^{-1}(D)$.

27. If $y \in f(A \backslash B)$, there is some x such that $y = f(x)$ with $x \in A$ and $x \notin B$. So $y \in f(A) \backslash f(B)$, and $f(A \backslash B) \subset f(A) \backslash f(B)$. Now suppose $y \in f(A) \backslash f(B)$. So there is some $x \in A$ such that $y = f(x)$. Since f is one-to-one, this is the only preimage for y, so $x \in A \backslash B$. Therefore, $f(A) \backslash f(B) \subset f(A \backslash B)$.

29. By Theorem 3 of Sec. A.2, $f(f^{-1}(C)) \subset C$. Let $y \in C$. Since f is onto, there is some x such that $y = f(x)$. So $x \in f^{-1}(C)$, and hence $y = f(x) \in f(f^{-1}(C))$. Therefore, $C \subset f(f^{-1}(C))$.

Section A.4

1. Base case: $n = 1 : 1^2 = \frac{1(2)(3)}{6}$
Inductive hypothesis: Assume the summation formula holds for the natural number n.
Consider

$$1^2 + 2^2 + 3^2 + \cdots + n^2 + (n + 1)^2$$
$$= \frac{n(n + 1)(2n + 1)}{6} + (n + 1)^2$$
$$= \frac{n + 1}{6}(2n^2 + 7n + 6)$$
$$= \frac{n + 1}{6}(2n + 3)(n + 2)$$
$$= \frac{(n + 1)(n + 2)(2n + 3)}{6}$$

3. Base case: $n = 1 : 1 = \frac{1(3-1)}{2}$
Inductive hypothesis: Assume the summation formula holds for the natural number n.
Consider

$$1 + 4 + 7 + \cdots + (3n - 2) + [3(n + 1) - 2]$$
$$= \frac{n(3n - 1)}{2} + (3n + 1)$$
$$= \frac{3n^2 + 5n + 2}{2}$$
$$= \frac{(n + 1)(3n + 2)}{2}$$

5. Base case: $n = 1 : 2 = \frac{1(4)}{2}$
Inductive hypothesis: Assume the summation formula holds for the natural number n.
Consider

$$2 + 5 + 8 + \cdots + (3n - 1) + [3(n + 1) - 1]$$
$$= \frac{1}{2}(3n^2 + 7n + 4)$$
$$= \frac{(n + 1)(3n + 4)}{2}$$
$$= \frac{(n + 1)(3(n + 1) + 1)}{2}$$

7. Base case: $n = 1 : 3 = \frac{3(2)}{2}$
Inductive hypothesis: Assume the summation formula holds for the natural number n.
Consider

$$3 + 6 + 9 + \cdots + 3n + 3(n + 1)$$
$$= \frac{1}{2}(3n^2 + 9n + 6)$$
$$= \frac{3}{2}(n^2 + 3n + 2)$$
$$= \frac{3(n + 1)(n + 2)}{2}$$

9. Base case: $n = 1 : 2^1 = 2^2 - 2$
Inductive hypothesis: Assume the summation formula holds for the natural number n.
Consider

$$\sum_{k=1}^{n+1} 2^k = \sum_{k=1}^{n} 2^k + 2^{n+1}$$
$$= 2^{n+1} - 2 + 2^{n+1}$$
$$= 2^{n+2} - 2$$

11. From the data in the table

n	$2 + 4 + \cdots + 2n$
1	$2 = 1(2)$
2	$6 = 2(3)$
3	$12 = 3(4)$
4	$40 = 4(5)$
5	$30 = 5(6)$

we make the conjecture that

$$2 + 4 + 6 + \cdots + (2n) = n(n + 1)$$

Base case: $n = 1 : 2 = 1(2)$
Inductive hypothesis: Assume the summation formula holds for the natural number n.
Consider

$$2 + 4 + 6 + \cdots + 2n + 2(n + 1)$$
$$= n(n + 1) + 2(n + 1)$$
$$= (n + 1)(n + 2)$$

13. Base case: $n = 5 : 32 = 2^5 > 25 = 5^2$
Inductive hypothesis: Assume $2^n > n^2$ holds for the natural number n.
Consider $2^{n+1} = 2(2^n) > 2n^2$. But since $2n^2 - (n + 1)^2 = n^2 - 2n - 1 = (n - 1)^2 - 2 > 0$, for all $n \geq 5$, we have $2^{n+1} > (n + 1)^2$.

15. Base case: $n = 1 : 1^2 + 1 = 2$, which is divisible by 2.
Inductive hypothesis: Assume $n^2 + n$ is divisible by 2.
Consider $(n + 1)^2 + (n + 1) = n^2 + n + 2n + 2$. By the inductive hypothesis, $n^2 + n$ is divisible by 2, so since both terms on the right are divisible by 2, then $(n + 1)^2 + (n + 1)$ is divisible by 2. Alternatively, observe that $n^2 + n = n(n + 1)$, which is the product of consecutive integers and is therefore even.

17. Base case: $n = 1 : 1 = \frac{r-1}{r-1}$
Inductive hypothesis: Assume the formula holds for the natural number n.
Consider

$$1 + r + r^2 + \cdots + r^{n-1} + r^n$$
$$= \frac{r^n - 1}{r - 1} + r^n$$
$$= \frac{r^n - 1 + r^n(r - 1)}{r - 1}$$
$$= \frac{r^{n+1} - 1}{r - 1}$$

19. Base case: $n = 2 : A \cap (B_1 \cup B_2) = (A \cap B_1) \cup (A \cap B_2)$, by Theorem 1 of Sec. A.1
Inductive hypothesis: Assume the formula holds for the natural number n.
Consider

$$A \cap (B_1 \cup B_2 \cup \cdots \cup B_n \cup B_{n+1})$$
$$= A \cap [(B_1 \cup B_2 \cup \cdots \cup B_n) \cup B_{n+1}]$$
$$= [A \cap (B_1 \cup B_2 \cup \cdots \cup B_n)] \cup (A \cap B_{n+1})$$
$$= (A \cap B_1) \cup (A \cap B_2) \cup \cdots \cup (A \cap B_n) \cup (A \cap B_{n+1})$$

21.

$$\binom{n}{r} = \frac{n!}{r!(n - r)!}$$
$$= \frac{n!}{(n - r)!(n - (n - r))!}$$
$$= \binom{n}{n - r}$$

23. By the binomial theorem,

$$2^n = (1 + 1)^n = \sum_{k=0}^{n} \binom{n}{k}$$

Index